Annual Session of Scientific Papers "IMT ORADEA 2019"

IOP Conference Series: Materials Science and Engineering
Volume 568

Oradea, Romania
30 - 31 May 2019

Editors:

Gavril Grebenisan
Alexandru-Viorel Pele

ISBN: 978-1-5108-9811-0
ISSN: 1757-8981

Printed from e-media with permission by:

Curran Associates, Inc.
57 Morehouse Lane
Red Hook, NY 12571

Some format issues inherent in the e-media version may also appear in this print version.

Institute of Physics
Dirac House, Temple Back
Bristol BS1 6BE UK

Phone: 44 1 17 929 7481
Fax: 44 1 17 920 0979

techtracking@iop.org

Additional copies of this publication are available from:

Curran Associates, Inc.
57 Morehouse Lane
Red Hook, NY 12571 USA
Phone: 845-758-0400
Fax: 845-758-2633
Email: curran@proceedings.com
Web: www.proceedings.com

Table of contents

Volume 568

Annual Session of Scientific Papers "IMT ORADEA 2019"

30–31 May 2019, Oradea, Felix SPA, Romania

Accepted papers received: 13 June 2019
Published online: 17 September 2019

Preface

Papers

IOP Conf. Series: Materials Science and Engineering **568** (2019) 011001 doi:10.1088/1757-899X/568/1/011001

Preface

The Annual International Session of Scientific Papers - IMT Oradea 2019 is an International Conference which held in Felix Spa, near Oradea, in the North-Western part of Romania. We are delighted, proud and honored to be your hosts every year at the scientific events of the Faculty of Managerial and Technological Engineering at the University of Oradea. The high value of the works presented at each of the Annual Session of the Scientific Works makes us the Editorial Board, The Scientific Committee and The Organizing Committee, trust that we have carried out intensive work dedicated to the progress and dissemination of scientific, multidisciplinary engineering information. We have proposed that the rate of acceptance of papers should always be a reflection of the scientific review process, the more demanding, the higher the scientific level, and the lower the rate of acceptance of the works, respectively. The high degree of exigency in the review process has been instrumental in accepting publication by IOP Conference Series: Materials Science and Engineering (MSE). In this regard, the papers presented at The Annual Session of Scientific Papers, IMT Oradea, have been indexed SCOPUS and since this year they will be indexed in Web Of Sciences (Clarivate Analytics, ISI). The road was not easy, nor straight, due to the many difficulties encountered, but the results are obviously positive. Our objectives, courageous and motivating, at the same time, were based on ethical criteria unequivocally. We have succeeded in promoting this scientific event to such an extent that it has been accepted as full membership in the respected and recognized of global profile organization, we named here The Committee on Publication Ethics (COPE). In the meantime, we did not give up on the proposed objective, namely that the papers presented at the IMT Oradea Conference should be free of any deviations from scientific ethics, thus discouraging plagiarism. We verify the degree of "overlap" of the texts of the works, with a maximum error of 5% being allowed, which are accepted at the level of writing References titles. All the papers presented at the IMT Oradea Conference are identified by the Digital Object Identifier (DOI) as we are full members of CROSSREF, which allows us to have access to any search engine as it enables robots and crawlers to recognize Crossref DOIs as URLs. There, it is significant for the scientific world because the DOI structure is built on the URL link system, but DOI survives even after the URL link of the paperwork has disappeared. We are very confident that we have always received full credit from you, the participants in this elegant event, and we recognize that you have always had a positive influence on the quality of our work, but especially on the outcome of this work: IMT Oradea.
Editorial Board

IOP Conf. Series: Materials Science and Engineering **568** (2019) 011001 doi:10.1088/1757-899X/568/1/011001

1) THE CONFERENCE'S COMMITTEES

- **Editors:**
  - **Gavril GREBENIŞAN,** University of Oradea, Romania
  - **Alexandru-Viorel PELE**, University of Oradea, Romania

- **Editorial Board:**
  - **Călin BĂBAN**, University of Oradea, Romania
  - **Horia BELEŞ**, University of Oradea, Romania
  - **Florin Sandu BLAGA**, University of Oradea, Romania
  - **Izabela DEMBINSKA**, University of Szczecin, Poland
  - **Gavril GREBENIŞAN**, University of Oradea, Romania
  - **Alexandru-Viorel PELE**, University of Oradea, Romania
  - **Mircea Teodor POP**, University of Oradea, Romania
  - **Delia POP**, University of Oradea, Romania

- **Scientific Committee:**
  - **Stelian ALACI**, Stefan cel Mare University, Suceava
  - **Catalin ALEXANDRU**, Transilvania University Brasov, Romania
  - **Silvia AVASILCĂI**, Technical University Gheorghe Asachi of Iaşi, Romania
  - **Calin BABAN**, University of Oradea, Romania
  - **Marius BABAN**, University of Oradea, Romania
  - **Miroslav BADIDA**, Technical University of Košice, Slovakia
  - **Pouya DERAKHSHAN-BARJOEI**, Islamic Azad University, Iran
  - **Horia BELEŞ**, University of Oradea, Romania
  - **Florin BLAGA**, University of Oradea, Romania
  - **Sanda BOGDAN**, University of Oradea, Romania
  - **Ioan BONDREA**, Lucian Blaga University of Sibiu, Romania
  - **Dan BRINDASU**, Lucian Blaga University of Sibiu, Romania
  - **Adriana BUJOR**, Technical University Gheorghe Asachi of Iaşi, Romania
  - **Constantin BUNGAU**, University of Oradea, Romania
  - **Nicolae BURNETE**, Technical University of Cluj-Napoca, Romania
  - **Samuel SÁNCHEZ CABALLERO** - Univeristat Politecnica de Valencia, Spain
  - **Miguel ANGEL SELLES CANTO** - Univeristat Politecnica de Valencia, Spain
  - **Marco CECCARELLI**, University of Cassino, Italy
  - **Florina-Carmen CIORNEI**, Stefan cel Mare University, Suceava, Romania

IOP Conf. Series: Materials Science and Engineering **568** (2019) 011001 doi:10.1088/1757-899X/568/1/011001

- **Alexandru RUS**, University of Oradea, Romania
- **Nazzal SALEM**, Zarqa University, Jordan
- **Dan SAVESCU**, Transilvania University Brasov, Romania
- **Laurentiu SLATINEANU**, Gheorghe Asachi Technical University of Iasi, Romania
- **Radu STANCIU**, Polytechnic University of Bucharest, Romania
- **Ivan NAGY SZOKOLAY**, Széchenyi István University, Hungary
- **Edit SZUCS**, University of Debrecen, Hungary
- **Ioan Constantin TARCA**, University of Oradea, Romania
- **Radu Catalin TARCA**, University of Oradea, Romania
- **Imre TIMAR**, University of Pannonia, Hungary
- **Marian TOLNAY**, Slovak University of Technology, Slovakia
- **Stefan VALCUHA**, Slovak University of Technology, Slovakia
- **Radu VELICU**, Transilvania University Brasov, Romania
- **Tiberiu VESSELENYI**, University of Oradea, Romania
- **Agostino VILLA**, Polytechnic University of Torino, Italy
- **Marian ZAHARIA**, Petroleum- Gas University of Ploiesti, Romania
- **Máté ZÖLDY**, Budapest University of Technology and Economics, Hungary
- **Sahin YILDIRIM**, Erciyes University, Turkey

- **Steering Committee:**
  - **Dan CHIRA**, University of Oradea, Romania
  - **Nicolae CRECAN**, University of Oradea, Romania
  - **Lehel Szabolcs CSOKMAI**, University of Oradea, Romania
  - **Anamaria MADURA**, University of Oradea, Romania
  - **Ovidiu MOLDOVAN**, University of Oradea, Romania
  - **Georgeta NICHITA**, University of Oradea, Romania
  - **Alin POP**, University of Oradea, Romania
  - **Marina ARTENIE**, University of Oradea, Romania

IOP Conf. Series: Materials Science and Engineering **568** (2019) 011002 doi:10.1088/1757-899X/568/1/011002

Peer review statement

All papers published in this volume of *IOP Conference Series: Materials Science and Engineering* have been peer reviewed through processes administered by the proceedings Editors. Reviews were conducted by expert referees to the professional and scientific standards expected of a proceedings journal published by IOP Publishing.

IOP Conf. Series: Materials Science and Engineering **568** (2019) 012001 doi:10.1088/1757-899X/568/1/012001

Multi-body system simulation of the sun trackers used for PV panels

C Alexandru

Transilvania University of Braşov, Romania

E-mail: calex@unitbv.ro

Abstract. The article approaches the design algorithm of the sun trackers used to improve the energy conversion performance of the photovoltaic systems. In this regard, three specific types of mechanical models are developed in an integrated concept, in the following sequence: kinematics - inverse dynamics - dynamics. The application is carried out by considering a dual-axis sun tracker that uses two linear actuators as driving sources for the diurnal and elevation movements. The MBS package ADAMS is used to model and simulate the behaviour of the sun tracker in virtual prototyping environment.

1. Introduction

The renewable energy is a very good alternative for avoiding the major problems that arise from the use of traditional energy sources, such as the limited resources of fossil fuels, the pollution, or the green-house effect. Improving the efficiency of the conversion devices of the renewable sources (e.g. sun, wind, or water) is a constant concern and challenge for the research, development and innovation (RDI) in the field. In the case of photovoltaic systems (PV), one of the viable solutions for this purpose is the implementation of sun tracking systems, which allow the capture of as much incident solar radiation as possible [1-4]. Most of the tracking systems are based on different structures of mechanical systems (more or less complex), whose drive is made by actuators (motors) that are commanded so that to allow a more accurate tracking of the sun [5-8]. Estimating the behavior of the tracking systems is crucial for their optimal design in accordance with the specific implementation area [9-12].

The literature reveals a variety of ways for evaluating the behaviour of the mechanical/mechatronic systems. The analytical methods are frequently used, being based on the determination of motion laws in the form of analytical functions. Although seemingly more difficult, the analytical approach offers a high accuracy of the results obtained, it can apply to all types of systems (planar, spatial and spherical), and is suitable for computer programming. The analysis can be accomplished with both in-house specialized programs, based on a programming language platform, as well as with MBS (Multi-Body Systems) commercial programs.

An important advantage of MBS programs is that the movement equations are formulated and then automatically resolved. Among the commonly used formalizing algorithms used by the commercial MBS software solutions, the following can be mentioned: Lagrange, Euler, Kane, or d'Alembert. From this point of view, there are two major MBS software groups: programs that form numerically the motion equations for each integration step; programs that symbolically formulate the motion equations, the user then establishing whether or not to integrate the equations, which can be carried out with symbolic computing programs (such as MATLAB or MAPLE). Complex virtual prototypes can

IOP Conf. Series: Materials Science and Engineering **568** (2019) 012001 doi:10.1088/1757-899X/568/1/012001

be created with the help of MBS software solutions, in order to obtain products that function in line with the high market demands [13-17]. Thus, it is possible to accurately model both the mechanical or mechatronic system itself and the environment in which it operates. This eliminates much of the experimental testing, which is an expensive and time consuming process. However, the experimental testing will still be necessary, but at a much lower level, for validating the virtual prototype.

In this regard, the work shows the analysis algorithm for PV tracking systems. The application is carried out by considering a bi-axial mechanism, which uses linear actuators as driving sources for the daily and elevation motions. The MBS package ADAMS of MSC.Software is used for conducting the tests in virtual environment.

2. Tracking systems designing algorithm

For the tracking systems design, the following mechanical models can be developed/conceived, in accordance with the flow-chart from Figure 1:

a. kinematic model - contains the bodies (parts), connected by geometric constraints; the actuating is modeled by motion restrictions that are applied in the rotational or translational joints of the driving elements (rotary or linear actuators), thus controlling their linear or angular position;
b. inverse dynamic model - includes the external and internal forces that act in the tracking mechanism (e.g. mass and inertia forces), the actuating being performed in a similar way with the kinematic model (by motion restrictions); the model is used to determine the motor torques or forces (depending on the type of actuating element - rotary or linear actuator) that generate the kinematically prescribed behavior of the mechanism;
c. dynamic model - its structure is similar to that of the inverse dynamic model, the distinction consisting in the way in which the action is modeled (by motor torque/force instead of movement restrictions); the dynamic model is useful for evaluationg the tracking system behavior under the action of external and interal forces/torques.

In this paper, for implementing the design algorithm, a bi-axial tracking mechanism is proposed, the two degrees of freedom of the system, namely the daily and elevation movements, being generated/conducted by linear actuators. Sun tracking for the both movements is realized in steps, the motion functions of the photovoltaic panel being designed with the purpose to enssure an optimal number of steps, with a minimal energy demand to carry out the movements, what will be found in increasing the system efficiency.

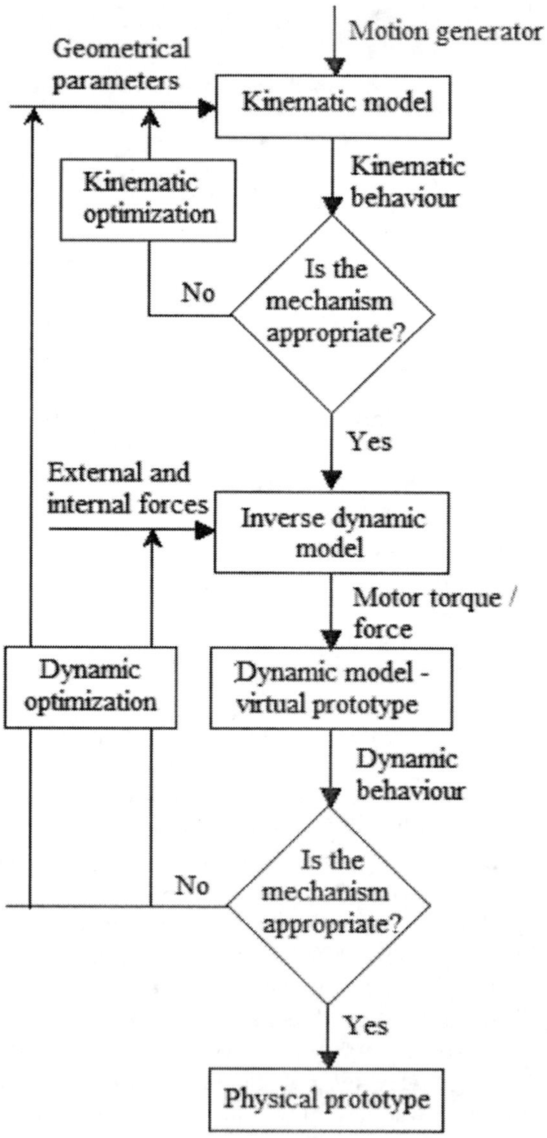

Figure 1. The designing algorithm.

IOP Conf. Series: Materials Science and Engineering **568** (2019) 012001 doi:10.1088/1757-899X/568/1/012001

3. Case study

The MBS model of the bi-axial sun tracker (Figure 2) contains two kinematic loops, corresponding to the diurnal (ABCD) and elevation (EFGH) movements. The geometric constraints in A, B, D, E, F, and H have been modeled by rotational joints, whereas the connections C and G (between the two components of the acuators) are translational joints. For the diurnal movement, the panel (1) rotates relative to the intermediary support (2) around the axis A-A'(the joint A' is kinematically passive). The elevation movement is performed by rotating the support relative to the fixed pillar (0), which is rigidly connected to ground.

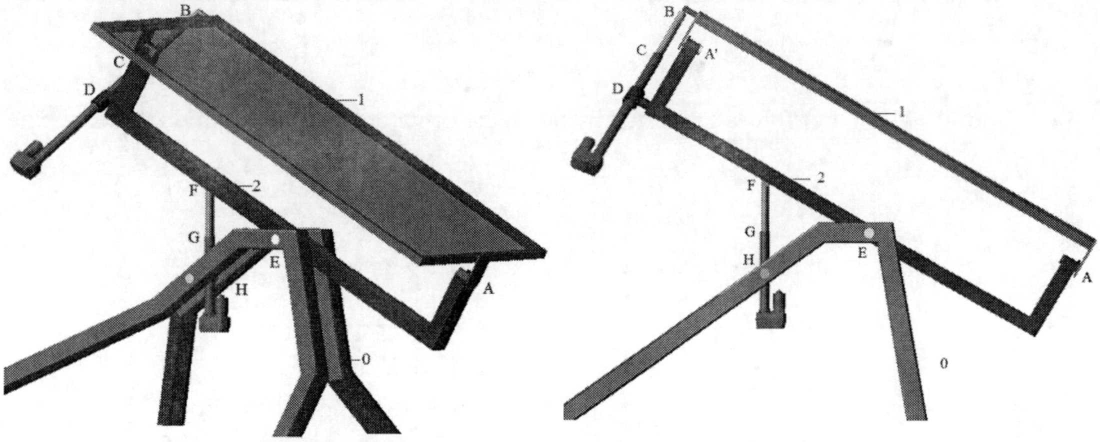

Figure 2. The MBS model of the bi-axial PV system.

Simulations performed for this work correspond to summer solstice day, and Braşov implementation area (sunrise at 4.26, sunset at 19.74, in solar time), which actually determines the orientation of the PV panel according to the sun position. In case of the daily motion, the angular domain of the panel is 120°, in the interval [+60°, -60°], relative to the zero position at solar noon. At sunset, the panel returns to the initial position. The adjustment of the elevation angle of the panel is performed with two motion steps (passing from 11° to 22.05°, and then back to 11°, the angle being measured relative to the horizontal axis). In the virtual model developed in ADAMS, the motion laws of the panel have been modeled by summing a series of STEP time functions, by using ADAMS/Function Builder. In these terms, the variation diagrams of the diurnal and elevation angles are shown in Figure 3. Afterwards, the inverse kinematic analysis was carried out with the aim to determine the corresponding linear displacements (strokes) in the two actuators (Figure 4).

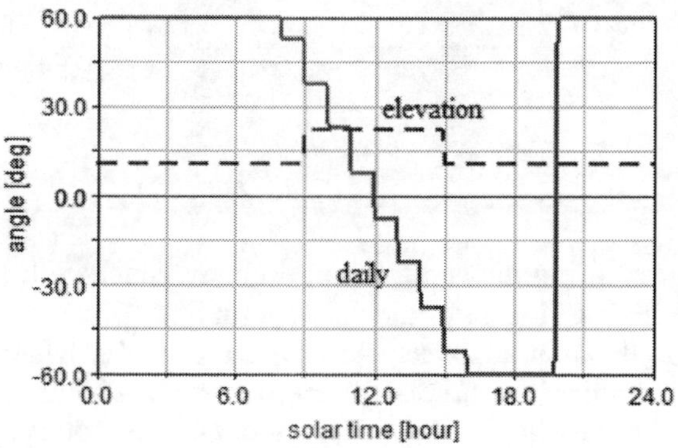

Figure 3. The angular motion functions.

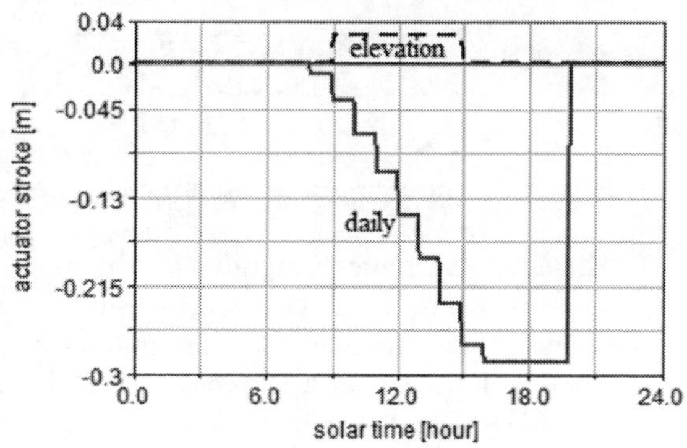

Figure 4. The linear displacements in actuators.

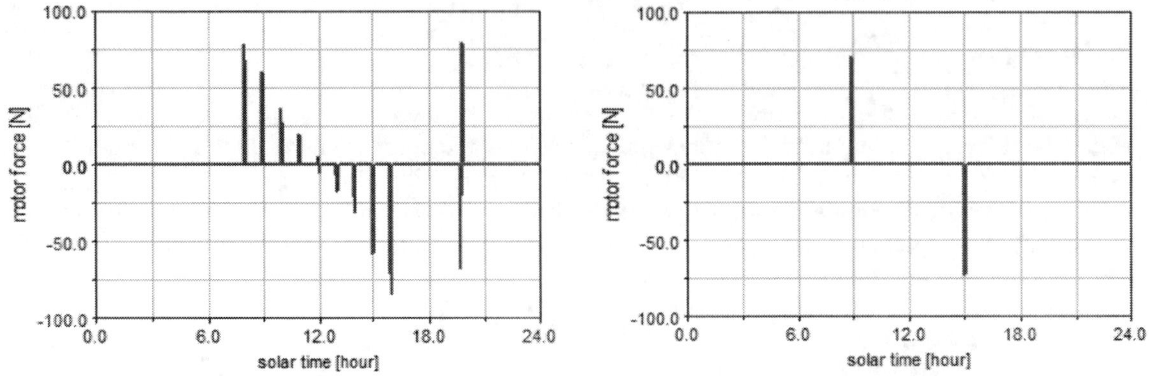

a. b.

Figure 5. The motor forces generated by the linear actuators (a - daily motion, b - elevation).

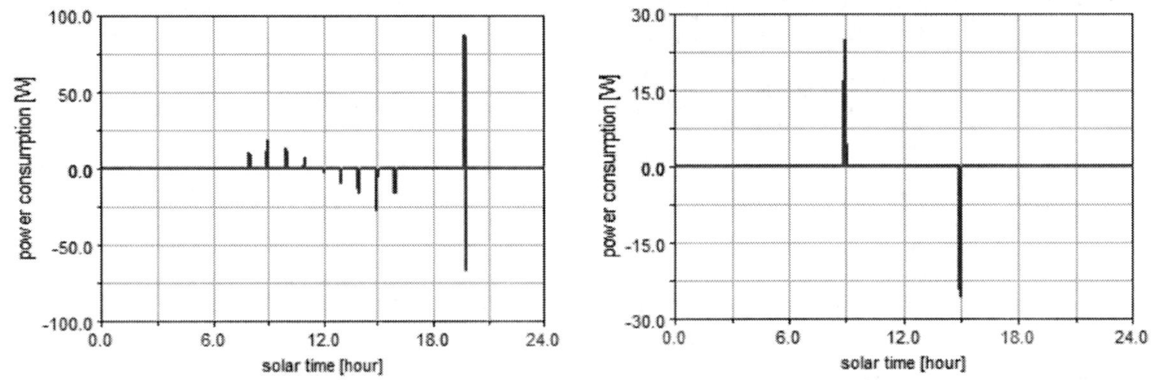

a. b.

Figure 6. The power consumptions for performing the tracking (a - daily motion, b - elevation).

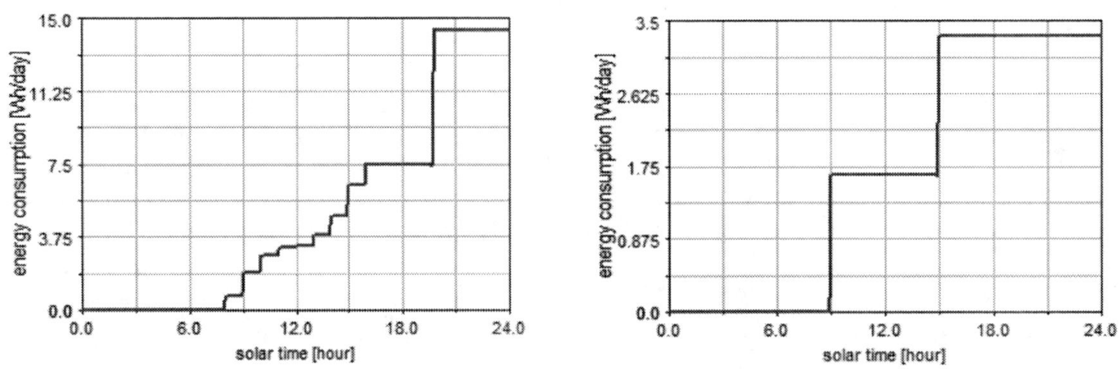

a. b.

Figure 7. The energy consumptions for performing the tracking (a - daily motion, b - elevation).

The actuating forces generated by the motors for assuring the sun tracking in compliance with the imposed motion laws have been obtained by the inverse dynamic analysis (Figure 5). Then, these forces have been applied as input parameters in the dynamic analysis. Among the specific dynamic parameters, Figure 6 shows the power consumptions required to carry out the two movements, through whose integration the corresponding energy consumptions have resulted (Figure 7). It is noticeable that low consumption is obtained, which influences positively the PV system efficiency.

IOP Conf. Series: Materials Science and Engineering **568** (2019) 012001 doi:10.1088/1757-899X/568/1/012001

4. Conclusions

This work aims to highlight the importance of the testing in virtual prototyping environment, given that the alternative for identifying the optimal solution would have been the development and testing of experimental models, which is an expensive and time consuming process. An important advantage of the testing in virtual prototyping environment consists in the ability to make measurements that are actually very difficult to be performed on physical prototypes. The paper was focused on the analysis process, following that in a forthcoming work to accomplish the optimization of the proposed sun tracker, thus addressing the whole design flow-chart in the manner in which it is shown in Figure 1.

References

[1] Alexandru C and Comşiţ M 2008 The energy balance of the photovoltaic tracking systems using virtual prototyping platform *Proceedings of the 5th IEEE International Conference on the European Electricity Market - EEM* pp 253-258

[2] Alexandru C 2009 The design and optimization of a photovoltaic tracking mechanism *Proceedings of the 2nd IEEE International Conference on Power Engineering, Energy and Electrical Drives - POWERENG* pp 436-441

[3] Flores-Hernández D A, Palomino-Resendiz S, Lozada-Castilloc N and Luviano-Juárez A 2017 Mechatronic design and implementation of a two axes sun tracking photovoltaic system driven by a robotic sensor *Mechatronics* **47** pp 148-159

[4] Nsengiyumva W, Chen S G, Hu L and Chen X 2018 Recent advancements and challenges in solar tracking systems (STS) *Renewable & Sustainable Energy Reviews* **81** pp 250-279

[5] Alexandru C and Tatu I N 2013 Optimal design of the solar tracker used for a photovoltaic string *Journal of Renewable and Sustainable Energy* **5** pp 1-16

[6] Çinar S M, Hocaoğlu F O and Orhun M 2014 A remotely accessible solar tracker system design *Journal of Renewable and Sustainable Energy* **6** pp 1-12

[7] Ioniţă M and Alexandru C 2012 Dynamic optimization of the tracking system for a pseudo-azimuthal photovoltaic platform *Journal of Renewable and Sustainable Energy* **4** pp 1-15

[8] Vieira R G, Guerra F K, Vale M R and Araujo M M 2016 Comparative performance analysis between static solar panels and single-axis tracking system on a hot climate region near to the equator *Renewable & Sustainable Energy Reviews* **64** pp 672-681

[9] Altuzarra O, Macho E, Aginaga J and Petuya V 2015 Design of a solar tracking parallel mechanism with low energy consumption *Proceedings of the Institution of Mechanical Engineers, Part C: Journal of Mechanical Engineering Science* **229** pp 566-579

[10] Seme S, Stumberger B and Hadziselimovic M 2016 A novel prediction algorithm for solar angles using second derivative of the energy for photovoltaic sun tracking purposes *Solar Energy* **137** pp 201-211

[11] Sharaf Eldin S A, Abd-Elhady M S and Kandil H A 2016 Feasibility of solar tracking systems for PV panels in hot and cold regions *Renewable Energy* **85** pp 228-233

[12] Sinha S and Chandel S S 2016 Analysis of fixed tilt and sun tracking photovoltaic-micro wind based hybrid power systems *Energy Conversion and Management* **115** pp 265-275

[13] Alexandru C and Pozna, C 2009 Dynamic modeling and control of the windshield wiper mechanisms *WSEAS Transactions on Systems* **8** pp 825-834

[14] Alexandru C and Alexandru, P 2010 The virtual prototype of a mechatronic suspension system with active force control *WSEAS Transactions on Systems* **9** pp 927-936

[15] Alexandru C and Alexandru, P 2011 A comparative analysis between the vehicles' passive and active suspensions *International Journal of Mechanics* **5** pp 371-378

[16] Alexandru P, Macaveiu D and Alexandru C 2012 Design and simulation of a steering gearbox with variable transmission ratio *Proceedings of the Institution of Mechanical Engineers, Part C: Journal of Mechanical Engineering Science* **226** pp 2538-2548

[17] Tarniţă D and Marghitu D 2013 Analysis of a hand arm system *Robotics and Computer-Integrated Manufacturing* **29**, pp 493-50

IOP Conf. Series: Materials Science and Engineering **568** (2019) 012002 doi:10.1088/1757-899X/568/1/012002

Metal coin error study using 3D modelling and FEM Analysis

C C Gavrila[1] and M T Lates[1]

[1]Product Design, Mechatronics and Environment Department, "Transilvania" University of Brasov, Brasov, Romania

E-mail: cgavrila@unitbv.ro

Abstract. Mints around the world produced a large amount of metal coins currency. Besides the coin basic design, in most of the situations, various amount of coins wearing different error types are released. The error type is depending on the coin metal, dimension and also by manufacturing step where it could appear. This coins and medals with mint errors are searched by collectors around the world. The paper presents some aspects about the cracked dies error, which appear at the coin striking tools or coin properly strike, because of some dies misalignments, failures and the reduced relief dimension. In the first part of the paper, some aspects about the cracked dies error are presented, with some particular examples. Then, the simplified dies 3D model is presented. Also, the finite element analysis on this model is realized and is achieved for different initial conditions. In the paper final part the analysis results and conclusions are presented.

1. Introduction

Metal coin currency manufacturing process is followed in most of the situations by a various amount of error types. The error type is depending on the coin manufacturing step where it could appear. There are known errors on the coin metal alloy, on the coin blank manufacture, at the striking tools manufacturing and, respectively, on the properly coin striking. The coins and medals with mint errors are most valuable on the collectors market: for increased spectacular error on coin or medal, the increased value for the subjected piece [1].

The metal coin is manufactured by pressing at high loads the coin blank with hardened steel negative dies, which forms a closed space to be filled by coin material. Generally, the model on the negative die is incused. After the coin striking, result the relief model on coin. The negative die is obtained before hardening, by pressing with a hardened steel positive die (also known as hub) with the relief model. A positive die is used to obtain a number of negative dies, used after hardening to strike the large amount of coins [2, 3]. If some errors or model changes appear on dies, the pressed coins will result with same error or model changes.

The cracked letters (or broken letters) can be observed as missing parts of some letters of the impressed inscription. The literature doesn't reveal exactly the causes of this error in the coins manufacturing process. Moreover, the Romanian pieces wearing this error are often considered variants from the basic design [4]. In figure 1, a is presented a Romanian 1 ban bronze coin, minted in year 1900, with cracked letters on the obverse inscription. In figure 1, b the same error type is presented at Romanian 2 bani bronze coin, minted in year 1879, with cracked letters on the reverse inscription, respectively a 2 bani bronze coin from year 1882 with cracked letters on the obverse inscription, presented in figure 1, c.

a. b. c.

Figure 1. The cracked letters at various types of bronze coins

There are known various other coins with cracked letters at inscriptions; well known are also 2 bani 1900 (bronze), 1 leu 1873 (silver), 3 lei 1966 (nickel clad steel) or 50 lei 1994 (brass clad steel) [4].

From figure 1 it can be observed that the font used for the upper case letters inscription presents some thin portions, as the top and bottom of O, C, R. Also, the letters are round shape, eventually closed, or open with sharp edges. The coin diameter is also important: the 1 ban coin has 15 millimetres and 2 bani coin, 20 millimetres. At this coin size, the details of royal effigy, country armour and inscription should be clearly distinguished.

a. b.

Figure 2. The straight cracked dies impressed on coins

The cracked negative dies error is known at various Romanian and foreign modern coins [1, 4]. The properly die crack could be observed at coins, as a straight or curved rib impressed over the coin relief. Usually, there is one crack impressed on one coin face, but multiple cracks on the same face or cracks on both coin faces should not be excluded. The coins showing this error are considered by literature as properly error coins. In figure 2 there are presented two examples of dies cracks, pressed on Venezuelan 5 bolivars obverse, issued in 1989 (nickel), and also on Romanian 20 lei obverse, issued in 1943 (zinc). If the damaged die is not replaced, the crack increase and finally lead to the loosing of a die portion. In figure 3 can be observed the die crack in two different stages, on Romanian 20 lei 1944 coin reverse (zinc): from simple rib on coin face, to the die missing part; in that area, the coin has no model pressed.

IOP Conf. Series: Materials Science and Engineering **568** (2019) 012002 doi:10.1088/1757-899X/568/1/012002

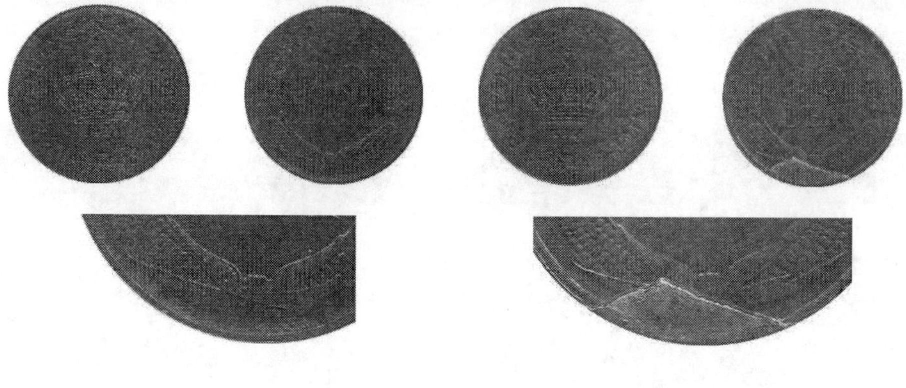

Figure 3. The crack evolution impressed on coins

2. The virtual model

As was presented, on a coin face there are multiple complex details. On a virtual model, these details cannot be faithfully reproduced. So, simplified models should computed for study. Were chosen two simple positive and also negative dies, each wearing a round closed letter O and also an open and sharp letter M. The CATIA software is used [7, 8].

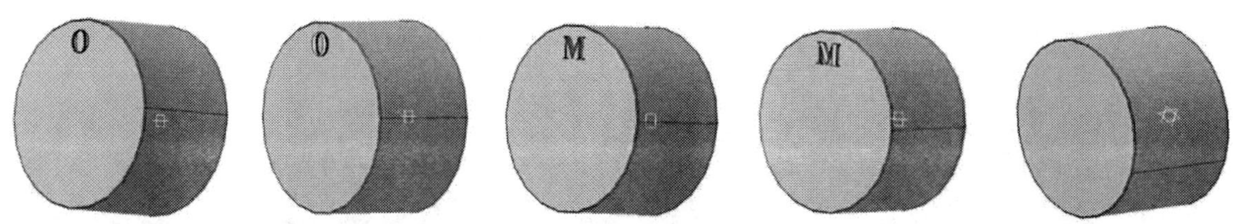

a. Negative die, letter O; b. Positive die, letter O; c. Negative die, letter M; d. Positive die, letter M; e. Blank cylinder

Figure 4. The 3D model

The each die simplified model consist in a cylinder with engraved letter, for negative die, respectively a cylinder with a relief letter, for positive die. After the sketching of the parts, it follows the each part's virtual model computing, using the Part Design module. To simulate the dies loads, each die will be pressed on a blank cylinder. Using the Assembly Design module, the device assembly is computed, following the corresponding constraints.The dies and blank model are presented in figure 4, a, to 4, e.

3. Finite element analysis, simulation and results.

For the analysis, the ANSYS software is used. The analysis objective is to determine the pressed dies ensemble behaviour under the load. For the analysis, the simplified version of the virtual model is used, as is presented in figure 5. The material properties, for dies material (steel) and coin material (aluminum), as Young modulus, Poisson coefficient, Tensile Yield Strength, Ultimate Strength should be defined [6, 8].

In the contact area it is chosen a smooth mesh with the minimum edge length equal with 0.001 mm. The applied normal force is equal with 650KN, in order to obtain high contact pressures, over the allowable stress – 1600 MPa [2, 8].

IOP Conf. Series: Materials Science and Engineering **568** (2019) 012002 doi:10.1088/1757-899X/568/1/012002

Figure 5. The finite element model

The each model of negative and positive dies will be pressed on the blank cylinder. For the dies models the considered material is steel. For the blank cylinder, the considered material may is steel. The results, presented in figures 6 to 9 and also in table 1, consist in the contact pressure maximum values and also the maximum values of the penetration in material.

Figure 6. The negative die with letter M pressed on steel

Figure 7. The negative die with letter O pressed on steel

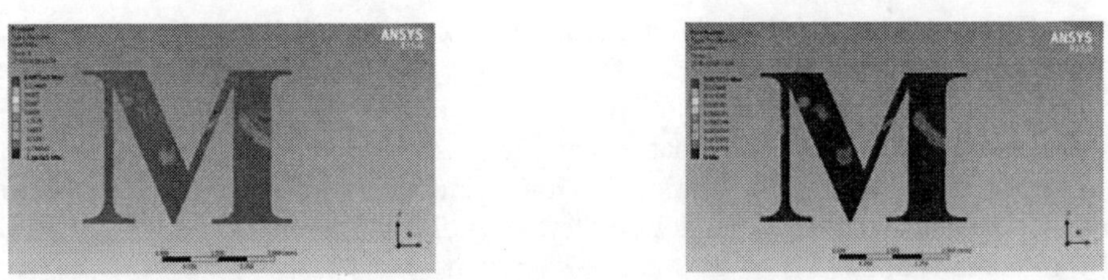

Figure 8. The positive die with letter M pressed on steel

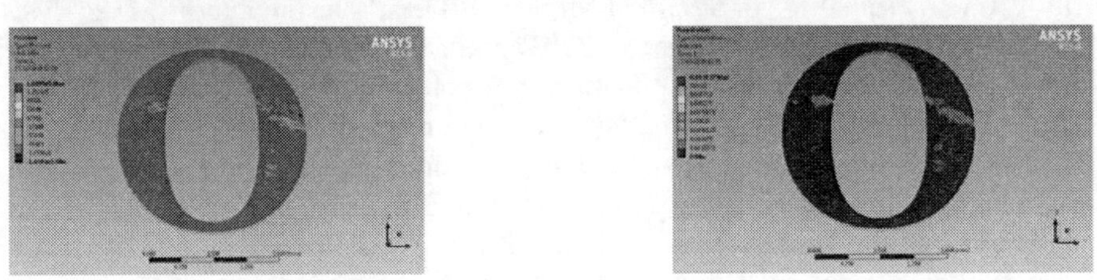

Figure 9. The positive die with letter O pressed on steel

IOP Conf. Series: Materials Science and Engineering **568** (2019) 012002 doi:10.1088/1757-899X/568/1/012002

4. Conclusions

When the negative die is pressed on the blank, the first contact appears on the full circle area, except the incused letter. If the letter shape is closed, as O, the interior portion of it should be added to the contact area. So, the contact pressure and the penetration in material are relatively reduced. For studied letters, the penetration is larger to the fat details than to the thin details. The contact could be reduced to a circular segment if appear some misalignments between involved parts. In this case, the properly die crack could appear, only to the negative die. Following this, the crack is pressed on the coin together with the entire model. If the negative die crack appears at the coin striking, a reduced number of coins with the same error are released. If this error appears when the positive die is pressed, it is transmitted from negative die to the produced positive dies; then, to the next number of negative dies and finally to a large number of coins pressed with the same error.

When the positive die is pressed on the blank, the first contact appears on the area defined by the small letter contour. This area is dramatically reduced and the contact pressure and the penetration in material are highly increased. The positive letter profile penetration in the material is deeper, but the thin details of the letter are exposed to be damaged. The letter shape has a direct influence on the contact area: in the studied case, the O letter is in a slightly better condition than M, but the damages could also appear. This led to the case of the letter crack on the positive die, which led to the negative pressed with the broken letter having a missing part, which is finally pressed on the coins. Also, the appearance of this error can be combined with some misalignments between involved parts, which led to worse conditions. The coins wearing this error could be considered properly error coins.

Table 1. The contact pressure and the penetration in material, maximum values

The dies combination	Contact pressure, MPa	Penetration in the material, mm
Negative die with letter M pressed on steel	1743.4	0.0034622
Negative die with letter O pressed on steel	3909.3	0.00032802
Positive die with letter M pressed on steel	168710	0.015153
Positive die with letter O pressed on steel	160990	0.013837

References

[1] *** 2007 Euro 4, Coins and Banknotes (Euro4, Monnaies et Billets), *Edition Les Chevau-Lègers* Paris France (in french) pp. 193 – 204.

[2] Iliescu C and Tureac O 1987 Cold Pressing Technology (Tehnologia presării la rece) *Editura Universității din Brașov* (in romanian) pp. 266 - 267.

[3] *** 1945 Finance Ministry, National Mint, Ten Years of Activity (Ministerul Finanțelor, Monetăria națională. Zece ani de activitate) *Monitorul oficial și Imprimeriile statului* București (in romanian) pp. 41 - 43.

[4] Buzdugan G Luchian O and Oprescu C C 1977 Romanian Coins and Banknotes (Monede și bancnote românești) *Editura Sport Turism* București (in romanian) pp 256 - 259.

[5] Gavrilă C C and Velicu R 2016 Virtual Modeling, Detail Design and FEM Analysis for a Testing Device *Current Solutions in Mechanical Engineering* Trans Tech Publications pp. 3 - 6.

[6] Ghionea I G 2009 CATIA v5. Application in Mechanical Engineering (CATIA v5. Aplicații în ingineria mecanică) *Editura Bren* București, (in romanian).

[7] Lateș M T 2008 Finite Element Method. Applications (Metoda elementelor finite. Aplicații, in romanian) *Editura Universității Transilvania Brașov.*

[8] Lee H H 2012 Finite Element Simulations with ANSYS Workbench 14. Theory, Applications, Case Studies. *Schroff Development Corporation* Kansas USA.

IOP Conf. Series: Materials Science and Engineering **568** (2019) 012003 doi:10.1088/1757-899X/568/1/012003

Pedestrian visibility at night: the influence of the pedestrian clothing and the defective headlamp of the car

B C Benea[1], D D Trusca[1], G R Toganel[1] and A I Radu[1]

[1]Department of Automotive and Transportation, Mechanical Engineering Faculty, Transilvania University of Brasov, Brasov, Romania

b.benea@unitbv.ro

Abstract. Low night vision is a factor contributing to the production of traffic accidents. In this work, we investigated the distance from which pedestrians can be seen by night, by a driver driving a car with a single functional headlamp. Six pedestrian clothing conditions were tested (black, blue, red, green, yellow and reflective vest). Pedestrians wearing dark clothing (black, blue) were only visible from a distance of 10 meters; the best visibility was for the pedestrian that wearing the reflective vest. The results demonstrate the importance of pedestrian clothing during night time on visibility while traveling on the public road.

1. Introduction

Road accident is an unpredictable event occurring on a public roadway where at least one moving vehicle and other traffic participants are involved, resulting in material damage, injury or death of the person. The cause of most accidents is the low visibility of car participants at night [1]. This is primarily due to the difficulty in reading the traffic information, the colour differences of other traffic cars, the ability to see the edges of the low road, and the effort to the human eye [2].

In the case of drivers, their reduced night-time visibility is due to several factors:

- Fatigue is a factor that influences visibility. This leads to a decrease in driving capacity. Typically, in the time range of 03:00 - 07:00 this tends to the maximum, as in the period when he normally sleeps [3].
- Alcohol is the next major contributor to road accidents. This, along with night driving, is the leading cause of death among young drivers [3,4].
- Age, being a major influence of night vision. The decrease in visual acuity begins at the age of 20 years. Over time, the amount of light that goes through the cornea decreases, so somewhere around the age of 60 it is needed 3 times more light than at the age of 20 to see the same details [5].

In addition to driving factors, there are external factors that influence their visibility, such as:

- Lighting with headlights, where the visible road sector is much lower than during the day. Darkness allows perceiving distances with great difficulty, so that headlights, especially poorly tuned, can create a blinding effect [6,7].
- Unlit or poorly illuminated roads, which affect the visibility of the drivers, becoming much lower and pedestrian sensing, will be much slower [7,8].
- Weather conditions (fog, rain, snow, etc.) are an unpredictable external factor and cause night vision [9].

IOP Conf. Series: Materials Science and Engineering **568** (2019) 012003 doi:10.1088/1757-899X/568/1/012003

Pedestrians are the most vulnerable traffickers due to lack of protection during a collision with a car. A proper visibility of the pedestrian can reduce the consequences of an accident. During night driving, there is the possibility that the objects placed in the vehicle paths are not detected in time [8,9].
The objective of the work is the influence of the colour of pedestrian clothes on the visibility in the night conditions, in the case of vehicles running with defects of the lighting system.

2. Method

To perform this experiment, the measurements were made on the test track, with a length of 50 m, without street lighting. On this test track marks were made at intervals of 10 meters, so the test track was divided into 5 intervals.

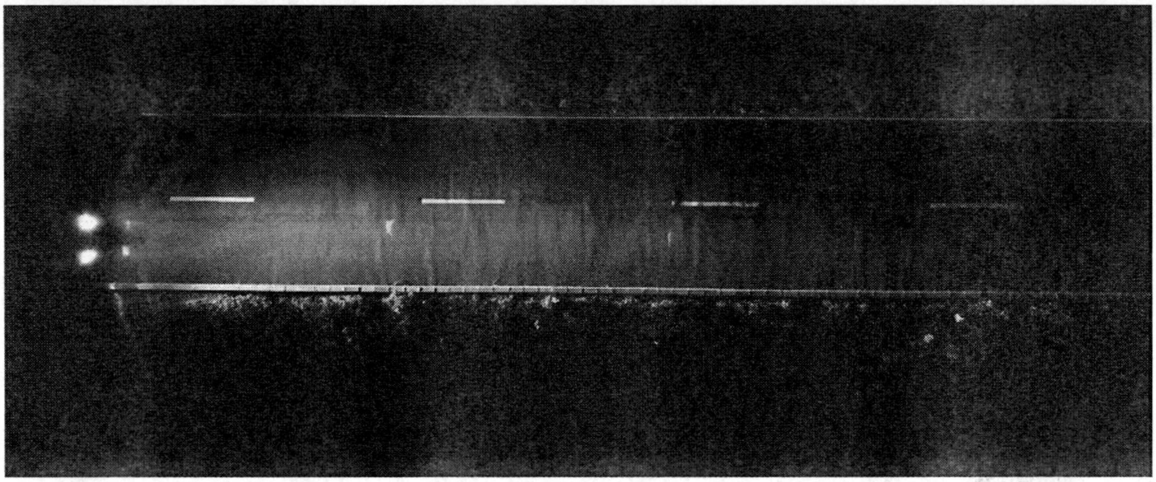

Figure 1. Test track.

Six pedestrians equipped with different colours clothes and a Dacia Logan were used to achieve the objectives. The pedestrian was located at a distance of 0.5 m from the right merge of the road, and the car was positioned at a distance of 10 to 50 m from the pedestrian. The tests were carried out using a single car headlight (simulating the failure of the other headlamp).
Pedestrians wore a black, blue, red, green, yellow blouses and black blouse with reflective vest.

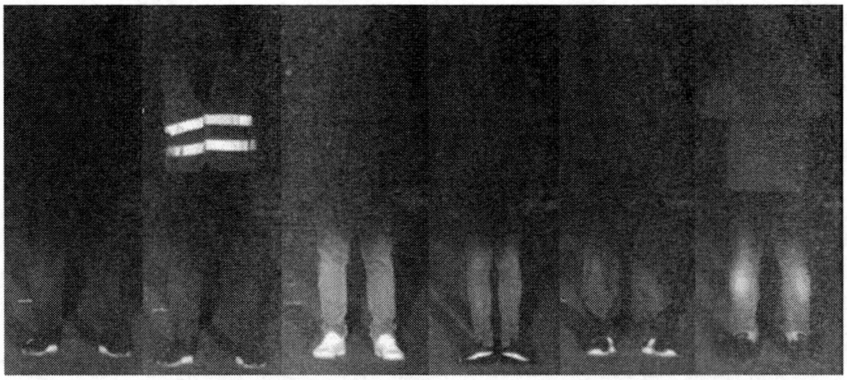

Figure 2. Pedestrians clothes.

A Fujifilm X-T1 camera was used to capture photos from the car. This camera allowed capturing photos at night without changing the focus.

For image processing was used the PC-RECT 4.2 software, developed by DSD Austria, specializing in photogrammetry. This software determines the brightness on different areas of the image and can calculate the contrast between the two analysed areas, so the software gives the visibility of the pedestrian by comparing the two luminance values and the contrast determination. The program uses Adrian Werner's method, which is the method of stability of the visibility of the object under different light conditions.
The analysed area of the pedestrian was the chest area, the wearing area of the reflective vest.
The program determines the value of the luminance of the object in the area of the pedestrian's chest (1) and compares it with the luminance of the environment (2), thus obtaining the contrast value.

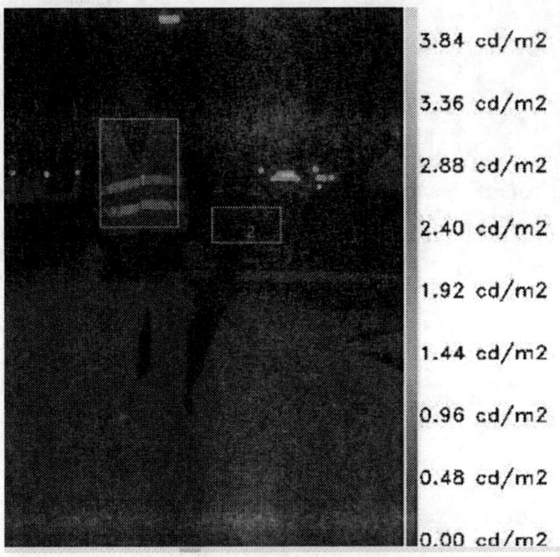

Figure 3. Photo processing to determine the visibility of the pedestrian.

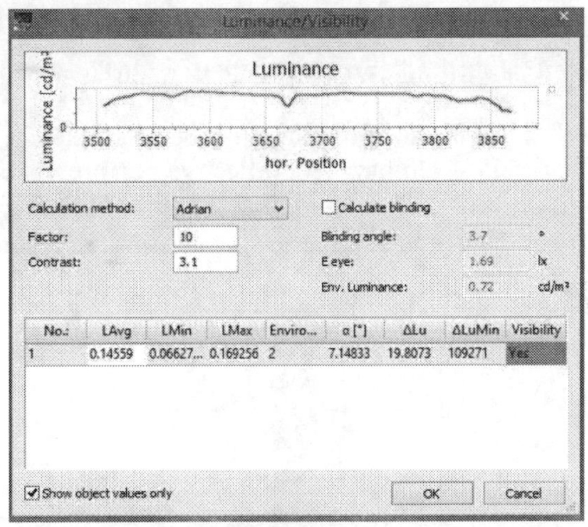

Figure 4. Luminance determination window from PC-RECT.

The software extracts all the data in a table showing the object's visibility based on the ΔLu parameter, which represents the difference between the two selected areas.
L_{avg} – average luminance; L_{Min}, L_{Max} – shows the minimum and maximum luminance of the respective area; α- object size; Environment – assigned background area; ΔL_u – luminance difference, it is the difference between LAvg of the defined area and LAvg of the background area;
ΔL – luminance difference threshold.

3. Results

The pedestrians wore a black (Bk), blue (B), red (R), green (G), yellow (Y) blouses and black blouse with reflective vest (BkRV).

In figure 5 are presented the results for the distance of 10 meters between the car and the pedestrian. The greatest contrast is for the pedestrian that wears the reflective vest, regardless of the functional headlamp.
The small contrast is found for the pedestrian that wears black clothing when the right headlamp works.
For the distance of 20 meters between the car and the pedestrian (figure 6), the biggest contrast is for the pedestrian that wear the reflective vest. At this distance, the pedestrians that wear black or blue clothing are not visible (they are confused with the environment). The pedestrians that wear green or red clothing are more visible when the left-hand headlight of the car works.

IOP Conf. Series: Materials Science and Engineering **568** (2019) 012003 doi:10.1088/1757-899X/568/1/012003

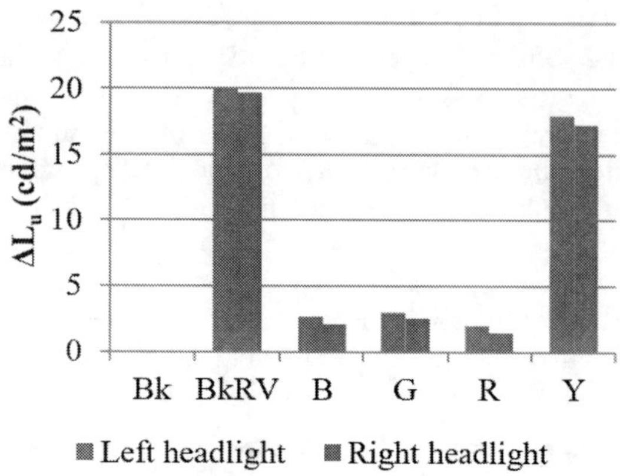

Figure 5. Contrast at 10 m distance.

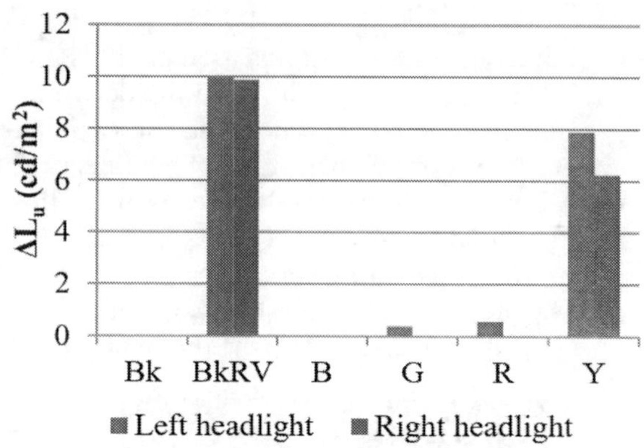

Figure 6. Contrast at 20m distance.

At a distance of 30 meters between the car and the pedestrian (figure 7), the pedestrian that wears yellow clothing or a reflective vest are visible, regardless of the headlight that works.

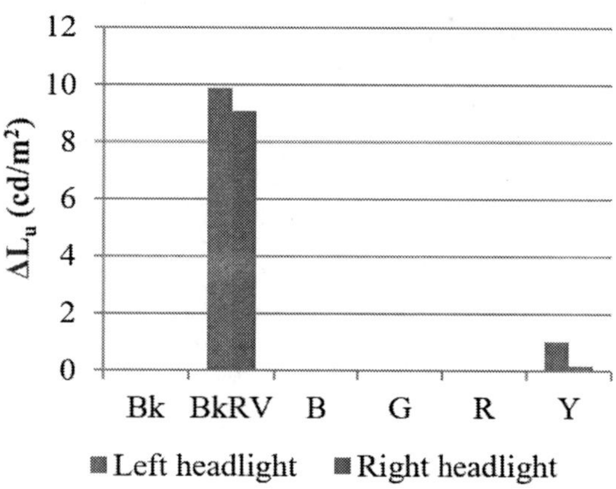

Figure 7. Contrast at 30m distance.

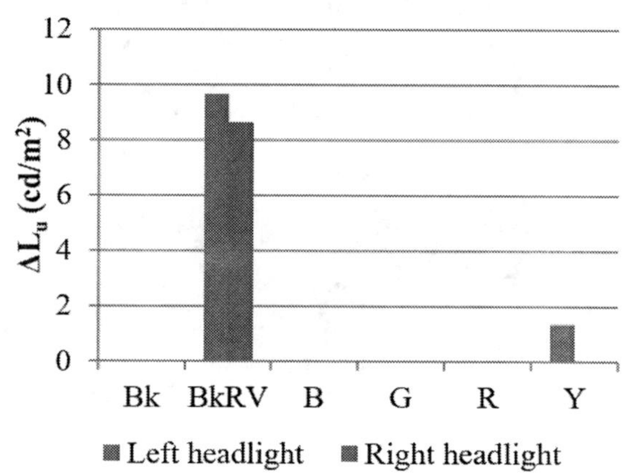

Figure 8. Contrast at 40m distance.

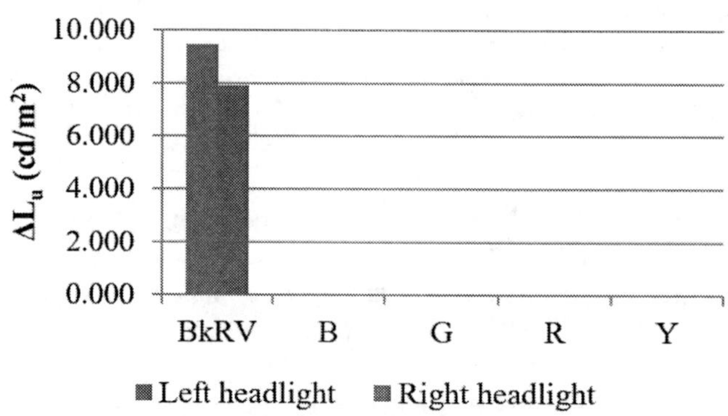

Figure 9. Contrast at 50m distance.

At a distance of 40 meters (figure 8), pedestrian with yellow clothing are not visible if the car has the left headlamp defective.

At a distance of 50 meters (figure 9), only pedestrians with a reflective vest are visible.

IOP Conf. Series: Materials Science and Engineering **568** (2019) 012003 doi:10.1088/1757-899X/568/1/012003

4. Conclusions

The present experiment sought to determine the visibility of pedestrians depending on their worn clothing, in the event of damage to the vehicle lighting system (a defective light bulb). The data demonstrate that pedestrian clothing and functional headlamp significantly affect both the probability that a driver will recognize the presence of a pedestrian on the road and the distance to which pedestrians are recognized in the night.

Pedestrian clothing had a powerful effect on pedestrian visibility. When the pedestrian wore the black coat he was hardly sizable of it at a distance of 10 m. For the other distances tested, the pedestrian in black was confused with the background (it was invisible). The moment of visibility of the pedestrian wearing black clothes occurs at a point that may not allow drivers from stopping in time to avoid a collision.

The visibility of pedestrians is improved by wearing light-coloured clothes (greater contrast with the environment). In the case of green or red blouses, the pedestrians could be seen from a distance of 20 m. Pedestrians who wear yellow blouses can be seen from a distance of 40 m. Pedestrians who wear a reflective vest can be seen from a distance of 50 m.

The functional headlamp had a significant effect on pedestrians' recognition. In the case of the operation of the left-hand headlamp, pedestrians are more easily recognized.

5. References

[1] Rumar K 1990 The basic driver error: late detection. *Ergonomics*, 32, **1281-90**

[2] Sullivan J M and Flannagan M J 2002 The role of ambient light level in fatal crashes: inferences from daylight saving time transitions. *Accident Analysis and Prevention*, 34, **487-498**

[3] Summala H and Mikkola T 1994 Fatal accidents among car and truck drivers: effects of fatigue, age, and alcohol consumption. *Hum. Factors* 36 (2), **315–326**

[4] Owens D A and Sivak M 1996 Differentiation of visibility and alcohol as contributors to twilight road fatalities. *Human Factors*, 38, **680-689**

[5] Chrysler S T, Danielson S M and Kirby VM 1996 Age differences in visual abilities in nighttime driving field conditions. *Proc Human Factors Ergonomics Soc* 40th Meeting, **923-927**

[6] Tolea B, Antonya C and Beles H 2015 Assessment of the injury severity of the pedestrian lower limbs at the collision with a vehicle. *Proc of the Annual Session of Scientific papers "IMT-Oradea 2015"* XIV, **189-192**

[7] Brooks J O, Tyrell R A and Olson H M 2001 Pedestrian visibility at night: evaluating a method of teaching young drivers to be safer pedestrians. *80th Annual Meeting of the Transportation Research Board, Washington*

[8] Gegenfurtner K R, Mayser H and Sharpe L T 1999 Seeing movement in the dark. *Nature* 398, **475-476**

[9] Mannering F L and Bhat C R 2014 Analytic methods in accident research: methodological frontier and future directions. *Anal. Methods Accid. Res.* 1, **1-227**

IOP Conf. Series: Materials Science and Engineering **568** (2019) 012004 doi:10.1088/1757-899X/568/1/012004

Focusing device based on overconstrained mechanism

L D Racila[1], I Dumitru[1*], D Tutunea[1], L Matei[1], I Geonea[1], O Otat[1], M Racila[2]

[1] Faculty of Mechanics, University of Craiova, Romania
[2] Department of Applied Mathematics, University of Craiova, Romania

*dumitru_ilie@yahoo.com

Abstract. The paper presents an application based on a very special 6R overconstrained mechanism. This mechanism is started from 6R Wohlhart symmetric mechanism, not in its classical spatial position but in a different one, with three non-adjacent joints constrained to remain in a fixed plane. This disposition allows the conception of an interesting device, a focusing table, capable to obtain a good accuracy.

1. Introduction

The paper analyzes a mechanism with six revolute joints in a closed loop, namely Bricard rectangular mechanism or Wohlhart symmetric mechanism. Overconstrained mechanisms with revolute joints are composed by four, five or six joints. Sarrus [1] presents in 1853 a 6R closed-loop mechanism with six revolute joints, transforming a rotation movement in a translation one. Raoul Bricard described in his papers [2,3] six types of overconstrained linkages, three mobile octahedral mechanisms and three symmetric mechanisms. J.E. Baker analyse in 1980 these mechanisms [4], giving the closure equations for all these mechanisms. Karl Wohlhart [5] presents in 1987 a new overconstrained mechanism with six revolute, considered as a generalization of Bricard rectangular (or threefold symmetric) mechanism. Other researchers have studied these types of mechanisms, like Bennett [6], Myard [7] or Goldberg [8], Waldron [9], Mavroidis and Roth [10] or Dietmaier [11].

Overconstrained mechanisms have been studied by researchers, initially for kinematical point of view and in the last years from their application point of view. Many authors, like Schatz [12], Gan and Pellegrino [13], Chen [14] and more recently Ding [15] or Song [16] have presented possible applications of these mechanisms.

A new possible application of overconstrained mechanisms have been presented by Racila [17, 18], based on 6R Wohlhart symmetric mechanism, obtaining a 6R translational device. A generalization of this translational device was made it by Zeng [19], with a new design based on general forms of spatial overconstrained linkages.

The 6R translational device have three non-adjacent joints in a fixed horizontal plane, while the other three joints will be forced by the new design to remain in a plane, parallel with the horizontal plane

2. The new spatial disposition

In its classical position (figure 2a), the Bricard rectangular mechanism is analyzed with the help of Denavit-Hartenberg parameters [20]. In figure 1 is shown three revolute joints using Denavit and Hartenberg notations, where θ_i is the kinematical variable and a_i, α_i and d_i are geometrics parameters.

IOP Conf. Series: Materials Science and Engineering **568** (2019) 012004 doi:10.1088/1757-899X/568/1/012004

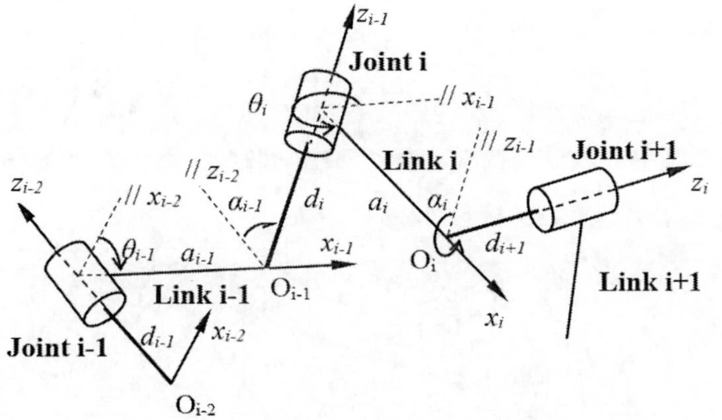

Figure 1. Denavit – Hartenberg parameters

The condition to obtain a closed loop mechanism is that the six transfer matrix product to be equal with the unity matrix:

$$^6Q_1 \cdot {}^1Q_2 \cdot {}^2Q_3 \cdot {}^3Q_4 \cdot {}^4Q_5 \cdot {}^5Q_6 = I \qquad (1)$$

Using D-H parameters, the input - output equation between the input angle θ and the output angle φ is [17]:

$$\cos\theta \cdot \cos\phi \cdot (1 + \cos^2\alpha) + (\cos\theta + \cos\phi) \cdot \sin^2\alpha - 2 \cdot \sin\theta \cdot \sin\phi \cdot \cos\alpha + \cos^2\alpha = 0 \qquad (2)$$

In equation (2) α is the twist angle.

In the new spatial disposition (figure 2b), three non-adjacent joints are constrained to have movements in a horizontal plane to the centre of their circumscribed circle. The horizontal plane is supposed to be a fixed one. The others three joints will describe a plane, parallel with the horizontal one.

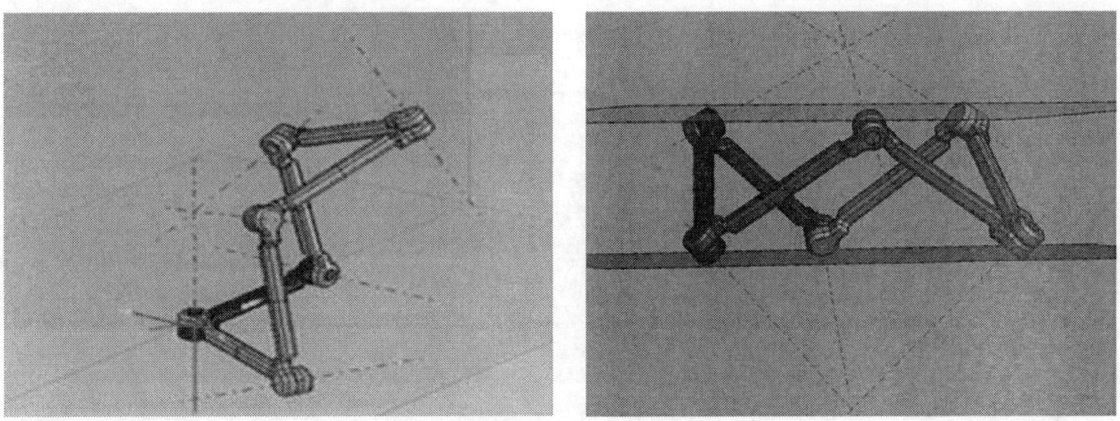

a) b)

Figure 2. The 6R mechanism in classical position (a) and the new spatial disposition (b)

The new design for the mechanism is considering (figure 3) and in the initial position all the six joints are in the horizontal plane and the input angle θ has zero degrees. The actuated joint is the revolute joint O_1, situated in the initial position in O_1^1. When O_1 is actuated, all the odd joints O_1, O_3 and O_5 make movements to the centre O of circumscribed circle of the triangle $O_1O_3O_5$. The even joints O_2, O_4 and O_6 will be situated in a parallel plane with the horizontal and fixed plane. The distance between the two planes is noted h and this distance depend to the input angle θ and the input distance b.

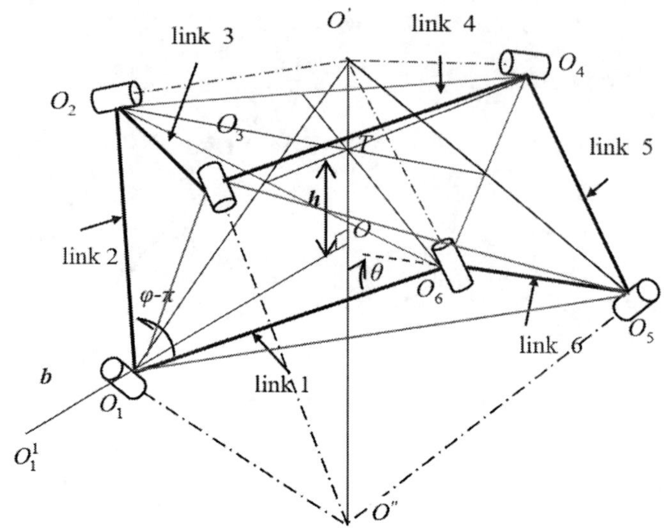

Figure 3. The the new spatial disposition in an intermediary position

This variation of the distance h between the horizontal plane (135) and the mobile plane (246) according to the input displacement b is [18]:

$$h = a \cdot \sqrt{\frac{4}{3} \cdot \left(1 - \frac{\left(2a - b\sqrt{3}\right)^2}{4a^2}\right) \cdot \left(1 - \frac{a^2}{\left(2a - b\sqrt{3}\right)^2}\right)} \tag{3}$$

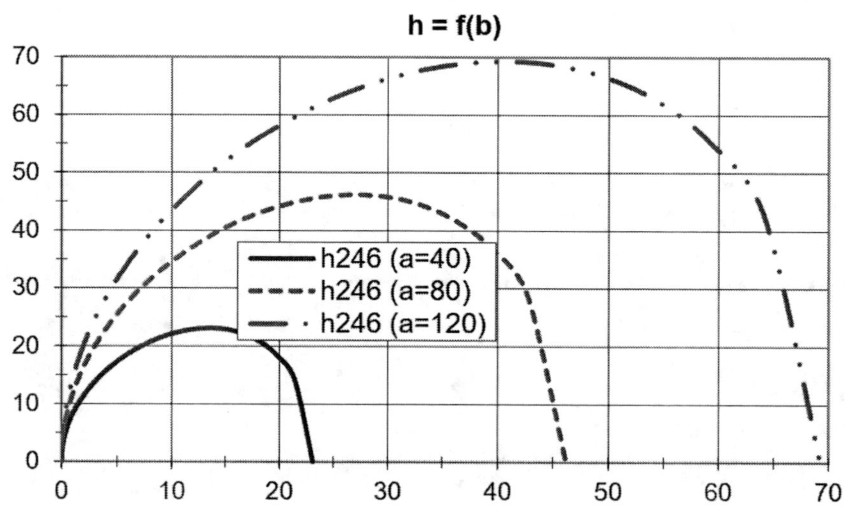

Figure 4. The variation of the height h according to the input distance b

This variation of the distance h between the horizontal plane and the mobile one, according to the input displacement b is given in figure 4 for three links length.

IOP Conf. Series: Materials Science and Engineering **568** (2019) 012004 doi:10.1088/1757-899X/568/1/012004

3. Focusing table

Starting from this new design, an interesting device can be obtained. The cross point of even joints axes noted O' and the cross point of odd joints axes, noted O", form a straight line O'O" always orthogonal to the two planes, the horizontal fixed plane and the mobile one.

For a twist angle $\alpha=\pi/2$, the initial position is defined by the all six links in the horizontal plane. In this position the point O' is to infinity because the axes of the odd joints are parallel. The O" point is situated in the same horizontal plane, with all six revolute joints. When the input parameter b increases the point O' will move to the horizontal plane and the point O" will move under the horizontal plane, arriving finally to infinity, when all six joints will be in the horizontal plane, but in a different position.

This property can be utilized in a opposite direction, when the input movement is the displacement of the cross point O" and the output parameter is the distance between the horizontal (135) and the mobile plane (246).

To obtain this dependence, we start from the input-output equation of Bricard rectangular mechanism, with a twist angle $\alpha=\pi/2$, a particular case of equation (2):

$$\cos\theta + \cos\phi + \cos\theta \cdot \cos\phi = 0 \quad (4)$$

We note by z the distance OO''. From figure 3 it can be written:

$$(OO'')^2 = z^2 = a^2 \cdot \cos^2\frac{\theta}{2} \cdot \left(\frac{1}{\sin^2\frac{\theta}{2}} - \frac{4}{3}\right)^2 \quad (5)$$

and finally:

$$\cos^2\frac{\theta}{2} = \frac{a^2 - 3\cdot z^2 + \sqrt{(a^2 - 3\cdot z^2)^2 + 48\cdot a^2\cdot z^2}}{8\cdot a^2} \quad (6)$$

We note m(z) the expression:

$$m(z) = a^2 - 3\cdot z^2 + \sqrt{(a^2 - 3\cdot z^2)^2 + 48\cdot a^2\cdot z^2} \quad (7)$$

The expression of the height h between the two planes according to the distance z is:

$$h = \sqrt{\frac{[m(z) - 2\cdot a^2]\cdot[8\cdot a^2 - m(z)]}{6\cdot m(z)}} \quad (8)$$

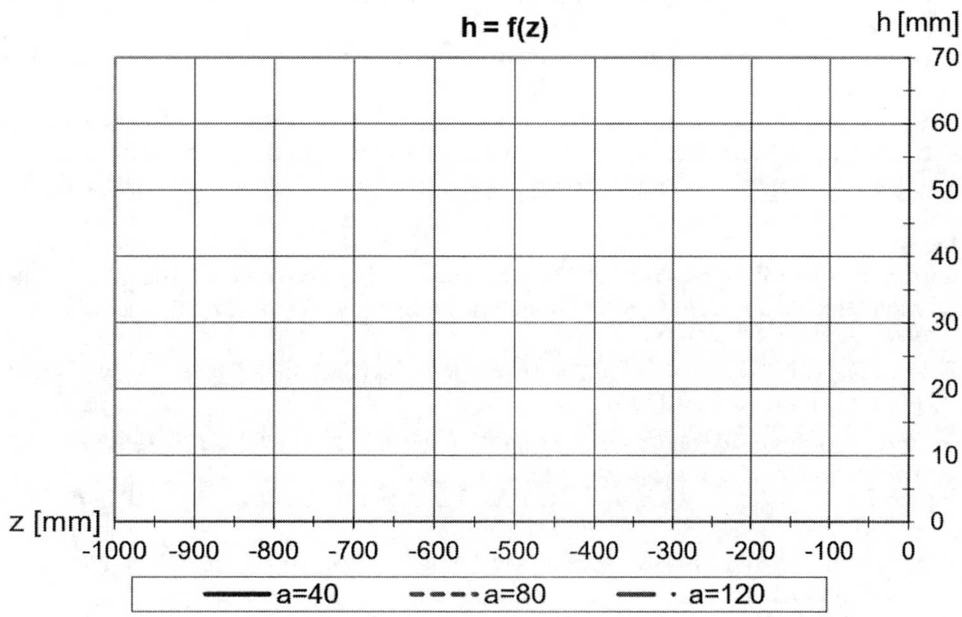

Figure 5. Variation of the height h according to z for three links length

IOP Conf. Series: Materials Science and Engineering **568** (2019) 012004 doi:10.1088/1757-899X/568/1/012004

In figure 5 is shown the variation of the distance OO", noted h, according to the input distance z, for three links length. A similar analyze can be obtained for different twist a variation twist angle, for example $\alpha=2\pi/3$ and $\alpha=5\pi/6$.

A virtual model of this device has been realized, confirming the accuracy of the analytical calculus. This accuracy can vary from 1:100 to 1:1000, depending of the links length and O"O variation domain.

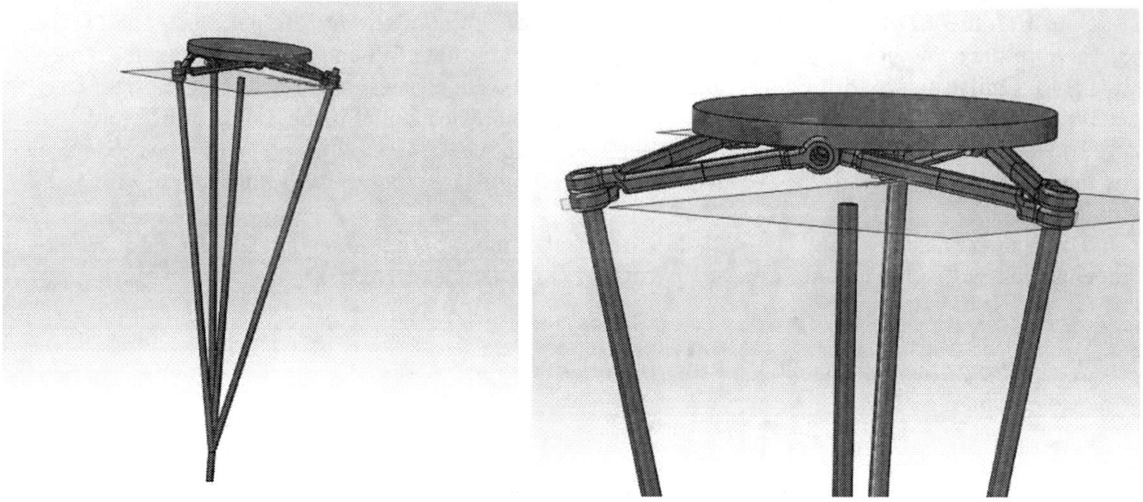

Figure 6. Virtual model of focusing device

4. Conclusion

The paper presents a 6R overconstrained closed-loop mechanism, not in its classical position, with one fixed element, but in a new spatial disposition, with no fixed elements. This special spatial position is used to obtain a new device, with three non adjacent joints imposed to make a particular movement in a fixed plane.

This device can be used to develop many applications, in the industrial area, or even in the entertainment domain. With the actuator placed in the cross point of the three nonadjacent joints of fixed plane, this device can be used, for example, to realize a focusing table for optical microscopes, or as plate table in rectifying machines, due to the its accuracy in this particular position. Some other devices are can be imagined, depending on the domain in which they are intended to be used.

References

[1] Sarrus P F 1853 Note sur la transformation des mouvements rectilignes alternatifs, en mouvements circulaires et réciproquement, *Comptes Rendus des Séances de l'Académie des Sciences*, vol 36, p 1036-1038

[2] Bricard R . 1897 Mémoire sur la théorie de l'octaèdre articulé, *Journal de Mathématiques pures et appliquées*, vol 3, p 113–148

[3] Bricard R. 1927 *Leçons de Cinématique, Tome II – Cinématique Appliquée*, ed Gauthier (Villars, Paris)

[4] Baker J E 1980 An Analysis of the Bricard Linkages, *Mechanism and Machine Theory*, vol 15, p. 267–286

[5] Wohlhart K 1987A New 6R Space Mechanism, *Proc. of the 7th IFToMM World Congress*, (Sevilla), p 193–198

[6] Bennett G T 1903 A new Mechanism, *Engineering* vol 76, p 777–778

[7] Myard F E 1931 Chaîne fermée à cinq couples rotoïdes, déformable au premier degré de liberté, *Comptes Rendus Hebdomadaires des Séances de l'Académie des Sciences*, vol 192, p

1352-1354
[8] Goldberg M 1943 New five–bar and six–bar linkages in three dimensions, *Transaction of the A.S.M.E.*, vol 65, p 649–663
[9] Waldron K J 1979 Overconstrained Linkages, *Environment and Planning B*, vol 6, p 393-402
[10] Mavroidis C and Roth B 1994 Analysis and Synthesis of Overconstrained Mechanisms, *Proc. of the ASME Design Technical Conferences*, vol. DE-70, p 115-133 (Minneapolis)
[11] Dietmaier P 1995 A new 6R space mechanism, *Proc of the 9th IFToMM World Congress*, (Milano) p. 52–56
[12] Schatz P 1975 *Rhytmusforschung und Technik* Ed Verlag Freies Geistesleben (Stuttgart)
[13] Gan W W and Pellegrino S 2003 Closed-loop deployable structures, *Proc. of 44th AIAA /ASME /ASCE /AHS /ASC Structures, Structural Dynamics, and Materials Conference*, (Norfolk)
[14] Chen Y, You Z and Tarnai T 2005 Threefold-symmetric Bricard Linkages for Deployable Structures, *International Journal of Solids and Structures*, vol 42 p 2287-2301
[15] Ding W, Qiang R, Yao Y 2016 Design and Locomotion Analysis of a Novel Deformable Mobile Robot with Two Spatial Reconfigurable platforms and Three Kinematic Chains *Proc. of the Institution of Mechanical Engineers, Part C Journal of Mechanical Engineering Science*, vol 231 issue 8, p 1481-1499
[16] Song X, Deng Z, Guo H, Liu R, Li L and Liu R 2017 Networking of Bennett linkages and its application on deployable parabolic cylindrical antenna, *Mechanism and Machine Theory*, vol 109, p 95–125
[17] Racila L and Dahan M 2010 Spatial properties of Wohlhart symmetric mechanism, *Meccanica*, vol 45, p 153-165
[18] Racila L and Dahan M 2011 6R Parallel Translational Device *Proc. of the 13th IFToMM World Congress*, (Guanajuato)
[19] Zeng Q and Ehmann K F 2014 Design of parallel hybrid-loop manipulators with kinematotropic property and deployability, *Mechanism and Machine Theory*, vol. 71, p 1-26
[20] Denavit J and Hartenberg R S 1955 A Kinematic Notation for Lower–Pair Mechanism Based on Matrices, *Journal of Applied Mechanics*, vol 22, p 215–221

IOP Conf. Series: Materials Science and Engineering **568** (2019) 012005 doi:10.1088/1757-899X/568/1/012005

Organizational factors that influence job satisfaction in a company from automotive industry

G M Moraru[1] and D Popa[1]

[1] "Lucian Blaga" University of Sibiu, Engineering Faculty, 4, Emil Cioran Street, Sibiu, 550025, Romania

E-mail: gina.moraru@ulbsibiu.ro

Abstract. The aim of this paper is to present the most important organizational factors that influence job satisfaction in many companies from automotive industry. Therefore, we have firstly analyzed several theoretical aspects in the specialty literature regarding the factors that influence employee satisfaction. Secondly, we have developed a research based on questionnaire in a branch plant of an international organization in automotive industry. There are many factors that place the automotive industry ahead of other areas of activity in terms of maintaining high work satisfaction among employees. We have found which of these factors have a great impact on job satisfaction in the studied branch plant: the lunch vouchers received by employees, the microclimate at work (light, silence, lack of emissions etc.), the relationship with hierarchical boss, the work schedule (duration and time interval), the organization's support in solving personal problems, the clarity of the received tasks and others.

1. Introduction

In the last half century, many specialists have focused on Herzberg's theory to explain work satisfaction [1-4]. Several surveys have concluded that intrinsic motivational factors are more strongly related to the concepts of satisfaction and dissatisfaction than the extrinsic motivational factors [2].

In the last decade, specialists have identified ten organizational variables that influence employee satisfaction [5]:

- Organization development;
- Policies of compensation and benefit;
- Promotion and career development;
- Job satisfaction;
- Job security;
- Working environment and condition;
- Relationship with the supervisor;
- Work group;
- Leadership styles;
- Other factors, like: feedback and encouragement, feeling of belonging to the group, use of high technology at work.

On one hand, we have found that, in part, these variables are interdependent. By example, besides financial rewards, there are many other factors that influence job satisfaction, like: good working conditions, relationship between supervisors and subordinates, availability of effective human resources structures, efficient dissemination of information from management to subordinates, and achieving challenging goals [6]. On the other hand, we have noticed that some specialists have emphasized the relationship between job satisfaction and organizational performances [7, 8], and other have been concerned on the employee psychological well-being, as emotional factor affecting his/her work performance [9, 10]. An important conclusion drawn from the specialty literature is that income and other material rewards are in most cases unimportant to happiness [11].

Taking into account all considerations above, we have developed our selective research based on simple questions and easy to understand by respondents, choosing only the most important factors that influence the job satisfaction in the automotive industry.

2. Research in a branch plant of an international organization in the automotive industry

We have conducted the research regarding the organizational factors that influence job satisfaction in a branch plant of an international organization in automotive industry in Sibiu County, Romania, in which the organizational culture is based on respect, fidelity and cooperation. Even this branch plant has been founded expressly for production activities, in the last years its top management has also developed here various research activities.

One of the reasons we have chosen this factory was the discover that we have made in a previous research [12] that its management has developed here a “culture of communication”, in the sense proposed by specialists: communication builds trust, stimulates employees to bring creative suggestions, helps management in decision making, stimulates engagement and better performance in the changing scenario [13].

2.1. Methodology

We have developed a research based on questionnaire. We have selected 100 respondents from all compartments, covering all jobs and occupational levels.

The most important group of questions was related to five groups of organizational factors affecting job satisfaction: general labor conditions, financial motivations, hierarchical relations, profession, and workplace climate. At each analyzed organizational factor, respondents have had to choose the variant corresponding to the importance of respective factor (“not important”, “minor importance”, “important”, or “very important”) and the variant corresponding to their own satisfaction with that factor (“not at all satisfied”, “slightly satisfied”, “satisfied”, or “very satisfied”).

We have noted the importance of a factor with I and have evaluated this importance using the following scale: 0 points for the variant “not important”, 1 point for “minor importance”, 2 points for “important” and 3 points for “very important”. The respondents' satisfaction was noted with S and was evaluated with a similar scale: 0 points for “not at all satisfied”, 1 point for “slightly satisfied”, 2 points for “satisfied” and 3 points for “very satisfied”.

The final score F obtained for a factor was determined with the relation:

$$F = I \times S \tag{1}$$

So the final score can take values between 0 and 9. In our statistical analysis, we have noted for each factor with $\bar{I}$, $\bar{S}$, respective $\bar{F}$ the mean of importance, of satisfaction and of final score. The variance of responses collected for a factor was noted with $Var(I)$ for importance, $Var(S)$ for satisfaction and $Var(F)$ for final score. Taking into account that I and S are independent variables, we have used the variance of final score the following relation [14]:

$$Var(F) = Var(I) \times Var(S) + Var(I) \times \bar{S}^2 + Var(S) \times \bar{I}^2 \tag{2}$$

2.2. Results

The first analyzed group of organizational factors that affect job satisfaction has included four general labor conditions: work schedule (duration and time interval), breaks during working hours, microclimate at work (light, silence, lack of emissions etc.), and opportunities for career development (figure 1).

IOP Conf. Series: Materials Science and Engineering **568** (2019) 012005 doi:10.1088/1757-899X/568/1/012005

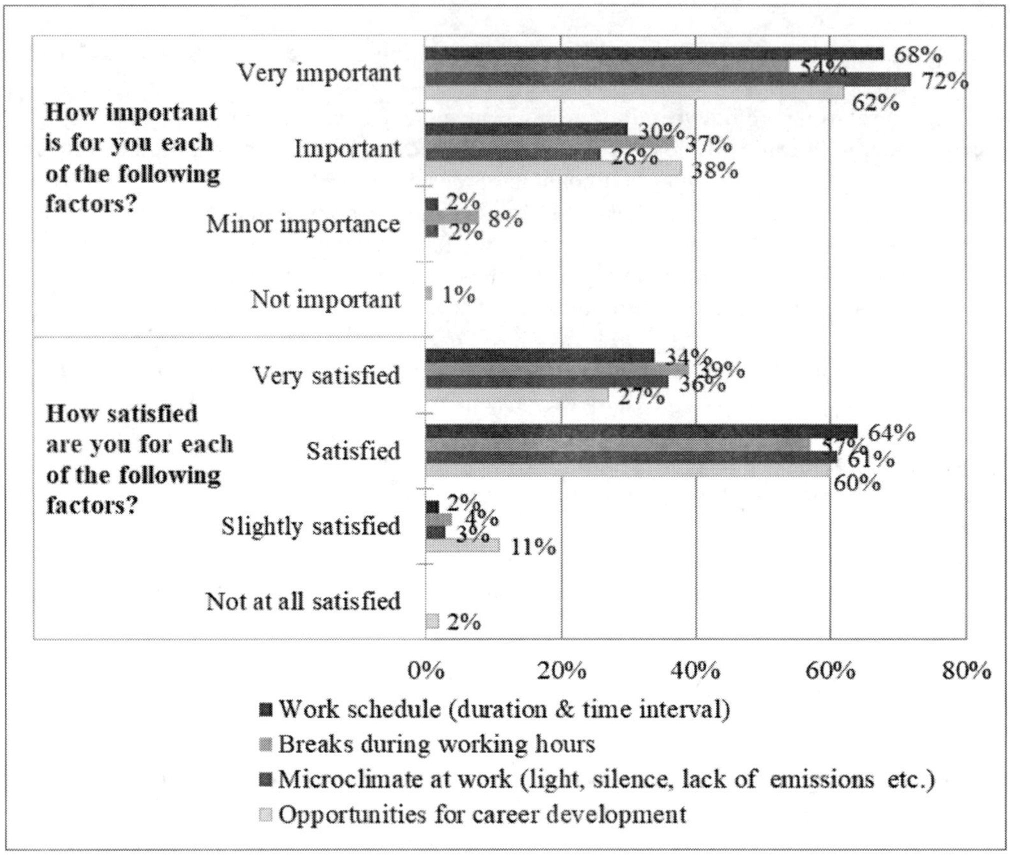

Figure 1. Importance of general labor conditions and employees' satisfaction related to them.

Table 1 contains statistical parameters obtained for the organizational factors from the group of general labor conditions. Responses regarding all these factors are homogenous, homogeneity coefficients being under 0.3 or a little over 0.3. We observe that microclimate at work and work schedule are real strengths for the automotive industry (mean 6.291, respective 6.171).

Table 1. Statistical parameters for the general labor conditions.

The factor	Importance *(I)*				Satisfaction *(S)*				Final score *(F)*			
	Mean	Variance	St. Dev.[a]	HC.[b]	Mean	Variance	St. Dev.[a]	HC.[b]	Mean	Variance	St. Dev.[a]	HC.[b]
Work schedule	2.66	0.264	0.514	0.193	2.32	0.258	0.508	0.219	6.171	3.314	1.820	0.295
Breaks	2.44	0.466	0.683	0.280	2.35	0.308	0.555	0.236	5.734	4.550	2.133	0.372
Microclimate	2.70	0.250	0.500	0.185	2.33	0.281	0.530	0.228	6.291	3.477	1.865	0.296
Career opportunities	2.62	0.236	0.485	0.185	2.12	0.446	0.668	0.315	5.554	4.223	2.055	0.370

[a] St. Dev. = Standard deviation
[b] HC. = Homogeneity coefficient

IOP Conf. Series: Materials Science and Engineering **568** (2019) 012005 doi:10.1088/1757-899X/568/1/012005

The second group of organizational factors that affect job satisfaction has included several financial motivations (figure 2). Looking at statistical analysis (table 2), we observe that lunch tickets /vouchers surprise us with the highest mean (almost 7) of the factors list.

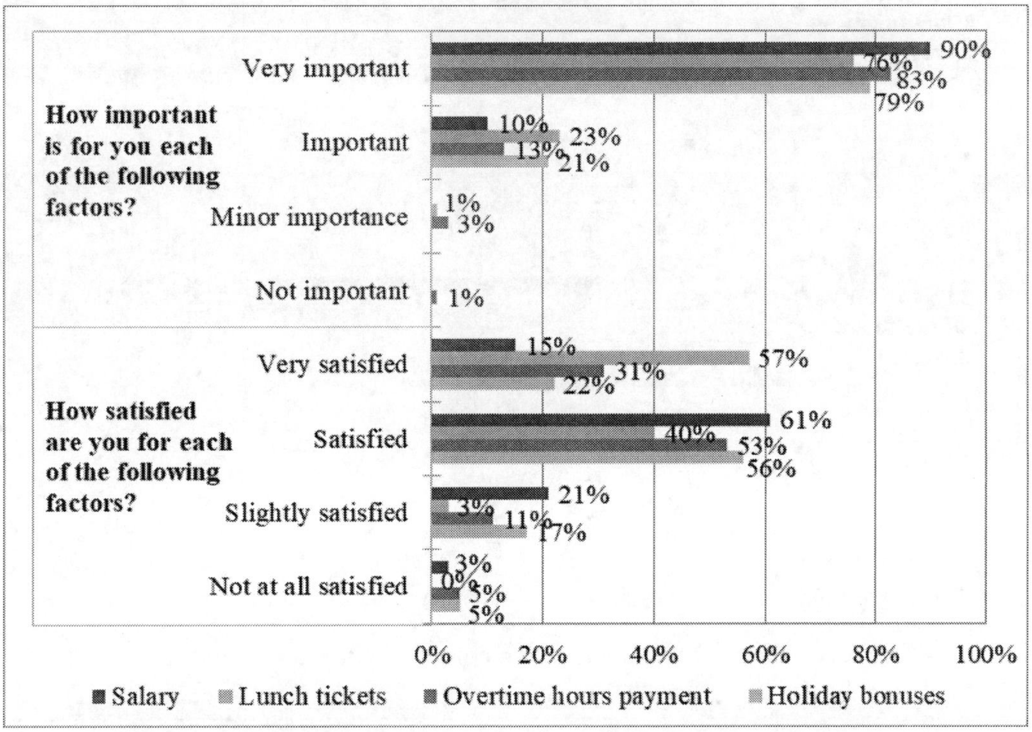

Figure 2. Financial motivations and employees' satisfaction related to them.

Table 2. Statistical parameters for financial motivations.

The factor	Importance *(I)*				Satisfaction *(S)*				Final score *(F)*			
	Mean	Variance	St. Dev.[a]	HC.[b]	Mean	Variance	St. Dev.[a]	HC.[b]	Mean	Variance	St. Dev.[a]	HC.[b]
Salary	2.90	0.090	0.300	0.103	1.88	0.466	0.682	0.363	5.452	4.276	2.068	0.379
Lunch tickets	2.75	0.208	0.456	0.166	2.54	0.308	0.555	0.219	6.985	3.735	1.933	0.277
Overtime hours payment	2.78	0.292	0.540	0.194	2.10	0.610	0.781	0.372	5.838	6.178	2.486	0.426
Holiday bonuses	2.79	0.166	0.407	0.146	1.95	0.588	0.766	0.393	5.441	5.301	2.302	0.423

[a] St. Dev. = Standard deviation
[b] HC. = Homogeneity coefficient

In a similar manner, we have analyzed four factors that define the hierarchical relations of an employee (figure 3) and four factors regarding the importance of profession for employees (figure 4). Looking at figures 3 and 4, we observe that most of respondents have appreciated these factors being very important, according to maximum on the evaluation scale (3 points), but they are only satisfied of these factors (2 points on the evaluation scale).

IOP Conf. Series: Materials Science and Engineering **568** (2019) 012005 doi:10.1088/1757-899X/568/1/012005

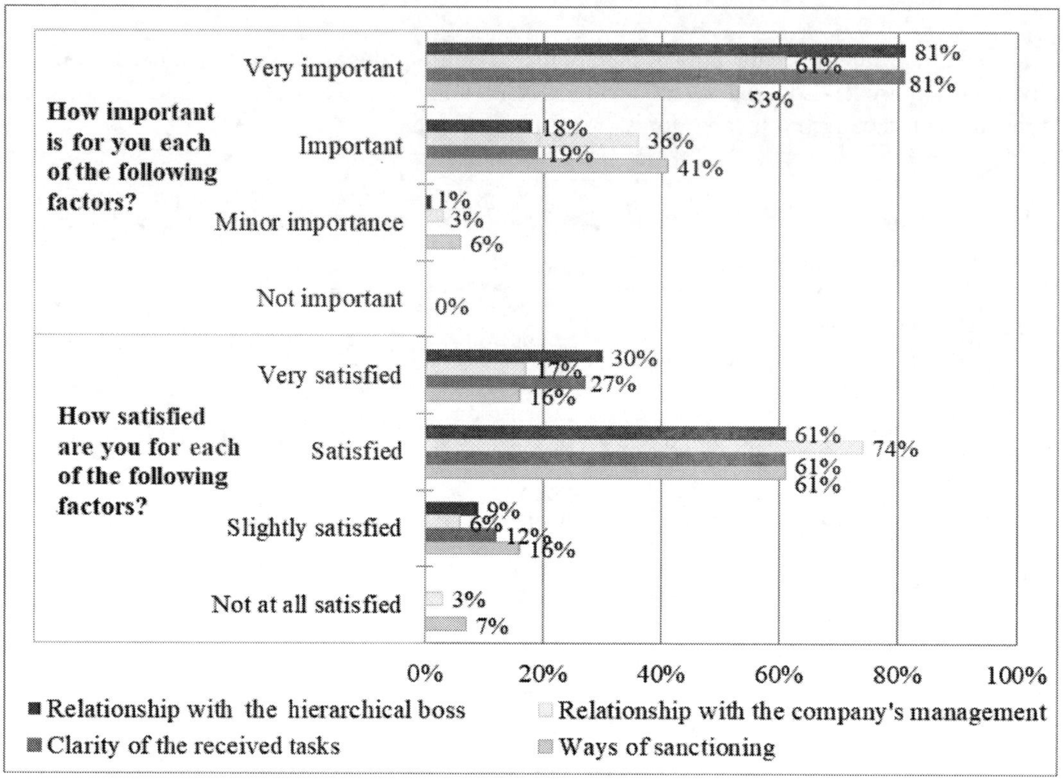

Figure 3. Hierarchical relations and employees' satisfaction related to them.

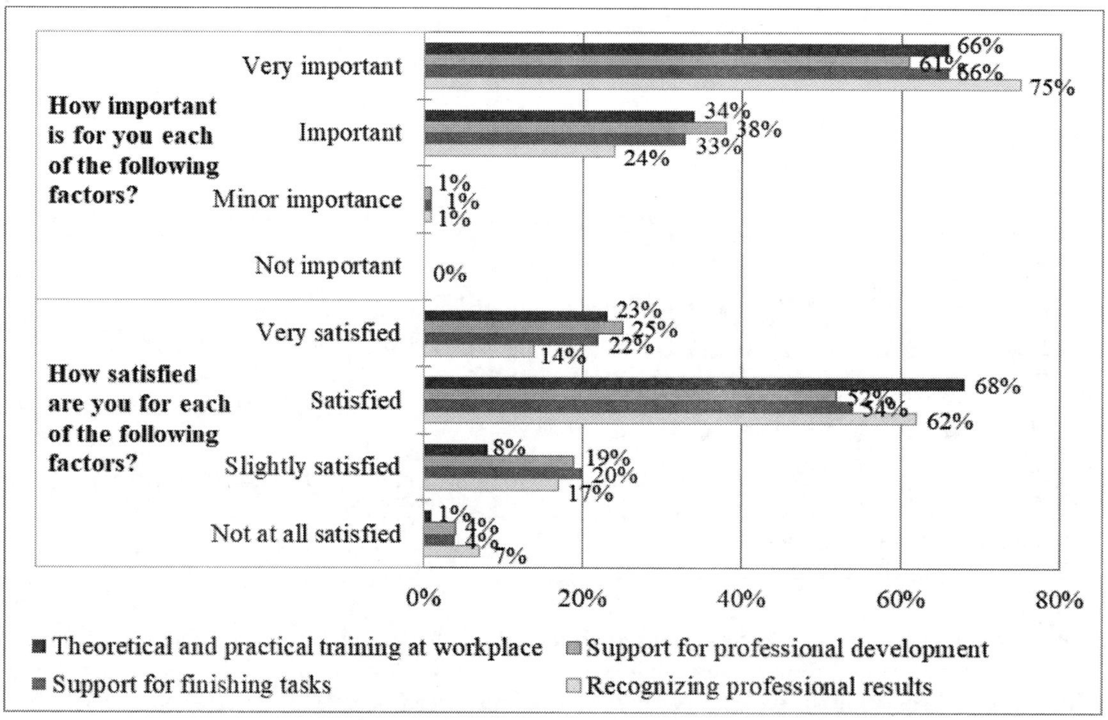

Figure 4. Factors regarding the importance of profession for employees.

IOP Conf. Series: Materials Science and Engineering **568** (2019) 012005 doi:10.1088/1757-899X/568/1/012005

The statistical analysis of hierarchical relations (table 3) shows that relationship with hierarchical boss is on the first place in this group of factors (mean of final score being 6.188), and clarity of the received tasks is on the second (mean 6.042). Therefore we affirm that a strong point of the branch plant is the way in which the right people are chosen as operational managers. It seems that these people have solid professional knowledge, high communication abilities and the ability to inspire trust in employees, items that are of great importance to employees.

Table 3. Statistical parameters for hierarchical relations.

The factor	Importance *(I)*				Satisfaction *(S)*				Final score *(F)*			
	Mean	Variance	St. Dev.[a]	HC.[b]	Mean	Variance	St. Dev.[a]	HC.[b]	Mean	Variance	St. Dev.[a]	HC.[b]
Relationship with the boss	2.80	0.180	0.424	0.152	2.21	0.346	0.588	0.266	6.188	3.653	1.911	0.309
Relationship with the top management	2.58	0.304	0.551	0.214	2.05	0.348	0.589	0.288	5.289	3.694	1.922	0.363
Clarity of the received tasks	2.81	0.154	0.392	0.140	2.15	0.368	0.606	0.282	6.042	3.670	1.916	0.317
Ways of sanctioning	2.47	0.369	0.608	0.246	1.86	0.580	0.762	0.410	4.594	5.032	2.243	0.488

[a] St. Dev. = Standard deviation
[b] HC. = Homogeneity coefficient

Table 4 summarizes the statistical analysis of the factors regarding the importance of profession for employees. The training system is the first at this category, but the mean of 5.666 is lower than means obtained by many factors from the other three categories above analyzed (see tables 1, 2, and 3).

Table 4. Statistical parameters for the factors regarding the importance of profession.

The factor	Importance *(I)*				Satisfaction *(S)*				Final score *(F)*			
	Mean	Variance	St. Dev.[a]	HC.[b]	Mean	Variance	St. Dev.[a]	HC.[b]	Mean	Variance	St. Dev.[a]	HC.[b]
Training	2.66	0.224	0.474	0.178	2.13	0.333	0.577	0.271	5.666	3.450	1.857	0.328
Support for finishing tasks	2.60	0.260	0.510	0.196	1.98	0.600	0.774	0.391	5.148	5.228	2.287	0.444
Support for development	2.65	0.248	0.497	0.188	1.94	0.576	0.759	0.391	5.141	5.122	2.263	0.440
Recognizing results	2.74	0.212	0.461	0.168	1.83	0.561	0.749	0.409	5.014	5.043	2.246	0.448

[a] St. Dev. = Standard deviation
[b] HC. = Homogeneity coefficient

Responses regarding the last group of organizational factors affecting job satisfaction – the workplace climate – are sketched in figure 5. The statistical analysis related to them is presented in table 5.

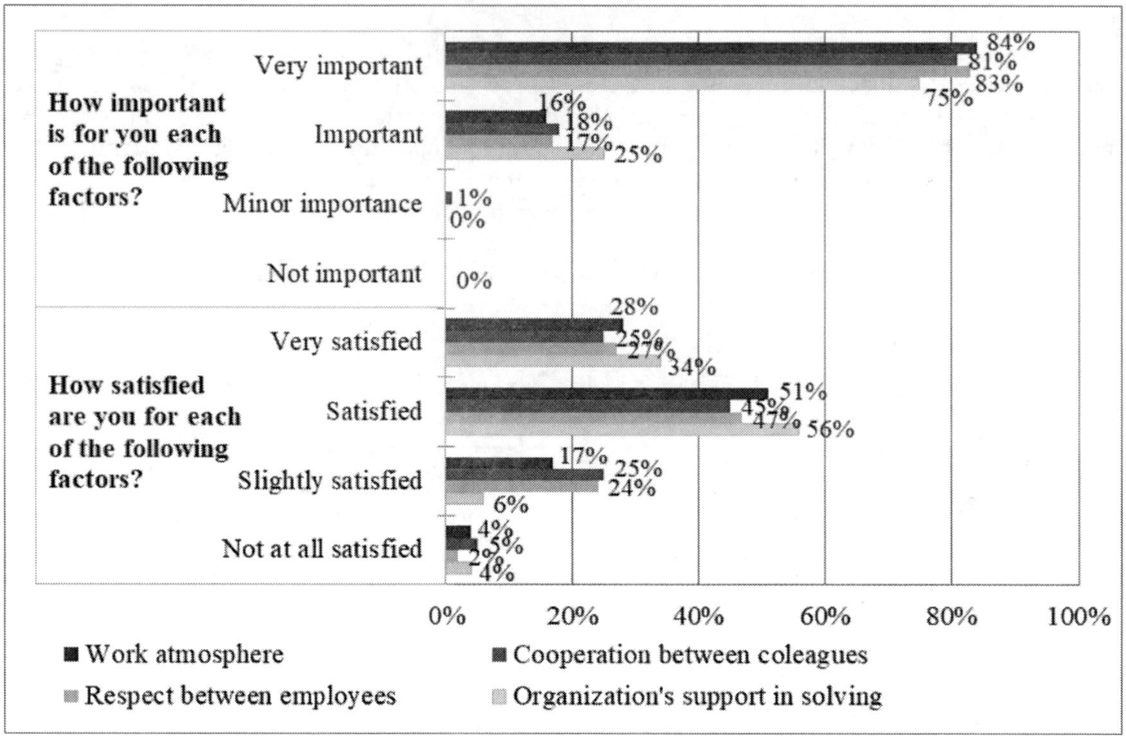

Figure 5. Workplace climate.

Table 5. Statistical parameters for factors related to the workplace climate.

The factor	Importance *(I)*				Satisfaction *(S)*				Final score *(F)*			
	Mean	Variance	St. Dev.[a]	HC.[b]	Mean	Variance	St. Dev.[a]	HC.[b]	Mean	Variance	St. Dev.[a]	HC.[b]
Work atmosphere	2.84	0.134	0.367	0.129	2.03	0.609	0.780	0.384	5.765	5.548	2.356	0.409
Cooperation	2.79	0.148	0.385	0.138	1.90	0.690	0.831	0.437	5.301	6.008	2.451	0.462
Respect between employees	2.83	0.141	0.376	0.133	1.99	0.590	0.768	0.386	5.632	5.366	2.317	0.411
Support in solving personal problems	2.75	0.188	0.433	0.157	2.20	0.520	0.721	0.328	6.050	4.938	2.222	0.367

[a] St. Dev. = Standard deviation
[b] HC. = Homogeneity coefficient

The support received by employees for solving personal problems has the highest score in this group (table 5). Compared to the other groups of factors above analyzed, the answers related to the workplace climate had a rather high degree of inhomogeneity, the homogeneity coefficient having values even above 0.4 (table 5), similar to the statistical results from the importance of profession (table 4).

IOP Conf. Series: Materials Science and Engineering **568** (2019) 012005 doi:10.1088/1757-899X/568/1/012005

3. Conclusions

The most of respondents have appreciated the organizational factors affecting job satisfaction as being very important, but they were only satisfied with them, and too few were very satisfied. So, one conclusion is the fact that managers knows to choose the proper factors for increase the job satisfaction, but they have still to work regarding the quality of these factors in their compartments.

The factors that have a great impact on job satisfaction in the studied branch plant, with a mean of final score above 6, are: the lunch vouchers received by employees on the first place, the microclimate at work on the second place, the relationship with hierarchical boss on the third place, the work schedule on the fourth place, the organization's support in solving personal problems on the fifth place, and the clarity of the received tasks on the sixth place (tables 1-5). We mention that we have taken into account at this top both the importance of the factors and the way in which those factors satisfy employees in the studied branch plant. Based on the study presented in this paper, we intend to expand our research to other companies in the automotive industry, but also to other industries.

References

[1] Bexheti L 2016 The Impact of Herzberg's Two Factor Theory and Efficiency at Work *European Journal of Multidisciplinary Studies* Vol 1 Nr 2 pp 379-386

[2] Ewen R B Smith P C Hulin C L 1966 An empirical test of the Herzberg two-factor theory *Journal of Applied Psychology* Vol 50(6) pp 544-550

[3] Graen G B 1966 Motivator and hygiene dimensions for research and development engineers *Journal of Applied Psychology* Vol 50(6) pp 563-566

[4] Pardee R L Motivation 1990 Theories of Maslow, Herzberg, McGregor & McClelland *A Literature Review of Selected Theories Dealing with Job Satisfaction and Motivation* (U S Department of Education: Educational Resources Information Center - ERIC)

[5] Sageer A Rafat S Agarwal P 2012 Identification of Variables Affecting Employee Satisfaction and Their Impact on the Organization *IOSR Journal of Business and Management* Vol 5 Issue 1 pp 32-39

[6] Sarvar S Abugre J 2013 The Influence of Rewards and Job Satisfaction on Employees in the Service Industry *The Business & Management Review* Vol 3 Number 2 Retrieved from http://www.abrmr.com/myfile/best_track/best_track_50317.pdf on April 10 2019

[7] Raziq A Maulabakhsh R 2015 Impact of Working Environment on Job Satisfaction *2nd Global Conf. on Business, Economics, Management and Tourism (Oct 2014)* Procedia Economics and Finance 23 (Prague: Czech Republic/ Elsevier) pp 717-725

[8] Siggins J A 1993 Job Satisfaction and Performance in a Changing Environment *Library Trends* vol 41 No 2 Fall 1992 pp 299-315 (The Board of Trustees: University of Illinois)

[9] Wright T A Cropanzano R 2000 Psychological Well-Being and Job Satisfaction as Predictors of Job Performance *Journal of Occupational Health Psychology* Vol 5 No 1 pp 84-94 (Educational Publishing Foundation

[10] Chartered Institute of Personnel and Development U K 2016 Moving the employee well-being agenda forward *A collection of thought pieces* Reference 7191 (CIDP)

[11] Judge T A Piccolo R F Podsakoff N P Shaw J C Rich B L 2010 The relationship between pay and job satisfaction: a meta-analysis of the literature *Journal of Vocational Behavior* 77 pp 157-167 (Elsevier)

[12] Limbăşan G (author) Moraru G M (coordinator) 2012 The organizational culture of a firm *Bachelor's degree in engineering (The economic part)* unpublished (Sibiu: Romania/"Lucian Blaga" University of Sibiu)

[13] Rajhans K 2012 Effective Organizational Communication: a Key to Employee Motivation and Performance *Interscience Management Review (IMR)* vol 2 Issue 2 pp 81-85

[14] Butănescu-Volanin R C 2015 *Probabilities and mathematical statistics with applications in economics* (Sibiu: Romania/"Lucian Blaga" University of Sibiu Pub. House)

IOP Conf. Series: Materials Science and Engineering **568** (2019) 012006 doi:10.1088/1757-899X/568/1/012006

Bush chain friction depending on speed and influence of centrifugal effect

R Velicu

Transilvania University of Brasov, Dept. of Product Design Mechatronics and Environment

E-mail: rvelicu@unitbv.ro

Abstract. Evaluation of friction in chain drives is the first step in the process of reducing the friction. This paper presents the results of experimental evaluation of friction in a bush chain considering only the friction in the joints of the chain (pin-bush joints and chain-sprocket joints) and not the friction in bearings. Friction is presented as friction torque, measured for a basic chain drive with transmission ratio equal to one, depending on speed. The influence of the centrifugal effect on friction is also analysed.

1. Introduction

The importance of reducing friction in automobiles transmission is as high as possible in the context of the regulations imposed by the 20/20/20 goals from the Directive for Energy Efficiency of the European Union (2012/27/UE), imposing targets for year 2020, like: 20% reduction of CO2 emissions compared to 1990 levels and 20% reduction of final energy consumptions. In the same directions have been moved the goals of automotive industry for year 2020. American standards CAFE impose a one third reduction of fuel consumption, relative to year 2008. European Commission set a goal for 2021 of average CO_2 emissions of 95 grams CO_2 / km for the whole fleet produced in 2020.

One of the main applications for bush chains is in the timing system of combustion engines, where tensioning and speed reach some of the extreme values of the field of use. Reducing friction in timing chains is an important goal since, according to [1, 2, 3], the part of the fuel energy dedicated to overcome friction in the engine system is about 15% (7–18%) of total losses.

Very few experimental researches on chain friction have been published. A procedure for measuring losses in the timing chain and the loss in the guides of an engine, using equipment consisting in a full engine is presented in [3]. The contribution of chain friction and guide friction on the global timing system friction are only presented for one oil temperature and one value of rotational speed. Friction between chain and guide is subject of several papers referring to friction coefficients [4], friction contribution [5] or theoretical approaches [6].

At the Tribology Centre of Transylvania University of Brasov, a research on friction in chain drives, used for timing system of combustion engines, has been developed with the finance of Schaeffler Group. The procedure for measuring chain friction is presented in [7]. Friction losses are measured as friction torque depending on speed, tensioning and oil temperature.

A global friction torque in chain and bearings is measured in a first stage. In a second stage, the friction torque in bearings (sum of friction in input bearing box and output bearing box, at the same rotational speed and load) is measured in the same conditions [7]. Friction in chain results by extracting bearing friction from global friction.

$$T_{f\ chain} = T_{f\ chain+bearings} - T_{f\ bearings}. \tag{1}$$

In order to correctly evaluate the friction in the chain drive, all the influences must be considered. Since the tensioning force along the chain, also acting in the pin-bush joint, has great importance for the friction, the influence of the centrifugal effect on this force must be carefully studied.

The theoretical approach [8, 9] on the centrifugal effect on chain drives presents a component F_c, which is equilibrating the centrifugal force. The force F_c is creating a supplementary tensioning force on the whole length of the chain and, in the same time, it creates a release of the initial tensioning of the transmission. The value of F_c force is

$$F_c = \frac{m}{l} v^2, \tag{2}$$

where (m/l) is the mass of chain per unit of length and v is the linear speed of the chain.

A study of the influence of the stiffness of supports and chain on the effect of centrifugal force is presented in [10]. It shows that the force F_c divides in two components: one, with a value of $2\kappa_S F_c$, goes on release of the shafts; the second component, with a value of $\kappa_C F_c$ determines supplementary tensioning of the chain. The coefficients indicating how much of the F_c force goes to the chain tensioning (κ_C) and how much goes to the shafts releasing (κ_S) are [10]:

$$\kappa_C = \frac{c_C}{c_S + c_C};\ \kappa_S = \frac{c_S}{c_S + c_C}, \tag{3}$$

where c_C, respectively c_S, are the stiffness of chain (2 parallel arms), respectively of the supports.

This paper presents the results of experimental evaluation of friction in a bush chain considering only the friction in the joints of the chain and not the friction in bearings. Friction is presented as friction torque, measured for a basic chain drive with transmission ratio equal to one, depending on speed. The influence of the centrifugal effect is also analysed.

2. Theoretical approach on chain friction

Horovitz [8] presents the calculus relation of the mechanical work loss by friction L_f in the case of one pin-bush joint, for entering or exiting a sprocket, along a pitch angle $\varphi = 2\pi/z$ rad

$$L_f = \frac{\mu \pi d_3 F}{z}, \tag{4}$$

where μ is the friction coefficient in joint, z is the number teeth of the sprocket, d_3 is the diameter of the pin and F is the force in joint. In order to calculate the friction losses as the friction torque for a basic chain drive (transmission ration equal to one), the sum of mechanical work losses, in the four pin-bush joints entering and exiting the two identical sprockets, must be divided by the pitch angle

$$T_f = \frac{2\mu\pi d_3 (F_1 + F_2)}{z} / \frac{2\pi}{z} = \mu d_3 F, \tag{5}$$

where F_1 and F_2 are the tensioning forces on the two arms of the chain and F is their sum. In the absence of the centrifugal effect and neglecting the weight of the chain, the force F is loading the supports. As it can be seen from equation (5), the friction torque should be depending only on the pin-bush friction coefficient, diameter of the pin and tensioning force.

3. Experimental results on bush chain friction

The tests have been developed on the chain rig described in [7], based on the procedures presented in the same reference. The subject of the test is a bush chain B8, with 8 mm pitch, 64 links working with identical 23 teeth sprockets. Tensioning is created by automatic enlarging the centre distance, ensuring a constant tensioning force F_0 of 1 kN. This is the force constantly loading the two bearing boxes. The transmission is vertical with transmission ratio equal to one.

Lubrication of the chain is made with two jets of oil oriented to the points of chain entering the two sprockets. 5W30 Castrol Edge synthetic oil is used with a controlled temperature of 90 °C. The measured kinematic viscosity at 90 °C is $27 \cdot 10^{-6}$ m^2/s, resulting a $23 \cdot 10^{-3}$ Pas dynamic viscosity.

IOP Conf. Series: Materials Science and Engineering **568** (2019) 012006 doi:10.1088/1757-899X/568/1/012006

Four identical chains have been tested in the same conditions, measuring friction torque for a series of values of rotational speed *n* {500, 1000, 1800, 3000, 5000} rpm. Each chain has been previously subject of running in for 50 hours. Friction in each chain results by extracting measured bearing friction from measured global friction. The diagram showing measured global friction for the first tested chain C1, measured friction in bearings and the resulted friction in the first tested chain C1 is presented in figure 1. The values of friction torque from figure 1 are presented as percentage of the minimum chain friction torque resulted for 1800 rpm ($Tf\% = T_f \cdot 100/T_{f\,\mathrm{chain\ 1800\ rpm}}$).

Figure 2 presents the results for the tests on four identical B8 chains (C1, C2, C3 and C4) and an average of the measurements. The values of friction torque from figure 2 are presented as percentage of the minimum average chain friction torque resulted for 1800 rpm ($Tf\% = T_f \cdot 100/T_{f\,\mathrm{C\_avg\ 1800\ rpm}}$).

The analyses of the results presented in figure 2 allow us to draw few conclusions:

- Even looking identical chains, there are differences in measured friction up to 20%;
- The trend of all the friction vs. rotational speed curves is the same;
- A comparison with the theoretical model shows that the influence of speed on chain friction may act on modification of friction coefficient and/or on chain tensioning;
- Instead of constant friction vs. speed, there is a friction change with maximum 35% at a 10 times increasing speed (from 500 to 5000 rpm);
- Friction tend to diminish with increasing rotational speed at small rotational speed; this could be explained by boundary or mixed friction in the joints, following Stribeck's curve;
- For high speed, friction tends to increase with increasing speed; this may partially be explain by the influence of increasing centrifugal effect.

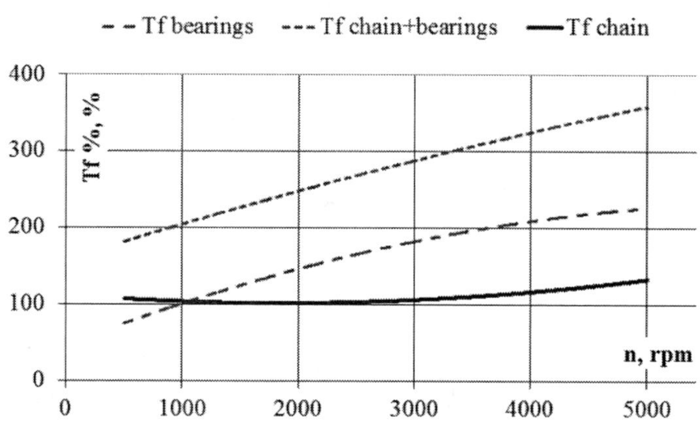

Figure 1. Global friction, bearing friction and resulted friction in chain C1.

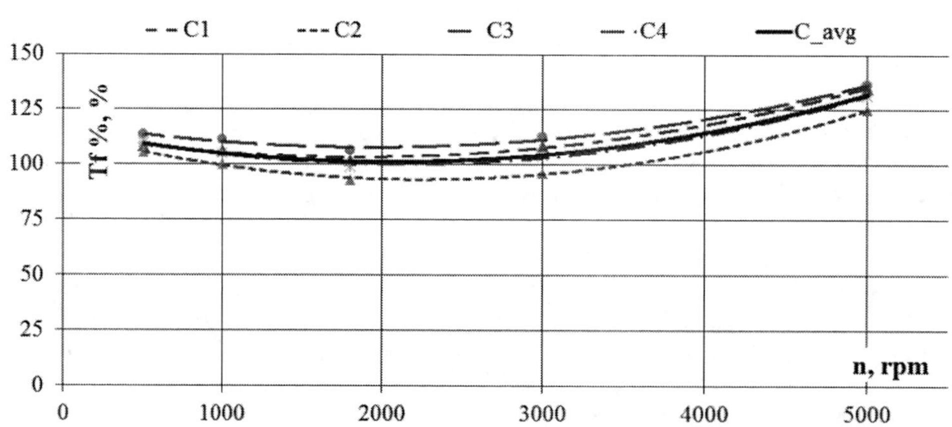

Figure 2. B8 chain friction depending on rotational speed.

4. Centrifugal effect on bush chain friction

All the tests have been made with constant F_0 tensioning force of 1 kN, with automatic control on tensioning.

Figure 3 presents three stages of loading the chain transmission:

- in the initial stage (figure 4, a), tensioning at rest, F_0 tensioning is acting on both support and the two arms of the chain together;
- if the two supports are blocked and chain is moving with linear speed v (figure 4, b), the component of the centrifugal force F_c (see equation (2)) is divided in order to release the supports and at the same time to supplementary tension the chain;
- in the case of our tests, with automatic tensioning, the load on the supports F_0 was all the time constant (figure 4, c), including the centrifugal effect; in this case the entire F_c component is supplementary tensioning the two arms of the chain.

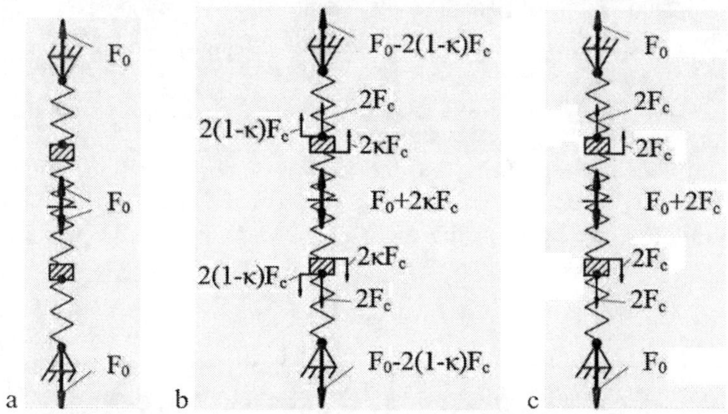

Figure 3. Forces on initial tensioning (a), repectively with centrifugal effect with blocked supports (b) and with automatic tensioning (c).

For a B8 bush chain ($m/l = 0.3383$ kg/m), figure 4 shows the change in chain tensioning with the increasing rotational speed, considering both arms of the chain $F = F_1 + F_2$.

Taking into account the measured value of friction torque at 1000 rpm rotational speed and considering the theoretical calculus relation for friction torque (equation (2)) with centrifugal influence on chain tensioning force F, and constant friction coefficient, figure 5 presents a comparison between this and the measured averaged friction torque. The differences may be explained by continuing decreasing friction coefficient up to 4000 rpm rotational speed and a small increase of friction coefficient at high rotational speed due to bad lubrication.

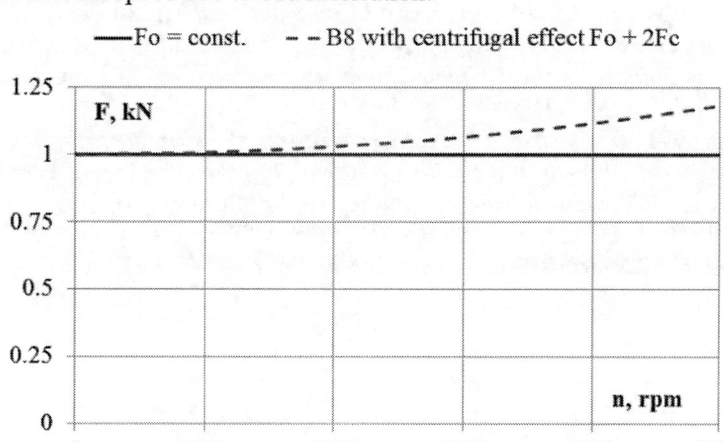

Figure 4. Chain tensioning with entire centrifugal effect.

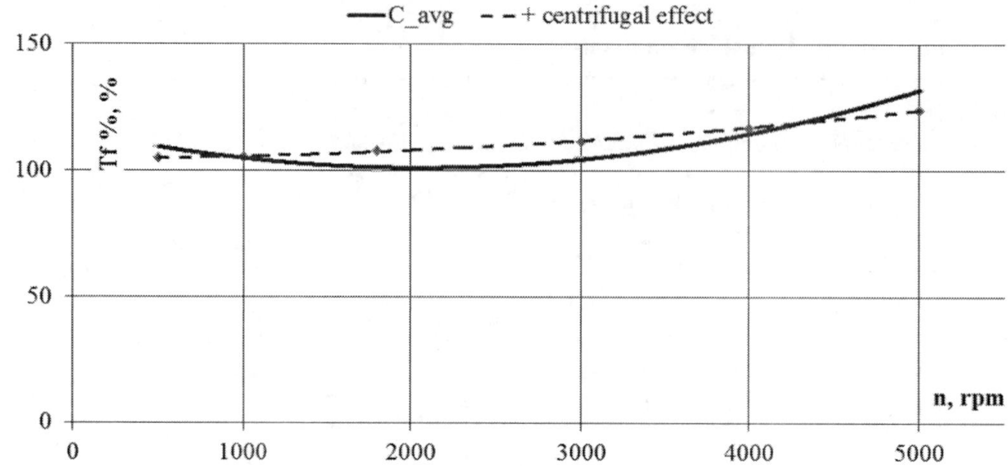

Figure 5. Comparison between experimental results and theoretical friction model with correction of centrifugal effect on tensioning.

5. Conclusion

Future research should focus on the influence of tensioning and lubrication on chain friction. Friction coefficients from the theoretical model should be evaluated. There is also needed an evaluation of other than pin-bush friction since sliding of the bush on the tooth of the sprocket has been clearly demonstrated.

References

[1] Tung S C and McMillan M L 2004 Automotive tribology overview of current advances and challenges for the future *Tribology International* **37 (7)** p 517-536

[2] Holmberg K, Andersson P and Erdemir A 2012 Global energy consumption due to friction in passenger cars *Tribology International* **47**

[3] Hyakutake T, Inagaki M, Matsuda M, Hakamada N and Teramachi Y 2001Measurement of friction in timing chain *JSAE Review* **22:5**

[4] Papuc R, Gavrila C and Lates M T 2017 Tribological tests of steel on polyamide 66, polyamide 46 type surface contacts *MATEC Web of Conferences* **126** 02003

[5] Papuc R, Gavrila C and Lates M T 2017 The study of chain link/guide contact in relative alternative translational motion *MATEC Web of Conferences* **126** 01003

[6] Lates M T, Velicu R G and Papuc R 2014 Multiscale modeling of chain-guide contact by using tests and FEM *11th World Congress on Computational Mechanics*

[7] Velicu R and Lates M 2015 On the Measurement Procedure for Testing Friction in Bearing Mountings *Annals of the Oradea University, Fascicle of Management and Technological Engineering* **XXIV (XIV)**

[8] Horovitz B 1971 *Transmisii si variatoare prin curele si lanturi Transmissions and variable transmissions with belts and chains (in Romanian)* (Bucuresti: Ed. Tehnica)

[9] Niemann G 1975 *Machinenelemente Band II* (Berlin: Springer Verlag)

[10] Saulescu R, Jurj L and Velicu R 2018 Centrifugal effect on shaft's release of chain transmissions *Materials Science and Engineering* **444 (5)** 052006

IOP Conf. Series: Materials Science and Engineering **568** (2019) 012007 doi:10.1088/1757-899X/568/1/012007

The dynamic capabilities framework implementation in innovation and creative ecosystems

E Avram[1], S Avasilcai[1], C A Hutu[1], A Bujor[1]

"Gheorghe Asachi" Technical University of Iasi, Blvd. D Mangeron no 29, Tex 2, Doctoral School, Iasi, Romania

egalateanu@tex.tuiasi.ro

Abstract. The arising concern for the inter-industry collaborations was concretised through the continuous development of new collaborative structures as it is the case of business ecosystems emergence. The use of open innovation in existing ecosystems became an essential trigger for the emergence, defining and development of dynamic capabilities. Consequently, these capabilities were used as the most relevant resources in value networks evolution as they defined and created new ways for opportunities exploration and exploitation especially within existing ecosystems. From this point of view the current research seeks to identify and define what are dynamic capabilities, to present the existing framework of capabilities analysis and to understand how it is used in practice within existing ecosystems. Also, this research seeks to uncover how collaborative efforts and the use of innovative technologies influence organisational growth.

1. Introduction

Business ecosystems emergence and development was triggered by the need to establish a comprehensive framework where collaborative relations were perceived as mechanisms for actors' co-evolution [1]. From this point of view, in order to ensure collaborative processes implementation each ecosystem's participant should design own business model so that it could provide and support the creation of required capabilities for ecosystems orchestration [2].

The arising technological challenges along with the increasing attention for continuous innovation and value creation represent weighting factors in business model innovation processes. As companies strive to adopt new technologies and to engage customers into critical internal processes, new innovation capabilities are acquired. Teece suggested the relevance of dynamic capabilities as a binding linkage between a strategic and entrepreneurial approach to the use of internal resources and firm's capability to establish collaborations with various external partners [3]. This approach requires a continuous analysis of arising opportunities in terms of their exploration and exploitation [3]. In terms of business ecosystem perspective, the development of dynamic capabilities first of all depends on internal innovation of each engaged actor (e.g. universities, complementors, financial agencies, etc.) [4]. Subsequent, based on the key aspects presented, there will be highlighted the dynamic capabilities framework and its relevance for ecosystems 'development within different industries.

2. The dynamic capabilities framework: an innovation ecosystem perspective

For the first time the dynamic capabilities concept was closely linked to the Resource Based View theory. However, according to the Wang and Ahmed this approach seems to be static and do not reflect

IOP Conf. Series: Materials Science and Engineering **568** (2019) 012007 doi:10.1088/1757-899X/568/1/012007

entirely the nature of dynamic capabilities. The main identified gap suggested that RBV reflects only the firm's perspective on the necessary internal resources [5]. In terms of business environments dynamics this approach is no longer valid It seems neglected the fact that dynamic capabilities should reflect not only organizational resources perspective but also strategical view by implementing specific processes (e.g. to establish new partnerships) or product development [6]. Also, they can be perceived as an important strategical tool as this concept emphasizes the relevance of organizational attributes in dynamic business environments [7]. From this point of view dynamic capabilities emphasize the relevance of organizational routines which are seen also as organizational capabilities [8]. According to Salvato and Vassolo the engagement of the human resource and continuous dialogue building contribute to the creation of the internal knowledge base necessary for opportunities exploration and exploitation [8]. More than that Bingham et al suggest that learning process can be achieved though learning processes of the past experiences [9].

According to Teece the emergence of innovation ecosystems rises also the concern for new type of resources and dynamic capabilities development in order to ensure the companies adaptability to the new market's requirements [4]. Thus, it is more about how to build collaboratively new capabilities for innovation rather than owning them. From this point of view, the building process of dynamic capabilities requires three main stages: sensing, seizing and transformation [4], presented in figure 1.

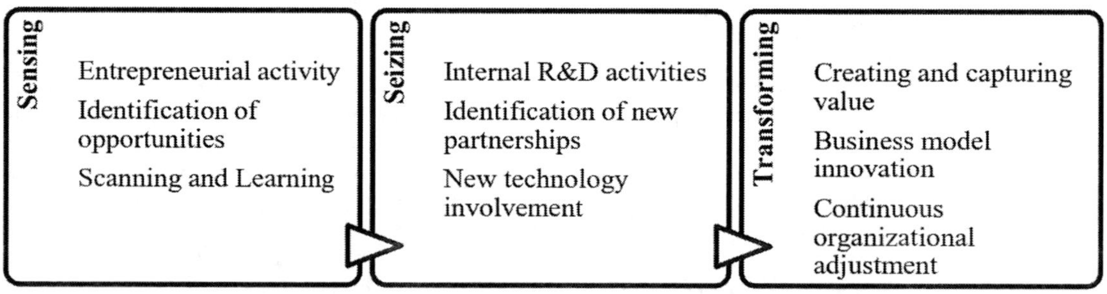

Figure 1. The dynamic capabilities framework according to Teece [4].

Williamson explored the dynamic capabilities framework in terms of creating innovation. According to the author the use of open innovation, social media and the creation of dedicated platforms contribute to the experience building by acquiring necessary knowledge in a dynamic way [10]. The major arising implications are linked to the dynamic response from the market, adaptability to the continuously changing customers' preferences, product modularity, the use of waterfall type of processes and the creation of virtual services [10]. Therefore, the dynamic capabilities framework emphasizes the relevance of building and adopting flexible and dynamic type of processes rather than static ones. Hence, the companies from innovation ecosystems should be able to adapt to the continuously evolving market requirements.

3. Appling the dynamic capabilities framework in innovation ecosystem versus creative one

The concepts of innovation and creative ecosystems by definition suggests the engagement of different type of stakeholders into co-creation processes. However, the main different lies in the type of resources used in order to acquire necessary knowledge, the product development processes and innovation capabilities built by actors. Innovation ecosystems, from the dynamic capabilities' perspective, can be perceived as a dynamic structure which combines knowledge base derived from external partners with various experiences, abilities and capacities [11, 12] and usually can be observed in the case of keystones actors [13]. In the case of creative ecosystems along with continuous innovation, the building of necessary knowledge base is perceived as a creative process which especially implies the engagement of the creative community [14]. In this type of ecosystems, the building of innovative capabilities is grounded to the symbiotic relations and dynamic interactions among different type of stakeholders [14]. In order to illustrate how dynamic capabilities framework can be applied in innovation and creative ecosystems there will be taken into consideration two cases: Electrolux innovation ecosystem and 99designs creative ecosystem.

3.1. The dynamic capabilities framework in Electrolux innovation ecosystem

In order to understand how dynamic capabilities are built within an innovation ecosystem there was selected the case of Electrolux, as the company itself is well known for a remarkable history of created innovations. Founded in 1919, Electrolux has known multiple transitions especially in terms of innovation implementation in product development [15]. According to their philosophy, innovation represents the main driver of organizational growth and success [15].

The Electrolux innovation ecosystem comprises a variety of collaborative relations with internal and external stakeholders: employees, suppliers, open source organization OSF, Google, start-ups, etc. [16]. By promoting and implementing the open innovation and co-creation processes the main concern of the company, as a part of an innovation ecosystem, is to build innovative dynamic capabilities. In table 1 is presented an analysis of the dynamic capabilities created by Electrolux based on Teece framework.

Table 1. The dynamic capabilities framework: the case of Electrolux Innovation Ecosystem [17-22]

Sensing	Seizing	Transforming
The use of open innovation	Created dedicated platforms (Design Lab, Ideas Lab)	Continuous innovation of product portfolio
The potential of connected products - Internet of Things ecosystems	Use employees creative potential through the brainstorming session iJam	Implemented three-steps development process focused on consumers' wellbeing experience
Shift to customer-centric production	Partners with different alliances (Allseen), Non Governmental Organisations (OSF) or ecosystems (Internet of Things, Drop)	Ensures the remote connectivity and interactivity between customer and appliences
Design new structure - Innovation Triangle	Implemented ethernet network for internal communications	Offering interactive and intelligent products
Use creative potential of design students	Use of the 5G technology in product development	
Promotes innovative organisational culture		

3.2. The dynamic capabilities framework in 99designs creative ecosystem

Established in 2008, 99designs represents an important key player within graphic design creative ecosystem [14]. According to the company beliefs 99designs seeks to promote creative potential of the young designers and there should not exist barriers between innovative ideas and their dissemination [14]. Concentrated entirely in virtual environment the creative ecosystem of 99designs offers an interactive meeting point for customers and potential designers, limited timeline for contest development and is capable to reach a large pool of stakeholders [14].

In terms of participating actors, 99designs is mainly concentrated on their suppliers (potential clients), graphic customers and graphic designers [14].

Table 2 The dynamic capabilities framework: the case of 99designs Creative Ecosystem [14,23, 24]

Sensing	Seizing	Transforming
Sense the gap between the young designers and potential clients Seek to use the potential of crowdsourcing modules Understood the relevance of the crowd as a creative community Use brainstorming sessions with own employees Rebranding process by using one of the own designers	Created dedicated platform Implemented one-to-one projects in order to create agency experience Use social media as an instrument for graphic promotion Ensures the flow of knowledge concerning the current trends Different revenue model Integrates API	Created a marketplace - recruiting and graphic designs Business model innovation by implementing the platform Ensures the trust building between ecosystem partners Emphasize the relevance of uniquness and creativity Offers continuous feedback and assistance

4. Findings and Conclusions

The dynamic capabilities seem to be the most relevant tools in opportunities seeking for companies. Based on Teece's framework applied in different ecosystems there can be found that each stage of the dynamic capabilities' development seeks to present different aspects especially from the innovation point of view. Sensing stage is relevant for companies in order to understand the dynamic changes in different markets. Companies are rather concentrated on acquiring new knowledge from the external environment or other relevant actors than on own resources and capabilities. It is relevant at this stage to explore the arising opportunities in ecosystems in terms of the technology development or other technological challenges. This type of knowledge represents an important trigger for organizational growth and transformation (e.g. rebranding process or new organizational structures development). Seizing stage represents the first step in building own capabilities. At this stage companies come forward with new technological solutions (e.g. platform development for co-creation or crowdsourcing). In innovation ecosystem this stage is concretized by adopting new communication technologies or 5G technology implementation in product development. However, in creative one, the case of 99designs, company adopted a creative approach by using social media and virtual environment only. The transformation stage comprises the results of companies' efforts in order to create own dynamic capabilities. It is more about how to ensure continuous value creation to the customers, to be able to respond to their needs and to create an interactive environment. However, as it can be observed in the first case it targets especially the value created by product innovation. In the case of 99designs there can be traced a remarkable attention for community involvement and value created by creative crowds. Consequently, the dynamic capabilities framework offers a variety of valuable insights on how companies are evolving in platform-based ecosystems and provides interesting findings in terms of innovation dynamic capabilities developed by different actors.

References

[1] Galateanu (Avram) E and Avasilcai S 2013 Business ecosystem architecture *Fas. of Manag. and Techn. Eng.* **XXII(XII) 1** 79-84
[2] Helfat C and Raubitschek R 2018 Dynamic and integrative capabilities for profiting from innovation in digital platform based ecosystems *Res Pol* **47(8)** 1391-1399
[3] Teece D 2014 A dynamic capabilities based entrepreneurial theory of the multinational entreprise *J. Int. Bus. St.* **45** 8-37
[4] Teece D 2010 Technological innovation and the theory of the firm: the role of enterprise-level

IOP Conf. Series: Materials Science and Engineering **568** (2019) 012007 doi:10.1088/1757-899X/568/1/012007

knowledge, complementarities, and (dynamic) capabilities in Hall B and Rosenberg N (eds) *Handbook of the Economics of Innovation* **1** 679-730

[5] Wang C and Ahmed P 2007 Dynamic capabilities: a review and research agenda *J. Manag. Rev.* **9(1)** 31-51

[6] Eisenhardt K and Martin J 2000 Dynamic capabilities: what they are? *Strat Manag J* **21** 1105-1121

[7] Winter S 2003 Understanding dynamic capabilities *Strat Mgmt J* **24** 991-995

[8] Salvato C and Vassolo R 2017 The sources of dynamism in dynamic capabilities *Strat Mgmt J* **39(6)** 1728-1752

[9] Bingham C, Heimeriks K, Schijven M and Gates S 2015 Concurrent learning: how firms develop multiple dynamic capabilities in parallel *Strat Mgmt J* **36(12)** 1802-1825

[10] Williamson P 2016 Building and leveraging dynamic capabilities: insights from accelerated innovation in China *Glob Strat J* **6** 197-210

[11] Jackson D 2011What is an Innovation Ecosystem? *National Science Foundation, Arlington, VA.* National Science Foundation, Arlington, VA. From http://ercassoc.org/sites/default/files/topics/policy_studies/DJackson_Innovation%20E

[12] Tejero A and Leon G 2016 The Relevance of Innovation Ecosystems Using Taxonomies from Ecological Sciences. *The ISPIM Innovation Summit, Kuala Lumpur* 1 - 13 Kuala Lumpur, Malaysia: ISPIM.

[13] Valkokari K 2015 Business, Innovation, and Knowledge Ecosystems: How They Differ and How to Survive and Thrive within Them *Techn Inn Mgmt Rev* **58** 17 - 24

[14] Galateanu (Avram) E and Avasilcai S 2017 Emerging creative ecosystems: platform development process *Fas. of Manag. and Techn. Eng.* **XXVI(XVI) 3** 5-10

[15] Galateanu (Avram) E and Avasilcai S 2014 Pro-active consumers' engagement as driver of co-innovation-the case of Electrolux (A) *Fas. of Manag. and Techn. Eng.* **XXIII(XIII) 3** 39-44

[16] Electrolux 2018 Electrolux expands smart appliences collaboration with Google retrieved from https://www.electroluxgroup.com/en/electrolux-expands-smart-appliances-collaboration-with-google-26049/

[17] Avram E 2016 Case Study: Electrolux in Avasilcai S and Hutu C A (eds) Innovation through co-creation in online era Performantica 116-146

[18] Electrolux 2018 Electrolux embeds digital experiences in its design strategy retrieved from https://www.electroluxgroup.com/en/electrolux-embeds-digital-experiences-in-its-design-strategy-25822

[19] Electrolux 2014 Electrolux joins AllSeen Alliance to enable seamlessly connected appliances retrieved from https://www.electroluxgroup.com/en/electrolux-joins-allseen-alliance-to-enable-seamlessly-connected-appliances-19555/

[20] Electrolux 2018 Electrolux partners with smart kitchen startup Drop retrieved from https://www.electroluxgroup.com/en/electrolux-partners-with-smart-kitchen-startup-drop-26180

[21] AB Electrolux 2018 Well positioned to create value *Annual Report* retrieved from https://www.electroluxgroup.com/annualreports/2018/files/elux-2018/Electrolux_AReview_2018_FINAL_low.pdf

[22] Electrolux 2016 Electrolux Ideas Lab retrieved from https://www.electroluxgroup.com/en/electrolux-ideas-lab-lets-bring-healthy-home-cooked-food-to-every-dinner-table-22752/

[23] Bujor A, Avasilcai S and Galateanu(Avram) E 2018 Creative ecosystem model innovation via crowdsourcing modules *Proceedings of 6th Review of Management and Economic Engineering 6th International Management Conference*: Performance Management or Management Performance, 20-22 septembrie, Cluj-Napoca, ISSN: 2247 – 8639, pp. 669-677

[24] 99designs 2018 Creativity. Now avaible via API retrieved from https://en.99designs.es/api

IOP Conf. Series: Materials Science and Engineering **568** (2019) 012008 doi:10.1088/1757-899X/568/1/012008

NI LabVIEW based camera system used for gait detection

D Salánki[1] and K Sarvajcz[1]

1University of Debrecen, Hungary, Ótemető street 2-4.

danisalanki@yahoo.com

Abstract. In these times, with the development of the world, biometric identification systems are becoming more and more widespread. Access control systems, but even the most mid-range smartphones have biometric authentication features, and even ID cards can include a person's fingerprint. The research group previously realized a rudimentary gait recognition system, which was upgraded to a multicamera system with high-resolution cameras and instead of reference points, the new version recognizes different templates. The program can compare and evaluate the functions that are matched to the reference curve and the current curve in a specific way, whether two walking images are identical. The comparison is decided by the definite integrals of the two suited functions. The self-developed gait recognition system was tested by the research team on several test subjects and according to the results, permission was never given to a strange person.

1. Introduction

There are several disciplines where motion tracking and motion capture devices play an important role. Examples include automation, robotics and computer animation. Different application areas may have different specific requirements, other needs may be important. There are places where it is enough to follow one or two independent points, elsewhere, e.g. at character animation, the following points form a complex transformation hierarchy where the complete rigid transformation of each element is required [1].

Compared to physiological biometric identification (fingerprint identification, iris and face identification, ear tag geometry identification), walking has the advantage of being able to identify human subjects with low-resolution or hidden camera images [2]. Physiological biometric methods are incapable of reliably recognizing non-cooperating individuals, especially at a distance, under changing environmental conditions in the real world [3]. However, the accuracy of gait recognition can be influenced by several factors, such as the angle of view, which changes the matching of visual characteristics with each other [4].

2. The Created Gait Recognition Software [5]

In 2017, the research team set up an initial walk detection system with a 640x480 resolution camera in LabVIEW [6]. The improved version will be presented below.

The main change of the newer version compared to the previous one is that the program can now handle two cameras instead of one, and instead of real time processing due to the higher system requirements, the walking is recorded, then the video file is analyzed by the program. The cameras were placed side by side to complement each other. A further modification compared to the old version is that instead of point recognition, pattern recognition is done, which can better filter out the false sensed

IOP Conf. Series: Materials Science and Engineering **568** (2019) 012008 doi:10.1088/1757-899X/568/1/012008

points, and that instead of comparing the coefficients, the definite integral values of the functions are compared.

2.1. *The Front Panel*

The front panel of the new version of the program is shown in Figure 1. The user can switch between 4 tabs using Tab Control. The first tab shows the recordings made by the two cameras. Under the displays can be selected the computer location of the video files. There is a larger green lamp that flashes when the test subject is recognized. There are 8 smaller lamps above the larger one, which indicate the logical values of 8 comparisons. In the bottom line, the locations of the templates can be selected, in this case 6, 3 per camera, because in each case 3 templates were placed on the test subject: one on the ulna proximalis (forearm), one on the fibula proximalis (knee) and one on half distance between the right fibula proximalis and the outer ankle (calf). These templates will be searched in the videos. This version no longer requires a Stop button because the program stops automatically after analyzing the videos.

On the second tab of the front panel, the coordinates of the detected templates are stored in arrays X and Y, respectively. On the third tab, the resulting walking curves are displayed separately for the first and second video, and the curve described by the template on the forearm is evaluated for both videos. During the evaluation, the program fits a function on the curve described by the template and calculates the definite integral of it between 0 and 1920.

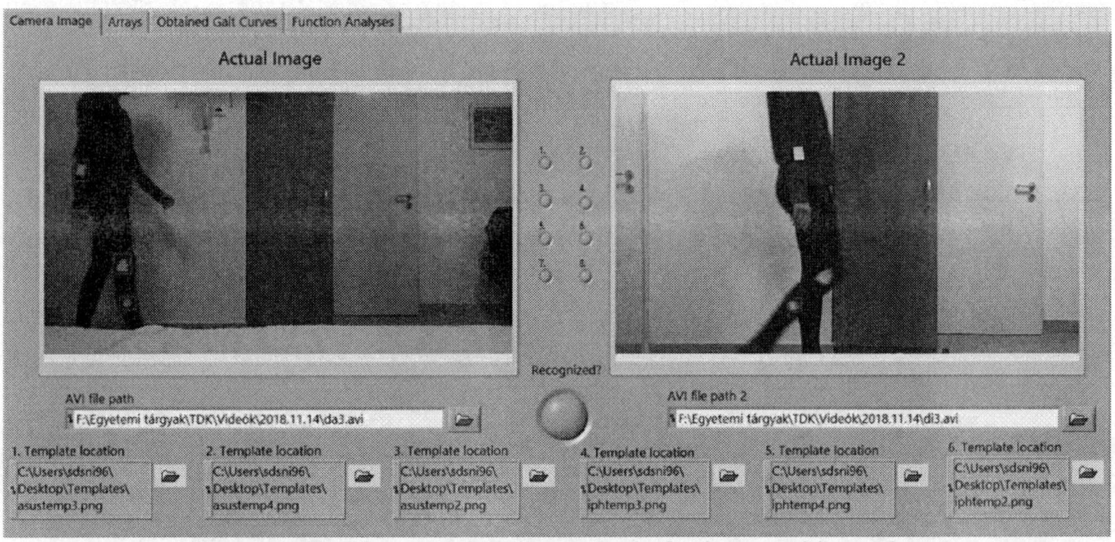

Figure 1. The first tab of the front panel of gait recognition software.

The fourth tab contains the evaluation of the other walking curves. The second graph shows the curves drawn by the template on the calf in the first video, as shown in Figure 2. The third and fourth graphs show the suited functions on the curves described by the knee's and the calf's template in the second video. For the forearm and knee-mounted templates, a second-degree alignment is performed, because these templates describe a path with a small curvature, so they can be considered straight with a good approximation, but because of better accuracy, the second-degree member was also considered. However, the calf template's curve cannot be written with a single function. Based on our observations, about 3 steps fit into the camera image, which can be described with 3 second-degree functions. The integration limits are the local maximums of the functions.

IOP Conf. Series: Materials Science and Engineering **568** (2019) 012008 doi:10.1088/1757-899X/568/1/012008

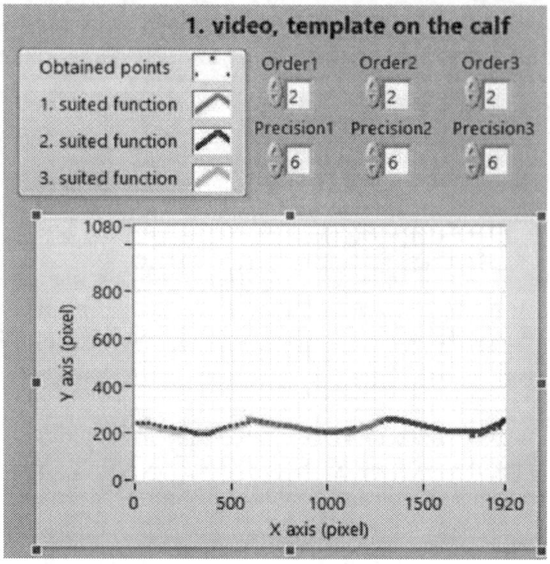

Figure 2. A part of the fourth tab. The figure shows curves drawn by the template on the calf in the first video and the functions suited on them.

2.1.1. The limits of the definite integral. Taking the average of several 10 measurements, the limits of the definite integrals of each function have been determined. If the difference from this limit is within 5%, or in some cases within 12.5%, then a logical 1 is given from that part of the program. In the case of the calf template, it was formed the sum of the values of the definite integrals since a smaller standard deviation was observed between the individual values, but their sum was approximately constant during the measurements. The program makes a total of 8 comparisons. If each one closes with a logical 1, the green light will flash, and the test subject will be noted as recognized.

3. Results [5]

		1. camera					2. camera					
	1+2+3	1. calf	2. calf	3. calf	Knee	Forearm	1. calf	2. calf	3. calf	Knee	Forearm	Lights
1.	404825	155900,9	151685,6	97238,51	689877,1	1505048	113513,2	131300,7	57626,29	657069,3	1582469	8
2.	404336,2	151084,5	137148,7	116103	684607,8	1508530	112636	125737,1	58156,21	671815,8	1588548	8
3.	407951	131239,4	135300,6	141411	678855	1497569	90849,29	121681,8	71900,5	645783	1572135	6
4.	410753,5	148262,4	141133,2	121357,9	684912	1492215	111130,2	121252,9	65424,52	673045,6	1565411	8
5.	412703,4	179398,4	136128	97177,02	684714	1500263	118396,4	124636,5	59458,87	633641,4	1572650	8
6.	404583,4	147187,2	143169,9	114226,4	679971,2	1501445	107789,5	122463,6	66480,57	685295,5	1573187	8
7.	408340,1	134015,8	141292,2	133032,1	680918,3	1486649	101269,1	120697,3	67942,65	683546,2	1555418	8
8.	406633,9	137548,8	156790	112295,1	680750,2	1488126	110686,4	120324,1	61604,29	647499,3	1563726	8
9.	411615	154885,7	138469,1	118260,3	679610,3	1486759	108561,6	127727,8	60319,06	612740,2	1559566	8
10.	391804,8	144159,9	138948	108696,9	676811,2	1497082	102893	124120,1	54797,94	657944,3	1566160	8
11.	399453,6	137030,4	158742,8	103680,4	679787,1	1496530	107543,7	128705,6	55152,24	604656,2	1566528	8
12.	406380,2	144129,1	140629,5	121621,6	674742,4	1492289	103795,4	120809,2	66398,06	665115,1	1573108	8
13.	402587,5	145807,6	150664,7	106115,2	676462,2	1484978	105895,2	128511,1	55181,08	668137,3	1557829	8
14.	392768,1	153320,8	136799,3	102648	675255,9	1481227	119166,1	117615,1	60369,81	674478	1549399	8
15.	399154,1	137491,9	141396,9	120265,2	677788,4	1474409	100297,6	128346,2	59830,2	624090	1546348	8

Figure 3. Gait data from a known test subject.

After setting up the camera system, the research team tested it with several subjects. The measurements were made in normal clothing (trousers and long-sleeve T-Shirt). The test subjects took the first step with their left foot. Fig. 3. illustrates the results of a test subject known by the program. The first grey column shows the sum of the values of the first 3 blue columns. The blue columns show the results of the first video. The values represent the definite integrals of the functions suited on the curves drawn by

IOP Conf. Series: Materials Science and Engineering **568** (2019) 012008 doi:10.1088/1757-899X/568/1/012008

the templates, the first, second, and third functions of the template placed on the calf, the function of the knee, and the curve of the template fitted to the forearm, respectively. The green columns show data obtained from the video produced by the second camera in a similar division to the first one. The last column shows the number of flashing small lights on the user interface.

The program correctly detected the known test subject 14 times out of 15 (Figure 3.). In this case, the false rejection rate (FRR) was 6.25% (correct recognition rate: 93.75%). In the case of a similar test subject to the known one, but unknown to the system, the program has never given the permission, so the False Acceptance Rate (FAR) was 0%.

4. Comparison with Other Gait Recognition Systems

Dacheng Tao, Xuelong Li, Xindong Wu, and Stephen J. Maybank have been tested a Gabor-based gait recognition system. Gabor filters can be used to amplify linear structures in a given direction. With this method, a detection rate of 92% was obtained from different views and in footwear, and 87% for test subjects with a briefcase [7].

The GEI-based (Gait Energy Image) system tested by Khalid Bashir, Tao Xiang, and Shaogang Gong has been run by multiple databases without the cooperation of test subjects. Based on the CASIASetA database, where there is no packet and no big jacket with test subjects, the recognition rate was 100%, however, for CASIASetB and CASIASetC, where the test subjects are with pack or wearing large jackets, the detection rate was 77.8%, and 43.1% [8].

Chin Poo Lee, Alan W.C. Tan and Shing Chiang Tan used the TBP - Transient Binary Patterns method to develop walking detection system and compared it with other methods. The TBP and GEI methods resulted in a 99% detection rate. The Local Binary Patterns from Three Orthogonal Planes (LBP-TOP) resulted in only a 49% detection rate [9].

Bence Gálai and Csaba Benedek investigated district-based personal identification and action recognition with LiDAR sensors (Light Detection and Ranging). Several people have been introduced to their systems at one time, with a 100% detection rate for up to 5 people, and then gradually reduced to 75% for 22 people after which it stagnates [10].

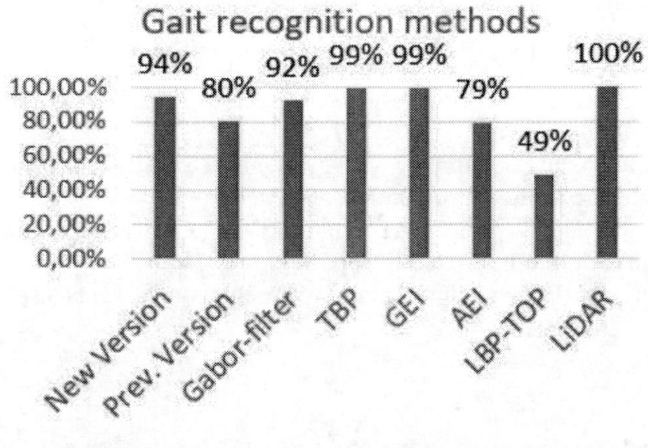

Figure 4. The research team's gait recognition system compared with other systems.

Figure 4 summarizes the accuracy of the current and previous systems of the research group and other systems. Based on this, we can say that the most accurate gait detection systems are the LiDAR (100%), TBP (99%) and GEI (99%). The system set up by the research team is also good with its 93.75% recognition rate, while the least efficient is the LBP-TOP system with 49%.

5. Conclusion

In summary, the research team has succeeded in creating a complex gait recognition system in NI LabVIEW environment that can detect multiple samples using multiple cameras and is able to fit predefined curve on the collected samples. The program is also able to compare the values of the reference curve and the current curve in a specific way and to evaluate if the image of the two curves is the same. The self-developed gait recognition system has been tested on several individuals, which never gave permission to a foreign person.

In the future, the processing speed of the program could be improved with a camera or a software that can be used to immediately save the video in AVI format. The 2014 LabVIEW version used in the research is only capable of post-processing AVI videos. Writing the program in other software (e.g. OpenCV), which is capable of handling MP4 files, or in which a high-quality real-time camcorder does not hold back the performance, can also improve the processing speed.

Acknowledgement

The work/publication is supported by the EFOP-3.6.1-16-2016-00022 project. The project is co-financed by the European Union and the European Social Fund.

References

[1] D. Viktor, „Optikai alapú Motion Capture rendszer," pp. 5, 7-9, 2011.
[2] Mark S. Nixon, T. N. Tan, R. Chellappa, „Human Identification Based on Gait," New York, Springer Science + Business Media, Inc., 2006., p. 11.
[3] Ju Han, Bir Bhanu, „Individual Recognition Using Gate Energy Image," IEEE transactions on pattern analysis and machine intelligence, %1. szám28.2, pp. 316-322, 2006.
[4] Wei Zeng, Cong Wang, „Neurocomputing - View-invariant gait recognition via deterministic learning," ,175, pp. 324-335, 2016.
[5] Salánki Dániel, Sarvajcz Kornél, Dr. Géza Husi: "Járásfelismerés fejlesztése NI LabVIEW környezetben", Debrecen, University of Debrecen, Mechatronical Engineering BSc, thesis, 2018.
[6] Salánki Dániel, „Járásfelismerés fejlesztése NI LabVIEW környezetben," University of Debrecen, Faculty of Engineering, National Council of Student Research Societies, Debrecen, 2017.
[7] Dacheng Tao, Xuelong Li, Xindong Wu, Stephen J. Maybank, „General Tensor Discriminant Analysis and Gabor Features for Gait Recognition," *IEEE transactions on pattern analysis and machine intelligence,* pp. 1-35, 2007.
[8] Khalid Bashir, Tao Xiang, Shaogang Gong, „Gait recognition without subject cooperation," Pattern Recognition Letters, pp. 2052-2060, 2010.
[9] Chin Poo Lee, Alan W.C. Tan, Shing Chiang Tan, „Gait recognition with Transient Binary Patterns," J. Vis. Commun. Image R., pp. 69-77, 2015.
[10] Gálai Bence, Benedek Csaba, „Járás alapú személyazonosítás és cselekvésfelismerés LiDAR szenzorokkal", 2017.

IOP Conf. Series: Materials Science and Engineering **568** (2019) 012009 doi:10.1088/1757-899X/568/1/012009

Optimization methods of the industrial robots' trajectory

M Ratiu[1], G Grebenisan[2], M A Prichici[1] and S Bogdan[1]

[1]Department of Mechanical Engineering and Automotive, University of Oradea, 410087 Oradea, Romania
[2]Department of Industrial Engineering, University of Oradea, 410087 Oradea, Romania

E-mail: mratiu@uoradea.ro

Abstract. In this paper, we present a part of our wider research regarding modeling, simulation and optimization of the trajectory of the industrial robots. After a short introduction on the importance of the industrial robots, the necessity and widely interest regarding the trajectory optimization, we present a classification of the optimization methods: numerical, computational and hybrid. Then, we make a brief presentation of the main optimization criteria approached in the last few years in several scientific papers, regarding the optimization of the industrial robots' trajectory.

1. Introduction

In the last decades, the industrial robots are more and more used across all industries, and across all types of companies, from the small and medium-sized companies to the large companies, in special in automated production processes, where the companies seeking agile production methods. Taking into consideration the important role of the robots in the present and future industrial processes, there are a lot of studies in robotics on increasing performance, precision and flexibility, and reducing the effort, time, energy and cost, through streamline and optimize the trajectories and the motions of the industrial robots [1].

Generally, optimization aims at establishing, based on one or more predetermined criteria, the best solution, called the optimal solution, for a given situation, when several variants are possible. In the current context, optimization of processes of any kind, in this case of engineering, is considered as of particular importance, as evidenced by a huge volume of research, publications and scientific events dedicated to this subject.

2. Optimization methods

Optimization methods can be categorized by several criteria. Thus, from the point of view of the restrictions imposed on the variables, there are problems of optimization without restrictions and with restrictions. After the kind of objective function, there are linear optimization problems, in which both the objective function and the restrictions are linear, and nonlinear optimization problems [2]. There are multimodal optimization for functions with multiple global or local optima, multicriteria optimization when there are multiple optimization criteria that are conflicting, optimization in dynamic or non-deterministic environments.

2.1. Numerical optimization methods

In a mathematical sense, optimization is a process of minimizing or maximizing an objective function, and at the same time satisfying some constraints or restrictions. There are standard optimization techniques and algorithms, which are found in most of the specialized literature. There are also some more modern approaches that combine mathematical rigor with engineering practice, such as, for example, linking the analytical character of an optimization problem expressed in terms of optimality, and analyzing numerical algorithms optimization used to solve the problem [3].

The numerical methods used in engineering [4] to resolve optimization problems are divided into two major classes: direct methods and indirect methods. Direct methods are most often used because they are easier to configure and resolve, but they do not lead to precise solutions. Indirect methods are more difficult but provide very good precision, which is why they are used in certain applications that require a precise solution. In solving the optimization problems are used: systems of nonlinear algebraic equations (on indirect methods), differential equations and integration of functions (both indirect and direct methods) and nonlinear programming methods, namely gradient methods and heuristic methods (to direct methods). Lately, in optimization issues are used combinations of direct and indirect methods in a so-called hybrid approach [5].

The main features of an optimization problem are:
- the decision variables, which can be selected from a certain set
- the objective function, which must be minimized or maximized
- the constraints or restrictions to be observed, by equality and/or inequality.

2.1.1. Optimization without constraints. To solve optimization problems without constraints, numerical methods of decreasing (relaxation) are used, namely, methods that lead to decreasing the value of the objective function at each iteration and which, depending on the algorithms used, are classified as follows:

- one-dimensional optimization methods, or zero-order optimization, that only require the calculation of the objective function values at the current point and possibly at some neighboring points and are based either on direct, cyclical search, by coordinate axis directions, or random (the golden section search method, the bisection method, the Fibonacci sequence method, the Newton-Raphson classical method and the secant method, Hooke-Jeeves, the Nedler-Mead simplex method), or on the approximation of the objective function with a polynom determined by interpolation using the function values and/or the derivatives of the objective function at certain points (quadratic interpolation in two or three points and cubic interpolation).

- first order numerical optimization methods or linear search methods, based on the information obtained from the valuation of the objective function and its gradient (gradient method or steepest descent method, conjugate direction method, variable metric methods). These linear search methods are very much used in practice because they make a favorable compromise between simplicity and efficiency.

- second-order numerical optimization methods or Newton type, more complex, because they use information about the curve of the objective function or the Hessian matrix (the Newton or Newton-Raphson method and the quasi-Newton method).

- numerical methods to optimize some objective functions with special structure, to the type of the least squares (Gauss-Newton method and Levenberg-Marquardt method).

2.1.2. Optimization with constraints. When solving optimization problems with constraints, are used:

- first- and second order methods for nonlinear programming with convex constraints (descending direction methods - conditional gradient method and projected gradient method - and the projected Newton method).

- optimization methods for nonlinear programming with equality constraints (methods for quadratic programming with equality constraints, Lagrange's first order method, Lagrange-Newton method, Newton method with equality constraints).

IOP Conf. Series: Materials Science and Engineering **568** (2019) 012009 doi:10.1088/1757-899X/568/1/012009

- optimization methods for general nonlinear programming (active-set methods, sequential quadratic method, penalty methods, barrier methods, interior point methods). Of these, interior point methods are considered as a modern alternative to others and have the advantage of replacing non-smooth conditions with a set of smooth conditions, which are an approximation of the original ones.

2.2. Computational optimization methods

Often, optimization in engineering is doing by default for relatively simple problems, on the base on own knowledge and experience, the engineer trying to find an optimal model, through several attempts, in an iterative process, using certain software packages for modeling and analysis. To solve more complex problems, are used algorithms and programs capable to automate optimization, and the computer is used both to evaluate the model and to find a better solution in a shorter timeframe. Solving a computational optimization problem requires the following steps: description of the problem, discussion of the problem, defining the optimization problem, model development, ordering the equations, coding the model, possibly debugging it, linking the model with the optimization software and checking, optimizing the model [6].

There are some difficult optimization problems for both people and computers, problems characterized by a large search space, processes that are carried out in complex or dynamic environments, etc., so-called "bad" problems. In these situations, are used calculus and optimization methods based on computational intelligence, "Computational Intelligence" or "Soft Computing". It is about the neural calculus, the evolutionary calculus, and the fuzzy calculus, which use the artificial neural networks, the evolutionary algorithms, and the fuzzy system as the working tools, inspired by the human brain, the processes of evolution in nature and the specifics of human reasoning and language natural. This new style of calculus is based on fuzzy logic and several artificial intelligence tools: Probabilistic Reasoning, Genetic Algorithm, Evolutionary Programming, Neural Networks, and Machine Learning [7, 8].

If artificial intelligence is considered the pioneer, computational intelligence is the successor. Thus, if in the artificial intelligence is considering that the expert is putt into the computer, in the case of computational intelligence there is a collaboration between the expert and the computer [9].

2.3. Hybrid optimization methods

As even some of the most modern methods may be inadequate in some situations, it has been tried and proven by studies that the shortcomings of a method can be removed by incorporating specific techniques of another method, complementary. Thus, by combining concepts and algorithms from two or more methods, can be created a new, hybrid method with superior behavior. An example of such a hybrid method is obtained by combining genetic algorithms that excel in exhaustive exploration of an area of a problem, but are deficient in local refining of promising solutions, with algorithms such as Pattern Search (PS) that offer a highly effective local optimization but they are not able to explore the whole field. The new hybrid method thus obtained has both global optimization capabilities provided by the genetic algorithms, and local optimization capabilities provided by the Pattern Search numerical algorithms [10].

3. The industrial robots' trajectory optimization

In engineering, a problem of optimization means to minimize or maximize a function, that is subject to constraints, that, mathematically, can be inequalities or equations, the optimization problem being called the nonlinear programming problem [11]. The aim of the industrial robots' trajectory optimization is to minimize or maximize at least one of the following objective functions:

- to minimize the execution time, respectively to maximize the robot productivity, taking into consideration the limits of the relative speeds of the actuators' elements
- to minimize the energy consumption or mechanical work needed for execution
- to minimize the maximum power necessary for operating the robot
- to minimize the maximum forces and moments necessary for actuation.

To get better results, many times is achieved multi-criteria optimization, when the objective functions that have to be minimized or maximized express simultaneously combinations of more criteria.

As presented in detail in [1], and illustrative in figure 1, the most common optimization criteria used for industrial robots' trajectory optimization, after a review of 40 scientific papers in this area, are:
- minimum time trajectory planning (7)
- minimum energy trajectory planning (12)
- minimum jerk trajectory planning (1)
- multi-criteria optimization (14): minimum time and minimum cost; minimum time and minimum energy; minimum time and continuous jerk; time-optimal and jerk-bounded; time-jerk optimal; weighting between the execution time, acceleration, and jerk; minimum transmission time and maximum load carried; minimum of the Euclidean distance between the target positions and minimum effort of the servomotors; optimal speed-trajectory length; optimization of the singularities, joint limits and collisions; optimization of six objective functions: weighted balance of traveling time, actuator torques, singularities avoidance, joint jerks, joint accelerations and gripping force.
- optimization of the robotic task sequencing (6).

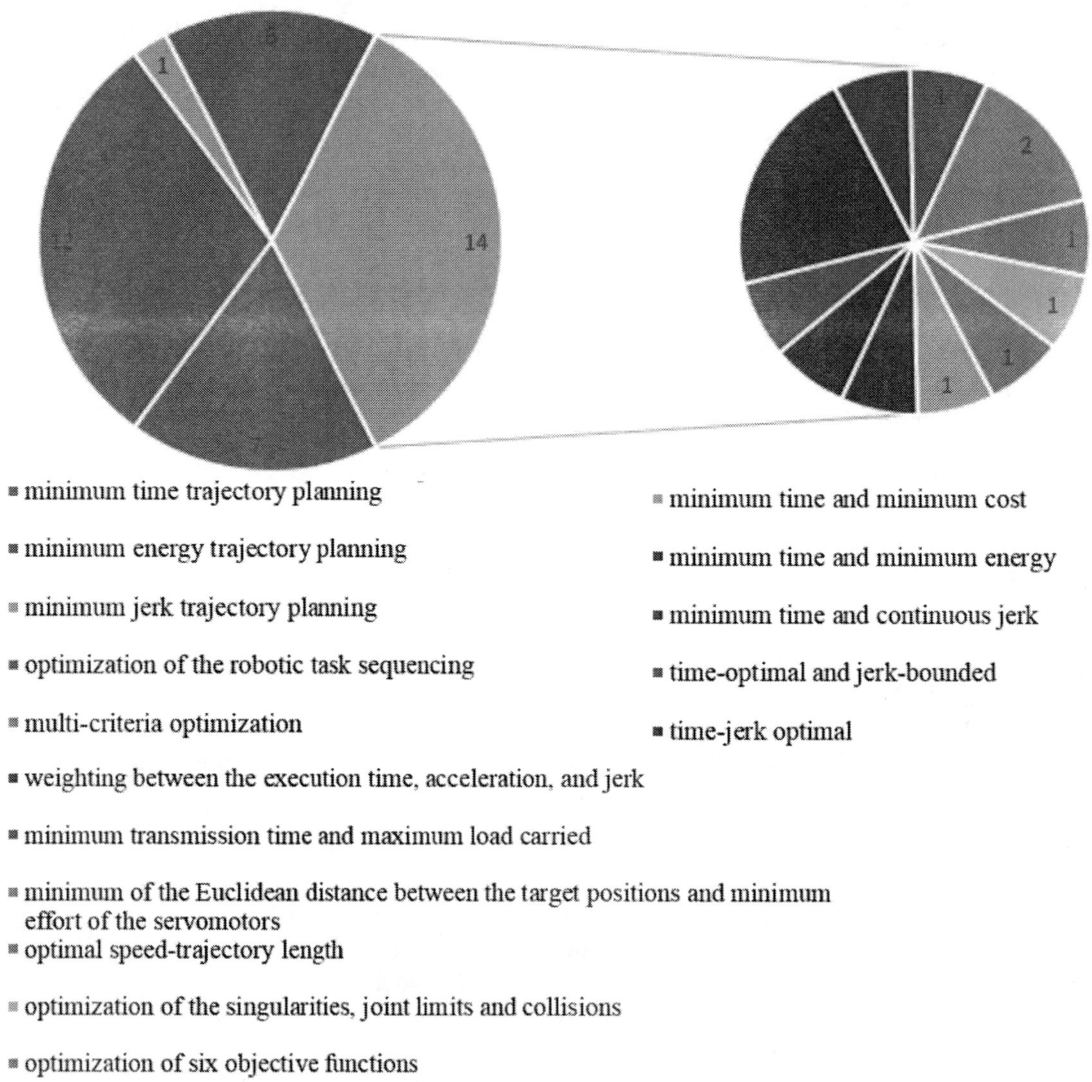

Figure 1. Optimization criteria of the robots' trajectory.

IOP Conf. Series: Materials Science and Engineering **568** (2019) 012009 doi:10.1088/1757-899X/568/1/012009

In the process of optimization in virtual modeling, an accurate parametric model of the mechanical system is very important [12], especially in the case of tasks control, the positioning accuracy, and the robot calibration [13].

4. Conclusion

The optimization of any kind of processes, including those in engineering, is today particular importance, this being demonstrated by a large interest dedicated to this topic. How, in the present, the industrial robots are used on a more and more wider scale, across all industries, searching the optimal trajectories is a complex and difficult task, assumed by researchers and specialized companies in many kinds of approaches. This work is integrated into this current, and represent a part of wider research developed by us regarding modeling, simulation, and optimization of the trajectory of the industrial robots.

References

[1] Ratiu M and Prichici M A 2017 Industrial robot trajectory optimization- a review *Annual Session of Scientific Papers IMT ORADEA 2017 (MATEC Web of Conferences)* **126**, 02005

[2] Berbente C, Mitran S and Zancu S 1998 *Numerical methods* (Bucharest: Technical Publishing House)

[3] Necoara I 2013 *Methods of numerical optimization* (Polytechnic University of Bucharest-Department of Automatic and System Engineering)

[4] Grebenisan G 2008 *Metode numerice în inginerie, folosind MATLAB* (Oradea University Press)

[5] Rao A V 2014 Trajectory optimization: a survey *Optimization and Optimal Control in Automotive Systems* (Springer) **455** pp 3–21

[6] Parkinson A, Balling R, Hedengren J 2013 *Optimization Methods for Engineering Design. Applications and Theory* (Brigham Young University)

[7] Dzițac I 2008 *Artificial Intelligence* (Arad: Aurel Vlaicu University Publishing House)

[8] Konar A 2005 *Computational Intelligence. Principles, Techniques and Applications* (Springer)

[9] Kordon A 2010 *Applying Computational Intelligence. How to create value* (Springer)

[10] Cazacu P 2015 *Contributions regarding the use of evolutionary algorithms to the constructive-technological optimization of prismatic parts in motor vehicles* (University of Oradea: doctoral thesis)

[11] Grebenisan G and Salem N 2017 The multi-objective genetic algorithm optimization, of a superplastic forming process, using ansys *Annual Session of Scientific Papers IMT ORADEA 2017 (MATEC Web of Conferences)* **126**, 03003

[12] Antonya C, Butnariu S and Beles H 2015 Parameter estimation from motion tracking data. *Lecture Notes in Computer Science* (Springer) **9185** pp 113-121

[13] Antonya C, Butnariu S and Beles H 2015 Geometric identification of a four-bar linkage from noisy tracking data *IFToMM World Congress Proceedings* IFToMM 2015 (Taipei, Taiwan) pp 827-832

IOP Conf. Series: Materials Science and Engineering **568** (2019) 012010 doi:10.1088/1757-899X/568/1/012010

Digital world of co-creation: strategy and process

A Bujor[1], S Avasilcăi[1], C A Hutu[1], E Avram[1]
Department of Engineering and Management „*Gheorghe Asachi" Technical University of Iasi*, Bd. D Mangeron, no.29, Iasi, Romania

adbujor@gmail.com

Abstract. The business world, irrespective of the type of industry, has become increasingly complex and amazingly competitive, and companies have to move towards innovation as one of the few sustainable sources of competitive advantage. Innovation is now considered to be one of the top priorities for most major companies in the world. Innovation has become a necessary process, widely accepted, but beyond their enthusiasm for bright ideas, most leaders know that in order to be successful in the long run, they need to develop a strong culture of open innovation and to accept that closed innovation is no longer enough to be competitive on the market. Thus, innovation challenges are considered to be one of the top priorities for most major companies in the world. From this point of view, this paper aims to highlight the strategic implications of the open innovation in co-creation processes, to present collaborative stages in terms of product development and to provide valuable examples of co-creation process and customers' engagement in various industries.

1. Introduction

Due to the fact that innovation has become more and more an open process as a result of companies' transition from closed business models to open-ended business models, as Chesbrough calls them, companies have also become more and more effective in creating and capturing values [1]. By practicing open innovation, which has become increasingly appealing due to a growing division of labor, companies are developing ideas that they do not immediately bring to the market but seek collaborators to put them into practice or partners to handle the marketing of these ideas. In this context, the digital world plays a significant role, as shown by recent research on digital marketing and virtual consumers. They have revealed the growing involvement of consumers/ stakeholders in all aspects of marketing processes, from product innovation to printed and broadcast advertising. Consequently, switching from a traditional business model to a business model based on open innovation is an important step taken by many companies that have grasped the importance of their ongoing business development, focusing in particular on 3C: co-creation, crowdsourcing and community.

Nowadays, more and more companies understand and agree that for the development of products and/ or services, communities, crowds, stakeholders represent the driving force, the main engine. In this respect, in order to achieve an interactive development of products and/ or services, companies delegate the main tasks to the crowd (empowers people), as stated by Bujor & Avasilcai [2]

2. Co-creation as a strategy for innovation performance

Among open innovation strategies, co-creation is an important economic strategy that has already been adopted by companies from different industries, from the traditional ones (e.g. automotive, consumer goods, food and beverages) to most emergent ones (creative industries, e.g. IT, fashion design, interior and ambient design, graphic design, architecture), highlighting thus its importance as key resources for innovation performance, and based on so-called Client Knowledge Management (CKM) [3].

Most of the co-creation definitions offered by the literature automatically lead to a collaboration, namely a creative collaboration. This type of collaboration invites both parties involved in co-creative initiatives to be prepared to take risks, open their hearts and minds to create something new or re-create something already existing. In this way, value is increasingly created by both barricades: customer and producer.

The most common definition of co-creation was provided by Prahalad and Venkat, according to which co-creation is "active, creative and social process, based on collaboration between producers and users that is initiated by the firm to generate value for customers" [4]. Value refers either to new concepts and solutions or new products and services, and the collaborative development of this new value takes place between the company and experts and/ or stakeholders such as customers, fans, suppliers, etc [5].

Co-creation, as a strategy of open innovation, generally addresses three main issues, according to Neumann:

- is based on the relationship between a company and a group of stakeholders, most of the time its customers;
- involves working with end-users of a product or service, sharing knowledge, experience and resources to finally offer a unique experience using the company's value proposition;
- is a way to involve customers directly, even in product development processes [6].

According to Belkahla W and Triki A, CKM, defined as the combination of the knowledge management and the client/ customer relationship management principles, increases two main roles that customers have during the co-creation process: that of co-creator, and that of co-producer, the last one being considered by Hilton and Hughes a specific component of the co-creation process, related to customer specific tasks that may occur anytime during the co-creation process [7].

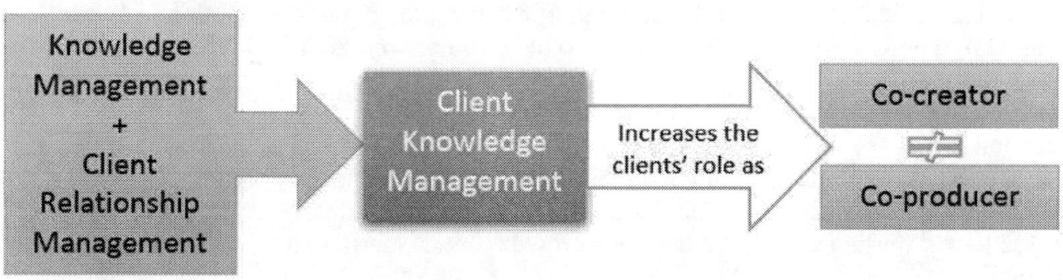

Figure 1 Client Knowledge Management approach, developed according to [7].

Starting with integrating innovation as a practice-based co-creation process, Russo-Spena & Mele developed the Five "Co-s" for innovation Model, which includes: co-ideation, co-evaluation, co-design, co-test and co-launch [8]. Each "Co-" represents a stage in the innovation process that results from dynamic and ongoing interactions between resources, actions and a group of actors that are interconnected through a strong and dense network.

In fact, on this occasion, Russo Spena and Mele make two proposals: the first proposal is that co-creation in the innovation process can be disrupted in several "Co-s", surrounded by value-generating processes from multiple stakeholders, while the second relates to the fact that co-creation of innovation can be seen as a set of practices [8].

According to the model proposed by Russo-Spena and Mele, each stage of innovation can be considered as a process of finding new ways to co-create value through more effective participation in the integration of resources from a multilateral perspective, as Mele et. al. highlight and Gammesson also points out [9].

IOP Conf. Series: Materials Science and Engineering **568** (2019) 012010 doi:10.1088/1757-899X/568/1/012010

The 5 Co-s for innovation - Innovating together

Co-ideation	Co-evaluation	Co-design	Co-test	Co-launch
The generation of ideas from an external network of stakeholders; Pro-active involvement of different actors, from lead users to consumers, customers, fans, intermediaries, partners, and professionals in the idea generation and shaping phase; Involved actors become part of a collaborative community.	Linked to the generation of ideas; Two main activities: commenting and voting. Stakeholders can vote without participating in the idea generation phase.	Encompasses a wide range of practices based on the involvement of many different actors; Participants are interconnected by a common context and interest, with a more specific purpose, the gap between identified ideas or needs and finding a solution being thus reduced.	Closely related to the launch of products and services on the market. Used to support product or service prototype improvement before being marketed; Often used to test the commercial character of a product or service.	The launch of the product or service created with the collaborative community based on the ideas generated and voted, then designed and tested.

Figure 2 Innovation stages in terms of collaborative engagement [8].

Each stage of the collaborative innovation process reflects a set of activities which are relevant in order to achieve the symbiosis between developed processes, necessary resources and customers' engagement.

From this point of view, it seems important for companies to continuously explore digital environment in order to respond and to adapt to the turbulent market requirements.

3. Co-creation initiatives: examples from the industry

Co-creation initiatives did not appear in the context of Web 2.0, but they managed to flourish because of the fact that these platforms included many forms of co-creation marketing. Since then, as Chaney stated, Web 2.0 platforms have been seen as social and consumer communities that have become "ambassadors", "buzz agents", "intelligent crowds", and "participants" that transform the product experience [10]

In light of the above, the authors of this paper have conducted an analysis of some of the companies belonging to different types of industries, from the most traditional to the most emerging, as is the case for creative industries that have developed dedicated instruments, as a specific meeting environment for stakeholders, to host innovation and to co-create together.

Consequently, there can be seen that companies seek to implicate different type of stakeholders, especially customers, into own product development process as valuable sources of knowledge and innovation. However, depending on the industry affiliation of the companies the innovation initiatives are concentrated more or less on the revealing of the stakeholders' creative potential. It seems relevant to enhance the collaborative relations in designing the product in order to achieve a high level of product customization.

Table 1 Examples of business that have developed websites in order to host open innovation, specifically co-creation with stakeholders [11-18]

Type of industry	Company/ *Project*	Co-creation process's stages	Engaged stakeholders	Key aspects
Automotive	BMW/ *Co-creation Lab*	Co-ideation Co-design	Employees Students Clients (BMW owners)	Develops new designs and functionalities; Ideas gathering from professionals; Brainstorming with employees.
Automotive	Volvo/ *Involve*	Co-ideation	Group experts Students Universities	Innovative ideas gathering based on specific topics; Creating concepts of future product; Establish long term collaboration and partnership.
Consumer Goods	Procter & Gamble/ *Connect and Develop*	Co-design	Experts Intermediaries Managers Partners Professionals	To generate and develop new solutions; To design prototypes.
Food & Beverage	Barilla/ *Barilla Factory*	Co-ideation Co-evaluation	Creatives Artists	To express themselves in different creative languages; Dialogue and create new landscapes and new stories together; Share the Barilla Factory projects with all friends; Vote on the attractiveness of the design and the best videos.
Food & Beverage	Heineken / *Ideas Brewery / Innovators Brewhouse*	Co-ideation	Entrepreneurs Inventors Digital innovators Consumers Suppliers Experts	Creating innovative packaging; Recycling and up cycling objectives; Creating consumers' consumption experience; Developing new technologies through stakeholders' collaboration.
Fashion Design	A Way to Mars	Co-ideation Co-evaluation Co-design	Designers Professionals Creatives	Share ideas by posting on the co-creation forum Evaluate ideas fairly and equally Provide feedback and help improve designs.
Music	Propellerhead	Co-ideation Co-test Co-launch	Passionate about music Producers Composers	Suggest new ideas and solutions for company software development; Test the software development project.
Graphic Design	99Designs	Co-ideation Co-evaluation Co-design	Clients (companies) Designers (graphic)	Require various graphic representations (e.g. logos); To opt for one-to-one help; To propose, provide graphical representations as required; Appreciate, vote on the projects provided; Set the reward method.
Graphic Design	Threadless	Co-ideation Co-evaluation Co-design, Co-test Co-launch	Consumers Fans Designers Collectors Artists Managers	To propose designs; To appreciate, evaluate the attractiveness of the design; Take responsibility for advertising.

4. Conclusions

The most successful companies around the world are deciding to adopt the new open innovation model due to the competitive advantages they gain by digitizing innovation and engaging customers in virtual communities where they can share ideas, bring new perspectives in terms of packaging, new technologies or new methods for designing products.

So did the companies that were the subject of the analysis presented in Table 2, who understood the importance of direct involvement of their own consumers in the co-creation of products that correspond to their wishes or needs. For some of them, co-creation is the chance that their dream will become reality, while for the companies, it represents the chance to improve their innovation performance. Consequently, co-creation is about creating value for both parties, and with the digitization of innovation all became even easier, the involvement of the consumers in the process of co-creation being possible irrespective of the geographic position of either party.

In terms of strategic implications, the co-creation process development became a keystone in value creation and capturing. The continuous engagement of stakeholders seems to have a strong impact on designing activities and ensures the product conformity with customers' expectation. As it can be

IOP Conf. Series: Materials Science and Engineering **568** (2019) 012010 doi:10.1088/1757-899X/568/1/012010

observed the co-ideation and co-design stages are widely explored by various companies from different industries.

By using virtual environments and open innovation techniques companies are increasing their own visibility on the market and profitability.

References

[1] Chesbrough H M 2007 *Why Companies Should Have an Open Business Model* MIT Sloan Manag Rev 48(2) 22

[2] Bujor A and Avasilcăi S 2018 Open innovation in creattive industries. Part I: Innovation and design *IOP Conf. Ser.: Mater. Sci. Eng.* 400 062007

[3] Belkahla W and Triki A 2011 Customer knowledge enabled innovation capability: proposing a measurement scale *J Know Manag* 15(4) 648-674

[4] Prahalad C.K. and Ramaswamy V 2000 Co-opting Customer Competence *Harv Bus Rev* 78 (1) 79-87

[5] Fronteer https://fronteer.com/what-is-co-creation/

[6] Neumann M 2017 Open innovation vs Crowdsourcing vs Co-creation https://www.wazoku.com/open-innovation-vs-crowdsourcing-vs-co-creation/

[7] Hilton T and Hughes T 2008 Co-production and co-creation using self service technology: the application of service-dominant logic, *Otago Forum 2 – Academic Papers*, Paper no 4

[8] Russo-Spena T and Mele C 2012 "Five Co-s" in innovating: a practice-based view *J Serv Manag* 23(4) 527 – 553

[9] Gummesson E 2008 Customer centricity: reality or a wild goose chase? *Euro Bus Rev* 20(4) 315-330

[10] Chaney D 2012 The Music Industry in the Digital Age: Consumer Participation in Value Creation *Int J Arts Manag* 15(1) 42-52

[11] Avram E 2016 Case Study: BMW in Avasilcai S and Hutu C A (eds) Innovation through co-creation in online era Performantica 147-182

[12] Avram E 2016 Case Study: Volvo in Avasilcai S and Hutu C A (eds) Innovation through co-creation in online era Performantica 183-218

[13] Agafitei I G 2016 Case Study: Procter & Gamble in Avasilcai S and Hutu C A (eds) Innovation through co-creation in online era Performantica 255-280

[14] Bujor A 2016 Case Study: Barilla in Avasilcai S and Hutu C A (eds) Innovation through co-creation in online era Performantica 309-336

[15] Rusu G 2016 Case Study: Heineken in Avasilcai S and Hutu C A (eds) Innovation through co-creation in online era Performantica 377-402

[16] Bujor A, Avasilcai S and Alexa L 2017 Co-creation in fashion industry: The case of AWAYTOMARS, *Fas. of Manag. and Techn. Eng.* XXVI(XVI) 3 22-25

[17] Galateanu (Avram) E and Avasilcai S 2017 Emerging creative ecosystems: platform development process *Fas. of Manag. and Techn. Eng.* XXVI(XVI) 3 5-10

[18] Propellerhead 2018 https://www.propellerheads.se

[19] Parmentier G and Mangemantin V 2014 Orchestrating innovation with user communities in the creative industries *Technological Forecasting and Social Change* 83 (1) 40-53

[20] Avasilcai S and Bujor A 2018 Open innovation in creative industries. Part II: The case of Threadless *IOP Conf. Ser.: Mater. Sci. Eng.* 400 062002

IOP Conf. Series: Materials Science and Engineering **568** (2019) 012011 doi:10.1088/1757-899X/568/1/012011

Modeling the mechanical and electrical characteristics of the synchronous motor for different variations of torque in time

C Cristea[1] and I Stroe[1]

[1] Centre "Advanced Research on Mechatronics", Transilvania University of Braşov, Romania

cosmin.cristea@unitbv.ro

Abstract. The paper presents the mechanical and electrical characteristics of the synchronous motor at different imposed variations of torque over time. Modelling the resistive torque allows the loading with different capacities vibrations. The experimental setup is designed and presented in such a way as to allow a servomotor to stop the synchronous motor in variable torque values. The resulting characteristics highlight the mechanical characteristics of the synchronous motor as well as the resistance to repeated sudden repeated breaking within a shorter or longer period, observing the evolution of the speed up to the point of de-synchronisation of the motor. It is presented the mechanical characteristics of the mechanical power, electrical current, rotational speed and mechanical torque in function of time observing the behaviour of the synchronous machine when applying sudden varying resistive torques and its de-synchronization. The determination of these characteristics is made at a constant value of the excitation current.

1. Introduction

The synchronous machine is used in most industrial applications as an motor, generator or synchronous compensator [1]. It is an AC machine at which the speed is constant no matter the torque value at the shaft, within normal limits and regardless of the stabilized operating regime. The speed is synchronous and is strongly linked to the frequency 'f' of the alternating current power grid from which the machine is supplied [2].

Just like the other rotating electric machines the synchronous machine consists of three main parts: the stator, the rotor and the construction parts, but it has a wide constructional variety.

Depending on the role played by the two armatures, there exist:

— synchronous machines in normal construction, with induced stator and inducting rotor;

— synchronous machines in reverse construction, with inducting stator and induced rotor.

In the paper the normal construction is considered, being the most widespread; the reverse construction is encountered in relatively low power machines, and especially in synchronous machines used as excitation sources for brushless excitation systems, consisting of the synchronous generator which outputs power on the rotating rectifier [3].

The basic operating mode of the synchronous machine is the electric generator mode, just as the motor mode is the basic mode for the asynchronous machine. The synchronous machine as a generator is the economic basis for electricity generation in all existing power plants. In this mode the synchronous machines reach the highest rated powers, being the largest man-made electrical machines [4].

In the generator mode, the machine converts the mechanical power received at the shaft from an auxiliary motor into electrical power, fed into an AC network. In the motor mode, the machine converts the electrical power received from an AC network into mechanical power, presented at the shaft and provided to a mechanical installation [2].

Three-phase synchronous generators are used to generate electricity in power plants. They can be driven into rotation by primary motors such as: hydraulic turbines, diesel engines, thermal turbines, etc. Three-phase synchronous motors are built for high power (hundreds of KW) and are used for the driving at constant speed of certain machines: pumps, compressors, fans, etc [4].

In the following, only three-phase synchronous machines will be referred [5].

2. The synchronous motor

The Synchronous motor mode is especially used due to the benefits of asynchronous motors (higher efficiency, the power factor tending towards unity, steady torque at varying speed, bigger air-gap). This was only possible after technology managed to successfully solve two serious deficiencies of the synchronous motor: the absence of the starting torque and the possibility of developing a pendulum-like behavior with the danger of exiting synchronism (loss of stability) [6].

In this operating mode, the synchronous machine is used in all drives requiring constant speed (compressors, turbochargers, irrigation pumps, etc.), replacing asynchronous motors (especially at high power where economic considerations predominate: efficiency, power factor) [2].

Due to their rigid characteristics, the synchronous motor is increasingly used on a large scale, especially in the mechatronic drives.

The main disadvantage of the synchronous motor is that it develops torque only at synchronous speed and additionally has a rigid mechanical characteristic until the resistive torque reaches the maximum value of the electromagnetic torque at which time the motor exist synchronism.

Other drawbacks are related to the oscillations that may occur at sudden load variations at the shaft, the more complicated construction due to the exciter and the higher cost. In case of reduced power drives, startup electrical schematics are more complex in the case of varying the imposed speed dependent on the technological process. Synchronous motors are currently preferred especially to drive work machines that require high power and constant working speed [5].

3. Mechanical and electrical characteristics of the synchronous motor

The synchronous motor has a rigid mechanical characteristic in relation to the load at its shaft [7]. In the following section we study the behavior of this machine at different variations of torque from time and amplitude point of view. The aim is to study the behavior of the synchronous machine to see how far it can work until it exists synchronism [8].

In order to determine the mechanical and electrical characteristics it consists of the following components:

– Universal power supply with thermal protection;
– Digital-analogue measuring device, voltmeter, ammeter;
– Varying transformer drive – to generate the excitation current and voltage;
– Servomotor controller;
– Servomotor with brake and controller;
– Active Servo dedicated software;
– Synchronous machine;
– General tripolar switch with 2 positions.

This was used to simulate an increasing value of the torque.

The stator is supplied with three-phase alternating current and the rotor is powered by a DC voltage called excitation voltage. Both voltages as well as the excitation current are monitored. A servomotor applies a resistive torque to a synchronous motor coupled to the shaft with it. A controller monitors the results and displays them as graphs [8].

IOP Conf. Series: Materials Science and Engineering **568** (2019) 012011 doi:10.1088/1757-899X/568/1/012011

4. Torque modelling for characteristics determination

The execution and monitoring system consists of a servomotor, a controller and dedicated software called ActiveServo.

Through ActiveServo we generate torque diagrams that will brake the motor. Torque charts can be built to act on the synchronous motor shaft for a longer or shorter time.

ActiveServo monitors the electrical and mechanical parameters of the motor via the brake servomotor, displaying the results.

5. The mechanical and electrical characteristics at different variations of the motor torque with time

In the experimental determinations performed to study the characteristics, a synchronous motor with the following characteristics was used: $P_n = 0.27kW$, $I_n = 1.5A$, $f = 50Hz$, $n = 1500rpm$, $U_{err} = 20Vcc$; $I_{err} = 4A$. The excitation source generated a fixed voltage at the excitation current value of 3A.

For the studied synchronous motor we will observe the time dependence of the following quantities: The current of the rotor circuit I [A], the mechanical power on the shaft P2 [W], the motor torque M [Nm], the speed n [rpm]

By imposing torque variations at the synchronous machine's shaft, we observed its behavior in terms of speed because any variation of torque at the shaft leads to variations of speed, current and torque. Four different torque characteristics were set-up to observe and determine these aspects.

In the first part of this study, we considered that variations are short-lived, and they occur for 10s, a leap of linear increasing or decreasing variation occurs for one second. The resulting torque characteristic is shown in figure 1.

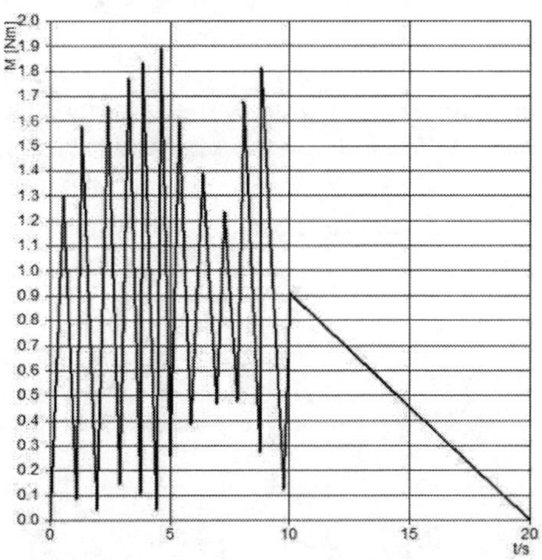

Figure 1. Variable imposed torque characteristic for I_{err}=3 A, in 10s

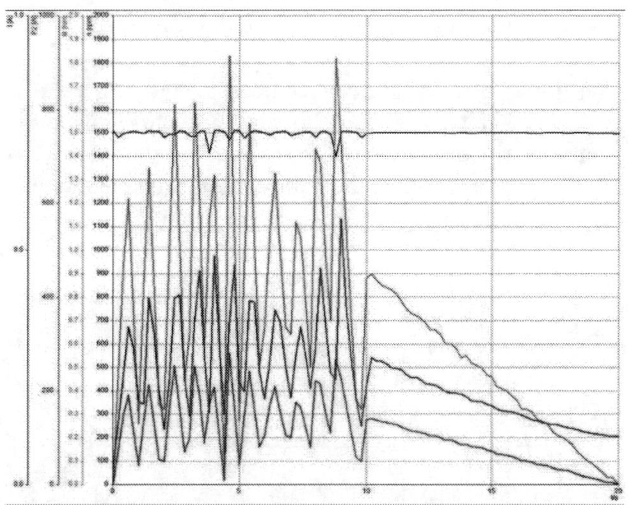

Figure 2. Mechanical and electrical characteristics for varying torque with time, at an excitation current of I_{err}=3 A, in 10s

From the analysis of the studied graph, in figure 2 it is observed that the speed has small variations at linear increasing and decreasing torque up to the value of 1.63Nm. From this value, the motor tends to become out of sync, especially when the motor torque drops. Speed is variable and drops to 1400 rpm. At variations that return to smaller torques, it returns to the synchronism speed, but then, the speed variation returns at a sudden increase of torque to 1.8Nm.

IOP Conf. Series: Materials Science and Engineering **568** (2019) 012011 doi:10.1088/1757-899X/568/1/012011

For this reason it was found appropriate to see what happens to the synchronous machine if it has the same torque variations, but over a 50s time period, shown in Figure 3.

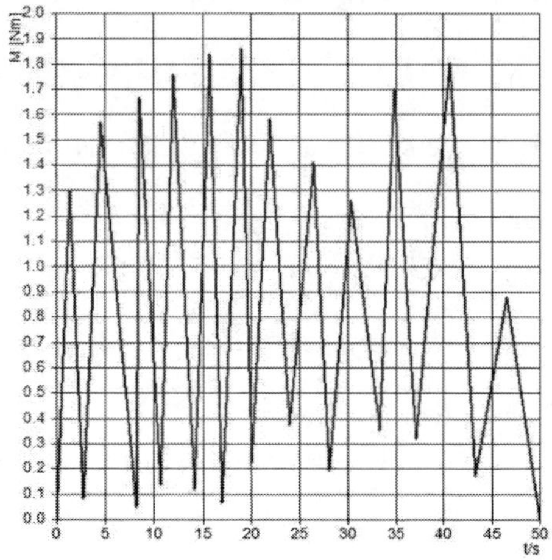

Figure 3. Variable imposed torque characteristic for I_{err}=3 A, in 50s

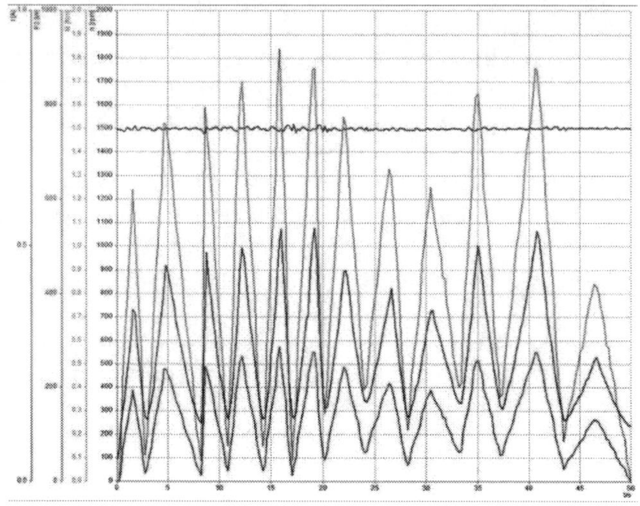

Figure 4. Mechanical and electrical characteristics for varying torque with time, at an excitation current of I_{err}=3 A, in 50s

Analyzing the figure 4, in which the torque curve was kept at the same motor torque values but with a run time of 50s, the following is observed:

The speed is kept relatively constant at linear increasing and decreasing variations such that it fits within a 10% variability. We need to study how the synchronous generator supports these variations.

Next, it was studied how the synchronous motor behaves at sudden torque variations because nature can be unpredictable and can accelerate or brake the motor's shaft.

We have imposed the following torque characteristic, which looks like in figure 5.

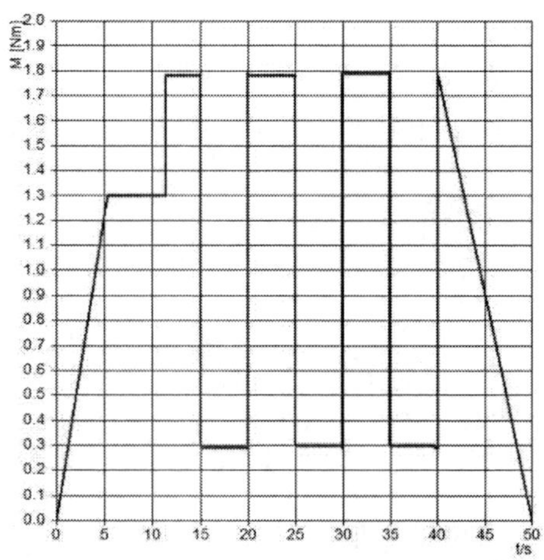

Figure 5. Variable imposed torque characteristic for I_{err}=3 A, in 50s

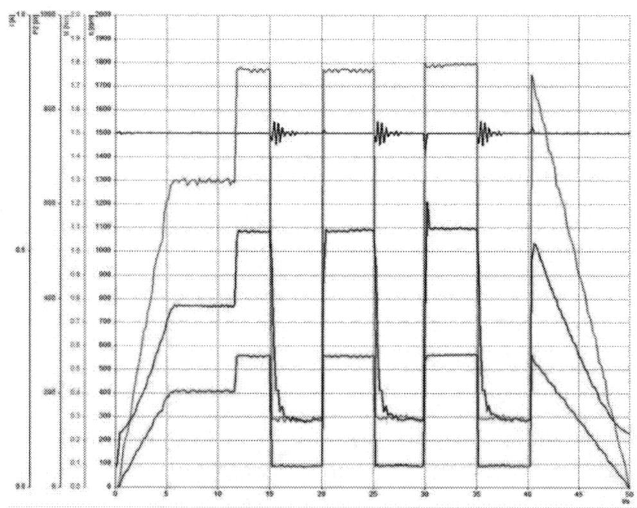

Figure 6. Mechanical and electrical characteristics for sudden varying torque with time, at an excitation current of I_{err}=3 A, in 50s

I considered the motor is starting and has a linear torque increase from 0 to 1.3Nm. It is kept at this value and after 7 seconds, the torque increases sharply to 1.8Nm.

After 4s it suddenly increases to 1.8Nm and after another 5 seconds it drops to 0.3Nm. After this value, the sudden increase and decrease of torque is maintained until 40s, after which it decreases linearly from 1.8Nm to 0Nm. The results in figure 6 were obtained.

From the graph analysis, figure 6, it is observed that the speed is kept constant at the increasing value up to 1.8Nm. However, there is a variation of the speed at the sudden decrease of the torque, which is repeated at each decrease. The electrical current and mechanical power keep the torque characteristic. In conclusion, the motor is slightly offset from synchronism by sudden torque drops, but it does not go out of synchronism.

For this reason it was considered necessary to observe what happens to the speed at variable values of increasing and decreasing values of the motor torque (figure 7).

From the graph analysis in figure 8, it is observed that the speed is kept constant at the increase value up to 1.8Nm. However, there is a variation of the speed at the sudden decrease of the torque, which is repeated at each decrease, but the variations are smaller, with the decrease of the torque value.

The electrical current and mechanical power keep the torque characteristic. In conclusion, the motor is slightly offset from synchronism at sudden torque drops, but it does not go out of synchronism.

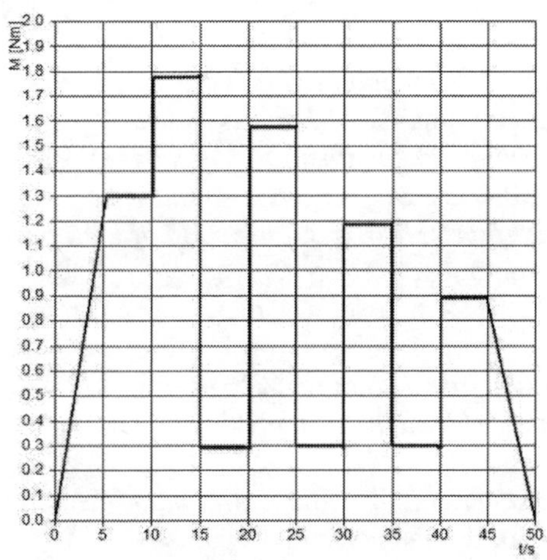

Figure 7. Imposed torque characteristic for I_{err}=3 A, in 50s

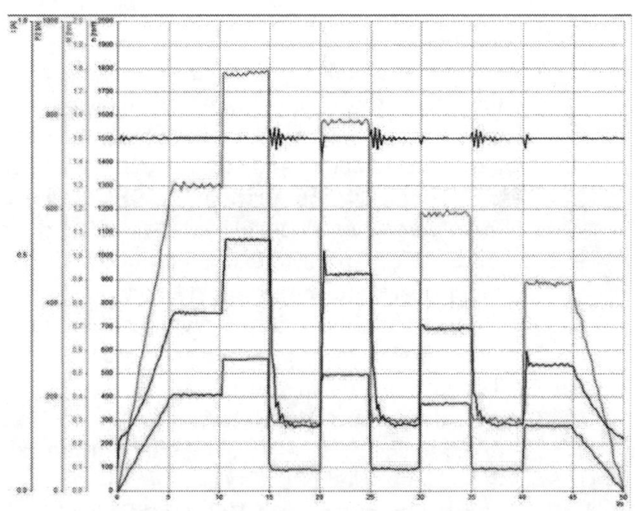

Figure 8. Mechanical and electrical characteristics for sudden varying torque with time, at an excitation current of I_{err}=3 A, in 50s

6. Conclusions

In the experimental determinations performed to study the characteristics, a synchronous motor with the following characteristics was used: Pn = 0.27kW, In = 1.5A, f = 50Hz, n = 1500rpm, Uerr = 20Vcc; Ierr = 4A. The excitation source generated a fixed voltage at the excitation current value of 3A.

For the studied synchronous motor we will observe the time dependence of the following parameters: The current of the rotor circuit I [A], the mechanical power on the shaft P2 [W], the motor torque M [Nm], the speed n [rpm].

Torque variations at the synchronous machine's shaft were imposed, and it was observed its behavior in terms of speed because any variation of torque at the shaft leads to variations of speed,

IOP Conf. Series: Materials Science and Engineering **568** (2019) 012011 doi:10.1088/1757-899X/568/1/012011

current and torque. Four different torque characteristics were set-up to observe and determine these aspects.

In the first part of this study, we considered that the variations are short-lived, and they occur for 10s, a leap of linear increasing or decreasing variation occurs for one second.

In the second part, it was considered appropriate to see what happens to the synchronous machine if it has the same torque variations, but over a 50s time period.

In the third part, it was studied how the synchronous motor behaves at sudden torque variations because nature can be unpredictable and can accelerate or brake the motor's shaft.

For this reason it was considered necessary to observe what happens with the speed at varying increasing and decreasing values of torque.

Analyzing the first two graphs, we can conclude that at linear increasing and decreasing variations, the synchronous motor keeps constant speed around the nominal synchronism value if the variations are over a period of time greater than 2s.

However, if the same variations occur in a shorter time, the motor has a fluctuating speed, but it does not exit synchronism.

From the last graphs, it is noticeable that at torque variations, the motor is slightly offset, the speed variability being higher at decreasing torque, which diminishes if the torque value on the motor shaft is lower.

7. References

[1] Nedelcu V 1968 *Electrical Machines,(Maşini electrice)* Editura Didactică şi Pedagogică, Bucureşti

[2] Calueanu D *Electrotechnics and Electrical Machines (Electrotehnica și Mașini Electrice)* Facultatea de Nave şi Inginerie Electrica, Universitatea Dunarea de Jos din Galaţi

[3] Cioc I, Nica C 1994 *Design of electric machines (Proiectarea mașinilor electrice)* Editura Didactică şi pedagogică R.A. Bucureşti, ISBN:973-30-2937-8

[4] Cioc I, Bichir N and Cristea N 1981 *Electrical Machines, Design Guide (Masini electrice, Indrumar de proiectare) Vol. I, II, III.* Ed. Scrisul Românesc , Craiova

[5] Engine Operating Modes Continuous Current and the synchronous motor (Regimuri de functionare ale motorului de current continuu si ale motorului sincron)[online] 10 March 2019. Available at: http://www.euedia.tuiasi.ro/uee/uee_files/Lucrarea%2014.pdf

[6] Teodoru E C and Gogu M 2000 *Electrical Machines,(Maşini electrice)*Editura Universităţii Tehnice Gh.Asachi , Iaşi

[7] Cernat M, Bidian D and Marinescu C 1998 *Electrical machinery and equipment (Masini si aparate electrice)* – Vol. II. Braşov, Editura Universitatii Transilvania din Brasov

[8] Cristea C and Stroe I 2017 *The study of the synchronous motor*, Bulletin of the Transilvania University of Braşov • Vol. 10 (59) No. 2 – 2017 Series I: Engineering Sciences, ISSN 2065-2119 (Print), ISSN 2065-2127 (CD-ROM)

IOP Conf. Series: Materials Science and Engineering **568** (2019) 012012 doi:10.1088/1757-899X/568/1/012012

Assembly project management of a suspended conveyor

F S Blaga[1], I Stănăşel[1], A Pop[1], V Hule[1] and D Craciun[1]
[1]University of Oradea, Faculty of Management and Technological Engineering, str.Universităţii nr.1, Oradea 410087, Romania

alinpop23@yahoo.com

Abstract. Modern manufacturing systems are characterized by the fact that they integrate various machines and work equipment, industrial handling robots, industrial robots performing technological operations (welding, riveting, painting, etc.), transfer systems (conveyors, conveyor belts, automated guided vehicle -AGV), storage systems and inspection stations. The paper presents the way in which the assembly project management of a suspended conveyor system was carried out. Critical path method was used. It has identified the assembly operations, the way they succeed in the technological process of assembly, the duration of the operations and the resources (labor force) necessary for their realization. This information has been processed through Microsoft Project resulting in critical activities (critical path), project duration, the ratio between the available and necessary resources (in the analyzed case a significant deficit of the machinist resource) and the corresponding costs. In order to solve the problem of resource requirements, two solutions were used: leveling resources and supplementing them in order to reduce the duration of critical activities. Finally, a version of the conveyor assembly project management resulted in a balance between costs and duration.

1. Introduction

The project management of the conveyor assembly will be done by applying the critical path analysis methods. Critical path analysis defines a scientific discipline for the implementation of projects using graphs theory. In the theory of the Critical Path Analysis Method, the project means a high intensity action or a complex process that has the objective a well known and defined result [1].

Critical path method is commonly used in the production management of unit type and not only Assembly processes can be managed through these methods. Thus, the paper [2] delves into the scheduling of complex assembly lines with workforce constraints. This paper presents an extended Critical Path Method (CPM) using a simulation approach to calculate probable Critical Paths of the projects. A new algorithm that obtains an approximation of the Critical Path in schedules generated using the disjunctive graph model that represents the Job Shop Scheduling Problem (JSSP) is presented in [3]. The study [4] aims to perform the best scheduling project of the empennage structure design of Indonesia's aircraft with limited resources. Critical Path Method is used to schedule the project based on activity and resource.

The paper [5] shows how the critical path method was used in the project management and execution of automatic detergent packaging equipment. The paper [6] proposes using the Critical Path Method for Making Process Layout of a T-Shirt within Earliest Finish Time.

This paper presents the way in which, using the critical path method, the assembly activity for a conveyor transfer system was managed.

IOP Conf. Series: Materials Science and Engineering **568** (2019) 012012 doi:10.1088/1757-899X/568/1/012012

2. Defining the activities of the conveyor assembly project

The conveyor is designed for the transfer of components on assembly lines in the automotive industry but can be adapted to carry tasks specific to other areas of activity. The designed and constructed conveyor is a suspended type and carries individual loads between the loading and unloading stations on the technological route. The track on which the conveyor works is a complex one, consisting of linear and curved segments, arranged horizontally or inclined at an angle of 30 ° and is a maximum length of 15m. In figure 1 is presented the 3D model of the conveyor that is the object of the work, and in figure 2 the conveyor made practically.

The main components of the conveyor are (figure 1 and figure 2): 1- metal frame; 2- drive subassembly; 3- drive gear; 4- trolleys rollaway, 5-transport trolleys rollaway; 6- transport trolleys 7- drive cable; 8- strut; 9- driven gear; 10- electric and command subassembly.

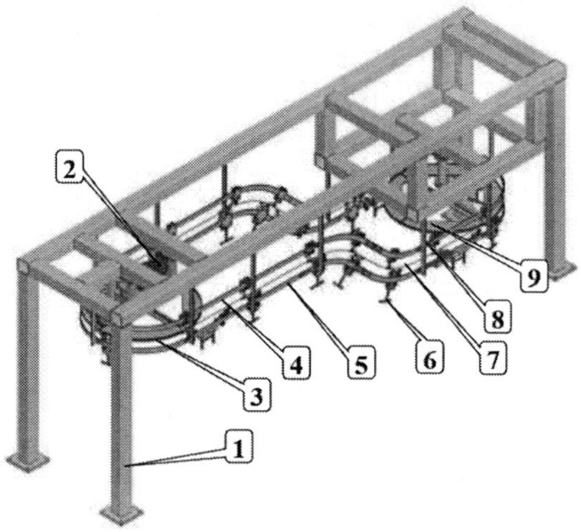

Figure 1. Conveyor - 3D model.

Figure 2. Conveyor made practically.

In table 1 is presented the activities of the conveyor assembly project.

Table 1. The activities of the conveyor assembly project.

Nr. crt.	Name	Time (h)	Predecessors	Resource	Necessary
1.	Frame (metal frame)	0	-	-	
2.	Support motor mounting	2	1	Machinist	2
3.	Side struts mounting	6	1	Machinist	2
4.	Upper Rollaway mounting	8	3	Machinist	3
5.	Lower Rollaway mounting	8	3	Machinist	3
6.	Gearmotor mounting	2	2	Machinist	2
7.	Vertical axis mounting	1	6	Machinist	2
8.	Support bush mounting bearing_1	2	7	Machinist	1
9.	Driver gear mounting	2	8	Machinist	3
10.	Support bush mounting bearing_2	2	2	Machinist	1
11.	Driven axis mounting	1	10	Machinist	2
12.	Driven gear mounting	2	11	Machinist	3
13.	Trolleys mounting	6	4	Machinist	2
14	Cable mounting	3	13	Machinist	3
15	Trolleys mounting	6	14	Machinist	2
16.	Control cam mounting	2	15	Machinist	2
17.	Mounting sensors	2	16	Electrician	1
18.	Electrical cabinet installation	2	17	Electrician	2
19.	Making electrical connections	4	18	Electrician	2
20.	Making pneumatic connections	2	19	Electrician	2

All activities take into account the following associated information: the duration in hours, the activities to be carried out for the conveyor assembly, technological dependencies between activities (predecessors) and the type of resource needed to carry out the activity, the resource requirements.

Table 2 provides information on the available workforce (resources) and on workforce costs.

Table 2. Workforce (resources) available and costs

Nr. crt.	Resource	Type	Available	Standard rate (lei/h)	Overtime Rate (lei/h)	Accrue
1.	Machinist	Work	3	20	40	Prorated
2.	Electrician	Work	2	20	40	Prorated

3. Project implementation in the *Microsoft Project®* program

The project will be implemented in Microsoft Project®. It allows a complex analysis of all aspects of project management: definition of activities, critical path determination, allocation of resources, cost analysis, etc.

The working hypotheses are: it was considered that the week has five working days: Monday, Tuesday, Wednesday Thursday and Friday; the working day has eight hours of work; the program starts at 8 am; the break is between 12: 00-13: 00; the end of the working day is at 1700.

3.1. First version

In a first version, the activities are transferred to the program according to the information in table 1. The Gantt chart is shown in figure 3. The same figure shows the activities that have a resource deficit - the necessary is greater than the available.

The synthesis information regarding the first version in the Microsoft Project® program: the project runs for 5.13 days; the time worked is 157 hours; the labor costs are 3.140 lei.

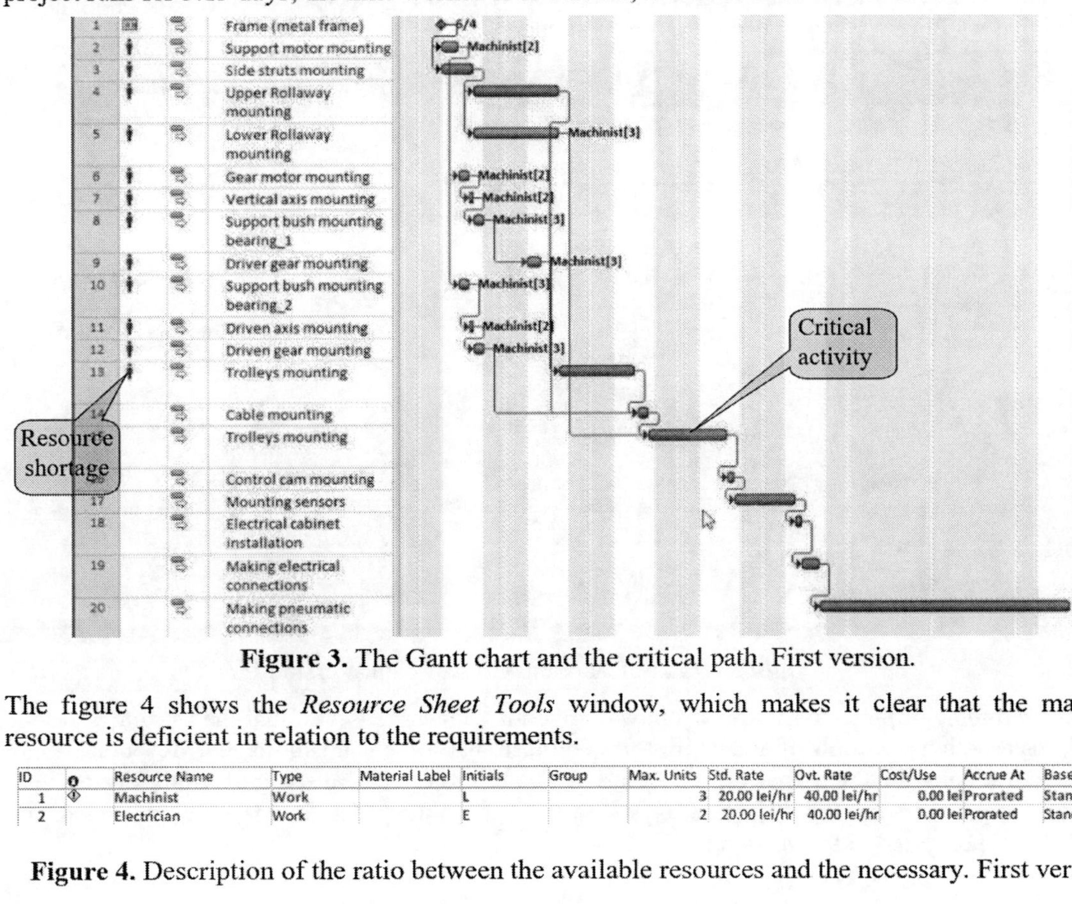

Figure 3. The Gantt chart and the critical path. First version.

The figure 4 shows the *Resource Sheet Tools* window, which makes it clear that the machinist resource is deficient in relation to the requirements.

ID		Resource Name	Type	Material Label	Initials	Group	Max. Units	Std. Rate	Ovt. Rate	Cost/Use	Accrue At	Base Calendar
1	◇	Machinist	Work		L		3	20.00 lei/hr	40.00 lei/hr	0.00 lei	Prorated	Standard
2		Electrician	Work		E		2	20.00 lei/hr	40.00 lei/hr	0.00 lei	Prorated	Standard

Figure 4. Description of the ratio between the available resources and the necessary. First version.

IOP Conf. Series: Materials Science and Engineering **568** (2019) 012012 doi:10.1088/1757-899X/568/1/012012

In the figure 5, through the Resource Graph Tools option, the ratio between necessary and available resource, for each day of the project is highlighted. Thus, it is found that there is a shortage of 9 machinist resource on Monday 3.06.2018, and 6 machinist resource on 4.06.2018. Figure 6 shows the cost of the machinist resource each day of the project runtime.

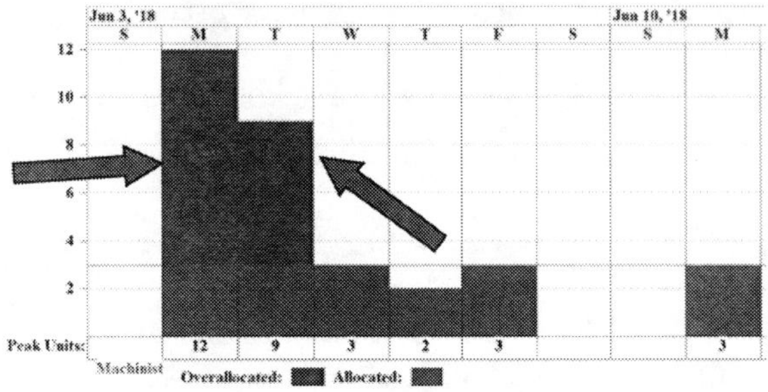

Figure 5. The description of the ratio between the available and the necessary machinist resource - in the Resource Graph Tools window. First version.

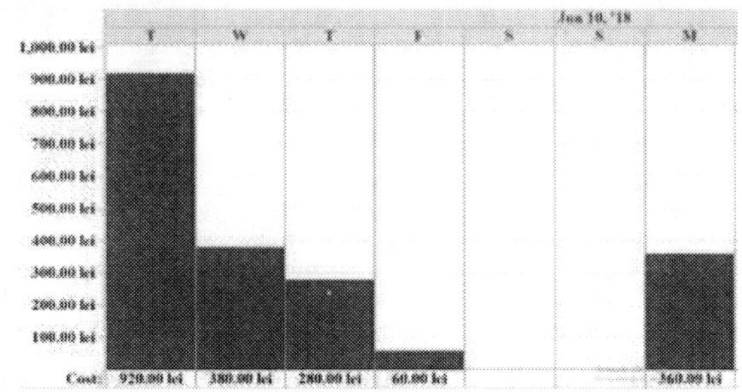

Figure 6. The cost chart with the machinist resource every day of the project runtime. First version.

3.2. Version II

To solve the problem of resource shortage (labor force) in the case of the conveyor assembly project, it can be use a Microsoft Project® feature. This feature is Level Resource. With this option, Microsoft Project® schedules activities so that the necessary resources are not greater than available. The deficit refers to the machinist resource. After applying this feature, it is noted that significant changes occur in the Gantt chart (figure 7). The deficit refers to the machinist resource. After applying this feature, it is noted that significant changes occur in the Gantt chart (figure 7).

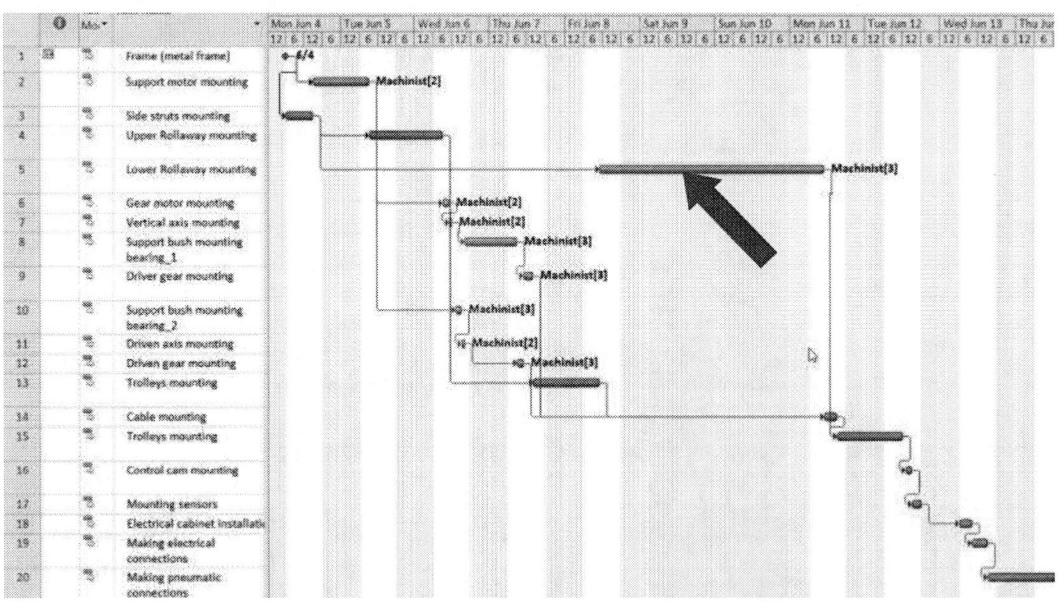

Figure 7. Gantt Chart and Critical Path. Version II.

The beginning of the activity lower rollaway mounting is was offset so that the machinist resource can be used within the limits of availability. It is noticed that there is no shortage of resources. Due to the reprogramming of the activities, the duration of the project changes significantly. Thus, in this version, the project would run for 7,88 days (7 days and 7 hours). The increase in project execution time is 35%, compared to the version I.

3.3. Version III

In order to solve the deficit problem and for a short duration of the project, the following measures are taken: the availability of the machinist resource is increased from 3 to 8; for some of the critical operations, additional resources will be allocated so that their execution time decreases and implicitly the execution time of the project will be shorter.

These operations are: upper rollaway mounting - in the version I 3 machinists are allocated, in version III are allocated 4 machinists; trolleys mounting- in the version I 2 machinists are allocated, in the version III are allocated 4 machinists.

Synthesis information regarding the version III project implementations in Microsoft Project ® are: the project runs for 4.13 days (4 days and 1 hour); the working time is 157 hours; the labor costs are 3.140 lei.

A one-day decrease (meaning 20%) in the duration of the project was found. The cost of the machinist resource is 2780 lei (88.53% of the total), and the costs with the electrician resource is 360 lei. The distribution of the working time on the two resource categories: 139 hours (88.53% of the total) are allocated to the machinist resource and 18 hours to the electrician.

4. Conclusion

Network programming, critical path determination methods, are effective tools for unique production management. These methods make it possible to optimize solutions by taking into account the available resources.

In the analyzed case study - the project management of a suspended conveyor assembly, the critical path method was implemented through *Microsoft Project®*. The first version showed the critical activities and implicitly the critical path that was defined to be the shortest time in which the project can be done. Also in this first variant there was found a significant deficit in terms of the machinist resource. This problem was solved by leveling the resources - programming the activities so that the resources are used within the limit of availability (version II).

However, in order to shorten the duration of the project implementation, the availability of the machinist resources was increased, the duration of the project execution decreased by 20% (version III). The *Microsoft Project®* program provides the decision maker (the project manager) a variety of information that allows him to take the most effective measures to balance the project's duration with costs.

5. References

[1] Wolfgang L 2007 *Project management* (*Managementul Proiectelor*- in Rumanian) Bucureşti: Editura ALL)

[2] Evangelos A, Rose O 2012 *Proceedings of the 2012 Industrial and Systems Engineering Research Conference,* eds G Lim and JW Herrmann

[3] Cruz-Chávez MA and Frausto-Solís J 2006 *Mexican International Conference on Artificial Intelligenc,* eds A Gelbukh and CA Reyes-Garcia (Apizaco, Mexico: Springer) p 450

[4] Wulandari A, Dachyar M and Farizal 2018 *4th Engineering Science and Technology International Conference* (ESTIC 2018) MATEC Web Conf.Volume 248, eds A. Hazmi, R.A. Hadiguna, H. Suherman and R. Desmiarti (Padang, West Sumatra, Indonesia) pp 5

[5] Blaga F, Stanasel I, Hule V and Crăciun D 2015, ed. G Grebenisan *ANNALS OF THE UNIVERSITY OF ORADEA, Fascicle of Management and Technological Engineering* ISSUE #3 (Oradea) p 13

[6] Takebira UM and Mohibullah ATM 2017 *J. of Tex. S & Eng.* **7 (5)** pp 5

Acknowledgments

The paper has been realized within the research contract 19/2015 - MONORAIL CONVEYOR SYSTEM FOR AUTOMATED PRODUCTION LINES. Program MANUNET-ERA-NET.

IOP Conf. Series: Materials Science and Engineering **568** (2019) 012013 doi:10.1088/1757-899X/568/1/012013

Computation of a torsion spring stabilizer bar rigidity and fatigue resistance

I Geonea[1], N Dumitru[1], A Rosca[1], N Craciunoiu[1], L Racila[1] and C Copilusi[1]
[1]Faculty of Mechanics, University of Craiova, Craiova, Romania.

igeonea@yahoo.com

Abstract. The stabilizing bar is the suspension element that reduces the sliding of the car by stopping any kind of uneven vertical movement between the wheels to which it is connected. The stabilizer bar is a piece designed to increase stability during curves, in the case of sharp turns or maneuvers, and to enhance vehicle maneuverability. In general, sports cars are equipped with such stabilizing bars both on the front and rear. In this paper we present research on the experimental determination of the characteristic and rigidity of the torsion bar springs used in suspensions of motor vehicles. The determination of the torsion bar spring characteristic is done by two methods: analytical and experimental. The results obtained by the two methods are similar. In the second part of the paper there is discussed a test bench used for the experimental determination of fatigue resistance of the torsion bar spring. Also, angular deformations of torsion bar with resistive strain gauge transducers are monitored during fatigue applications. The results of the fatigue application are also studied by numerical simulation in the ANSYS FEA program.

1. Introduction

Stabilizers serve to reduce rolling oscillations and increase vehicle stability in turn. The most widespread stabilizers are U-shaped, with the heads attached, directly or through intermediate rods, to the car body or suspension arms, and the middle part hinged to the body.

The main purpose is to obtain a proper design of the stabilizer bar and also to obtain proper torsion stiffness for this component [1]. Because the stabilizing bar is subjected to alternating twisting stresses, the material and its construction must be safe to operate in order to prevent fatigue breakage. For this purpose, a series of achievements are presented in the literature showing the study of fatigue resistance of the stabilizing bar, proposing optimal constructive variants [2, 3].

Most fatigue studies use the finite element method, especially the ANSYS program [4]. The influence of factors of the production process as well as of the material on the fatigue resistance is presented in [5]. Different methods of finite element modeling are presented in [6]. In [7] the construction of a stand for fatigue testing of stabilizing bars is presented. In paper [8] it is considered a crank and connecting rod mechanism which performs rotation fatigue loads upon a stabilizer bar.

The main purpose of the present research study is the experimental computation of the characteristic and rigidity of torsion bar spring. Also another purpose is to compare the experimental results with theoretical results, and determination of fatigue resistance of torsion bar spring, for torsion solicitation, with an alternant symmetric cycle.

2. Theoretical considerations

The curve that represents the dependence between the load that is acting upon the spring and the elastic deformation of the spring produced by the load, measured on the load direction, is called spring characteristic.

For metallic springs, that respects the Hooke law and they are without interior friction, the characteristic is linear (with constant rigidity), as presented in Figure 1. For these springs, based on the Figure 1, is defined the rigidity as in equation (1):

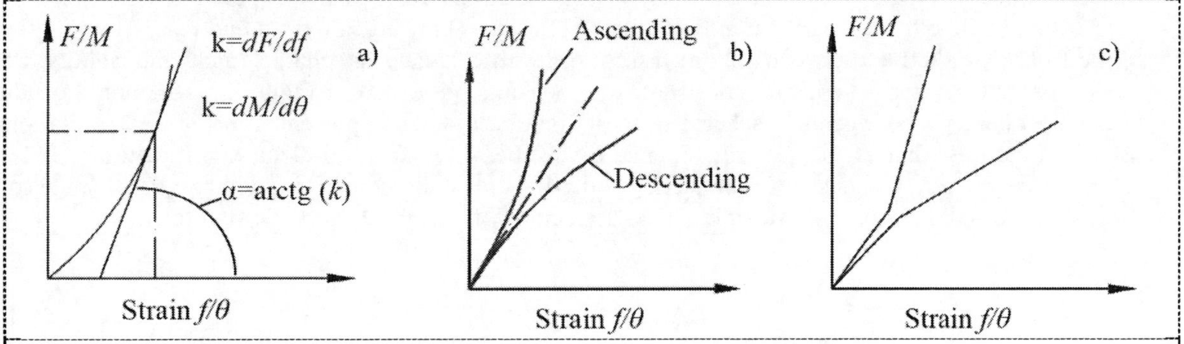

Figure 1. Torsion bar spring characteristic: a) -nonlinear characteristics; b)-progressive and regressive; c) - in steps.

$$c = \frac{F_i}{f} = tg\alpha = \frac{F}{f};\ c' = \frac{T_i}{\varphi_i} = tg\alpha' = \frac{T}{\varphi} \quad (1)$$

where: c is the spring rigidity; f is the bending displacement, in (mm); φ is the rotation, in (radians).

Under the torsion-only load, based on the calculation scheme in Figure 2, we can write the relationships for stresses and angular deformation as in equation (2).

$$\tau_{t\max} = \frac{T}{W_p} = \frac{16Fa}{\pi \cdot d^3} \le \tau_{at};\ \varphi = \frac{Tl}{GI_p} = \frac{Fal}{G\dfrac{\pi d^2}{32}} = \frac{32Fal}{G\pi d^4} = \frac{2\tau_t l}{Gd} \quad (2)$$

where: T is the applied torsion, in (N·mm); F- applied force, in (N); *a*- cantilever length of the spring, in (mm); *d* is the bar diameter; *l* is the spring length (mm); G is the material transversal elasticity module, in (MPa); φ –is the angular deformation of the spring, in (mm); τ_t – is the tangential stress, in (MPa).

To design, from the load is established the spring diameter *d*, and from the deformation the length *l*. The clamping of these levers is done by adjusting the shape of the corresponding ends of the springs: eccentric, flattened, hexagonal, square, triangular, grooves. To increase the fatigue strength, the shape from Figure 2, a) is recommended.

3. Experimental test rig

In Figure 2, a) is the drawing of a stabilizer at the rear axle of a car. The front stabilizer consists of a U-shaped bar which is transversely mounted to the longitudinal axis of the car in two rubber bushings fixed by the frame. The bar ends are connected by means of the rods to the support plates of the springs. If the deformations of the two springs are equal, the torsion bar will rotate freely in the rubber sleeves without getting into action. At turns, when the body is tilted laterally and one of the springs will compress more, the bar will work as a torsion bar and opposes a transverse tilt resistance of the car. With the help of stabilizers, the lateral tilting of the body can be reduced by 20 ... 30% without changing the suspension characteristics. Stabilizers are fitted to most cars as well as to some buses. The rigidity of the stabilizer can be determined with the equation (3):

IOP Conf. Series: Materials Science and Engineering **568** (2019) 012013 doi:10.1088/1757-899X/568/1/012013

$$k_{st} = G_c \cdot h\left(\frac{Y}{\alpha G_a} \pm 1\right) - k'_g \tag{3}$$

in which: G_c - is the weight of the car body; h - the distance from the center of gravity to the roll axis; Y - lateral force; α - the roll angle in radians; k'_g - the overall angularity of the front and rear suspension without a stabilizer. It is recommended that the stabilizer mounts only if the roll angle is greater than 7(°). If the stabilizer is mounted only on the front axle the car will be under the turn. If the car is fitted with both front and rear axle stabilizers, the ratio between the front and rear axle rigidities of the front axle shall be between 1.2 and 1.6. In order to calculate the characteristic and rigidity, the torsion bar spring, with shape and constructive dimensions indicated in Figure 2, is mounted on the fatigue test bench. The kinematic scheme of the fatigue test bench is presented in Figure 3. Notations used in Figure 3, that is the test rig kinematic scheme, have the following significations: 1, 2 – actuation gearboxes; 3 – rocker; 4- connecting rods; 5 – links with connecting devices; 6 – vehicle roll rods; 7 –front stabilizer bar; 8 – flexible blocks for stabilizer bar; 9- rear vehicle stabilizer bar.

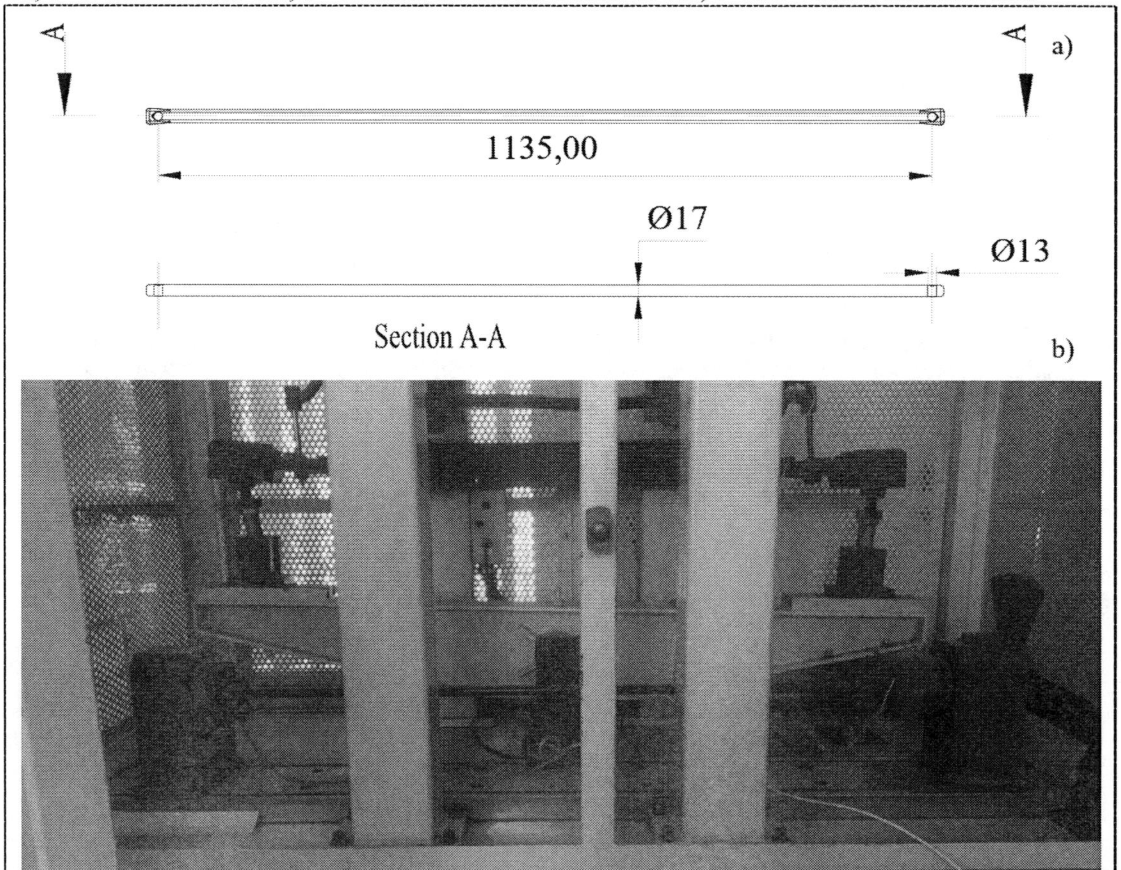

Figure 2. Geometric dimensions of the torsion bar spring a), and mounting of test rig b).

The control panel of the fatigue test rig of the torsion bar springs is presented in Figure 4, a). The control panel allows the following actions and features: manual mode start or automatic mode start of the gear motor, necessary for the torsion spring fatigue mechanism; emergency stop button; indicator for number of cycles of solicitations (number of cycles performed and the number of maximum specified cycles); the electric motor rpm indicator; deformations indicator; general key for start – stop.

IOP Conf. Series: Materials Science and Engineering **568** (2019) 012013 doi:10.1088/1757-899X/568/1/012013

The torsion bar spring, with the shape and dimensions indicated in Figure 2, is mounted on the torsion fatigue test rig. The torsion solicitation is achieved with crank rod rocker mechanism. The rod length can be adjusted with a screw-nut mechanism. In this way is modified the angular amplitude of torsion solicitation. The measurement equipment mounted on the test rig, consist in strain gauges, which measures the deformations produced by the torsion. In Figure 4, b) is observed the strain gauge transducer that measures the angular deformations of the torsion spring stabilizer.

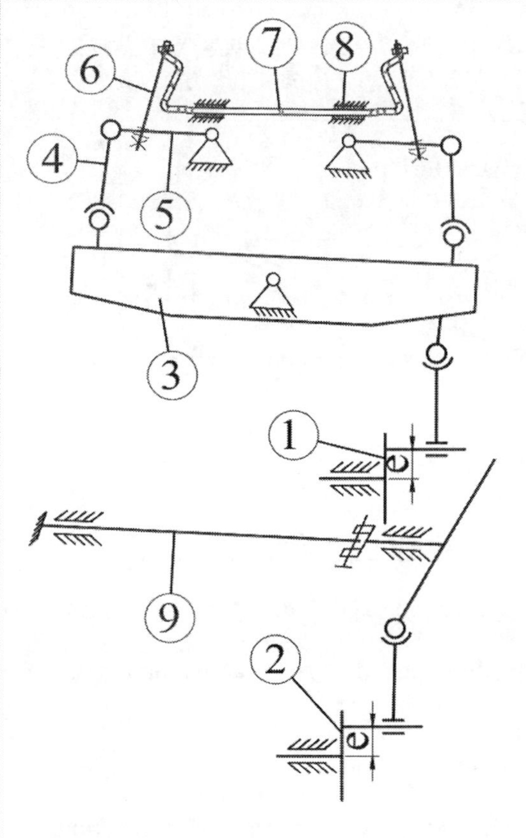

Figure 3. Test bench for torsion spring bar testing.

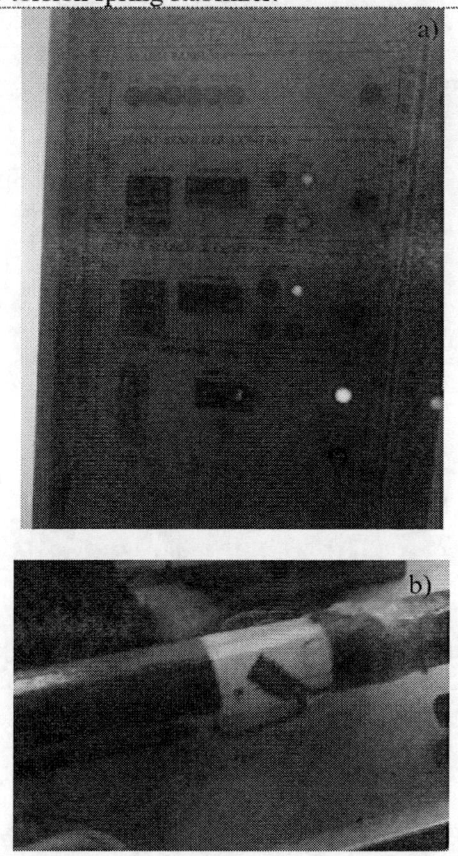

Figure 4. Control panel of test rig and. strain gauge transducer.

4. Torsion bar characteristic determination

The torsion bar spring is subjected to torsion by applying to one head of a torsion torque and the other side is fixed, as is observed in Figure 3. The torsion torque is applied by rotating of the actuation mechanism rocker.

It aims to establish the torsion bar spring rigidity, namely the dependence from the torsion torque applied to the spring head and produced angular deformation θ, in (radians) or (degrees).

$$k = \frac{dM}{d\theta} = \frac{GI_P}{l} \tag{4}$$

where: G –is the transversal elasticity modulus of the spring material, in (MPa).

Ip – is the polar moment of inertia of spring transversal section (mm^4); *d*- torsion spring diameter (mm).

Based on the geometrical dimensions of the torsion spring, l=1135 (mm), d=17 (mm), depicted in Figure 2, is computed $k_{\text{theoretic}}$ = 577,6 (N·mm/rad).

For experimental determination of the torsion spring constant, is dissembled the rod of the actuation mechanism. With a screw-nut mechanism mounted on the rocker side, is applied on the rocker 3-5 progressive loads, measured with a force transducer. Corresponding to each load is read on the pointer the angular rotation of the torsion spring bar. Taking into account the length between the point of force application and the point of rocker articulation, the experimental torsion torque is determined. The torsion torque applied to the spring, as well the angular rotation allows to draw the spring characteristic diagram.

Based on this experimental data, is plotted the spring experimental characteristic, respectively the dependence between the torsion torque, T in (Nm) and the spring angular rotation θ (°). The characteristic shows a linear dependence between the torsion torque and the angular rotation, as presented in Figure 5.

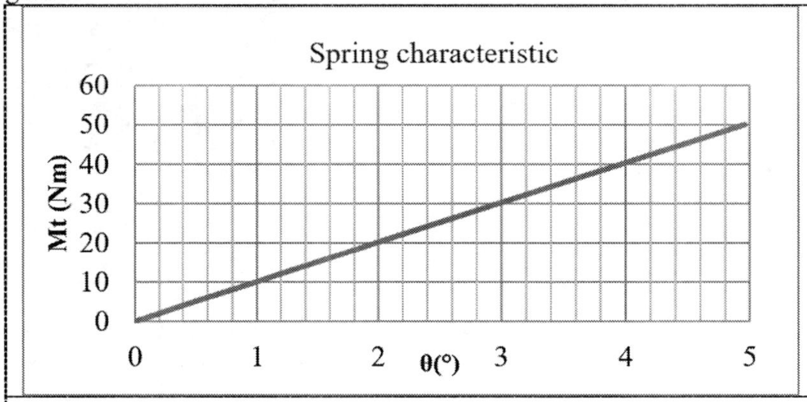

Figure 5. Experimental characteristic of torsion bar.

Based on the experimental characteristic, is computed the experimental constant, of the torsion bar spring, respectively: k_{exp}=585,872 (N·m/rad).

Is compared the spring computed constant, with that obtained experimentally, and is concluded that they have similar values.

5. Analysis of fatigue behavior in ANSYS of rear stabilizer bar

To this paragraph is studied the fatigue behavior of vehicle rear suspension stabilizer bar, with ANSYS finite element analysis software. The CAD model of rear stabilizer bar is imported in ANSYS database. The stabilizer bar is meshed in finite elements, are specified the contour conditions and loads.

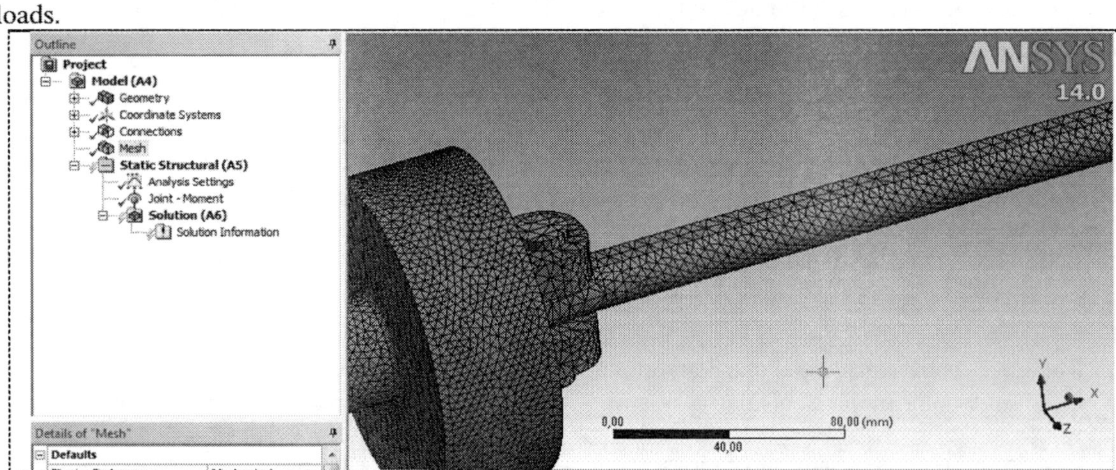

Figure 6. Stabilizer bar meshed in finite elements.

In the next step are defined the connecting bolts of the stabilizer bar with the revolute joint and fixed joint elements. To the stabilizer bar ends, where the stabilizer bar is fixed with bolts to the vehicle wheel spindle, are imposed the following conditions. One side of the stabilizer bar is fixed, as is presented previously and to the other side a torsion torque is imposed. This situation corresponds to the hypothesis when one of the wheels passes to an obstacle. The obtained results of the finite element static structural analysis, respectively the stress for present load and safety coefficient are presented in Figure 7.

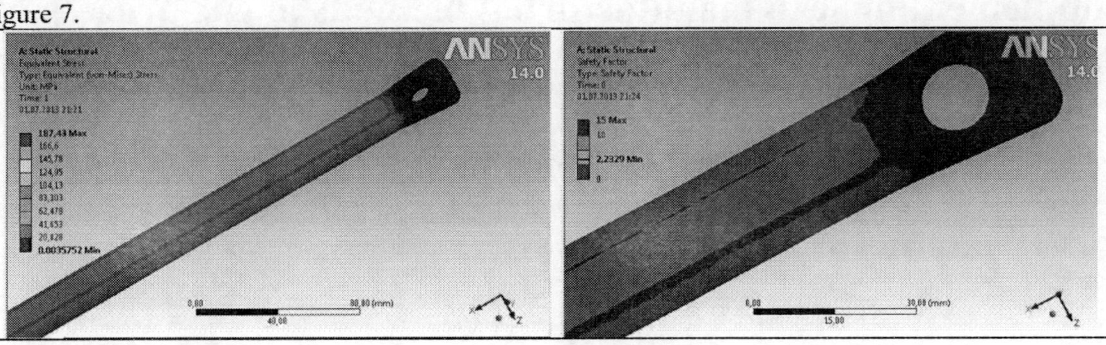

Figure 7. Equivalent von Misses stress distribution and fatigue safety coefficient.

6. Conclusions

The paper presents the theoretical and experimental determination of the characteristic and stiffness of a spring torsion spring from the suspension. The experimental analysis was performed on a fatigue test bench which stresses a rear stabilizer bar through an alternant symmetric load cycle. The characteristic diagram obtained for the torsion bar spring is linear, and the stiffness of the spring calculated theoretically and experimentally is of the same order of magnitude. Also the stabilizer bar model is subjected to a stress analysis with ANSYs software. The maximum equivalent stress obtained is 166 (MPa). The minimum safety coefficient for fatigue breakage is 2.23.

References

[1] Nikravesh P E and Gim G 1990 Ride and stability analysis of a sports car using multibody dynamic simulation Math. *Comput. Model. J* **14** pp. 953–958

[2] Marzbanrad J and Yadollahi A 2012 Fatigue life of an anti-roll bar of a passenger vehicle *World Acad. Sci. Eng. Technol. Int. J. Mech. Aerosp. Ind. Mechatron. Manuf. Eng.* **6(2)** pp. 407–413

[3] Palma ES 2002 Fatigue damage analysis in an automobile stabilizer bar *Proceedings of the Institution of Mechanical Engineers, Part D-Journal of Automobile Eng.* **216** pp. 865-871

[4] Liu S H and Li F 2012 Fatigue life analysis of the stabilizer anti-roll bar using ANSYS *Adv. Mater. Res.* **383–390** pp. 5894–5898

[5] Wittek A M, Burdzik R, Folega P, Konieczny L and Lazarz B 2015 Influence of production process and material factors on fatigue strength of tubular stabilizer bar *Arch. Metall. Mater.* **60** pp. 2485–2492

[6] Zhang L, Liu H, Xu Y, Wu S 2009 Study on Modelling Method of Anti-roll Bar Using One Dimensional Beam Element *SAE Technical Paper 2009-01-1454*

[7] Rosca A S, Craciunoiu N, Geonea I and Ploscaru C 2018 Design of a Test Rig for Vehicle Stabilizer Bar Fatigue *Study Applied Mechanics and Materials 880* (Trans Tech Publications) pp. 226-231

[8] Dumitru N, Copilusi C, Geonea I and Margine A 2016 Kinematic and Dynamic Study of a Mechanism for a Vehicle Front and Rear Stabilizer Bars *CONAT* (Brasov) pp.161-169

Lead time reduction and increasing productivity by implementing lean manufacturing methods in cnc processing center

I C Gherghea[1], C Bungau[1] and D C Negrau[1]

[1] Engineering and Management Department from Doctoral School of Science Engineering, University of Oradea, Universitatii street 1, Romania

cosmin.gherghea@csud.oradea.ro

Abstract. Growing competitiveness of manufacturing companies has led companies board to look for new innovative solutions to increase the productive efficiency. More and more companies are looking to maximize profit through different ways, some of them use classical methods that are focused on increasing production capacity, expanding the product portfolio, manufacturing and marketing campaigns, other operational activities, are implementing modern techniques to reduce losses. This paper propose to present the way that productivity is increased by reducing lead time through Lean Manufacturing method implementation. For this purpose a piece will be manufactured using a CNC processing center, which will contribute significantly to processing the part at a higher precision, and at the same time eliminating the down times. To highlight the contribution and the efficiency of the CNC processing center, will be present the classic production method using a classical milling machine, respectively FUS 22, the method that was also implemented, according to this case study. To analyze and graphically illustrate these two production methods, respectively classic and modern methods, VSM will be built, then the 2 methods will be compared.

1. Introduction

With increasing the competitiveness of manufacturing companies, more and more companies are investing massively in developing flexible production systems, acquiring latest technologies. Along with the new technologies and flexible manufacturing systems, it is also necessary to implement innovative techniques and methods to increase production efficiency. To this end, manufacturing companies and not only, build their strategy by using different methods and techniques for streamlining processes, and Lean manufacturing is one of the most known methods for continuous improvement that these companies have implemented [1].

Lean Manufacturing is a concept developed by the Japanese from the Toyota Production System (TPS) after the second World War (1950), when the car industry and not only were looking for new solutions to reduces production costs, and also increases production capacity [2].

Thereby, this concept of Lean Manufacturing has been developed which is a long - term strategy, having as objective identifying, respectively reducing / eliminating waste, with a focus on continuous improvement of all operational processes by implementing Lean Manufacturing specific techniques such as: Kaizen, Value Stream Mapping, just in time, Poka Yoke, 5S, Kanban, Single minute exchange of die (SMED), Pull System, Gemba, etc. [3, 4].

According to some analyzed case studies, Lean manufacturing has proven to be one of the most effective methodologies applied in the manufacturing industry, contributing significantly to the elimination of waste due to techniques and tools for continuous improvement [5, 6, 7, 8].

One of the most popular methods for identifying waste in Lean manufacturing is Value Stream Mapping (VSM), which aims to graphically visualize the entire operational flow (material flow and informational flow) [9, 10].

Once the waste from the operational process are identified, different innovative solutions are proposed through which these waste are eliminated.

In Lean manufacturing, the are called muda (in Japanese), and these waste can identified in different forms, and the waste that most affect production are transportation, waiting time and unnecessary motion [11].

Eliminating these waste (those activities that do not create value) lead to a reduction in lead time, respectively productivity growth, which is the main objective of manufacturing companies [12].

The implementation of various Lean tools within the CNC processing centers leads to optimization the production process and the elimination of dead times [13].

2. Methodology

The research methodology is based on the use of a Lean manufacturing technique, respectively VSM, with the help of which it is exemplified graphical the operational process for manufacturing a mobile plate. Data analyzed in this case study, were obtained from a production company, a company that executed this part using both production methods, respectively modern methods, using a CNC processing center and classical method using a classic machine tool. With the help of this method will be presented in parallel the benefits of production of mobile plate on a CNC processing center, respectively the classic production on a conventional machine tool. The results of the research highlight the identification and elimination of some waste in the production on CNC processing center, compared to the classic production. It reduces and eliminates certain down times, which leads to reducing lead time and increasing productivity.

3. Application of VSM and lead time reduction – a case study

In the below figures are presented the execution drawing of a mobile plate, which was executed on a CNC processing center (HAAS VF-2) and also on a conventional machine tool (FUS 22).

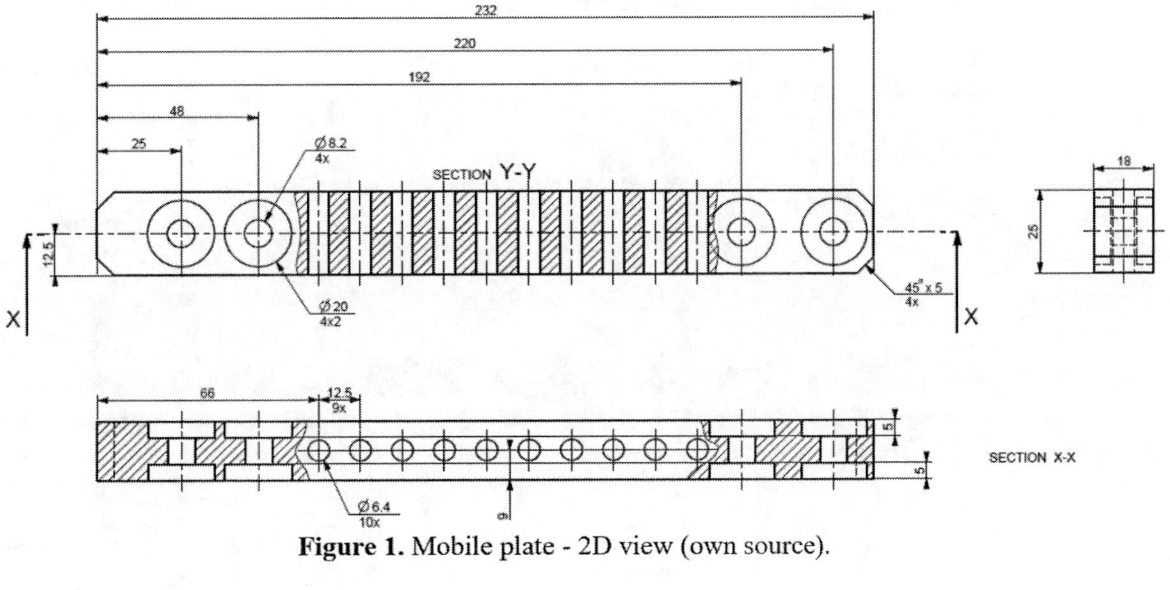

Figure 1. Mobile plate - 2D view (own source).

IOP Conf. Series: Materials Science and Engineering **568** (2019) 012014 doi:10.1088/1757-899X/568/1/012014

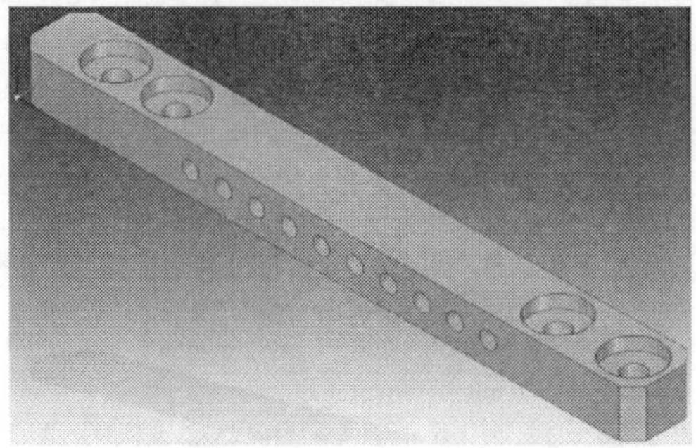

Figure 2. Mobile plate - 3D view (own source).

Part name: mobile plate
Material OLC 45 (C45)
Processing of this piece requires 3 fasteners:
1. Front Centering Φ 2, Drilling Φ 8,2, . Lamar Φ 20 plus chamfering holes outline;
2. Lamar Φ 20 (the other side) plus chamfering holes outline;
3. Front Centering Φ 2, Drilling Φ 6,4 on the front area, chamfering holes.

For processing the workpiece, the mobile plate requires the following cutting tools:
1. Drilling center Φ 2;
2. Drill Φ 8,2;
3. Milling Φ 20 with 2 teeth;
4. Hole chamfer Φ 20 at 90^0;
5. Drill Φ 6,4.

In order to execute this piece, the necessary times for the three clamps were summed up, respectively: time for setting coordinates according to the 2D execution drawing, tool changing time, tool setting time, cycle time and interphase measurement time.

After identifying the operations and the time required for the execution of the workpiece, the VSM was built, exemplifying both operational flow, respectively classic production and CNC production (according to the two figures below).

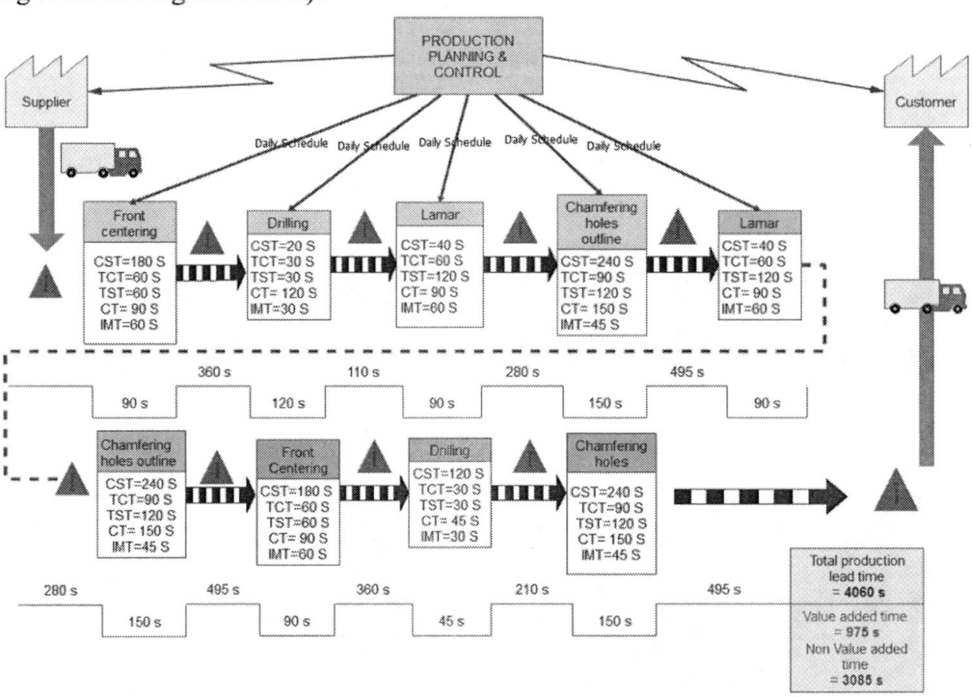

Figure 3. VSM - classical production on FUS 22 (own source).

IOP Conf. Series: Materials Science and Engineering **568** (2019) 012014 doi:10.1088/1757-899X/568/1/012014

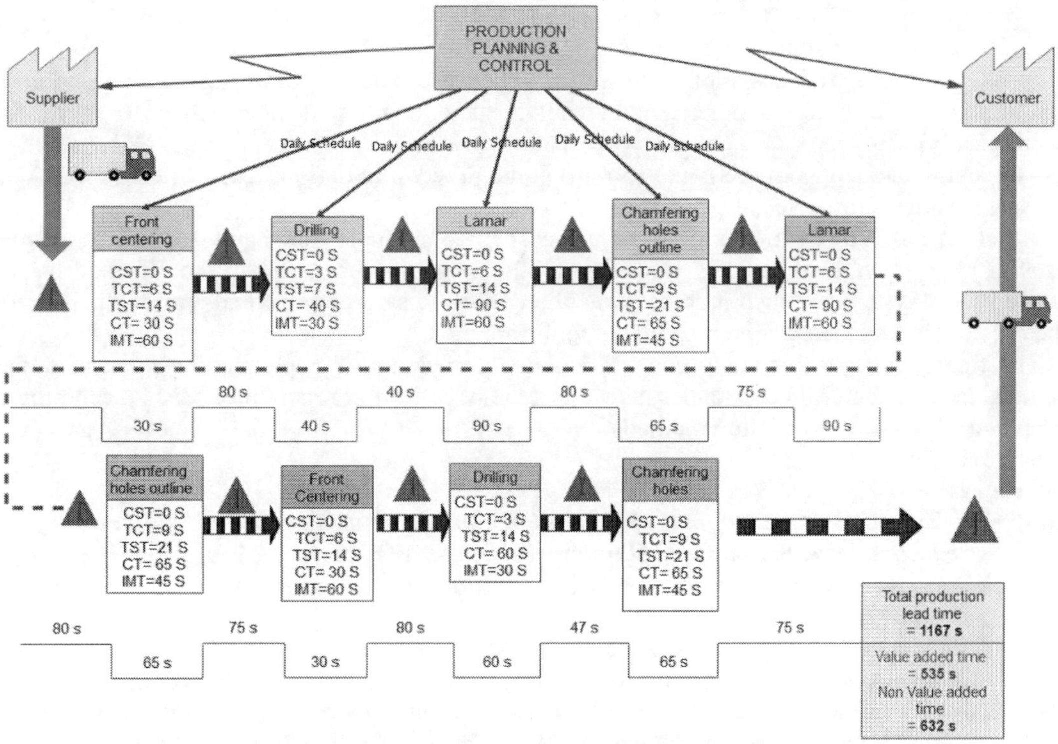

Figure 4. VSM - CNC production on HAAS VF-2 (own source).

4. Results and conclusion

The fundamental purpose of this research is to identify and eliminate the waste, respectively those activities that do not create value in the manufacturing process and also to increase productivity by reducing the lead time. Following the completion of this research, through the implementation of VSM, the percentage of down time in the classic production process was identified, as compared to the percentage of down time in the manufacturing process at the CNC processing center.

In the case of classical production, the total production lead time is 4060 seconds for a single piece, of which the value-added time (VAT) is 975 seconds and the non-value added time (NVAT) is 3085 seconds, which in the Lean production are identified as waste.

In the case of production on CNC processing center, the total production lead time is 1167 seconds for a single piece, of which the VAT is 535 seconds and the NVAT is 632 seconds (figure 5).

Applying VSM and production on CNC processing center the lead time were reduced by 2893 seconds, respectively by 71.25%.

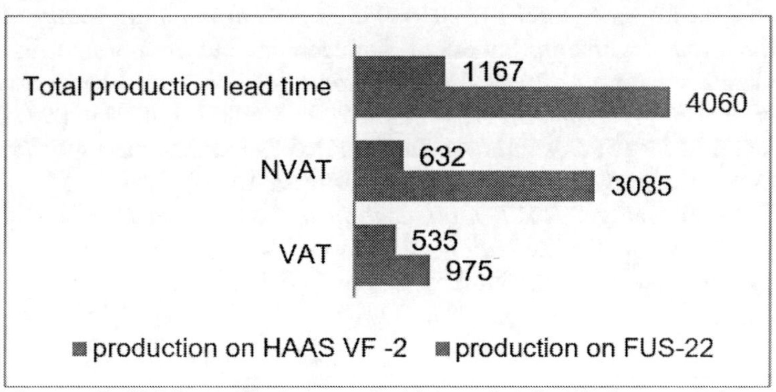

Figure 5. CNC production vs classic production (own source).

Here are some of the conclusions that was reached after comparing the two methods of production:

- processing on FUS 22 involves many down times (working with individual tools, no tool magazine, operator must move the machine table by following the coordinates)
- CNC processing reduces a lot of down times (because it has a tool magazine, displacement at the coordinates of the holes is done automatically, calculating the distances from 0 "zero" is done automatically)
- Checking the depths of the lamar (5 mm) at the processing on FUS 22, while at CNC processing center this value is respected at a precision of 0.01 mm
- the most of down time is at the change sharpening tools by hand and adjust them to the benchmark part, while at the CNC it is insert all the necessary tools, adjusts the offset of the part with the Renishaw system and then it starts without any need to stop the machine until the part is finished. The only down time being is picking up the tool from the CNC magazine.

The final results of this study confirm again the benefits of implementing Lean manufacturing methods in manufacturing companies, as well as the benefit of production on CNC machining centers compared to classical production methods.

References

Philipp H., Thomas B. 2016 *Optimal configuration of manufacturing cells for high flexibility and cost reduction by compo-nent substitution*, Procedia CIRP **41**: 111 – 116.

Womack, J.P., Jones, D.T., Ross, D. 1991, *The Machine that Changed the World: The Story of Lean Production*, Harper Collins: New York, NY, USA.

Sundara R., Balajib A.N., Satheesh Kumar R.M. C 2014, *A Review on Lean Manufacturing Implementation Techniques*, Procedia Engineering **97**: 1875 – 1885.

Gherghea I. C., Bungau C. 2018, *Poka Yoke application synthesis in manufacturing engineering*, Proceedings of the 6th Review of Management and Economic Engineering International Management Conference **6**, pp 564-571.

William M. G., Samson M., Alphonce M. 2011, *Use of the Value Stream Mapping tool for waste reduction in manufacturing. case study for bread manufacturing in Zimbabwe*, Proceedings of the International Conference on Industrial Engineering and Operations Management 236 – 241.

Bungau C., Blaga F., Gherghea I. C. 2011, *Method of analysis and audit, used to implement 5S in operational management*, Review of Management and Economic Engineering, 36-47.

Bungau C., Groza M., Buidos T., Gherghea I. C. 2011, *Measurement system used in the implementation of 5S*, Proceedings of Conference: International scientific assembly ES-NBE Economic Science in Function of Creating a New Business Environment, 667-679.

Bungau C., Blaga F., Gherghea C. 2012, *Continuous improvement of processes in cutting operations*, Review of Management and Economic Engineering, 208-216.

Manjunath M., Shivaprasad H. C., Keerthesh Kumar K. S., Deepa Puthran 2014, *Value Stream Mapping as a Tool for Lean Implementation: A Case Study*, International Journal of Innovative Research and Development, **3**, Issue 5, 477-481.

Bungau C., Gherghea I. C., Prichici M. 2010, *Value stream mapping analysis, efficency methods of operational management*, Review of Management and Economic Engineering, 188-198.

Arunagiria P., Gnanavelbabub A. 2014, *Identification of Major Lean Production Waste in Automobile Industries using Weighted Average Method*, Procedia Engineering **97**, 2167 – 2175.

Amir A, Thulasi a/p M. 2015, *Designing a Future Value Stream Mapping to Reduce Lead Time using SMED-A Case Study*, Procedia Manufacturing **2**, 153 – 158. 83.

Ramakrishnan, V. Nallusamy S 2017, *Optimization of Production Process and Processing Time in CNC Cell through the Execution of Different Lean Tools*, International Journal of Applied Engineering Research, **12**, Number 23, pp. 13295-13302.

IOP Conf. Series: Materials Science and Engineering **568** (2019) 012015 doi:10.1088/1757-899X/568/1/012015

Manufacturing based on feature recognition using NX

I Stănăşel[1], F S Blaga[1], T Buidoş[1]

Department of Industrial Engineering, Faculty of Managerial and Technological Engineering, University of Oradea, Romania

stanasel@uoradea.ro

Abstract. Competition in the market causes companies to update continuously their products both in terms of design and performance and in terms of manufacturing technologies to ensure high performance, low cost and manufacturing time. The integration of Computer Aided Design (CAD) applications with Computer Aided Manufacturing (CAM) applications has led different researchers to study the concept of recognition of features in as many activities as possible that contribute to the achievement of a product, starting from design, manufacturing, production planning, assembly, quality management. In order to make the production process more efficient, it was approached the automatic realization of the programs for numerically controlled (CNC) machine tools based on the characteristics of the features that make up the part. The paper presents a case study on how to perform machining programs on CNC machine tools using the facilities offered by Siemens NX application based on feature recognition.

1. Introduction

The evolution of the competitive environment determines the companies in the market to update the requirements related to product design, complexity, quality, durability, environmental protection, delivery deadlines. This development has made it necessary to identify some performant technologies which allow, once a faster development of products, and on the other hand, to ensure a manufacturing process faster and of superior quality. By the development of computer technology they have appeared and have developed applications to assist the whole life cycle of products from conception to end of use and its recycling.

The computer can process very large amounts of data in a short time without error, so reducing the intellectual effort of the designer, who in this way does not have to do a routine work, but to focus on the innovative aspects of the product.

In the design stage, the use of the applications of CAD (Computer Aided Design) allows the designer to alter the data quickly and to realize multiple variants of design of the same product from which to be identified the optimal variant.

Computer Aided Manufacturing (CAM) can have several interpretations. Computer Aided Manufacturing refers both to the obtaining the G code programs to control the CNC machine tools, flexible manufacturing systems, and the automatic design of production processes in which all aspects of the production process are controlled. Currently, however, computer-aided manufacturing involves the use of applications which allow the management and monitoring of manufacturing equipment such as CNC machine tools and industrial robots and flexible manufacturing systems.

Using CAM allows shortening the time of product development, increases the efficiency of the manufacturing process and also allows increasing the quality of processing by checking the CNC program performed by simulating tool paths.

Integration of CAD applications with CAM is a necessity, because it allows reducing errors which occur when using neutral formats, increases productivity, reduces product development time, and ensures greater flexibility [1]. The development of applications in the field of engineering has led various researchers to study the concept of the feature recognition in as many activities as possible, which contribute to the achievement of a product from design, manufacturing, production planning, assembly, quality management. For the purpose of making the production process more efficiently, it was approached the automatic realization of the CNC programs, starting from the characteristics of the features which make up the piece.

In the paper [2] is presented a method and a case study regarding the execution of the CNC programs for different features (slots, holes, etc.). The CNC program is based on reading and recognizing these forms and transposing of the repetitive commands of these shapes into macros.

In article [3] is approached the realization of parametric CNC programs based on recognition of the features of the 3D model and is proposed a new model for Feature-based Machining.

The researchers [4] analyze the role of modeling and the characteristics of the product features in its manufacturing process.

The authors [5] approach the development of a manufacturing model based on features recognition in terms of productivity and cost. Experiments are performed to compare the model made with a commercial system.

The article [6] proposes a way of sequencing of a machining process based on both knowledge-based rules and geometric reasoning rules. A set of rules is developed to perform the machining sequences which are validated by a case study.

Taking into account of these considerations, the integration of CAD with CAM applications at the current stage of development is a necessity, features having a very important role in the integration process, because the design and manufacturing data can be associated with features that contain the information which can be interpreted to achieve sequencing of technological process. The features with common characteristics can be grouped in order to reduce the processing time.

From the studies carried out, it was found that there are concerns regarding the development of algorithms for automating the process of automate manufacturing through the recognition the features of 3D model, with the aim of reducing the time for developing the CNC programs.

The paper shows how to make the G code programs for numerically controlled machines using feature recognition facilities provided by the Siemens NX application.

2. Considerations about Feature-Based Machining in NX CAM

At the present stage, most systems CAM may import the 3D model in a neutral format of the parts. Traditionally, achieving the code G for processing on the CNC machines tools using a CAM application complies, in principle, the steps mentioned in figure 1.

Siemens NX is one of the most powerful integrated CAD /CAM solutions, offering efficient 3D modeling tools as well as advanced tools for programming numerically controlled machines up to five axes, high speed machining, hybrid manufacturing, programming of robots for manufacturing operations [7]. NX CAM has also the possibility to automate the process of making the programs for CNC machine tools by using specific manufacturing templates for different types of features (hole, pockets, steps, threads, boss, etc.).

The feature recognition is possible for both native 3D models made on the NX platform, as well as 3D models made in other CAD systems and imported in a neutral format. Once the entities have been recognized, the information is processed and, in accordance with the rules and filters defined in Machine Knowledge, are identified the processing operations, tools, cutting parameter, and the path of each tool is generated. The block diagram for automating the CNC automation process is presented in figure 2..

Machining Knowledge Editor (MKE) and Machining Knowledge Library (MKL) are the core for automating machining process. MKE is the tool that enables the manufacturing technology based on the data provided by MKL. MKL contains rules regarding on the features that can be recognized, which manufacturing operations can be adopted, what type of tool can be used and which size of tool can be selected for the considered manufacturing operations.

IOP Conf. Series: Materials Science and Engineering **568** (2019) 012015 doi:10.1088/1757-899X/568/1/012015

Recognized Features are checked based on the rules defined in MKE. These rules can achieve automatic manufacturing process, select the operations and the necessary tools, and if the material of the workpiece and the tool are known, the cutting parameters are established, and finally it will be determined the optimal tool paths for each tool used to process a recognized feature.

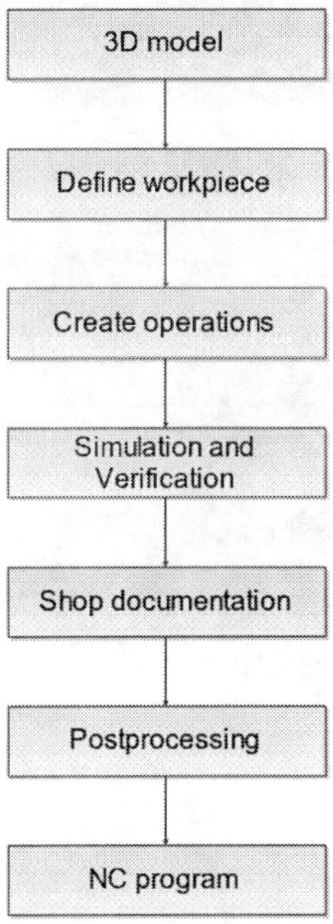

Figure 1. Main steps for getting G code using CAM application

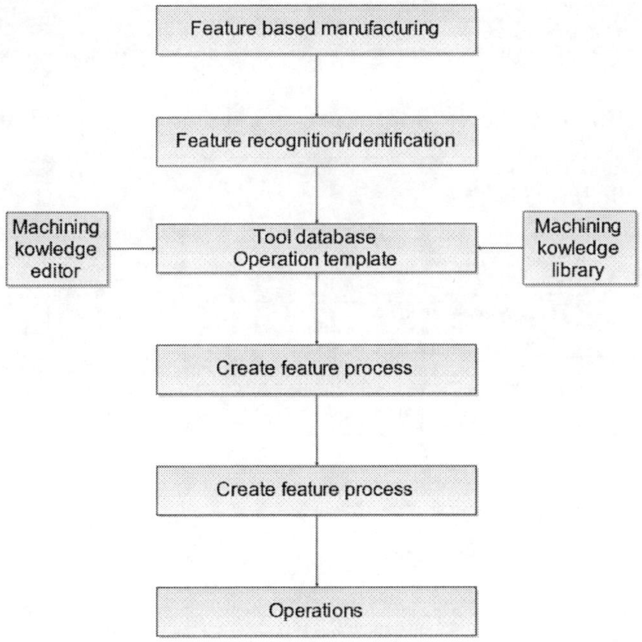

Figure 2. Block diagram for automatic generation of CNC programms

The features which can be recognized and processed automatically refer to the holes, slots and pockets, and irregular feature created by the user. Siemens NX provides more options for the recognition of features like as: feature identification, parametric recognition, legacy hole recognition, legacy making, pocket recognition, and manual recognition. In the above methods, feature identification and parametric recognition (figure 3) are more commonly used and NX recommends that users use the second way of parametric recognition [7].

The NX feature-based machining module can recognize and create processing operations for a wide range of manufacturing processes such as turning, milling, wire EDM.

Regarding the milling operations, feature-based machining module can create different operations such as: Hole Making, Floor /Wall Milling, Cavity Milling, Thread Milling, and Hole Milling, Planar Milling, Plunge Milling, Z -Level Milling, Fixed Axis Surface Contouring, Variable Axis Z -Level Milling, Variable Axis Surface Contouring. In the process of feature recognizing by NX, the user does not need to do any intervention [7]..

3. Example of use Feature-Based Machining in NX CAM

The workpiece that is processed is shown in figure 4. It is a prismatic one, with circular pockets, holes, counterbore holes and thread holes. Some of these features are chamfered. The material of the workpiece is aluminum alloy 7075.

After defining the coordinate system used in the manufacturing and at the dimensions of the blank, the features are recognized. Recognition can be done automatically, or you can specify the types of features to be recognized (figure 5). You can introduce some search restrictions, by indicating a specific area where features to be recognized are located, or by specifying the access direction of the tool during processing. In figure 6 are shown recognized features and their position on the piece.

IOP Conf. Series: Materials Science and Engineering **568** (2019) 012015 doi:10.1088/1757-899X/568/1/012015

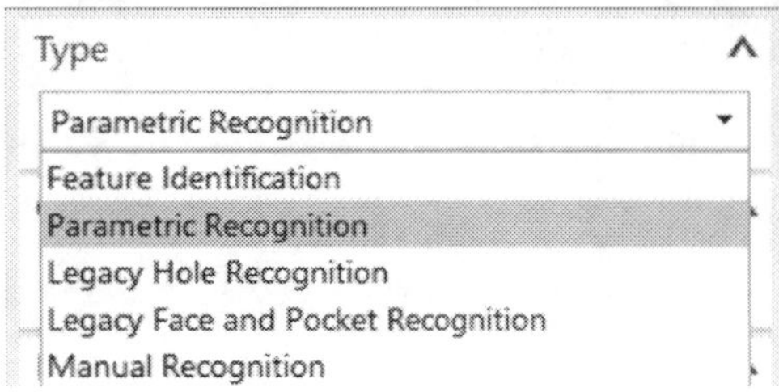

Figure 3. Types of feature recognition.

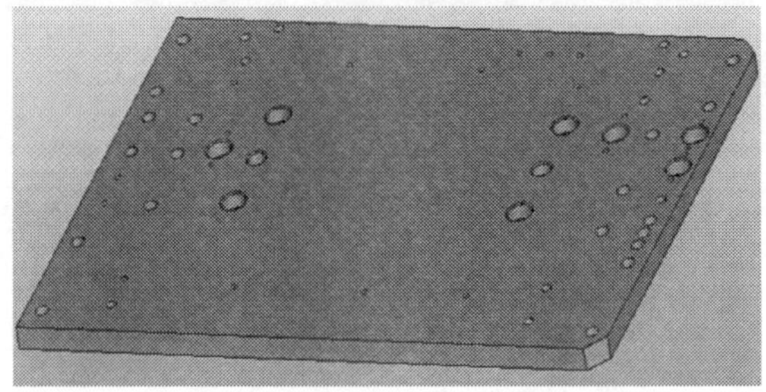

Figure 4. The workpiece to be processed.

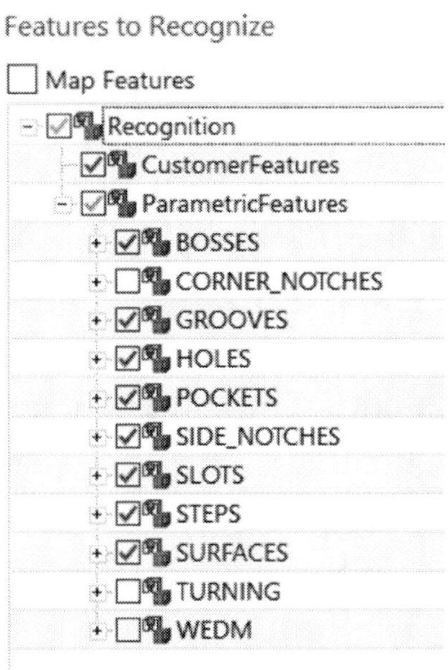

Figure 5. Features to be recognized.

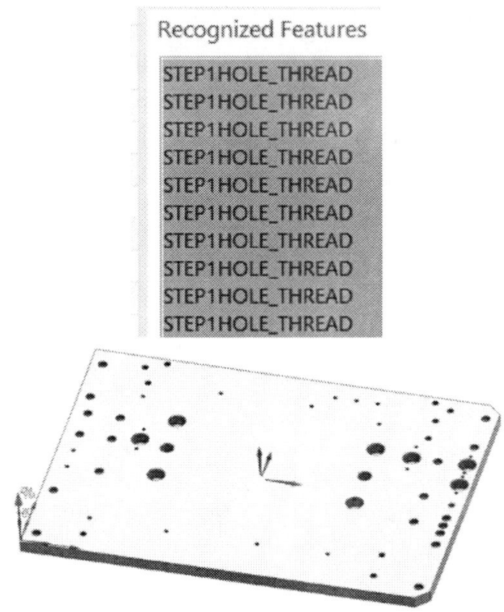

Figure 6. Recognized features and their position on the workpiece.

The next step consists to create feature process. Knowing the workpiece material and the dimensions of the recognized features (figure 7), based on the data in the knowledge libraries are identified the types of manufacturing operations (figure 8), the types of tools, and are chosen their dimensions, are determined the cutting parameters, and it is performed sequencing of operations. The tool paths for each tool is automatically generated (Figure 9) and their 3D simulation is shown in Figure 10.

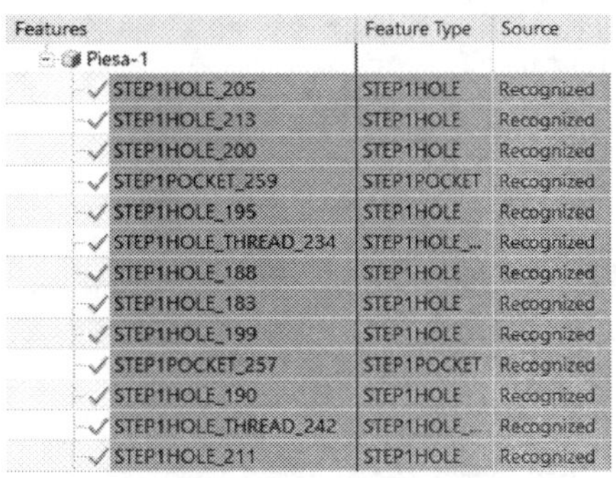

Figure 7. Feature navigator.

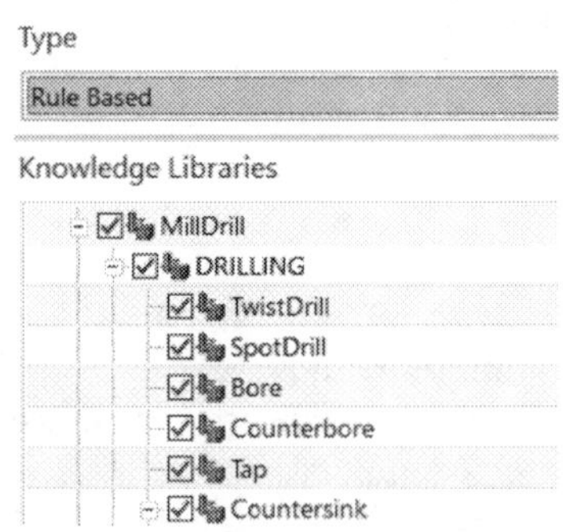

Figure 8. The types of processing operations.

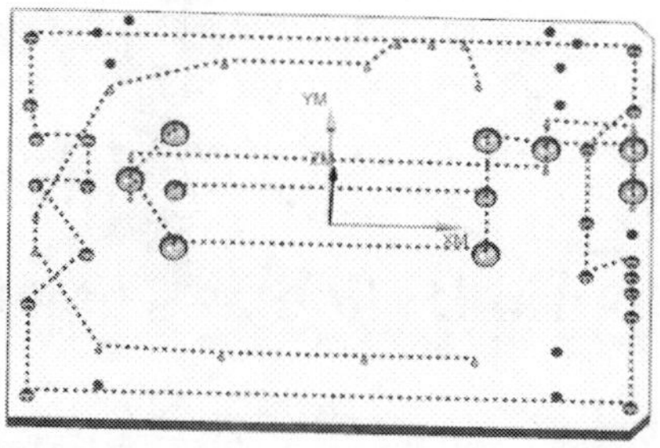

Figure 9. The cutting paths for each tool.

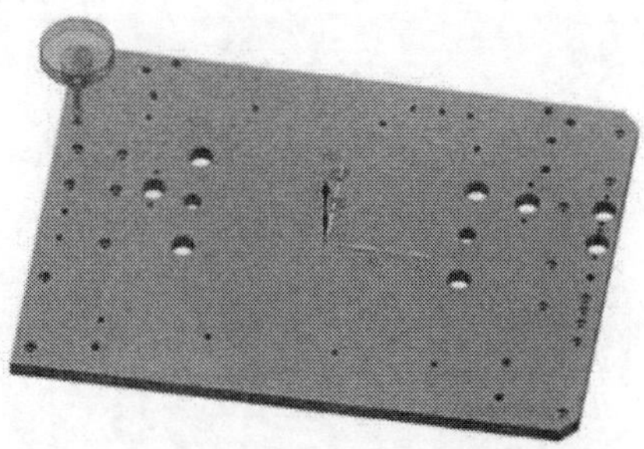

Figure 10. The 3D simulation of the cutting paths.

The CNC program was done by post-processing for a HAAS CNC machining center. Before running on machining center, the program was tested using a backplotter. The machined part is shown in Figure 11.

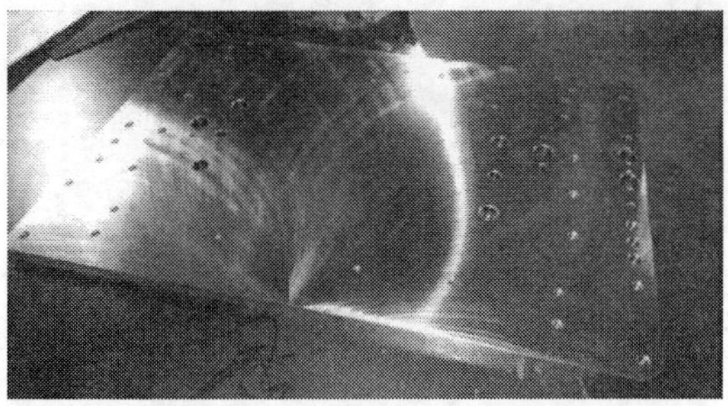

Figure 11. The obtained part.

4. Conclusions

Feature-Based Machining reduces the time required to achieve CNC programs. They are used standardized manufacturing processes which contribute to the decreasing of programming errors. Feature-Based Machining is realized automatically, in this way being eliminated the repetitive work of the programmer. Feature-Based Machining can be used for parts from various fields (mold, machinery, automotive, etc.) dominant features being pockets, holes, slots.

References

[1] Gideon H 2001 *Handbook of Production Management Methods* Butterworth-Heinemann

[2] Razak M., Jusoh A., Zakaria A., 2012 Feature-Based Machining using Macro, *International Journal of Mathematical, Computational, Physical, Electrical and Computer Engineering* Vol:6, No:8 World Academy of Science, Engineering and Technology

[3] Razak M.A., Ibrahim M.R., Sulaiman S., Jusoh A., Zakaria A. 2013 Parametric Programming in Feature-Based Machining *International Journal of Engineering & Technology Sciences* (IJETS) 1(4): 218-225 ISSN 2289-4152 Academic Research Online Publisher.

[4] Mäntylä M., Nau D., Shah J. 1996 Challenges în feature-based manufacturing research, *Communications of the Association for Computing Machinery*, Volume 39 Issue 2, Feb. 1996, pp 77-85

[5] Shukla A, Jani S.S, Acharya G.D. 2016 Developing a Cam Estimator for feature based machining time estimation în VB enviomment *International Journal of Advance Research and Innovative Ideas în Education*, Vol-2 Issue-3 2016, pp 527-535.

[6] Zhenkai L, Lihui W 2007 Sequencing of interacting prismatic machining features for process planning *Computers în Industry* Volume 58, Issue 4 (2007) pp 295–303, Sciencce Direct Elsevier

[7] Tom van 't Erve 2015 *NX CAM Feature-Based Machining Introduction* Siemens PLM Software

Considering constructive-functional optimization as a stage of product design in industrial practice

V. F. Banciu[1,3], E. Pămînțaș[1] and D. I. Ioaniciu[2]

[1] Materials and Manufacturing Engineering Department, Mechanical Faculty, Politehnica University Timişoara, Bv. Mihai Viteazu No. 1, 300222, Timişoara, Romania
[2] Continental Automotive AG, Timişoara, Romania

[3]felicia.banciu@upt.ro

Abstract. This paper is part of a product optimization study, conducted at the request and together with an important national and international economic agent in the automotive industry. The presented results are focused on highlighting how the design activity is influenced by technical and technological progress, by the specific knowledge of the design engineer and by the way they integrate into new generations of products on the one hand having constructive and functional features and on the other hand, minimized costs.

1. Introduction

The development of a product requires the determination of customer requirements, their "translation" against the perspective of the company, the development of the product concept as a result of the marketing study, the development of the technical specifications of the product, the design of the product and the manufacturing technologies, the distribution and the service for the products sold, and ultimately the disposal of the product. All of this is based on management, communication and decision-making processes for organizing and integrating product life cycle phases, including financial, legal, strategic planning, research-development, etc. [1, 2]

Regarding the product development (PD) process, in literature exists exist several approaches. The descriptive and prescriptive approaches to product development [3] result from investigation into actual design practice the descriptive one and the second approach recommend or prescribe guiding principles, steps/phases or techniques that, in theory, are thought to improve specific aspects of PD.
The prescriptive approaches have at basis methods, tools, approaches that refer to product design and development – [4, 5, 6]. They offer a general orientation to products development (design) offering a problem solving approach using tasks and decision making to a design team rather to indicate what tools to use. They say what is to be done but they must be adapted to specific of PD project. The product development process reaches its objective through knowledge accumulated passing from abstract to detail.

The detailed design provides all other data missing from the exhaustive description of the new product, such as materials, installation / adjustment rates, etc. It is necessary to complete iterative steps in order to solve these conception / optimization problems and to analyse several variants using knowledge about means such as statistical methods and tools, manufacturing process [6,7], software,

3

hardware and systems implementation, product reliability and qualification, physical analysis methods, computer-aided design and simulation programs [6,7], specific technology etc.

This paper is part of a product optimizing study, product that is a part of a larger project, conducted at the request and with an important national and international company in automotive horizontal industry, highlighting differences from the theoretical approach. A major company in the Eurasian market (which we call Original Equipment Manufacturer with the acronym OEM) produces a car (called the new CAR), which incorporates a series of sensors (called SZs) with various functions and size (overall dimensions) features.

OEM requires to a company from the horizontal automotive industry (acronym MCP) that besides the actual SZ, to deliver also the support (hereafter referred to as SS) which allow the sensor to be embedded and also, allow mounting of the subassembly in the indicated location of the CAR. OEM expresses the requirements for this SS in the form of functions it desires to be fulfilled. Functions are translated in univocal technical features, in the form of tables, with engineering language, so that their reading should not allow interpretation. To design the SS, the engineer must meet all OEM requirements and embed them in a single product, even if the number of technical specifications can reach up to several thousand.

2. Fulfillments of functional conditions

The SS that embed the SZ must match the OEM 's CAR "environment". For this, investigations of the machine's "environment" for about 240 hours have been made, during which an intense exchange of information between OEM engineers and those of MCP that will produce SS has been carried out to determine which changes are possible and if there is a solution, among those found to be agreed on both sides, so that the design can continue based on it.

After the investigation period, the design team of the MCP sent the OEM first result, in the form of the V1 support concept. This concept is a "box" type with a wall thickness of 2.5 mm. The support incorporates 2 interfaces:

A. Interface with SZ (figure 1) - consisting of the 5 geometric structures numbered in the figure as follows:

1. Two M4 internal threads that are inserted into the mould prior to injection and fasten SZ through 4 screws. Fixing is done with screws; 2. Locating (set up) surface 2 consisting of 4 surfaces, which together represent the horizontal plane of SZ location (set up); 3. Four surfaces denoted with 3, which are distributed 2 by 2 on each side of SS, are designed so as to provide a small SZ guide in SS and left-right positioning;

4. Two geometric elements in the form of "fins" which come in contact with the side of the SZ and block the movement on the Z axis. The element 4, together with the elements 2 and 1, constrains the SZ on the Z axis and pre-stresses it so that it is fixed; 5. Two SZ contact surfaces at the top. This element together with the element 1 positions the SZ in the front and rear direction.

B. Interface with CAR (figure 2) - This interface is customer-conditioned and constraints in the machine environment, so in this case, we have the following elements:

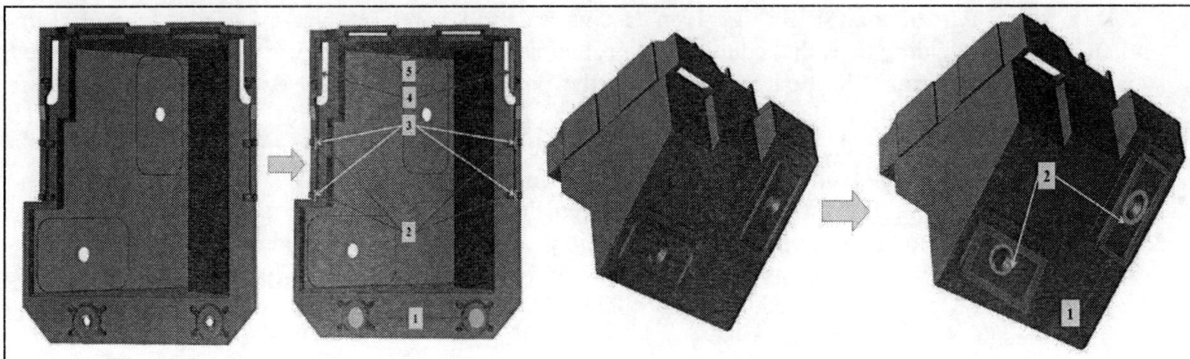

IOP Conf. Series: Materials Science and Engineering **568** (2019) 012016 doi:10.1088/1757-899X/568/1/012016

Figure 1. SZ –SS interface	.**Figure 2.** CAR-SS Interface..

1. The joint surface created by 2 ribs that come in contact with the CAR plate on which the SZ is positioned;
2. Two passage holes for 2 screws, which will then assemble 2 elastic metal clips that pass through the metallic CAR metal sheet, and once past, pretension and make enough tension to position the SZ and fix it;

3. Product constructive optimization

3.1. Optimizing product weight

The imperative desire to achieve the best products determines customers to collaborate with other companies to get more expertise (opinions) for the same product. For the previously described product, because OEM
later introduced the 20% reduction in CAR weight, it sent variant 1 of the SS product to a third company to conduct a study and provide solutions to reduce its weight.

The finality of third-party studies ended with a report containing several weight reduction suggestions and recommendations that could be grouped into two broad categories:

- Making cuts in the SS walls
- Adding stiffening ribs after a certain logic

Taking into account the suggestions above, SS solutions development are thought so as to embed this suggestions. The initial SS variation has undergone numerous changes by which it was mainly intended to optimize the amount of material used in the SS in relation to its mechanical strength. After the multiple variants and finite element analysis, they came to a variant with removed walls but without the addition of ribs, as it was found that they only needlessly load the model, and there is no real need for additional resistance to the given force loads - variant 2 of figure 3.

It is noteworthy that, following the stress tests, it was found that the SS in the V2 variant is even more resistant than the previous variant, given that all characteristics of the injection material were preserved as V1 (PA6.6 GF30 with a density of 1360 kg / m3) and the weight decreased from 86 g in V1 to 55.8 g. Therefore, a reduction of 30.2 g was obtained, i.e. V2 is lighter by about 26%. Both the OEM and the third-party consultant were pleasantly surprised that without adding ribs but merely modifying the geometry, an easier product and better mechanical properties were obtained.

3.2. Optimizing the SS manufacturing process

For SS in V2, the model details were made taking into account the tool making process and were looking for the suppliers to deliver this part.

In parallel with this stage of preparation of the SS manufacturing in V2 - although the analysis of injected material flow into the mould cavities is an optional one and often avoided the design team has also decided to carry out an analysis of this type using the Mold Flow Analysis software [8]. The results of this type of analysis for the given case were as follows (table 1). As a result of material flow, there are three welding lines that could cause problems for the piece. In figure 4, the location where SS is maximally deformed, located for on one of the positioning legs (C2). By virtue of a responsible design activity, it has been decided to put the injecting tool preparation into stand-by and to continue the design process, also taking into account the capabilities of the injection process as outlined above.

After several models and trials, the V3 variant for SS has come to life, distinguished by a new geometry that allows a better material flow. Given the localization of the main issues at V2, it was concluded that by merging the two legs of SS, the problem can be considered as solved, as indicated by the new Mold Flow analysis for V3 (see table 2). It can be seen in figure 5 that the place of injection point has been modified (C3), and this has greatly contributed to avoiding the welding lines; the final result, SS in V3 variant is show in figure 6.

IOP Conf. Series: Materials Science and Engineering **568** (2019) 012016 doi:10.1088/1757-899X/568/1/012016

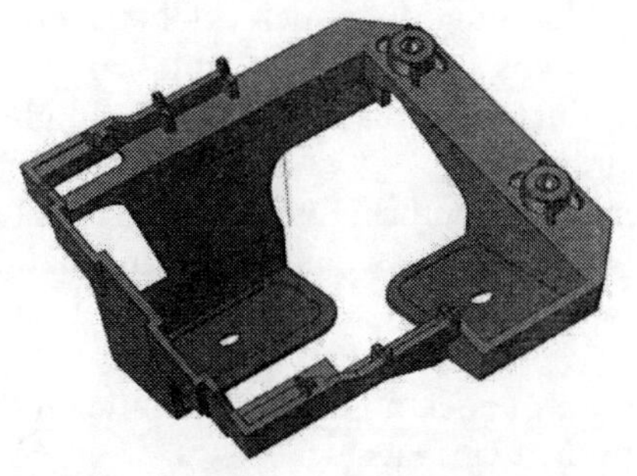

Figure 3. SS-variant 2

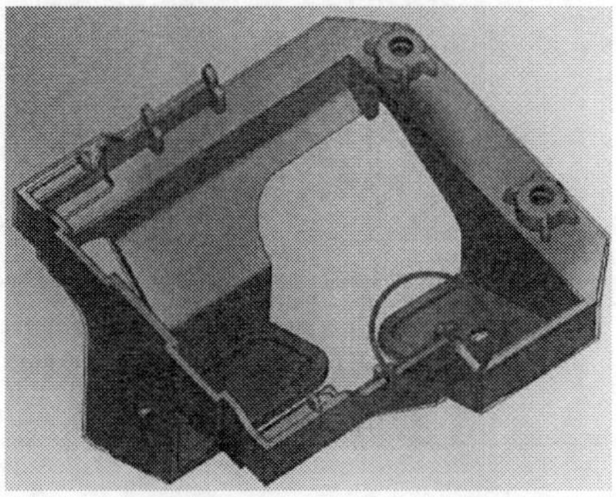

Figure 4. Maximum deform. place-C2

Table 1. Mold Flow Analysis Results for SS in V2 variant

Variant	Injection time [s]	Wieight[g]	Inj. Point. Temp. [°C]	Pressure [MPa]	Vol. Contract [%]	Estimated contraction [mm]	Deflections [mm]			
							X	Y	Z	T
SS/V2	1.48	55.8	291.4	40.84	14.61	1.08	1.02	0.96	0.8	1.35

Table 2. Mold Flow Analysis Results for SS in V3 variant

Variant	Injection Time[s]	Wieight[g]	Inj. Point. Temp. [°C]	Pressure [MPa]	Volume contract[%]	Estimated contraction [mm]	Deflections [mm]			
							X	Y	Z	T
V3	1.24	59	290.3	27.5	15.15	0.67	0.41	0.72	0.5	0.87

In figure 7 we can see the places C4 where the largest deformations are located: the two corners of the piece, areas that do not have functional importance.

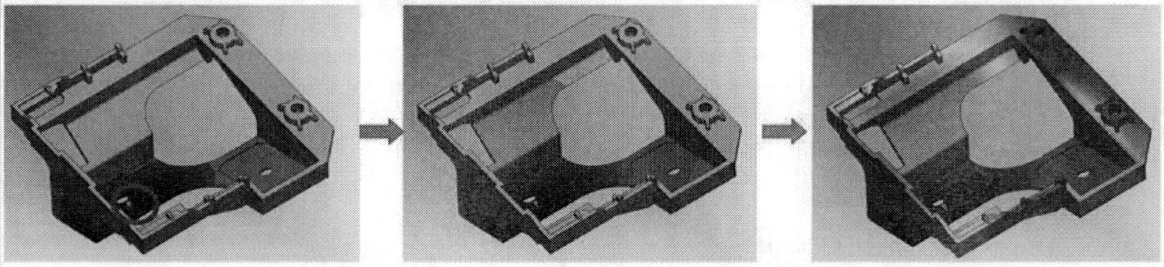

(a) (b) (c).

Figure 5. Cavity filling and injection point

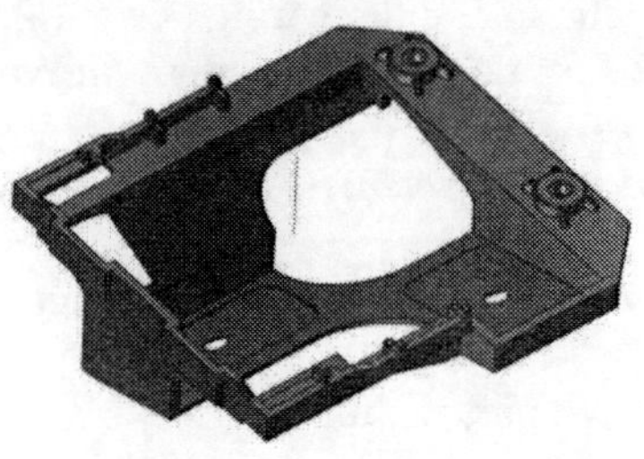

Figure 6. SS in V3 variant.

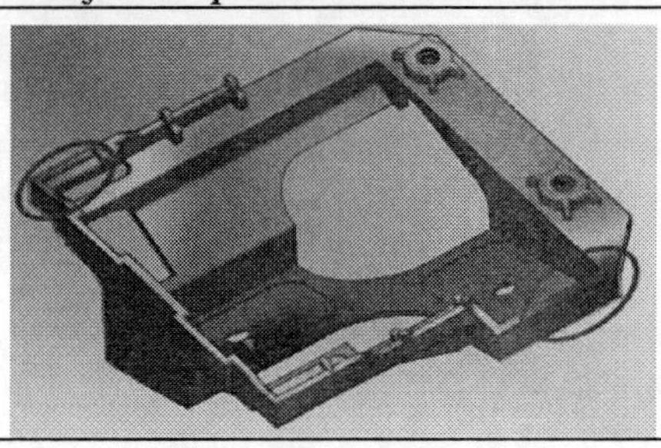

Figure 7.Places with maximum deformations -C4

4. Conclusions

The design phase of this project, by reaching the V3 variant of the SS, can be considered as completed, but the project is far from complete in terms of production, which offers the opportunity to meet new challenges from which can result in better solutions or can be examples of good practices that are included in new industrial product design procedures. In this sense, we highlight at least 3 issues that should complement the development of new projects:

1. Consideration of the type and conditions of the manufacturing process must be a mandatory phase of analysis among the design stages for example, if the product is one of the injected types, then Mold Flow Analysis should be mandatory from the early stage of a design, even in a simplified version.

2. Reiterating the design process with the inclusion of each additional specification should be considered as an essential step in generating improved product variants.

In the analysed case, the V2 variant of the SS brought a weight reduction of 26%, and in V3, even if it required the addition of material (the introduction of the bridge between the two legs added 3.2 g of material), it is easier than V1 by 23%.

3. The maturing of the implementation of new product design of the first two aspects above represents an important step towards the realization of a technological ideal generally pursued by companies: obtaining products that best meet the basic requirements, ie light and resistances.

The design case presented showed that in the working mode that embraced the "tips" above, the easiest piece was obtained according to the customer's requirements, provided that its gauge is not the smallest, but corresponding to the conjugated piece in which will be assembled.

Moreover, the mechanical properties of the piece have improved, which makes us affirm that optimization has been achieved at least in two directions. The data in table 3 is an argument in favour of this statement: only the weight is a negative aspect for SS variant 3 while the other parameters favour its choice, sufficient reasons to decide that the project will be continued in this variant.

Table 3. Comparative Mold Flow Analysis for V2 and V3

SS Variant	Timp de umplere[s]	Weight [g]	Pressure [MPa]	Estimated Contractions [mm]	Deflections [mm]
V2	1.48	55.8	40.84	1.08	1.35
V3	1.24	59	27.5	0.67	0.87

More than the realization of the design experience by completing the procedures, it is to notice that in the case of a responsible design activity, the research of the optimization study carried out by the competing third party is inevitable.

5. References

[1] Armstrong Stephen C. 2001 *Engineering and Product Development Management*, (Cambridge University Press, ISBN-13 978-0-521-01774-9)

[2] Dym L., Brown D. C. 2012 *Engineering design: representation and reasoning* /. – 2nd ed., ISBN 978-0-521-51429-3

[3] Finger, S. and J. R. Dixon 1989 A review of research in mechanical engineering design. Part I: Descriptive, prescriptive, and computer-based models of design processes. *Research in Engineering Design* **1** (1):51-67

[4] Ulrich, K. T., and S. D. Eppinger 2008 *Product Design and Development*. (Fourth ed. New York: McGraw-Hill International Edition)

[5] G. Pahl et al. 2007 *Engineering Design - A Systematic Approach*, (Third Edition, Springer, ISBN 978-1-84628-318-5, e-ISBN 978-1-84628-319-2)

[6] https://www.improve-your-injection-molding.com/mold-flow-analysis.html

[7] https://www.autodesk.com/products/moldflow/overview

[8] Gligor Liviu, "*Mold Flow Analysis for U6XX*", PPT Presentation, April 2017, Timisoara

Kinematics analysis of the swash plate mechanism

S Alaci[1], R D Pentiuc[1], F-C Ciornei[1], F Buium[2] and O T Rusu[1]
[1]"Stefan cel Mare" University, Suceava, Romania
[2]"Gheorghe Asachi" Technical University, Iasi, Romania

stelian.alaci@usm.ro

Abstract. The swash plate mechanism is one of the most exploited mechanisms from technical applications due to its constrictive simplicity. The kinematical analysis supposes establishing the motion of the plunger as function of the motion of the plate and of the constructive characteristics of the mechanism. The work presents the kinematical analysis of the mechanism performed by two methods. In the first way, the mechanism is replaced by a mechanism with lower pairs to which the Hartenberg-Denavit procedure is applied. In the second way, the definition conditions of the pair of first class, plunger-plate, are used. The results are obviously identical for the two methods but for the second approach, the labour required for the analysis is significantly reduced. The main outcome of the paper refers to the scheme of kinematical solving of a mechanism. The type of pairs should be a criterion and for the higher pairs, the use of direct condition of the structural definition is recommended.

1. Introduction

The swash plate mechanism consists in a driving element, made from a rotating shaft with an attached plane plate and a driven element, linked to the ground via a translational pair, as shown in figure 1. The motion is transmitted using a kinematical pair of class I (sphere-plane). The above description attests that the mechanism corresponds to the definition given by Duca [1] for a cam mechanism: "the mechanism that transmits the motion from a driving element to a driven element by means of a higher pair". The contact stresses developed between the plate and the spherical tip of the plunger produces wear and in order to reduce it, an intermediate part that makes spherical pairs both with the end of the plunger and with the plane plate, is interposed [2]. This operation increases the degree of mobility of the mechanism with one unit, fact that is explained by the presence of a passive motion – the rotation of the part about its axis of symmetry. The kinematical analysis of the mechanism assumes finding the dependency between the displacement of the driven element and the rotation of the driving element.

2. Kinematics analysis using the replacing mechanism

Hartenberg and Denavit [3] proposed a general method for kinematics analysis of spatial mechanisms named the method of homogenous operators, based on the transformation of the coordinates of a point when these are changed from one reference system to another one.

$$\begin{bmatrix} x_1 \\ 1 \end{bmatrix} = \begin{bmatrix} \multicolumn{3}{c}{R_{12}} & d_{12} \\ 0 & 0 & 0 & 1 \end{bmatrix} \begin{bmatrix} x_2 \\ 1 \end{bmatrix} \tag{1}$$

The position vectors of the point in the two coordinate systems have the column matrices $x_{1,2}$, the vector describing the displacement necessary to overlap the origin of the system 1 over the origin of the system 2 is d_{12} and the rotation matrix that superposes the axes of the system 1 over the axes of system 2 is R_{12}, with the columns as the projections of the versors of system 2 on the axis of system 1.

Figure 1. The swash plate mechanism

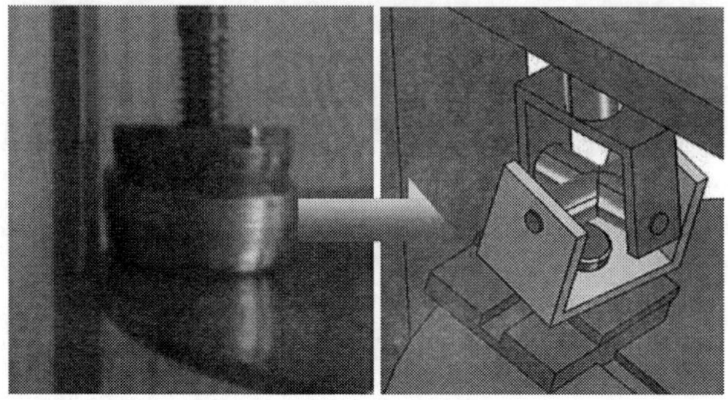

Figure 2. The sphere-plane pair replaced by the Oldham and Cardan clutches

The equation (1) can be expressed in concentrated manner:

$$\boldsymbol{r}_1 = \boldsymbol{T}_{12}\boldsymbol{r}_2 \tag{2}$$

Next, a closed contour made by the elements of a mechanism is considered and the convention that the $"n+1"$-th element coincides to the first element $"1"$. The consequence of relation (2) is:

$$\boldsymbol{r}_1 = \boldsymbol{T}_{12}\boldsymbol{r}_2 = \boldsymbol{T}_{12}\boldsymbol{T}_{23}\boldsymbol{r}_3 = \ldots = \boldsymbol{T}_{12}\boldsymbol{T}_{23}\ldots.\boldsymbol{T}_{n-1,n}\boldsymbol{r}_n = \boldsymbol{T}_{12}\boldsymbol{T}_{23}\ldots.\boldsymbol{T}_{n-1,n}\boldsymbol{T}_{n,1}\boldsymbol{r}_1 \tag{3}$$

From equations (3) it results:

$$\boldsymbol{T}_{12}\boldsymbol{T}_{23}\ldots\boldsymbol{T}_{n-1,n}\boldsymbol{T}_{n,1} = \boldsymbol{I}_4 \tag{4}$$

Hartenberg and Denavit demonstrated [3] that when the structure of the kinematic chain contains only cylindrical pairs (with particular cases of rotation or translation), the relative position of two systems attached respectively to two neighbouring elements necessitates only four parameters for describing it instead of six, as in the general case. To this end, the axes of cylindrical pairs play the part of $"z"$ axes while the $"x"$ axes will be represented by the common normals of the neighbouring $"z"$ axes. The displacement of the current frame $"k"$ over the next frame, $"k+1"$ is accomplished by two rototranslations, the first along the $"z"$ axis, characterized by the angle θ_k and the translation s_k followed by a rototranslation about the $x_{k,k+1}$ axis (which is the common normal for the axes $"z_k"$ and $"z_{k+1}"$) with the angle $\alpha_{k,k+1}$ and translation $a_{k,k+1}$. One can write:

$$\boldsymbol{T}_{k,k+1} = \boldsymbol{Z}(\theta_k, s_k)\boldsymbol{X}(\alpha_{k,k+1}, a_{k,k+1}) \tag{5}$$

where:

$$\boldsymbol{Z}(\theta,s) = \begin{bmatrix} \cos\theta & -\sin\theta & 0 & 0 \\ \sin\theta & \cos\theta & 0 & 0 \\ 0 & 0 & 1 & s \\ 0 & 0 & 0 & 1 \end{bmatrix}; \quad \boldsymbol{X}(\alpha,a) = \begin{bmatrix} 1 & 0 & 0 & a \\ 0 & \cos\alpha & -\sin\alpha & 0 \\ 0 & \sin\alpha & \cos\alpha & 0 \\ 0 & 0 & 0 & 1 \end{bmatrix} \tag{6}$$

Therefore, the wobble plate mechanism should be replaced by an equivalent mechanism containing in its structure only cylindrical pairs. Taking into account the fact that the higher pair of first class cancels only the translation along the common normal in the contact point, it results that the simple relative motions from the sphere-plane contact are two translations along two normal directions from the plane of the plate and three rotations performed about three axes reciprocally perpendicular. Figure 2 presents one of the possible replacements of the higher pair. It can be noticed that the substitution was made using an Oldham clutch and a Cardan clutch, linked between a rotational pair. The mechanism that substitutes the swash plate mechanism is presented in figure 3.

$$\begin{aligned}&\boldsymbol{Z}(\theta_1,0)\boldsymbol{X}(\pi/2+\alpha,)\boldsymbol{Z}(\pi/2,s_2)\boldsymbol{X}(\pi/2,0)\boldsymbol{Z}(\pi/2,s_3)\boldsymbol{X}(\pi/2,0)\boldsymbol{Z}(\theta_4,0)\boldsymbol{X}(\pi/2,0)\cdot\\&\cdot\boldsymbol{Z}(\theta_5,0)\boldsymbol{X}(\pi/2,)\boldsymbol{Z}(\theta_6,0)\boldsymbol{X}(\pi/2,0)\boldsymbol{Z}(\pi/2,s_7)\boldsymbol{X}(0,d)=\boldsymbol{I}_4\end{aligned} \tag{7}$$

The unknowns from the equation (7) are the translations from the pairs 2, 3 and 7 and the rotations from the pairs 4, 5 and 6. McCarthy specifies [4] that the use of the closing equation under this form conducts to complex calculus and the equation must be rewritten by decomposing the left member into two factors. For the present case, the equation was rewritten as:

$$\begin{aligned}&\boldsymbol{Z}(\theta_1,0)\boldsymbol{X}(\pi/2+\alpha,)\boldsymbol{Z}(\pi/2,s_2)\boldsymbol{X}(\pi/2,0)\boldsymbol{Z}(\pi/2,s_3)\boldsymbol{X}(\pi/2,0)\boldsymbol{Z}(\theta_4,0)\boldsymbol{X}(\pi/2,0)=\\&[\boldsymbol{Z}(\theta_5,0)\boldsymbol{X}(\pi/2,0)\boldsymbol{Z}(\theta_6,0)\boldsymbol{X}(\pi/2,0)\boldsymbol{Z}(\pi/2,s_7)\boldsymbol{X}(0,d)]^{-1}\end{aligned} \tag{8}$$

Considering the identity $(\boldsymbol{AB})^{-1}=\boldsymbol{B}^{-1}\boldsymbol{A}^{-1}$, the equation (8) can be amended as:

$$\begin{aligned}&\boldsymbol{Z}(\theta_1,0)\boldsymbol{X}(\pi/2+\alpha,)\boldsymbol{Z}(\pi/2,s_2)\boldsymbol{X}(\pi/2,0)\boldsymbol{Z}(\pi/2,s_3)\boldsymbol{X}(\pi/2,0)\boldsymbol{Z}(\theta_4,0)\boldsymbol{X}(\pi/2,0)=\\&\boldsymbol{X}(0,d)^{-1}\boldsymbol{Z}(\pi/2,s_7)^{-1}\boldsymbol{X}(\pi/2,0)^{-1}\boldsymbol{Z}(\theta_6,0)^{-1}\boldsymbol{X}(\pi/2,0)^{-1}\boldsymbol{Z}(\theta_5,0)^{-1}\end{aligned} \tag{9}$$

A second remark that can reduce considerably the time for calculus comes from Uicker [5] who stipulates that the inverse of any of the operators $\boldsymbol{Z}(\theta,s)$ or $\boldsymbol{X}(\alpha,a)$ is obtained by changing the signs of the two scalar parameters from the definition relations. The final form of the relation (9) becomes:

$$\begin{aligned}&\boldsymbol{Z}(\theta_1,0)\boldsymbol{X}(\pi/2+\alpha,)\boldsymbol{Z}(\pi/2,s_2)\boldsymbol{X}(\pi/2,0)\boldsymbol{Z}(\pi/2,s_3)\boldsymbol{X}(\pi/2,0)\boldsymbol{Z}(\theta_4,0)\boldsymbol{X}(\pi/2,0)=\\&\boldsymbol{X}(0,-d)\boldsymbol{Z}(-\pi/2,-s_7)\boldsymbol{X}(-\pi/2,0)\boldsymbol{Z}(-\theta_6,0)\boldsymbol{X}(-\pi/2,0)\boldsymbol{Z}(-\theta_5,0)\end{aligned} \tag{10}$$

$$\begin{bmatrix}-\sin\theta_1\cos\alpha\cos\theta_4-\cos\theta_1\sin\theta_4 & -\sin\theta_1\sin\alpha & -\sin\theta_1\cos\alpha\sin\theta_4+\cos\theta_1\cos\theta_4 & s_3\cos\theta_1+s_2\cos\alpha\sin\theta_1\\ \cos\theta_1\cos\alpha\cos\theta_4-\sin\theta_1\sin\theta_4 & \cos\theta_1\sin\alpha & \cos\theta_1\cos\alpha\sin\theta_4+\sin\theta_1\cos\theta_4 & s_3\cos\theta_1-s_2\cos\alpha\cos\theta_1\\ \sin\alpha\cos\alpha & -\cos\alpha & \sin\alpha\sin\theta_4 & -s_2\sin\alpha\\ 0&0&0&1\end{bmatrix}= \\ =\begin{bmatrix}\sin\theta_5 & -\cos\theta_5 & 0 & -d\\ -\cos\theta_5\cos\theta_6 & -\sin\theta_5\cos\theta_6 & -\sin\theta_6 & 0\\ \cos\theta_5\sin\theta_6 & \sin\theta_5\sin\theta_6 & -\cos\theta_6 & -s_7\\ 0&0&0&1\end{bmatrix} \tag{11}$$

From equation (11) it can be remarked that the information on the translations occur in the last column of the matrices. By equating the corresponding elements of the last columns, it results:

$$\begin{cases}s_3\cos\theta_1+s_2\cos\alpha\sin\theta_1=-d\\ s_3\cos\theta_1-s_2\cos\alpha\cos\theta_1\\ -s_2\sin\alpha=-s_7\end{cases} \tag{12}$$

The solutions of the system are immediately obtained:

$$s_2(\theta_1)=-d\sin\theta_1/\cos\alpha;\quad s_3(\theta_1)=-d\cos\theta_1;\quad s_7(\theta_1)=-d\tan\alpha\sin\theta_1 \tag{13}$$

IOP Conf. Series: Materials Science and Engineering **568** (2019) 012017 doi:10.1088/1757-899X/568/1/012017

The first two relations can be regarded as the parametric equations of the trajectory of the contact point and represent an ellipse having the semi-axes $d / cos\,\alpha$ and d . An interesting aspect can be noticed from figure 4: the trajectory of the contact point is actually an ellipse but it is eccentric with respect to the circular plate. This fact is produced by the spherical surface from the end of the plunger. The plate and the plunger in the extreme positions are presented in figure 5.

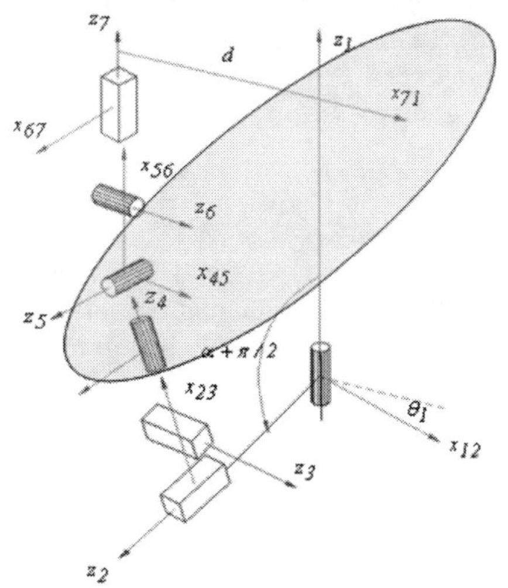

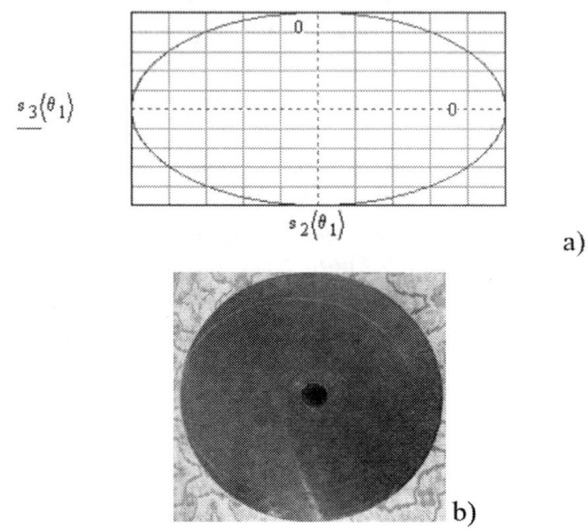

Figure 3. The mechanism that replaces the swash plate mechanism

Figure 4. The trajectory of the contact point: a) model; b) experimental trace

For this arrangement of the plate, the angle α between the normal to the plate and the rotation axis of the main shaft is seen in its actual size. The points $T_{1,2}$ correspond to the contact points between the plate and the plunger when the ball radius is zero and are placed at the distance $d / cos\,\alpha$ with respect to the axis of rotation of the plate, in agreement to the remarks based on the relations (13). When a ball exists at the end of the plunger, the contact points move from $T_{1,2}$ in the positions $C_{1,2}$, and therefore the elliptical trajectory "ascends" on the plate with the distance $R\,tan\,\alpha$.

3. The direct kinematical analysis using the constraint equation of the higher pair

The law of motion of the plunger was found after replacing the pair of first class between the plate and the plunger and using the equivalent mechanism. The drawback of the method consists in the large amount of calculus required for establishing the motion in all the pairs of the replacing chain (part of them not being necessary). When the structure of the mechanism contains pairs of first or second class, it is convenient to perform the analysis on the actual mechanism directly, because the small number of constrains imposed by those pairs require a reduced number of equations of condition. For the actual case, the equation of condition obliges that the tip of the plunger always is contained in the plane of the plate. The current position of the normal to the plane of the plate, after a plate rotation of θ angle about the vertical of versor $\boldsymbol{k}$, shown in figure 6, is, according to Angeles [6]:

$$\boldsymbol{n} = [\boldsymbol{k}\boldsymbol{k}^T + (\boldsymbol{I}_3 - \boldsymbol{k}\boldsymbol{k}^T)\cos\theta + \tilde{\boldsymbol{k}}\sin\theta]\boldsymbol{n}_o \tag{14}$$

where $\boldsymbol{n}_0$ is the versor of the plane of the plate for the initial position $\theta = 0$.

$$\boldsymbol{k} = \begin{bmatrix} 0 & 0 & 1 \end{bmatrix}^T ; \quad \boldsymbol{n}_o = \begin{bmatrix} sin\,\alpha & 0 & cos\,\alpha \end{bmatrix}^T \tag{15}$$

After the calculus is made, it results:

$$\boldsymbol{n} = \begin{bmatrix} sin\,\alpha\, cos\,\theta & sin\,\alpha\, sin\,\theta & cos\,\alpha \end{bmatrix}^T \tag{16}$$

The equation of the plane of the plate for a current position is:

$$\boldsymbol{n} \cdot \boldsymbol{r} = 0 \Leftrightarrow x \sin\alpha \cos\theta + y \sin\alpha \cos\theta + z \cos\alpha = 0 \tag{17}$$

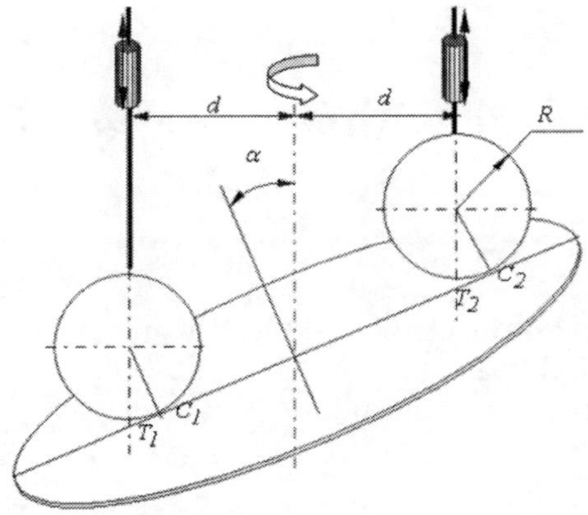

Figure 5. The plate and the plunger in the extreme positions

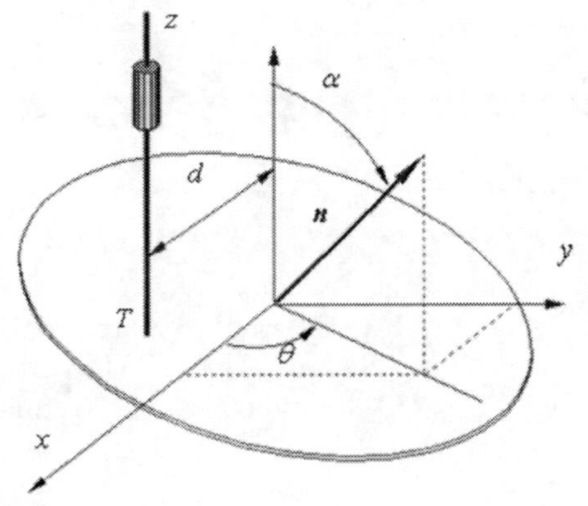

Figure 6. The normal to the plate for a current position

The equations of the direction of motion of the plunger are added:

$$x = d; \quad y = 0; \quad z = -d \tan\alpha \cos\theta \tag{18}$$

Equation (13) and equation (18), express identical harmonic motions with the amplitude $d \tan\alpha$, but with different origins for the angle of rotation of the plate.

4. Conclusions

The paper presents the kinematical analysis of the swash plate mechanism, for which two methods are considered. The first one assumes replacing the pair between the plunger and the plate with a kinematical chain, structurally equivalent, consisting only in rotational pairs and translational pairs to which the Hartenberg-Denavit method can be applied. The method has the advantage of an increased degree of algorithmization but also has the disadvantage of large calculus required for finding the motions from all the pairs of the kinematical chain of the replacing mechanism, from which some are not necessary of even don't have a physical significance. The second method consists in using directly the actual mechanism. Since the number of constraints imposed by the higher pair is reduced, only one for the present case, the equations of condition are simply to express, that is to state mathematically the fact that the tip of the plunger is always placed in the plane of the plate. The second method is more expedite than the first one but requires the identification and formulation of the accurate mathematical expressions of the condition equations for kinematical pairs.

Acknowledgment
„This work was supported by a grant of the Romanian Ministry of Research and Innovation, CCCDI – UEFISCDI, project number PN-III-P1-1.2-PCCDI-2017-0404 / 31PCCDI/2018, within PNCDI III".

References

[1] Duca C 1982 *Mechanisms* (*Mecanisme*, in Romanian) (I P Iasi)
[2] Alaci S, Ciornei FC, Amarandei D, Filote C, Cerlinca D and Irimescu L 2010 *Fiability and Durability* **2**(6) 79
[3] Hartenberg R S and Denavit J 1964 *Kinematic synthesis of linkages* (McGraw-Hill)
[4] McCarthy J M and Soh G S 2011 *Geometrical design of linkages* (Springer)
[5] Uicker J J, Ravani B and Sheth P N 2013 *Matrix Methods in Design Analysis of Mechanisms and Multi-body Systems* (Cambridge University Press)
[6] Angeles J 1988 *Rational Kinematics* (Springer New York)

IOP Conf. Series: Materials Science and Engineering **568** (2019) 012018 doi:10.1088/1757-899X/568/1/012018

The study of the motion of a gyroscope without rotational symmetry

F-C Ciornei[1], S Alaci[1], R D Pentiuc[1], C Bujoreanu[2] and O T Rusu[1]
[1]"Stefan cel Mare" University, Suceava, Romania
[2]"Gheorghe Asachi" Technical University, Iasi, Romania

florina.ciornei@usm.ro

Abstract. In rigid dynamics theory the gyroscope has a special role, as one of the bodies with the most complex motion, having three degrees of freedom. As consequence of the complexity of the motion is the fact that only for three situations, the equations of motion can be analytically integrated. For all the three situations, it is assumed that the principal moments of inertia (PMI) with respect to the principal axes that are normal to the rotation axes, are equal, condition fulfilled by the shape of body of revolution of the gyroscope. The paper analyses the case when the two principal moments of inertia are unequal. The equations of motion are numerically integrated for different values of the ratio between the two PMI. The effect of the disparity between the two PMI upon the motion of the gyroscope is proved by comparison with the motion of the axisymmetric gyroscope, both obtained by numerical integration.

1. Introduction

In figure 1, the motion of the rigid body is referred to the immobile Cartesian coordinate system $Oxyz$ of versors $\boldsymbol{i}$, $\boldsymbol{j}$ and $\boldsymbol{k}$. The frame attached to the rigid body is also a tri-orthogonal coordinate system $Ox'y'z'$ with versors $\boldsymbol{i}'$, $\boldsymbol{j}'$ and $\boldsymbol{k}'$. The position vector of the point $\boldsymbol{M}$ is expressed by the relation:

$$\boldsymbol{r} = \boldsymbol{r}_0 + \boldsymbol{r}' \tag{1}$$

where $\boldsymbol{r}_0$ is the position vector of the origin of the mobile frame. Under the form (1) the relation cannot be applied since the vectors from the right member are expressed in different coordinate systems. The explicit relation is:

$$x\boldsymbol{i} + y\boldsymbol{j} + z\boldsymbol{k} = x_0\boldsymbol{i} + y_0\boldsymbol{j} + z_0\boldsymbol{k} + x'\boldsymbol{i}' + y'\boldsymbol{j}' + z'\boldsymbol{j}' \tag{2}$$

Applying the scalar product for the equation (2) and the versors of the immobile frame, the following matrix form is obtained:

$$[x \;\; y \;\; z]^T = [x_O \;\; y_O \;\; z_O]^T + A[x' \;\; y' \;\; z']^T \tag{3}$$

In a concentrated manner, the equation (3) is written as:

$$\boldsymbol{r} = \boldsymbol{r}_0 + \boldsymbol{A}\boldsymbol{r}' \tag{4}$$

IOP Conf. Series: Materials Science and Engineering **568** (2019) 012018 doi:10.1088/1757-899X/568/1/012018

where A is an orthogonal matrix $A^T A = AA^T = I_3$, $A_{ij}A_{jk} = \delta_{ik}$, [1] describing the position of the axes (x', y', z') with respect to (x, y, z) $x', y', z')$, where δ_{ik} is the Kroneker symbol.

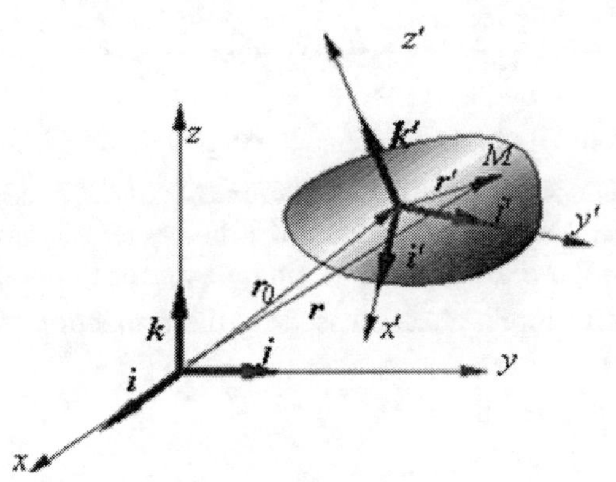

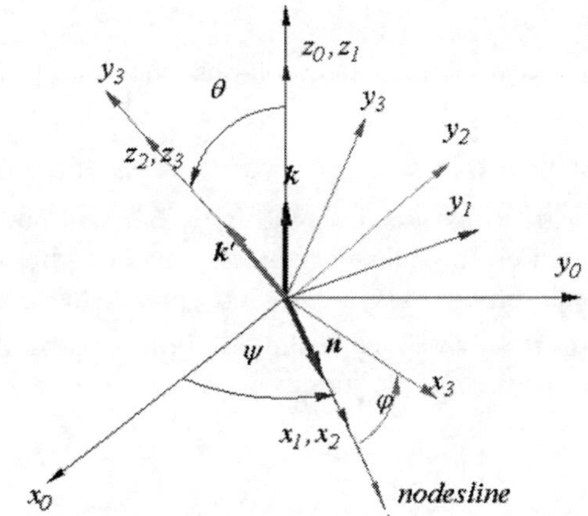

Figure 1. Position of a point of a rigid body **Figure 2.** Over posing two coordinate systems

As a conclusion, *six parameters are required* for stipulation of the position of the rigid: three needed for specifying the position of the origin of the mobile system and the other three for specifying the orientation of the axes of the mobile system with respect to the axes of the fixed frame. Regarding the first three parameters, they are the projections of the $\boldsymbol{r}_0$ vector on the axes of the fixed system, but the identification of the last three parameters is more difficult. In order to overlap two systems having the same origin, there are required three finite rotations about three well précised axes, as shown by Euler [2]. Thus, the A matrix may be obtained as the product of three matrices, corresponding to the three rotations [3]. According to Euler [2], the line of the nodes is the intersection between the planes x_0y_0 and x_1y_1. The three rotations are: rotation (*precession*) around z_0 axis of $\boldsymbol{k}$ versor, with the ψ angle, of the system $''0''$ about the system $''1''$; rotation (*nutation*) about Ox_1 axis, of $\boldsymbol{n}$ versor, with the ψ angle and rotation (*self-rotation*) about Oz_2 axis, of $\boldsymbol{k}'$ versor. After applying the three successive rotations, the 0 system will finally get into the 3 position. By denoting with Ω the infinitesimal vector obtained as the sum of the three infinitesimal rotations:

$$d\Omega = \boldsymbol{k}d\psi + \boldsymbol{n}d\theta + \boldsymbol{k}'d\varphi \tag{5}$$

it will be on its turn, an infinitesimal rotation which applied to a vector will produce a vector placed in a plane normal to the direction of $d\,\Omega$ and the rotation angle is given by the modulus of the vector (5). By dividing the relation (5) to dt, a ω vector is obtained:

$$\omega = \dot{\psi}\boldsymbol{k} + \dot{\theta}\boldsymbol{n} + \dot{\varphi}\boldsymbol{k}' \tag{6}$$

With the product used in finding the A matrix, there can be obtained the projections of any versor of an axis on the axes of another system.

2. The equations of motion for the rigid body with a fixed point

The rigid body with a fixed point (spherical joint) is represented in figure 3. The unknowns of the problem are the expressed using the three Euler angles and the components of the reaction from the spherical pair. The position of the rigid are obtained from Euler equations, where the mobile axes are the principal axes of inertia, passing through the fix point:

$$\begin{bmatrix} J'_x & 0 & 0 \\ 0 & J'_y & 0 \\ 0 & 0 & J'_z \end{bmatrix}\begin{bmatrix} \dot\omega'_x \\ \dot\omega'_y \\ \dot\omega'_z \end{bmatrix}+\begin{bmatrix} 0 & -\omega'_z & \omega'_y \\ \omega'_z & 0 & -\omega'_x \\ -\omega'_y & \omega'_x & 0 \end{bmatrix}\begin{bmatrix} J'_x & 0 & 0 \\ 0 & J'_y & 0 \\ 0 & 0 & J'_z \end{bmatrix}\begin{bmatrix} \omega'_x \\ \omega'_y \\ \omega'_z \end{bmatrix}=\begin{bmatrix} M'_{Ox} \\ M'_{Oy} \\ M'_{Oz} \end{bmatrix} \Leftrightarrow \begin{matrix} J'_x\dot\omega'_x+(J'_z-J'_y)\omega'_y\omega'_z=M'_{Ox} \\ J'_y\dot\omega'_y+(J'_x-J'_z)\omega'_x\omega'_z=M'_{Oy} \\ J'_z\dot\omega'_z+(J'_y-J'_x)\omega'_x\omega'_y=M'_{Oz} \end{matrix} \tag{7}$$

The dynamics monographs show that the system (7) can be analytically integrated in three situations: 1) the Euler Poinsot case $\boldsymbol{M}'_a=0$; 2) the Lagrange Poisson case $J'_x=J'_y$; 3) the Sofia Kovalevskaia case, the center of mass of the rigid should be in the plane $Ox'y'$; $J'_x=J'_y=2J_3$.

An axi-symmetric rigid body but without rotational symmetry is considered, having a fixed point placed on the axis of symmetry, as in figure 4 actuated by its own weight $\boldsymbol{G}$; the center of gravity is placed on the axis of symmetry at a distance ζ from the fixed point. The single exterior torque acting upon the gyroscope is the one produced by the gravity force $\boldsymbol{G}$, expressed by the projections on the axes of the mobile frame:

$$\boldsymbol{M}_O=Mg\zeta\sin\theta\begin{bmatrix}\cos\varphi & \sin\varphi & 0\end{bmatrix}^T \tag{8}$$

3. Numerical integration of the equations of motion

The values of the angles ψ, θ, φ and their derivatives must be stipulated for the initial instant, in order to integrate numerically the system (7).

For the derivatives calculus, the relation (6) of definition of angular velocity is recalled. The versors $\boldsymbol{k}$ and $\boldsymbol{n}$ are expressed in the mobile frame:

$$\begin{bmatrix} \omega'_x \\ \omega'_y \\ \omega'_z \end{bmatrix}=\begin{bmatrix} \sin\theta\sin\varphi \\ \sin\theta\cos\varphi \\ \cos\theta \end{bmatrix}\dot\psi+\begin{bmatrix} \cos\psi \\ -\sin\psi \\ 0 \end{bmatrix}\dot\theta+\begin{bmatrix} 0 \\ 0 \\ 0 \end{bmatrix}\dot\varphi \tag{9}$$

Using the expressions of the angular velocity components expressed as functions of Euler's angles, their derivatives with respect to time and applying the relations (7), a system of three differential equations of second degree is obtained. In order to solve this type of system, the introduction of an auxiliary variable is required for obtaining a system of six linear differential equations of first order. The equations (9) permit writing the derivatives of Euler's angles as functions of the components of the angular velocity on the axes of the mobile system.

$$\begin{bmatrix} \dot\psi \\ \dot\theta \\ \dot\varphi \end{bmatrix}=\begin{bmatrix} \sin\varphi/\sin\theta & \cos\varphi/\sin\theta & 0 \\ \cos\varphi & -\sin\varphi & 0 \\ -\sin\varphi\cos\theta/\sin\theta & -\cos\varphi\cos\theta/\sin\theta & 1 \end{bmatrix}\begin{bmatrix} \omega'_x \\ \omega'_y \\ \omega'_z \end{bmatrix} \tag{10}$$

From relations (8) and (9) it is obtained:

$$\begin{bmatrix} \dot\omega'_x \\ \dot\omega'_y \\ \dot\omega'_z \end{bmatrix}=\begin{bmatrix} \omega'_y\omega'_z(J_y-J_z)/J_x \\ \omega'_z\omega'_x(J_z-J_x)/J_y \\ \omega'_y\omega'_y(J_x-J_y)/J_z \end{bmatrix}+Mg\zeta\sin\theta\begin{bmatrix} \cos\varphi \\ \sin\varphi \\ 0 \end{bmatrix} \tag{11}$$

The equations (10) and (11) can now be regarded as a system of six differential nonlinear equations of unknowns $\psi, \theta, \varphi, \omega'_x, \omega'_y$ and ω'_z. A numerical method must be applied for the integration of this system [4] and the stipulation of the values of the six unknowns for the initial moment is required.

IOP Conf. Series: Materials Science and Engineering **568** (2019) 012018 doi:10.1088/1757-899X/568/1/012018

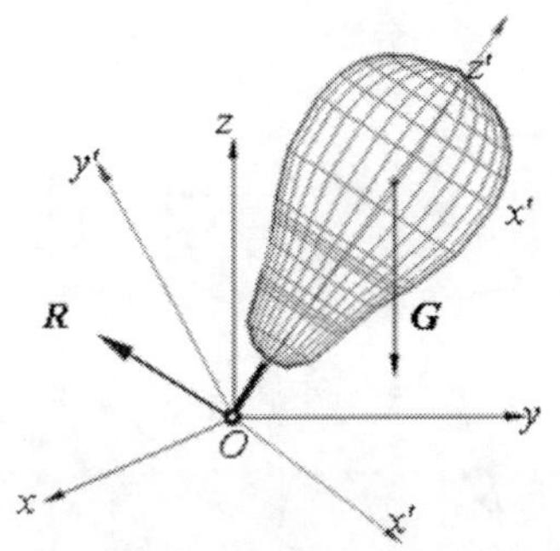

Figure 3. The rigid body, with a fixed point, with rotational symmetry

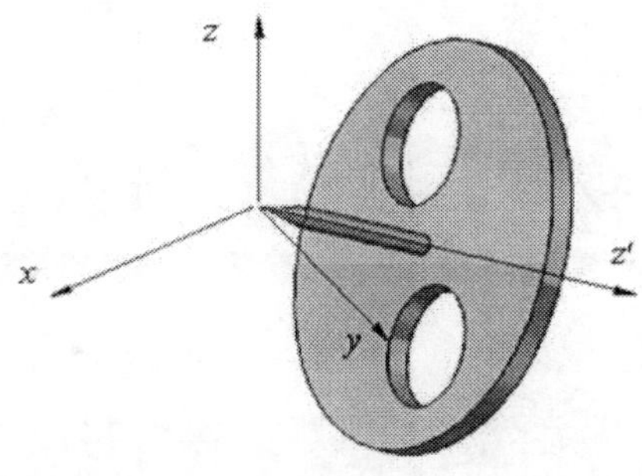

Figure 4. Gyroscope with axial symmetry without rotational symmetry

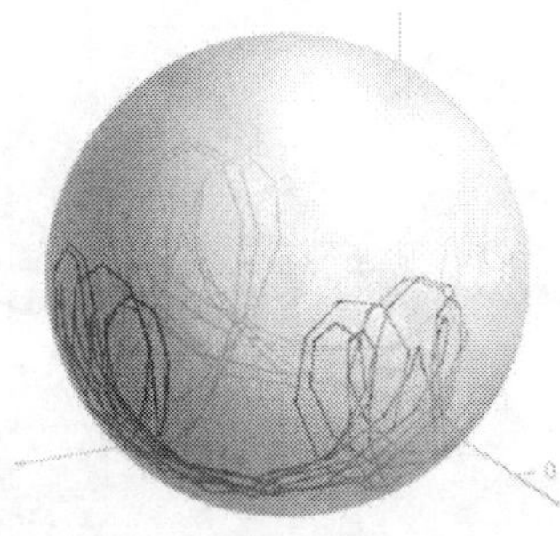

Figure 5. Spatial trajectory of the centre of mass of the gyroscope

$$\begin{bmatrix} \dot{\psi} \\ \dot{\theta} \\ \dot{\varphi} \\ \dot{\omega}'_x \\ \dot{\omega}'_y \\ \dot{\omega}'_z \end{bmatrix} = \begin{bmatrix} \omega'_x \sin\varphi / \sin\theta + \omega'_y \cos\varphi / \sin\theta \\ \omega'_x \cos\varphi - \omega'_y \sin\varphi \\ -\omega'_x \sin\varphi \cos\theta / \sin\theta - \omega'_y \cos\varphi \cos\theta / \sin\theta + \omega'_z \\ \omega'_y \omega'_z \left(J_y - J_z\right) / J_x + Mg\zeta \sin\theta \cos\varphi \\ \omega'_z \omega'_x \left(J_z - J_x\right) / J_y + Mg\zeta \sin\theta \sin\varphi \\ \omega'_x \omega'_y \left(J_x - J_y\right) / J_z \end{bmatrix} \tag{12}$$

The system (12) is the general system of equations which describes the motion of any type of gyroscope. Next, a gyroscope that has the Oz' axis as principal axis of inertia is considered, with the particularity that the moments of inertia with respect to the other principal axes are not equal anymore, as in the cases of integrability. Such a gyroscope can be obtained practically from a gyroscope with axial rotational symmetry, by removing some material symmetrically with respect to the axis of symmetry, as shown in figure 5. As it can be noticed from the relation (12), a first consequence of the asymmetry $J_x \neq J_y$ is the fact that the rotation motion of the gyroscope about the axis of symmetry is no longer a rotation with constant velocity. In figure 5 is presented the position of the centre of mass of the gyroscope.

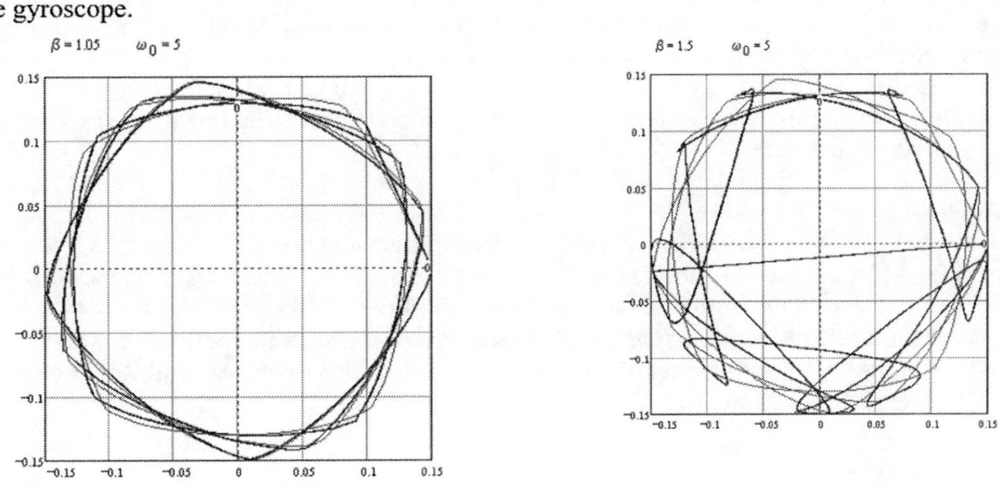

Figure 6. The effect of the geometrical asymmetry β upon the trajectory of the centre of mass

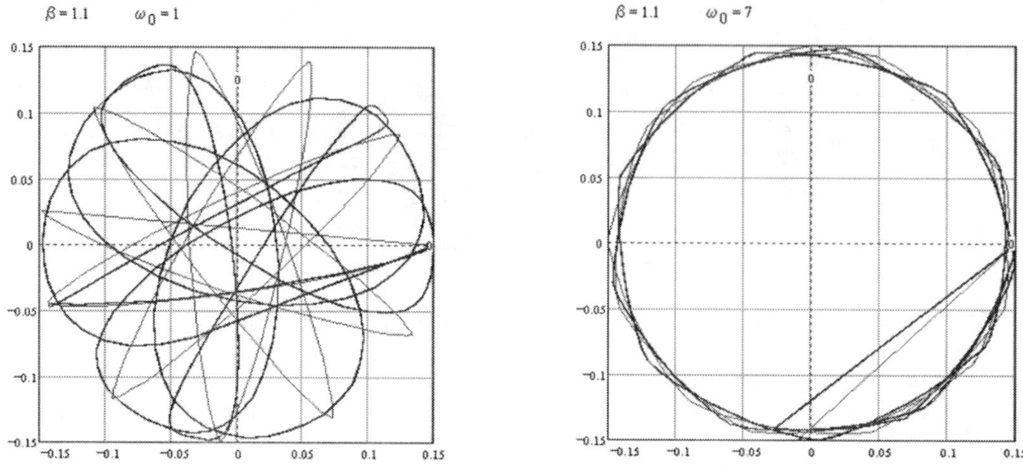

Figure 7. The effect of the own initial angular velocity upon the trajectory of the centre of mass

In order to emphasize the effect of lack of rotational symmetry, the traces of the trajectory of the center of mass on the fixed plane Oxy are presented for different values of the asymmetry coefficient $\beta = J_y / J_x$ in figure 6 and as function of the own initial angular velocity, in figure 7, respectively. The plots from figure 6 and figure 7 present comparatively the graphs for the gyroscope with rotational symmetry (red traces) and for the gyroscope without rotational symmetry (blue traces).

4. Conclusions

The paper presents the method of obtaining the finite equations for a symmetrical gyroscope but without rotational symmetry. First, the system of differential equations of the gyroscope under the form of a nonlinear system of six differential equations of first order is obtained; the unknowns of the system are the derivatives of the Euler's angles and the derivatives of the projections of angular velocity on the axes of the coordinate frame attached to the gyroscope.

The system obtained is valid for the motion of any type of gyroscope. In order to illustrate the effect of asymmetry upon the motion of the gyroscope, the projections of the trajectories on the horizontal plane for the gyroscope with rotational symmetry and for the one without rotational symmetry were presented comparatively. The graphical representations were made firstly for different values of the coefficient of asymmetry but for the same initial conditions and secondly, for the same asymmetry but for different values of the own angular velocity.

As anticipating, the differences diminish with increasing angular velocity, fact easy to perceive qualitatively. The paper permits achieving quantitative aspects concerning the differences between the motions from the two cases.

5. References

[1] Brandt L 1947 *Vector and Tensor Analysis* (John Willey and Son)
[2] Voinea R, Voiculescu D and Simion F P 1989 *Introduction to Solid Mechanics (Introducere în mecanica solidului* in Romanian) Ed. Academiei Romane Bucuresti)
[3] McCarthy J M 1990 *Introduction to Theoretical Kinematics* (MIT Press)
[4] Hamming R W 1987 *Numerical Methods for Scientists and Engineers* 2nd Revised ed. (Dover Books on Mathematics)

Acknowledgement

This work was supported by a grant of the Romanian Ministry of Research and Innovation, CCCDI – UEFISCDI, project number PN-III-P1-1.2-PCCDI-2017-0404 / 31PCCDI/2018, within PNCDI III.

IOP Conf. Series: Materials Science and Engineering **568** (2019) 012019 doi:10.1088/1757-899X/568/1/012019

Dynamic simulation of the adaptive sun tracking system used for an electric unmanned ground vehicle

C Alexandru

Transilvania University of Braşov, Romania

E-mail: calex@unitbv.ro

Abstract. In this article, the modeling and simulation of the adaptive sun tracking system used for an electric unmanned ground vehicle is carried out. The power supply of the vehicle is an electric one, supported by a photovoltaic (PV) panel with solar tracker, which is positioned on the upper platform of the vehicle (practically, the tracking system thus defined is one with mobile base). The main feature of the adaptive tracking mechanism consists in its ability to correlate the predefined tracking law (which it is designed as system would be one with fixed base) with the random movement of the vehicle. The mechatronic model contains both mechanical and electronic (control) elements, being conceived and analyzed in virtual prototyping environment by using the MBS software package ADAMS.

1. Introduction

The unmanned ground vehicles (UGV) are autonomous robotic systems, without human operator on board, which are frequently used for applications where human presence is not possible or necessary (for example, in agriculture or military applications), for various purposes, such as detecting and/or removing objects, collecting data about the environment, and others [1-3]. These vehicles are usually electrically driven, the energy being stored in batteries fitted to the vehicle. For long-term operation, photovoltaic panels can be attached to the vehicle, to provide supplementary energy or to charge the batteries. Solar tracking mechanisms can improve the conversion system efficiency by maximizing the incident solar radiation on the panel surface [4-8]. The existing solar trackers are predominantly mechatronic systems, which integrates mechanical and control components, whose drive is made by actuators (motors) that are commanded so that to allow a more accurate sun tracking.

The literature reveals many types of solar trackers (more or less complex), in mono-axial and bi-axial configurations (with one or two degrees of mobility). Usually, the mono-axial systems are used to track the sun diurnal movement, while the bi-axial systems are able to assure the both movements in the Earth - Sun system, namely the diurnal and elevation (altitudinal) motions [9-14]. Unlike the most existing tracking systems, whose base (support) is fixed connected to ground, the tracking mechanism proposed in this work has mobile base (namely the platform of the unmanned ground vehicle), making it necessary to adapt (correlate) the predefined tracking law (which it is designed as system would be one with fixed base) with the random movement of the vehicle. The software package ADAMS has been used to model and simulate the adaptive sun tracker behavior. Such a virtual prototyping solution provides important benefits, as pointed out in [15-19]. For this study, only the yaw movement of the vehicle was taken into account, the PV tracking system being approached in the mono-axial version, which is derived from an azimuthally system, the axis of the diurnal motion being vertically oriented, parallel to the axis around which the vehicle yaw occurs (by turning to the left - right).

IOP Conf. Series: Materials Science and Engineering **568** (2019) 012019 doi:10.1088/1757-899X/568/1/012019

2. Tracking system modeling

The design algorithm (flow-chart) of the sun trackers is based on building and testing the kinematic, inverse dynamic, and direct dynamic models, as presented in [20]. By going through this algorithm, the sun tracking mechanism for this work (which is a mono-axial azimuthally system) has been designed, the virtual model conceived in ADAMS being shown in Figure 1.

The tracking system contains the vertical pillar (1), on which the fixed part of the drive motor (based on worm gear reducer) is mounted, the moving part of the motor (i.e. the rotor) being coupled to the support frame (2) of the PV panel (3). The diurnal movement is carried out by rotating the panel (along with the support on which it is mounted) around the vertical axis - x.

In the classical variant of tracking system with fixed base, the pillar is rigidly connected to ground. The diurnal movement is performed according to a stepping orientation function, developed so as to ensure an optimal number of motion steps, with a minimal energy demand to carry out the sun tracking, what will be found in increasing the PV system efficiency. The application for this article has been conceived considering the summer solstice day, and the Braşov implementation area, which actually determines the orientation path.

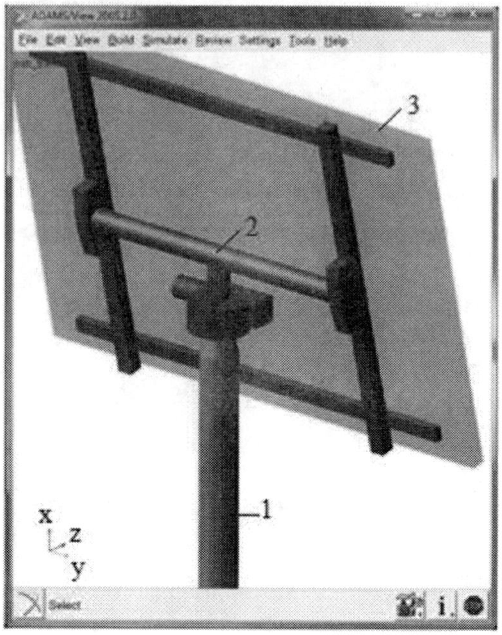

Figure 1. The MBS model of the mono-axial azimutally sun tracker.

The angular domain of the diurnal motion is 180°, in the interval $\psi^* \in [90°, -90°]$, with the following actuating timing (in local time): 9.819 ($\Delta\psi^*$=21°), 11.439 ($\Delta\psi^*$=30°), 12.769 ($\Delta\psi^*$=39°), 13.869 ($\Delta\psi^*$=39°), 15.199 ($\Delta\psi$=30°), 16.819 ($\Delta\psi$=21°), 21.166 ($\Delta\psi$= -180° - return to the initial position, after sunset). In ADAMS, the motion law has been modeled by a sum of STEP functions, with the following format STEP(time, begin at, initial value, end at, final value), in which the time is the independent parameter, corresponding to the number of motion steps, in respect with the previosly presented values, as follows (where *d* indicates that the value is in degrees):

$$\begin{gathered}F = \mathrm{STEP}(time, 9.82, 0.0d, 9.92, 21.0d) + \mathrm{STEP}(time, \\ 11.44, 0.0d, 11.54, 30.0d) + \mathrm{STEP}(time, 12.77, 0.0d, \\ 12.87, 39.0d) + \mathrm{STEP}(time, 13.87, 0.0d, 13.97, 39.0d) \\ + STEP(time, 15.20, 0.0d, 15.30, 30.0d) + \mathrm{STEP}(time, \\ 16.82, 0.0d, 16.92, 21.0d) + \mathrm{STEP}(time, 21.17, 0.0d, \\ 21.27, -180d)\end{gathered} \tag{1}$$

It should be mentioned that the final value in a STEP function is expressed by reference to the initial one, which is actually the final one from the previous motion (tracking) step. For example, Figure 2 shows the modeling of the first tracking step, which occurs during the time interval 9.82-9.92.

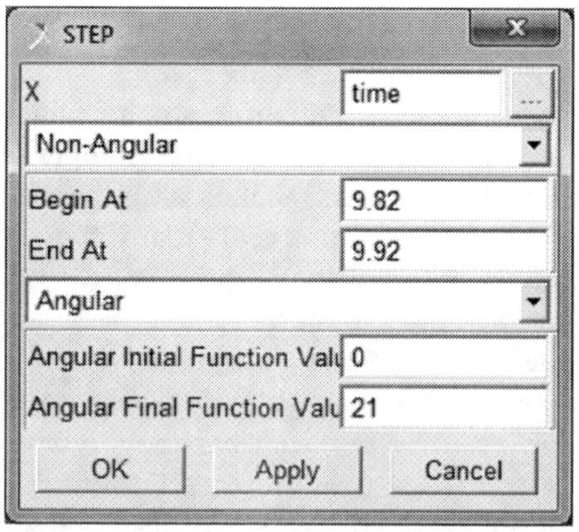

Figure 2. The STEP function modeling.

Under these circumstances, the imposed variation diagram of the diurnal (azimuthallly) angle of the PV panel is shown in Figure 3.a, with the corresponding angular velocity in Figure 3.b.

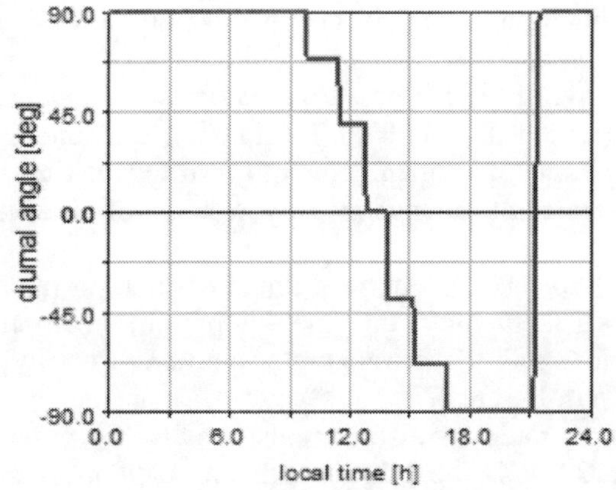

a.

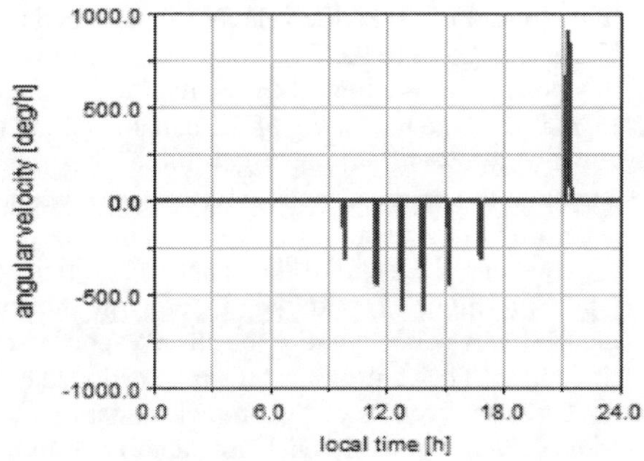

b.

Figure 3. The tracking law for the PV system with fixed base.

Afterwards, the so-designed sun tracking mechanism was implemented (mounted) on an unmanned ground vehicle, whose virtual model is the one in Figure 4, the tracking mechanism base (i.e. the vertical pillar) being rigidly connected to the upper platform of the vehicle. For this work, only the yaw movement of the vehicle, which is generated by the vehicle turning to the left or to the right, has been considered. Consequently, the adaptive tracking mechanism must be able to correlate its predefined tracking law (shown in Figure 3), which it is designed as the PV system would be one with fixed base, with the random turning movement of the vehicle, so as to maintain the optimal incidence of solar radiation on the PV panel.

In the virtual model, a sensor is used to measure the yaw angle of the vehicle (in reality, this movement can be measured by using a gyroscope or an accelerometer), the value read by the sensor being then used in the form of a state variable for changing/adapting the tracking law in the control system of the rotary motor for the diurnal motion, as follows:

$$\begin{aligned} F = {} & \mathrm{STEP}(time, 9.82, 0.0d, 9.92, 21.0d) + \mathrm{STEP}(time, \\ & 11.44, 0.0d, 11.54, 30.0d) + \mathrm{STEP}(time, 12.77, 0.0d, \\ & 12.87, 39.0d) + \mathrm{STEP}(time, 13.87, 0.0d, 13.97, 39.0d) \\ & + STEP(time, 15.20, 0.0d, 15.30, 30.0d) + \mathrm{STEP}(time, \\ & 16.82, 0.0d, 16.92, 21.0d) + \mathrm{STEP}(time, 21.17, 0.0d, \\ & 21.27, \text{-}180d) + \mathrm{VARVAL}(VARIABLE\_1) \end{aligned} \tag{2}$$

VARVAL being an ADAMS/View predefined function that is used to provide the current value of the variable mentioned in the argument. The variable is defined by using the displacement function YAW, with the following format: YAW(vehicle.MARKER_101, ground.MARKER_102). This function returns the first angle of rotation from one coordinate system marker (MARKER_101, which belongs to the vehicle) to another (MARKER_102 - belonging to the ground part). In the initial (modeling) position, the two coordinate system markers are coincidental (as location/position, and orientation).

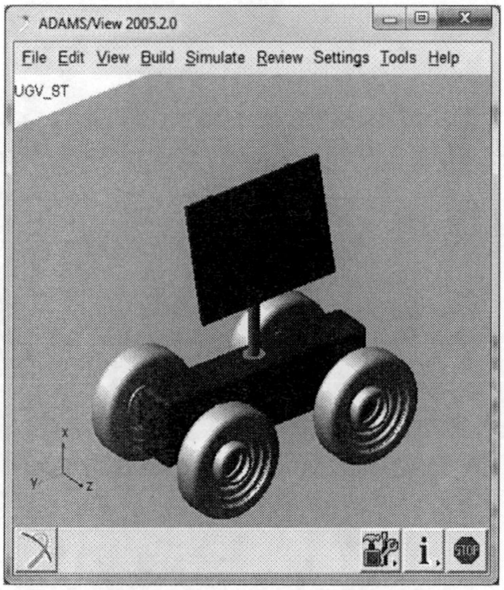

Figure 4. The virtual model of the UGV equipped with PV solar tracker.

IOP Conf. Series: Materials Science and Engineering **568** (2019) 012019 doi:10.1088/1757-899X/568/1/012019

3. Results and conclusions

In the initial position, the vehicle is positioned so that its longitudinal axis (y) is directed on the South-North axis, so as to the PV panel is oriented towards the East. If the vehicle would remain stationary in this position throughout the day, then the orientation law of the PV panel would be the one shown in Figure 3. At the beginning of the dynamic simulation, which starts at 8:00 (local hour), the vehicle will move to the South, in a straight line in the first 30 minutes, and then begins to turn left or right (for example, to explore a specific workspace), which generates the change in the yaw angle of the vehicle according to the diagram shown in Figure 5.

The control system of the adaptive sun tracker will correlate of the predefined orientation function (shown in Figure 3) with the random yaw movement of the vehicle, thus making possible the optimal incidence of solar radiation on the PV panel at all times, as if the tracking system were one with a fixed base. The control law thus obtained is that shown in Figure 6.

The effectiveness of the proposed adaptive tracking strategy is also demonstrated by the graphical frames from Figure 7, which are captured at midday (13.206 in local time), at this time the mechanism assuring the best orientation of the PV panel, namely facing South, on y-axis direction (Figure 7,a), which cannot be assured if the system would be a non-adaptive one (Figure 7,b), meaning that it is not able to correlate the tracking law with the vehicle trajectory.

In a forthcoming work, the adaptability of the PV tracking system will be extended so that to consider other movements of the vehicle, such as the pitch, which occurs when the vehicle goes down-up a slope/ramp or it passes over obstacles.

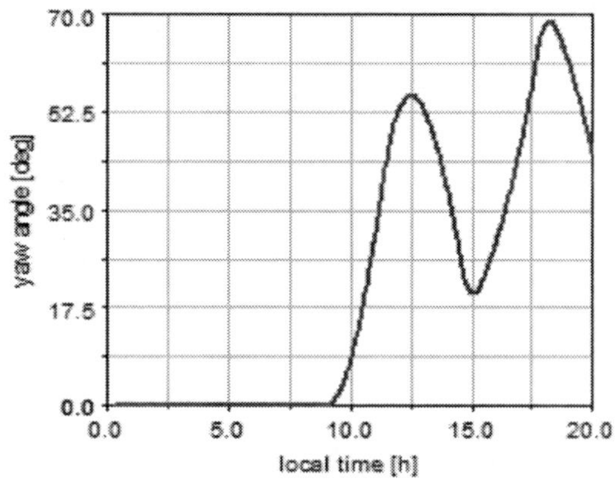

Figure 5. The yaw angle of the vehicle.

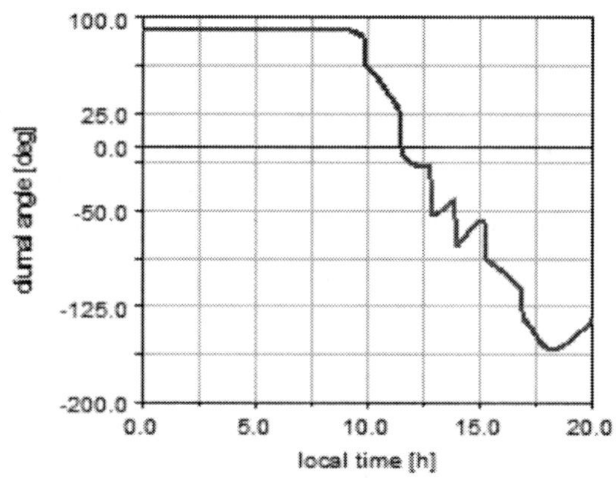

Figure 6. The control law for the diurnal motion.

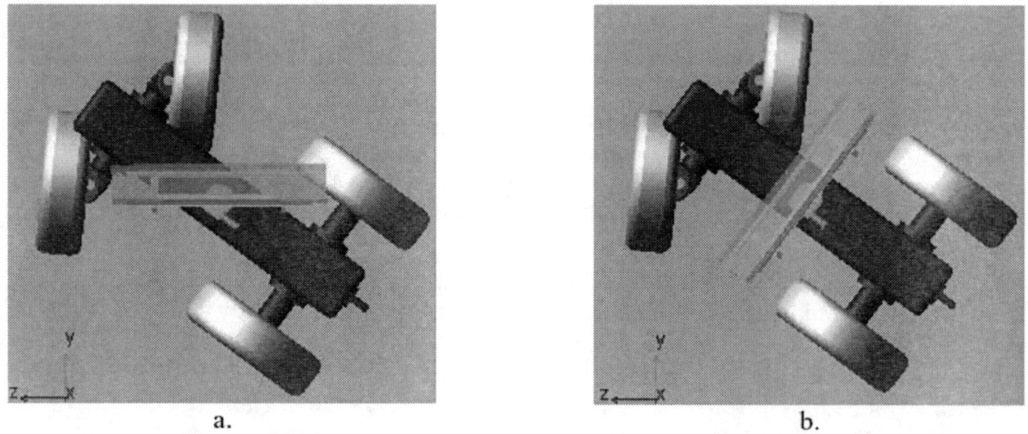

Figure 7. Simulation frames for the adaptive (a) and non-adaptive (b) tracking system.

References

[1] Durst P, Goodin C, Anderson D and Bethel C 2017 A reference autonomous mobility model *Proceedings of the IEEE Winter Simulation Conference - WSC* pp 4026-4037

[2] Gage D 1995 A brief history of unmanned ground vehicle (UGV) development efforts *Unmanned Systems Magazine* **13** pp 1-10

[3] Hassan S, Alam M, Siddiqui N A, Siddiqui A A and Qadri T 2017 Designing and control of autonomous unmanned ground vehicle *Proceedings of the International Conference on Innovations in Electrical Engineering and Computational Technologies - ICIEECT* pp 1-5

[4] Alexandru C and Comşiţ M 2008 The energy balance of the photovoltaic tracking systems using virtual prototyping platform *Proceedings of the 5th IEEE International Conference on the European Electricity Market - EEM* pp 253-258

[5] Alexandru C 2009 The design and optimization of a photovoltaic tracking mechanism *Proceedings of the 2nd IEEE International Conference on Power Engineering, Energy and Electrical Drives - POWERENG* pp 436-441

[6] Nsengiyumva W, Chen S G, Hu L and Chen X 2018 Recent advancements and challenges in solar tracking systems (STS): A review *Renewable & Sustainable Energy Reviews* **81** pp 250-279

[7] Seme S, Stumberger B, Hadziselimovic M 2016 A novel prediction algorithm for solar angles using second derivative of the energy for photovoltaic sun tracking purposes *Solar Energy* **137** pp 201-211

[8] Sharaf Eldin S A, Abd-Elhady M S and Kandil H A 2016 Feasibility of solar tracking systems for PV panels in hot and cold regions *Renewable Energy* **85** pp 228-233

[9] Alexandru C and Tatu I N 2013 Optimal design of the solar tracker used for a photovoltaic string *Journal of Renewable and Sustainable Energy* **5** pp 1-16

[10] Altuzarra O, Macho E, Aginaga J and Petuya V 2015 Design of a solar tracking parallel mechanism with low energy consumption *Proceedings of the Institution of Mechanical Engineers, Part C: Journal of Mechanical Engineering Science* **229** pp 566-579

[11] Ioniţă M and Alexandru C 2012 Dynamic optimization of the tracking system for a pseudo-azimuthal photovoltaic platform *Journal of Renewable and Sustainable Energy* **4** pp 1-15

[12] Flores-Hernández D A, Palomino-Resendiz S, Lozada-Castilloc N and Luviano-Juárez A 2017 Mechatronic design and implementation of a two axes sun tracking photovoltaic system driven by a robotic sensor *Mechatronics* **47** pp 148-159

[13] Hoffmann F M, Molz R F, Kothe J V, Nara E O B, Pablo L and Tedesco C 2018 Monthly profile analysis based on a two-axis solar tracker proposal for photovoltaic panels *Renewable Energy* **115** pp 750-759

[14] Vieira R G, Guerra F K, Vale M R and Araujo M M 2016 Comparative performance analysis between static solar panels and single-axis tracking system on a hot climate region near to the equator *Renewable & Sustainable Energy Reviews* **64** pp 672-681

[15] Alexandru C and Alexandru, P 2010 The virtual prototype of a mechatronic suspension system with active force control *WSEAS Transactions on Systems* **9** pp 927-936

[16] Alexandru C and Alexandru, P 2011 A comparative analysis between the vehicles' passive and active suspensions *International Journal of Mechanics* **5** pp 371-378

[17] Alexandru P, Macaveiu D and Alexandru C 2012 Design and simulation of a steering gearbox with variable transmission ratio *Proceedings of the Institution of Mechanical Engineers, Part C: Journal of Mechanical Engineering Science* **226** pp 2538-2548

[18] Bernard A 2005 Virtual engineering: methods and tools *Journal of Engineering Manufacture* **219** pp 413-421

[19] Tarniţă D, Marghitu D 2013 Analysis of a hand arm system *Robotics and Computer-Integrated Manufacturing* **29**, pp 493-50

[20] Alexandru C 2019 Multi-body system simulation of the sun trackers used for PV panels *Submitted to IMT'2019*

IOP Conf. Series: Materials Science and Engineering **568** (2019) 012020 doi:10.1088/1757-899X/568/1/012020

8D complaint solving method in an automotive component processing company

E S Muncut[1], L I Culda[2] G M Erdodi[3] and G Sima[4]

[1,2,3,4] Department of Automation, Industrial Engineering, Textiles and Transport, Aurel Vlaicu University of Arad, RO, Arad, Romania

muncutstela@yahoo.com

Abstract. The 8D Report is a method used by engineers and other specialists to solve product or process problems. The purpose of the report is to identify, correct and eliminate recurring problems. The 8D report sets out permanent corrective actions based on the statistical analysis of the problem as well as the origin of the problem. This type of report has been developed by Ford Motor Company and has become a standard in the automotive industry, especially used to solve customer complaints about non-compliant products. The name of the 8D report comes from "8 Disciplines". These 8 disciplines are actually 8 points that delimit the steps you take if a problem or defect is observed in a process or product. Solution stages are: choice 8D team identification and responsibilities, problem description, defining of the control actions, cause analysis, defining the corrective actions and demonstration of efficacy, the implementation of corrective actions and demonstration of effectiveness, Prevention of recurrence of nonconformities and completion and signing the final report 8D.

1. Introduction

The name of the 8D report comes from "8 Disciplines". These 8 disciplines are actually 8 points that delimit the steps you take if a problem or defect is observed in a process or product. The 8D complaint-handling method in a component manufacturing company for automotive industry has the following steps [1],[2].

2. Working method

2.1. D1 - 8D team identification and responsibilities

This stage involves nominating the team members with the necessary skills and experience. This is done on the basis of the 8D ratio of figure 1.

This report contains the following: date of issue - start date of the report, final report - date of last signature on the 8D report, interim report - date on which the 8D report is sent to the client in the interim phase, the date of the complaint – when the complaint was received from the customer (complaint number, description of the non-compliance, etc.), figure 2.

The report also contains a data set related to the product: supplier name, material, client material code, series etc.

The piece we are going to analyses is shown in figure 3 and it is made of EN 10294-1 19MnVS6. Steel with the following characteristics: higher strength, better ductility, better machinability, good weldability and even hardenability and dimensional stability.

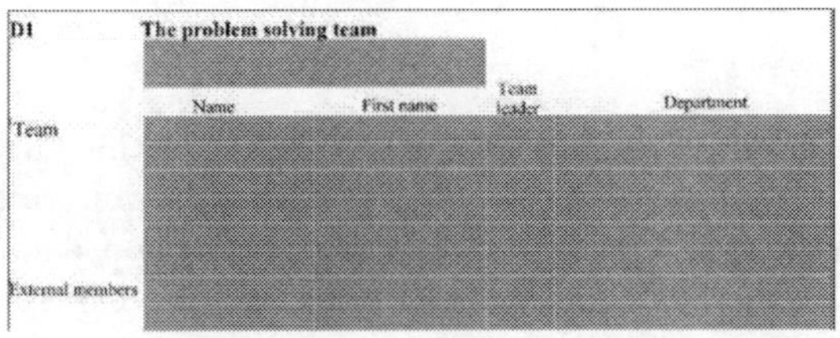

Figure 1 Report of naming of the 8D team and responsibilities

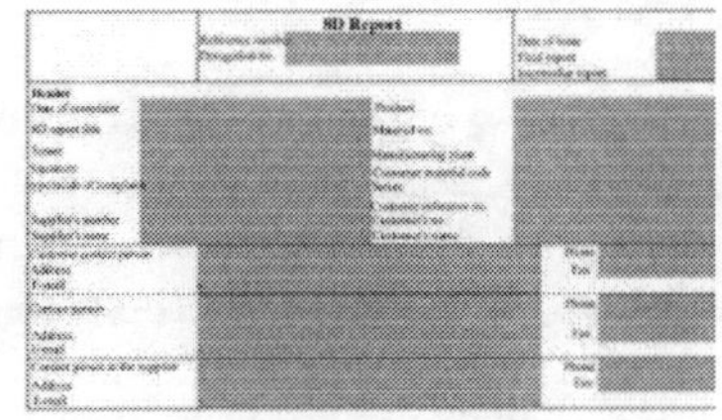

Figure 2. Model of the 8D Report

This piece requires a large number of machining operations and high precision according to the processing plan shown in figure 4. Choosing the right people to be part of the team is critical to solving the problem. The team has to be a multidisciplinary team, to have one member in each department to look at the problem from different perspectives.

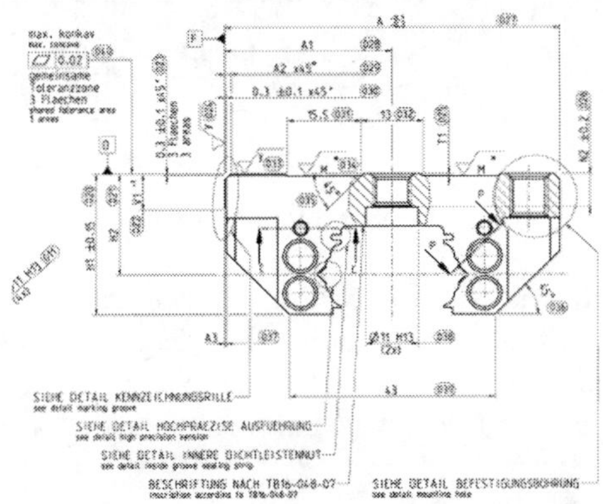

Figure 3. Technologically analyzed piece

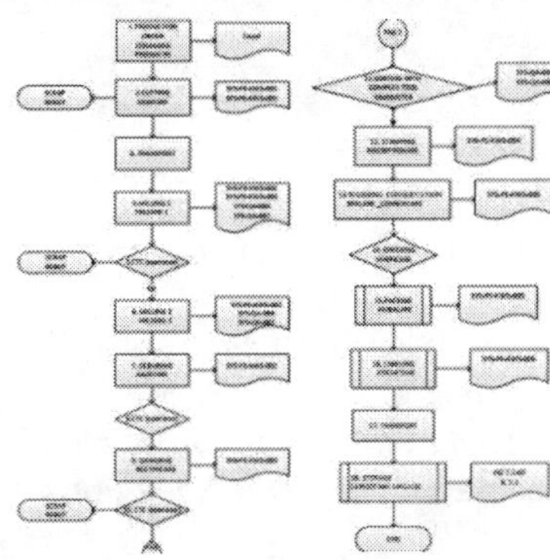

Figure 4. Processing stages

Every person will have a well-established role in this team. The role of the quality engineer is the link between the customer and the company. He / she needs to know the client's requirements very well to be able to present to the team both the problem and what is to be done in this 8D report.

2.2. D2 - Problem description

Describing the issue clearly and concisely. Collecting information - the issue is addressed from two points of view, namely: the customer and the manufacturer. Results: A clear and complete description of the problem, as shown in figure 5.

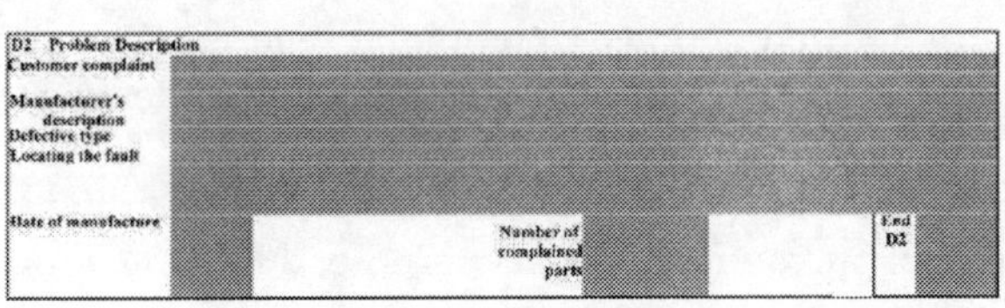

Figure 5 Report to present the problem

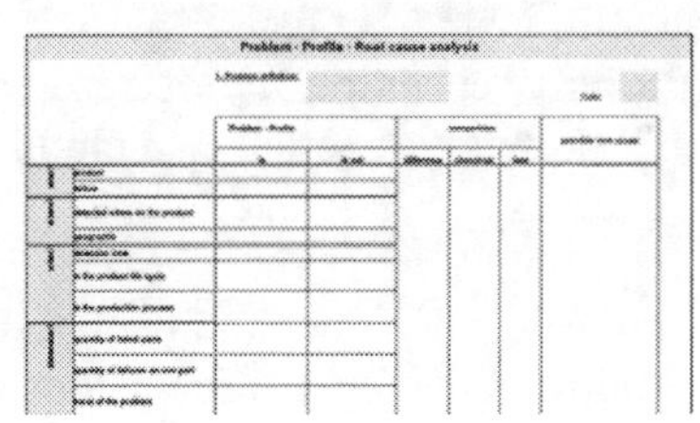

Figure 6 Report on the analysis of the causes

The customer's complaint must contain a description of the non-compliance with all relevant customer-specific influences from the time it was discovered.

The manufacturer's description contains the non-conformity from the manufacturer's point of view with all relevant data, figure 6.

2.3. D3 – Defining of the control actions

The manufacturer has a catalog of 37 possible defects only for the processing of this piece, which details the ones that are accepted or rejected. This analysis is made according to the consumer's requirements.

Initialisation of the most appropriate actions to isolate the problems. Isolation actions consist of the following steps, figure 7: securing current production (egg taking into account the use of other lines or other locations, blocking stock, applying warning signs, informing employees); Checking the storage stock / stock in circulation - on transport; Checking the stock at the customer (egg customer's warehouse, items being in transport to the customer, other affected customers?).

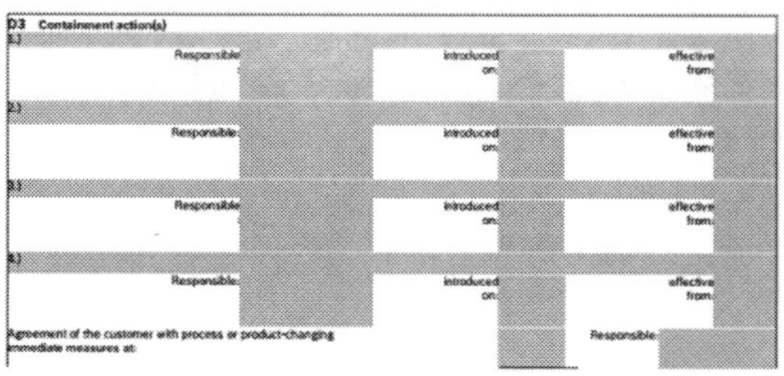

Figure 7 Report of control actions

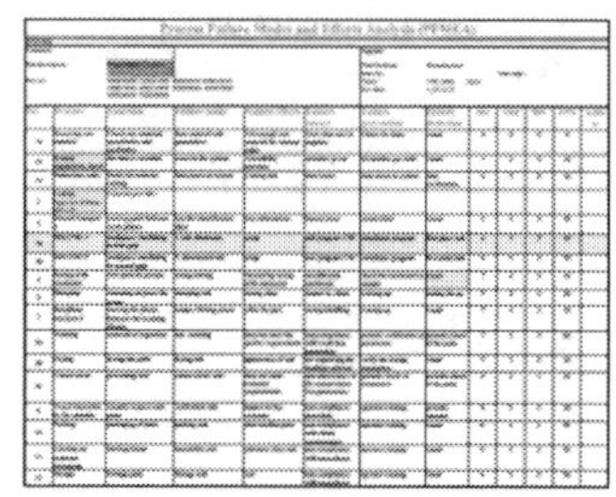

Figure 8 Control Plan

The control plan for the part under consideration is shown in figure 8. Compliance with the control plan during the manufacturing process makes it possible to find the noncompliance stage and to remedy very quickly. After completion of step D3, the 8D Report will be sent to the customer. The customer will reject or approve the taken actions to stop the production of defective parts.

2.4. D4 Cause analysis

The objective of step D4 is to determine the root cause of occurrence and non-detection of nonconformity. Analysis of the non-compliance in terms of appearance "Why did the defect occur?" and from the non-detection point of view "Why the fault was not detected?";

The analysis of the occurrence of the defect is studied from the point of view of the MRC managerial root cause and TRC technique, figure 8.

After determining the causes and the risk analysis, the 8D report is updated and the control actions implemented in step D3 are reviewed to determine their effectiveness and their need for further development. The client is informed about the status of the 8D report.

2.5. D5-Defining the corrective actions and demonstration of efficacy

Defining corrective actions and demonstrating effectiveness are aimed of creating a list of all actions that can eliminate the cause of nonconformity, actions for the occurrence and non-detection of nonconformity, checking the effectiveness of the identified actions, determining and confirming the "optimal" corrective actions, including steps and responsibilities for implementing actions and communicating corrective actions to customers, figure 9.

An important part is the size verification module during the processing process that follows the dimensions of the designer and the possibility of defect prevention. For this purpose, during the process, all machining and related dimensions will be included in the measurement plan of the dimensions resulting from the technological processing (presented in part) containing 107 operations, figure 10.

After this stage, all possible corrective actions, regardless of their costs, will be mentioned; The basic principle is, "Preventing errors is more important than eliminating defects."

IOP Conf. Series: Materials Science and Engineering **568** (2019) 012020 doi:10.1088/1757-899X/568/1/012020

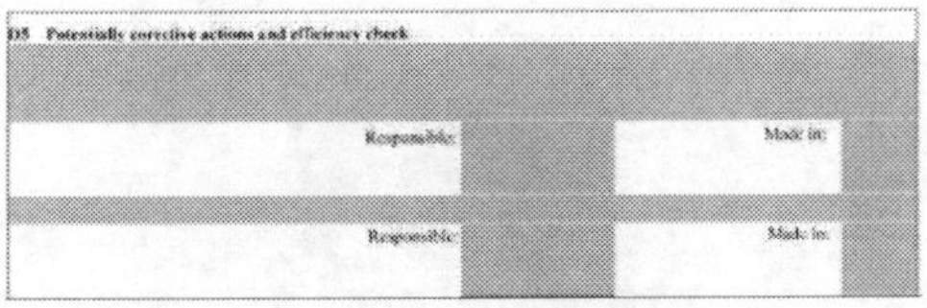

Figure 9 Defining the corrective actions

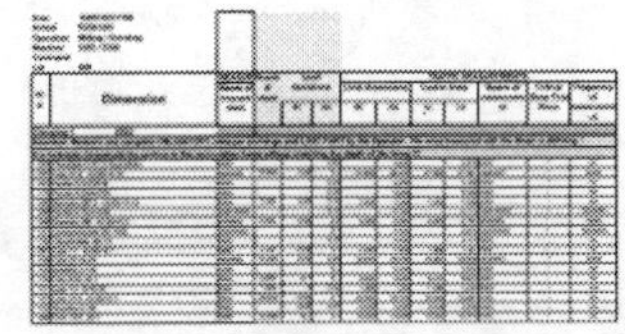

Figure 10 Measurement plan during processing

Are all actions traced - for appearance and non-detection, TRC and MRC have been taken into account? Are the proofs of the effectiveness of the actions attached and approved? Is there a relationship between the root cause and the measures taken at D4?

2.6. D6 The implementation of corrective actions and demonstration of effectiveness

The objective of Step D6 is to implement the actions defined in Step D5.

The activities of this step are: implementing corrective actions, confirming the effectiveness of the implemented actions, eliminating the emergency actions implemented in step D5.

The corrective actions chosen from those determined in Step D5 are being implemented, the measures must be both to avoid and to detect the nonconformity. The way how the problem was solved is verified and the actions are monitored, figure 11.

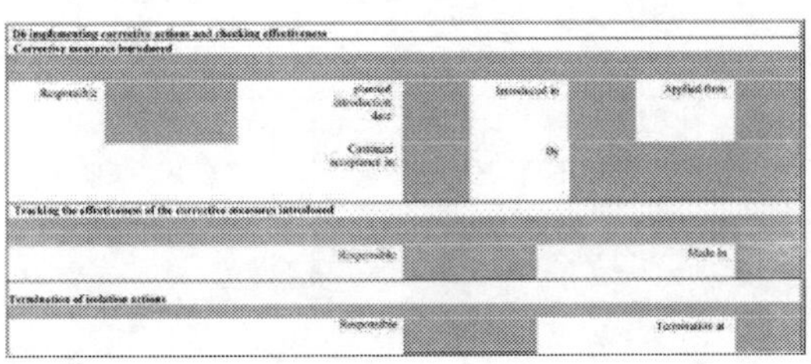

Figure 11 Corrective actions report

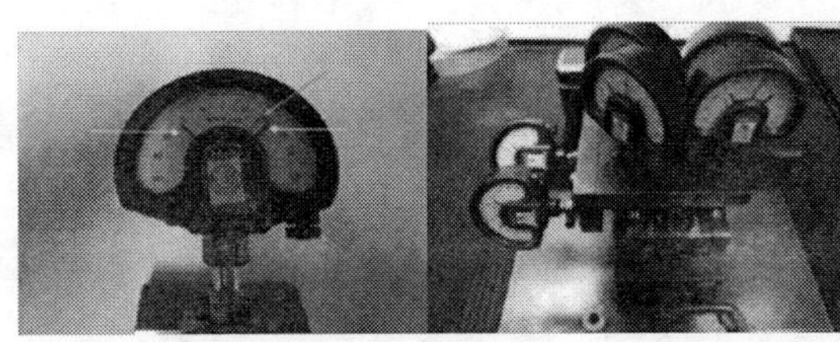

Figure 12 Internal measurement procedure with micrometer

The control actions implemented in step D3 are removed only after the effectiveness of the corrective actions has been demonstrated - the justification for cancelling the control measures is documented.

After processing the parts on the grinding machine, it will be measured with the dial comparator to determine the precision class in which the piece conforms to the internal standard. The measurement method with the comparator is described in: Internal measurement procedure with micrometer the QA-004, figure 12.

If it is necessary to implement actions that bring changes to the product / process, the client's consent is required to implement the corrective actions defined in step D5 and document this.

2.7. D7-Prevention of recurrence of nonconformities

The objective of step D7 is to identify the measures necessary to avoid the recurrence of defect or similar defects. A plan for implementation, monitoring of the effectiveness and evaluation of preventive actions is established. The defined actions are implemented and documented, and if there are actions that are not implemented, the reason will be justified, figure 13.

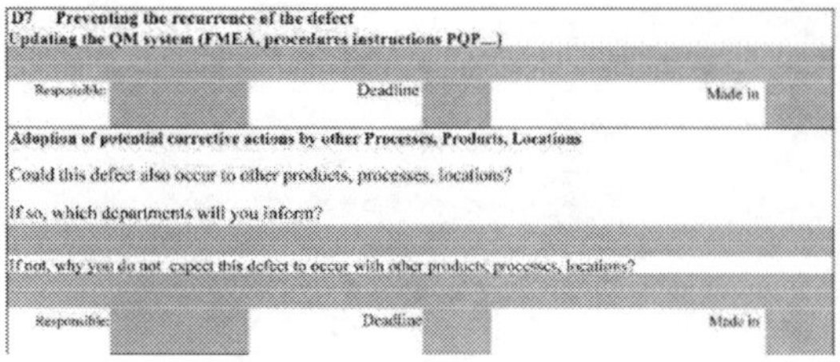

D7 Preventing the recurrence of the defect
Updating the QM system (FMEA, procedures instructions PQP...)
Responsible: Deadline Made in
Adoption of potential corrective actions by other Processes, Products, Locations
Could this defect also occur to other products, processes, locations?
If so, which departments will you inform?
If not, why you do not expect this defect to occur with other products, processes, locations?
Responsible: Deadline Made in

Figure 13 Implementing preventive actions to improve the process

Figure 14 Positioning the piece in the vies of the imprinting machine

Updating the QM system (FMEA, Instructions-Procedures, PQP) - Upgrading the FMEA, details of IPC changes, the frame control plans and other relevant documents are very important in the event of customer noncompliance.

2.8. *D8 – Completion and signing the report*

The objective of step D8 is to successfully complete the process of resolving nonconformities. A meeting is organized with all those involved in the 8D report, for the evaluation of steps D1-D7 and closure of the report. Conclusions and completion of the 8D Report in figure 15.

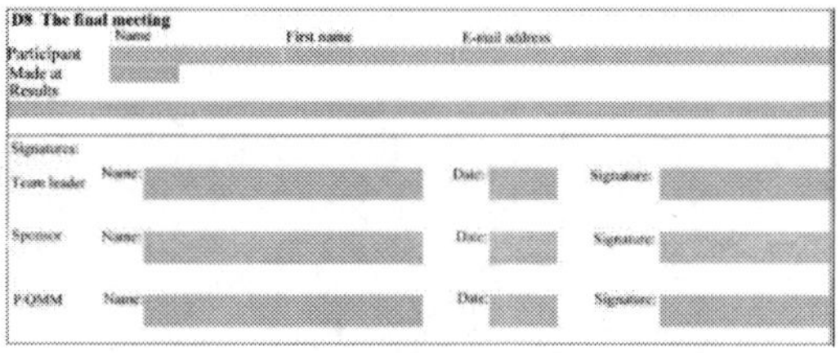

D8 The final meeting
Name First name E-mail address
Participant
Made at
Results
Signatures:
Team leader Name: Date: Signature:
Sponsor Name: Date: Signature:
P-QMM Name: Date: Signature:

Figure 15 The 8D final report

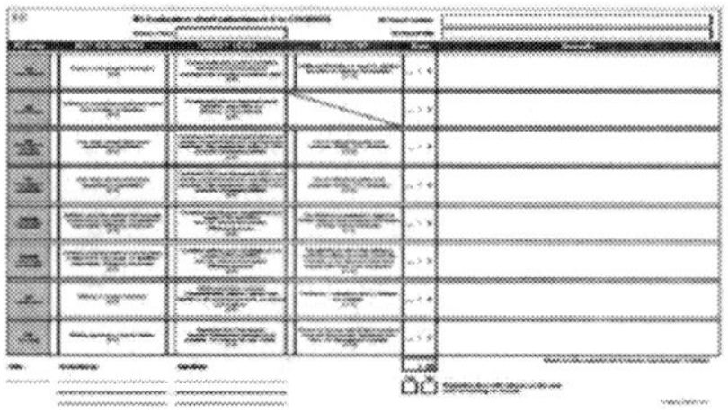

Figure 16 The final evaluation of the method 8D

A final meeting is taking place with members of the 8D Reporting Team, presenting what was good in solving the 8D Report and what needs to be improved. The results are documented by the final format assessment called "FR 08958-2", figure 16.

3. Conclusions

This way of dealing with nonconformities can be applied when technology flow can be verified. Namely, the company has clear and interconnected procedures. The control plans are sustainable, there are measurement and control equipment and skilled people who can do it. All of this data is stored in a company database that makes it possible to access it from anywhere.

An important step is the corrective action by which the company must eliminate the cause of the non-compliance in order to prevent its recurrence. As a final idea, the action includes: customer complaint and defect analysis, determination of the causes of non-compliance, undertaking actions to prevent recurrence of non-compliance, implementation of necessary actions and execution of the final 8D report.

References

[1] Miclaus I.M, Quality management (Managementul calității), Ed. Dacia, Cluj, 2008;
[2] Gheorghe Sima, Ramona Lile, Glavan Dan, Elena Muncut, Management, Traceability and Control of Industrial Processes, Soft Computing Applications pp 971-980, 03 November 2015, DOI: 10.1007, ISBN 978-3-319-18415-9, Online ISBN 978-3-319-18416-6;

IOP Conf. Series: Materials Science and Engineering **568** (2019) 012021 doi:10.1088/1757-899X/568/1/012021

Kinematic modeling of a 6R industrial robot

M Ratiu[1], A Rus[1] and M L Balas[2]

[1] Department of Mechanical Engineering and Automotive, University of Oradea, 410087 Oradea, Romania
[2] Department of Product Design, Mechatronics and Environment, Transilvania University of Brasov, 500068 Brasov, Romania

E-mail: mratiu@uoradea.ro

Abstract. In this paper, we present the process of kinematic modeling and simulation, in ADAMS MBS of MSC software, of an articulated robot with six revolute joints, through direct and inverse kinematics. First, by using the STEP function, we define a spatial trajectory of the end effector, and, through inverse kinematics, we determine the motion laws of the six revolute joints. Then, we apply the inverse process on another virtual model of the same robot, by imposing the motion laws to the joints, to obtain the desired trajectory of the end effector, through direct kinematics. This work is a small part of our research regarding modeling and optimization of the industrial robots' motion.

1. Introduction

Nowadays, in the context of using the industrial robots at a wide scale, across all industries, and due to the rapid and significant progress of computer designing, virtual prototyping is more and more used for analyzing and optimization of the robots' motion. As presented in [1], there are a lot of studies and several complex software packages on the modeling of industrial robots.

In our work, we use ADAMS (Automatic Dynamic Analysis of Mechanical Systems) from MSC software, which is one of the most widely used Multi-Body System (MBS) software for the modeling, analysis, and simulation of the complex mechanical systems' motion.

In paper [2] is presented the process of the modeling in ADAMS of the mechanical system of a 6DOF (degree of freedom) articulated robot, with six revolute joints, starting from the geometric 3D CAD model of the robot ABB IRB 1600-6/1.2, in PARASOLID format, available on the manufacturer's website [3].

In continuation of our work, in this paper, we present the process of kinematic modeling and simulation of the robot, through direct and inverse kinematics, in ADAMS/View, by following the steps outlined in [4, 5]. The kinematic model of the robot contains the kinematic elements of the robot, connected through that six revolute joints, the geometric parameters specific of the mechanism, represented by the locations of the joints, and the kinematic restrictions - the motion generators [6].

Following the procedure proposed in [7], first, by using the STEP function, we define a spatial trajectory of the end effector, and, through inverse kinematics, we determine the motion laws of the six revolute joints. Then, we apply the inverse process on another virtual model of the same robot, by imposing the motion laws to the joints, to obtain the desired trajectory of the end effector, through direct kinematics.

IOP Conf. Series: Materials Science and Engineering **568** (2019) 012021 doi:10.1088/1757-899X/568/1/012021

2. Kinematic analysis of the robot

The kinematic analysis of a robot means calculation of the kinematic parameters of the mechanical elements of the robot, without considering their masses, or the forces and torques that cause their motion.

The kinematic analysis of any robotic manipulator can be done by direct (forward) and inverse kinematic analysis. The goal of the direct kinematics is the determination of the robot's end effector position and orientation when all robot joints variables are known. In contrast, the goal of the inverse kinematics is the determination of the joints variables, when are known the position and orientation of the end effector.

While for the direct kinematic analysis the Denavit-Hartenberg convention is well known and widely used, the inverse kinematic analysis is complex and time-consuming because of the lack of a unique solution. The inverse kinematics is the basis for the motion planning and control of the robot trajectory and forms the basis for dynamic analysis.

3. Inverse kinematic analysis of the robot

The problem of inverse kinematics for the robot consists of determining the motion laws of the six revolute joints for a particular trajectory of the end-effector.

In order to impose a spatial trajectory of the end effector, for example, UO-shaped, are created eight points, with the coordinates presented in table 1 and illustrated in figure 1:

Table 1. Defining points on the trajectory.

Point number	Coordinates (mm)		
	X	Y	Z
7	600	-100	1000
8	500	100	600
9	700	150	500
10	800	-50	900
11	1000	0	800
12	900	200	400
13	1100	250	300
14	1200	50	700

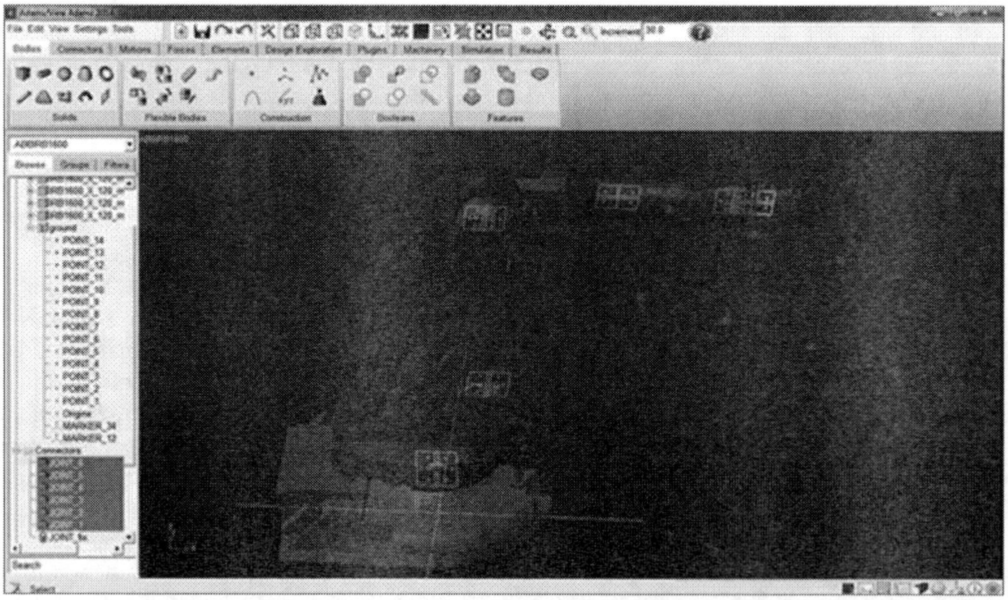

Figure 1. Creation in ADAMS the points for the imposed trajectory.

IOP Conf. Series: Materials Science and Engineering **568** (2019) 012021 doi:10.1088/1757-899X/568/1/012021

In ADAMS / View, with General Motion, using the STEP function, the end-effector displacements on X, Y, and Z are defined and presented in table 2:

Table 2. Defining STEP functions.

Displacement on X	Displacement on Y	Displacement on Z
STEP(time, 0, 0, 0.001, -290) +	STEP(time, 0, 0, 2.001, -100) +	STEP(time, 0, 0, 5, 38.5) +
STEP(time, 4.999, 0, 5, -100) +	STEP(time, 6.999, 0, 7, 200) +	STEP(time, 5, 0, 5.001, -400) +
STEP(time, 6, 0, 6.001, 200) +	STEP(time, 8, 0, 8.001, 50) +	STEP(time, 5.999, 0, 6, -100) +
STEP(time, 10.999, 0, 11, 100) +	STEP(time, 12.999, 0, 13, -200) +	STEP(time, 11, 0, 11.001, 400) +
STEP(time, 12, 0, 12.001, 200) +	STEP(time, 14, 0, 14.001, 50) +	STEP(time, 11.999, 0, 12, -100) +
STEP(time, 16.999, 0, 17, -100)+	STEP(time, 18.999, 0, 19, 200) +	STEP(time, 17, 0, 17.001, -400)+
STEP(time, 18, 0, 18.001, 200) +	STEP(time, 20, 0, 20.001, 50) +	STEP(time, 17.999, 0, 18, -100) +
STEP(time, 22.999, 0, 23, 100) +	STEP(time, 24.999, 0, 25, -200) +	STEP(time, 23, 0, 23.001, 400)+
STEP(time, 24, 0, 24.001, -200)	STEP(time, 26, 0, 26.001, -50)	STEP(time, 23.999, 0, 24, 100)

In ADAMS, the STEP function approximates an ideal mathematical STEP function (but without discontinuities). A STEP function is used when a value has to be changed from one constant to another.

An animation of the robot motion, for 27 s and 200 steps, and the trajectory followed by the end effector, are presented in figure 2:

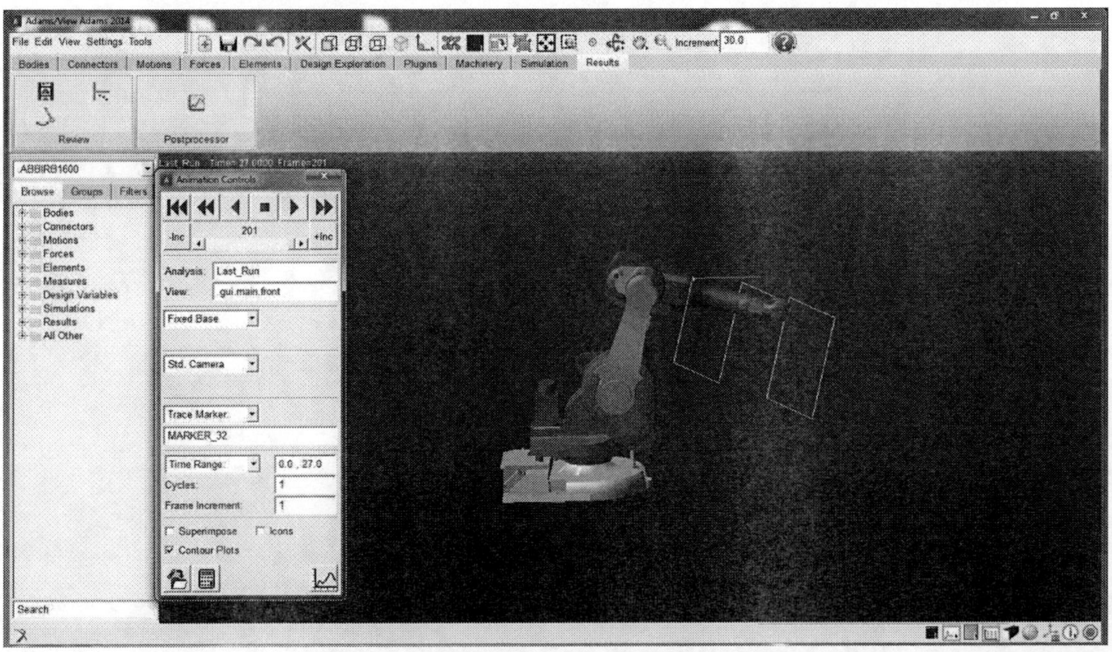

Figure 2. Simulation of imposed trajectory.

By using Adams / Postprocessor, the variation of the main kinematic parameters (position and velocity) of the end effector mass center, represented by Marker 32, for a simulation of 27 s and 200 steps are obtained. In figure 3a can be seen the displacement curves of the end effector on X, Y and Z. The curves that represent the variation of the end effector' velocity on X, Y, and Z, are presented in figure 3b.

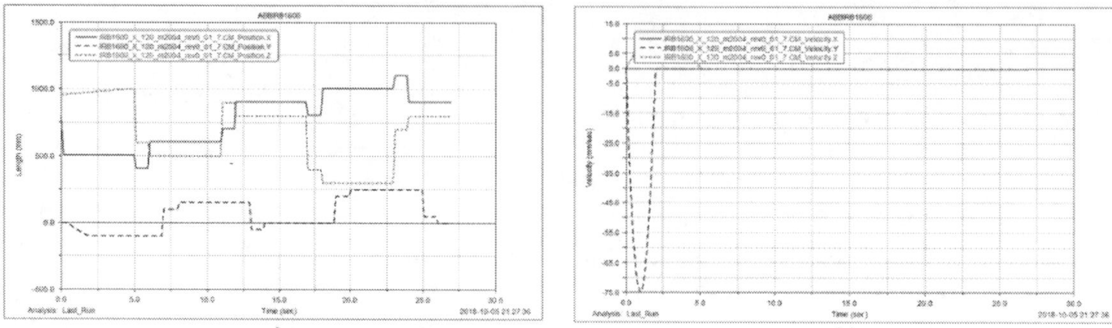

a) b)

Figure 3. The variation of the end-effector position (a) and velocity (b).

The motion laws of the six revolute joints of the robot, 1, 2, 3, 4, 5, and 6 (illustrated in figure 1), are presented in figure 4. They represent the variation curves of the angular displacements in each revolute joint, during the displacement of the end effector along the imposed trajectory.

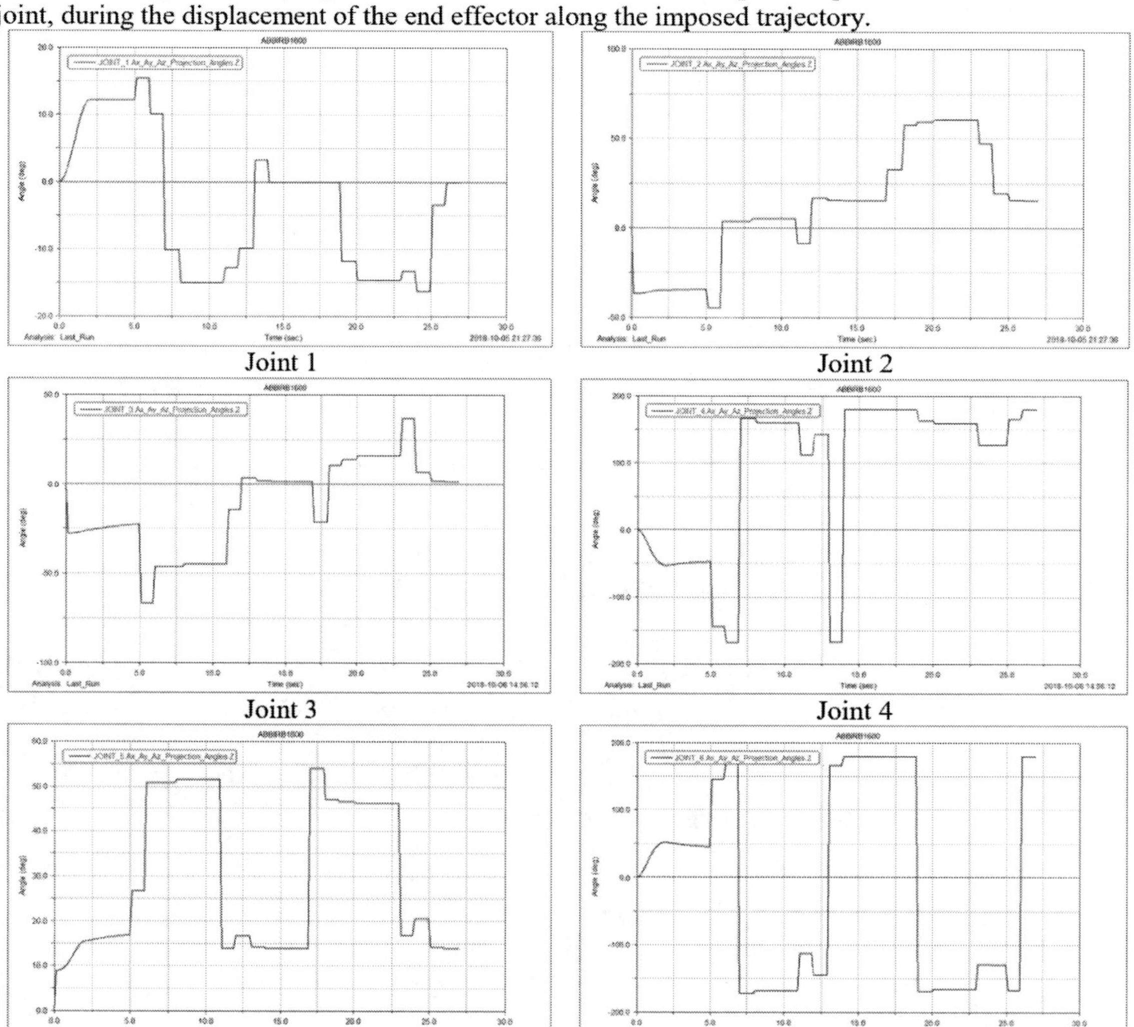

Joint 1 Joint 2

Joint 3 Joint 4

Joint 5 Joint 6

Figure 4. The motion laws of the six revolute joints, 1, 2, 3, 4, 5, and 6.

IOP Conf. Series: Materials Science and Engineering **568** (2019) 012021 doi:10.1088/1757-899X/568/1/012021

4. Direct kinematic analysis of the robot

The problem of direct kinematics in ADAMS consists in imposing the motion laws of the six revolute joints, in obtaining a certain trajectory of the end effector, by going through the following steps:
- measures for the angular displacements of the six revolute joints are created;
- the motion laws of angular displacement of the revolute joints are accessed from the simulation results of the inverse kinematics;
- spline curves are created;
- the motion laws of the six revolute joints are imposed/modified. The analysis of the kinematic model of the robot, presented in figure 5, will be the next step in our research.

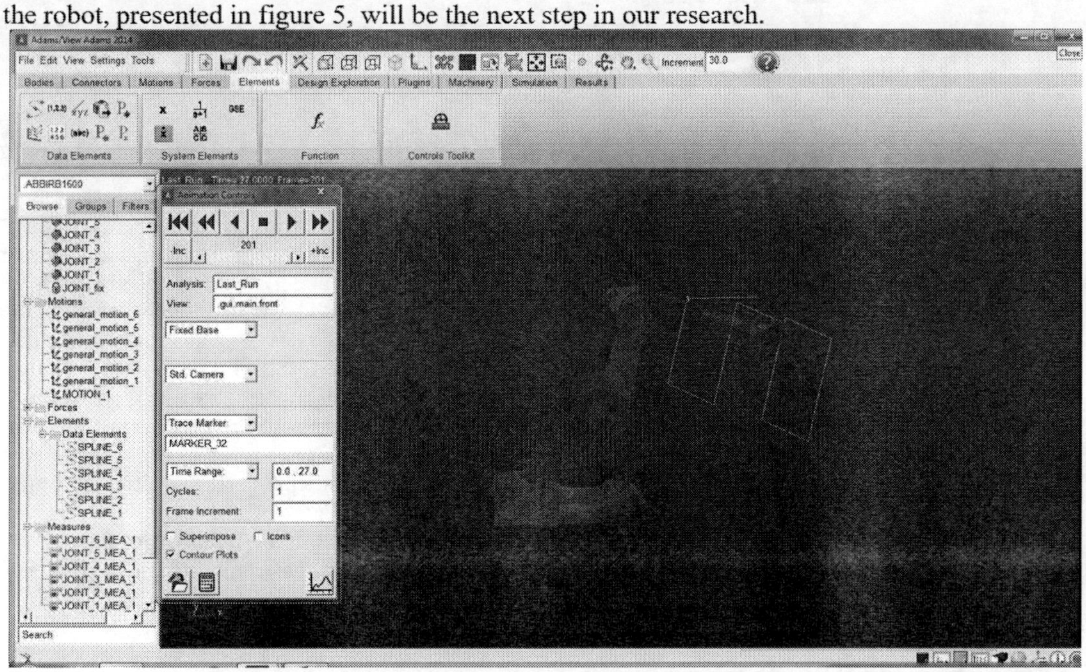

Figure 5. Kinematic model of the robot in ADAMS.

5. Conclusion

We presented this paper, which represents a small step in our larger research regarding modeling and optimization of the industrial robots' motion, in ADAMS MBS software, by hoping that it will be useful to the students or other interested people who start learning to model such robots in ADAMS.

References

[1] Ratiu M and Rus A 2017 Modeling of the trajectory-generating equipments *IEEE publication 14th International Conference on Engineering of Modern Electric Systems (EMES)* DOI: 10.1109/EMES.2017.7980418

[2] Ratiu M, Rus A and Balas M L 2018 Modeling in ADAMS of a 6R industrial robot, *Annual Session of Scientific Papers IMT ORADEA 2018 (MATEC Web of Conferences)* **184**, 02006

[3] www.abb.com/robotics accessed in July 2017

[4] *Getting started using ADAMS/View* 2014 retrieved from mscsoftware.com on September 2015

[5] *Adams Tutorial Kit for Mechanical Engineering Courses (Third Edition)* 2015 retrieved from www.mscsoftware.com on July 2017

[6] Alexandru C 2006 Modeling in mechatronic concept and virtual testing of orientation mechanisms used in solar panels *Mechanism and Manipulators Magazine* **2/2** pp 49-54

[7] Enescu M and Alexandru C 2011 Modeling and simulation of a 6 DOF robot *Proceedings of 2011 International Conference on Optimization of the Robots and Manipulators* (Sinaia, Romania: University Politehnica of Bucharest)

IOP Conf. Series: Materials Science and Engineering **568** (2019) 012022 doi:10.1088/1757-899X/568/1/012022

Analysis of the movement of an exoskeleton for the upper limb

I A Tătulea[1], C N Drugă[1]

[1]Design of Product, Mechatronics and Environment Department, Transylvania University of Brasov, Brasov, Romania

irinatatulea23@yahoo.com

Abstract. The theme approached in this paper falls within the general lines of an interdisciplinary research field, strategies provided by the European health insurance programs: Social Challenges (Health, Demographic Change and Wellbeing) and respectively in the national classification: Health (Fundamental and Border Research). The main objective of the research is the identification and knowledge of the biomechanical analysis elements that contribute to the increase of the mobility of the upper limbs in the persons with different motor dysfunctions (recoverable and rehabilitated dysfunctions), using customized exoskeleton-type systems together with opto-electronic systems to diversify the scope. In the first part of the paper are presented general aspects related to variants and fields of use of types of exoskeleton for upper limbs. The biomechanical analysis of upper limb movements, performed for the purpose of determining the action limits, is presented in the second part of the paper. The results of the analyses and the conclusions of this research are presented in the final part of the paper, along with the description of the proposed constructive variant of the exoskeleton.

1. Introduction

The design of the exoskeleton systems requires a systemic approach, starting from the anthropometric dimensions, personalized features of each subject, continuing with the identification and knowledge of motor dysfunctions (limits of anatomical and physiological functioning), simulating then computerized or experimental, certain situations of action, to follow the way of working in rehabilitation, training and recovery procedures, considering the movements of upper limb shown in figure 1. These mechanisms of action and the fundamental design aspects form the theoretical basis of research and represent a necessary step in the design of the structure of the proposed variant. The constructive variants on the market of these products are made of different materials, such as: metal, casted or moulded plastic, powders or composite alloys.

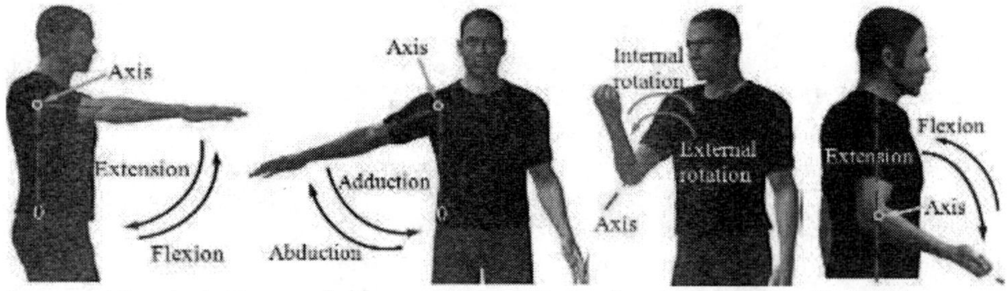

Figure 1. Types of movements and limit of action of upper limb [1]

2. Theoretical basis of upper limb movement

In order to carry out a generalized analysis of the movements of the arm-forearm assembly, it starts from the Denavit-Hartenberg Convention (D-H Convention), a form of calculation of the links that exist between the segments of an articulated system consisting of two bars, a coupling joint between them and an attachment joints with other elements of the human trunk. As mentioned in various papers [2-4] coordinate frames are attached to the joints between two links such that one transformation is associated with the joint, [Z], and the second is associated with the link [X]. The coordinate transformations along a serial robot consisting of n links form the kinematics equations of the robot:

$$[T] = [Z1]\,[X1]\,[Z2]\,[X2]\,\ldots\ldots\,[Zn]\,[Xn] \tag{1}$$

where *[T]* is the transformation locating the end-link [3].

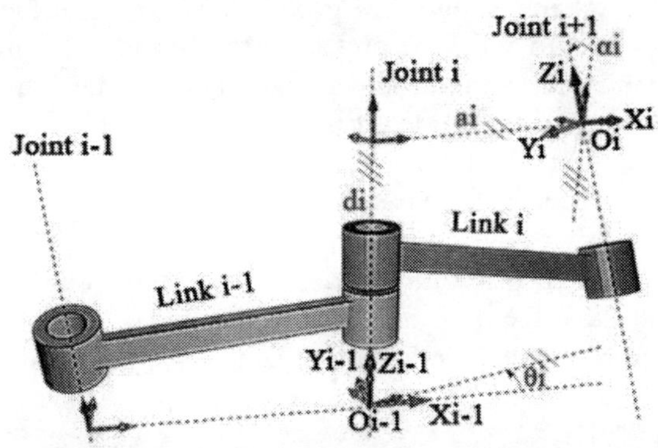

Figure 2. The four parameters of classic D-H Convention [3]

The parameters of the D-H Convention which will be used in the description of movements at the level of the upper limb are as it follows: *d*:offset along previous *z* to the common normal; θ: angle about previous *z*, from old *x* to new *x* axis; *r*: length of the common normal; α: angle about common normal, from old *z* axis to new *z* axis [3].

By using the D-H Convention method, a coordinate axis system will be attached to each kinematic chain and at the end it will be obtained, according to figure 2, transformation matrices 4x4 type between the two successive links (link i-1 and link i). The D-H Convention also allows the writing of kinematic equations (velocities, accelerations) or dynamic (linear and angular momentum, forces and couples) by using those forms of transformation matrices specific to determinations.

In addition to this, the D-H convention parameters can be modified, the difference between classical parameters as in equation (2) and modified parameters as in equation (3) consisting in moving the coordinate system from i+1 junction (classical convention) to i-1 junction (modified convention) [3, 5, 6]:

$$^{n-1}M_n = \left[\begin{array}{ccc|c} \cos\theta_n & -\sin\theta_n\cos\alpha_n & \sin\theta_n\sin\alpha_n & r_n\cos\theta_n \\ \sin\theta_n & \cos\theta_n\cos\alpha_n & -\cos\theta_n\sin\alpha_n & r_n\sin\theta_n \\ 0 & \sin\alpha_n & \cos\alpha_n & d_n \\ \hline 0 & 0 & 0 & 1 \end{array}\right] = \left[\begin{array}{ccc|c} & R & & T \\ \hline 0 & 0 & 0 & 1 \end{array}\right] \tag{2}$$

IOP Conf. Series: Materials Science and Engineering **568** (2019) 012022 doi:10.1088/1757-899X/568/1/012022

$$^{n-1}M_n = \left[\begin{array}{ccc|c} \cos\theta_n & -\sin\theta_n & 0 & a_{n-1} \\ \sin\theta_n \cos\alpha_{n-1} & \cos\theta_n \cos\alpha_{n-1} & -\sin\alpha_{n-1} & -d_n \sin\alpha_{n-1} \\ \sin\theta_n \sin\alpha_{n-1} & \cos\theta_n \sin\alpha_{n-1} & \cos\alpha_{n-1} & d_n \cos\alpha_{n-1} \\ \hline 0 & 0 & 0 & 1 \end{array}\right] \tag{3}$$

where *R is* submatrix 3x3 which describes the rotation movement and *T* is submatrix 3x1 which represents the translation movement.

To study human arm behaviour, it must be taken into account that it is assimilated to a complex structure, but usually in biomechanical calculations, the joints and segments of a human arm are simplified in a 7-DOF kinematic chain. The arm joint with the shoulder structure can be modelled as a 3-DOF spherical joint and the elbow joint as a cylindrical joint. In the same way, the wrist joint can be modelled as a 3-DOF joint which assembles with the forearm structure [7].

Using the modified D-H convention, it will be possible to model the motion of the proposed exoskeleton according to the scheme shown in figure 3, where a1 represents the length of the arm, a_2 represents the length of the forearm and J_1, J_2 and J_3 represent the joints of the entire upper limb in the following order: shoulder joint, elbow joint (between arm and forearm) and wrist joint.

As shown in the paper [8] the trajectories used in this research correspond to passive flexion/extension movements for shoulder, elbow and wrist joints used in passive rehabilitation exercises or in the evaluation processes. The initial values of the system are selected as $q(0) = [0,\pi/2,0]^T$, $\dot{q}(0) = [0, 0, 0]^T$.

The parameters of the sliding surface function are $p = 7$, $q = 9$, and the diagonal gain matrices are $\alpha = 3\ I$, $\beta = I$. For the control law, the gain matrix $\lambda = 10^3 diag([0.1,\ 0.3,\ 3])$. The initial value of the adaptation term $\hat{A}$ is equal to zero and the adaptation gain is $\Gamma = 10^2\ I$.

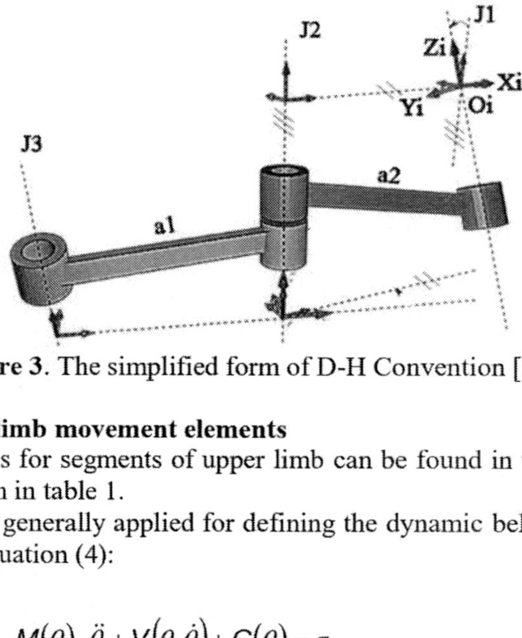

Figure 3. The simplified form of D-H Convention [3]

3. Determination of upper limb movement elements

Values of mass characteristics for segments of upper limb can be found in various papers [9]. Values for range of motion are shown in table 1.

The basic relationship that is generally applied for defining the dynamic behaviour of the exoskeleton structure is presented as in equation (4):

$$M(\theta)\cdot\ddot{\theta} + V(\theta,\dot{\theta}) + G(\theta) = \tau_{joints} \tag{4}$$

where: θ is the joint variables vector, τ is the generalized torques vector, *M* (*θ*) is the inertia matrix, $V(\theta,\dot{\theta})$ is the Coriolis /centrifugal vector, *G* (*θ*) is the gravity vector [9].

IOP Conf. Series: Materials Science and Engineering **568** (2019) 012022 doi:10.1088/1757-899X/568/1/012022

Table 1. Range of motion [9]

Type of motion	Anatomical range [9, 10]				
	Source 1[^{0}]	Source 2 [^{0}]	Source 3 [^{0}]	Source 4 [^{0}]	Source 5 [^{0}]
			Elbow & Forearm		
Flexion	140	140	145	120	116
Extension	0	0	5-15	0	0
Pronation	80	90	80	80	70
Supination	80	85	90	80	85

The joints noted with Ji-1, Ji and Ji + 1 represent the joint of the shoulder (assimilated to a cylindrical joint due to the proposed exoskeleton model), joint of elbow and joint of the wrist, which are linked by segments, their length being noted with a1, respectively a2.

In mechanics, a joint is an assembly, a linking element, based on a material shaft that allows angular displacement of the assembled parts. According to the paper [10], the elbow joint represents the link that forces the rigid solid to stay with one of its axis in permanent contact with a fixed axis in space. The position of a rigid linked through a cylindrical joint, which provides a fixed axis in space, will be determined by a single scalar parameter – the angle of rotation around this axis. Consequently, it will have only one degree of freedom (1 DOF), resulting in the fact that joint blocks five degrees of freedom (5 DOF). The torques of active mechanical stresses and reactions in J2 joint will have the following forms:

$$\vec{\tau}_0\left(\vec{F}_i\right)=\begin{Bmatrix}\sum\limits_i \vec{F}_i\\ \sum\limits_i \vec{M}_0\left(\vec{F}_i\right)\end{Bmatrix}=\begin{Bmatrix}\sum\limits_i X_i\\ \sum\limits_i Y_i\\ \sum\limits_i Z_i\\ \sum\limits_i M_{xi}\\ \sum\limits_i M_{y_i}\\ 0\end{Bmatrix}+\begin{Bmatrix}0\\0\\0\\0\\0\\ \sum\limits_i M_{zi}\end{Bmatrix}=\vec{\tau}_b+\vec{\tau}_l \tag{5}$$

In this case the torque of these mechanical stresses may be written in the following form as in equation (6), showing that the joint can be replaced with a single action having two components. Determining degrees of mobility (M), is based on the Gruebler & Kutzbach equation, which takes into account the number of connections in the system (L - Links), number of joints (J - Joints) and respectively the number of grounded links (G – Grounded links) as in equation (7).

$$\vec{\tau}_0\left(\vec{F}_i\right)=\begin{Bmatrix}\sum\limits_i \vec{F}_i\\ \sum\limits_i \vec{M}_0\left(\vec{F}_i\right)\end{Bmatrix}=\begin{Bmatrix}\sum\limits_i X_i\\ \sum\limits_i Y_i\\ 0\\ 0\\ 0\\ 0\end{Bmatrix}+\begin{Bmatrix}0\\0\\0\\0\\0\\ \sum\limits_i M_{zi}\end{Bmatrix}=\vec{\tau}_b+\vec{\tau}_l \tag{6}$$

IOP Conf. Series: Materials Science and Engineering **568** (2019) 012022 doi:10.1088/1757-899X/568/1/012022

$$M = 3L - 2J - 3G \tag{7}$$

In the case of any real mechanism, there is at least one grounded link, therefore G has the value 1 and the Gruebler & Kutzbach equation becomes:

$$M = 3(L-1) - 2J \tag{8}$$

4. Conclusions

According to these observations, the structure of an exoskeleton robot for the upper limb can assure the rehabilitation process in the circumstances that the whole arm leans in the elbow, on a hard surface and the shoulder movement is considered simplified as a movement that appears in a cylindrical joint.
The exoskeleton designed for training and rehabilitation procedures performs limited elbow movement without involving the movement of the shoulder or wrist. The last two movements are taken into account both for the initial positioning and for establishing the coordinate systems, taken as a reference in the elbow movement analysing. The anatomical elements involved in this analysis have been estimated according to the upper limb properties of a typical adult.

Acknowledgements
Practical realization of this project was made possible thanks to funding received in the| My License Project-2019 competition held in the Transylvania University of Brasov.

References:

[1] R. A. R. C. Gopura et al. 2009 *Mechanical Designs of Active Upper-Limb Exoskeleton Robots State-of-the-Art and Design Difficulties*, IEEE 11th International Conference on Rehabilitation Robotics Kyoto International Conference Center, Japan, June 23-26
[2] Kiran Y. Jadhav, Sangeetha Prasanna Ram 2017 *Biomechanical Analysis of Human Upper Limb*, International Journal of Advanced Research in Electrical, Electronics and Instrumentation Engineering, Vol. 6, Issue 1
[3] Wikipedia The Free Encyclopedia, *Article Denavit-Hartenberg parameters*, link accessed on: 15.04.2019, https://en.wikipedia.org/wiki/Denavit%E2%80%93Hartenberg_parameters
[4] I. Petre, Denavit Hartenberg 2011 *Analysis Of A Rehabilitation Equipment*, International Conference on Economic Engineering and Manufacturing Systems, Braşov
[5] Lipkin Harvey 2005 *A Note on Denavit–Hartenberg Notation in Robotics*, 29th Mechanisms and Robotics Conference, Vol.7, pp. 921-926, doi:10.1115/DETC2005-85460
[6] K. Waldron, J. Schmiedeler 2008 *Kinematics*, Springer Handbook of Robotics, pp. 9–33, doi:10.1007/978-3-540-30301-5_2
[7] Xin Wang, Qiuzhi Song, Xiaoguang Wang, Pengzhan Liu 2018 *Kinematics and Dynamics Analysis of a 3-DOF Upper-Limb Exoskeleton with an Internally Rotated Elbow Joint*, Appl. Sci., 8, 464; doi:10.3390/app8030464
[8] A. Riani, T. Madani, A. Benallegue, K. Djouani 2018 *Adaptive integral terminal sliding mode control for upper-limb rehabilitation exoskeleton*, Control Engineering Practice 75, 108–117
[9] M.H. Rahman, M. Saad, J.P. Kenné, P.S. Archambault 2010 *Exoskeleton Robot for Rehabilitation of Elbow and Forearm Movements*, 18th Mediterranean Conference on Control & Automation Congress Palace Hotel, Marrakech, Morocco, June 23-25
[10] I. C. Rosca, I. Serban 2015, Fundamentals of Biomechanics (*Fundamente de biomecanică),* Ed.Universitatii Transilvania Brasov

IOP Conf. Series: Materials Science and Engineering **568** (2019) 012023 doi:10.1088/1757-899X/568/1/012023

Surface roughness measuring in case of electro-erosion processed work pieces

D Popescu[1] and I Geonea[1]

[1] University of Craiova, Craiova, RO

daniel.popescu1119@yahoo.com

Abstract. Electroerosion is a manufacturing process for working difficult to machine materials. The paper describes a miniature punching die for perforation and cutting obtained from X15 using electro-erosion. This type of material cannot be processed using regular technologies because of its properties such as high hardness and complexity as well as protection against superficial layer damage. The manufacturing process provides a high quality and dimensional precision of the finished product. It is presented a digital surface roughness meter and the functional schematic of the PGI system. Also there are presented different roughness measurement variants as well as the graphical representation and the bearing area curve. There are emphasized the variations of the parameters in the 4 measurement points as well as the variation of the c coefficient. Finally the Ra, Rq and Rz parameters are analysed.

1. Introduction

The electro-erosion manufacturing process provides high precision material removal using controlled thermal energy of electric discharges which remove material through melting and vaporization [1], [2].

Electro-erosion allows processing of materials that are very complicated or even impossible to process using traditional technologies:

- ferrous alloys (steel, cast iron, etc.), hard alloys/exotic materials, metallic carbides, non-ferrous metallic materials (copper, aluminum, bronze, etc.), electricity conducting composite materials and so on;
- plastic or ceramic materials cannot be processed;
- the resulting pieces can have a wide range of shapes and dimensions;
- the work piece is not deformed during processing since there is no direct contact with the electrode and hence there are no pushing forces; the temperature influence in the work zone is reduced by use of dielectric liquid in which the work piece is immersed ;
- since material removal is performed through melting and vaporization, the surface finish is very good, without burrs [3], [4];
- complex shapes can be obtained since the machines are computer-controlled;
- repeatability is very high and precision is within the order of microns;

The presence of irregularities on the work piece surface exposes a series of disadvantages, especially in severe functioning conditions: decrease of actual contact surfaces, increased friction, reduced resistance to alternate tensions, reduced sealing and modifications in the actual work piece dimensions [5]. On the other hand, the absence of these irregularities affects the oil grease on the contact surface [6]. Under these assumptions, the surface roughness is determined as function of processing speed, contact surface size, load size and characteristics, precision and geometric shape. Another factor is the work piece design, which will be the scope of the main surface roughness parameters analysis in case of electro-erosion processing [7].

2. Theoretical aspects

Surface roughness is measured using surface roughness test meters that can provide both analogue and digital readings; they convert the vertical displacement of the touch sensor into proportional changes of the electrical output which requires amplification up to 2,000,000 in order to measure 0.01μm displacements.

Depending on the functioning principle, there are two types of analogue transducers:

- position based: provide displacement proportional signal even when the sensor is stationary; the output depends on the vertical sensor position; the advantage provided is accurate shape;
- movement based: provide output signal only when sensor is moving (depending on speed); zero when stationary;

For digital transducers the PGI system was used in order to allow reduced dimensions and increased measurement interval than HeNe laser interferometer measurement. The PGI system is comprised of an optical fascicle separation unit positioned on the pivoting arm as mobile interferometer element [8], [9]. The wavelength is the measuring reference, depending on ambient conditions. The interference pattern is detected by means of four photodiodes, which allows interpolation of output signal with resolution of 0.8nm in 12.5mm domain.

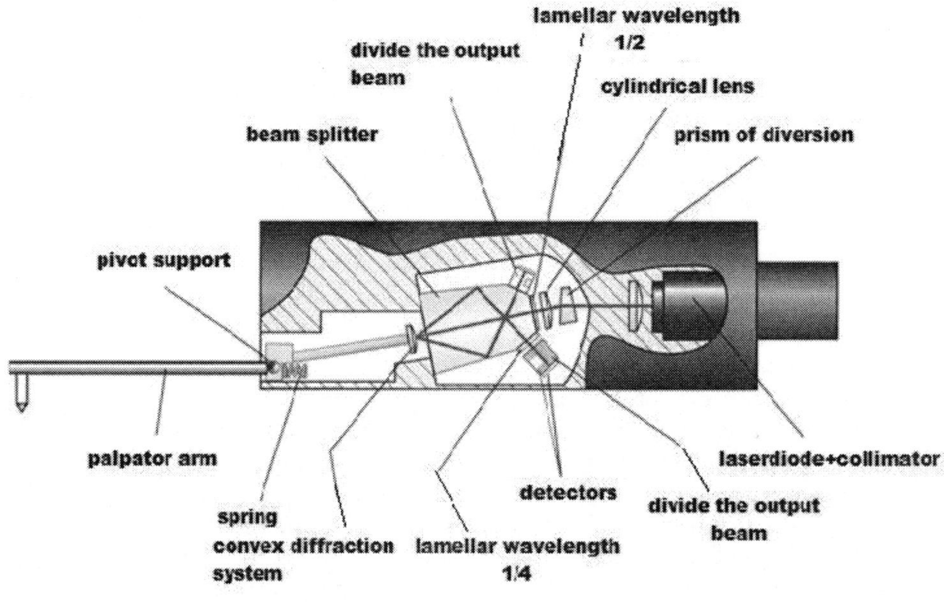

Figure 1. PGI measurement system functional schematic

The output electrical signal is the combination of the touch sensor and arm displacement from the surface level. In order to achieve the correct output signal for the analysed profile the system should travel a parallel path to the surface [10], [11].

3. Roughness measurement. Conclusions

A punching die for perforation and cutting of Aluminum rolling profiles (presented in the following figures) was used in surface roughness measurements.

IOP Conf. Series: Materials Science and Engineering **568** (2019) 012023 doi:10.1088/1757-899X/568/1/012023

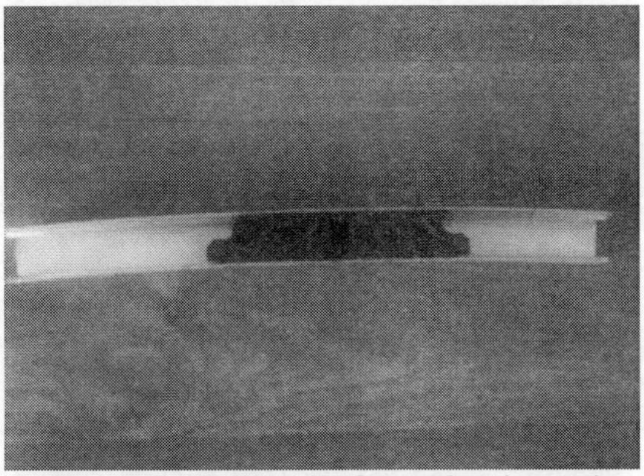

Figure 2. Complex die for perforation and cutting of Aluminum profiles

Figure 3. Shape of the cut in the Aluminum profile

The die is manufactured from X15 through the electro-erosion technological process.

For measuring the roughness the following measurement surfaces and directions were used depending on the surface roughness tester type and measurement performances [12]:

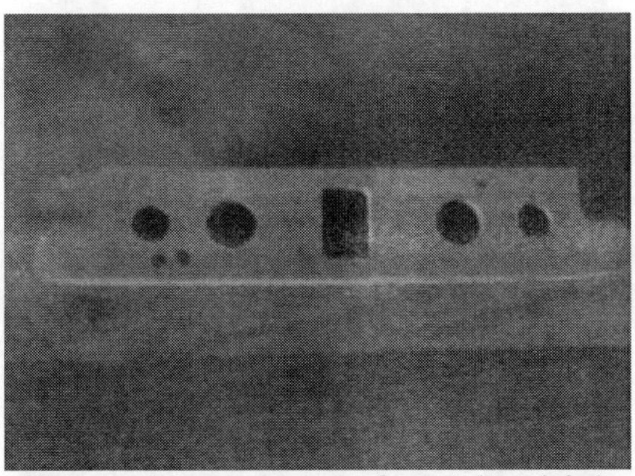

Figure 4. Measurement surfaces 1 (1 red dot) and 2 (2 red dots)

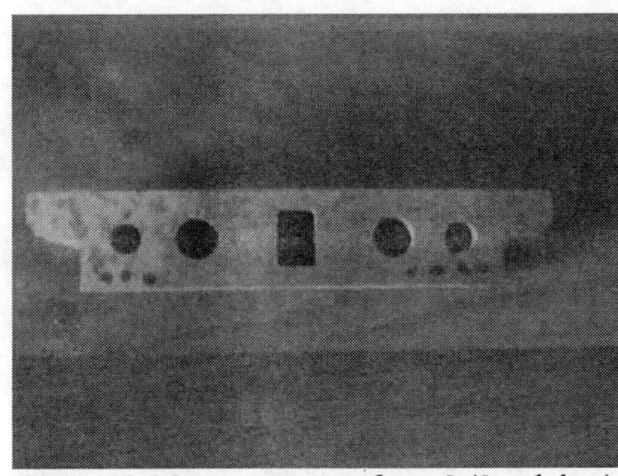

Figure 5. Measurement surfaces 3 (3 red dots) and 4 (4 red dots)

The resulting measurement curves are:

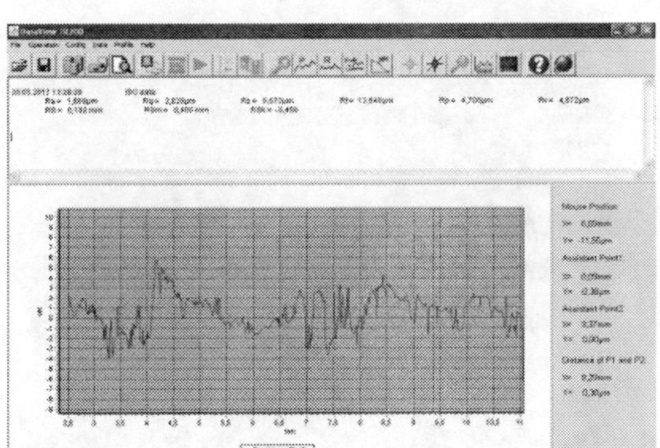

Figure 6. Surface 1 (1 red dot) measurement

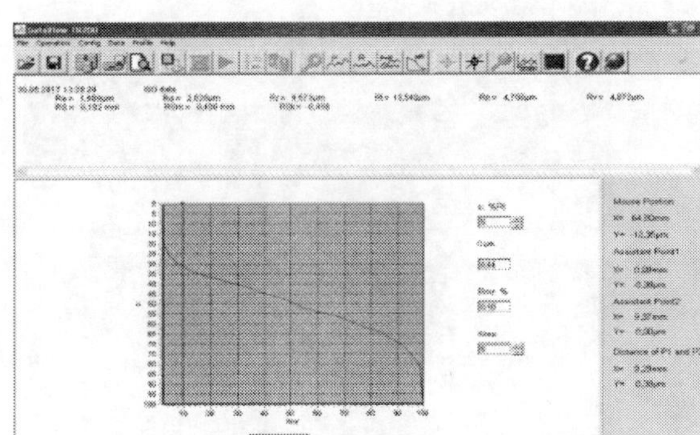

Figure 7. Surface 1 bearing area curve

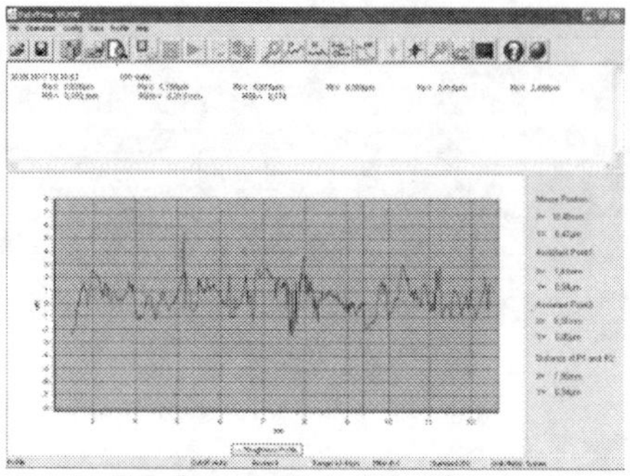

Figure 8. Surface 2 (2 red dots) measurement

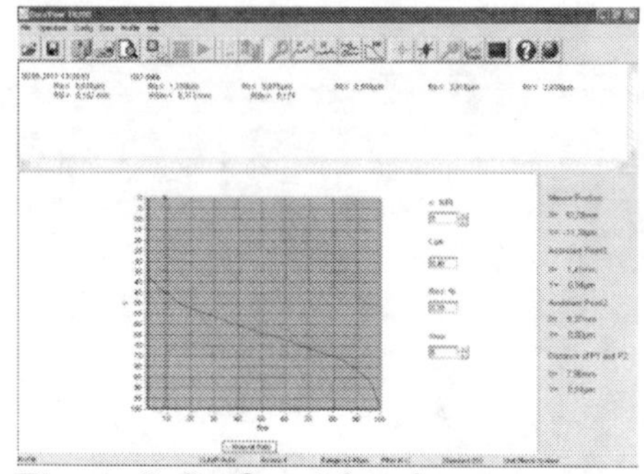

Figure 9. Surface 2 bearing area curve

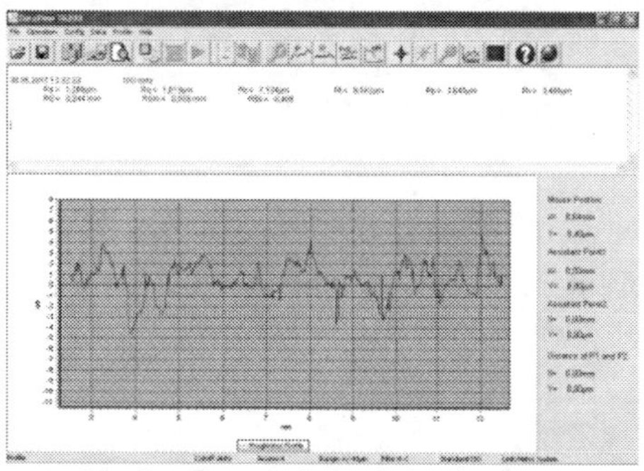

Figure 10. Surface 3 (3 red dots) measurement

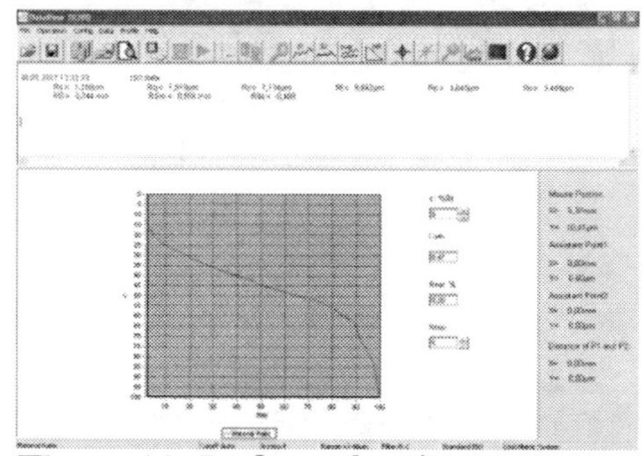

Figure 11. Surface 3 bearing area curve

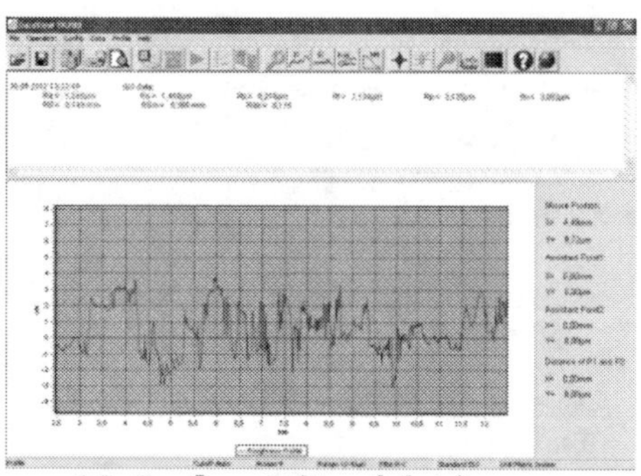

Figure 12. Surface 4 (4 red dots) measurement

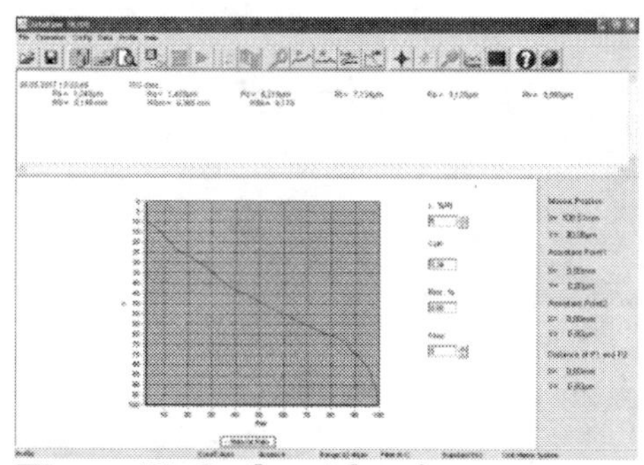

Figure 13. Surface 3 bearing area curve

The digital surface roughness tester and the setup for measurement of the given product are shown in the following pictures:

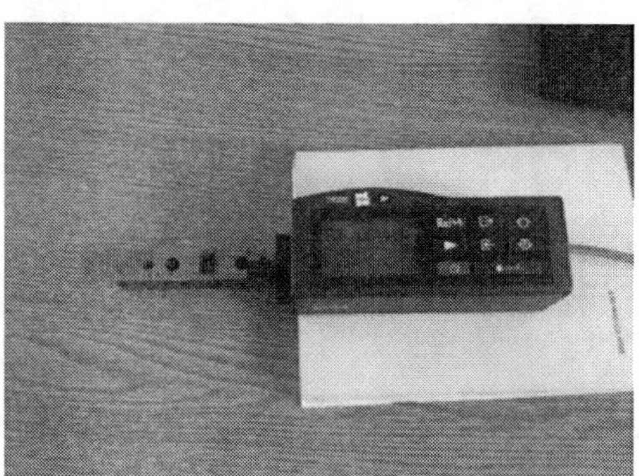

Figure 14. Surface roughness tester

Figure 15. Measurement setup [13]

The variation of Ra, Rq, Rz in the 4 measurement points is presented in the following diagrams:

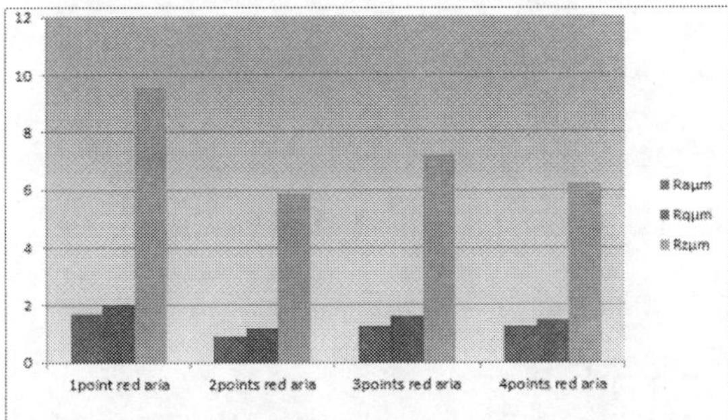

Figure 16. Variation of Ra, Rq, Rz parameters

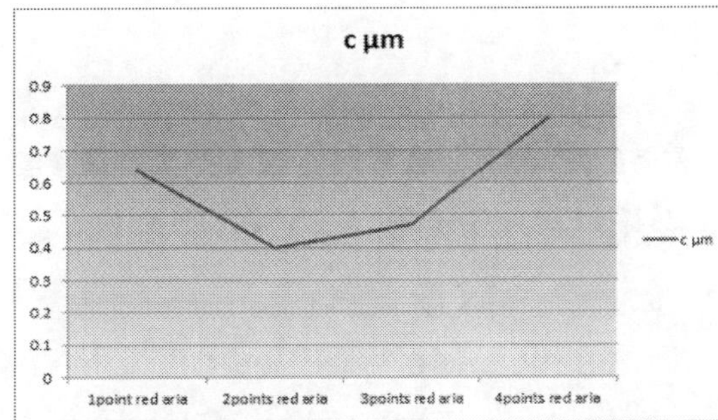

Figure 17. Variation of c coefficient

The variation of Ra, Rq, Rz parameters is within the limits presented in relevant literature. Also the variation of the bearing curve coefficient is within the limits indicated by the electro-erosion technological process specifications [14], [15].

References

[1] Balc, N., Berce, P., Gyenge, C., 2006, *Non-conventional Technologies*, Ed. Alma Mater, Cluj Napoca,.
[2] Balc, N., 2001,*Non-conventional Technologies*, Ed., Dacia
[3] Balc, N., Berce, P., Ancău, M., 2000, *Non-conventional Technologies.*
[4] Picoş,C., Coman, C., Dobre, N., 1979, *Non-conventional Technologies.*
[5] Taylor Hobson Ltd, Exploring Surface Texture – *A Fundamental Guide To The Measurement Of Surface Finish.*
[6] Taylor Hobson Ltd, Basic Surface Finish Filters - *Basic Surface Finish Parameters.*
[7] Micro – Top Consulting Engineering & Service S.R.L., *Course In Applied Metrology.*
[8] J. Paulo Davim, 2010, *Surface Integrity In Machining, Springer*-Verlag, London Ltd,pp.45-46.
[9] Whitehouse, David, 2012, *Surfaces And Their Measurement*, Boston, Butterworth-Heinemann. ISBN 978-0080972015.
[10] Degarmo, E. Paul; Black, J.; Kohser, Ronald A.,2003, *Materials And Processes In Manufacturing* (ed. 9th), Wiley, p. 223, ISBN 0-471-65653-4.
[11] Wen-Ruey Chang; S. Matz, Safety Science, October, 2000, *The Effect Of Filtering Processes On Surface Roughness Parameters And Their Correlation With The Measured Friction*, Part I: Quarry Tiles Volume 36, Issue 1, , Pages 19–33.
[12] Roughness Tester PCE-RT 1200, https://www.pce-instruments.com/english/measuring-instruments/test-meters/roughness-tester-pce-instruments-roughness-tester-pce-rt-1200-det_56329.htm
[13] Munteanu, A., Cioata, F., 2017, *Modern Measuring Devices And Systems*, Ed. Performantica, Iasi, , ISBN 978 -606-685-548-8.
[14] Nanu, A., 2003, *Non-conventional Technologies*, vol.I, Timişoara, Ed. Augusta,.
[15] Nagâţ, Gh., 1998, *Compendium of Non-conventional Technologies*, Iaşi, Ed. Rotaprint.

IOP Conf. Series: Materials Science and Engineering **568** (2019) 012024 doi:10.1088/1757-899X/568/1/012024

Estimation of braking capability for two distinct auxiliary engine brake systems paired with an exhaust brake

I C Manolache-Rusu, M Glovnea and C Suciu

Stefan cel Mare University, Faculty of Mechanical Engineering, Mechatronics and Management, Department of Mechanics and Technologies, 13 University Street, 720229, Suceava, Romania

E-mail: cozmin.manolache@usm.ro

Abstract Nowadays vehicles achieved an important increase in energy efficiency. Though highly desirable, in the case of heavy transport vehicles, this leads to a pronounced deficiency in their native deceleration capacity. This deficiency is more pronounced when traveling down slopes with important inclination, or when the traveling mode is with multiple starts and stops, as in the case of city travel. In such operating conditions, using only the conventional braking system can induce important thermal overloads, reducing braking efficiency, or even rendering the system useless. In order to limit this deficiency, auxiliary braking systems are often provided for heavy transport vehicles. Such systems serve to obtain a negative torque at the engine's crankshaft thus actively slowing the vehicle. The present paper aims to simulate the braking capacity obtained by using a combination of two distinct engine brake systems: the Jake brake and the Bleeder brake. The obtained results provide information regarding the evolution of thermodynamic parameters over the engine working cycle, with applications to optimize the deceleration capacity of vehicles that use diesel engines. Based on the presented data, it can be determined which of the analyzed auxiliary braking systems may be best suited, for implementation on particular road vehicles.

1. Introduction

By making use of the latest discoveries in the automotive industry, with regards to: injection systems, supercharging technologies, vehicle aerodynamics, new materials used in the construction of internal combustion engines and of vehicle bodies respectively, an increased vehicle efficiency was reached by reducing friction losses and by increasing engine power. In the case of heavy transport vehicles a lower curb weight is thus attained, leading to a higher carrying capacity. Although the aspects listed above represent vehicles advantages, in the case of heavy-duty vehicles there remains an important deficiency in the reduction of travel speed, especially when descending slopes with a high degree of inclination or when traveling through towns. In the case of frequent stopping maneuvers, said action can be regarded as a continuous and lengthy operation of the classic braking system. Since the classic braking system is a mechanical one based on direct contact between two elements, the friction surfaces are dimensionally limited due to the necessity of heat dissipation on the one hand and wheel dimensions on the other hand. For this reason, auxiliary braking systems such as the engine brake are often employed in heavy transport vehicles. These include various assemblies mounted on the engine at the level of the valve actuation system. The auxiliary engine brake systems dealt with in this article are the Jake brake and the Bleeder Brake.

IOP Conf. Series: Materials Science and Engineering **568** (2019) 012024 doi:10.1088/1757-899X/568/1/012024

The advantage of these two auxiliary systems consists in a high braking efficiency, a relatively low production price, and an easy implementation, even for vehicles not designed for this purpose, according to [1]. With these types of auxiliary systems, the sequence and the duration of the processes taking place during a motor cycle are modified with the intent of extracting as much energy as possible from the kinetic energy of the vehicle.

2. Model setup

In order to accomplish the proposed study, a computational program consisting of two distinct parts was developed. The first part deals with the evolution of thermo-gas-dynamic parameters for each process of the engine cycle, while the second part represents a dynamic engine model. The computational program is developed in the Mathcad environment and was validated in [2]. This program consists of a series of fixed pitch iterative loops that use differential equations characteristic for each process.

The engine braking process is characterized in the computation program by a gas exchange process. In this case, in addition to the gas state equation, the differential equation of temperature, mechanical work, heat exchanged with the walls, and others, a supplemental series of equations describing the flow through pipes are used. By means of several variables implemented in the calculation code, or by using the conditioning functions, the regime and direction of the gas flow are estimated. Depending on the momentary values of these variables, the calculation code automatically determines whether critical or subcritical regime equations found in [3], must be used.

At the end of each iteration step, the computational program stores the instantaneous values of the thermo-gas-dynamic parameters of the process within a matrix. These values are used as input data for the next process. The computational program allows the process evolution to be customized by imposing the characteristic values for every studied case. Thus, for the implementation of the exhaust brake regime, the value of $dcol_{ev}$ parameter that characterized the exhaust gas pipe diameter will be set equal to 5 mm, different from 50 mm in case of operation without the action of exhaust manifold butterfly valve.

This study is carried out by introducing in the computational model, the catalog data corresponding to a Lombardini 6LD400 mono-cylinder, summarized in Table 1.

Table 1. Constructive parameters for the engine.

Cylinder diameter	86 mm
Stroke	68 mm
Engine displacement	395 cm^3
Maximum power speed	3600 rpm
Engine power	8 HP
Compression ratio	18
Rod length	112 mm

For engine operation in Jake or Bleeder braking mode, it is necessary to impose a motion law for exhaust valve lifting within the calculation code. This is implemented by using a cubic interpolation function, for which it is necessary to indicate the maximum lifting height, the angle values of the opening and closing of the exhaust valve, in order to achieve braking. The value of the gap between the exhaust valve and the valve seat is considered to be 0.4 mm for the Bleeder brake action.

The exhaust valve first opens at a crank angle of 5 degrees ahead of the top dead center firing (TDCF), when the engine is running in the Jake brake mode. This value was adopted for this study, based on previous research performed by the authors in [4].

In order to evaluate the breaking capacity, the dynamic engine model developed in Mathcad was used. This calculation uses the values of the thermo-dynamic parameters within the matrices as input data in order to determine the instantaneous engine torque. The evolution of the engine torque is imposed by the values of the tangential force component of the crank mechanism, denoted by F_t, at every step. According to [3], this force is directly influenced by the in cylinder gas pressure p_{cil}, as.

$$F_t = \left[\left(P_{cil} - P_a \right) \cdot A_p + F_j \right] \cdot \frac{sin(\alpha + \beta)}{cos(\beta)} \tag{1}$$

where, P_a - atmospheric pressure, A_p - area of the piston head, F_j - inertia forces of the masses in translation, α - crank shaft angle, β - angular displacement of the connecting rod.

Engine operation in braking mode is achieved by suppressing fuel injection for all study cases. For this reason, in the computational code the combustion process is completely replaced by a gas exchange process. Also, when using the exhaust brake, the release process will be replaced by a compression one due to the high pressure set at the exhaust valve gate as a result of decreasing gas passage section at the level of the exhaust manifold. Other studies on auxiliary braking systems are made by the authors in [5, 6, 7], but these do not treat the influence of the Bleeder braking system on the vehicle's deceleration capacity.

3. Results

This paper aimed to evaluate the braking torque evolution obtained at the Lombardini 6LD400 engine shaft when it is used in different engine brake modes. The results obtained are related to the evolution of the in cylinder pressure, with applications to optimize the deceleration capacity of vehicles equipped with diesel engines.

Five cases were studied, aiming to evaluate the braking performance of the mono-cylinder engine during operation in three distinct auxiliary braking modes: Jake brake, Bleeder brake and exhaust brake. In all of these situations, the fuel injection is suppressed. It is also desirable to evaluate the braking capacity for combining each type of engine brake with the exhaust brake. Actuation of the exhaust brake assumes complete closure of the exhaust butterfly valve for all study cases. The study is performed by imposing a constant engine speed, considered at 3600 rpm. The braking power can be evaluated based on simulated results of the average torque obtained at the crankshaft.

Figure 1 shows the evolution of in cylinder pressure for all five study cases.

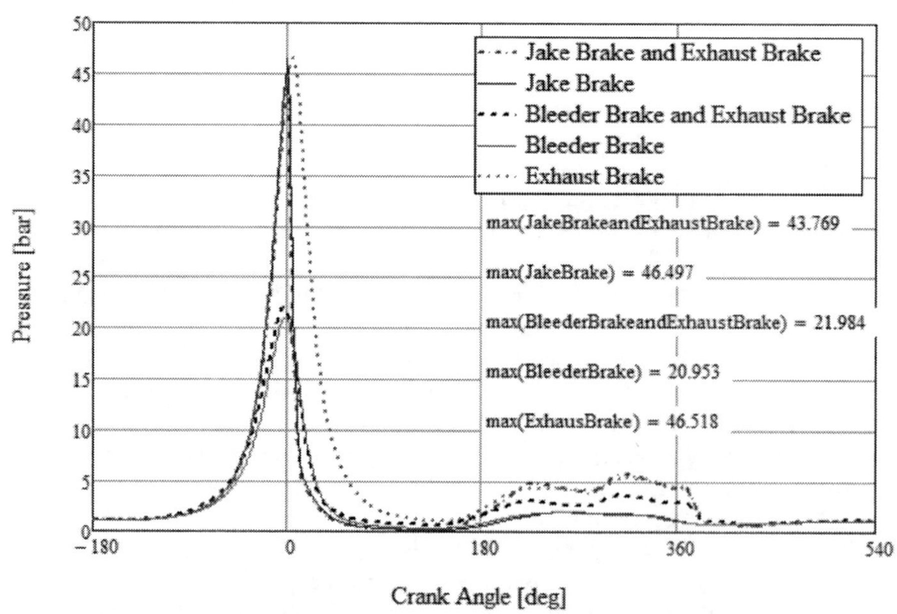

Figure 1. In cylinder pressure evolution.

The graph shows that the in cylinder pressure records high values when the engine is used in Jake Brake and Exhaust Brake modes. A maximum value of 46.518 bars is reached in the case of Exhaust Brake. Also, the use of the engine in this brake mode leads to an increase in cylinder pressure during the forced evacuation process. During the discussed process case in cylinder pressure value exceeded 5. On the other hand when the engine is running in the Bleeder Brake mode, the maximum cylinder pressure is around half from the peak pressures reached in the previously discussed cases.

In Jake braking mode, a sudden drop in cylinder pressure near the TDCF is observed, as a result of lifting the exhaust valve from its seat. When combining the action of the engine brake system with the exhaust brake, it is observed that the characteristic pressure evolutions for the two distinct braking modes are preserved.

Using the proposed dynamic engine model, which uses the cylinder pressure matrix, it is possible to evaluate the engine torque. The evolution of the engine torque and its average values per cycle are illustrated in Figure 2.

The lowest mean retarding torque value of -8.9 Nm is obtained in the case of the exclusive exhaust brake action, although the in cylinder pressure value in this operation case records a maximum value.

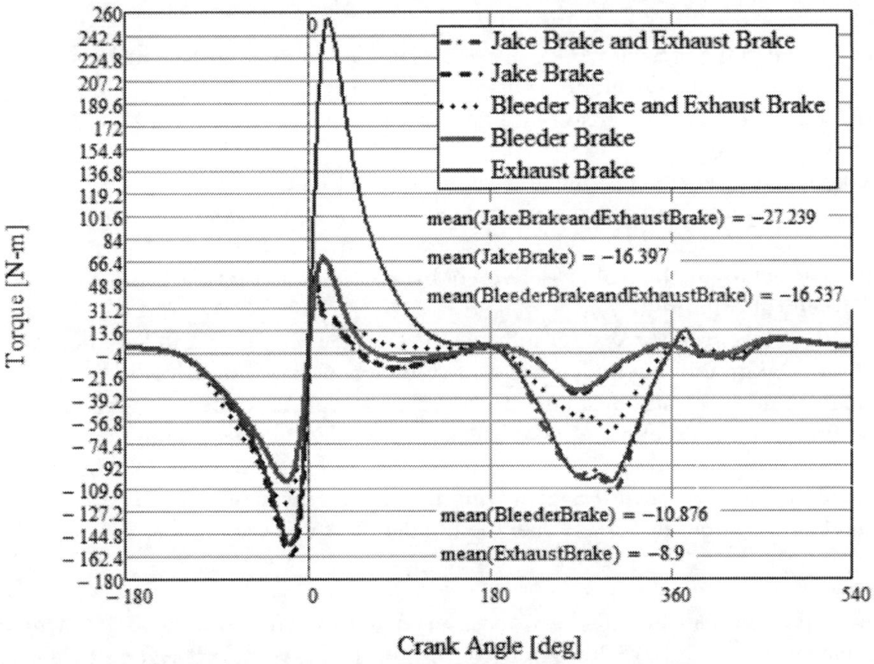

Figure 2. Instantaneous engine torque at the shaft.

Also in this operation mode the biggest difference between the minimum and maximum values of the instantaneous engine torque can be observed. This fact leads to major mechanical stress for the shaft. A higher average resistive torque value is obtained in the Bleeder mode, reaching -10,876 Nm and even higher in Jake brake mode for which the average value is -16,397Nm. This value is approaching the maximum generated engine power at 3600 rpm, but reversed. According to simulated torque values, it is obvious that the in Jake brake mode offers the best braking capacity. However, this system is more expensive and harder to implement than the others because synchronization with the distribution mechanism is required. Moreover, this braking system, according to [8], develops a strong noise during operation, this being the reason why its use was restricted in some countries.

The simulated results show that the simultaneous use of a Bleeder brake system with an exhaust restrictor lead to values of engine torque similar to those obtained in the case of Jake brake operation. This result shows that this constructive version can be used to replace the Jake brake system.

By attaching an exhaust brake system to the Jake brake system the mean value of the resistive engine torque is much higher, reaching -27,239 Nm. The braking power corresponding to this obtained value of the engine torque at the maximum power speed is -13,771hp, which is 1.7 times greater than the power that can be generated by the engine in case of normal operation. Table 2 summarizes the peak to peak values of engine torque expressed in Nm obtained for the five studied cases.

Table 2. Peak to peak engine torque values [Nm].

Jake Brake and Exhaust Brake	Jake Brake	Bleeder Brake and Exhaust Brake	Bleeder Brake	Exhaust Brake
203,466	217,009	188,819	173,848	407,841

It can be noticed that when using the engine in Bleeder mode, the peak to peak value is the lowest one. Use of the engine in this brake mode leads to minimal mechanical stress for the engine components while developing a braking power approximately equal to the one generated by the engine in combustion operation.

4. Conclusion

The results obtained in the present work can be summarized by several conclusions as further described. The Bleeder brake auxiliary system used in combination with an exhaust brake can easily replace the Jake Brake system as it would offer similar braking performance. In this way, the noise produced by the Jake system can be avoided.

In all studied cases, the mean engine torque value was found to be more important for the combined use of engine brake system and an exhaust restrictor than in the case of individual use.

The braking power developed by the Bleeder brake system in combination with an exhaust restrictor is approximately equal to the value delivered by the Lombardini 6LD400 single-cylinder engine operating normally at full throttle.

The use of the Bleeder brake system subjects the engine components to less mechanical stresses.

Although the maximum in cylinder pressure value when using the Bleeder braking mode is around half from the maximum pressure from the use of exhaust brake, the mean engine torque is significantly higher in the first case due to elimination of the spring effect of gases during the relaxation process.

Using an exhaust brake simultaneously with a Jake brake system in the case of Lombardini 6LD400 engine at maximum power speed it is achieving the brake power of -13,771hp, 1.7 times greater than the maximum engine output in normal operation under the same conditions.

5. References

[1] Maxwell F C and Gurdijan E 1996 Accessory brakes for motorhomes *FMC Magazine* **86-95**
[2] Manolache-Rusu I C 2015 Research concerning the Jake brake system in diesel engines *PhD Thesis* Stefan cel Mare University Suceava Romania
[3] Berthold G 1980 *Theory, calculus and construction of engines for road vehicles (Teoria calculul si constructia motoarelor pentru autovehicule rutiere)* (Bucharest: Ed. Did. & Ped.)
[4] Manolache-Rusu I C, Suciu C, Patuleanu L and Andronic F 2014 *J. Engineering Studies and Research* **20** 45-51
[5] Manolache-Rusu I C, Suciu C, Andronic F and L Patuleanu 2014 *J. Engineering Studies and Research* **20** No.1 52-59
[6] Manolache-Rusu I C, Patuleanu L, Andronic F and Radion I 2013 *J. Tehnomus* **20** 270-75
[7] Manolache-Rusu I C, Severin T L, Patuleanu L and Andronic F 2013 *J. Tehnomus* **20** 369-75
[8] Jacobs Vehicle Systems Inc. 2000 Vehicle Noise Levels and Compression Release Engine Braking P/N 028307 Rev B

IOP Conf. Series: Materials Science and Engineering **568** (2019) 012025 doi:10.1088/1757-899X/568/1/012025

Sustainable mobility as a result of peoples' awareness on environmental problems generated by transport activity

G Mitran[1], S Ilie[1 2], S V Igret[3], S Mihailescu[4]

[1] University of Pitesti, Pitesti, Targul din Vale Street, No. 1, Romania
[3] University Aurel Vlaicu Arad, Arad, Revolutiei Boulevard, No. 77, Romania
[4] University of Petrosani, Petrosani, Universitatii Street, No. 20, Romania

[2] sorin_ilie_dis@yahoo.com

Abstract. The field of mobility and transport has undergone major changes in recent decades, being a close link between it and the general economic development. Taking into consideration that urban areas concentrate the most of inhabitants and socio-economic activities, the negative effects generated by the increased road traffic are present mainly in the cities, especially in their central areas. People's involvement in awareness-raising actions on environmental issues caused by transport activity is a starting point for achieving a sustainable urban mobility system. Information and awareness of people about the link between travel behavior that they adopt and negative effects on the environment is a first step towards educating them in order to shift to sustainable mobility patterns. The paper highlights the awareness and participation component of people in the development and implementation of sustainable urban mobility plans, as an educational action for environmental protection and sustainable mobility. This concept of involving citizens is detailed within a case study.

1. Introduction

According to the latest statistical data, at the level of the European Union 70% of the population lives in the urban environment, which generates about 80% of the total GDP. These factors, in addition to the high usage levels of personal passenger vehicles powered by fuels from non-renewable sources have led to more difficulties in the smooth running of transport activities. One of the objectives of European Commision is to reduce the greenhouse gas emissions generated by the transport sector by 60% relative to the 1990 levels until 2050. Taking into account these target values and the travel characteristic of the urban environment – a high percentage of short-distance travels [1, 2], the cities/ towns have a high potential of shifting towards a low carbon emission transport, which can be achieved through modal relocation, increasing pedestrian travel, travelling by public transport, using non-motorized means of transport or those means of transport powered by energy from renewable sources. Knowing that the three pillars which support sustainable development are the society, the environment and the economy, the citizens, as representatives of the society, play an important role in all the stages of the process of sustainable transport development – diagnosing the current situation, identifying solutions, devising development plans and implementing them. The effective involvement of the society in sustainable transport development is influenced by their knowledge of the field and the degree of understanding of the phenomena affecting the transport activities and the associated externalities [3, 4, 5], which can be achieved through environmental education.

IOP Conf. Series: Materials Science and Engineering **568** (2019) 012025 doi:10.1088/1757-899X/568/1/012025

Environmental education is the process by which individuals acquire knowledge and skills that help them become aware of and assess the negative effects caused to the environment by natural and anthropic activities. Thus, it brings about the determination and motivation for the participation of each individual and of the collectivity on the whole in solving the environmental problems of the locality in which they live. Environmental education also facilitates the decision making process ensuring certainty with regard to the effects on the environment and the adoption of responsible measures both at the individual and the community level. The higher the number of informed individuals, the greater the efficiency of promoting and achieving sustainable development goals. In Romania the ratio between the number of subjects that interacts with knowledge in educational environment the total population is very small. Subsequently, in addition to the educational means the population must be informed using seminars, workshops, brochures, leaflets, informative guides, etc.

Paper emphasizes the way in which the inhabitants of a Romanian town are involved in devising and implementing the sustainable urban mobility plan. The case study was conducted in Mioveni, the second town in Romania that started devising a sustainable urban mobility plan, document that entails the approach of participatory process in transport and sustainable mobility planning.

2. Methodology

The interaction between the transport system and the social, political and economic environment implies a complex interdisciplinary approach, which integrates aspects related to engineering, management, economy and last but not least, environmental protection. All these specific features contribute to the classification of the transport system as a large technical system. The specific spatial, temporal and structural features of the transport demand require a permanent adaptation of the offer and of the technologies involved to serve it, which makes transport become a very dynamic system. This dynamic character generates the difficulties encountered by the decision factors in transport planning. Their decisions have to ensure the effective running of the transport system with the lowest consumption of resources, minimum negative external effects and a degree of performance in conformity with the demand of the beneficiaries. Under these circumstances, the participation of the public in the decision making process in due time is essential for the success of any transport plan.

Kemp and Stephani presented 40 good practices in the field of transport planning; these were extracted from case studies conducted in 33 U.S. states and in 19 states from other continents. The following conclusion is underlined among the good practices suggested in the study: *"Planning and public participation are crucial ingredients to the success of a city's eco-transportation plan"* [6]. The idea was first brought into discussion by Sherry Arnstein in 1969. He represented the degree of citizen participation in the decision making process by analogy with the steps of a flight of stairs. The maximum power was reached by those participation typologies placed in the upper part: "partnership", "delegated power" and "citizen control" [7]. Booth and Richardson argue the need for public participation in transport planning in the UK, in the context of the institutional changes that took place in the late 1990s [8]. They propose a conceptual framework which integrates the aspects that must be taken into account when planning the public involvement plans in the field of transport planning, offering answers to the following questions: "Why involve the public?", "What is negotiable and what is non-negotiable?", "How will the public be involved?", "Who should be involved?", "When should the public be involved and where?".

The methodology proposed in this paper and meant to be used in the citizen education process, whose aim is to achieve sustainable transport, is based on public participation (citizens, representatives of the civil society (NGOs), representatives of public institutions) at all the stages of the sustainable urban mobility plan (figure 1): diagnosing the current situation; establishing the development goals, the course of action and the measures necessary for achieving the goals; action plan and its implementation, respectively.

The participation of the interested factors at all these stages allows them to collect information on the interdependence between the transport system and all the aspects of the socio-economic life.

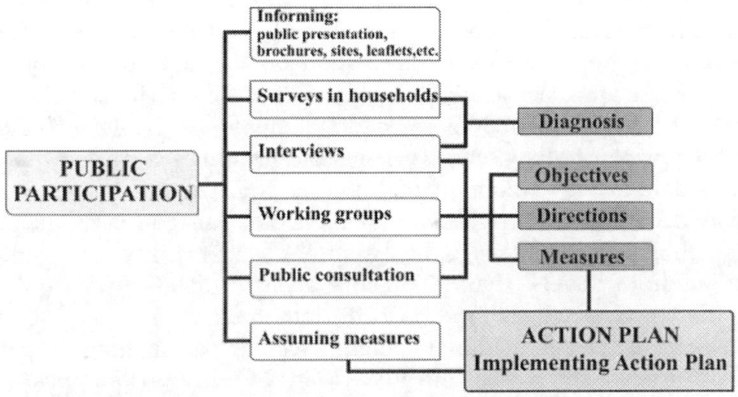

Figure 1. Public participation in devising the sustainable urban mobility plan.

Within the first part, dedicated to characterizing the current situation and diagnosing current mobility, the public is familiarized with the fact that the environment, economic prosperity, and social characteristics specific to a region depend on the transport system that serves the given/ respective territory and on its characteristics, such as: safety, accessibility, effectiveness, and impact on the environment (noise and air pollution, greenhouse gases). This stage represents the most consistent part of the plan, as a good substantiation for diagnosing mobility issues enables the realistic identification of the malfunctions in this field. In addition to the documenting process performed by the expert team, whose members come from interdisciplinary fields (transport engineering, urbanism, sociology, and economy), it is neccessary to involve the citizens as a target group for surveys and interviews which will offer information about the mobility behaviour at the level of the area under study.

It is important to mention that in order to collect data more effectively it is necessary to promote these activities through all the media and to inform the people about the notion of "sustainable transport". Thus, it is recommended that through the environmental education provided within working groups and through public consultation to present/ explain the principles underlying the concept of sustainable transport – economy (saving of costs), social equity, environmental protection, which should be completed by the adequate institutional framework intended to support integration, transparency, assumption of responsibility, performance, effectiveness, long-term planning, international cooperation, innovation, best use of limited resources, and education.

Applying these principles knowledgeably to establish the objectives, the courses of action and the proposed measures will increase the effectiveness of the action plan implementation process by taking responsibility, an activity that can be regarded as a significant challenge taking into account the fact that the shift towards sustainable transport implies the adoption of some radical measures, with a major impact on all the users of the transport system.

3. Case study. Results and discussion

The case study in which the environmental education methodology in the field of transport was applied has been conducted within the sustainable urban mobility plan for city of Mioveni, a heavily industrialized Romanian town with high mobility values. The motorisation rate (number of passenger cars per 1,000 inhabitants) in the City of Mioveni was 33% higher than the average national value. Between 2009 and 2017, this indicator has grown from 220 to 350 cars / 1,000 inhabitants.

The participation of the population was materialized in informing activities, household surveys regarding the mobility behaviour, interviews with the main socio-economic factors operating in the area, working groups with the representatives of the public institutions, etc. The aim of these activities was to involve all the social entities in promoting sustainable mobility and environmental protection in a conscious manner.

IOP Conf. Series: Materials Science and Engineering **568** (2019) 012025 doi:10.1088/1757-899X/568/1/012025

The most complex action was that of identifying the mobility behaviour based on household surveys. The surveys were conducted for 567 persons from 200 households of the total of 11,164 recorded in the area, thus ensuring a representative sample of 1.8% [9]. The interview is divided into three main parts referring to: (i) general information on the size of the household, including the number of persons, passenger cars available, income level, etc.; (ii) specific information on each member of the household, such as the age, gender, occupation, whether or not they have a driving licence, workplace or place of study, etc.; (iii) specific information on the routes travelled by each member of the household the day before, in a 24-hour period. The information includes the origin and destination of travel, departure and arrival times, the mode of transport used, the purpose of travel, etc.

An indicator which is very important for analyzing the sustainability that results from these surveys is the modal distribution of travels (figure 2). The analysis of data shows a positive aspect, which is the fact that pedestrian travel accounts for 43% of the total travels.

The negative aspects are represented by the high percentage accounting for personal transport (car: 32%) and the low values reached by the collective means of transport (minibus: 11%, Bus - employee shuttle: 13%). Since pedestrian travel covers a significant percentage, another extensive action was the interviewing of the pedestrians regarding the existing facilities and those they would like to have at their disposal for non-motorized transport – the technical state of the infrastructure, the existence of pedestrian and bicycle tracks between the main points of interest, bicycle rentals, and accessibility. These interviews were conducted in the busy pedestrian areas (figure 3), on a sample of 500 people. As in the case of the household surveys, the first part of the discussion consisted of the introduction into the context of the study, each interviewee being given an informative brochure. After the data had been processed, it was found that 70% of the respondents would like to have bicycle lanes.

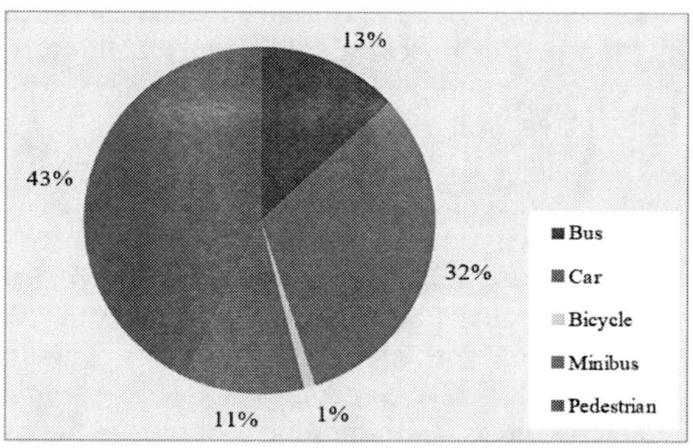

Figure 2. Modal distribution of travels.

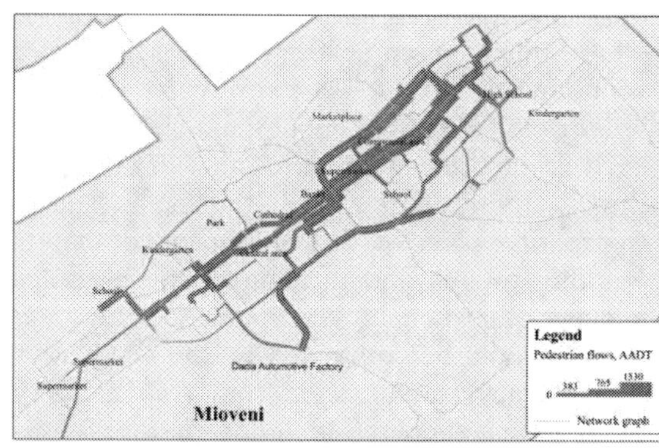

Figure 3. Pedestrian flows, Daily Traffic.

These topics, as well as other aspects related to the travel behaviour for working purposes were found in the survey forms distributed to the employees domiciled in the town of Mioveni and working in the main economic companies located in the area under study. A total of 542 persons were interviewed. The results obtained after processing the primary data reveal that most employees travel to their workplace on foot or by mass transport systems provided by the employer (buses – employee shuttles). All the distributed questionnaires were accompanied by informative brochures and instructions on how to fill them in. Furthermore, informative brochures were also distributed to groups of employees who were not part of the selected sample. This activity was performed with the help of the representatives of the companies involved in the action, who also participated in the working groups established for this purpose. The working groups were organized separately, for transport of people and transport of goods, respectively. Different organisations interacting directly with the two transport subsystems, both at local and county level were co-opted, thus creating a common framework which helped them identify the current internal mobility problems and find possible solutions to remedy and reduce them. The presentation of the conclusion reached by analyzing the data from these surveys and from the documentation work on the current mobility situation conducted by the experts involved in designing the plan was made at the first public consultation, attended by 92 people (citizens, representatives of the civil society - NGOs, and mass-media). The second public consultation was meant to consolidate the proposed projects, taking into account the opinions of the public interested in this issue. 150 people attended the meeting, most of them belonging to that category of population with high levels of mobility (pupils/ university students) and persons with special needs (elderly persons). During the period in which the action was in progress, information relating to the project was published on the website of the local administration and informative materials were distributed among the people attending local events organized by the local administration (about 500 brochures).

4. Concusions

The participation of all the interested parties in all the design stages of the sustainable urban mobility plan, first entails providing them with adequate information on the impact of the transport activity on the environment, which is a component of environmental education.

The current case study involved approximately 2000 people, which was a major effort, taking into account the fact that the participatory process in the field of tranport planning in Romania is still at its initiation stage, because civic awareness is not properly developed. Once they were involved in this process, the citizens discovered that their opinion was important, investing feelings and intellectual capacity to propose and support those projects whose implementation should contribute to ensuring the sustainability of the transport system. This raised their awareness, which contributed to adapting the mobility behaviour of each citizen, so that their actions might minimize the negative effects on the environment. The agreement of the interested parties on the action plan creates the premise which supports the implementation stage of the plan, the stage in which the role of the concerned actors (citizens and organisations) is crucial for the acceptance of the proposed measures, as some of them involve the radical shift of the travel habits towards environmentally friendly means of transport – pedestrian and public transport, car-sharring, car-pooling and a reduction in the use of passenger cars. The education process based on participation in sustainable planning will continue in the stages monitoring the degree of the plan implementation.

References

[1] Stead D and Marshall S 2001 The relationships between urban form and travel patterns. An international review and evaluation *European Journal of Transport and Infrastructure Research* **2**, 113 - 141
[2] Naess P 2006 *Urban Structure Matters: Residential Location, Car Dependence and Travel Behaviour* (London-New York: Taylor & Francis Group)
[3] Van H T, Fujii S, Nakamura F and Emori H 2007 Educational methods to change the attitudes of transport planners towards environmentally sustainable transportation systems in developing countries *IATSS Research* 2 74-83
[4] Laing R, Tait E and Gray D 2012 Public engagement and participation in sustainable transport issues *Risc Cobra Conference Las Vegas*
[5] Schiller P, Bruun E and Kenworthy J 2010 *An Introduction to Sustainable Transportation: Policy, Planning, and Implementation*, (London: Earthscan)
[6] Kemp R and Stephani C 2015 *Urban Transportation Innovations Worldwide: A Handbook of Best Practices Outside the United States* (North Carolina: McFarland Company)
[7] Arnstein S 1969 A ladder of citizen participation *AIP Journal* 216-224
[8] Booth C, Richardson T 2001 Placing the public in integrated transport planning *Transport Policy* **8** 141-149
[9] Atlanta Regional Commission 2011 *Regional travel survey - final report*

IOP Conf. Series: Materials Science and Engineering **568** (2019) 012026 doi:10.1088/1757-899X/568/1/012026

Tribological properties of the PA46 and Teflon added PA46 polyamides

M T Lates[1]

[1]Product Design, Mechatronics and Environment Department, "Transilvania" University of Brasov, Brasov, Romania

E-mail: latesmt@unitbv.ro

Abstract. In order to increase the efficiency of the mechanical transmissions, one of the actually research directions is oriented to develop new materials characterised by high endurance, small friction coefficients and internal structure stability with the temperature. One of these materials is represented by the polyamides which are used in mechanical industry, food industry and medicine. The paper provides the results of tests performed on a tribometer for a PA46 and a Teflon added PA46 polyamide materials being on contact with a PA46 polyamide disk, in lubricated conditions. The tests are achieved for a set of temperatures of the oil bath, a set of rotational speeds of the PA46 polyamide disk and a continuous variation of the local pressure. During the running-in process there are measured the evolution of the dynamic friction coefficient and of the wear. After this, it is studied the variation of the dynamic friction coefficient with the contact pressure, the rotational speed and the temperature. Finally there are given conclusions regarding the tribological behaviour of the tested materials and regarding their applications.

1. Introduction

Increasing the efficiency of the mechanical systems represents a new actually trend in their optimization process. The increase of mechanical systems efficiency is oriented in two main directions: finding new solutions in the case of the mechanical systems structures and using new materials with low-frictional properties.

Applications in the first direction are referring on a wide area of mechanical systems as: robotics where the friction of a new class of robots with flexible joints it conducts to energy loses [1]; mechanical systems as speed increasers used in the wind systems, where the power coefficient can be improved by increasing the mechanical efficiency of the speed increasers [2, 3, 4]; solar tracking systems where the energy consumption used to orient the solar panels or the solar collectors can be reduced by designing new mechanical trackers [5, 6].

The second direction of optimization the mechanical systems efficiency is using new materials with low-frictional properties. In this case, the researchers are looking for materials characterised by low friction, low wear and stabile structures (depending on temperature). A class of these materials is represented by polyamides. The literature shows the high endurance and small deformation of the polyamides in the case of loadings [7, 8, 9]. Their mechanical properties, with good performances at high temperatures are presented in [10, 11, 12, 13, 14, 15]. The frictionless behaviour of polyamides is tested under different testing conditions in [16] for sliding-rolling contacts; even in dry conditions [17] the polyamides shows their slow frictional behaviour and their reduced wear.

The limits of polyamides applications in the construction of the mechanical systems are represented mainly by the heavy industry. In medicine, food industry, fine mechanics their usage is wider and wider. Due to this, the paper investigates the tribological properties of two polyamide type materials being in contact: PA46 with PA46 and Teflon added PA46 with PA46.

2. The tests

The tests are performed on a tribometer which is connected thought an acquisition plate to a computer as it can be observed in figure 1. The tribometer is equipped with a two dimensional force sensor which allows force measurements in an interval of 0.1 ... 1000 N with the resolution up to 50 mN [18]; the measured forces are oriented about the vertical axis and about a direction tangent to the rotational motion of the lower disk. On the force sensor is mounted a suspension in order to damp the oscillations. On the base plate of the force sensor is mounted a holder which allows the mounting of different pieces as ball, pins, plates. The vertical stroke is equal with 150 mm and it can be reached with a speed situated in an interval of 0.001 ... 10 m/s.

Figure 1. The tribometer.

On the tribometer is mounted a slider which allows motions in a direction parallel with the horizontal plane. The lateral positioning stroke is equal with 75 mm and can be reached with a speed in the interval of 0.01 ... 10 m/s; the resolution of the lateral positioning motion is equal with 2 μm. The heater allows the control of the temperature in the oil bath up to 150 °C. The wear can be measured with an accuracy of 50 nm [18].

The tribometer is equipped with three modules: the rotary module, the reciprocating module and the block on ring module; in the paper the tests are performed by using the rotary module. The disk of the rotary module allows rotations about two directions with a rotational speed between 0.001 ... 5000 rpm and a resolution equal with 1 μm [18]. The maximum radius for the contact between the disk and the upper holder is 75 mm [18].

The testing procedure begin with a running-in process which is performed for a period of 2 hours, at a room temperature of 22 °C, a normal pressure of 10 MPa and the rotational speed of the disk equal with 500 rpm.

In the next step, the tests continue with a set of rotational speeds equals with: 50, 200, 500, 1000 and 2000 rpm; the local mechanical pressure has a linear variation between 5 and 10 MPa. All the tests are performed for a period of 15 minutes and for the following temperatures of the oil bath: 22 °C, 90 °C and 120 °C.

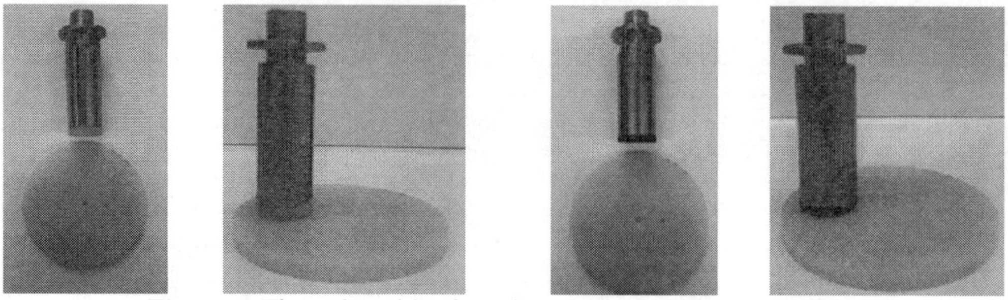

Figure 2. The polyamide plates in contact with the PA46 disk.

3. Results and conclusions

The results present the variation of the friction coefficient with the test parameters: the rotational speed of the PA46 polyamide disk, the normal pressure and the temperature of the oil bath.

The evolution of the wear during the running-in process is presented in figure 3; the value of the wear, for the PA46 polyamide is stabilised after approximately 35 minute at a value around 0.06 mm. The wear for the Teflon added PA46 polyamide is higher (about 0.09 mm) and it is stabilised after 70 minutes.

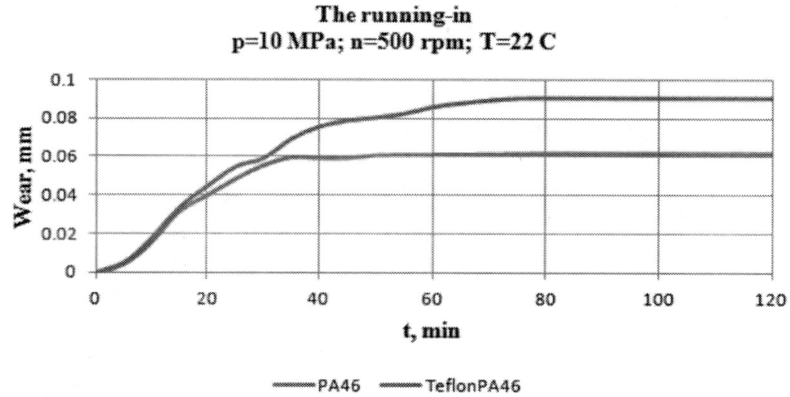

Figure 3. The evolution of the wear during the running-in process.

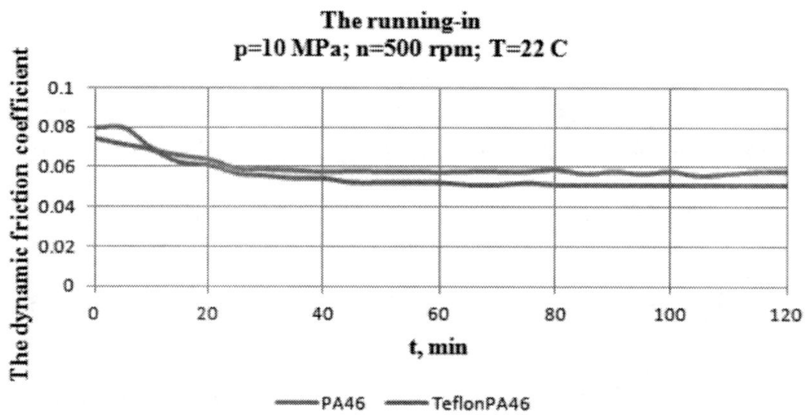

Figure 4. The dynamic friction coefficient.

The variation of the dynamic friction coefficient, during the running-in process, is presented in figure 4; in the case of the PA46 polyamide, after 25 minutes of tests, its value it is stabilised at a value of 0.06. The value of the dynamic friction coefficient is smaller – about 0.05.

IOP Conf. Series: Materials Science and Engineering **568** (2019) 012026 doi:10.1088/1757-899X/568/1/012026

Figure 5 presents the variation of the dynamic friction coefficient at a contact pressure equal with 10 MPa depending on the rotational speed and the temperature. The value of the friction coefficients increases with the increasing of the temperature and decreases with the increasing of the rotational speed. In all the cases, the value of the friction coefficient for the Teflon added PA46 is smaller than the value of the friction coefficient for the PA46 polyamide; with the increasing of the temperature, the difference between the friction coefficients of the two polyamide materials decreases.

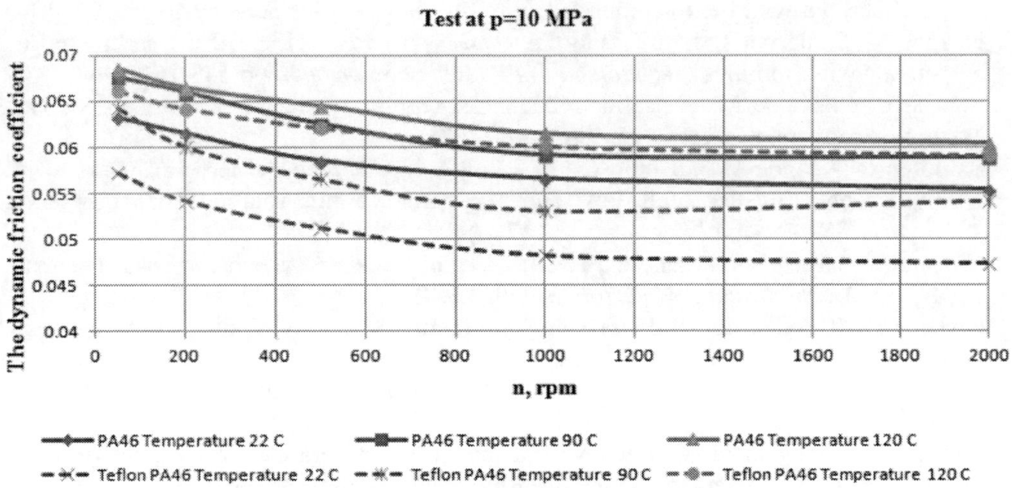

Figure 5. The dynamic friction coefficient.

Figure 6 presents the variation of the dynamic friction coefficient with the contact pressure and the temperature. The value of the dynamic friction coefficient, for the two polyamide type materials, increases with the increasing of the temperature and decreases with the increasing of the pressure. In all the cases, the dynamic friction coefficient for the Teflon added PA46 polyamides has smaller values than the dynamic friction coefficient of the PA46 polyamide; the difference between the two sets of friction coefficients decreases with the increasing of the temperature.

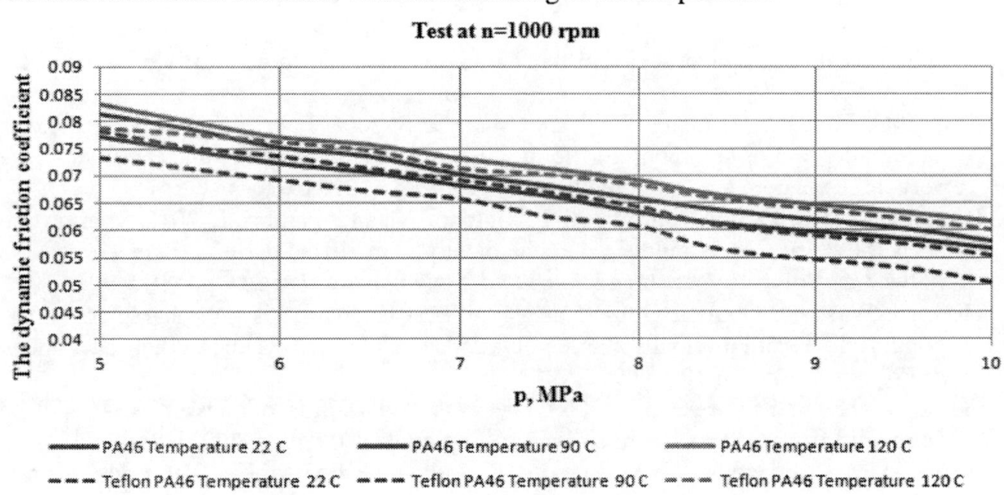

Figure 6. The dynamic friction coefficient.

As a general conclusion, the Teflon added PA46 polyamide has a higher wear than the PA46 polyamide; from this point of view, in the case of high endurance and small wear conditions, the PA46 polyamide it is preferable to be used. From the dynamic friction coefficient point of view, in all the tested cases, the friction coefficient of the Teflon added PA46 polyamide has smaller values than the friction coefficient of the PA46 polyamide; the difference between the friction coefficient of the Teflon added PA46 polyamide and the friction coefficient of the PA46 polyamide decreases with the increasing of the temperature.

References

[1] Cretescu N, Neagoe M and Saulescu R 2016 Kinematic and dynamic analysis of a 4DOF parallel robot with flexible links *Mechanisms and Machine Science* **46** pp 473-481

[2] Neagoe M, Saulescu R, Jaliu C and Cretescu N 2016 Novel speed increaser used in counter-rotating wind turbines *Mechanisms and Machine Science* **46** pp 143-151

[3] Saulescu R, Neagoe M and Jaliu C 2018 Conceptual synthesis of speed increasers for wind turbine conversion systems *Energies* **11-2257**

[4] Saulescu R, Neagoe M and Jaliu C 2016 Improving the energy performance of wind turbines implemented in the built environment using counter-rotating planetary transmissions *Materials Science and Engineering* **147 (1)**

[5] Ciobanu D, Jaliu, C and Saulescu, R 2014 Chain Tracking System for Solar Thermal Collector *Applied Mechanics and Materials* 658 pp 35-40

[6] Ciobanu D 2011 Conceptual design of a solar thermal system with dish solar collector *Journal of Environment Engineering and Management* **10 (8)**

[7] Fink T and Bodestein H 2011 Friction reduction potentials in chain drives *MTZ Worldwide Edition* **72** pp 46-51

[8] Hyakutake T, Inagaki M, Matduda M, Hakamada N and Teramaki Y 2001 Measuring of friction in timing chain *Society of Automotive Engineers of Japan* **22** pp 343-347

[9] War X and Dwyer-Joyce R S Model experiments on automotive chain drive systems *Elsevier Science* **39** pp 851-861

[10] Mouhmid B, Imad A, Benseddiq N and Lecompte D 2009 An experimental analysis of fracture mechanics of short glass fibre reinforced polyamide 6.6 (SGFR-PA66) *Journal of Composite Science and Technology* **69** pp 2521-2526

[11] De Monte M, Moosbruger E and Quaresimin M 2010 Influence of temperatura and thickness on the off-axis behavior of short glass fiber reinforced polyamide 6.6 – quasi-static loading *Journal of Composites* **41** pp 859-871

[12] Chiu F-C and Kao G-F 2012 Polyamide 46/multi-walled carbon nanotube nanocomposites with enhanced thermal, electrical and mechanical properties *Journal of Composites* **43** pp 208-218.

[13] Chiu F-C, Fu S W, Chuang W T and Sheu H S 2008 Fabrication and characterization of polyamide 6.6/organomontmorillonite nanocomposites with and without a maleated polyolefin elastomer as a toughener *Polymer* **49** pp 1015-1026

[14] Buccella M, Dorigato A, Pasqualini E, Caldara M and Fambri L 2012 Thermo-mechanical properties of Polyamide 6 chemically modified by chain extension with Polyamide/Polycarbonate blend *Journal of Polymer Research* **19** pp 9935-9944

[15] Hsiao S, Wang H, Chou J, Guo W and Tsai T 2012 Synthesis and characterisation of novel organosoluble and thermally stable polyamides bearing triptycene in their backbones *Journal of Polymer Research* **19** pp 9902-9911

[16] Gordon D H and Kukureka S N 2009 The wear and friction of polyamide 46 and polyamide 46/aramid-fibre composites in sliding–rolling contact *Wear* pp 669-678

[17] Akiyama M, Yamaguchi T, Matsumoto K and Hokkirigawa K 2010 Friction and wear of polyamide 66 composites filled with RB ceramics particles under dry condition *Tribology Online* **5** vol. 2 pp. 87-91

[18] *** 2009 UMT Multi-Specimen Test System. Hardware Installation & Application Manual. Center for Tribology, Dell Ave, Campbell, USA.

IOP Conf. Series: Materials Science and Engineering **568** (2019) 012027 doi:10.1088/1757-899X/568/1/012027

Experimental research on Diesel Particle Filter (DPF) in relation to the fuel type

B M Jurchiş[1], **A I Borzan**[1] **and D L Băldean**[1,2]
[1] Automotive and Transportation Department, Technical University of Cluj-Napoca, Cluj, Muncii 103-105, Romania

[2] doru.baldean@auto.utcluj.ro

Abstract. The experimental research is developed to validate or to dismiss theoretical suppositions considering the efficiency of particulate matter filters for internal combustion engines. This paper is pointing out the ideas of a practical test regarding soot impact upon the diesel particle filter from after treatment system, supplied with different fuels in operation mode. Measurements are recorded with after treatment system scanner and subsequently the experimental data are processed and graphically presented using digital module. Experience and know-how concerning the after-treatment system applied to the exhaust pipe of a Diesel engine is extremely useful in modelling the protocol for sustainable technical design of the automotive powertrains. The present paper outlines an applied process of controlling and analysing at least four parameters defining the clogging level of an actual diesel particle filter in K9K792 euro 5 internal combustion engine in relation to the fuel type. Engine supplied with Diesel standard has a wide range of clogging, but when changing to more blended fuels containing biodiesel from soy methyl ester and ethanol, the after treatment process is influenced significantly. The experimental testing is designed to reiterate for each fuel type in real time and after a distance of one thousand kilometres – in engine speed regime of two thousand revolutions per minute.

1. Introduction

The necessity for approaching the after treatment of the exhaust gases from car engines is based on the researchers and engineers awareness and efforts to diminish the pollution impact upon the environment by creating more sustainable solutions in combustion and in motorized mobility of the future, both on our planet and beyond if space travel would be further developed [1]. The chemical and physical pollution affect through nitrous oxides, particulate matter and aerosols climatic changes, biodiversity, oceanic acidity, ozone layer thickness and the life quality in general for all living beings. In the last decades, the after treatment systems have been considered and developed for different types of exhaust architectures in automotive sector, but in order to keep up engine's performances and to maintain power value when configuring particle filter and catalyst the system has to be further analysed and electronically controlled [2]. Potential of after treatment system is yet to be highlighted and further studied through the spectrum of modern equipment in sooth combustion control and exhaust gas dynamics [3].

The complex transport operations lead to higher particulate matter emissions, thus polluting the environment, with different sizes of particles and residues. Recent developed technologies are available today for reducing traffic and transport particulate products and to lower pollution [4]. Researching and measuring the particulate filters and their efficiency is now possible with digital assisted equipment, both in laboratory and road operation of cars [5]. Main objective of this research is

IOP Conf. Series: Materials Science and Engineering **568** (2019) 012027 doi:10.1088/1757-899X/568/1/012027

to outline an innovative solution for particulate matter (PM) filter in K9K792 Diesel engine operation. The specific objectives are the tests with four types of fuels and mixtures in order to point out the viable experimental operating regime and also to evaluate the sensors data in working state. The integrated system for differential pressure measurement and soot impact, based on complete handling of Diesel engine, allows the creation of a regeneration map. The important achievement will be the creation a specific model available for testing and applicable on large scale production, if possible. Other specific steps are: development of a working plan for particulate matter monitoring and an electronic control phase for actual components in the after treatment system destined to evaluate the soot impact, temperature, voltage and differential pressure.

PM stands for particulate matter (also tagged or labelled as particle pollution): the expression for a mixture of solid particles and liquid droplets appearing in the atmosphere [6]. Solid or material particles, such as dirt, soot, dust and smoke, are large enough to be observed with the eye. Some other particulate matter appears so small they can only be monitored using electronic equipment or digital micro-scope. The particulate matter exists in many sizes and shapes and may be generated from multiple and various chemicals. Some PMs are generated directly from a specific source, such as a construction site, unpaved road, fields, smoke stacks or fires. Most PMs from the earth's atmosphere were generated by complex reactions of many chemicals such as sulphur dioxide and nitrogen oxides. These groups of substances are pollutants emitted by power plants, industries and automobiles [7]. Pollution with particulate matter includes: PM10: small particulate matter, with medium diameter of 10 micrometres; PM2.5: smaller and fine inhalable particles, with diameters of 2.5 micrometres or lower. Particulate matter consists of microscopic solid compounds and liquid droplets which are small enough to be inhaled and cause cancer or other health conditions [8]. PMs have less than 10 micrometres in size, posing the highest risk, due to the fact that they can travel deep inside the lungs. Through this mechanism some of the particulate matter will get in the blood and other organs. Small particles as PM2.5 are the important causal factor for lowered visibility (haze) in polluted zones, some parks and other areas [9].

2. Methodology and material

By using the ubiquitous networking and computer processing power of today in applied science and engineering regarding even the PM research, there may be designed a specialized tool or a method for particulate matter control in a complex after treatment system from automotive field. Approaching the subject of particulate matter emission impact correlated with automotive and transportation allows the control and proper management of these pollutant compounds, materials and substances which harm the environment and living beings. The present research paper is based on a method of experimental testing in four distinct scenarios when supplying the engine with four different types of fuels. The research method is developed on a compared analyse of three solutions with the standard recipe when supplying the engine with Euro 5 commercial Diesel (Standard Diesel). A specialized automotive engineering methodology for after treatment monitoring and control is put into practice. Methodological solution applied in this case makes use of computer aided engine scanning and automotive expertize for PMs control in the Diesel Particle filter 14, according to figure 1. In this graphical representation of the tested power train there are shown the main parts of the applied equipment, such as the internal combustion engine 1, clutch 2, gear box 3, transaxle 4, traction wheels 5, engine speed sensor 6, fuel pressure regulator 7, common rail 8, fuel pressure sensor 9, temperature sensor 10, electronic control unit 11, air temperature sensor 12, exhaust gas manifold 13, after treatment equipment 14 and exhaust pollutant analyser / smoke meter.

The research methodology consists in comparative analyse of the three strategies of supplying the engine with mixtures of Diesel Standard (DS) with Bio Diesel (BD) and Ethanol (E) in different ratios with the reference supply mode with DS only. The schematic strategy of the testing protocol is shown in figure 2, which describes the structure of the dual branch evolution in the performed experimental research. First branch consists in specific and individual tests with each fuel type, starting with the reference fuel (Diesel Standard). After recording the actual values with DS as reference fuel, the testing has been developed with alternative fuel recipes such as: Diesel Standard 95% + Bio Diesel 5%

IOP Conf. Series: Materials Science and Engineering **568** (2019) 012027 doi:10.1088/1757-899X/568/1/012027

- DS95BD5 (numerical indexing expressing the volumetric participation of each fuel component); Diesel Standard 90%+ Bio Diesel 10% - DS90BD10; Diesel Standard 85% + Bio Diesel 10% + Bio Ethanol 5% - DS85BD10BE5. The second direction or branch of the research consists in a compared analyse of the collected actual values. This study makes a parallel graphic comparison of numerical data received from electronic control unit monitoring the after treatment system with DPF.

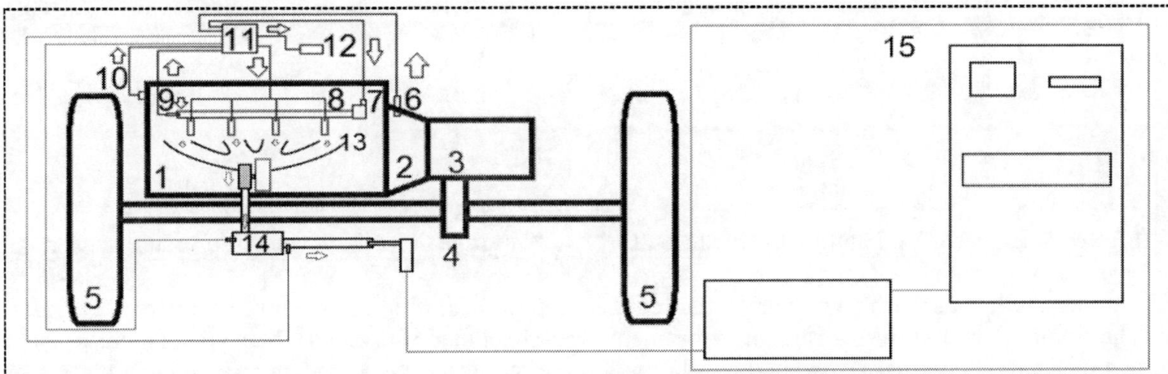

Figure 1. The main components of the tested power train with after treatment system and analyser.

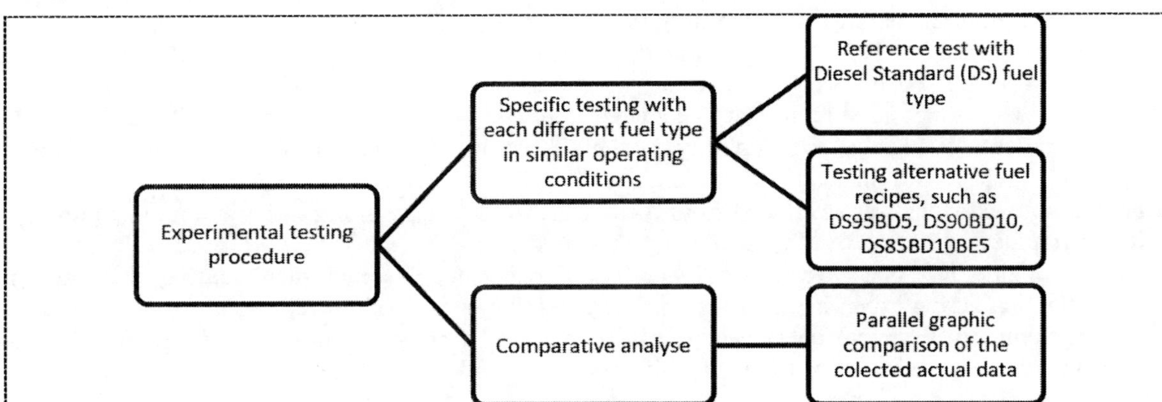

Figure 2. The simplified schematic of the experimental testing strategy put in the research practice..

3. Experimental results and discussions

The first analysed data come from the particular tests with each fuel type. The second part of analyse is concentrated upon the overall comparison of the actual values, as it is shown in figure 3 for the soot impact, in figure 4 the differential pressure, exhaust temperature in figure 5 and voltage in figure 6.

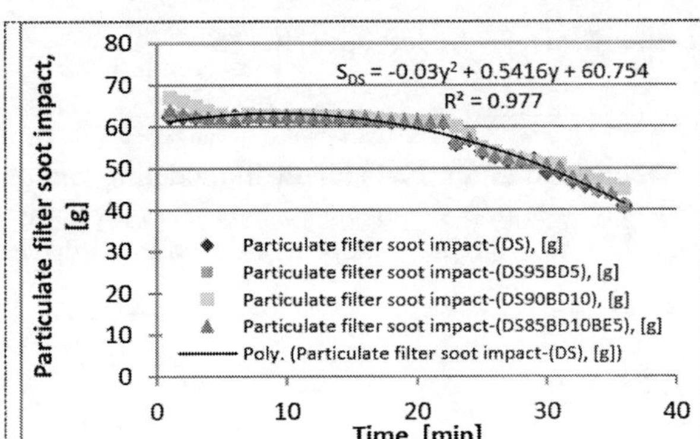

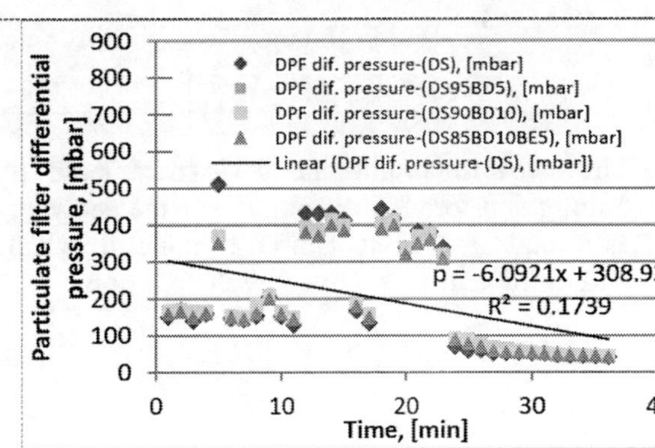

Figure 3. Soot mass in DPF for tested scenarios.

Figure 4. Differential pressure before DPF.

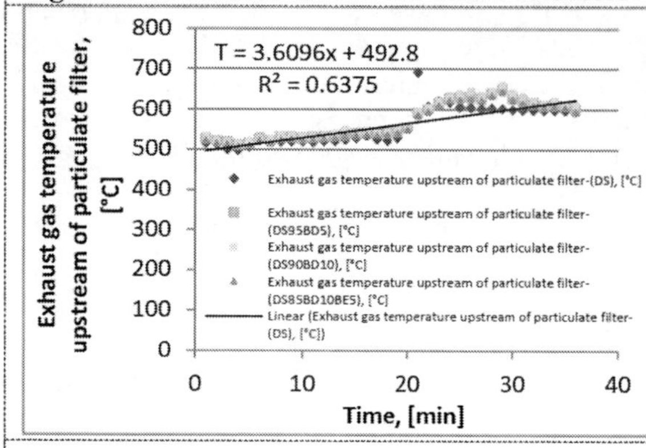

Figure 5. Exhaust Temperature upstream DPF.

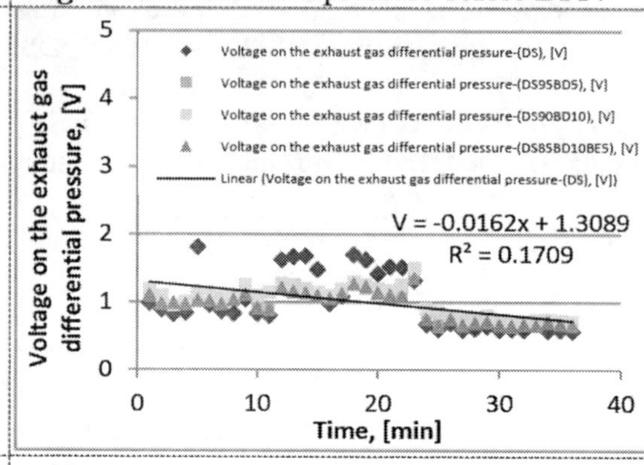

Figure 6. Voltage on diff. pressure sensor.

The measurements were taken during the regeneration procedure in an overall interval of 40 minutes. The first and the last two extreme measurements were excluded. In statistical analyse, an outlier value is a measured point that is deviated in relation to other collected values. Such an outlier in the present case is due to variability in the (initiating and ending) after treatment monitoring procedure and it will indicate experimental error; thus to avoid the latter mentioned errors the problematic extremes were excluded from the actual data sheet, resulting in only 36 points presented in the graphics.

The measurements were taken during the regeneration procedure in an overall interval of 40 minutes. The first and the last two extreme measurements were excluded. In statistical analyse, an outlier value is a measured point that is deviated in relation to other collected values. Such an outlier in the present case is due to variability in the (initiating and ending) after treatment monitoring procedure and it will indicate experimental error; thus to avoid the latter mentioned errors the problematic extremes were excluded from the actual data sheet, resulting in only 36 points presented in the graphics.

The engine has been wormed up and made ready for testing when starting the practical determination during experimental research. The trend line was traced for reference measurement.

The summary technical data sheet is given in table 1, showing the targeted operation during experimental tests, exhaust gas flow in manifold and some other significant values.

Table 1. Technical data of the Diesel Particulate Filter after treatment architecture.

Parameter	K9K792 Diesel Particle Filter
Target process	Particle capture (PM 10 & 2,5)
Exhaust gas flow, [m³/h]	100
Temperature, [°C]	600
Targeted Pollutant	Particulate matter (PM), Hydrocarbons (HC)
Efficiency, [%]	98

The most important and basic parameter, that is considered for mathematical modeling and is highlighted for a more comprehensive analyze, consists in soot impact correlated to regeneration time, as a total mass of soot in DPF monolith when K9K792 tested engine is kept in full load and driven with third gear ratio, given by the relations:

$$S_{DS} = -0.57x + 67.8 \quad (1)$$

where x is the time variable considered in minutes of operation in regeneration mode, for a linear progression of the soot combustion model.

$$S_{DS} = -0.03y^2 + 0.54y + 60.8 \quad (2)$$

where y is the time variable considered for a polynomial progression of the soot mass model.

4. Conclusions

In this research paper, were studied the parameters of after treatment process with multiple fuel recipes, taking as a reference fuel type the Diesel Standard Commercial euro 5, with the detailed specification given at https://www.petrom.ro/services/downloads/00/petrom.ro/1522164440041/sf-43-motorin-standard.

The highest soot impact factor was recorded in the test made with DS90BD10 fuel type in the beginning of the procedure. Most elevated differential pressure was for point 5 with DS at 512 mbar. Peak temperature was at 692 °C for measurement point 21 with DS.

Acknowledgments

This work was supported by a grant of the Romanian Ministry of research and Innovation, CCCDI-UEFISCDI, project number PN-III-P2-2.1-CI-2018-1227, within PNCDI III

References

[1] Băldean D et al 2018 *Contributions on experimental research of fuel consumption and management system in K9K892 Diesel engine from Dacia-Renault* (Annual Session of Scientific Papers IMT ORADEA 2018 Section Mech Eng and Automotive MATEC Web of Conferences 184 01018) DOI https://doi.org/10.1051/matecconf/201818401018

[2] Băldean D et al 2017 *Contributions in Experimental Research Concerning Diesel Fuel Supply and Lubrication in the Case of Comparative Study Between Euro V and IV Common Rail Engines* (In Chiru A Ispas N (eds) CONAT 2016 Internat Congress of Automotive and Transport Engineering CONAT 2016 Springer Cham) Online ISBN 978-3-319-45447-4

[3] Băldean D 2017 *Researching the Dysfunctional aspects concerning Diesel Particle Filter (DPF) in case of Dacia Logan E5 1.5dCI Engine* (In Science and Engineering / Ştiinţă şi Inginerie AGIR Press) eISSN 2359 – 828X

[4] Borza EV et al 2018 *Research Concerning Fuel Economy Coefficient and Carbon Foot Print in Various Conditions for a City Compact Size Vehicle with Digital Control for a Green Solution and Method at Technical University from Cluj-Napoca* (In: Burnete N Varga B (eds) Proceedings of the 4th Internat Congress of Automotive and Transport Eng (AMMA 2018) AMMA2018 Proceed in Automotive Engineering Springer Cham) Online ISBN 978-3-319-94409-8 DOI https://doi.org/10.1007/978-3-319-94409-8_22

[5] Borzan AI et al 2018 *Research contributions through virtual modeling of an innovative particle filter for alternative application in automotive field using advanced engineering methods* (In: The International Congress of Automotive and Transport Engineering, AMMA 2018) Online Last modified 2018-07-11 DOI http://amma2018.ro/index.php/amma/2018/paper/view/28

[6] Borza EV et al 2005 *Functional characteristics of compression ignition engines with improved equipment which work with rape oil* (In: J Acta Technica Napocensis, Publishing date 2005) pp 67-73

[7] Burnete N et al 2008 *Diesel engines and biofuels for urban transport* (in ro *Motoare Diesel şi biocombustibili pentru transportul urban* Cluj-N Mediamira ISBN 978-973-713-217-8)

[8] Hodor AI 2012 *Studies and research on implementation of flexible molds for small and medium manufacturing series* (Technical University of Cluj-Napoca Publishing date 2012/12/15 Publisher UTPress)

IOP Conf. Series: Materials Science and Engineering **568** (2019) 012027 doi:10.1088/1757-899X/568/1/012027

[9] Odenie VS et al 2017 *Experimental researches concerning the clogging level of diesel particle filter (DPF) after 45000 km in case of k9k892 e5 1.5 dci engine* (Bucharest AGIR Publishing Press)

IOP Conf. Series: Materials Science and Engineering **568** (2019) 012028 doi:10.1088/1757-899X/568/1/012028

Reducing NOx emissions from diesel engines supercharged with pressure wave compressor driven by the electric motor

C I Leahu[1,*] and S Radu[1]

[1] Automotive and Transport Engineering Department, Transilvania University of Brasov, 1 Politehnicii Street, Brasov 500024, Romania

E-mail: leahu.cristian@unitbv.ro

Abstract. The overwhelming parts of the modern Diesel engines for vehicles are supercharged. However, the supercharging process lead to generating a significant amount of NOx emissions. Therefore, the supercharging intensity may be intentionally limited in order to avoid increasing NOx emissions excessively, as these emissions are drastically limited by the Euro 6 pollution standard. By moving from Euro 5 to Euro 6, the NOx emission's permissible limits have been reduced by up to 80%. These reductions of NOx emissions were made possible mainly due to the utilization of the some conventional equipment and procedures presented in the paper. Diesel engine's supercharging could be realized efficiently also with pressure wave compressor (PWC). The paper highlights the possibility of realizing the EGR process directly through the PWC. The concept of supercharging with PWC leads to a significant reduction of NOx, for a significant increase of both the pressure and temperature of the intake air.

1. The current context of reducing NOx emissions

1.1. Supercharged diesel engines

The overwhelming parts of the modern Diesel engines for vehicles are supercharged. More and more often, the concept of downsizing is applied (Figure 1). Thus, by increasing the supercharging pressure, the engine's specific power also increases [1].

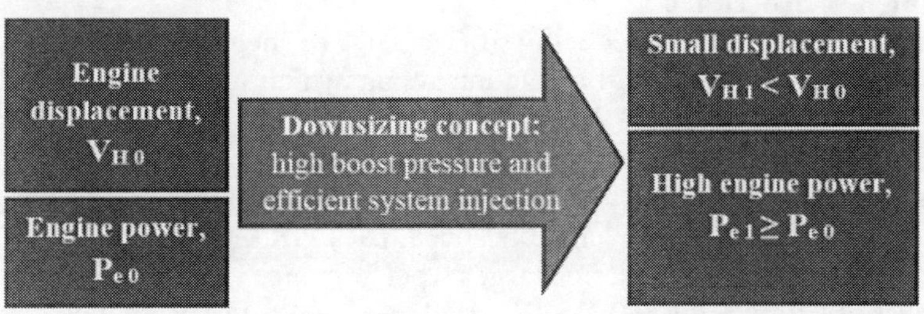

Figure 1. Downsizing concept.

A downsizing example is presented in Table 1, according with [2]. However, the supercharging process increases the level of pressure and temperature of the engine cycles [3]. That means that a poor air-fuel mixture (with oxygen in excess) it is formed in cylinders, which burns at high temperature.

IOP Conf. Series: Materials Science and Engineering **568** (2019) 012028 doi:10.1088/1757-899X/568/1/012028

These two conditions lead to generating a significant amount of NOx emissions – which makes more difficult to meet the Euro 6 pollution standard [4,5,6].

Table 1. Downsized diesel engine example.

Parameter	Before downsizing	After downsizing	Difference
Engine displacement	2000 cmc	1600 cmc	-20 %
Supercharger pressure	2.5 bar	3 bar	20 %
Maximum cylinder pressure	170 bar	200 bar	18 %
Specific power	64 kW/l	80 kW/l	25 %

1.2. The Euro 6 emissions standard

NOx emissions of the exhaust gas for Euro 6 Diesel engines must not exceed the concentration levels from Table 2. The values are valid for the vehicle's operating conditions according with the cycle specified in table: New European Driving Cycle (NEDC), World Harmonized Stationary Cycle (WHSC) and World Harmonized Transient Cycle (WHTC).

Table 2. Euro 6 standard emissions – NOx limit values.

Type vehicles	Cycle	NOx emissions limit values, g/km	Reduction from Euro 5 standard, %
Passenger cars	NEDC	0.08	56
Light vehicles	NEDC	Class I: 0.08	56
	NEDC	Class II: 0.105	64
	NEDC	Class III: 0.125	55
Heavy duty vehicles	WHSC	0.40	77
	WHTC	0.46	80

It could be noticed that by passing from the pollution standard Euro 5 to Euro 6 one, the NOx emissions have been significantly reduced, the differences between the two standards reaching values up to 80%. These reductions of NOx emissions were made possible mainly due to the utilization of the following equipment and procedures: exhaust gas recirculation (EGR), Diesel Oxidation Catalyst (DOC), Lean NOx Trap (LNT) and Selective Catalytic Reduction Technology (SCRT) [5,7,8].

Regarding NOx emission, the Euro 5 standard could be attained largely through utilizing the EGR system alone. In case of Euro 6 standard, depending on the engine displacement, a LNT or SCRT usage is supplementary needed - beside that of EGR.

2. The experiment's procedure

The EGR procedure consists in introducing of 6 to 10% of the exhaust gases back into the cylinders. Essentially, exhaust gas replace a part of the intake air, which reduces the oxygen quantity within the combustion chamber. In addition, by introducing triatomic molecules that have high specific temperature – such as CO_2 or H_2O – the maximum temperature of the cycle is also reduced [7].

Three types of EGR systems can equip the supercharged engines: low pressure EGR, high pressure EGR or both of them (depending on engine's speed, LP EGR or HP EGR is used). At low speeds, LP EGR is used to avoid affecting the turbocharger's performance. At low-medium speeds (when there is sufficient exhaust gas flow rate), HP EGR is used – having a faster response, while not affecting the reliability of intake air's systems. At medium-high speeds, the EGR system does not work.

Generally conventional EGR contains the components from Table 3 [7,8]. There are two main disadvantages of using an EGR system: the increase of complexity of the engine due to the amount of additional equipment used exclusively for recirculating the exhaust gas and the decrease of engine's reliability due to the frequent obstruction of EGR's valve and throttle.

Table 3. The role of the main components of the conventional exhaust gas recirculation.

EGR components	The components role
EGR valve	Control of the volume of recirculated exhaust gas
EGR cooler and temperature sensor	Reducing the temperature of the recirculated exhaust gas and finally the cycle temperature.
EGR throttle	Adjust intake air or exhaust gas to supercharged engines.
EGR valve and throttle control unit	On the basis of information on sensor (temperature, pressure, flow, speed, position and gas composition) controlling the EGR adjusting

The main advantages of recirculation the exhaust gas when using PWS are: increasing the reliability of the engine and reducing its cost because it eliminates the components of the conventional EGR that are used exclusively for the recirculation and improving the air filling of the cylinders by reducing pumping losses - as the intake or exhaust manifold throttle are removed.

The previously described internal EGR can be successfully used thanks to the innovative PWC driving system. This system consists in driving PWC with an electric motor having a variable or sequential speed. On the other hand, the classic solution involves driving the PWC at a speed proportional to the engine's one. Due to the fixed ratio between the engine speed and the PWC, it is difficult to provide simultaneously high supercharging pressures and a satisfactory flow rate of EGR [3,10]. According to the new PWC driving solution, its speed is independent of that of the internal combustion engine (Figure 3). The experimental research can identify the optimal speed with which PWS should be driven for each of the engine operating conditions.

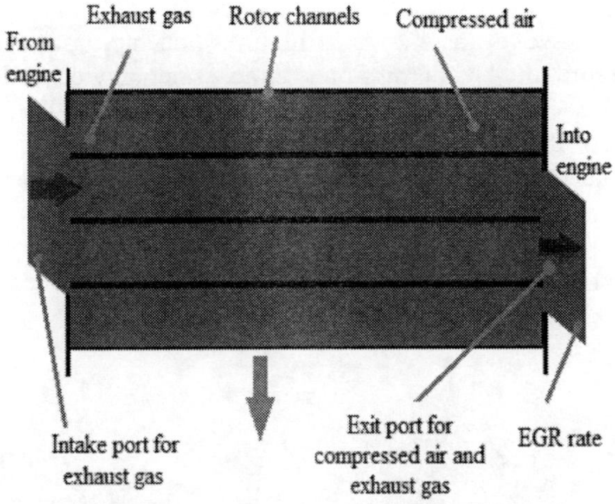

Figure 2. Scheme of the internal EGR trough PWC.

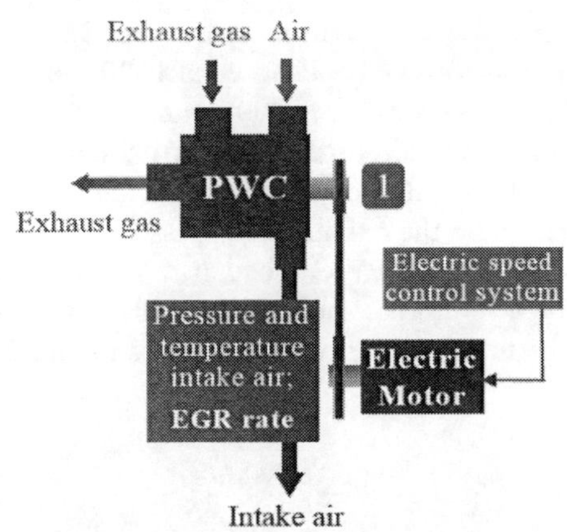

Figure 3. Supercharging system with PWC driven by the electric motor.

When choosing the PWC optimal speed could be taken into account of engine performances or more precisely of specific fuel consumption and emissions pollutants including EGR rate.

3. Interpretation of experimental results

Experimental researches were performed on a supercharging Diesel engine. During experimental research the engine was equipped with a turbocharger (original variant) and Comprex type PWC, model CX-93 (modified variant). The PWC was driven by a variable speed from an electric motor. As a result of research show the efficiency of the PWC in comparison with turbocharger (TC).

The experimental results presented in Figures 4, 5 and 6 were obtained for the engine supercharged with TC and PWC. The evolution of the parameters varies with the engine speed. Engine load and PWC speed have been kept at the same values throughout this research. Figure 4 shows that the fuel consumption is reduced if the engine is supercharged with PWC. At the same time, the air consumption increases due to the higher pressure of the supercharging. Figure 5 shows that the supercharging pressure generated by PWC is greater than the one achieved by the TC. The increase of the supercharging pressure takes place along with a significant increase of the intake air temperature.

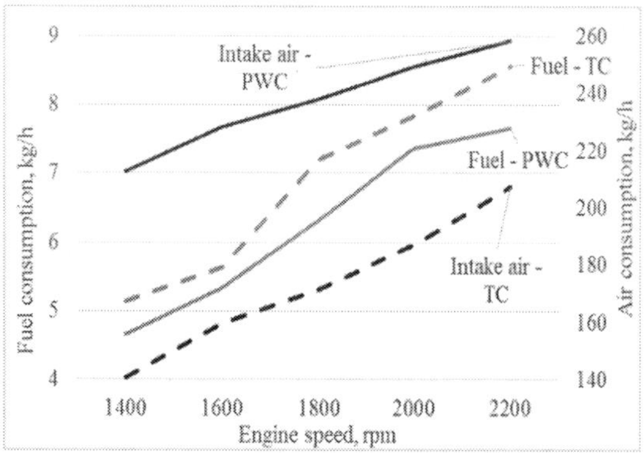

Figure 4. Comparing the fuel and intake air consumption obtained during supercharged with TC and PWC.

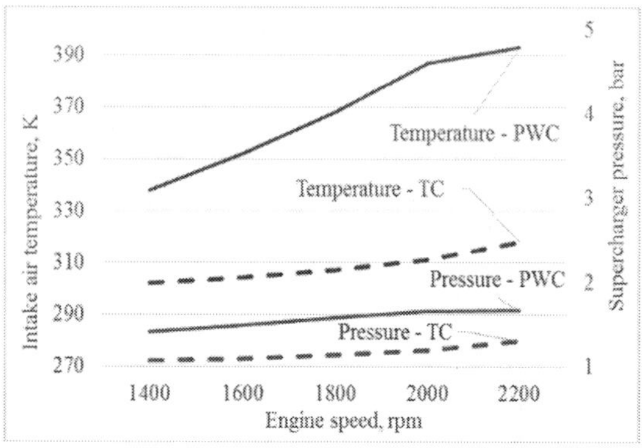

Figure 5. Comparing temperature and pressure intake air obtained during supercharger with TC and PWC.

The supercharging with PWC led to a reduced amount of diesel fuel injected per cycle, an increased amount of air entering the cylinders and a significant raise of the intake air temperature. These aspects normally lead to an increase in NOx emissions. However, as shown in Figure 6, an increase in NOx emissions is only observed at 1600 rpm. At 1800 rpm, the NOx emission is almost unchanged, while for the speeds of 1400, 2000 and 2200 rpm, the NOx emission is, against the expectations, lower. This situation could be explained by the internal EGR of the PWC.

According to [3], the time when the exhaust gas reaches the end of the channel depends on the PWC speed, on the exhaust gas pressure and on the pressure wave speed. Therefore, considering Figures 4-6, for a constant PWC speed, it can be stated that the rate of EGR depends on the last two factors mentioned above. Figure 7 shows the results of the experiments for a constant engine speed. PWC was driven with 3 different speeds (S1pwc, S2pwc and S3pwc).

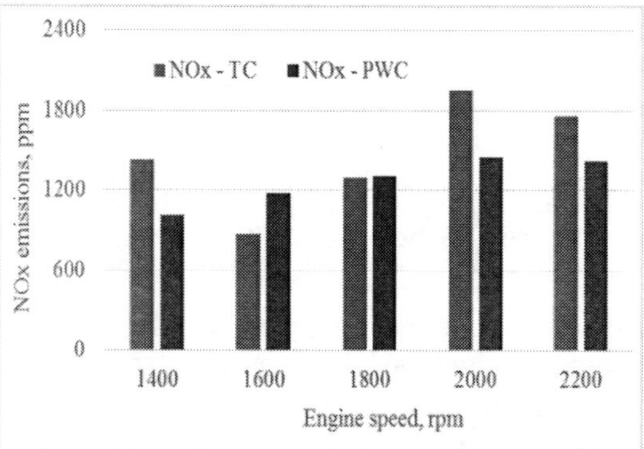

Figure 6. Comparing NOx emissions from exhaust gases obtained during supercharged with TC and PWC.

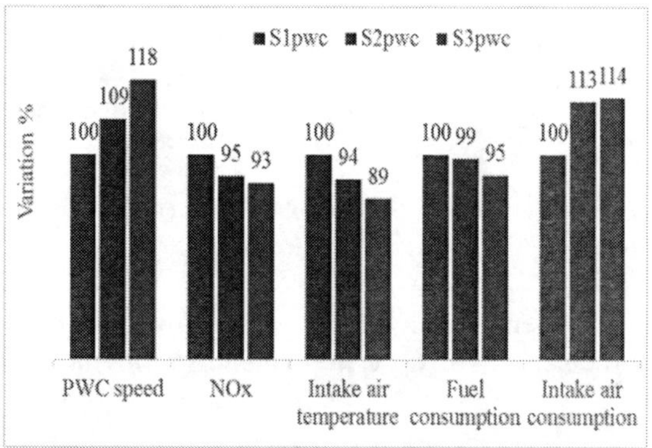

Figure 7. Comparing parameters obtained with the PWC driven at different speeds - at the same speed and engine load.

The reference values (100%) are the ones measured during driving PWC with the S1pwc speed. By consecutively increasing the PWC speed by 9% (S2pwc) and 18% (S3pwc), NOx emissions decreased by 5% and 7%, respectively. After analyzing the chart, it can be seen that NOx have been reduced due to a 6% (11%) decrease of the intake air temperature and to a 1% (5%) decrease of the fuel consumption.

4. Conclusions

When comparing the measured values for NOx emissions during PWC supercharging with those measured during turbocharging, the emission reduction is a substantial one.

The concept of supercharging with PWC (i.e. direct contact between exhaust gas and intake air) leads to a significant reduction of NOx, for a significant increase of both the pressure and temperature of the intake air.

The recirculation of exhaust gases through PWC represents another advantage of this compressor that justifies supercharging Diesel engines with it. In addition the driving with variable or sequential speed successfully enables the internal EGR process.

References

[1] Busch H, Adolph D, Genz T, Korfer T, Kolbeck A, Lamping M, Pischinger S 2007 Downsizing of HSDI Diesel engines – Potential and Challenges *6th Symp – Towards Clean Diesel Engines* 20-22 June 2007 Ischia

[2] Stan C and Cipolla G 2008 *Alternative propulsion systems for automobiles* vol 2 (Wankelstr: Expert Verlag)

[3] Leahu C I 2018 Improvement of exhaust gas pressure's utilization for compressing the intake air in diesel engine's cylinders *Int. J. of Automotive Technol.* **16** p 913

[4] Leahu C I and Tarulescu S 2016 The influence of thermal regime on gasoline direct injection engine performance and emissions *IOP Conf. Series: Materials Science and Engineering* **147**

[5] Leahu C I, Tarulescu S and Tarulescu R 2018 The exhaust gas temperature control through an adequate thermal management of the engine *IOP Conf. Series: Materials Science and Engineering* **444**

[6] Dragomir G, Beles H, Blaga V, Stanasel I, Negrea V D and Blaga F S 2008 Diesel Engines pollution and functional-constructive performances compromise optimization, using fuzzy sets *Annals of DAAAM for 2008 & Proc of the 19th Int. DAAAM Symp.* (Vienna: DAAAM Int Vienna) p 437

[7] Ivan F, Liţă D and Buşoi A 2014 *Methods and means of depolution of the engines for vehicles (Metode şi mijloace de depoluare a motoarelor pentru automobile, in Romanian)* (Bucharest: MATRIX ROM) p 222

[8] Basshuysen R and Schäfer F 2002 *Internal combustion engine handbook: basics, components, systems, and perspectives* (Warrendale: SAE International)

[9] Weber F 2001 *Mean Value* Modelling of Wave Pressure Supercharger Including Exhaust Gas Recirculation Effects *Swiss Federal Institute of Technology* Zürich

[10] Leahu C I, Dogariu D M and Chiru A 2015 Researches on the influence of pressure wave compressor on the intake air temperature at the supercharged engines *Bulletin of the Transilvania University of Brasov Series I: Engineering Sciences* **8** p 7

IOP Conf. Series: Materials Science and Engineering **568** (2019) 012029 doi:10.1088/1757-899X/568/1/012029

Methods of project management used in small and medium enterprises

G M Moraru[1] **and D Popa**[1]

[1] "Lucian Blaga" University of Sibiu, Engineering Faculty, 4, Emil Cioran Street, Sibiu, 550025, Romania

gina.moraru@ulbsibiu.ro

Abstract. The paper presents the most important methods of project management and general management used in small and medium enterprises in the central area of Romania, as well as the degree of their use. Therefore, we have developed a research based on questionnaire in small and medium enterprises from various fields, in order to find how their managers combines these methods and which ones are used more frequently. Unfortunately, our research has found poor or mediocre use of project management methods, as well as most of the general management methods. Training employees of enterprises in project management field, hiring experts in management, or developing post-graduate courses in this field by universities are the most affordable solutions at this time for Romanian small and medium enterprises.

1. Introduction

Project management is planning, coordination, management and control of a project during its life cycle in order to achieve its objectives, keeping the deadlines, cost and the quality set [1].

The specialty literature shows that the use of project management in an organization increases the efficiency of its resources and helps its sustainable development, with positive effects in the state economy [1-3]. In literature is also mentioned the functional organizational type, where the project is a part of functional enterprise organization [4]. In this context, the aim of the paper is to discover the most important methods of project management and general management used in small and medium enterprises (SMEs) in the central area of Romania, as well as the degree of their use.

2. Research in small and medium enterprises

We have conducted an extended research based on questionnaires in small and medium enterprises from different fields in the central area of Romania. Between the objectives of this research, was that to determine what methods of project management and general management use the SMEs' managers, and how often they use those methods.

2.1. Methodology of the research

We have distributed 130 questionnaires as follows: 49% in micro-enterprises with less than 10 employees, 29% in small enterprises with a number of 10 – 49 employees, and 22% in medium enterprises with a number of 50 – 249 employees. 43% of respondents were from firms with a turnover below 100,000 euro, 29% were from SMEs with a turnover between 100,000 and 500,000 euro, and 28% were from firms with a turnover over 500,000 euro. Other demographic characteristics of respondents are shown in figure 1. We mention that respondents who have studies other than technical or economic, are trained in the legal, psychological, medical or other field.

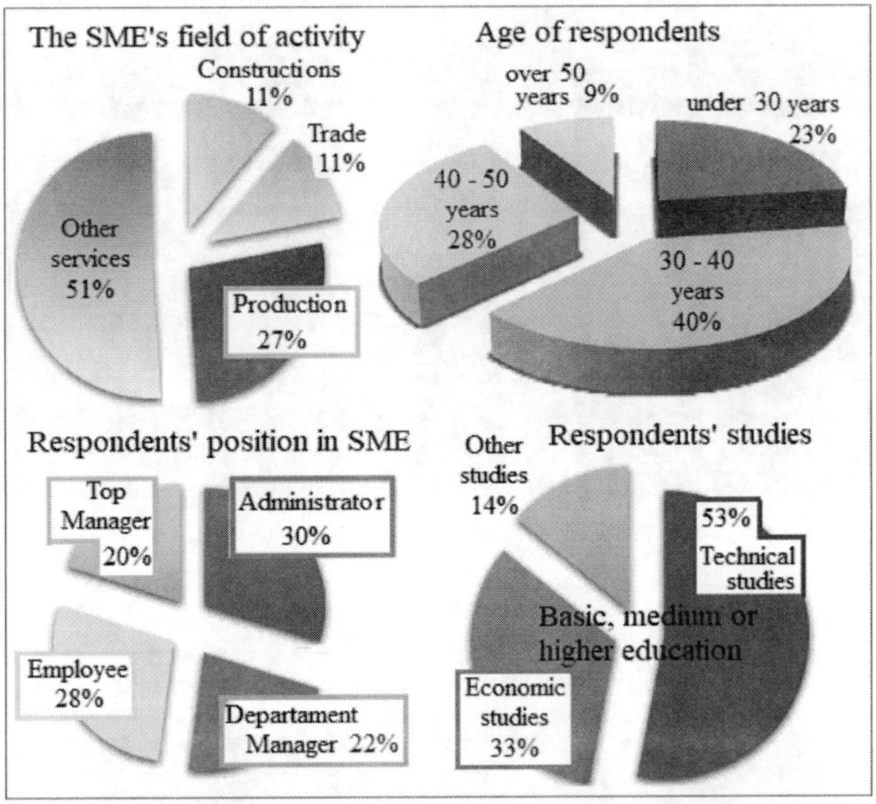

Figure 1. Several demographic characteristics of respondents

2.2. Main results

We have presented to respondents a list with some of the most common methods of project management and general management, known in the literature [3, 5-7]. They were asked to choose one option from the following list, without knowing the score given by us to the chosen variant: "I have never heard of this method" – 0 points, "I know this method, but I do not use it" – 1 point, "I use this method very rarely" – 2 points, "I use rarely this method" – 3 points, "I use sometimes this method" – 4 points, "I use often this method" – 5 points, and "I use this method very often" – 6 points.

The analyzed methods of general management were (figure 2): SWOT analysis (Strengths, Weaknesses, Opportunities, and Threats), PEST analysis (Political, Economic, Socio-cultural, Technological factors), Pareto chart, cause-effect method, forecast budget, brainstorming and critical road method. The forecast budget and SWOT analysis are in the top of respondents' preferences.

The list of project management method has included (figure 3): PERT (Program Evaluation Review Technique), Gantt technique, problem tree method, logical framework matrix, analysis of interest factors, and CPM (Critical Path Method). According to respondents' opinions, the analysis of interest factors and Gantt graphics are preferred from the list above.

Using the soft PASW18 and realizing various crosstabs between all methods in our database, we have found that respondents' preferences have focused on one of the following two groups of methods: the first group includes SWOT and PEST analysis, PERT and Gantt techniques; the second group includes the forecast budget, analysis of interest factors, problem tree method, cause - effect analysis and brainstorming. So, respondents who use methods from the first group do not use those from the second group, and vice versa.

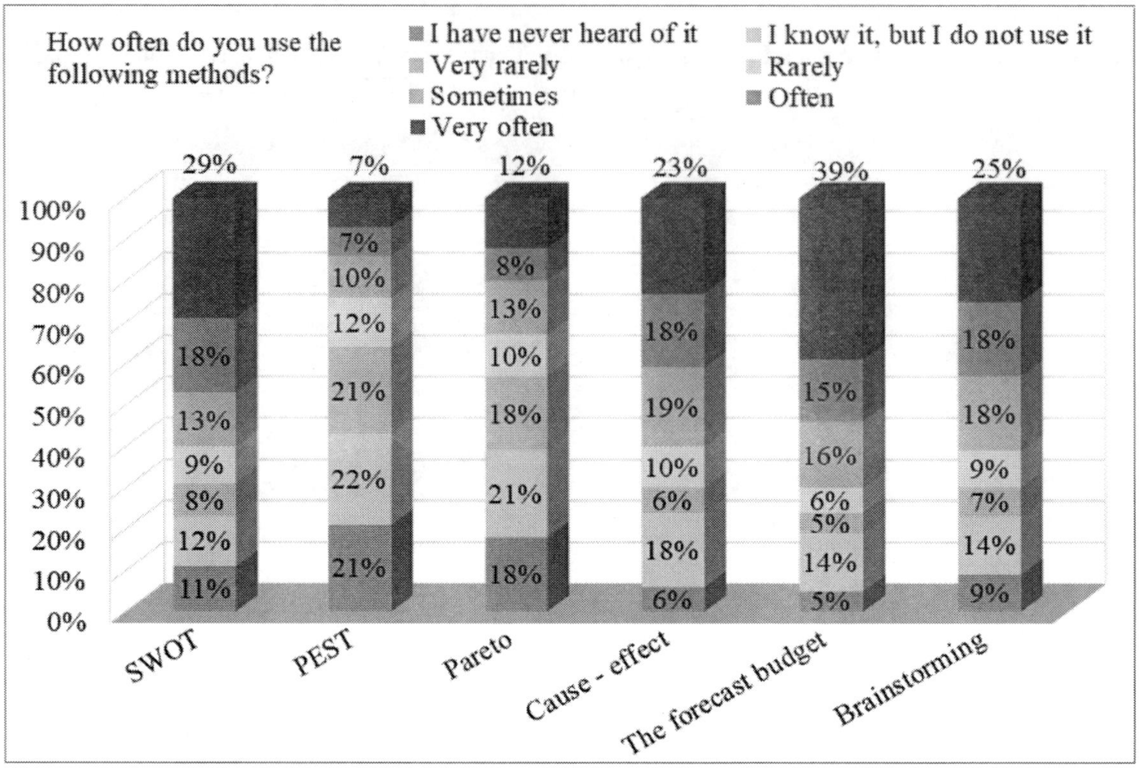

Figure 2. The use of general management methods in the studied SMEs

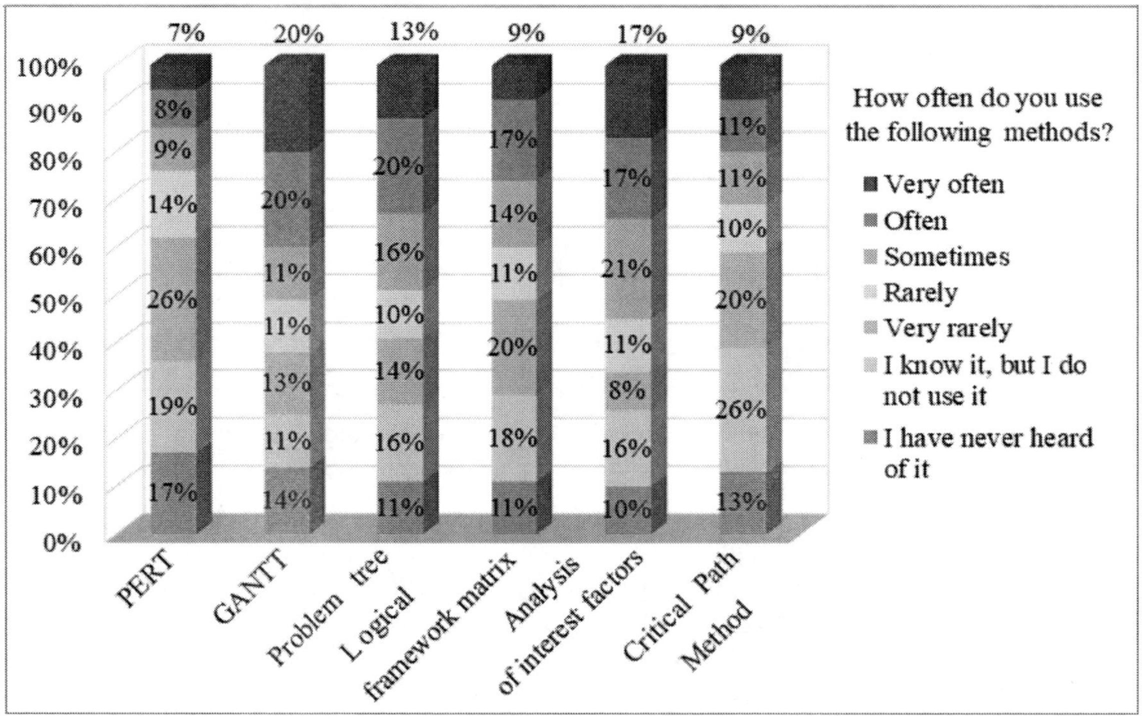

Figure 3. The use of project management methods in the studied SMEs

However, we observe on the whole a weaker use of project-specific management methods (figure 3) than general management methods (figure 2). This is also reflected in the statistical analysis of the answers (table 1). On one hand, excepting the forecast budget method, all other methods have a homogeneity coefficient above 0.5 (table 1). This fact shows that the means are not representative for the sample. On the other hand, taking into account that the median of our evaluation scale is 3, the average values of the use of the various management methods should raise questions about the sustainable development of SMEs in Romania. Only 7 of the 12 methods analyzed have obtained averages above the median value of the scale. In order to improve this situation, respondents have proposed to an open question the following solutions: training employees of enterprises in project management field, hiring experts in management, and developing post-graduate courses in this field by universities. The most used management methods in the studied SMEs are (table 1): the forecast budget, SWOT analysis, the brainstorming, and the cause - effect analysis.

Table 1. Statistical analysis: the use of various management methods in SMEs

Group of methods	The method/technique	Mean	Variance	Standard deviation	Homogeneity coefficient
General management methods	SWOT	3.71	4.506	2.123	0.572
	PEST	2.17	3.301	1.817	0.837
	Pareto	2.51	3.930	1.982	0.790
	Cause - effect	3.64	3.890	1.972	0.542
	The forecast budget	4.15	4.008	2.002	0.482
	Brainstorming	3.67	4.141	2.035	0.554
Project management methods	PERT	2.31	3.114	1.765	0.764
	GANTT	3.34	4.424	2.103	0.630
	Problem tree	3.16	3.874	1.968	0.623
	Logical framework matrix	2.86	3.520	1.876	0.656
	Analysis of interest factors	3.36	3.910	1.977	0.589
	Critical Path Method	2.49	3.510	1.873	0.752

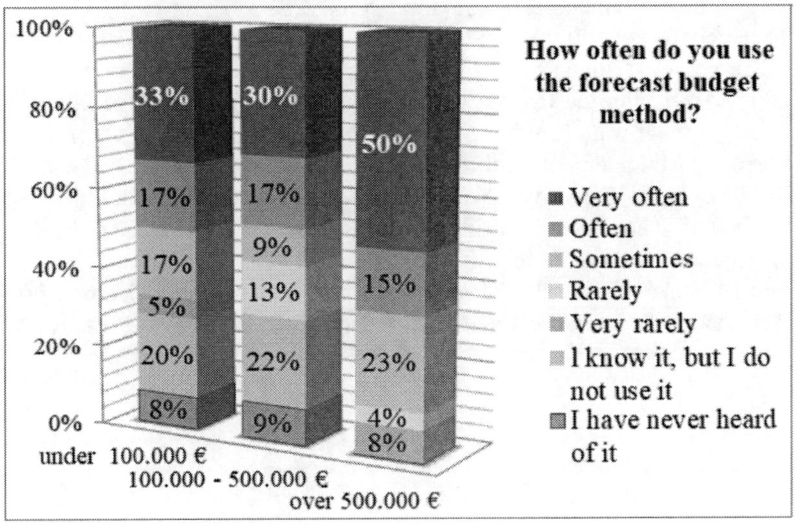

Figure 4. The use of the forecast budget based on SMEs' turnover

IOP Conf. Series: Materials Science and Engineering **568** (2019) 012029 doi:10.1088/1757-899X/568/1/012029

Analyzing the use of the forecast budget based on SMEs' turnover, we notice that it is used more frequently by those with a turnover of more than 500,000 euros (figure 4).

We have also found that project management methods (PERT, Gantt, problem tree and so on) are mainly used by firms with a turnover of more than 100,000 euros. Some of the SME managers with a turnover below 100,000 euros even have said they did not take the risk of applying these methods. An explanation of this may be a cause already reported by some specialists: even the use of project management within organizations increases, there is an attendant poor rate of success among these projects [8]. It is often found in practice a misalignment between the intended benefit of project management methodologies at the strategic level and the reported benefits by project managers at the project level [9]. However, Romanian SMEs should make more investments in project management, as experts say, "especially through strengthening people and organization competence" [10].

3. Conclusions

Most of the project management methods are known by the majority of the respondents in the analyzed SMEs but the degree of their use is still low. The most used management methods in the process of project planning and implementation in the analyzed SMEs are: forecast budget, SWOT analysis, brainstorming, and cause - effect analysis.

Training employees of enterprises in project management field, hiring experts in management, or developing post-graduate courses in this field by universities are the most affordable solutions at this time for Romanian small and medium enterprises.

References

[1] Popa D 2015 The knowledge of concept of project management in small and medium enterprises *Revista Economică* Vol 67 (Issue Supplement) ed D Popescu and L Mihăescu pp 351-362 http://economice.ulbsibiu.ro/revista.economica/archive/suplimente/67S32popa.pdf

[2] Dumitraşcu D Pascu R V 2005 *Project Management* (Sibiu: Romania/"Lucian Blaga" University of Sibiu Pub. House)

[3] Vîrvorea (married Popa) D 2012 Performing project management – essential solution for optimum management of SME's *PhD Thesis* unpublished (Sibiu: Romania/"Lucian Blaga" University of Sibiu)

[4] Dusan B 2008 Project Management Organization *Management Information Systems* Vol 3/2008 No 1 pp 3-9 UDC 005.8

[5] Lock D 2000 *Project Management* Romanian version (Bucharest: Romania/CODECS)

[6] Turner J R Stephen J S (eds.) 2004 *Gower Handbook of Project Management* Romanian version translator S Nistor (Bucharest: Romania/CODECS)

[7] Boddy D Buchanan D 2000 *Take the Lead! Interpersonal Skills for Project Managers* Romanian version (Bucharest: Romania/CODECS)

[8] Mark I 2005 Identifying the Contextual Elements of Project Management within Organizations and their Impact on Project Success *Project Management Journal* Vol 36 Issue 1 (USA/Project Management Institute) pp 37-50

[9] Wells H 2012 How Effective Are Project Management Methodologies? An Explorative Evaluation of Their Benefits in Practice *Project Management Journal* Vol 43 Issue 6 (USA/Project Management Institute) pp 43–58

[10] Radujković M Sjekavica M 2017 Project Management Success Factors *Procedia Engineering* 196 (Primosten: Croatia/Creative Construction Conference 2017/ Elsevier) pp 605-615

IOP Conf. Series: Materials Science and Engineering **568** (2019) 012030 doi:10.1088/1757-899X/568/1/012030

Performance evaluation of the support bush manufacturing line by modeling and simulation

F S Blaga[1], I Stănăşel[1], A Pop[1], V Hule[1] and T Buidos[1]
[1]University of Oradea, Faculty of Management and Technological Engineering, str.Universităţii nr.1, Oradea 410087, Romania

alinpop23@yahoo.com

Abstract. Improving the performance of manufacturing systems is the objective of management activities. The paper presents how this objective can be achieved by modeling and simulation. Base on the manufacturing process of the part support bush it is modeled the production line. Modeling is done using the TECNOMATIX PLANT SIMULATION program. Using the first model, by simulation, it was found that the resulting production after eight hours was relatively reduced: 28 pieces. Under these conditions, an additional welding post was introduced in the second model. It results in double the production and also in narrow places. The third model is designed in such a way that bottleneck is eliminated, and there is also an increase in production.

1. Introduction

Modeling and simulation are tools commonly used in designing, realization and improving the performance of manufacturing systems. Thus, in [1] is a synthesis of how the modeling and simulation of flexible manufacturing systems were presented in the specialty literature. The paper focused on the literature survey of the use of flexible manufacturing system design and operation problems on the basis of simulation tools and their methodology which have been widely used for manufacturing system design and analysis.

The paper [2] proposes a new approach to model and simulate manufacturing systems in order to rapidly respond to the changes of the manufacturing environment, including the enhancement of the manufacturing processing, reconfiguration of the manufacturing resources and the dynamic control. In [3] the authors discuss how simulation is used to design new manufacturing systems and to improve the performance of existing ones.

In the paper [4] is presented the practical application of the modeling and simulation in managing and planning of machines' setup. There are described principles and stages of the production system improvement with the theory of constraints and illustrated by the example of the analysis of the whole production line, which aim was the bottleneck determining, and the analysis of operations, which are realized in the bottleneck. The problem of the bottleneck is also discussed in [5].

This paper shows how using modelling and simulation procedures can put in evidence the bottleneck in the manufacturing flow. Using modeling and simulation (TECNOMATIX PLANT SIMULATION-software) it is also possible to identify the solutions to allow equilibration of the lines of the manufacturing process.

IOP Conf. Series: Materials Science and Engineering **568** (2019) 012030 doi:10.1088/1757-899X/568/1/012030

2. Presentation of support bush

The piece whose manufacturing line will be modeled is the *support bush*. The figure 1 shows the part drawing. In the figure is presented the support bush in the conveyor assembly.

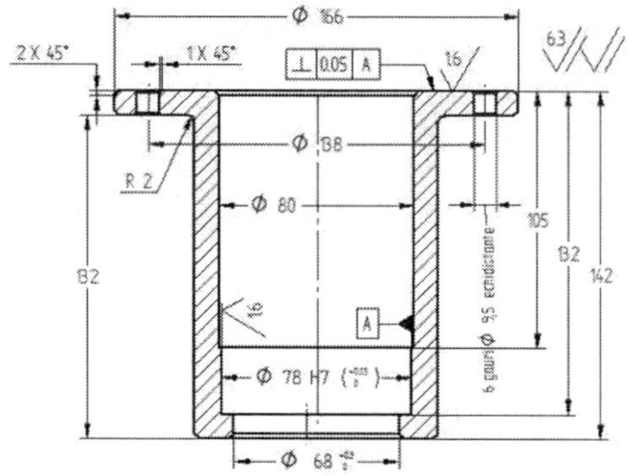

Figure 1. *Support bush* part drawing.

Figure 2. *Support bush* part in conveyor assembly.

Table 1 presents the process sheet of the support bush with the corresponding operations and times.

Table 1. Process sheet of the *support bush*

Nr. op.	Operations order	Machine	Number of parts made simultaneously	Time (min)	Symbol model
10	Welding		1	15	Welding_1(2)
20	Facing	Parallel lathe	1	4	Facing
30	Rough turning OD_	Parallel lathe	1	1	R_OD_166
40	Rough turning ID_	Parallel lathe	1	6	R_ID_67
50	Rough turning ID_	Parallel lathe	1	5	R_ID_76
60	Rough turning ID_	Parallel lathe	1	4	R_ID_78
70	Finish turning ID_ Ø78.4	Parallel lathe	1	7	F_ID_78
80	Finish turning ID_ Ø79.8	Parallel lathe	1	2	F_ID_79
90	Chamfer	Parallel lathe	1	1	Chamfer
100	Facing 142	Parallel lathe	1	2	F_142
110	Drilling Ø9.5x6	Drilling machine	1	4	Dr_9x6
120	Grinding ID_ Ø78H7	Round grinding machine	4	8	Gr_ID_78
130	Grinding ID_ Ø80H7	Round grinding machine	1	3	Gr_ID_80

3. Modeling and simulation of manufacturing line for the support bush using Plant Simulation

Plant Simulation, Siemens Industry Software's simulation software solution treats events in a discreet, step by step manner. The application is object-oriented, allowing greater flexibility and ease of modifying and maintaining complex models. The model is built from a collection of objects that interact with each other. Each object can receive, process, and send data. Also, each object can be seen as an independent machine. Simple objects can be used as blocks to build more complex objects.

Plant Simulation uses an object-oriented interface similar to other Windows applications, which means that the object or objects you want to work must be selected before the relevant options become available.

3.1. Version I

In a first model version, every single operation in the manufacturing process was associated with a SingleProc block. Each operation is done at one workstation. Figure 3 shows how the manufacturing line model is made. For each operation, it was set the duration and other workstation specific features.

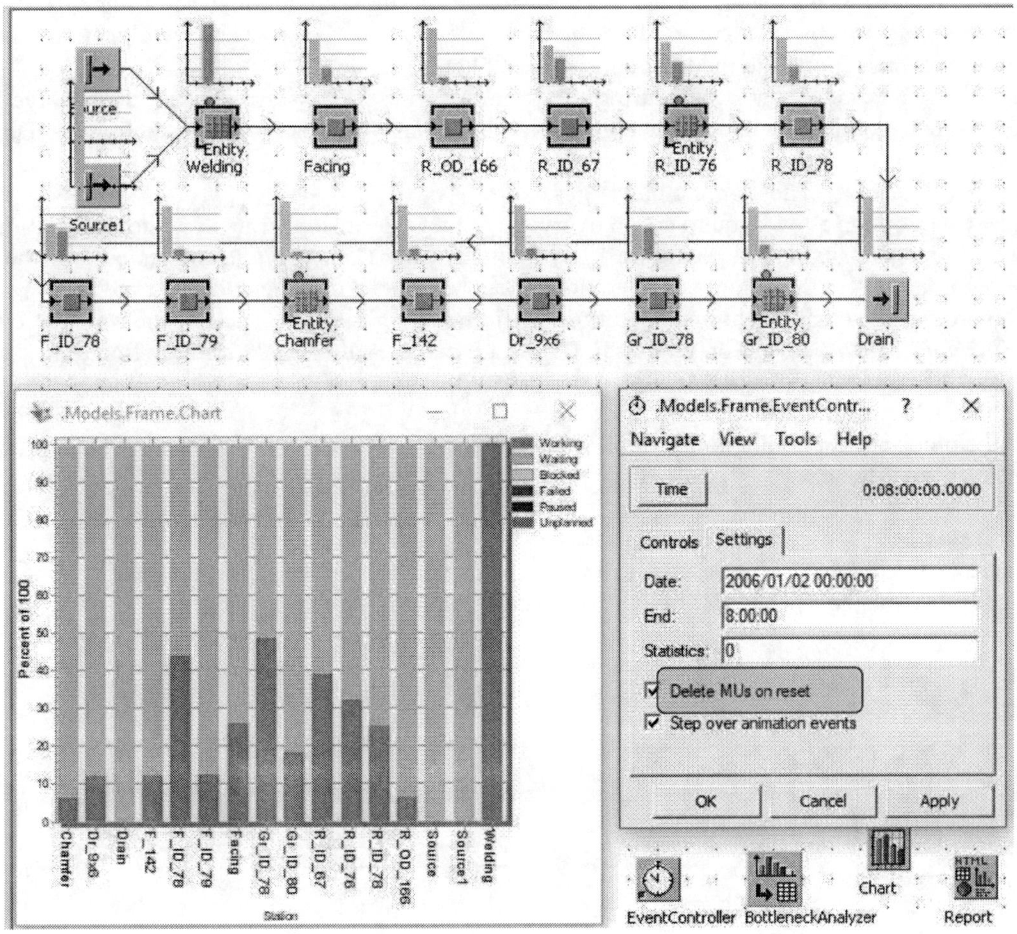

Figure 3. The model of the manufacturing line for the part support bush in TECNOMATIX PLANT SIMULATION after simulation. Version 1

System inputs are modeled through the Source block (figure 4). The interval between two successive entries is expected to be 15 minutes.

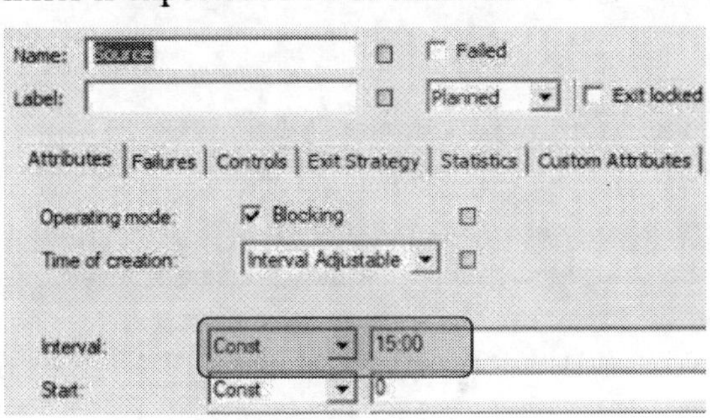

Figure 4. Source block - input modeling

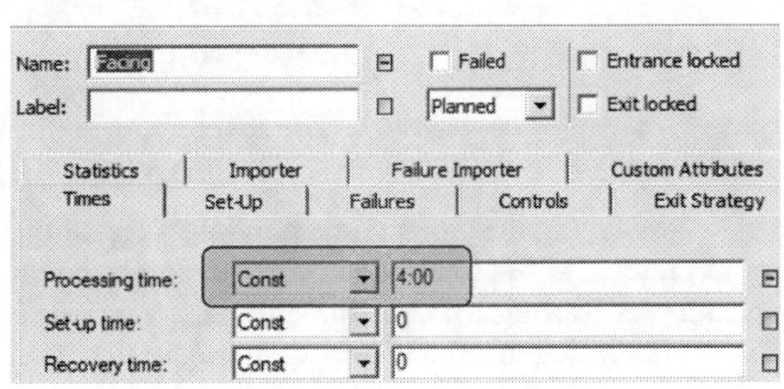

Figure 5. Block for *Facing operation*

Regarding the actual processing operations, their modeling is exemplified by the way in which the settings for the *Facing* operation are made. The figure 5 shows the window corresponding to this

IOP Conf. Series: Materials Science and Engineering **568** (2019) 012030 doi:10.1088/1757-899X/568/1/012030

operation (workstation): Name – *Facing*; Processing time- 4 minute. The processing time is considered constant.

The figure 3 shows the manufacturing system after the simulation (the version I). Using the *BottleneckAnalyzer* option, it can be described the states that characterize each workstation (operation), and their weights reported to the time considered (8 hours).

Thus, the blockages that occur during the manufacturing process are marked with the yellow bar. Also, normal working hours (green color), waiting times (gray color) are highlighted. During the reference time, workstations may be broken (red), may be in pause (blue color), or not in the work schedule (blue color). There are no blockages.

The Chart option allows you to get an overview of the states that characterize the workstations of the system during the simulation. In figure 3, by activating the *Chart option*, the statuses of the stations were recorded during the 8 hours of operation. It can be noticed that much of the time they are waiting.

Regarding the actual machining operations, their modeling is exemplified by the way the settings for the *Finishing Turning* operation Ø78.4- F_ID_78 are made. After simulation it is found that out of the 8 hours - the reference time - it functioned normally 43.75% and it was 56.25% waiting. Also 30 bushes have been processed.

If the block corresponding to the drain is analyzed, it is noted that 28 bushes have been processed in the system for 8 hours.

3.2. Version II

To increase the number of bushes made, it is supplemented the production line with a welding post. The model of this version (Version 2) of the manufacturing line is shown in figure 6.

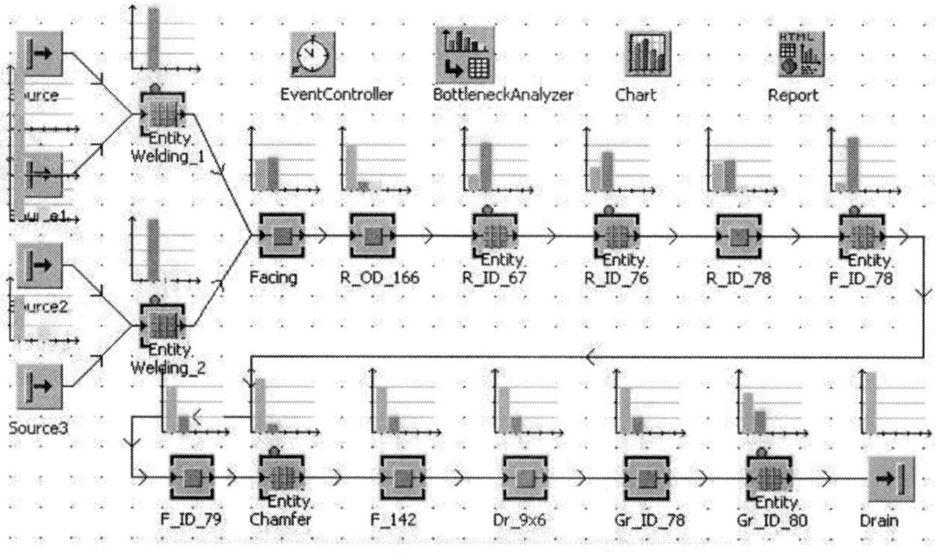

Figure 6. The manufacturing line model for the part support bush in TECNOMATIX PLANT SIMULATION after simulation. Version 2

Analyzing the block corresponding to this *Finish turning operation Ø78.4*- F_ID_78 after simulation (Version 2), it is found that out of the eight hours - the reference time - it functioned normally 86.67% and was waiting 13.33%. In addition, 59 bushes were processed.

Analyzing the block corresponding to the drainage system (Drain), it is found that 57 units have been processed in the system during 8 hours (version 2).

The Chart option allows you to get an overview of the states that characterize the workstations of the system during the simulation. It can be noticed that the workstations *Rough turning* OD_ Ø166 and *Rough turning* ID_ Ø78.5 are blocked (yellow color) for part of the reference time.

3.3. Version 3

To remove bottleneck, more workstations will be allocated for operations with longer processing times than previous operations. Thus, the operations: *Rough turning* ID_ Ø67.3 (R_ID_67) and *Finish turning* ID_ Ø78.4 (F_ID_78) will be executed at two positions. *ParallelProc blocks* are used to model this situation.

Analyzing the *ParallelProc* block corresponding to this *Finish turning operation Ø78.4*- F_ID_78 the block modeling the *Finish turning* operation ID_ Ø78.4 (F_ID_78), it is found that the two posts work 74.58% of the reference time, and wait for 25.42% of the 8 hours.

The figure 7 shows the result of the simulation of the manufacturing system in version 3. It is noticed that there are no longer any blockages in the process operations. Analyzing the block corresponding to the drainage system (*Drain*), it is noticed that 58 units were processed during the 8 hours (figure 7).

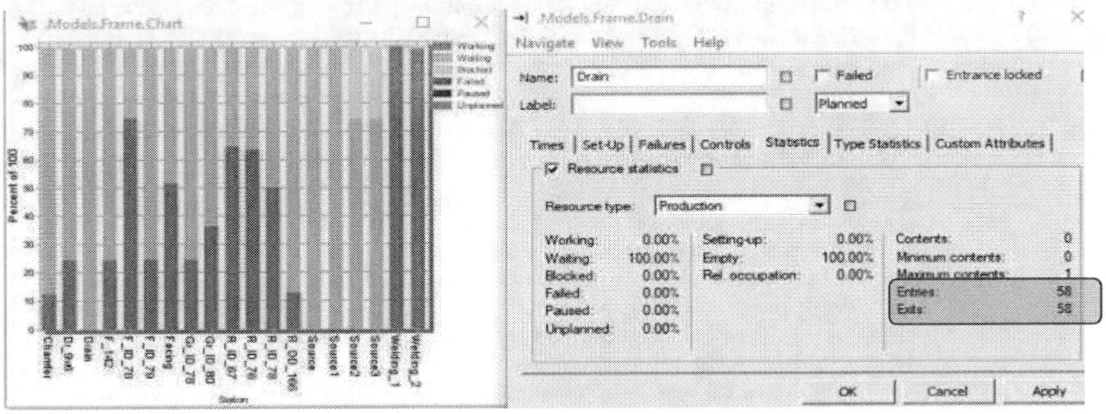

Figure 7. The result of the simulation of the manufacturing system. Version 3.

4. Conclusion

The dynamics of the manufacturing systems development is highly accentuated. The decision makers responsible for the management of these systems have to take decisions in a short time about how these systems are going to evolve. The evolution of manufacturing systems must be guided so as to meet the requirements of the market. Under these conditions, modeling and simulation can be useful.

The paper presented is a concrete example of solving the problems that appear in the manufacturing of a component part (*support bush*) from a complex product: a conveyor transfer system. Through an iterative process of model improvement, these being validated by simulation, the best structure of the fabrication system was made to make the part support bush.

Acknowledgments

The paper has been realized within the research contract 19/2015 - MONORAIL CONVEYOR SYSTEM FOR AUTOMATED PRODUCTION LINES. Program MANUNET-ERA-NET.

5. References

[1] Saren SK and Vesselenyi T 2016, ed. Grebenisan *ANNALS OF THE UNIVERSITY OF ORADEA, Fascicle of Management and Technological Engineering* ISSUE #1 (Oradea) p 113
[2] Xiaofeng H and Ruxiao A 2011 *Proceedings of the World Congress on Engineering,* eds. Ao SI, Gelman L, Hukins DWL, Hunter A and Korsunsky AM (London, UK), p 601
[3] Law AM and McComas MG 1998, *Winter Simulation Conference. Proceedings*, vol.1 (Washington DC, USA: IEEE), p 49
[4] Gregor M, Matuszek J and Plinta D 2013 *Adv. in M. Sc. and Tech.* **37** (1), p 7
[5] Blaga F, Stanăşel I, Hule V and Pop A, 2017 *MATEC Web of Conferences 112, 06012, IManE&E 2017*, eds Slătineanu L, Nagit G, Dodun O, (Iasi, Romania: EDP Sciences), pp 7

IOP Conf. Series: Materials Science and Engineering **568** (2019) 012031 doi:10.1088/1757-899X/568/1/012031

Concepts and tools used to manage industrial manufacturing processes optimization

F. V. Banciu[1,2], E. Pămîntaş[1]

[1] Materials and Manufacturing Engineering Department, Mechanical Faculty, Politehnica University Timişoara, Bv. Mihai Viteazu No. 1, 300222, Timişoara, Romania

[2] felicia.banciu@upt.ro

Abstract. This work is the result of a study on the production process of a company that manufactures safety features for vehicles. Repeated problems encountered in the production process significantly affect the company's performance and reputation. The aim is to provide an overview of the problems of production, allowing identifying the causes of faults to remedy them. Once a fault has occurred and the product was delivered to the client it will be very difficult to recover both reputation and costs, and to replace the damaged stock. The paper contains arguments for the need to optimize the production processes as a whole, describing the concepts and tools used to analyse and solve a concrete case of improving the production system for a complex and sensitive product. The information obtained from the detailed analysis and improvements in the quality level, the decrease of the production costs and the keeping the delivery rate to the client are presented.

1. Arguments for process optimization approach

Due to a very high risk factor brought by an Airbag Curtain Lateral Project (ACL or P32S), it was decided to start an extensive workshop to improve the production process. A brief analysis reveals that an improvement area is a huge one because the production process was very difficult and brought the company's weakest results in 2014:

- Poor quality level - of the total of 20,000,000 airbags sold, ACL contributed only with 600,000 units, generating 19 of the total of 27 customer complaints (figure 1);

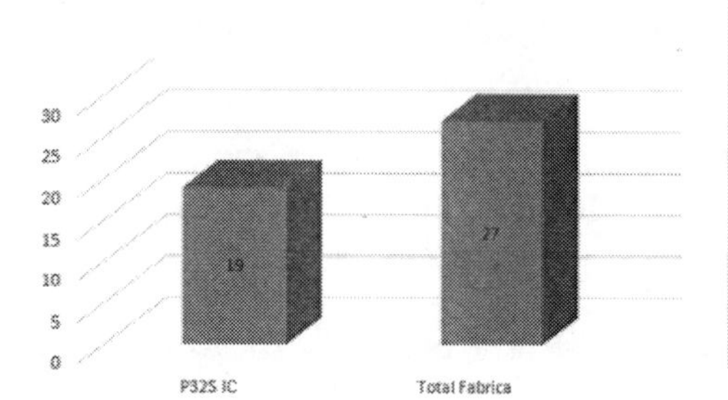

Figure 1. Number of customer complaints.

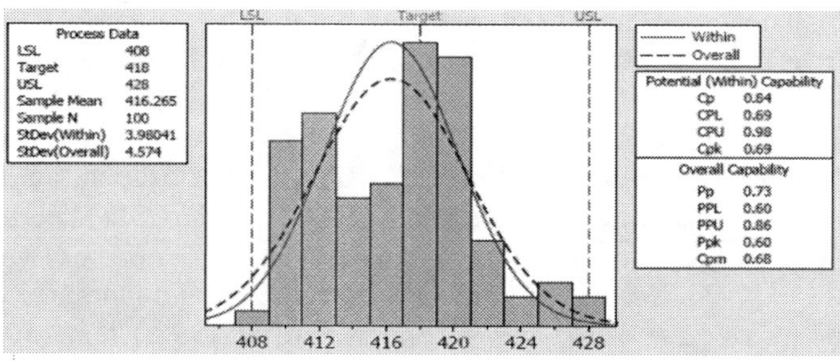

Figure 2. Gauss's curve on the Cp Process

IOP Conf. Series: Materials Science and Engineering **568** (2019) 012031 doi:10.1088/1757-899X/568/1/012031

- The ACL product defect rate was 3% and the factory target was 1.2%;
- The OEE for the ACL product was 65% and the factory target was 85%;
- The manufacturing process was far from 6sigma, PPK and CPK for most requirements was below the target of 1.33 / 1.67 of a production series, - (figure 2);

2. Tools used to analyze the production process

Lean Manufacturing Principles (5 in number) according to which all activities that do not add value to the client's eyes (but only generate costs) are identified as waste [1, 2, 3].

- One Piece Flow Concept (OPF) - which assures the customer what he wants, in the amount he / she wants, when he / she wishes. OPF is the production and movement of a single piece in a certain amount of time through a series of processes, as continuously as possible, in order to adjust itself to the best of tact time, at each stage achieving only what is necessary to proceed to the next operation. Implementation of the OPF requires: adequate number of operators with appropriate mentality, elimination of Muda, standardized work, production line balancing and in return a number of benefits in several areas: quality, inventory, productivity and safety [4];
- Value Stream Mapping (VSM) - both the Current State Map (graphical representation of the value flow in the existing situation) and the "Map of the Future State" (the improved value flow - by applying all Lean tools) are constructed by mapping the value flow based on process data (inventory, cycle time, change over time, up-time, default rate, etc.) [5, 6];
- "Zero defects" concept - quality defects have associated significant costs: materials, resources, time and reputation [7]. Debugging programs can be very expensive and can consume a lot of resources and time. Does the solution consist of eliminating defects without taking into account the necessary cost, or accepting a small percentage of defects? So far, Philip Crosby's solution has been superior, "everything has to be done first time and every time" [8].

3. Example of good practice in using concepts and tools to optimize the product manufacturing

The management study with optimization of ACL product manufacturing has been aimed at significantly reducing the presented difficulties and satisfying an increased demand from the customer.

3.1. Intensive Workshop Lean Manufacturing: „One Piece Flow"

First, they defined steps / stages that facilitate the implementation of OPF, planning necessary time for each step/stage, during 10 working days, as shown in Table1.

Table 1. Planning and scheduling of the intensive OPF workshop steps.

Step no.	Day	Step description	Time [min]
1	1st	Choosing the Team	10
2	1st	Choosing the Line (OPFWS Plan)	10
3	1st	Management challenge (Laminar flow, OPF	10
4	1st	Session Training (all team)	120
5	1st	GEMBA walk: data collection + Current Layout (all team)	210
6	1st	Scope /objective definition (all team)	60
7	2nd	Process Analyse and data collection form	480
8	3rd	Current layout drawing and attaching to magnet board	30
9	3rd	SWC coloured sheet for manual or automatized time	60
10	3rd	Compute CT/operator	120
11	3rd	Time introduction in SWCS form	120
12	3rd	Sequencing: update SWCS layout (Synchronicity)	120
13	4th	Takt time for 3 months	60
14	4th	Machine Improvements: updates SWCS layout	180
15	4th	Number of Operators = Manual time/TT*115% (the highest)	30

16	4th	Eliminating Muda	60
17	4th	Line balancing on board (it is used the smallest takt time)	150
18	5th	Redesign for line layout (reducing the space inside the line)	120
19	5th	Line adjusting/new layout	120
20	5th	SWI : SWS or STD work element	240
21	6th	Moving the line in choose configuration	300
22	6th	Re-measuring line CT	180
23	7th	Line documentation Update with SWS	240
24	7th	All the procedures from above will be repeated for all takt time configurations	240
25	8th	Operators Training	240
26	8th	New process release	30
27	9th	Line start	480
28	10th	Closing ceremony	30

3.2. Mapping „Value Stream Design".

Due to a very high requirement from the customer and the very low logistics contract, there was a great risk of entering into a capacity shortage and no longer able to honour orders to the client. Only 3 production lines were available for the ACL and with the increase of the requirements, it was necessary to validate / qualify the product on a 4th production line also used for the same customer. Deliveries to customers were at a disappointing level, a 85% for the logistics contract reported under the logistics contract at a minimum of 95% and Lead time was very high, generating both large stocks and delivery delays. To solve these new problems, a Success Story has been used following a Best Practice.

The Improvement Workshop started with the formation of a team that, after attending a training session on flow, ergonomics and regular delivery themes, went to the macro-planning of activities to identify and analyze the causes and finalize the workshop agenda and the current Value Stream Mapping (see figure 3 - before).

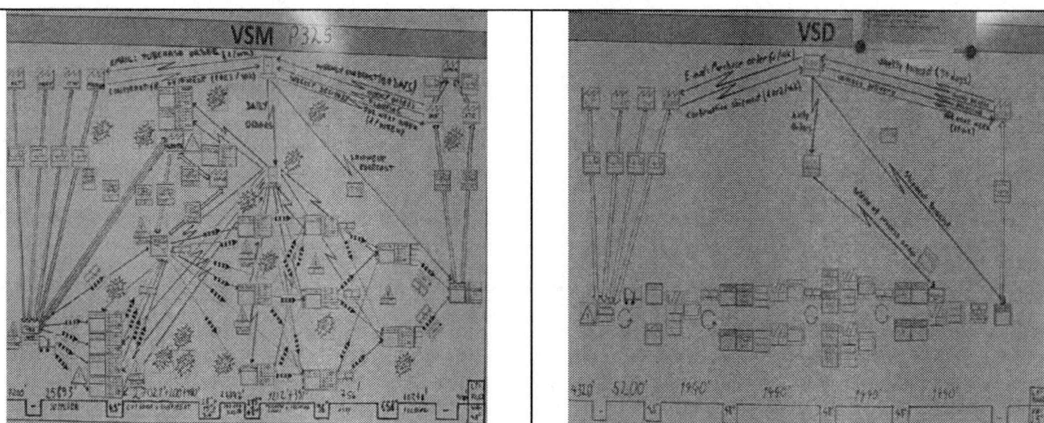

(a) (b)

Figure 3. Value Stream Design - (a)before (left) and (b) after workshop (right).

In order to implement the "Value papers" concept, it was necessary to set goals:

1. Production without variations: Time Takt, Logistic contract; Sharing and levelling customer orders.

2. Getting the shortest delivery time: Continuous flow, Laminar flow, One piece per stream, No isolated operators, No unnecessary objects at the work point, Separation of the operator from the work point and Creating standardized work instructions containing the work items ;

3. "Pull" flow (Kanban system);

IOP Conf. Series: Materials Science and Engineering **568** (2019) 012031 doi:10.1088/1757-899X/568/1/012031

4. Balancing production lines;
5. Standardized work.

Removing "non-value add" elements and optimizing the logistics operation process (line transfer, assembly line transformation, line transfer, merging operations, material changes, etc.) resulted in a Value Stream Design that reduced Lead time from 72.6 days to 10.6 days.

In addition to substantially reducing the lead time for the ACL product, following the use of this concept, three more important insights have been gained:

1. A new "Lessons Learned": Simplify processes by meeting the production plant's objectives regarding material flow and production organization.
2. Open new ways: a workshop on identifying components and material flow and then physically relocating equipment.
3. Clear and SMART objectives to be pursued in the future:

- Implementing the best actions on all production lines: "Yokoten Ergonomics and Performance" - to save about 50square meters, 3 operators and 11% Labor Minutes Per Unit;
- SMED (Single Minute Die Exchange [9]) on assembly lines - for shorter project change time from 43 to 10 min;
- SMED on Laser cutting lines - shortening project change time from 21 to 10 min;
- Continuous Improvement.

3.3. „Zero Defects" Workshop

It starts from the defect definition, on the production line is created the standard notice board as a visual management factor, and then the 8D - Problem Solving method is distributed successively in 5 steps:

1.*Prepare t*he objective of this preparatory work is to ensure that the workshop bases are set, that all documentation is ready and that the product specifications coincide with the reality of the prod. line;

2.Macro Workshop Planning - This step comprises the first two disciplines *(D1:* Forming a team for 8D Analysis [11] – it must be as a result of a thought process to get the best possible result under the given conditions. It has to go through all 8D analyse stages should be done following Category A complaints (highest severity), *D2:* Apply the Go & See method to find evidence);

3. *Troubleshoot current issues* - *t*he objectives of this stage are to eliminate all the problems one by one in turn and to ask for help from the management to solve the systematic problems or to support the engineering department to solve the technical problems. In this stage, the whole team is focused on problem solving *(D3:* Implementation of problem-limiting actions);

4. *Anticipating Potential Problems-t*he objectives are to identify potential problems using the experience gained in conducting a PFMEA Go & See [10], (D3: Implementation of problem-limiting actions as a first customer protection response, D4: Identify the main cause);

5. Yokoten. - *t*he objective of this step is to avoid recurrence of a problem in other areas as well *5. (D5.* Choosing and implementing corective actions. *D6. Evaluation of results, D7. Preventing the problem,* D8. Congratulations to the team).

4. Conclusions

In addition to the benefits already mentioned in sections 3.1 and 3.2, among the many improvements brought about by the application of the concepts and tools discussed, we only mention a few:

- Improving tool management in the production line, full check of inline tools and dedication of a special backup toolbox, implementing weekly verification and saving as an excel file in the database, implementation of stainless steel counter top to prevent friction and improve flow of tools, improvement of component detection system, Process Flow Modification and line layout integrating the One Piece Flow concept;
- A series of systems have been implemented which, if the deflector is positioned incorrectly sends an electric impulse and ignites a bulb, respectively, on a production line a Yokoten system and on the other two lines delivering directly to the customer instead of lighting a the bulb is set to stop the line.

As an overview of "how it was before" and "how is it after" the application of leadership concepts with optimization of the manufacturing processes of the ACL product in the considered company, we present two graphical suggestions regarding the evolution of the number of defects and their category (figure 4) and the number of customer complaints for this product (figure 5).

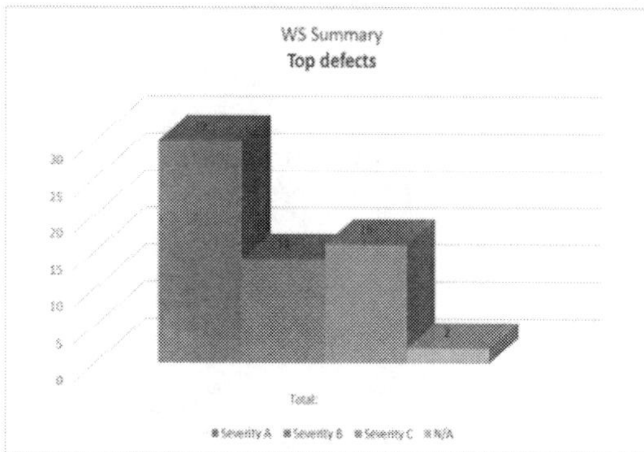

Figure 4. Defects classification.

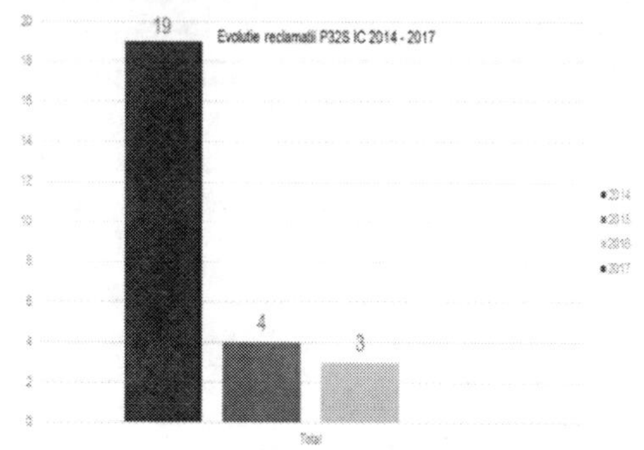

Figure 5. Evolution of customer complaints after improvements to the production line.

Finally, we note that the most important gain is that it was possible to prove that the "theory" of the concepts and tools of organizing and optimizing production is of great help in the case of rigorous application in practice and that already without them a firm no longer can survive nowadays.

References

[1] Liker J K and Hoseus M 2008, *Toyota Culture: The Heart and Soul of The Toyota Way*, (McGraw-Hill, New York, ISBN 978-0-07-149217-1)

[2] Pascal D 2015 *Lean Production Simplified, Third Edition* (Routledge Publisher, ISBN-13: 978-1498708876)

[3] Levinson W A and Rerick R A 2002 *Lean Enterprise: A Synergistic Approach to Minimizing Waste* (Milwaukee, WI: Quality Press)

[4] Kenichi Sekine 1992 *One-Piece Flow: Cell Design for Transforming the Production Process* (Productivity Press, ISBN 978-0915299331)

[5] Rother M and Shook J 1999 *Learning to See: value-stream mapping to create value and eliminate muda* (Brookline, Massachusetts: Lean Enterprise Institute. ISBN 0-9667843-0-8)

[6] http://leanmanufacturingtools.org/551/creating-a-value-stream-map/

[7] Harwood W B 1993 *Zero Defects Was Invented Here. Raise Heaven and Earth: The Story of Martin Marietta People and Their Pioneering Achievements* (New York: Simon & Schuster, ISBN 9780671749989)

[8] Crosby P 1970 *Quality is Free* (New York: McGraw-Hill, ISBN0-07-014512-1)

[9] Moreira A C and Campos Silva Pais 2011 G *Single Minute Exchange of Die:-A Case Study Implementation* (Journal of Technology Management and Innoviation) pp. 29-46

[10] Kmenta S and Ishii K 2004 *Scenario-Based Failure Modes and Effects Analysis Using Expected Cost J. of Mechanical Design*. **126** (6): 1027, doi:10.1115/1.1799614

[11] Chris S and VisserP 2017 *8D problem solving explained: Turning operational failures into knowledge to drive your strategic and competitive advantages* (in2quality publishing, Eindhoven, Netherlands, Printed in USA, ISBN-13: 978-1543000931)

Acknowledgements: This work has been developed with the help of the Autoliv Company and the documentation provided by it. Data processing, implementation and technical advice to solve production problems is the contribution of the authors.

Use of dual numbers in kinematical analysis of spatial mechanisms. Part II: applying the method for the generalised Cardan mechanism

S Alaci[1], R D Pentiuc[1], I Doroftei[2] and F C Ciornei[1]
[1]"Stefan cel Mare" University, Suceava, Romania
[2]"Gheorghe Asachi" Technical University, Iasi, Romania

stelian.alaci@usm.ro

Abstract. The equations which permit obtaining the displacements from kinematical pairs were deduced in the first part of the paper. An actual case is presented in the second part of the work, the method being applied for the generalized Cardan mechanism. In a first stage, the parameters required for the completion of the dual matrices from the dual matrix equation of closing the kinematical chain are identified. Next, the matrix equation which allow for finding the rotations and the translations, respectively, from the pairs, are identified. A special attention is assumed on establishing the rotations from the pairs. The dual matrix equation is split into two real matrix equations, corresponding to rotations and to the translational motions, respectively. The equation corresponding to the translations presents a complex form, but it is simplified by using some characteristics of the matrices. Additionally, this matrix equation is written under a form which allows for algorithmization. Finally, there are obtained the equations which describe the position of the mechanism - the same as in the classical approach.

1. Introduction

The second part of the paper presents an application of the theoretical results deduced in the first part of the work. In figure 1 is presented the image of an actual Cardan joint. The Cardan joint is a spherical mechanism used for transmitting the rotational motion between concurrent axes. It consist in a driving element 1, an intermediate element (Cardanic cross) 2, a driven element 3 and an immobile element 4, the ground. All the joints of the mechanism are rotational pairs. The Hartenberg-Denavit [1] method requires denoting the axes of the four pairs by $z_k, k=1..4$ and then defining the common normals of the couples of axes from the same element. To be mentioned that for spherical mechanisms the length of the common normal for any two axes is null and thus it will be defined as the straight line normal to the plane formed by the two axes, passing through the intersection point. Additionally, in the matrix closure equation, all the vectors corresponding to translations must be identical zero. For the Cardan mechanism from figure 1, the matrix closure equation is written as:

$$z(\theta_1)x(\alpha_{12})z(\theta_2)x(\alpha_{23})z(\theta_3)x(\alpha_{34})z(\theta_4)x(\alpha_{41})=\boldsymbol{I}_3 \tag{1}$$

In equation (1), the position angle of the driving element is θ_1 and the angles θ_2, θ_3 and θ_4 corresponding to the motions from the other pairs must be found. The arguments of the x matrices represent the constant constructive parameters, with the values:

$$\alpha_{12} = \pi/2; \quad \alpha_{23} = \pi/2; \quad \alpha_{34} = \pi/2; \quad \alpha_{41} = \pi - \alpha \tag{2}$$

In the relation (2), α is the acute angle between the driven and driving axes. When the rotational motion ought to be transmitted between two crossed axes, a versatile solution consists in constructing a mechanism similar to the Cardanic transmission.

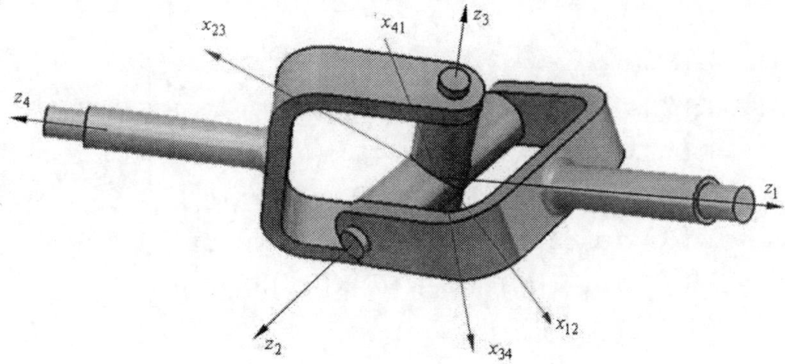

Figure 1. The Cardan mechanism

The family of the mechanism with crossed axes has the value 1, the mobility of the mechanism is:

$$M_0 = 6 \cdot n - 4c_5 = 6 \cdot 3 - 4 \cdot 5 = -2 \tag{3}$$

and therefore the Cardanic transmission cannot be directly used because the mechanism is blocked.

2. Use of dual matrices for obtaining the closure equation of the generalised Cardan mechanism

According to the transfer principle due to Kotelnikov [2], all the relations from spherical trigonometry are valid when the real angles are replaced by dual angles. From this replacement, it results that all rotational motions from the spherical mechanisms become helicoidally motions. Another consequence is the transformation of the spherical mechanisms into spatial ones. This remark led to the idea of studying the kinematics of the generalized Cardan mechanism based on the kinematics of Cardan transmission. The simplest way to "relax" the mechanism is by transforming three of the rotational pairs into cylindrical pairs, the mechanism becoming a RCCC one. The kinematical analysis of the RCCC mechanism for random values of the constructive parameters was made by Yang [3], using the method of dual quaternions. The test rig presented in figure 2 is considered for applying the method.

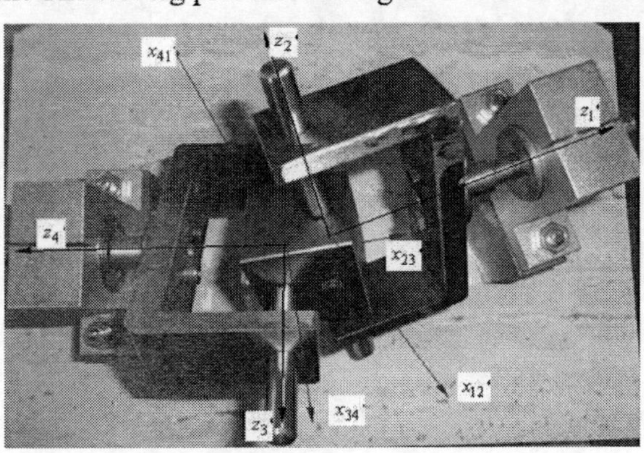

Figure 2. Generalized Cardan mechanism

The closure equation for the contour is obtained by transforming the angles from the equation (1) into dual angles, according the equations deduced in the first part of the work [4]. The closure equation is:

$$z(\hat{\theta}_1)x(\hat{\alpha}_{12})z(\hat{\theta}_2)x(\hat{\alpha}_{23})z(\hat{\theta}_3)x[\hat{\alpha}_{34}]z(\hat{\theta}_4)x(\hat{\alpha}_{41})=\hat{I}_3 \tag{4}$$

The form:

$$\begin{gathered}z(\theta_1)x(\alpha_{12})z(\theta_2)x(\alpha_{23})z(\theta_3)x(\alpha_{34})z(\theta_4)x(\alpha_{41})+\\ \varepsilon\begin{bmatrix}z(\theta_1)(s_1Q_\theta)x(\alpha_{12})z(\theta_2)x(\alpha_{23})z(\theta_3)x(\alpha_{34})z(\theta_4)x(\alpha_{41})+\\ z(\theta_1)x(\alpha_{12})(a_{12}Q_\alpha)z(\theta_2)x(\alpha_{23})z(\theta_3)x(\alpha_{34})z(\theta_4)x(\alpha_{41})+\\ z(\theta_1)x(\alpha_{12})z(\theta_2)(s_2Q_\theta)x(\alpha_{23})z(\theta_3)x(\alpha_{34})z(\theta_4)x(\alpha_{41})+\\ z(\theta_1)x(\alpha_{12})z(\theta_2)x(\alpha_{23})(a_{23}Q_\alpha)z(\theta_3)x(\alpha_{34})z(\theta_4)x(\alpha_{41})+\\ z(\theta_1)x(\alpha_{12})z(\theta_2)x(\alpha_{23})z(\theta_3)(s_3Q_\theta)x(\alpha_{34})z(\theta_4)x(\alpha_{41})+\\ z(\theta_1)x(\alpha_{12})z(\theta_2)x(\alpha_{23})z(\theta_3)x(\alpha_{34})(a_{34}Q_\alpha)z(\theta_4)x(\alpha_{41})+\\ z(\theta_1)x(\alpha_{12})z(\theta_2)x(\alpha_{23})z(\theta_3)x(\alpha_{34})z(\theta_4)(s_4Q_\theta)x(\alpha_{41})+\\ z(\theta_1)x(\alpha_{12})z(\theta_2)x(\alpha_{23})z(\theta_3)x(\alpha_{34})z(\theta_4)x(\alpha_{41})(a_{41}Q_\alpha)\end{bmatrix}=\begin{bmatrix}1&0&0\\0&1&0\\0&0&1\end{bmatrix}+\varepsilon\begin{bmatrix}0&0&0\\0&0&0\\0&0&0\end{bmatrix}\end{gathered} \tag{5}$$

is obtained as it follows: the left member of the equation (4) is expanded and contains the real part and the dual part; the dual part is identical to the left member of the equation of closure (1) the spherical mechanism; the dual part is a sum with the number of terms equal to the number of factors of the real part; each of these terms is obtained by introducing in each term of the dual part the matrix sQ_θ or aQ_α, depending on the type of the preceding matrix, $z(\theta)$ or $x(\alpha)$. The equation was written under the assumption that all the pairs of the mechanism were transformed into cylindrical pairs, in order to present a higher degree of generality for the solution. The equation (5) is equivalent to two equations obtained by equating the real and dual parts, respectively, from the members of equation (5), from which one is the equation (1) itself. The second equation seems complicated at first sight but it can be simplified based on equation (1). Following the remark that any term of the dual part is split by the Q_α or Q_θ matrix into two factors, any of these factor is the transpose of the other one, and the next equation is obtained:

$$\left.\begin{aligned}&z(\theta_1)(s_1Q_\theta)[z(\theta_1)]^T+\\ &z(\theta_1)x(\alpha_{12})(a_{12}Q_\alpha)[z(\theta_1)x(\alpha_{12})]^T+\\ &z(\theta_1)x(\alpha_{12})z(\theta_2)(s_2Q_\theta)[z(\theta_1)x(\alpha_{12})z(\theta_2)0]^T+\\ &z(\theta_1)x(\alpha_{12})z(\theta_2)x(\alpha_{23})(a_{23}Q_\alpha)[z(\theta_1)x(\alpha_{12})z(\theta_2)x(\alpha_{23})]^T+\\ &[x(\alpha_{34})z(\theta_4)x(\alpha_{41})]^T(s_3Q_\theta)x(\alpha_{34})z(\theta_4)x(\alpha_{41})+\\ &[z(\theta_4)x(\alpha_{41})]^T(a_{34}Q_\alpha)z(\theta_4)x(\alpha_{41})+\\ &[x(\alpha_{41})]^T(s_4Q_\theta)x(\alpha_{41})+\\ &I_3(a_{41}Q_\alpha)\end{aligned}\right.=O_3 \tag{6}$$

Considering the equations obtained in the part I of the paper [4]:

$$z(\theta)Q_\theta z(\theta)^T=Q_\theta;\qquad x(\alpha)Q_\alpha x(\alpha)^T=Q_\alpha \tag{7}$$

The equation (6) can now be simplified:

IOP Conf. Series: Materials Science and Engineering **568** (2019) 012032 doi:10.1088/1757-899X/568/1/012032

$$\begin{array}{l} s_1\boldsymbol{Q}_\theta + \\ z(\theta_1)(a_{12}\boldsymbol{Q}_\alpha)[z(\theta_1)]^T + \\ z(\theta_1)x(\alpha_{12})(s_2\boldsymbol{Q}_\theta)[z(\theta_1)x(\alpha_{12})]^T + \\ z(\theta_1)x(\alpha_{12})z(\theta_2)(a_{23}\boldsymbol{Q}_\alpha)[z(\theta_1)x(\alpha_{12})z(\theta_2)]^T + \\ [x(\alpha_{34})z(\theta_4)x(\alpha_{41})]^T(s_3\boldsymbol{Q}_\theta)x(\alpha_{34})z(\theta_4)x(\alpha_{41}) + \\ [z(\theta_4)x(\alpha_{41})]^T(a_{34}\boldsymbol{Q}_\alpha)z(\theta_4)x(\alpha_{41}) + \\ [x(\alpha_{41})]^T(s_4\boldsymbol{Q}_\theta)x(\alpha_{41}) + \\ a_{41}\boldsymbol{Q}_\alpha \end{array} = \boldsymbol{O}_3 \tag{8}$$

To be underlined that for the calculus of the dual part of the matrix closure equation (5) there are necessary $n=64$ matrix products while for the equivalent form, relation (8), only $n=2(2+4+6)=24$ matrix products are required, meaning 37.5% from the initial number of products, thus considerably reducing the time for calculus. As shown in the part I of the paper [4], the matrices resulting from products of $\boldsymbol{MQ}_\alpha\boldsymbol{M}^T$, $\boldsymbol{MQ}_\theta\boldsymbol{M}^T$ type are antisymmetric, and it results that the left member of equation (8) is a sum of antisymmetric matrices and it will by an antisymmetric matrix, too.

In conclusion, the equation (8) will provide three independent scalar equations. The angles of rotation from cylindrical pairs are present in these equations and thus the equation (1) must first be solved in order to obtain these angles and after that the translations from the pairs of the mechanism will be found.

3. The solutions of dual matrix closure equation. Observations and discussions

For the present case, three from the six unknowns are rotations (and three are translations) which can be found using only the matrix equation (1). But, as McCarthy shows [5], complicated calculus occurs when using the matrix equation under the form (1). To avoid this aspect, the equality is re-written in a convenient form, as the equality between two matrices. For the actual case, the matrix equation (1) is re-written as:

$$z(\theta_1)x(\alpha_{12})z(\theta_2)x(\alpha_{23})z(\theta_3)x(\alpha_{34})z(\theta_4)x(\alpha_{41}) = \boldsymbol{I}_3 \,|\, z(\theta_1) \tag{9}$$

$$[z(\theta_1)]^T \,\Big|\, z(\theta_1)x(\alpha_{12})z(\theta_2)x(\alpha_{23})z(\theta_3)x(\alpha_{34})z(\theta_4)x(\alpha_{41})z(\theta_1) = z(\theta_1) \tag{10}$$

and it results:

$$x(\alpha_{12})z(\theta_2)x(\alpha_{23})z(\theta_3)x(\alpha_{34})z(\theta_4)x(\alpha_{41})z(\theta_1) = \boldsymbol{I}_3 \tag{11}$$

The equation (11) is expressed now as:

$$x(\alpha_{12})z(\theta_2)x(\alpha_{23})z(\theta_3) = [x(\alpha_{34})z(\theta_4)x(\alpha_{41})z(\theta_1)]^T \tag{12}$$

The particular values $\alpha_{12}=\pi/2$, $\alpha_{23}=\pi/2$ and $\alpha_{34}=\pi/2$ lead to the explicit form:

$$\begin{bmatrix} \cos\theta_2\cos\theta_3 & -\cos\theta_2\sin\theta_3 & \sin\theta_2 \\ -\sin\theta_3 & -\cos\theta_3 & 0 \\ \sin\theta_2\cos\theta_3 & -\sin\theta_2\sin\theta_3 & -\cos\theta_2 \end{bmatrix} = \\ = \begin{bmatrix} \cos\theta_1\cos\theta_4 - \cos\alpha_{41}\sin\theta_1\sin\theta_4 & -\sin\theta_{41}\sin\theta_1 & \cos\theta_1\sin\theta_4 + \cos\alpha_{41}\sin\theta_1\cos\theta_4 \\ -\sin\theta_1\sin\theta_4 - \cos\theta_1\cos\theta_{41}\sin\theta_4 & -\sin\theta_{41}\cos\theta_1 & -\cos\theta_1\sin\theta_4 + \cos\alpha_{41}\cos\theta_1\cos\theta_4 \\ \sin\alpha_{41}\sin\theta_4 & -\cos\alpha_{41} & -\sin\alpha_{41}\cos\theta_4 \end{bmatrix} \tag{13}$$

The matrix equation (13) allows for finding the rotations from the three cylindrical pairs of the mechanism. Details upon the methodology are exhaustively given in [6], using special inverse trigonometrically functions [7-8].

4. Conclusions

The paper presents an example of applying the method of dual numbers in the kinematical analysis of spatial mechanisms. Initially, the Cardan mechanism that is a spherical mechanism, is considered and the closure matrix equation is written using rotation matrices of (3,3) type with real elements.

For the spatial mechanism obtained from the Cardan mechanism after the rotation pairs were transformed into cylindrical pairs, the closure matrix equation is obtained, according to Kotelnikov's principle, by the plain transformation of the actual angles from the closure equation of the spherical mechanism, into dual angles.

The dual equation is written according to the results obtained in Part I of the paper and leads to two matrix equations, resulting after equating the real parts and imaginary parts of the closing matrix equation. The first equation allows for finding the rotations from the pairs of the mechanism and the second equation helps finding the translations from the pairs. Some remarks concerning the method of finding the rotations from the pairs with the avoidance of strange solutions are made.

5. References

[1] Hartenberg R S and Denavit J 1965 *Kinematic synthesis of linkages* (McGraw-Hill series in mechanical engineering, New York McGraw-Hill)

[2] Kotelnikov A P 1950 Theory of Vectors and Complex Numbers, in Several Applications of Lobachevskii's Ideas in Mechanics and Physics (Collected Papers of Gostekhizdat Moscow) 7-47

[3] Yang A T and Freudenstein F 1964 Application of dual number quaternions algebra to the analysis of spatial mechanisms *ASME J. Eng. Ind.* **86** 300–308

[4] Ciornei F C, Alaci S, Pentiuc R D and Doroftei I 2019 Use of dual numbers in kinematical analysis of spatial mechanisms. Part I: principle of the method *IMT Oradea Ann. Ses. Scient. Papers*

[5] McCarthy J M 2000 *Geometric Design of Linkages* (Springer Berlin)

[6] Ciornei F C, Alaci S, Romanu I C, Mihai I and Tibu G A 2019 Kinematical analysis of a generalized Cardanic joint *IOP Conf. Ser.: Mater. Sci. Eng.* 477 012037

[7] Hahn B and Valentine D 2019 *Essential MATLAB for Engineers and Scientists* 7th Edition, (Academic Press)

[8] Maxfield B 2009*Essential Mathcad for Engineering* Second Edition (Academic Press)

Acknowledgement

This work was supported by a grant of the Romanian Ministry of Research and Innovation, CCCDI – UEFISCDI, project number PN-III-P1-1.2-PCCDI-2017-0404 / 31PCCDI/2018, within PNCDI III.

IOP Conf. Series: Materials Science and Engineering **568** (2019) 012033 doi:10.1088/1757-899X/568/1/012033

Use of dual numbers in kinematical analysis of spatial mechanisms. Part I: principle of the method

F C Ciornei[1], S Alaci[1], R D Pentiuc[1] and I Doroftei[2]
[1]"Stefan cel Mare" University, Suceava, Romania
[2]"Gheorghe Asachi" Technical University, Iasi, Romania

florina.ciornei@usm.ro

Abstract. The general methodology of solving a problem of kinematical analysis of a spatial mechanism is presented. The method is based on the system proposed by Hartenberg and Denavit, concerning the notations of the coordinate frames attached to the elements of the mechanism and on the closure matrix equation of a kinematical chain. Carrying out the closure equation for the case of spherical mechanisms and then applying the transfer principle of Kotelnikov, the equations which allow for finding the unknown parameters of the kinematical chain are obtained. Starting from the typical form of closure equation based on homogenous operators Hartenberg-Denavit, the closure matrix equation is obtained in dual format. By identifying two special matrices it is shown that the coefficient of the dual part from the closure equation is in fact a sum in which each term is attained from the real part of the closure equation by performing simple operations involving the special matrices, that reduce the calculus volume.

1. Introduction

The dual numbers were introduced by Clifford [1] related to the research conducted in the study of the properties of quaternions. The dual numbers occurred in abstract mathematical researches but after a short time they proved their usefulness in applied sciences, and the work of Ball [2] should be mentioned. In the field of theory of mechanisms Dimentberg [3] and Kotelnikov [4] are essential references. In the domain of theory of mechanisms, Yang [5] obtained important results applying the dual quaternions for the kinematical study of spatial mechanisms; thus, he obtained the analytical relations for the displacements from the pairs of the RCCC mechanism. The dual numbers have a structure similar to the complex numbers from elementary algebra, being expressed under the form:

$$\hat{a} = a + \varepsilon a_0 \tag{1}$$

where ε is the dual unit with the property $\varepsilon^2 = 0$ with the immediate consequence $\varepsilon^n = 0, n > 1$. The matrix method proved to be an extremely useful instrument in completing the kinematical analysis of spatial mechanisms. The matrix denoted $\boldsymbol{d}_{12}$ corresponds to the translation of the origin of the frame $"1"$ over the origin of frame $"2"$ and the matrix related to the rotation $\boldsymbol{R}_{12}$ gets the axes of the system $"1"$ parallel to the axes of the system $"2"$. The transformation relation has the general form:

$$\begin{bmatrix} x_1 & y_1 & z_1 \end{bmatrix}^T = \boldsymbol{d}_{12} + \boldsymbol{R}_{12} \begin{bmatrix} x_2 & y_2 & z_2 \end{bmatrix}^T \tag{2}$$

Or, in a concentrated manner:

IOP Conf. Series: Materials Science and Engineering **568** (2019) 012033 doi:10.1088/1757-899X/568/1/012033

$$\boldsymbol{x}_1 = \boldsymbol{d}_{12} + \boldsymbol{R}_{12}\boldsymbol{x}_2 \tag{3}$$

The drawback of this expressing style is evident when it is aimed the equivalent expression of some successive transformations, written as a unique transformation. Thus, writing:

$$\boldsymbol{x}_1 = \boldsymbol{d}_{12} + \boldsymbol{R}_{12}\boldsymbol{x}_2 = \boldsymbol{d}_{12} + \boldsymbol{R}_{12}(\boldsymbol{d}_{23} + \boldsymbol{R}_{23}\boldsymbol{x}_3) = \boldsymbol{d}_{13} + \boldsymbol{R}_{13}\boldsymbol{x}_3 \tag{4}$$

it results:

$$\boldsymbol{d}_{13} = \boldsymbol{d}_{12} + \boldsymbol{R}_{12}\boldsymbol{d}_{23}; \qquad \boldsymbol{R}_{13} = \boldsymbol{R}_{12}\boldsymbol{R}_{23} \tag{5}$$

Hartenberg and Denavit [6] proposed the explanation of the displacement of a system using a (4×4) type matrix, including in its structure both the rotation matrix and the displacement vector:

$$\begin{bmatrix} x_1 & y_1 & z_1 & 1\end{bmatrix}^T = \begin{bmatrix} \boldsymbol{R}_{12} & & & \boldsymbol{d}_{12} \\ 0 & 0 & 0 & 1 \end{bmatrix} \begin{bmatrix} x_2 & y_2 & z_2 & 1\end{bmatrix}^T \tag{6}$$

or, in concentrated form:

$$\boldsymbol{X}_1 = \boldsymbol{T}_{12}\boldsymbol{X}_2 \tag{7}$$

$$\boldsymbol{X}_1 = \begin{bmatrix} x_1 & y_1 & z_1 & 1\end{bmatrix}^T, \boldsymbol{X}_2 = \begin{bmatrix} x_2 & y_2 & z_2 & 1\end{bmatrix}^T, \boldsymbol{T}_{12} = \begin{bmatrix} \boldsymbol{R}_{12} & & & \boldsymbol{d}_{12} \\ 0 & 0 & 0 & 1 \end{bmatrix} \tag{8}$$

The advantage of expressing the transformation relation under the form (5) is clear for the situation of composing multiple displacements:

$$\boldsymbol{X}_1 = \boldsymbol{T}_{12}\boldsymbol{X}_2 = \boldsymbol{T}_{12}(\boldsymbol{T}_{23}\boldsymbol{X}_3) = (\boldsymbol{T}_{12}\boldsymbol{T}_{23})\boldsymbol{X}_3 = \boldsymbol{T}_{13}\boldsymbol{X}_3 \tag{9}$$

The outcome is:

$$\boldsymbol{T}_{13} = \boldsymbol{T}_{12}\boldsymbol{T}_{23} \tag{10}$$

This expression attests the homogenous nature of transformation relations and thus justifying the name "method of homogenous operators" under which the method is known. In the case of a closed contour the equation of closure of the kinematical chain is obtained:

$$\boldsymbol{T}_{12}\boldsymbol{T}_{23}\ldots\boldsymbol{T}_{n-1,n}\boldsymbol{T}_{n,1} = \boldsymbol{I}_4 \tag{11}$$

Hartenberg and Denavit [6] showed that the displacement of the $"k"$ system over the $"k+1"$ system will be obtained as the result of two helical motions: the first motion, performed about Oz_k axis, with the angle θ_k and the translation s_k and the second, about the $Ox_{k,k+1}$ axis, with the angle $\alpha_{k,k+1}$ and the translation $a_{k,k+1}$. The two above displacement can be completely characterized by the dual numbers:

$$\hat{\theta}_k = \theta_k + \varepsilon s_k; \quad \hat{\alpha}_{k,k+1} = \alpha_{k,k+1} + \varepsilon a_{k,k+1} \tag{12}$$

2. The closure equation of the contour in dual form

The two dual angles are represented in figure 1 and figure 2. The two displacements, the roto-translation with respect to Oz_k and the roto-translation with respect to $Ox_{k,k+1}$, will be expressed according to the transfer principle of Kotelnikov [4], using the matrices reflecting the rotation about the axes Oz and Ox respectively, where the actual angles should be replaced by dual angles. In order to obtain the explicit expressions of $z(\theta)$ and $x(\alpha)$ the form of the rotation matrix is particularized:

$$\boldsymbol{R}(\boldsymbol{u},\varphi) = \boldsymbol{u}\boldsymbol{u}^T + \left(\boldsymbol{I}_3 + \boldsymbol{u}\boldsymbol{u}^T\right)\cos\varphi + \tilde{\boldsymbol{u}}\sin\varphi \tag{13}$$

where $\boldsymbol{I}_3$ is the unit matrix of third order and $\tilde{\boldsymbol{u}}$ is the antisymmetric matrix attached to the $\boldsymbol{u}$ vector.

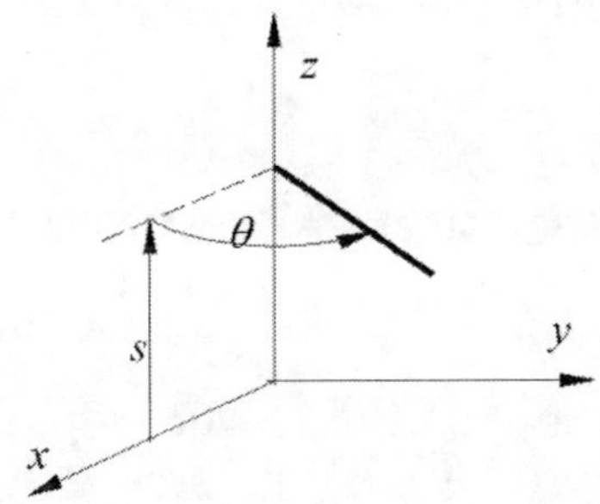

Figure 1. The dual number $\hat{\theta} = \theta + \varepsilon s$

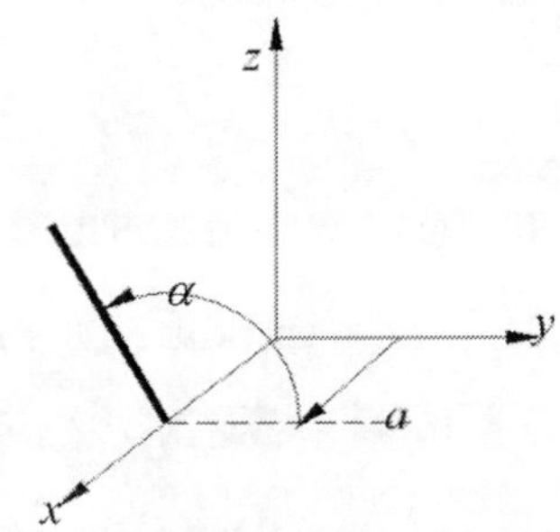

Figure 2. The dual number $\hat{\alpha} = \alpha + \varepsilon a$

For $\boldsymbol{u} = \boldsymbol{k}$ and $\varphi = \theta$ and for $\boldsymbol{u} = \boldsymbol{i}$ and $\varphi = \alpha$ it is obtained:

$$z(\theta) = \begin{bmatrix} \cos\theta & -\sin\theta & 0 \\ \sin\theta & \cos\theta & 0 \\ 0 & 0 & 1 \end{bmatrix};\ x(\alpha) = \begin{bmatrix} 1 & 0 & 0 \\ 0 & \cos\alpha & -\sin\alpha \\ 0 & \sin\alpha & \cos\alpha \end{bmatrix} \tag{14}$$

The following dual transformation corresponds to the $\boldsymbol{T}_{k,k+1}$ homogenous transformation:

$$\hat{\boldsymbol{t}}_{k,k+1} = z(\hat{\theta}_k)x(\hat{\alpha}_{k,k+1}) \tag{15}$$

The dual equation that corresponds to the closure equation (11) has the form:

$$\hat{\boldsymbol{t}}_{12}\hat{\boldsymbol{t}}_{23}\ldots\ldots\hat{\boldsymbol{t}}_{n-1,n}\hat{\boldsymbol{t}}_{n,1} = \hat{\boldsymbol{I}}_3 \tag{16}$$

where $\hat{\boldsymbol{I}}_3$ is the dual matrix of third order defined as:

$$\hat{\boldsymbol{I}}_3 = \begin{bmatrix} 1 & 0 & 0 \\ 0 & 1 & 0 \\ 0 & 0 & 1 \end{bmatrix} + \varepsilon \begin{bmatrix} 0 & 0 & 0 \\ 0 & 0 & 0 \\ 0 & 0 & 0 \end{bmatrix} \tag{17}$$

Recalling the definition of equality of two matrices and of two dual numbers respectively, a general equation of form (16) would give a number of $2\cdot(3\times 3) = 18$ scalar equations. Subsequently it is shown that for the case of a contour from the structure of a mechanism, the relation (16) generates 6 independent equations. In equation (14) the argument θ is replaced by the dual angle $\hat{\theta}$ and it results:

$$z(\hat{\theta}) = z(\theta)(\boldsymbol{I}_3 + \varepsilon s\boldsymbol{Q}_\theta);\ \boldsymbol{Q}_\theta = \begin{bmatrix} 0 & -1 & 0 \\ 1 & 0 & 0 \\ 0 & 0 & 0 \end{bmatrix} \tag{18}$$

$$\boldsymbol{x}(\alpha) = \boldsymbol{x}(\alpha)(\boldsymbol{I}_3 + \varepsilon a\boldsymbol{Q}_\alpha);\ \boldsymbol{Q}_\alpha = \begin{bmatrix} 0 & 0 & 0 \\ 0 & 0 & -1 \\ 0 & 1 & 0 \end{bmatrix} \tag{19}$$

The closure equation, written in dual form using the notations (14) and (16) take the form:

$$\prod_{k=1}^{n} z(\theta_k)\left[\boldsymbol{I}_3 + \varepsilon s_k\boldsymbol{Q}_\theta\right]\boldsymbol{x}(\alpha_{k,k+1})\left[\boldsymbol{I}_3 + \varepsilon a_{k,k+1}\boldsymbol{Q}_\alpha\right] = \hat{\boldsymbol{I}}_3 \tag{20}$$

The product from the left member of equation (20) is expanded and a polynomial of $2n$ degree with respect to variable ε is obtained, with matrix coefficients. Applying the definition of dual unit, it results that all the terms of second or higher degree must be zero. The equation (20) takes the form:

$$\prod_{k=1}^{n} z(\theta_k)x(\alpha_{k,k+1}) + \varepsilon \sum_{k=1}^{n} (\boldsymbol{P}_{\theta_k} + \boldsymbol{P}_{\alpha_k}) = \hat{\boldsymbol{I}}_3 \quad (21)$$

where $\boldsymbol{P}_{\theta_k}$ and $\boldsymbol{P}_{\alpha_k}$ are matrices obtained by replacing $z(\theta_k)$ by $z_k s_k \boldsymbol{Q}_\theta$, and $x(\alpha_{k,k+1})$ by $x(\alpha_{k,k+1})a_{k,k+1}\boldsymbol{Q}_a$ respectively, in the product $\prod_{k}^{n} z(\theta_k)x(\alpha_{k,k+1})$.

$$\boldsymbol{P}_{\theta k} = z(\theta_1)\boldsymbol{x}(\alpha_{12}).....z(\theta_{k-1})\boldsymbol{x}(\alpha_{k-1,k})z(\theta_k)s_k\boldsymbol{Q}_\theta\boldsymbol{x}(\alpha_{k,k+1})z(\theta_{k+1}).....z(\theta_n)\boldsymbol{x}(\alpha_{n,1}) \quad (22)$$

$$\boldsymbol{P}\alpha_k = z(\theta_1)\boldsymbol{x}(\alpha_{12}).....z(\theta_{k-1})\boldsymbol{x}(\alpha_{k-1,k})z(\theta_k)\boldsymbol{x}(\alpha_{k,k+1})a_{k,k+1}\boldsymbol{Q}_\alpha z(\theta_{k+1}).....z(\theta_n)\boldsymbol{x}(\alpha_{n,1}) \quad (23)$$

The equation (18) is equivalent to the following real equations:

$$\prod_{k=1}^{n} z(\theta_k)x(\alpha_{k,k+1}) = \boldsymbol{I}_3 \quad (24)$$

$$\sum_{k=1}^{n} (\boldsymbol{P}_{\theta_k} + \boldsymbol{P}_{\alpha_k}) = \boldsymbol{O}_3 \quad (25)$$

3. Solving the resultant matrix equations. Observations and discussions

The left member of equation (24) is the product of $2n$ orthogonal matrices, thus being an orthogonal matrix. Considering the physical significance of the elements of an orthogonal matrix, namely that the elements of a $"p"$ column are the projections of the $\boldsymbol{i}_p$ versor on the coordinate axes of the reference frame, it can be concluded that from the nine elements of the rotation matrix only three are independent because the orthonormation conditions must be satisfied:

$$\boldsymbol{i}_p \cdot \boldsymbol{i}_q = \delta_{pq} \quad (26)$$

From here it results the conclusion that only three equations from the nine provided by the matrix equality (24) are independent. Another property of an orthogonal matrix $\boldsymbol{R}$ is:

$$\boldsymbol{R}^{-1} = \boldsymbol{R}^T \quad (27)$$

From equation (24) it is obtained:

$$\boldsymbol{P}_{\theta k} = z(\theta_1)\boldsymbol{x}(\alpha_{12}).....z(\theta_{k-1})\boldsymbol{x}(\alpha_{k-1,k})z(\theta_k)s_k\boldsymbol{Q}_\theta\left[z(\theta_1)\boldsymbol{x}(\alpha_{12}).....z(\theta_{k-1})\boldsymbol{x}(\alpha_{k-1,k})z(\theta_k)\right]^T \quad (28)$$

$$\boldsymbol{P}_{\theta k} = \left[\boldsymbol{x}(\alpha_{k,k+1})z(\theta_{k+1}).....z(\theta_n)\boldsymbol{x}(\alpha_{n,1})\right]^T s_k\boldsymbol{Q}_\theta\boldsymbol{x}(\alpha_{k,k+1})z(\theta_{k+1}).....z(\theta_n)\boldsymbol{x}(\alpha_{n,1}) \quad (29)$$

$$\boldsymbol{P}_{\alpha_k} = z(\theta_1)\boldsymbol{x}(\alpha_{12})....z(\theta_k)\boldsymbol{x}(\alpha_{k,k+1})a_{k,k+1}\boldsymbol{Q}_\alpha\left[z(\theta_1)\boldsymbol{x}(\alpha_{12})....z(\theta_k)\boldsymbol{x}(\alpha_{k,k+1})\right]^T \quad (30)$$

$$\boldsymbol{P}_{\alpha_k} = \left[z(\theta_{k+1}).....z(\theta_n)\boldsymbol{x}(\alpha_{n,1})\right]^T a_{k,k+1}\boldsymbol{Q}_\alpha z(\theta_{k+1}).....z(\theta_n)\boldsymbol{x}(\alpha_{n,1}) \quad (31)$$

Next, the following assertion is necessary: Theorem 1: If $\boldsymbol{Q}$ is an antisymmetric matrix, $\boldsymbol{Q}+\boldsymbol{Q}^T=\boldsymbol{0}$., then for any $\boldsymbol{M}$ matrix, the $\boldsymbol{MQM}^T$ matrix is antisymmetric. Indeed:

$$\boldsymbol{MQM}^T + \left[\boldsymbol{MQM}^T\right]^T = \boldsymbol{MQM}^T + \left[\boldsymbol{M}^T\right]^T \boldsymbol{Q}^T\boldsymbol{M}^T = \boldsymbol{MQM}^T + \boldsymbol{M}(-\boldsymbol{Q})\boldsymbol{M}^T = \boldsymbol{O} \quad (32)$$

The relations (22) and (23), together with equation (24) lead to:

IOP Conf. Series: Materials Science and Engineering **568** (2019) 012033 doi:10.1088/1757-899X/568/1/012033

$$\boldsymbol{P}_{\theta k} = z(\theta_1)\boldsymbol{x}(\alpha_{12}).....z(\theta_{k-1})\boldsymbol{x}(\alpha_{k-1,k})z(\theta_k)s_k\boldsymbol{Q}_\theta\left[z(\theta_1)\boldsymbol{x}(\alpha_{12}).....z(\theta_{k-1})\boldsymbol{x}(\alpha_{k-1,k})z(\theta_k)\right]^T \quad (33)$$

$$\boldsymbol{P}_{\alpha_k} = z(\theta_1)\boldsymbol{x}(\alpha_{12}).....z(\theta_k)\boldsymbol{x}(\alpha_{k,k+1})a_{k,k+1}\boldsymbol{Q}_\alpha\left[z(\theta_1)\boldsymbol{x}(\alpha_{12}).....z(\theta_k)\boldsymbol{x}(\alpha_{k,k+1})\right]^T \quad (34)$$

The relations (18) and (19) show that the matrices $\boldsymbol{Q}_\theta$ and $\boldsymbol{Q}_\alpha$ are antisymmetric; based on the relations (33), (34) and Theorem 1, one can state that $\boldsymbol{P}_{\theta_k}$ and $\boldsymbol{P}_{\alpha_k}$ are antisymmetric matrices. As a consequence, the left member of the equation (21) is an antisymmetric matrix and it is completely defined by three scalar parameters; therefore, the matrix equation (21) provides three scalar equations.

The equations (33) and (34) can be simplified following the remark:

$$z(\theta)\boldsymbol{Q}_\theta z(\theta)^T = \boldsymbol{x}(\alpha)\boldsymbol{Q}_\alpha \boldsymbol{x}(\alpha)^T = \boldsymbol{I}_3 \quad (35)$$

The above relations are verified by direct calculus. Based on the relations (35) it is obtained:

$$\boldsymbol{P}_{\theta k} = z(\theta_1)\boldsymbol{x}(\alpha_{12}).....z(\theta_{k-1})\boldsymbol{x}(\alpha_{k-1,k})s_k\boldsymbol{Q}_\theta\left[z(\theta_1)\boldsymbol{x}(\alpha_{12}).....z(\theta_{k-1})\boldsymbol{x}(\alpha_{k-1,k})\right]^T \quad (36)$$

$$\boldsymbol{P}_{\alpha_k} = z(\theta_1)\boldsymbol{x}(\alpha_{12}).....\boldsymbol{x}(\alpha_{k-1,k})z(\theta_k)a_{k,k+1}\boldsymbol{Q}_\alpha\left[z(\theta_1)\boldsymbol{x}(\alpha_{12}).....\boldsymbol{x}(\alpha_{k-1,k})z(\theta_k)\right]^T \quad (37)$$

4. Conclusions

The paper presents the methodology of obtaining the closure equation in dual form based on the standard form of the closure equation for a spatial kinematical chain proposed by Hartenberg and Denavit. The identification of the real part and of the coefficients of dual part leads to two matrix equations and finding the solutions of these equations depends on the ratio between the number of translational and the number of rotational pairs. Thus, maximum three translational pairs may exist in the structure of the chain and in this situation, the two systems can be solved independently: first, the one that allows finding the rotations and next, the one that help obtaining the translations. If the number of translational pairs is less than three, the two systems must be considered altogether as a single system that gives simultaneously the solutions both for rotations and translations. By applying certain matrix properties for the matrices occurring in the structure of the dual part of the closure equation, the form of this equation is much simplified, the calculus volume is reduced significantly and an algorithm for finding the equations required for the positional analysis of a spatial mechanism can be formulated.

Acknowledgement
This work was supported by a grant of the Romanian Ministry of Research and Innovation, CCCDI – UEFISCDI, project number PN-III-P1-1.2-PCCDI-2017-0404 / 31PCCDI/2018, within PNCDI III.

5. References

[1] Clifford W K 1873 Preliminary sketch of biquaternions *Proc. Lond. Math. Soc.* **4**(64) 381–395
[2] Ball R S 1900 *Theory of Screws* (Cambridge University Press)
[3] Dimentberg F M 1968 *The screw calculus and its applications* Tech. Rep. AD 680993, (Clearinghouse for Federal and Scientific Technical Information VA USA)
[4] Kotelnikov A P 1950 Theory of Vectors and Complex Numbers, in *Several Applications of Lobachevskii's Ideas in Mechanics and Physics, Coll. Papers* (Gostekhizdat, Moscow) 7-47
[5] Yang A T and Freudenstein F 1964 Application of dual number quaternions algebra to the analysis of spatial mechanisms *ASME J. Eng. Ind.* **86** 300–308
[6] Hartenberg R S and Denavit J 1965 *Kinematic synthesis of linkages* (McGraw-Hill series in mechanical engineering, New York: McGraw-Hill)

IOP Conf. Series: Materials Science and Engineering **568** (2019) 012034 doi:10.1088/1757-899X/568/1/012034

Mechanical design and 3D printing process of an exoskeleton used for elbow joint rehabilitation

I A Tătulea[1], C N Drugă[1], I Şerban[1]

[1]Design of Product, Mechatronics and Environment Department, Transylvania University of Brasov, Brasov, Romania

irinatatulea23@yahoo.com

Abstract. This paper presents the design and 3D printing process of the components which form an exoskeleton robot used in elbow joint rehabilitation. In the first part of the paper are presented a series of exoskeleton models used to rehabilitate the upper limb (4 DOF and 7 DOF). The second part is entirely dedicated to the mechanical design process of the components which form the device. In the last part of the paper are presented a series of aspects related to the physical achievement of the device using rapid prototyping technology (3D printers, filaments of material, execution time, assembly of components etc.). In this case was chosen the natural PVA filament from Ultimaker that dissolves in water. The dissolving process of the support material was analysed over a 24 hours period, in which the pieces were immersed in an aquarium. This paper represents a part of a much wider project of designing, creating and testing a bio mechatronic device used in rehabilitation of elbow joint.

1. Introduction

Exoskeleton robots for upper limb have been used for various purposes, such as assisting joint movements, rehabilitating and treating joint dysfunctions, muscle strength augmentation, and haptic interface. In the case of therapy and rehabilitation of joint dysfunctions, most of these devices have less than seven degrees of freedom (7 DOF) of active joints [1].

The main features of 4DOF exoskeleton robot are the following: developed by the University of Saga; segments of application: shoulder, elbow and forearm; the degrees of freedom of the active joints: 4 DOF; type of actuators: actuators type I, powered by DC servo motors; power transmission methods: cable and gear drives; purpose of the application: power assist device.

The main components of the 4 DOF exoskeleton are: a shoulder motion support part, an elbow motion support part, a forearm motion support part, a wrist force sensor, and a mobile wheel chair. The mobile wheel chair itself has 2 DOF, which provide independent movement for physically weak persons. The device has the role to assist the following movements: horizontal flexion-extension of the shoulder, vertical flexion-extension of the shoulder, flexion-extension of the elbow and supination-pronation movement of the forearm.

The main advantage of the robot is the ability to adjust the distance between the arm holder and the centre of rotation of the shoulder joint of the robot, in accordance with the shoulder motion, in order to cancel out the ill effects caused by the position difference between the centre of rotation of the robot shoulder and the human shoulder. The control method of the robot is based on EMG signals generated by human muscles [1].

The main features of 7DOF exoskeleton robot are the following: developed by the University of Washington; segments of application: shoulder, elbow, forearm and wrist; the degrees of freedom of the active joints: 7 DOF; type of actuators: actuators type I, powered by brushed motors; power transmission methods: cable drives; purpose of the application: rehabilitation device for upper limb, virtual reality simulation, power assist device (active assist and passive assist) [1, 2].

7 DOF exoskeleton robot (CADEN-7) generates the following movements: flexion-extension of the shoulder, abduction-adduction of the shoulder, internal-external rotation of the shoulder, flexion-extension of the elbow, supination-pronation movement of the forearm, flexion-extension of the wrist. In addition to this movements, the robot generates the radial-ulnar deviation [1]. The mechanism is controlled by a PID controller (proportional–integral–derivative controller) and EMG sensors [2].

2. Mechanical design process

For designing and 3D printing of an exoskeleton used for elbow joint rehabilitation it was required to go through the following steps: effectuating the anthropometric measurements of the upper limb circumferences; performing the mechanical design process; choosing the manufacturing technology (3D printing, CNC etc.) and the materials that will be used; choosing the actuation technology: electric actuation (stepper motors, DC motors, CC motors, servo motors etc.), pneumatic actuation (pneumatic cylinder, pneumatic muscles etc.), hydraulic actuation etc.; choosing the mechanical fasteners (self-tapping screws, nuts etc.); making the gripping wristbands of the exoskeleton on the upper limb segments (upper arm and forearm); sizing the CAD model of the exoskeleton; validation of mechanical design dimensions; performing the technological process that has been chosen (3D printing); assembling component parts; making the command and control circuit of the stepper motor; programming the stepper motor– source code; testing the final device.

2.1. Effectuating the anthropometric measurements of the upper limb circumferences

To establish the suitable designing dimensions have been effectuated a number of 5 anthropometric measurements of the upper limb for a number of 20 person (10 women and 10 men), with ages between 22 and 27 years. It was taken into account the measurements of the following anatomical areas, shown in figure 1, as it follows: wrist circumference, noted with C1; forearm circumference, noted C2; the distal zone of the upper arm circumference, noted with C3; the length of the forearm segment, situated between the wrist joint and elbow joint, noted with L1; the length of the upper arm segment, situated between elbow joint and shoulder joint, noted with L2.

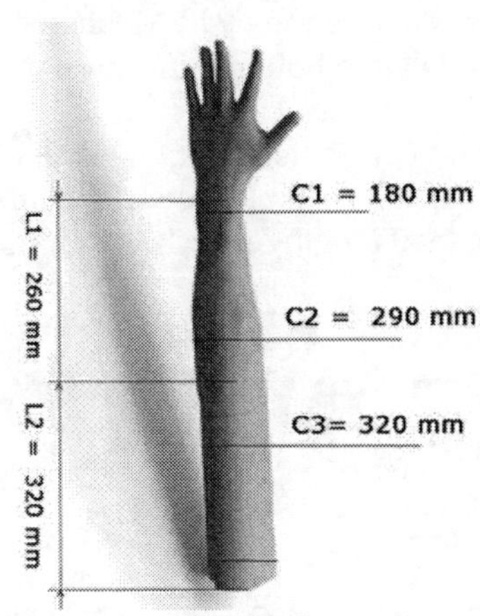

Figure 1. Anthropometric dimensions adopted

IOP Conf. Series: Materials Science and Engineering **568** (2019) 012034 doi:10.1088/1757-899X/568/1/012034

2.2. Performing the mechanical design process

The CAD model of the exoskeleton robot has been designed using the Solid Edge Academic Edition - Siemens PLM Software. The final model is made up of 24 pieces of which only 7 of these were obtained using 3D printing technology. The components of the CAD model are the following: cuffs for supporting the forearm C1-C2, shown in figure 3.a; cuff for supporting the upper arm C3, shown in figure 3.b; passive joints assembly; the holder; active joint; the bolted joint; three gripping wristbands; thirteen gripping screws; NEMA 17 stepper motor, shown in figure 4.a.

Designing process involves the succession of 7 main design stages. The cuff for supporting the wrist (C1) was designed with a radius of 29 mm and a wall thickness of 5 mm. Its width was generated using the Extrude command at a value of 25 mm. The cuff for supporting the forearm (C2) was designed with a radius of 46,5 mm and a wall thickness of 5 mm. Its width was generated using the Extrude command at a value of 30 mm. The distance between the cuffs (C1 and C2) is 160 mm. The cuff for supporting the upper arm (C3) was designed with a radius of 51 mm and a wall thickness of 5 mm. Its width was generated using the Extrude command at a value of 30 mm.

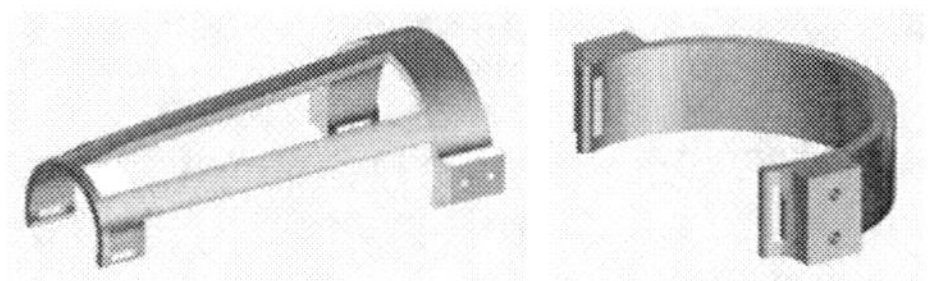

a) b)

Figure 2. The CAD model for: a) Cuffs for supporting the forearm C1-C2; and b) Cuff for supporting the upper arm C3

The joining elements creates a mechanical link between the cuffs. These joints have a length roughly around 84,5 mm respectively 82 mm and a maximum width of 10 mm. The central hole, shown in figure 3, represents the place where the bolted joint will be assembled. It has a outer diameter of 9,5 mm and a inner diameter of 7,5 mm.

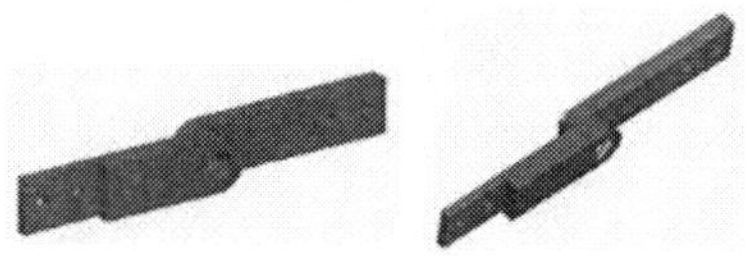

a) b)

Figure 3. The CAD model for: a) Passive joints; b) Mechanical link between the passive joints created by positioning the bolted joint through the existing holes

The bolted joint, shown in figure 4, represents the essential element which creates mobility of the passive subassembly within the biomechanical system. The outer diameter has a value of 6,6 mm and the inner diameter a value of 4,8 mm. The total length of the bolted joint is around 13 mm.

a) b)

Figure 4. The CAD model of the bolted joint; a) Sideway view; a) Bolted joint assembly;

IOP Conf. Series: Materials Science and Engineering **568** (2019) 012034 doi:10.1088/1757-899X/568/1/012034

The assembly process of all the components presented previously has been performed by using the ISO Metric Assembly module of the Solid Edge Academic Edition – Siemens PLM Software. The render image has been achieved using Key Shot for Solid Edge software. The final CAD model of exoskeleton robot is shown in figure 5.

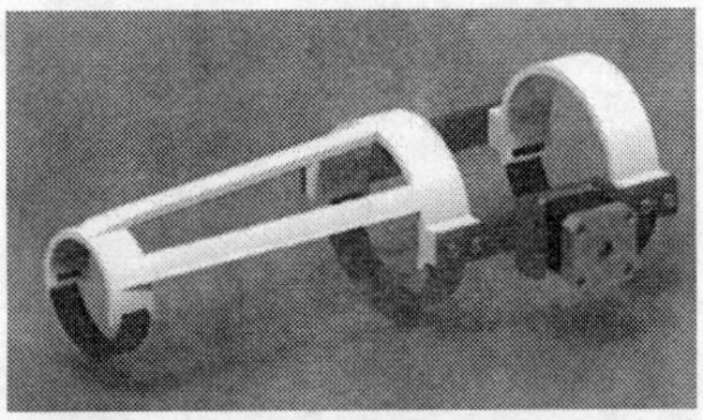

Figure 5. The final CAD model of exoskeleton robot

In the assembly were used a number of eight 4,2x13 self-tapping screws. These were used to perform the assembly between the joining elements and cuffs for supporting the upper arm (C3) and forearm (C1-C2). Four M3X10 screws have been used to mount the stepper motor housing. To perform the joint was used one 4,8x13 self-tapping screw.

To actuate the exoskeleton robot was chosen to be used a NEMA 17 stepper motor, shown in figure 6.a, which has the role of creating simulation of the gradual flexion-extension movement in the elbow joint. Technical characteristics of the stepper motor can be found in reference [3]. It is commanded and powered by means of TB 6600 stepper motor driver, shown in figure 6.b. This is a professional two-phase stepper motor driver. It supports speed and direction control. Technical characteristics of the stepper motor can be found in reference [4].

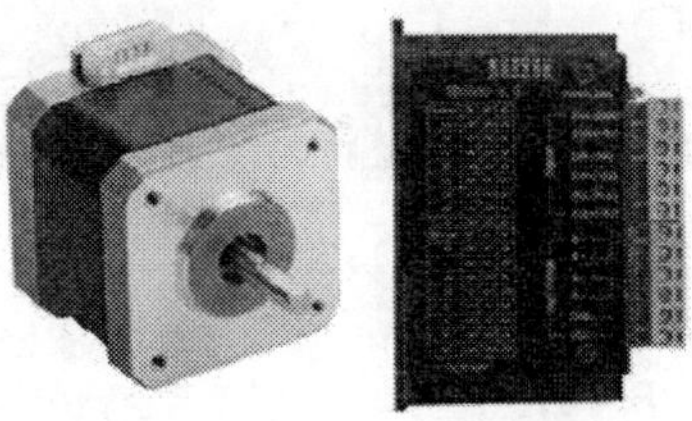

a) b)

Figure 6. a) NEMA 17 stepper motor; b) TB6600 stepper motor driver [3, 4]

3. 3D printing process

The 3D printing process of the final product (bio mechatronic exoskeleton for rehabilitation of elbow joint), was performed by using two models of 3D Ultimaker printers: Ultimaker 3 and Ultimaker 3 Extended. As work material, was chosen the CPE (copolyester) filament from Ultimaker, due to its remarkable mechanical, thermal and electrical properties. Mechanical, thermal and electrical properties of this material can be found in the reference [5]. The execution time of parts totalized a number of 35 hours of 3D printing out of which 22 hours have been allocated for printing the cuffs for supporting the forearm C1-C2. Due to the complex geometric configuration of the printing parts, it was required to use a support material. In this case was chosen the natural PVA filament from Ultimaker that dissolves in water. The dissolving process of the support material, shown in figure 7, took 24 hours. The pieces were placed inside of a transparent aquarium full of water so as to the dissolving process of the support material to be easily monitored. To generate the G-code was used Ultimaker Cura software. The final product achieved by means of rapid prototyping technology is shown in figure 8.

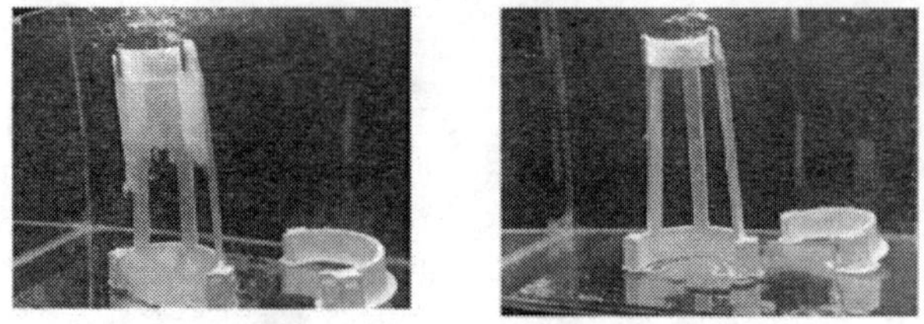

a) b)

Figure 7. Dissolving process of the support material: a) After 12 hours; b) After 24 hours

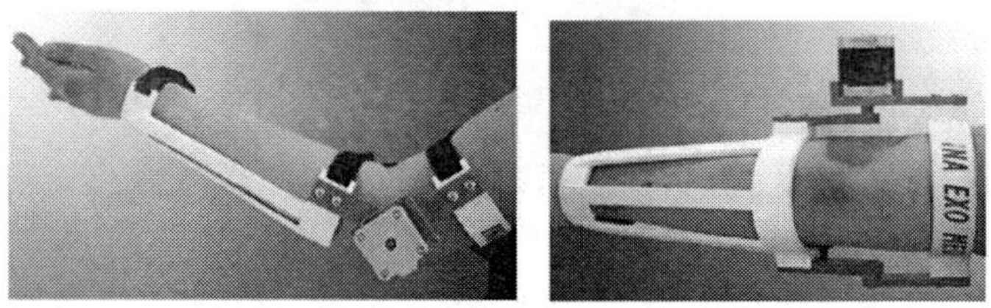

a) b)

Figure 8. Attaching the bio mechatronic exoskeleton for rehabilitation of elbow joint on the user's arm: a) In a flexion position; b) In extension position

4. Conclusions

Although the constructive solution chosen for this project is not very complex, the exoskeleton robot achieved by means of rapid prototyping technology presents a series of advantages: it's easy to put into operation and to be used, the amplitude and frequency of movements can be easily modified from the software, the cost price is low, can be used for didactic purposes in the Medical Engineering Laboratory for a number of specialized disciplines, maintenance is easy to carry out etc.

Acknowledgements

Practical realization of this project was made possible thanks to funding received in the| My License Project-2019 competition held in the Transylvania University of Brasov. I accord special thanks to Alexandru Dobre, the manager of 3D DOT Braşov, for providing the 3D printing equipment and for all the support given during the course of the project.

References:

[1] R. A. R. C. Gopura et al. 2009, *Development and Control of Upper-Limb Exoskeleton Robots*, University of Moratuwa, Sri Lanka

[2] Md Rasedul Islam, C. Spiewak, M. H. Rahman and R. Fareh 2017, *A Brief Review on Robotic Exoskeletons for Upper Extremity Rehabilitation to Find the Gap between Research Porotype and Commercial Type*, Adv Robot Autom Journal, DOI: 10.4172/2168-9695.100017

[3] 3D Modular Stsyems, *NEMA-17-KS42STH40 Data sheet,* link accessed on: 20.04.2019, http://3dmodularsystems.com/en/motors/271-nema-17-ks42sth40-stepper-motor-for-scalar-family-3d-printers.html

[4] DF Robot, *TB6600 Stepper Motor Driver User Guide,* link accessed on: 21.04.2019, https://forum.arduino.cc/index.php?action=dlattach;topic=558523.0;attach=266385
Ultimaker *2018, Technical data sheet CPE - UM180821 TDS CPE RB V11*, link accessed on: 22.04.2019,https://ultimaker.com/download/74973/UM180821%20TDS%20CPE%20RB%20V11.pdf

IOP Conf. Series: Materials Science and Engineering **568** (2019) 012035 doi:10.1088/1757-899X/568/1/012035

Study of the mechanical and electrical characteristics of the synchronous motor with varying resistive torque

C Cristea[1] and I Stroe[1]

[1] Centre "Advanced Research on Mechatronics", Transilvania University of Braşov, Romania

cosmin.cristea@unitbv.ro

Abstract. The paper presents the mechanical and electrical characteristics of the synchronous motor for imposed moment variations at the shaft. The test setup is presented, used to determine the electrical and mechanical characteristics of the synchronous motor. Modelling the resistive torque allows the loading with different torsion vibrations. It is presented the used experimental workbench in order to determine the mechanical and electrical characteristics of synchronous motors in such conditions. The engine torque model allows the shaft to be loaded with variable resistive torque values generated by an electromechanical brake. The characteristics presented show the mechanical characteristic of the synchronous motor. Mechanical rotation speed characteristics are presented based on the variation of the resistive torque for 40 seconds, taking note of the value of the resistance when the motor exits synchronism. The determination of these characteristics are made based on the variation of the excitation current applied to the rotor winding of the synchronous motor.

1. Introduction

The synchronous machine is reversible from an energetic and point of view.

The inducing magnetic-field for synchronous machines can be generated depending on the machine's construction. The frequency of the induced voltage is dependent on the rotational speed of the rotor and the number of poles of the machine. The current generated by the windings has the same frequency as the induced voltage. The magnetic field induced by symmetrical poly-phase machines is circular and has a rotational speed related to this frequency.

Between the two magnetic fields, inducing and induced, in relative resting state, there is an interaction specific to the stationary regime. Depending on the frequency, the rotational speed of the rotor is equal to the rotation speed of the rotating magnetic field generating the synchronicity state of the synchronous machine. The excitation current in the inductor winding is generated by the continuous current supplying this winding [1].

Depending on its location, the excitation winding in the machine can generate certain powers. Inner-pole machines where the excitation winding is located in the rotor generate significant power. Outer-pole machines where the excitation winding is located in the stator generate powers up to 20 kVA [2].

Cylindrical poles are used for rotational speeds of 1500 rpm and 3000 rpm at a frequency of 50Hz. Salient poles are used for rotational speeds of up to 1500 rpm. The shape of the rotor, dependent on the type of poles, may define the use of motors for different rotational speeds. The synchronous machine has a great use as a generating machine. If a synchronous machine is driven by a steam turbine, the synchronous generator is a turbo-generator with rotational speeds of over 1000 rpm. If the synchronous machine is of the hydraulic type, the synchronous generator is a hydro generator [1].

2. The synchronous motor

The synchronous machine has a rigid characteristic. The rotational speed does not depend on the load [3]. The engine keeps its rotational speed constant. The slope of the characteristic is linear and is given by expression [4].

$$M = \frac{d\Omega}{dM} \tag{1}$$

3. Mechanical and electrical characteristics of the synchronous motor

The choice of the type of engine depending on the application is determined by the specific power and rotational speed of the application, its rational use, the possibility of speed adjustment etc., require the study of the mechanical and electric characteristics of the motor. Each type of motor has a multitude of characteristics [5].

For determining the mechanical and electrical characteristics of the motors an experimental test setup is used comprised of a tripolar switch with two positions, a power supply with thermal overload protection, a digital/analog multimmeter, an adjustable excitation autotransformer, a controller and a brake servomotor and a synchronous machine.

The excitation winding is fed through an autotransformer. This voltage, as well as the three-phase supply voltage of stator winding, connected in the star configuration is monitored by two digital multimeters. In the Activeservo dedicated software, the controller monitors and induces braking at predefined moments of the engine torque by means of a servomotor [5].

4. Torque modelling for characteristics determination

The operation of the synchronous motor is determined by the variations in torque. The test program enables the experimental-setup to generate those application-specific resistive torques. The most important part of this system is the controller which, through the ActiveServo software, transmits the servomotor predefined torques for testing [6].

The Active Servo software is a tool for monitoring and registering the machine's characteristic in order to determine the static and dynamic operating points [5].

The controller generates resistive torques imposed on the studied characteristic.

The servomotor is in close-connection with the controller and is a monitoring and braking system. The servomotor has multiple characteristics, which are related to the speed regulation within very wide limits, the mechanical characteristics generated are linear, is protected against self-start and a high overload capacity [6].

5. Mechanical and electrical characteristics over time for variable torque values.

For the experimental determinations performed to study the characteristics, a synchronous motor with the following characteristics was used: P_n = 0.27kW, I_n = 1.5A, f = 50Hz, n = 1500rpm, U_{err} = 20Vcc; I_{err} = 4A. The excitation source generated a voltage in steps of 0-20 V_{cc} at imposed excitation current values of 0.5A, 1A, 2A, 3A.

For the studied synchronous motor we will observe the time dependence of the following quantities: The current of the rotor circuit, I [A]; Mechanical power at the shaft P2 [W]; Motor torque M [Nm] and Speed n [rpm].

The engine is driven at synchronism speed for 6 excitation current values, between 0.5 and 3A, in 0.5A steps. Via programming we have imposed a torque characteristic at which for 100 seconds we plotted the torque so that for variable values forced from 0 to the maximum synchronicity value we reduced and alternated with mechanical shocks up to the return value to the nominal values.

5.1. For I_{err}=0.5A

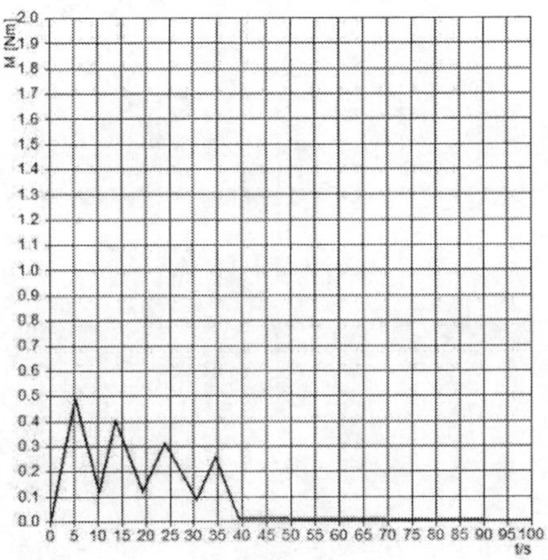

Figure 1. Imposed torque characteristic for I_{err}=0.5A

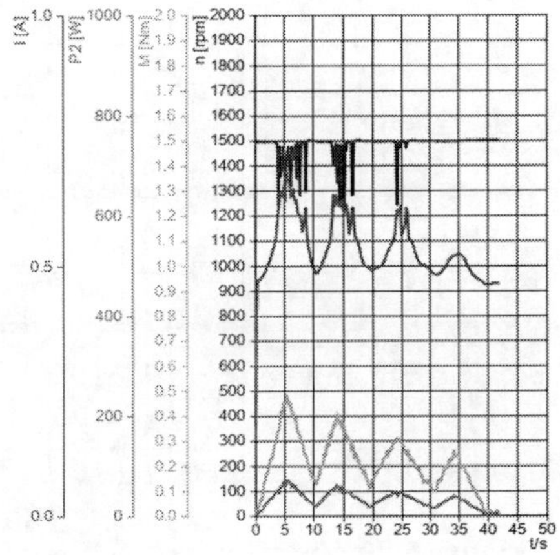

Figure 2. Mechanical and electrical characteristics for time-varying torques for an excitation current of I_{err}=0.5A

By analyzing the previous chart, it is noticed that the engine has a constant speed at increasing torque values, but tends to de-synchronize at higher torque values, however does not exit synchronism. Speed returns to a nominal value along with decreasing torque values. The electrical current in the stator circuit maintains the required torque curve, as well as the mechanical power at the motor shaft.

5.2. For I_{err} =1A

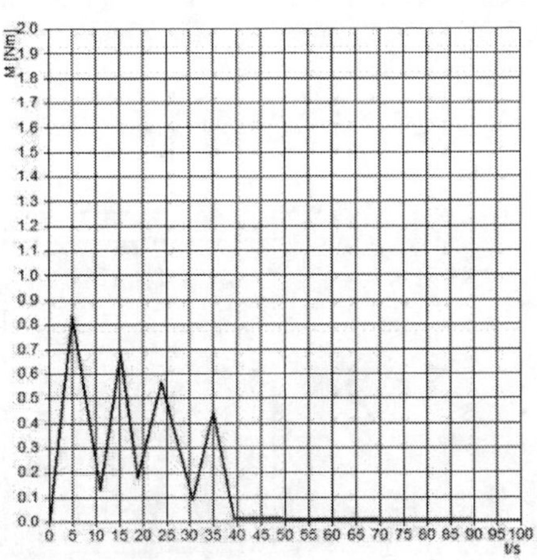

Figure 3. Imposed torque characteristic for I_{err} =1A

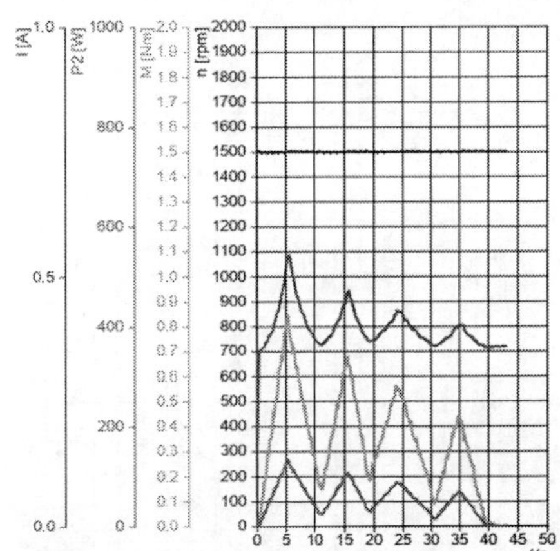

Figure 4. Mechanical and electrical characteristics for time-varying torques for an excitation current of I_{err}=1A

By analyzing the previous graph, it is noticed that the speed is kept constant at alternating variations of torque, if the torque increase or decrease is done linearly increasing within 5s limit. The electric current in the stator circuit maintains the imposed torque curve, as well as the mechanical power at the engine shaft.

5.3. For I_{err} =1,5A

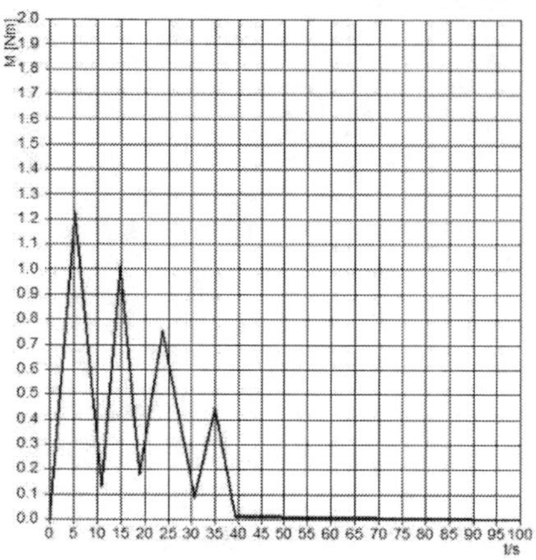

Figure 5. Imposed torque characteristic for I_{err}=1.5 A

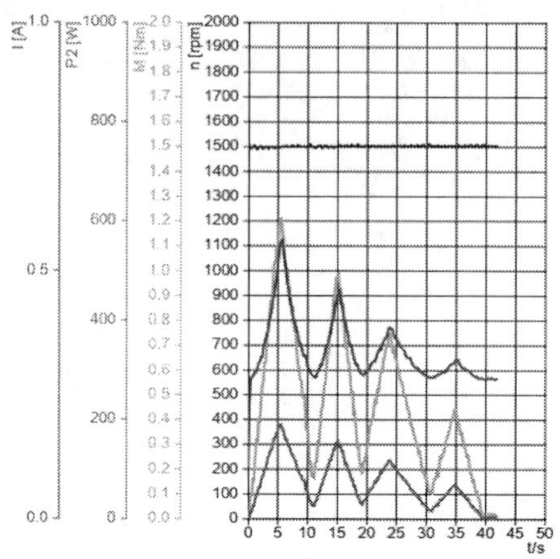

Figure 6. Mechanical and electrical characteristics for time-varying torques for an excitation current of I_{err}=1.5A

When analyzing the torque variation, we increased the maximum torque to 1.2Nm so that it came close to the exit-value of synchronism, we increased the other torque variables, but making sure the studied allure is kept.

From the previous graph, it is noticed that the speed is kept constant at alternating variations of torque, if the torque increase or decrease is done linearly increasing within 5s limit. The electric current in the stator circuit maintains the imposed torque curve, as well as the mechanical power at the engine shaft.

5.4. For I_{err} =2A

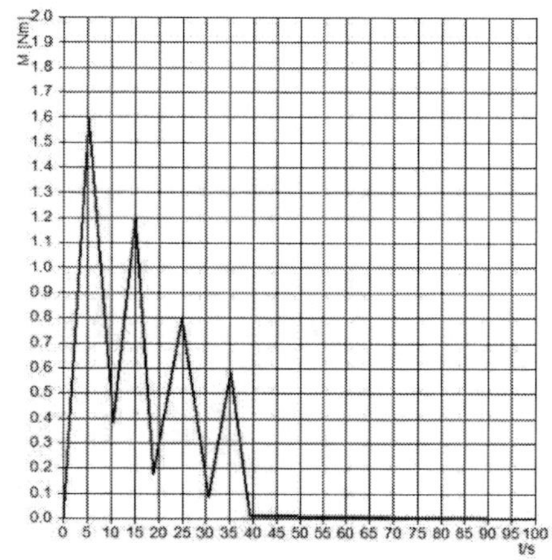

Figure 7. Imposed torque characteristic for Ierr =2 A

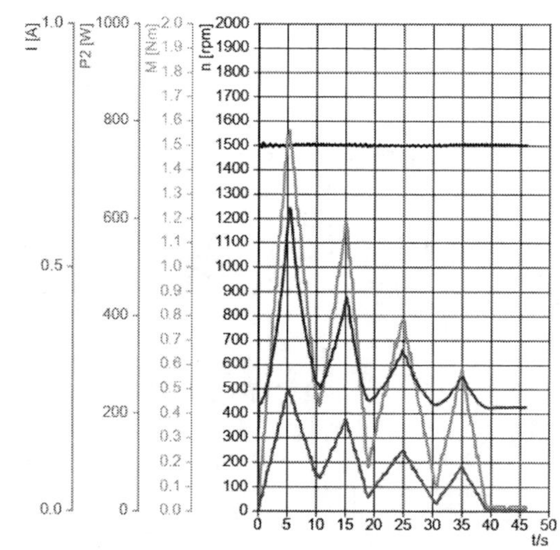

Figure 8. Mechanical and electrical characteristics for time-varying torques for an excitation current of I_{err}=2A

IOP Conf. Series: Materials Science and Engineering **568** (2019) 012035 doi:10.1088/1757-899X/568/1/012035

When analyzing the torque variation, we increased the maximum torque to 1.6Nm so that it came close to the exit-value of synchronism, we increased the other torque variables, but making sure the studied allure is kept.

From the previous graph, it is noticed that the speed is kept constant at alternating variations of torque, if the torque increase or decrease is done linearly increasing within 5s limit. Also, when increasing the excitation current, the speed variations are flattened.

The electric current in the stator circuit maintains the imposed torque curve, as well as the mechanical power at the engine shaft. From observation, if we increase the excitation current value, the engine resists larger torque shocks

5.5. For I_{err} =2.5A

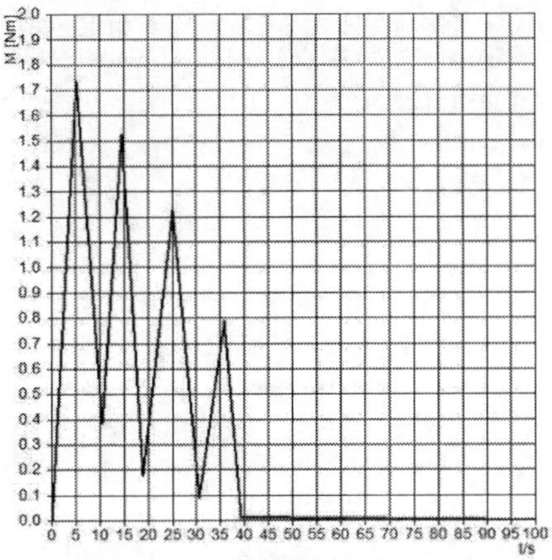

Figure 9. Imposed torque characteristic for I_{err} =2.5 A

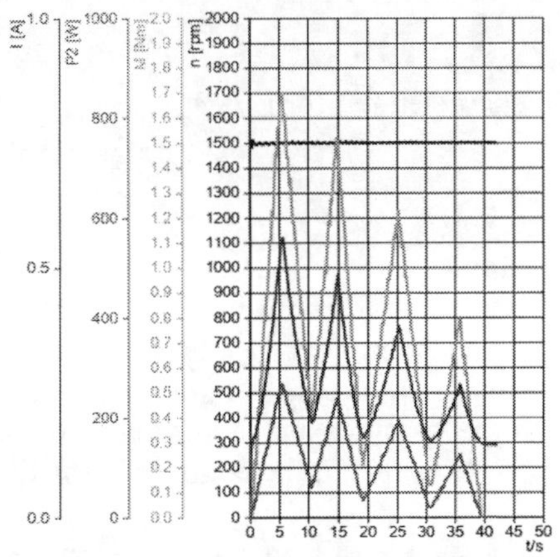

Figure 10. Mechanical and electrical characteristics for time-varying torques for an excitation current of I_{err}=2.5A

When analyzing the torque variation, we increased the maximum torque to 1.75Nm so that it came close to the exit-value of synchronism, we increased the other torque variables, but making sure the studied allure is kept.

From the previous graph, it is noticed that the speed is kept constant at alternating variations of torque, if the torque increase or decrease is done linearly increasing within 5s limit. Also, when increasing the excitation current, the speed variations are flattened.

The electric current in the stator circuit maintains the imposed torque curve, as well as the mechanical power at the engine shaft.

5.6. For I_{err} =3A

When analyzing the torque variation, we increased the maximum torque to 1.95Nm so that it came close to the exit-value of synchronism, we increased the other torque variables, but making sure the studied allure is kept.

From the graph presented in figure 12, it is noticed that the speed is kept constant at alternating variations of torque, if the torque increase or decrease is done linearly increasing within 5s limit. Also, when increasing the excitation current, the speed variations are flattened.

The electric current in the stator circuit maintains the imposed torque curve, as well as the mechanical power at the engine shaft.

From observation, if we increase the excitation current value, the engine resists larger torque shocks

IOP Conf. Series: Materials Science and Engineering **568** (2019) 012035 doi:10.1088/1757-899X/568/1/012035

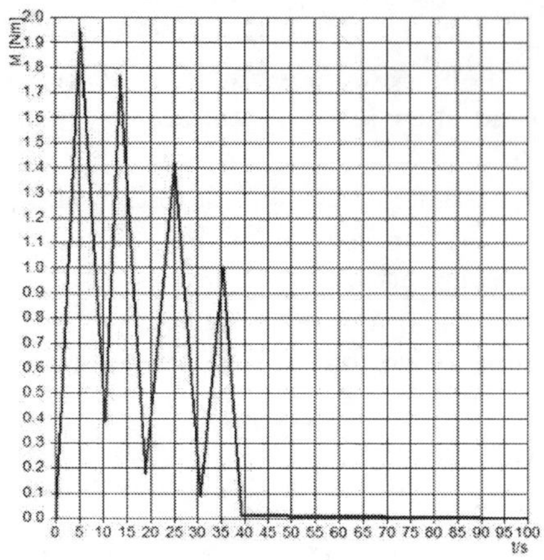

Figure 11. Imposed torque characteristic for I_{err} =3 A

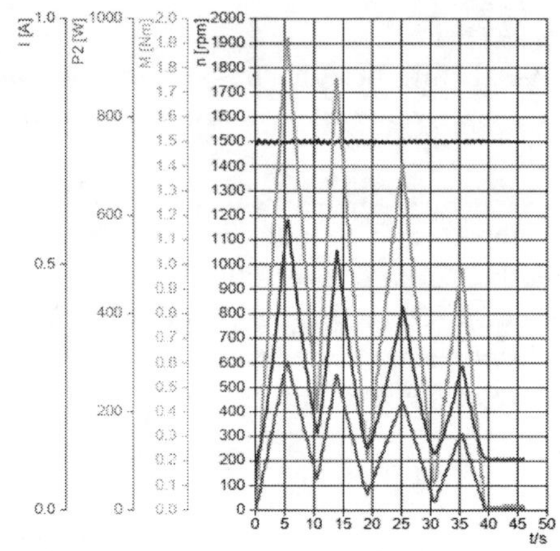

Figure 12. Mechanical and electrical characteristics for time-varying torques for an excitation current of I_{err}=3A

6. Conclusions

From the comparative analysis of the graphs, it is observed the influence of the excitation current on the synchronous motor when it is subjected to different values of resistive torque, which comes close to the maximum torque of the motor near the moment of de-synchronization.

Thus, if the synchronous motor tends to de-synchronize at a 0.5A excitation current, we notice that this does not happen and the engine returns to synchronous speed. If we increase the excitation current value, the engine resists larger torque shocks but does not de-synchronize and maintains its speed relatively constant within an acceptable range of variation.

The value of the excitation current is determined to be decisive and its variability influences keeping the engine's speed constant with varying resistive torque at the shaft.

7. References

[1] Toma B 2002 *Electrical Machines, *Theory (Maşini electrice, *Teorie)* Editura ASAB, Bucureşti, ISBN 973-86010-0-2

[2] Nedelcu V 1968 *Electrical Machines,(Maşini electrice)* Editura Didactică şi Pedagogică, Bucureşti

[3] Teodoru E C and Gogu M 2000 *Electrical Machines,(Maşini electrice)* Editura Universităţii Tehnice Gh.Asachi , Iaşi

[4] Engine Operating Modes Continuous Current and the synchronous motor (Regimuri de functionare ale motorului de current continuu si ale motorului sincron)[online] 10 March 2019. Available at: http://www.euedia.tuiasi.ro/uee/uee_files/Lucrarea%2014.pdf

[5] Cristea C and Stroe I 2017 *The study of the synchronous motor*, Bulletin of the Transilvania University of Braşov • Vol. 10 (59) No. 2 – 2017 Series I: Engineering Sciences, ISSN 2065-2119 (Print), ISSN 2065-2127 (CD-ROM)

[6] L@bsoft Multimedia Learning Environment,Version 6.05 - LUCAS NULLE GmbH 2000-2011, *Synchronous electric machines and electric machines with sliding contacts (Maşini electrice sincrone şi maşini electrice cu contacte alunecătoare)*

IOP Conf. Series: Materials Science and Engineering **568** (2019) 012036 doi:10.1088/1757-899X/568/1/012036

Influence of the silent chain link geometry on the link / chain guide contact

M T Lates[1] and C C Gavrila[1]

[1]Product Design, Mechatronics and Environment Department, "Transilvania" University of Brasov, Brasov, Romania

latesmt@unitbv.ro

Abstract. The mechanical behaviour of the chain drive transmissions is influenced of the contacts between elements being in relative motions, as: sprockets and links, links and pins and links and tensioning guides. These contacts have a complex behaviour and due to that it is required to be analysed separately, in order to find out their influences on the dynamical behaviour and on the tribological properties of the entire transmission. The article presents the analysis of the influence in the case of silent chain link geometry on the link/chain guide contact. It is studied the variation of the local normal maximum pressures and the variation of the contact area's width with the main radius of the chain link and with the normal applied force. The analysis is based on the finite element modelling. Finally there are given conclusions regarding the practical applications of the obtained results.

1. Introduction

The efficiency of the mechanical transmissions is influenced mainly by the lubricants used in order to reduce the friction between the elements being in contact, by the materials used in the construction of the mechanical parts and by the local geometry (dimensions and shapes) of the contact zone between the mechanical elements being in relative motions [1, 2, 3, 4, 5, 6].

Chain drives are characterised by high loading capacities. Due to this characteristic, the chain drive transmissions are used widely in the automotives industry and in the construction of the mechanical systems as industrial robots, machine tools and machines use in agriculture and mining industry. The main problems are represented by their efficiency and the noises developed during their functioning [7, 8, 9, 10]. The literature offers directions in order to reduce the friction and the noises [11, 12, 13]. One solution is represented by the silent chains; in this case the chains links have teeth with are gearing with the sprocket and, due to this, a reduction of the noise is obtained.

Generally, the friction developed in a chain drive transmission can be split in the friction in the chain links, the friction between the chain links and the sprockets and the chain links and the tensioning guide [7, 1, 12, 14]. Regarding the tensioning guide, the active part of it is made by polyamides which are characterised by small friction coefficients in combination with steel elements, small wear, high endurance and good mechanical behaviour at high loadings and high temperatures [15, 16, 17, 18, 19, 20].

The frictional behaviour of the chain link – tensioning guide contact (the mechanical efficiency, the wear and the level of the noises) is influenced by the geometrical characteristics of the contact area [8, 10, 12]. According to this, the paper studies the influence of the toothed chain link geometry on the contact with the tensioning guide.

IOP Conf. Series: Materials Science and Engineering **568** (2019) 012036 doi:10.1088/1757-899X/568/1/012036

2. The model

The theoretical analysis is achieved by using the finite element method based software Ansys 15.0; the geometrical model is made in Catia V5R18 – figure 1. The toothed chain link is in contact with a tensioning guide made by a polyamide type material which has the radius $\boldsymbol{R}$=120 mm.

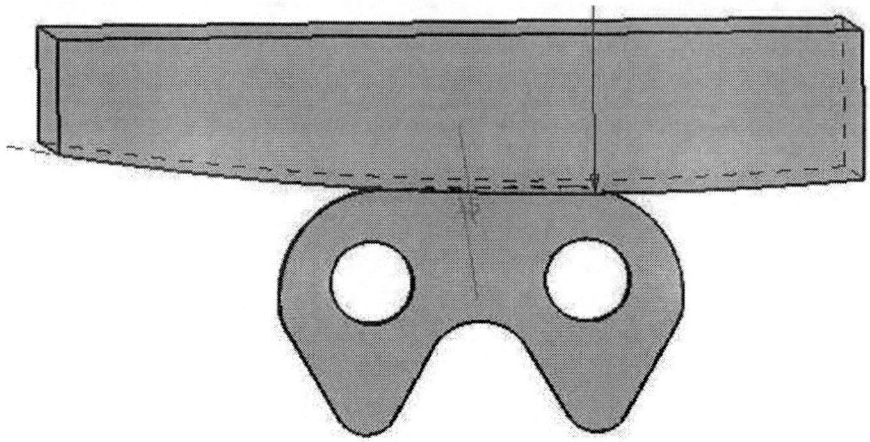

Figure 1. The geometrical model.

The geometry and the dimensions of the toothed chain link are presented in figure 2. For the finite element based analysis only the radius $\boldsymbol{R}$=4 mm is considered variable with values in the range of: 3.5; 3.75; 4; 4.25; 4.5 – figure 3. Other main dimensions are kept frozen in order to keep unmodified the values of the chains pitch, the height, the length and the width of the link, the diameters of the pins.

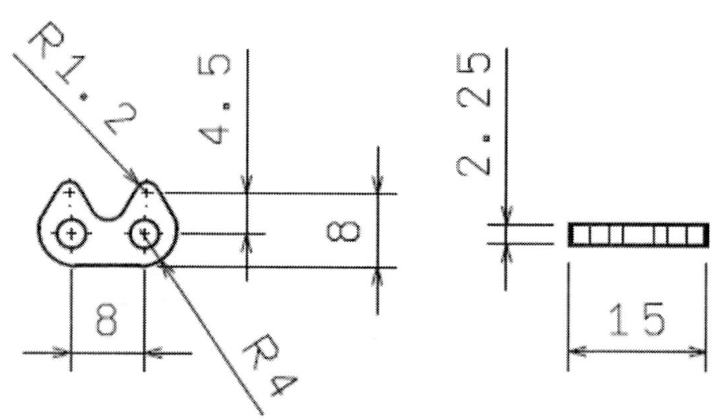

Figure 2. The toothed chain link.

By reducing the radius, the length of the contact zone between the link and the guide increases; high value of the radius produce a decrease of the length for the contact zone between the link and the guide.

Figure 4 presents the finite model of the link/guide contact. The tensioning force which is applied on the guide has the values of: 10 N, 15 N, 20 N, 25 N, 30 N; the holes where the pins are mounted are clamped. The links are made by steel and the guide is made by polyamide. The material data are: E_1, E_2 represent the Young's modulus for the skate and the link and ν_1, ν_2 represent the Poisson's coefficient of the two materials (for steel – the link, $E_1=2\cdot10^5$ MPa, $\nu_1=0.266$; for plastic – the skate, $E_2=2.2\cdot10^3$ MPa, $\nu_2=0.38$ [21]).

IOP Conf. Series: Materials Science and Engineering **568** (2019) 012036 doi:10.1088/1757-899X/568/1/012036

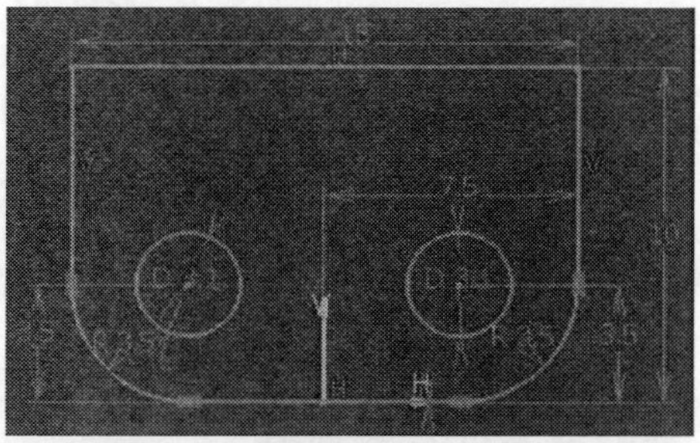

R=3.5 mm

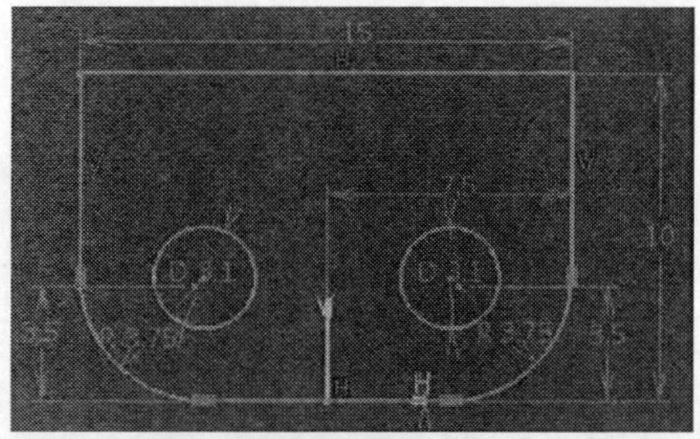

R=3.75 mm

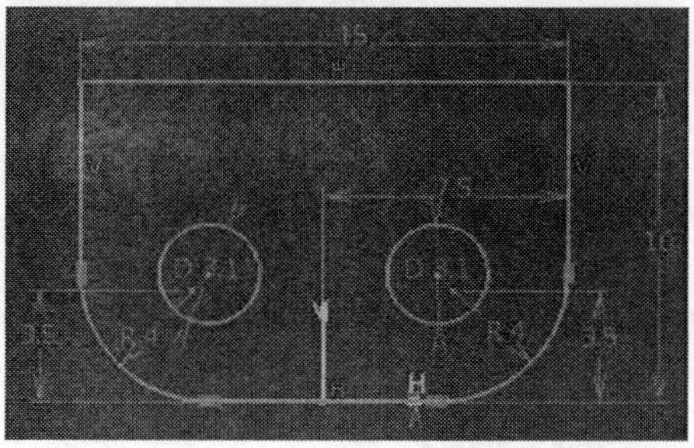

R=4 mm

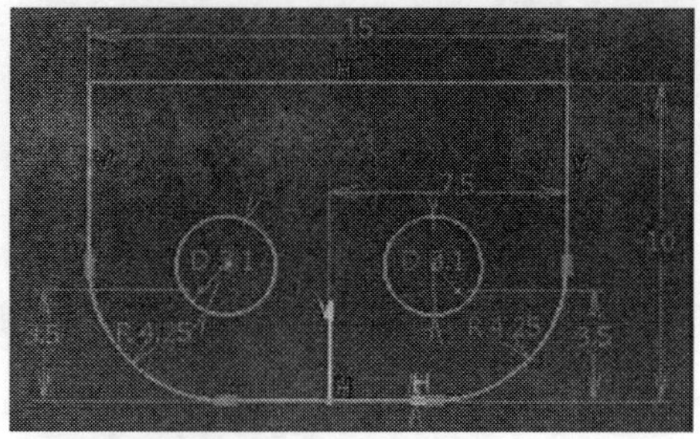

R=4.25 mm

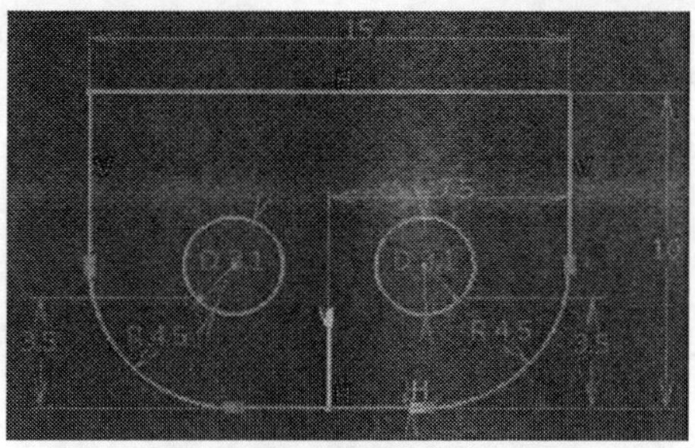

R=4.5 mm

Figure 3. The link shapes.

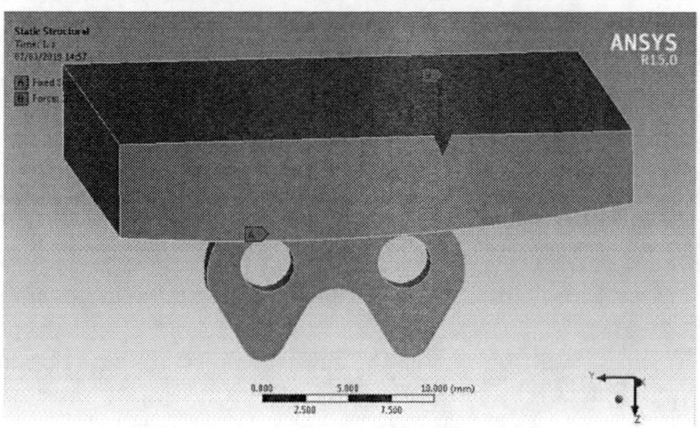

Figure 4. The finite element model.

3. Results and conclusions

The results of the finite element analysis presents as graphs, the variation of the contact pressure and of the contact areas half width with the normal tensioning force and the radius of the chain link.

IOP Conf. Series: Materials Science and Engineering **568** (2019) 012036 doi:10.1088/1757-899X/568/1/012036

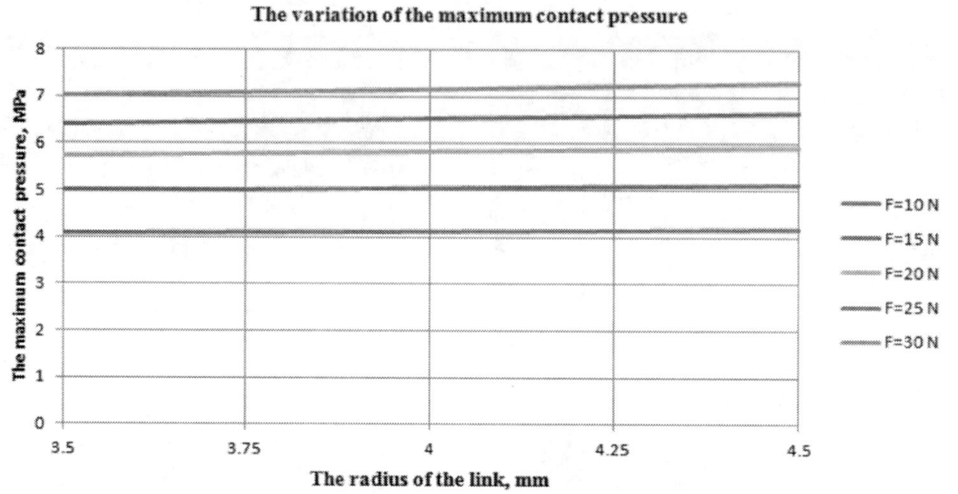

Figure 5. The variation of the maximum contact pressure.

Figure 5 presents the variation of the maximum contact pressure with the normal tensioning force which is applied on the guide and the radius of the chain link. The value of the maximum contact pressure increases with the increasing of the normal force and increases slightly with the increasing of the chain link's radius; higher values for the increasing of the contact pressure with the chain link's radius are obtained for higher normal forces.

The variation of the width of the contact area is presented in figure 6. The value of the width increases with the increasing of the normal force but is decreases slowly with the increasing of the chain link's radius; higher values for the decreasing of the width with the chain link's radius are obtained for higher normal forces.

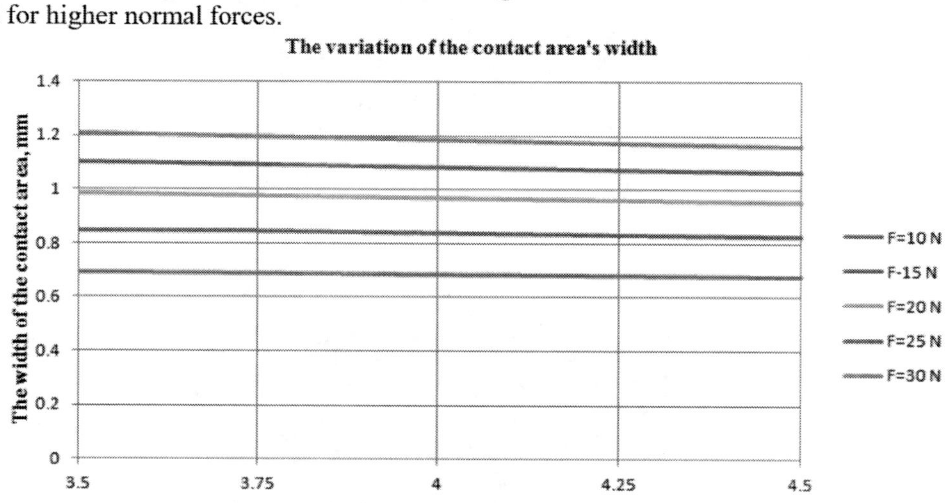

Figure 6. The variation of the width of the contact area.

According to the results, small values of the contact pressures are obtained in the case of small values of the chain links radiuses and these results offers the conditions for good lubrication of the contacts between the chain links and the guide; also small values of the width of the contact area are obtained in the same conditions. The minimum values of the chain links radius could be chosen depending on technological and dynamics behaviour constraints, while the variation of the radius is influencing the inertial characteristics if the chain.

IOP Conf. Series: Materials Science and Engineering **568** (2019) 012036 doi:10.1088/1757-899X/568/1/012036

References

[1] Cretescu N, Neagoe M and Saulescu R 2016 Kinematic and dynamic analysis of a 4DOF parallel robot with flexible links *Mechanisms and Machine Science* **46** pp 473-481

[2] Neagoe M, Saulescu R, Jaliu C and Cretescu N 2016 Novel speed increaser used in counter-rotating wind turbines *Mechanisms and Machine Science* **46** pp 143-151

[3] Saulescu R, Neagoe M and Jaliu C 2018 Conceptual synthesis of speed increasers for wind turbine conversion systems *Energies* **11-2257**

[4] Saulescu R, Neagoe M and Jaliu C 2016 Improving the energy performance of wind turbines implemented in the built environment using counter-rotating planetary transmissions *Materials Science and Engineering* **147 (1)**

[5] Ciobanu D, Jaliu, C and Saulescu, R 2014 Chain Tracking System for Solar Thermal Collector *Applied Mechanics and Materials* 658 pp 35-40

[6] Ciobanu D 2011 Conceptual design of a solar thermal system with dish solar collector *Journal of Environment Engineering and Management* **10 (8)**

[7] War X and Dwyer-Joyce R S Model experiments on automotive chain drive systems *Elsevier Science* **39** pp 851-861

[8] Pedersen S L 2005 Model of contact between rollers and sprockets in chain-drive systems *Archive of Applied Mechanics* **74** pp 489-508

[9] Metilkov S A, Berezhnoi S B and Yunin V V 2008 Wear of hinges in roller drive chain *Russian Engineering Research* **28-9** pp 839-844

[10] Metilkov S A and Yunin V V 2008 Influence of wear of a roller drive chain on transmission fitness *Russian Engineering Research* **28-8** pp 741-745

[11] Hyakutake T, Inagaki M, Matduda M, Hakamada N and Teramaki Y 2001 Measuring of friction in timing chain *Society of Automotive Engineers of Japan* **22** pp 343-347

[12] Fink T and Bodestein H 2011 Friction reduction potentials in chain drives *MTZ Worldwide Edition* **72** pp 46-51

[13] Sergeev S A and Moskalev D V 2009 Parametric Optimization of Chain-Transmission Sprockets *Russian Engineering Research* **29-5** pp 452-455

[14] J van Ruiten van J, Proost R and Meuwissen M 2012 How the choise of the polyamide type in timing chains tensioning systems affects the CO_2 emission and fuel economy of internal combustion engines *Presentation at VDI Veetiltrieb un Zylinderkopf*

[15] Mouhmid B, Imad A, Benseddiq N and Lecompte D 2009 An experimental analysis of fracture mechanics of short glass fibre reinforced polyamide 6.6 (SGFR-PA66) *Journal of Composite Science and Technology* **69** pp 2521-2526

[16] De Monte M, Moosbruger E and Quaresimin M 2010 Influence of temperatura and thickness on the off-axis behavior of short glass fiber reinforced polyamide 6.6 – quasi-static loading *Journal of Composites* **41** pp 859-871

[17] Chiu F-C and Kao G-F 2012 Polyamide 46/multi-walled carbon nanotube nanocomposites with enhanced thermal, electrical and mechanical properties *Journal of Composites* **43** pp 208-218.

[18] Gordon D H and Kukureka S N 2009 The wear and friction of polyamide 46 and polyamide 46/aramid-fibre composites in sliding–rolling contact *Wear* pp 669-678

[19] Buccella M, Dorigato A, Pasqualini E, Caldara M and Fambri L 2012 Thermo-mechanical properties of Polyamide 6 chemically modified by chain extension with Polyamide/Polycarbonate blend *Journal of Polymer Research* **19** pp 9935-9944

[20] Akiyama M, Yamaguchi T, Matsumoto K and Hokkirigawa K 2010 Friction and wear of polyamide 66 composites filled with RB ceramics particles under dry condition *Tribology Online* **5 (2)** pp 87-91

[21] Pereira C M, Ramalho A M and Ambrosio J A 2011 A critical overview of internal and external cylinder contact force models *Nonlinear Dynamics* **63** pp 681-697

IOP Conf. Series: Materials Science and Engineering **568** (2019) 012037 doi:10.1088/1757-899X/568/1/012037

Vibrations influence on the operation of gears

V Ionica[1], M Stanescu[1] , C Miritoiu[1] and I Geonea[1]

[1]Faculty of Mechanics, University of Craiova, Craiova, Romania.

e-mail: igeonea@yahoo.com

Abstract. The purpose to optimize gears requires a very good understanding of their dynamic behaviour. Very important is the influence of vibrations on the operation of the gears. As a result of environmental constraints such as acoustics, gears are increasingly subject to stringent requirements in terms of load capacity, efficiency, noise and vibrations. The deformations of the gears involve deformations, in particular, located near the contacts that require great finesse. Gearboxes are equally sources of current vibrations as they are commonly used in industry. They are the subject of numerous studies to characterize their kinematic and dynamic behaviour. At present, research activities focus on the development of multidisciplinary experimental, theoretical and numerical skills that are put into play when designing structures, machine elements or, in general, mechanical systems. In this paper, we have proposed to find the dynamic response of the gearbox to the action of external disturbing factors. To this end, we have called for the use of integral transformations which, algebraizing the problem, greatly simplifies calculations.

1. Introduction

The control and estimation of vibrations of mechanical systems is an important part of the design of new products [1]. Numerous sources of vibration may exist in a system such as friction and shocks [2, 3, 4, 5]. Gearboxes are equally sources of current vibrations as they are commonly used in industry. They are the subject of numerous studies to characterize their cinematic and dynamic behavior [6, 7, 8]. There are various approaches to modeling vibrations and gear noise [9, 10, 11]. The literature highlights the need to integrate both the elasticity of gear components and the descriptions of its vibrations in a three-dimensional dynamic behavior model [12, 13]. The deformations of the gears involve deformations, in particular, located near the contacts that require great finesse. As a result, we are encouraged to move towards a simplified model with respect to the three-dimensional finite element models, yet still sufficiently precise to study the elasticity of the engaging elements while permitting the integration of relatively complex geometries of the deformable wheels [14, 15]. As we have already shown, engagement appears to be the main source of excitation in power transmission components. The instantaneous movements of each wheel are represented by six degrees of freedom (three translations and three rotations). In [16] there is presented, in a multidisciplinary approach, the simplified mechanical model for the vibro-acoustic study of a gear, the elements of the gearing behaving like rigid solids. The objective of this approach is to optimize a mechatronic product. At present, research activities focus on the development of multidisciplinary experimental, theoretical and numerical skills that are put into play when designing structures, machine elements or, in general, mechanical systems [17, 18].

The aim of the research is to improve knowledge of the behavior of materials and structures, develop models and tools useful in designing structures and machines, and harness a technical culture of analytical, designing and manufacturing methodologies. This research is based on materials science, nonlinear mechanics of solids, fluid and coupled systems, acoustics, training and processing techniques, experimental measurement methods and numerical modeling. The appearance of simulation times is the capital for this type of analysis where the number of iterations can be significant. For this reason, it was decided to model the reducer by approaching rigid bodies. At the same time, the need to obtain vibrations on the outer surface of the gear requires the finite element method to be applied to the latter. The first part of the article shows the modeling obtained by coupling multiple rigid bodies with the finite elements of an epicycloidal reducer. The second part illustrates the multi physical modeling with Dymola. It is here to model an epicycloids reducer with straight teeth, proposing for this purpose a hybrid model of multiple rigid bodies / finite elements [19, 20]. A more complete hybrid model is found in the paper [21]. In this paper, we have proposed to find the dynamic response of the gearbox to the action of external disturbing factors. To this end, we have called for the use of integral transformations that, algebraizing the problem, greatly simplifies calculations.

2. Mechanical and mathematical models

In Figure 1, simulation of gearing is managed through the mechanical model that contains a spring and a shock absorber in parallel.

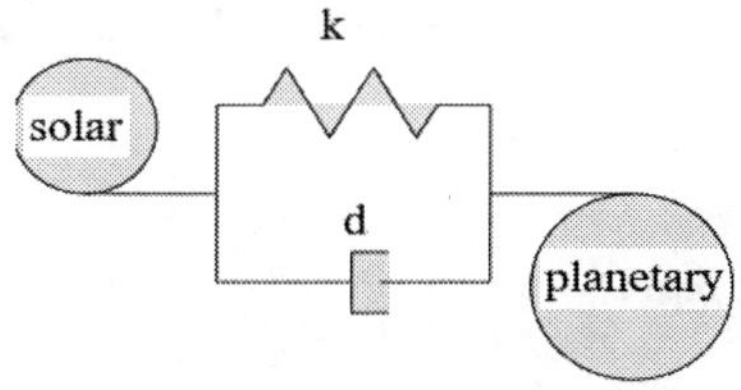

Figure 1. Mechanical model.

The mathematical model of the movement could be written in matrix form:

$$M\ddot{q}+D\dot{q}+Kq=F \tag{1}$$

The matrix of masses M, the damping matrix D, the stiffness matrix K and the excitation matrix F having the following representations, respectively:

$$M=\begin{bmatrix} M_s & 0 & 0 & 0 & 0 \\ 0 & M_{p1} & 0 & 0 & 0 \\ 0 & 0 & M_{p2} & 0 & 0 \\ 0 & 0 & 0 & M_{p3} & 0 \\ 0 & 0 & 0 & 0 & M_{p4} \end{bmatrix} \tag{2}$$

$$D=\begin{bmatrix} d_{sp} & -d_{sp1} & -d_{sp2} & -d_{sp3} & -d_{sp} \\ 0 & d_{sp1}+d_{rp1} & 0 & 0 & d_{sp1}-d_{rp1} \\ 0 & 0 & d_{sp2}+d_{rp2} & 0 & d_{sp2}-d_{rp2} \\ 0 & 0 & 0 & d_{sp3}+d_{rp3} & d_{sp3}-d_{rp3} \\ 0 & 0 & 0 & 0 & d_{sp}+d_{rp} \end{bmatrix} \tag{3}$$

$$K=\begin{bmatrix} k_{sp} & -k_{sp1} & -k_{sp2} & -k_{sp3} & -k_{sp} \\ 0 & k_{sp1}+k_{rp1} & 0 & 0 & k_{sp1}-k_{rp1} \\ 0 & 0 & k_{sp2}+k_{rp2} & 0 & k_{sp2}-k_{rp2} \\ 0 & 0 & 0 & k_{sp3}+k_{rp3} & k_{sp3}-k_{rp3} \\ 0 & 0 & 0 & 0 & k_{sp}+k_{rp} \end{bmatrix};\ F=\begin{bmatrix} F_1 \\ F_2 \\ F_3 \\ F_4 \\ F_5 \end{bmatrix}. \tag{4}$$

In which:

$$k_{sp}=\sum_{i=1}^{3}k_{spi},\ d_{sp}=\sum_{i=1}^{3}d_{spi},\ k_{rp}=\sum_{i=1}^{3}k_{rpi},\ d_{rp}=\sum_{i=1}^{3}d_{rpi} \tag{5}$$

Is defined also the angle $\psi_i=\phi_i-\alpha$.

The meaning of the indices is:

- ✓ *s*, p1, p2, p3 respectively represent the sun, the three planets and the satellite port;
- ✓ spi represents the contact between solar and planetary *i*;
- ✓ rpi represents the contact between the crown and the planetary *i*;
- ✓ the angle ϕ_i positions the center of the planet and the associated satellite port map;
- ✓ α is the pressure angle.

3. Dynamic response

We consider the system not amortized:

$$M\ddot{q}+Kq=F \tag{6}$$

Applying the transformed Laplace linear system (6) under homogeneous initial conditions:

$$q_i(0)=0,\ \dot{q}_i(0)\ \ i=\overline{1,5} \tag{7}$$

Results in an algebraic system, which has as unknown the Laplace coordinates $q_i(s),\ i=\overline{1,5}$ of the coordinates $q_i(t),\ i=\overline{1,5}$. By basically solving this system, we get the Laplace coordinates $q_i(s),\ i=\overline{1,5}$ of the coordinates $q_i(t),\ i=\overline{1,5}$, as follows:

$$q_1(s)=\Bigg[F_1-\frac{b}{M_{p1}s^2+f}\left(F_2-\frac{gF_5}{M_{p4}s^2+1}\right)-\frac{c}{M_{p2}s^2+h}\left(F_3-\frac{iF_5}{M_{p4}s^2+1}\right)-\frac{d}{M_{p3}s^2+j}\left(F_4-\frac{kF_5}{M_{p4}s^2+1}\right)- \\ -\frac{eF_5}{M_{p4}s^2+1}\Bigg]\bullet\frac{1}{M_s s^2+a}; \tag{8}$$

$$q_2(s)=\frac{1}{M_{p1}s^2+f}\left(F_2-\frac{gF_5}{M_{p4}s^2+1}\right);\ q_3(s)=\frac{1}{M_{p2}s^2+h}\left(F_3-\frac{iF_5}{M_{p4}s^2+1}\right); \tag{9}$$

$$q_4(s)=\frac{1}{M_{p3}s^2+j}\left(F_4-\frac{kF_5}{M_{p4}s^2+l}\right);q_5(s)=\frac{F_5}{M_{p4}s^2+l} \tag{10}$$

Reversing Laplace transformations of Lagrange coordinates, we obtain below the time functions $q_i(t),\ i=\overline{1,5}$, which represent the dynamic response of the gear unit to disturbing external stresses.

$$\begin{aligned}q_1(t)=&\frac{F_1}{\sqrt{M_s a}}\sin\left(\sqrt{\frac{a}{M_s}}\,t\right)-\frac{bF_2}{\sqrt{M_{p1}f}}\sin\left(\sqrt{\frac{f}{M_{p1}}}\,t\right)+\\&+\frac{bgF_5}{lM_{p1}-M_{p4}f}\left[\sqrt{\frac{M_{p1}}{f}}\sin\left(\sqrt{\frac{f}{M_{p1}}}\,t\right)-\sqrt{\frac{M_{p4}}{l}}\sin\left(\sqrt{\frac{l}{M_{p4}}}\,t\right)\right]-\frac{cF_3}{\sqrt{M_{p2}h}}\sin\left(\sqrt{\frac{h}{M_{p2}}}\,t\right)+\\&+\frac{ciF_5}{lM_{p2}-M_{p4}h}\left[\sqrt{\frac{M_{p2}}{h}}\sin\left(\sqrt{\frac{h}{M_{p2}}}\,t\right)-\sqrt{\frac{M_{p4}}{l}}\sin\left(\sqrt{\frac{l}{M_{p4}}}\,t\right)\right]-\frac{dF_4}{\sqrt{M_{p3}j}}\sin\left(\sqrt{\frac{j}{M_{p3}}}\,t\right)+\\&+\frac{dkF_5}{lM_{p3}-M_{p4}j}\left[\sqrt{\frac{M_{p3}}{j}}\sin\left(\sqrt{\frac{j}{M_{p3}}}\,t\right)-\sqrt{\frac{M_{p4}}{l}}\sin\left(\sqrt{\frac{l}{M_{p4}}}\,t\right)\right]-\\&-\frac{eF_5}{aM_{p4}-M_s l}\left[\sqrt{\frac{M_{p4}}{l}}\sin\left(\sqrt{\frac{l}{M_{p4}}}\,t\right)-\sqrt{\frac{M_s}{a}}\sin\left(\sqrt{\frac{a}{M_s}}\,t\right)\right];\end{aligned} \tag{11}$$

$$\begin{aligned}q_2(t)=&\frac{F_2}{\sqrt{lM_{p4}}}\sin\left(\sqrt{\frac{l}{M_{p4}}}\,t\right)-\\&-\frac{gF_5}{lM_{p1}-M_{p4}f}\left[\sqrt{\frac{M_{p1}}{f}}\sin\left(\sqrt{\frac{f}{M_{p1}}}\,t\right)-\sqrt{\frac{M_{p4}}{l}}\sin\left(\sqrt{\frac{l}{M_{p4}}}\,t\right)\right];\end{aligned} \tag{12}$$

$$\begin{aligned}q_3(t)=&\frac{F_3}{\sqrt{hM_{p2}}}\sin\left(\sqrt{\frac{h}{M_{p2}}}\,t\right)-\\&-\frac{iF_5}{hM_{p1}-M_{p2}l}\left[\sqrt{\frac{M_{p4}}{l}}\sin\left(\sqrt{\frac{l}{M_{p4}}}\,t\right)-\sqrt{\frac{M_{p2}}{h}}\sin\left(\sqrt{\frac{h}{M_{p2}}}\,t\right)\right];\end{aligned} \tag{13}$$

$$\begin{aligned}q_4(t)=&\frac{F_4}{\sqrt{jM_{p3}}}\sin\left(\sqrt{\frac{j}{M_{p3}}}\,t\right)-\\&-\frac{kF_5}{jM_{p4}-M_{p3}l}\left[\sqrt{\frac{M_{p4}}{l}}\sin\left(\sqrt{\frac{l}{M_{p4}}}\,t\right)-\sqrt{\frac{M_{p3}}{j}}\sin\left(\sqrt{\frac{j}{M_{p3}}}\,t\right)\right];\end{aligned} \tag{14}$$

$$q_5(t) = \frac{F_5}{\sqrt{IM_{p4}}} \sin\left(\sqrt{\frac{1}{M_{p4}}}\; t\right) \tag{15}$$

Where:

$$\begin{aligned} &a = k_{sp}; b = -k_{sp1}; c = -k_{sp2}; d = -k_{sp3}; e = -k_{sp}; f = k_{sp1} + k_{rp1}; g = k_{sp1} - k_{rp1};\\ &h = k_{sp2} + k_{rp2};\; i = k_{sp2} - k_{sp1}; j = k_{sp3} + k_{rp3}; k = k_{sp3} - k_{rp3}\, l = k_{sp} + k_{rp} \end{aligned} \tag{16}$$

4. Conclusions

Lately, as we know, the interest in using computers has increased considerably, including in the area of their application to various transformations and symbolic calculations. This is due, among other things, to the development of scheduled means that reduce the need to develop its own programs to solve various problems. Electronic analytical calculation systems, such as Mathematica, MatLab, Maple, or Mathcad, include thousands of commands included and library functions as well as visualization of the plotting process. Applying the unilateral Laplace transform in relation to time had the purpose in this paper to highlight its role in simplifying integrative operations, if we compare it with traditional methods, the role of mathematics being essential in the studies made by the researcher. Thus, it is unlikely that there could be serious scientific projects that do not have mathematical relevance and were completed without the use of electronic computers. A high level of security of calculations can ensure the use of several systems of computational mathematics. Many of the computational methods used are based on matrix formalism, easy to implement on the computer. Much of the solutions of differential equations, which describe, for example, the behavior of the material systems at the action of the internal and external factors acting on them, can be obtained with the help of the computer. To validate mechanical models an important role surely has dynamic tests. Thus, an important role is played by the modal experimental analysis problems, as well as the way in which the calculation-test correspondence can be done, as well as the updating of the model. Often used in mechanics and, in particular, in vibration mechanics are inverse problems. Among the applications and methods approached in this respect we can cite the inverse problems of the sources, the optimization of the shape, the active control of the fluid transport devices or, finally, the methods of the integral equations which are widely used in mechanics and acoustics. Such a research project can be conceived only in relation to other scientific concerns of the researcher or scientist, requiring the development of a series of competences such as:

- ✓ determination of the dynamic behaviour of complex mechanical systems in the design phase;
- ✓ development and evaluation of fast and solid techniques for the study of the stability of delayed systems (time simulation, multi-frequency methods, root clustering techniques);
- ✓ modelling of specific phenomena (cutting of metals) and interactions in the system in the presence of vibrations;
- ✓ development of techniques for active control of vibration phenomena in mechanical processing (delay control systems);
- ✓ taking into account the dynamic constraints in the design phase of the machines;
- ✓ integrating and managing models on different scales.

At present, research activities focus on the development of multidisciplinary experimental, theoretical and numerical skills that are put into play when designing structures, machine elements or, in general, mechanical systems. The aim of the research is to improve knowledge of the behaviour of materials and structures, develop models and tools useful in designing structures and machines, and harness a technical culture of analytical, designing and manufacturing methodologies.

References

IOP Conf. Series: Materials Science and Engineering **568** (2019) 012037 doi:10.1088/1757-899X/568/1/012037

[1] Abousleiman V and Velex P 2006 A hybrid 3D finite element/lumped parameter model for quasi-static and dynamic analyses of planetary/epicyclic gear sets *Mechanism and Machine Theory*, **41(6)** pp. 725–748
[2] Băgnaru D Gh, Stănescu M M and Cuţă P 2008 Vibrations of the tool and tool holder unit for a shaper device *8th International Conference, Research and Development in Mechanical Industry* RaDMI 2008 (Užice, Serbia) pp. 202-205
[3] Băgnaru D Gh 1997 Sur l'analyse des vibrations des barres viscoelsatique, les elements constitutifs des mecanismes plans *SYROM 97* (Bucureşti) pp. 25-30
[4] Băgnaru D Gh, Hadăr A ,Bolcu D, Stănescu M M and Cuţă P 2009 The vibrations influence on the field of accelerations of the linear-elastic connecting rod of a mechanism connecting rod lug *Annals of DAAAM for 2009 & Proceedings* **20(1)** pp. 833-835
[5] Băgnaru D Gh, Stănescu M M, Bolcu D and Cuţă P 2008 Approximate method in order to determine the displacement fields for the kinematic elements standing in vibration *Congress SMAT* (Craiova) pp. 9-12
[6] Blankenship G W and Singh R 1992 A comparative study of selected gear mesh interface dynamic models *International Power Transmission and Gearing Conference* (ASME Vol. 1) pp. 137–146
[7] Christophe Bard 1995 Modelisation du comportement dynamique des transmissions par engrenages *L'institut National des Sciences Appliquees de Lyon* (Lyon)
[8] Craig R R and Bampton M C C 1968 Coupling of substructures for dynamics analysis *AIAA Journal* **6(7)** pp. 1313–1319
[9] Gafiţanu M, Focşa V, Merticaru V and Biborosch, L 1980 "Vibrations and noises" ("Vibratii si zgomote", in Romanian) *Junimea Publishing House* (Iaşi)
[10] Kahraman A 1994 Planetary gear train dynamics *Journal of Mech. Des.* **116(3)** pp. 713–720
[11] Lin J and Parker R G 1999 Analytical characterization of the unique properties of planetary gear free vibration *Journal of vibration and acoustics* **121(3)** pp. 316–321
[12] Ohji K, Ogura K, Kubo S and Katada Y 1978 A study on governing mechanical parameters of creep crack growth rate by using notched round bar specimens with proportional geometry *Proc. 21st Japan Congr. Materials Res. Soc. Materials Sci.* (Japan) pp. 99-104
[13] Özguven H N and Houser D R 1988 Mathematical models used in gear dynamics –A review *Journal of Sound and Vibrations* **121** pp. 383-411
[14] Palaisi D, Guilbault R L, Thomas M, Lakis A and Mureithi N 2009 Simulations numériques du comportement vibratoire d'engrenages endommagés *Proceedings of the 27th seminar on machinery vibration* CMVA (Vancouver)
[15] Parker R G, Agashe V and Vijayakar S M 2000 Dynamic response of a planetary gear system using a finite element/contact mechanics model *Journal of Mech. Design* **122(3)** pp. 304–310
[16] Thouviot S, Omrani A and Tawfiq I 2009 Modele dynamique simplifie pour l'etude vibroacoustique d'un systeme d'engrenage, approche multidisciplinaire *19eme Congres Francais de Mecanique* pp. 1-6
[17] Stănescu M M, Băgnaru D Gh and Cuţă P 2008 The motion parameters of vibration for a linear viscoelastic plane plate *8th International Conference, Research and Development in Mechanical Industry* RaDMI 2008, 14-17 (Užice, Serbia) pp. 247-252
[18] Umezawa K, Sato T and Ishikawa J 1984 Simulation on rotational vibration of spur gear *Buletin of JSME* **27(223)** pp.102-109
[19] Velex P 1988 Contribution à l'analyse du comportement dynamique de réducteurs à engrenages à axes parallèles *PhD INSA Lyon* (Lyon)
[20] Verlinden O 1994 Simulation du comportement dynamique de systemes multicorps flexibles comportant des membrures de forme complexe *PhD thesis Faculté Polytehnique de Mons*
[21] Wilk O 1993 Simulation du comportement dynamique d'une automobile avec couplage aérodynamique *Thèse de l'Ecole Centrale de Paris*

IOP Conf. Series: Materials Science and Engineering **568** (2019) 012038 doi:10.1088/1757-899X/568/1/012038

Experimental design for *Diesel supply control* in order to improve fuel efficiency

A I Borzan[1] **and D L Baldean**[1,2]
[1] Automotive and Transportation Department, Technical University of Cluj-Napoca, Cluj, CJ 40064, Muncii 103-105, Romania

[2]doru.baldean@auto.utcluj.ro.

Abstract. The experimental testing and development of a system integrated in a vineyard farm in order to improve fuel efficiency of transport vehicles is a valuable asset and a way to go in the current research effort. The present paper is aiming to study and to present the practical application of equipment and ideas for automatic monitoring of fuel level inside the vehicle's tank. The system is created by digital hardware connected to the vehicles in question and controlled with proper software application in order record and offer an applied image of fuel level status or evolution. Actual data are recorded with the software module. This research points to the automation technology in automotive engineering and transportation field for a better fuel consumption management. Vehicle's engines are supplied with Diesel fuel via a volumetric sensor. From the fuel tank to the common-rail system, particles of the Diesel fuel are monitored and data is sent to digital control station via internet cloud. The experimental research is developed to create an environment for precise Diesel supply control to power trains in order to generate the basis of improving the fuel efficiency.

1. Introduction

Fuel loses and improper usage leads to lower efficiency when it comes to consider the energy management. The necessity for a proper fuel consumption management and control resides in significantly high costs of energy. Through a complex approach of the automation procedure in fuel level control during operation and even in park mode, the present research shows a design in which vehicles are strictly controlled regarding fuel supply management. Other researches were conducted on more electro-mechanical level of approach in small engine applications [1]. Some of the previous studies were developed analysing the link between fuel supply and lubrication [2], on one hand, or engine operation and Diesel particle filter [3], on the other hand. Fuel consumption influences the carbon foot print [4], thus it is reasonable to study these relations and their further potential. Residual particles [5], fuel consumption management and pollution evolve with a certain proportion characteristic [6] leading to the necessity of redefining the transport vehicles and powertrains [7].

Control and actuation system are thoroughly studied nowadays in other to develop more adaptable robotic and automatic applications and systems, with increased energy efficiency [8]. Highly complex transportation activities generate complicated measures to be taken for improving operating conditions [9], with multiple parameters to be considered [10]. Recent developed studies are pointing out to latest technologies and various possibilities in management of the energy resources [11], as well as to the numerical investigation of some factors [12]. Engine starting procedure and cold temperature operation of the entire power train lead to the possibility of influencing both the fuel consumption and the exhaust pollution [13][14]. Main objective of the paper is to present the implemented technology regarding the fuel consumption control with automatic reporting application on seven vehicles in

IOP Conf. Series: Materials Science and Engineering **568** (2019) 012038 doi:10.1088/1757-899X/568/1/012038

Jidvei area. As a part of the strategy to reach this stipulated target are nine specific objectives through which particular sequences are reached. These specific objectives are as follows: installation of the Titan Farmis application on the Jidvei digital platform; connecting the electronic platform with the mobile equipment; drawing the connections of the volumetric sensor with the proxy equipment; defining the experimental design for Diesel supply management; practical intervention and actual placement of the sensors and equipment on the studied vehicles (connections and adjustments); defining and applying the proper methodology for networking the vehicle and the server; location monitor and vehicle units' identification; vehicle speed control via the automatic reporting application; graphically representation of fuel tank level and control of supply. Thus it may be useful integrating some features and function based models into de application for experimental design of a diesel supply system with control possibilities in order to improve fuel efficiency.

Using Titan Farmis application it may contribute to a strictly monitoring the vehicle operational features and characteristics. This application is acquired together with other equipment necessary for this experimental endeavour.

2. Methodology and material

Through application of digital processing and automation capabilities available today in computational sector and internet, there may be defined an engineering method with specific instruments for Diesel fuel supply control in one digitally integrated system. In order to control the engine there is necessary to locate the fuel tank access for placing the fuel volumetric sensor. Specialized connection is linking the electronic control unit for the volumetric sensor with computing station, according to figure 1. Separated wires and digital unit were individually matched and connected in order to establish communication. Taking into consideration the problem of fuel control with vehicle powertrain allows the researcher to define a wide range of features that may be implemented in order to positively impact the energy efficiency and diminish carbon footprint. The present research use as material the power train as fuel consumer, the volumetric sensor for fuel tank level, electronic equipment, a network connection and software programs. The methodology consists in experimental application and set up of digital devices which offer the control possibilities regarding fuel level variation in real time and remote digital recording of the real values.

Multiple fuel level sensors are tightly placed inside the fuel tank of each one of the seven vehicles involved in the present study. The measuring pipe of the fuel level sensors must be permanently placed in the fluid inside the fuel tank. Digital unit of the sensor is placed in the upper part on the surface of the fuel tank of each vehicle considered in experimental part.

Connections and relations between the system parts that participate in the overall operation are sketched out in figure 1, in which may be observed the intricate influences of each factor.

IOP Conf. Series: Materials Science and Engineering **568** (2019) 012038 doi:10.1088/1757-899X/568/1/012038

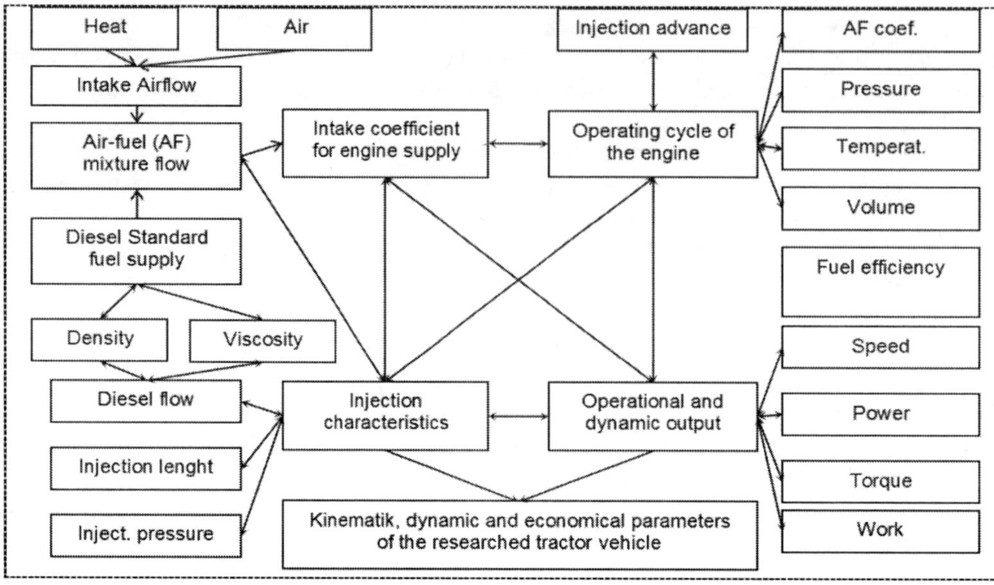

Figure 1. Schematics of the experimental designed methodology applied for Diesel fuel supply control and management during the practical part of the conducted research at Jidvei vineyard

3. Experimental results and discussions

With materials and by applying the specified methodology, the experimental part develops as planned and results are recorded.

In order to properly test in experimental conditions the equipment acquired and placed on all the vehicles took into consideration for the present research a finite small track was set up, as shown in figure 2, on which the vehicles were moved incrementally for the calibration procedure of the fuel control system.

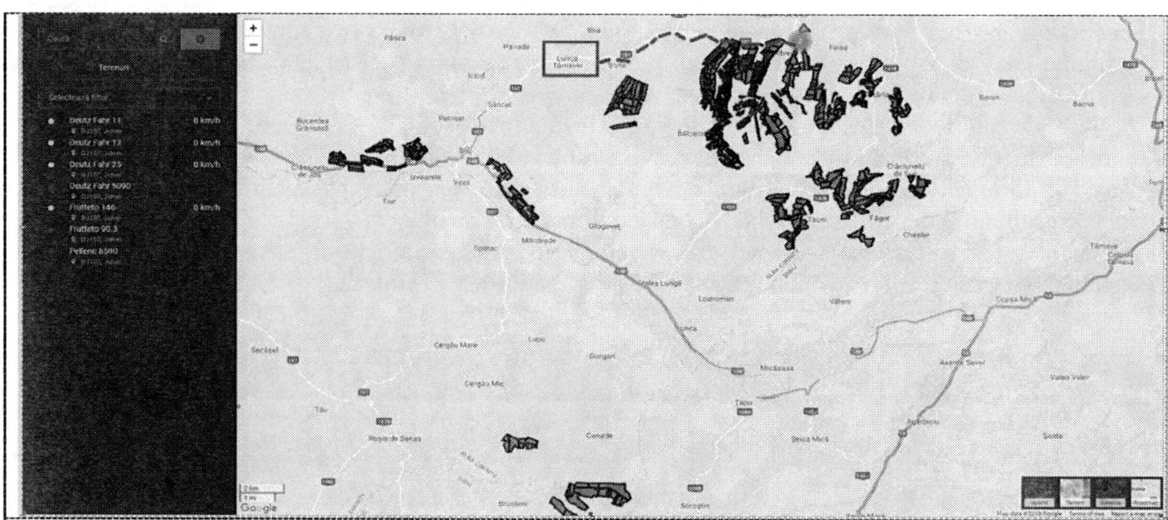

Figure 2. Geographical definition of the track between two locations in the experimental research log.

In the left side of the control panel are placed and numbered all the seven vehicles which were under the observation during the present experimental research development. Deutz Fahr, Frutteto and Pellenc vehicles were considered and studied. The actual track represented with red interrupted curve up side of the map is connecting the Jidvei city with Lunca Târnavei, both linked by 107 DJ road.

First thing before complete measurements and testing there was put in action the calibration on site procedure for all the vehicles took into study, thus some manoeuvres back and forth were made, as captured in experimental part, showing an increase of speed and a full stop afterwards. All data were recorded for further study.

Following the calibration step there was the measuring procedure with all the vehicles fully equipped and data transfer process in complete operation. Graphical and numerical representations are recorded and stored in digital programme installed on the Jidvei software platform.

The simplified data sheet with some technical references is given in table 1, showing the accuracy level, followed standard and available / used devices.

Table 1. Technical specifications of the control station for exhaust emissions recordings and study.

Parameter	Exhaust control station
Accuracy deviation level	0
Compliance standard	OIML R99
Tachometer	BDM (B+/B-)
Smoke meter	RTM 430
Gas analyse	CO, CO2, HC, O

The application and the systems were in full operation and completely automatized receiving and transmitting data between interconnected equipment in real time, and offering at request valuable reports on fuel efficiency on each individual vehicle took into consideration. By accessing the interface presented in figure 3 all the digital reports, both individually compiled or complex simultaneous comparisons are available. Thus specialized experimental design for Diesel supply control system was implemented to further improve fuel efficiency in practice of everyday operation of Jidvei farm tractors and machineries.

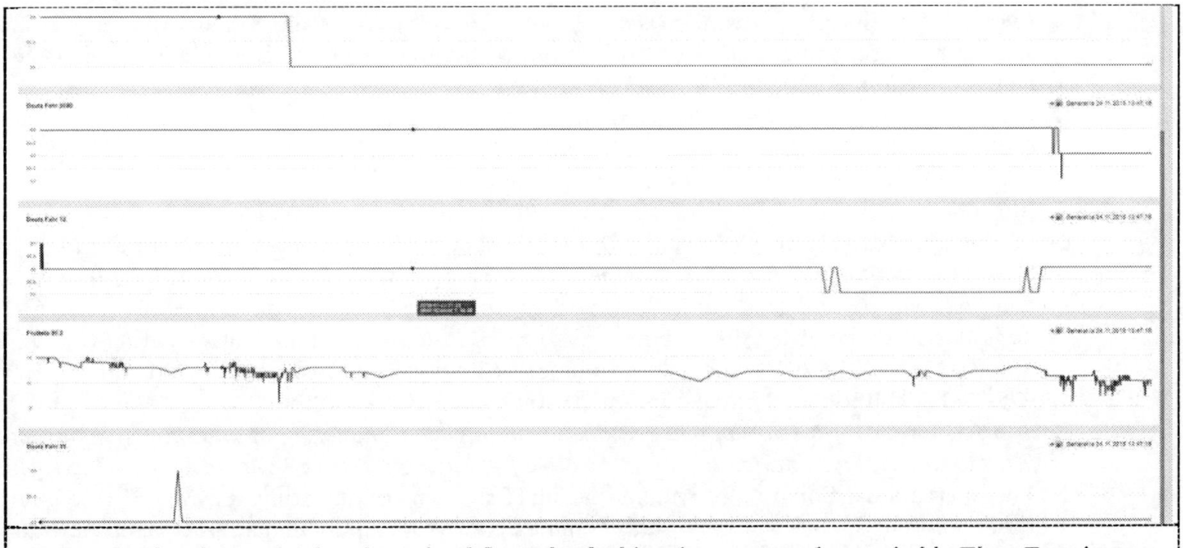

Figure 3. The electronic signal received from the fuel level sensors and recorded in Titan Farmis app.

Most fundamental parameter, that is important for consideration in mathematical modeling of the fuel flow and consumption efficiency, consists in *fuel hourly consumption*, given by the relation [7]:

$$F_{HC} = 3600 \bullet F_{inj.vol} / \tau, \quad (1)$$

where F_{HC} is hourly consumption of the fuel in [cm^3/h]; $F_{inj.vol}$ - fuel injected volume measured in [g]; τ - time in which there was injected the fuel in combustion chamber in [sec.].

4. Conclusions

In the presents research, were studied the possibilities and actual steps of implementation in Jidvei Vine Farm for a complete digitalised and automated system for fuel level control and energy consumption management in order to further improve the power train efficiency and lower the carbon footprint. Although the overall fuel efficiency improvement is a highly complex result after many hours of operation, which was not the main target of the present work, it may be still be assessed that by proper use of the automatic system there will be such an improvement in energy supply and work.

The highest variation in fuel level recorded on the platform was from the Frutteto 90.3 mobile unit. There were also induced oscillations in fuel level inside the tank in order to simulate and record abnormal activities on some of the other vehicles in question. System worked as expected and returned result in practice. Experimental research is still applicable on other aspects in order to assess more definitely the consumption for each operation in particular in order to map properly the pollution.

Acknowledgments
This work was supported by a grant of the Romanian Ministry of research and Innovation, CCCDI-UEFISCDI, project number PN-III-P2-2.1-CI-2018-1227, within PNCDI III

References

[1] Băldean D et al 2018 *Contributions on experimental research of fuel consumption and management system in K9K892 Diesel engine from Dacia-Renault* (Annual Session of Scientific Papers IMT ORADEA 2018 Section Mech Eng and Automotive MATEC Web of Conferences 184 01018) DOI https://doi.org/10.1051/matecconf/201818401018

[2] Băldean D et al 2017 *Contributions in Experimental Research Concerning Diesel Fuel Supply and Lubrication in the Case of Comparative Study Between Euro V and IV Common Rail Engines* (In: Chiru A., Ispas N. (eds) CONAT 2016 Internat Congress of Automotive and Transport Engineering CONAT 2016 Springer Cham) Online ISBN 978-3-319-45447-4, DOI: https://doi.org/10.1007/978-3-319-45447-4_29

[3] Băldean D 2017 *Researching the Dysfunctional aspects concerning Diesel Particle Filter (DPF) in case of Dacia Logan E5 1.5dCI Engine* AGIR Press eISSN 2359 – 828X http://stiintasiinginerie.ro/31-79-cercetarea-aspectelor-disfunctionale-legate-de-filtrul-de-particule-diesel-dpf-in-cazul-motorului-dacia-logan-e5-1-5dci/

[4] Borza EV et al 2018 *Research Concerning Fuel Economy Coefficient and Carbon Foot Print in Various Conditions for a City Compact Size Vehicle with Digital Control for a Green Solution and Method at Technical University from Cluj-Napoca* (In: Burnete N., Varga B. (eds) Proceedings of the 4th Internat Congress of Automotive and Transport Eng (AMMA 2018) AMMA2018 2018 Proceed in Automotive Engineering Springer Cham) Online ISBN 978-3-319-94409-8 DOI https://doi.org/10.1007/978-3-319-94409-8_22

[5] Borzan AI et al 2018 *Research contributions through virtual modeling of an innovative particle filter for alternative application in automotive field using advanced engineering methods* (In: The International Congress of Automotive and Transport Engineering, AMMA 2018) Online Last modified: 2018-07-11, DOI: http://amma2018.ro/index.php/amma/2018/paper/view/28

[6] Borza EV et al 2005 *Functional characteristics of compression ignition engines with improved equipment which work with rape oil* (In J Acta Technica Napocensis) pp 67-73

[7] Burnete N et al 2008 *Motoare Diesel şi biocombustibili pentru transportul urban* (Cluj-Napoca Mediamira ISBN 978-973-713-217-8)

[8] Covaciu F 2019 *Control and actuation system of a six degrees of freedom robotic arm* (in Acta Tehnica Napocensis Series Applied Mathematics and Mechanics vol 62 Issue I ISSN 1221-5872 Cluj-Napoca România) pp 99-106

[9] Crişan MA et al 2014 *Evaluation of laboratory analysis of lubricants and lubricating processes in compression ignited engines' mechanisms depending on biodiesel content in fuel* (in Multidisciplinary International Conference–"Professor Dorin Pavel – founder of Romanian hydro-energetics" in Science and Engineering XIV 26 ISSN 2067-7138) / *(Evaluarea analizelor de laborator ale lubrifianţilor şi proceselor de lubrifiere în mecanismele motoarelor cu aprindere prin comprimare în funcţie de conţinutul de biodiesel în combustibil* în A XIV-a Conferinţă Internaţională Multidisciplinară–"Profesorul Dorin Pavel – fondatorul hidroenergeticii româneşti" în Ştiinţă şi Inginerie An XIV Vol 26 ISSN 2067-7138 in Romanian)

[10] Crişan MA et al 2010 *Analysis of some operating parameters of S.I.E. from Seat Leon in different road conditions using a new testing technology* (in CONAT International congress on automotive and transport engineering 27-29 Oct Vol 1 p 8 ISSN 2069-0401 Bv Ro 2010)

[11] Mariasiu F 2012 *Energy sources management and future automotive technologies: Environmental impact* (Internat. Journal of Energy Economics and Policy 2 (4)) pp 342-347

[12] Mariasiu F 2013 *Numerical investigation of the effects of biofuel characteristics on the injector nozzle erosion process* (Tribology Transactions 56 (2) 10402004.2012.709918) pp 161-168

[13] Mariasiu F 2013 *Possibilities for reducing tractor engine friction losses at cold start using an ultrasonic irradiation technique* (Turk J Agric For 37 4) pp 623-631

[14] Moldovanu D et al 2018 *Influence of swirl and tumble motion inside the combustion chamber of a compression ignited engine on vertices formation* (MATEC Web of Conf 184 01022)

IOP Conf. Series: Materials Science and Engineering **568** (2019) 012039 doi:10.1088/1757-899X/568/1/012039

A brief assessment of outliers and malfunctions detecting techniques with an application on lubricant condition monitoring

G Grebenişan[1,3], S Bogdan[1], N Salem[2], D C Negrău[1]

[1]University of Oradea, Romania
[2]Zarqa University, Jordan

[3]grebe@uoradea.ro

Abstract. Condition monitoring and machine status classification are of great practical importance in the manufacturing industry as it provides online updates on the state of the machine, avoiding loss of production and minimizing the probability of generating catastrophic damage to the machine. In this paper, the classification of conditions is based on the processing of information using wavelets based on the results of the monitoring and the data collected during such an action, measuring the characteristics of the lubricating oil over some time sufficient to produce a time series of results. In this paper, the classification system is tested and validated using observation sequences based on the maximum wavelet distribution obtained from the collected signals, monitoring the state of the lubricating oil, to define and diagnose singularities in time series.

1. Introduction

In recent years, the rapid development of industrial automation has prompted the need for smarter and more reliable processing systems.

To minimize the losses caused by production interruption and the high cost due to machine and plant faults, it is necessary to monitor the status of the on-line machine using an effective condition monitoring system to provide timely information for faulty decision making, of any type, and diagnosing these defects promptly. In general, monitoring conditions involves observing machine condition using periodic dynamic response measurements, such as vibration signals obtained from multiple transducers, or measurable features for lubricating oils. Vibration measurements are taken from the machine usually contain lots of useful information but also noise components that should be removed from the signal before the information is used for machine condition classification and maintenance planning. Measurements made for oils refer to detections of the measurable components of the matrix of physical, chemical, electrical, properties. All these measurements, relating to the properties of the oils, are based on the use of performance sensors, [1].

2. Detecting and diagnosing the malfunction

When a process error occurs, it should be detected as soon as possible. The fault and malfunction detection system must indicate that something is wrong in the process. After detection, fault diagnosis is performed, the fault is isolated, and the cause of the malfunction is detected. Typically, the techniques used to detect and diagnose defects are divided into two broad categories: estimation methods and pattern recognition methods [2].

2.1 Methods of estimation
Estimation methods require mathematical models of the processes, not very complicated, to represent the real process satisfactorily, and solving the mathematical model is not excessively time-consuming. Detecting malfunctions based on state variables also involves the risk of a significant number of non-measurable variables that must be estimated. For directly estimation, a dynamic process model is linearized around an operating point. Estimation can be done using different methods depending on how stochastic the model is, then the residue evaluation, i.e., the differences between estimated and actual measured variables. This approach, based on mathematical models, requires relatively accurate knowledge of the linearized model parameters, with estimation errors generating model errors and validation errors. The detection of defects based on parameter estimation requires the existence of a known mathematical relationship, to be able to estimate their behavior by the physical parameters of the process. This requires specific knowledge of measurable process parameters as well as their variability. Because not all process parameters are directly measurable, their changes are calculated using the estimated process parameter. This is why the relationship between model parameters and process coefficients must be unique and, preferably, known precisely, a very rarely fulfilled condition.

2.2 Methods of pattern recognition.
These methods do not require the mathematical models of the process, the creation of neural network architecture, an algorithm for fault detection and diagnosis. The idea is that the process is classified according to measured data. From spatial representation, this classification is, in fact, a transposition of the measurement space into the decision-making area. The development of an algorithm for pattern recognition and pattern classification can usually be focused on three steps: measurement collection, extraction of features, and classification. As an example of the development of these algorithms, in the case of a real process, in a first step, the measured data is collected, then a characteristic vector is calculated, the redundant data is removed by the extraction, respectively the situations in which the data are missing (missing data), creating, at this extraction stage, the prerequisites for generating decision areas. In the last step, the characteristic vector is classified into one or more classes. These classes depend on the purpose of the algorithm created. If the target algorithm refers to error detection and diagnosis, the classes could be, for example, regular operation, number of A-type faults, number of B-type faults, etc. Any neural network or pattern recognition pattern or pattern classification algorithm performs the classification and recognition of patterns based on a complex overlapping operation, as is evident, of the measurement space in the decision making space. A human being has an amazing ability to recognize patterns, and often uses a very complex logic in pattern recognition and classification, but can not often define laws and rules for doing these operations. When a classification is done with neural networks, the entire mapping from the measurement space to the decision space is done at the same time, and the classification scheme is learned through examples extracted from the classified data collection. Thus, it is evident that methods of model recognition and classification do not require mathematical, analytical models but need representative data for training.

An outlier can be defined as a "data point" in a series of times that is very important difference from the rest of the data points. Outliers are important observations that affect data analysis and should, therefore, be treated with caution. There are different types of exceptional values that can occur in a series of times: a) -additive outlier (AO) is a measurement error at time T, $1 \leq T \leq N$, due to factors outside the system (e.g., o machine malfunction or a human error in data recording could be called outlier additions, an outlier additive does not affect the trend of a process); b) - Another type of outlier is innovative outlier (IO), which is caused by a certain change in a process/system. The main difference between an AO and an IO is that it indicates the beginning of a new trend in the process, which may eventually return to normal. Sometimes, changes in process/ system characteristics may involve a permanent change of process/system status (a change from stationary to non-stationary). The two main aspects associated with outliers are the detection of outliers, and the decision to be taken after the outliers have been detected. Detecting outliers involves identifying the occurrence time T, which may not be known, as well as recognizing the outlier type.

IOP Conf. Series: Materials Science and Engineering **568** (2019) 012039 doi:10.1088/1757-899X/568/1/012039

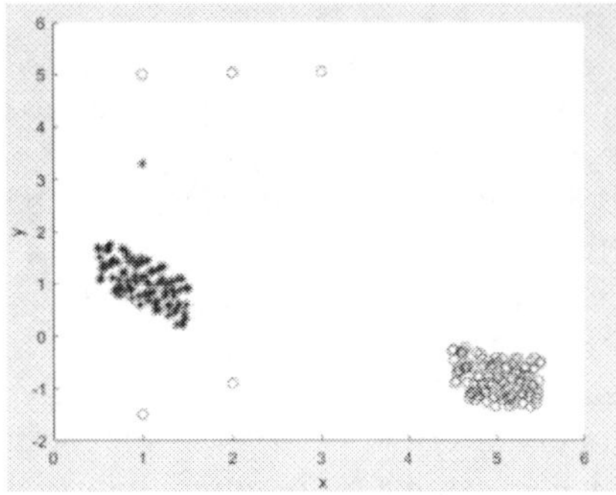

Figure 1- Normal regions and anomalies data

Figure 1 illustrates the dispersion of some anomalies in a simple set of data represented in a two-dimensional plot. Two distinct regions are distinguished, in which the data are grouped, each representing the normal behavior of a system or systems, while scattered, discordant data sufficiently distanced by the two typical regions are singularities, or anomalies of this data set, [4]. It is essential that these anomalies are not treated as "noises". Noises are generally defined as obstacles in the work of analyzing phenomena and processes, and should be treated as components of the mathematical model of the process, [5]. Considering a simple linear model that expresses the relationship between the time series of inputs, X_t, and the time series of outputs, Y_t, in the form:

$$\mathbf{Y_t = \upsilon(B) X_t + N_t} \qquad (1)$$

where, $\mathbf{\upsilon(B) = \upsilon_0 + \upsilon_1 B + \upsilon_2 B^2 + \ldots}$ is the transfer function, expressed as polynomial, $\mathbf{\upsilon(B) = \frac{\Omega(B)}{\delta(B)} = \delta^{-1}(B)\Omega(B)}$ respectively, represented as the ratio of mathematical operators, and which denotes the dynamic relation between outputs and inputs, where **B** is a backward shift operator (a set of time z_t observable with the components: : $z_1, z_2, \cdots, z_N$ then $Bz_t = z_{t-1}, B^m z_t = z_{t-m}$), while N_t represents the filtered noise component superimposed over the input signal, a_t (converts) the z_t component as follows:

$$\mathbf{z_t = \mu + a_t + \psi_1 a_{t-1} + \psi_2 a_{t-2} + \ldots = \mu + \psi(B) a_t} \qquad (2)$$

where μ is a parameter that generally expresses the "level" of z_t, and a_t, is a sequence consisting of randomly weighted components, after the operator $\mathbf{\psi(B) = 1 + \psi_1 B + \psi_2 B^2 + \ldots}$, called the filter transfer function (it is different, as a definition and mathematical model of anomalies, singularities, etc.: $\mathbf{N_t = \psi(B) a_t}$, so that analysis of a process or a system behavior requires the determination of both the transfer function $\mathbf{\upsilon(B)}$, and the filter transfer function, $\mathbf{\psi(B)}$, [5], [6], [7]. An iterative numerical procedure is presented in [7], starting from the analysis of two classes of outliers generated by dynamic models of exceptional interventions at unknown times: innovational outlier (IO) and additive outlier AO) Starting from a stochastic process model, x_t, following an autoregressive-integrated-moving average (ARIMA) model (possibly with the characteristics: **p**- a positive integer indicating the degree of the nonseasonal autoregressive polynomial, **d**-a nonnegative integer indicating the degree of nonseasonal integration in the linear time series, **q**-a positive integer indicating the degree of the known nonseasonal moving average polynomial), [8], [9], [10].

In [6], it is analyzed the case where the moment **T** of the occurrence and the unfolding of the exceptional intervention in the model of the process, and finally the appearance of the outliers, is known, and in [7] there is presented a survey including a practical procedure for the development of the analysis a stochastic process in which the time of intervention is unknown, it can be estimated statistically, or using other techniques, [10].

3. Experimental setup

The experiment was conducted in several phases, being part of a Research Program, supported by private companies in Romania and Italy. A first test phase of an experimental model that took place at Mecoil (Italy), in the company's lab, on a demo system with simulated parameters. The elements that were tested (physicochemical parameters) were evaluated in the first stage, on a single level. Data processing and computational work have yielded positive results in gaining the necessary data points to advance in the actual monitoring of a physical system. The overcoming of this early phase has enabled the decision to install the hydraulic system containing sensors and transducers sets, designed and physically installed in the industrial installations (pilot systems) at EMSIL Techtrans SA Oradea-Romania, as well as in four other locations in Italy, to the partners Mecoil [11], [12], [13].

4. Results and discussions

Status monitoring and diagnosis, respectively, predicting the estimation of some output parameters for a technical system can be framed in the "extrapolation" chapter as an immediate mathematical operation [14], [15]. Of course, the time series for which mathematical operations of this kind can be applied are subject to errors due to the statistical nature of these mathematical classes. The example prepares an LSTM network for two case studies: one to predict oil moisture and the other to predict the oil temperature using neural network analysis [13] based on the retrieved records by the sensors located in the hydraulic system designed to simulate the operation of an oiling plant in an automatic plant, figure 2. Prediction is understood as an estimate of the values of a time function based on values of a time series, values that can be, or can not be, affected by random errors. For example, a prediction problem could be expressed as follows: Given a series of time, S (t), which consists of a set of values, and a random set of disturbing signals assimilated to a set of noise, Z (t), it is proposed to estimate a future value, a prediction, therefore, P $(t + \tau)$, where τ is a positive constant, the prediction is also a continuous function of time. This case study shows how to predict data from time series using long short term memory (LSTM) network, [9]. To predict the values of the next steps of a sequence, a sequence regression LSTM network will be used, where the responses are the training sequences with values changed over a time step, figure 3. This means that at each step of the input sequence, the LSTM network learns to predict the output value at the next time step. To forecast the multiples of time values in the future, the predictedAndUpdateState function in Matlab was used to predict the time steps one by one, and to update the LSTM network state for each prediction, figures 4 and 5. This experiment uses the data set collected within the National Program PN II project, ERA MANUNET: NR 13081221 / 13.08.2013. The example prepares an LSTM network for two case studies: one to predict oil moisture and the other to predict the oil temperature using neural network analysis [13] based on the retrieved records by the sensors located in the hydraulic system designed to simulate the operation of an oiling plant in an automatic plant. The data set is partitioned as follows: 90% of the data volume will be used for network training, and the remaining 10% of the collected data set is used to test the network. The working procedure states that the response values of the system are the network training data sequence. An LSTM regression network was used for which it was specified to have an architecture of 200 hidden units, and train for 50 epochs. This will prevent the gradients from exploding the gradient threshold. The initial value of learning rate was set to 0.005 and provided to drop the learning rate after 25 epochs to be multiplied by a factor of 0.2. The network is initialized when the predictive training data is made. Next should be the first prediction using the last time step of the training response. The network function will be useful when prediction and outputs of the network are updated sequentially using predicted data. The phase of training will be monitored permanently by the calculated root-mean-square error (RMSE), which is the measure of errors induced by the network, figure 6.

IOP Conf. Series: Materials Science and Engineering **568** (2019) 012039 doi:10.1088/1757-899X/568/1/012039

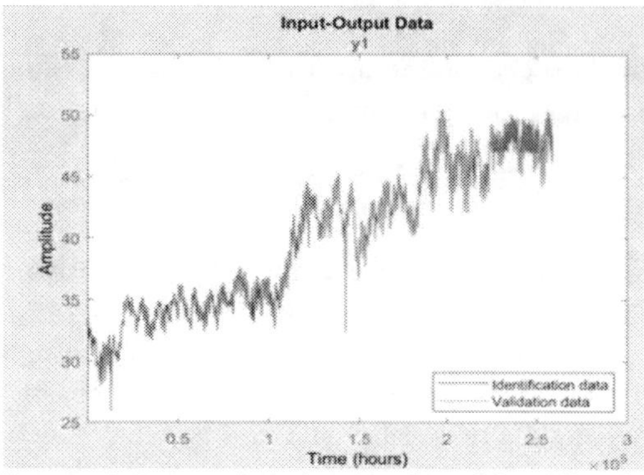

Figure 2 Separating the data into identification and a validation segment(time's scale in seconds, instead of hours)

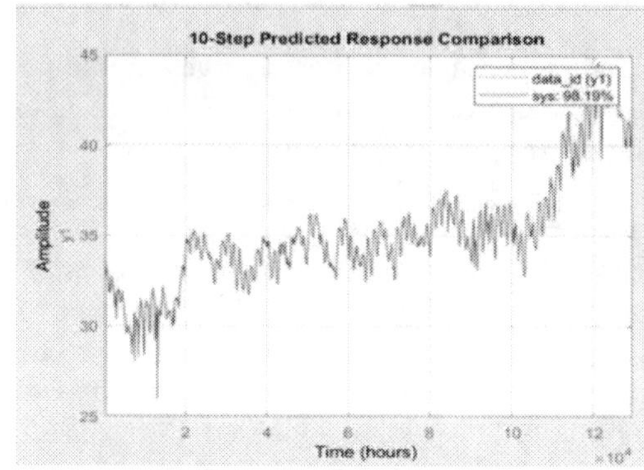

Figure 3 Using the 10 steps ahead predictor for the identification data and the independent validation data(time's scale in seconds, instead of hours)

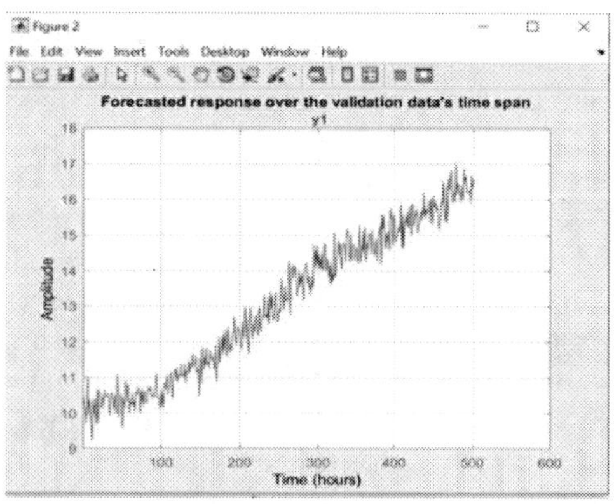

Figure4 The forecasting procedure uses the y_i, $i=1\div n$, measured data recorded, to compute the model state at time step n

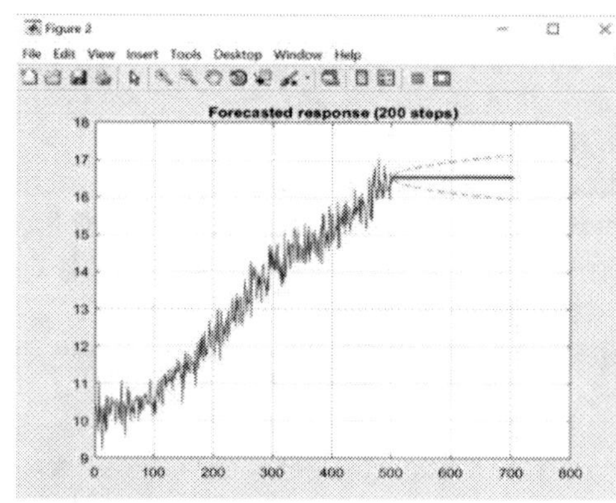

Figure 5 The forecasting results after 200 steps, shows not significant variance

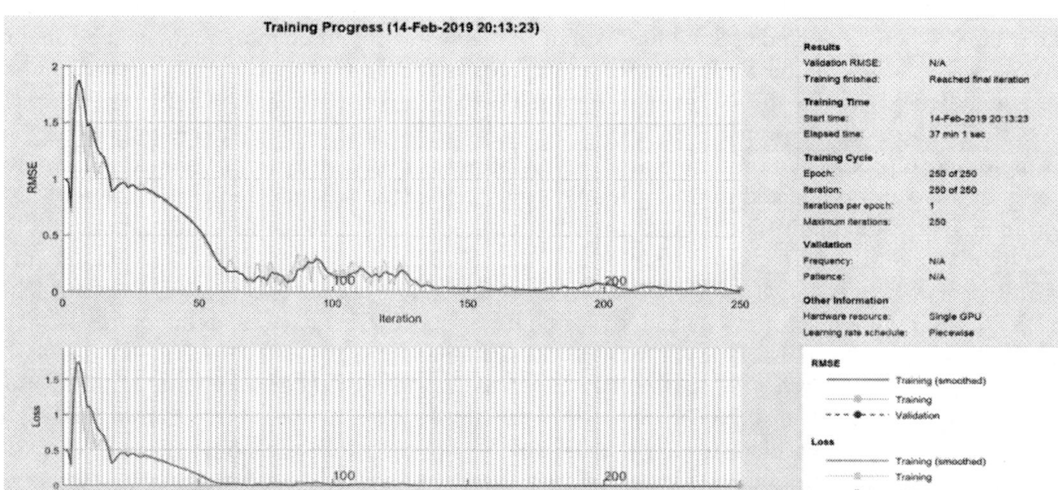

Figure 6 The plot of the training time series with the forecasted values, in terms of RMSE (root-mean-square-error) and LOSSE parameter

Acknowledgment
This work is supported by Machine Intelligence for Diagnosis Automation (MIDA), Research Program, National Program PN II, ERA MANUNET: NR 13081221/13.08.2013

References
[1] Sorsa T., Koivo H. N., and Koivisto H. 1991 *Neural Networks in Process Fault Diagnosis*, IEEE Transactions On Systems, Man, And Cybernetics, **Vol. 21**, no. 4, DOI: 10.1109/21.108299
[2] Teng H. S., Chen K. and Lu S. C-Y 1990 *Adaptive Real-time Anomaly Detection Using Inductively Generated Sequential Patterns* Conference Proceedings on Research in Security and Privacy IEEE Xplore DOI: 10.1109/RISP.1990.63857
[3] Chandola V., Banerjee A., and Kumar V. 2009 *Anomaly detection: A survey*. ACM Comput. Surv., 58 pages. http://doi.acm.org/10.1145/1541880.1541882
[4] Hodge V. J. and Austin J. 2004 *A Survey of Outlier Detection Methodologies*, Artificial Intelligence Review **22**: pp. 85–126, Kluwer Academic Publishers, doi: 10.1023/B: AIRE.0000045502.10941.a9
[5] Box G. E. P., Jenkins G. M., Reinsel G. C., Ljung G. M. 2015 *Time Series Analysis: Forecasting and Control*, Fifth Edition, John Wiley & Sons.
[6] Box G. E. P. & Tiao G. C. 1975 *Intervention Analysis with Applications to Economic and Environmental Problems*, Journal of the American Statistical Association, **70:349**, pp. 70-79, DOI: 10.1080/01621459.1975.10480264.
[7] Chang I., Tiao G. C., and Chen C. 1998 *Estimation of Time Series Parameters in the Presence of Outliers, Technometrics*30, **no. 2** 193-204. doi:10.2307/1270165.
[8] Abraham B., Box G. E. P. 1979 *Bayesian Analysis of Some Outlier Problems in Time Series* Biometrika **66**(2):229-236, DOI: 10.1093/biomet/66.2.229
[9] Statistics and Machine Learning Toolbox™ User's Guide, https://www.mathworks.com/help/releases/R2018a/pdf_doc/stats/stats.pdf, Accessed in February 2019.
[10] Rousseeuw P. J., Leroy A. M. 1987 *Robust Regression and Outlier Detection*, John Wiley & Sons.
[11] Adriani G., Campatelli M., Paoli M., 2013 *Monitoil® - Online sensors for efficient and cost-saving oil conditions' monitoring*, Mecoil Diagnosi Meccaniche s.r.l., Florence, Italy, http://www.mecoil.net/wp-content/uploads/2013/04/20120914-Articolo-colonne_Monitoil_eng.pdf
[12] Grebenişan G., Salem N., and Bogdan S. 2017 *The lubricants parameters monitoring and data collecting,* MATEC Web of Conferences, **Volume** 126, 2017, https://doi.org/10.1051/matecconf/201818403008.
[13] Grebenişan G., Salem N., and Bogdan S. 2018 *An approach of classification and parameters estimation using the neural network, for lubricant degradation diagnosis*, MATEC Web of Conferences, **Volume** 184 https://doi.org/10.1051/matecconf/201818403009.
[14] Wiener N. 1949 Extrapolation, *Interpolation, and Smoothing Of Stationary Time Series With Engineering Applications* The M.I.I. Press Massachusetts Institute Of Technology Cambridge, Massachusetts
[15] Zadeh L. A. and Ragazzini J. R. 1950 *An Extension of Wiener's Theory of Prediction* IEEE Xplore, Journal of Applied Physics **21**(7) pp. 645 – 655, DOI:10.1063/1.1699725

IOP Conf. Series: Materials Science and Engineering **568** (2019) 012040 doi:10.1088/1757-899X/568/1/012040

Analysis of the dynamics of road accidents in Sibiu county, Romania, in the period 2010-2018

L Tarnu[1] and C Deac[1]

[1] Department of Industrial Engineering and Management, Lucian Blaga University of Sibiu, Emil Cioran str., 4, Sibiu 550025, Romania

cristian.deac@ulbsibiu.ro

Abstract. Sibiu county is located in the very center of Romania. The county, and especially its main city, Sibiu, is an important industrial and trade hub. Also, it is located at the crossroads of major roads heading towards the south, the east, the northeast, and the northwest. All these factors result in dense road traffic, which, combined with the conditions encountered on some of the roads, leads in turn to a relatively high number of road accidents. In the current paper, the authors analyze the evolution of the number of road accidents, but also its more severe consequences (fatalities and serious injuries) in Sibiu county from 2010 to 2018 and try to identify road safety measures that could help to limit these numbers. The accidents are analyzed from the type of road, but also from the time of their occurrence (day of the week and hour, respectively).

1. Introduction

Road safety is a very complex problem of modern society, involving several widely different categories of factors: human participants in the road traffic (car drivers, pedestrians, bicycle riders etc,), vehicles (trucks, cars, motorcycles, bicycles etc.), transportation infrastructure (roads, bridges, road signals etc.), controlling authorities (police), weather conditions, disturbances (road accidents, pollution) etc. Therefore, to have the best possible picture of the road safety situation, but also to be able to determine risks and to anticipate and counteract problems, it is necessary to have as much data as possible on all of these elements.

Road accidents are a significant source of social and economic problems and have to be recorded and analyzed in all their aspects to better understand them and work towards preventing them and minimizing their consequences. On the other hand, their economic implications mean that road accident have to be viewed as work accidents. Therefore road safety-related risks can be assessed using a methodology similar to that used for determining the risks at the workplace [1, 2].

Continuous innovation – be it in the area of renewable fuel sources [3, 4] or the area of road infrastructure improvements – can help to reduce the risk of road accidents. However, it is the active conscientization of this risk by all participants in the road traffic that can help most in reducing it.

At European level, the European Commission has tried to spur interest in road safety by setting, even starting with the year 2001, the ambitious goal of reducing in 10 years the number of road fatalities by 50%. While this target was not fully reached in 2010, the goal was renewed in 2011 for the next period of 10 years. In 2017 there had been recorded a reduction of the number of fatalities in road accidents by 20% compared to 2011 (from 31500 to 25260) [5, 6, 7].

IOP Conf. Series: Materials Science and Engineering **568** (2019) 012040 doi:10.1088/1757-899X/568/1/012040

The Romanian Government has adopted in 2016 the National Road Safety Strategy 2016-2020 [8], reaffirming the country's commitment to the reduction by 50% of the number of road deaths between 2010 and 2020, as well as several other measures meant to improve many elements related to the management of road safety [9, 10].

Since the number of road accidents is generally about the density of road traffic, road accidents are problematic, especially in counties containing roads with significant road traffic.

Sibiu county is crossed by several essential roads that usually see a high volume of road traffic. Therefore, road safety is critical in this area, and the analysis of the number of road accidents and of the conditions in which they occur is a significant step in applying the right road safety measures.

In the current paper, the authors analyze the number of accidents and the consequences of these accidents (fatalities, severe injuries or light injuries) produced in Sibiu county between 2010 and 2018 from several points of view.

2. Sibiu County and its road network

Sibiu county is an administrative division of Romania, located in the center of the country.

The main road transportation arteries of the county are (figure 1 [11]):

Figure 1. The main roads in Sibiu county [11].

- The national road DN1 Bucharest-Borş enters the county from the east, coming from Braşov, passes through the city of Sibiu and leaves the county in the west, towards Sebeş and Alba Iulia;
- The national road DN7 Bucharest-Nădlac enters the county from the south, through the Olt River Gorge, passes through the city of Sibiu and then leaves the county in the west, going towards Sebeş; from Sibiu to the border between Sibiu county and Alba county DN1 and DN7 share the same route;
- A segment of the A1 motorway starts from Sibiu and goes on towards the border between Sibiu county and Alba county, doubling the DN1/DN7 national road in that area.
- The national road DN14 links the city of Sibiu with the city of Mediaş and the city of Sghişoara (the latter located in Mureş county).

IOP Conf. Series: Materials Science and Engineering **568** (2019) 012040 doi:10.1088/1757-899X/568/1/012040

All of these roads are heavily traveled, so, despite efforts to improve the road infrastructure and the traffic controls, accidents happen regularly.

3. Analysis of the number of accidents and their consequences

The authors have analyzed the number of accidents produced in Sibiu county from January 1st, 2010 until December 31st, 2018 based on data from the National Car Accidents Registry and from the General Inspectorate of the Romanian Police [12, 13].

The analysis has revealed that in Sibiu county, the number of accidents has known an ascending trend, from 434 accidents in 2010 to 542 in 2018. The additional 108 accidents represent an increase of almost 25%, which is a rather steep increase. Of these, 196 (36%) were severe accidents, and 346 (64%) were light accidents.

Figure 2 presents the evolution of the number of serious car accidents and light car accidents, respectively, on the roads of Sibiu county, from 2010 to 2018.

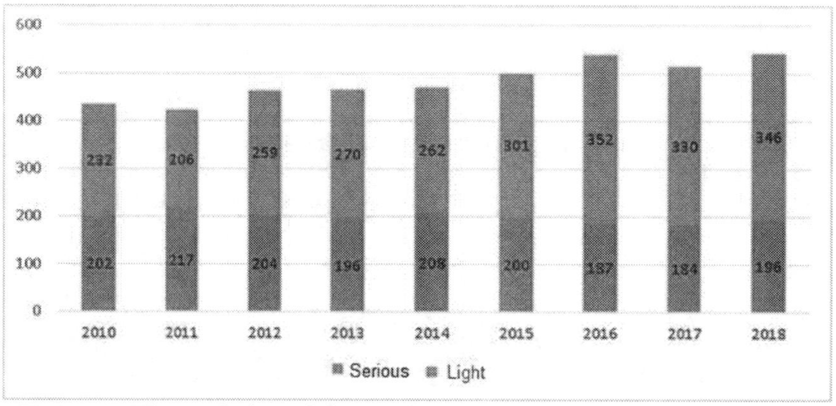

Figure 2. The evolution of the number of serious car accidents and light car accidents in Sibiu county from 2010 to 2018.

The consequences of the road accidents that occurred between 2010 and 2018 can be seen comparatively in figure 3.

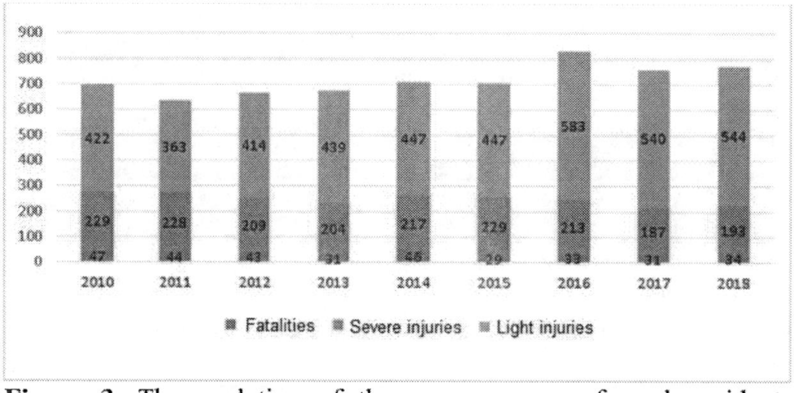

Figure 3. The evolution of the consequences of road accidents produced in Sibiu county from 2010 to 2018.

The road accidents produced in 2018 in Sibiu county resulted in the deaths of 34 persons (an increase by 3 compared to 2017 but a decrease by 13 compared to 2010), in severe injuries in 193 persons (again an increase of 6 compared to 2017 but a decrease by 36 compared to 2010) and in light injuries in 544 persons (an increase of 4 compared to 2017 and of 122 compared to 2010).

In this case, while there can be again noticed a generally ascending trend, there need to be emphasized the exceptions represented by the last two years, when the total number of accidents and the number of severe and light injuries has decreased compared to 2016. This can be attributed to the introduction, on two of the main road segments – DN1 from the border with Vâlcea county to Sibiu and DN7 from the border with Braşov county, at the end of 2016, of several measures for the increase of road safety and also to the opening of the A1 motorway segment from Sibiu to the border with Alba county.

Figure 4 presents the evolution of the number of accidents by road type. As can be seen, most accidents happened on the national roads, these being also the roads with the most substantial road traffic in the county. Also, 83% of all fatalities and severe injuries were recorded on national roads and streets within the limits of settlements.

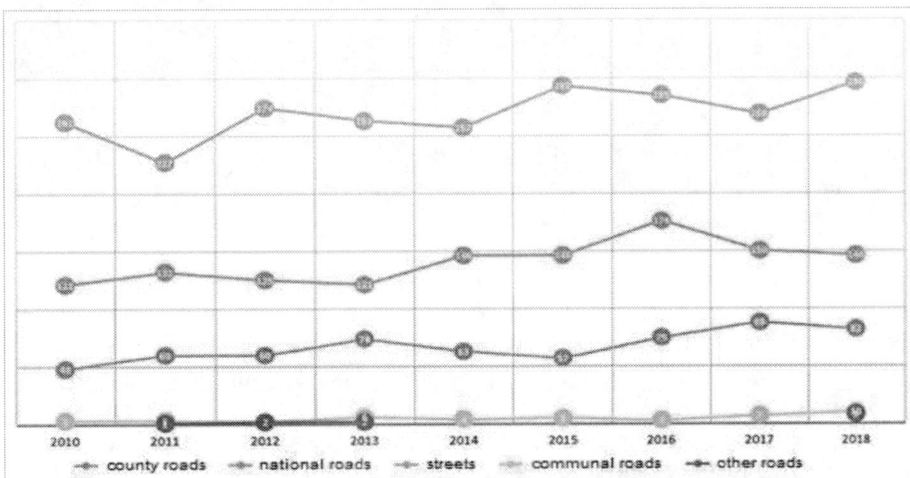

Figure 4. Evolution of the number of accidents by road type.

The distribution of road accidents in Sibiu county by the day of the week is indicated in figure 5.

It can be noticed that most accidents happened at the beginning of the weekend (Friday) but also at the beginning of the week (Monday and Tuesday). This is valid also for the city/village streets and the national roads, indicating an intensification of activity in the localities and a tendency towards traveling over longer distances on these days. On the other hand, Sundays see fewer accidents on the streets, but a peak of incidents on the county-level roads.

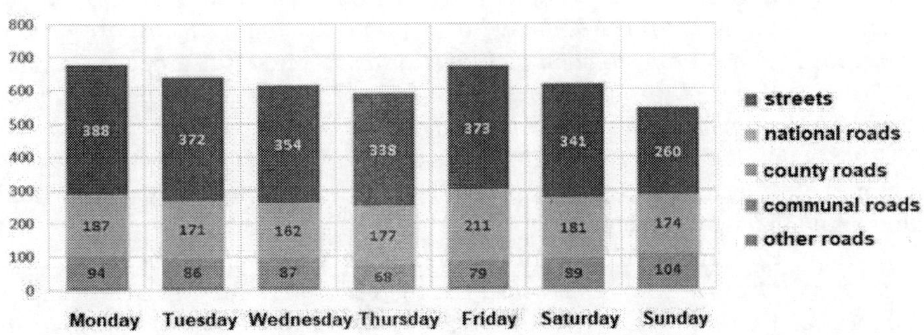

Figure 5. Distribution of road accidents in Sibiu county by the day of the week.

Figure 6 shows the hourly distribution of accidents in Sibiu county for the period from 2010 to 2018. As can be seen, most accidents produced in Sibiu county between 2010 and 2018 happened between 14:00 and 19:00. This can be explained by the end of the workday, by the tiredness of people after work but also by the eagerness of people to return home.

IOP Conf. Series: Materials Science and Engineering **568** (2019) 012040 doi:10.1088/1757-899X/568/1/012040

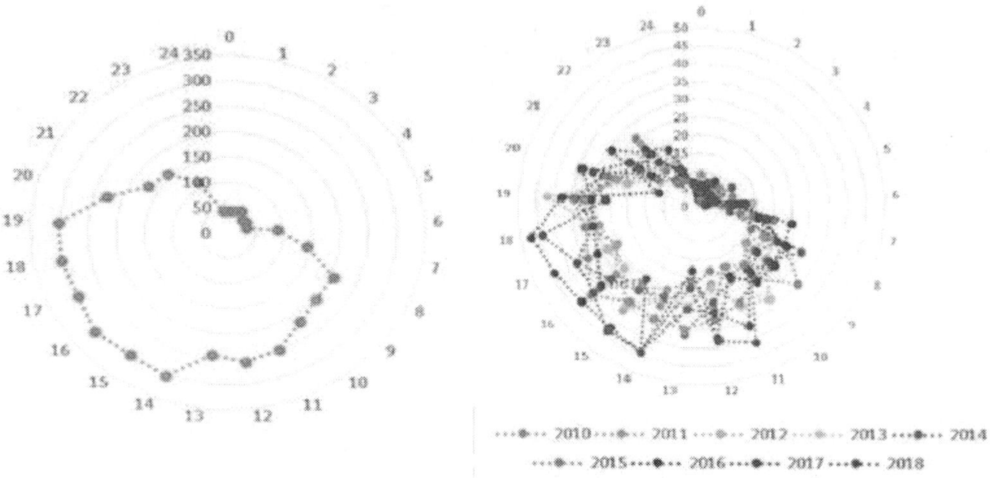

a) b)

Figure 6. The hourly distribution of road accidents in Sibiu county between 2010 and 2018: a) total number of accidents over the 9 years; b) number of accidents for each year

4. Conclusions

The current paper has analyzed the dynamics of the road accidents produced in Sibiu county between the years 2010 and 2018.

It can be concluded that the majority of accidents that occurred on the roads of Sibiu county in the period 2010-2018 had as a main favoring factor the human factor, which is the most important, but also the most difficult to manage.

This is especially significant when because 98% of the road accidents occurred on flat road segments and 77% of the road accidents occurred on dry roads.

In the future, the authors intend to expand their studies about road safety management to include a more in-depth analysis of the causes of road accidents and their correlation.

References

[1] Gligor A 2015 *Scientific Bulletin of the Petroleum-Gas University of Ploiesti, Technical Series*, **LXVI** 4 55-60
[2] Gligor A and Petrescu V 2018 *Matec Web Conf.* **184** 03005
[3] Dumitru M, Stanciu M, and Bibu M 2013 *Lucr. St. Manag. Agr.* **15** 156-159
[4] Bibu M and Dumitru M 2014 *Sci. Pap. Manag. Ec. Eng. Agri. Rural Dev.* **14** 35-38
[5] European Commission 2011 *Roadmap to a single European transport area – towards a competitive and resource efficient transport system* (Brussels: European Commission)
[6] ETSC 2018 *Ranking EU progress on road safety. 12th road safety performance index report* (Brussels: European Transport Safety Council)
[7] ERSO 2018 *Annual accident report 2018* (Brussels, European Commission)
[8] Government of Romania 2016 *National road safety strategy for the period 2016-2020* (Bucharest: Monitorul Oficial al Romaniei)
[9] Government of Romania 2012 *Law 265/2008 regarding the management of the safety of traffic on the road infrastructure, republished* (Bucharest: Monitorul Oficial al României)
[10] Government of Romania and AECOM 2015 *Final revised variant of the report on the master plan on short, medium and long term* (Bucharest: AECOM)
[11] Google 2018 *Map of Sibiu county* (Mountainview, California: Google)
[12] Inspectorate-General of the Romanian Police 2019 *EAC Registry of road accidents* (Bucharest: Romanian Police)
[13] Inspectorate-General of the Romanian Police 2019 *Statistics* (Bucharest: Romanian Police)

IOP Conf. Series: Materials Science and Engineering **568** (2019) 012041 doi:10.1088/1757-899X/568/1/012041

Experimental research regarding some dynamic parameters determination for sandwich bars reinforced with carbon fiber

C M Miriţoiu [1*], C O Burada [1], D F Tărâţă [1], A Bolcu [1], V Ionică [1]

[1]Faculty of Mechanics, University of Craiova, 165 Calea Bucureşti, 200620, Craiova, Romania.

*miritoiucosmin@yahoo.com

Abstract. For this study, we have built some composite sandwich bars with polypropylene honeycomb core reinforced with carbon fiber. We have clamped the bars at one end and we left them to vibrate freely. From the free vibrations we have determined the damping factors per unit mass and length, loss factor, dynamic Young modulus, dynamic stiffness and the frequency of the first eigenmode. The general conclusions obtained from the vibrations results show tat the sample width can influence the damping coefficient, by the fact that it determines the surface in which the air friction acts on the sample. The sample mass or specific linear mass influence the damping factor by the fact that the samples with higher mass and width, the deformation energy which is stored in the sample through the initial deformation, is dissipated in a larger quantity of material. An influence may occur due to the sample rigidity, explained by the fact that a force initially applied to the sample produces a smaller deformation if the stiffness is higher.

1. Introduction

Although carbon fiber is the most common reinforcement used in composite materials, many types of research still refer to this material. For example, in [1], the relationship between the property variables of the carbon fiber monofilament and the macroscopic parameters of the composites is established by using a regression tree, a type of decisions tree model, in machine learning. After a series of parameters adjustments and model selection, it was established a model with a better-generalized performance. Carbon fiber reinforced plastic is a multi-phase solid material which is a compound of two or more materials with different physical or chemical properties usually used in aerospace engineering [2]. In [3] a computational micromechanics model based on a representative volume element is used to predict the failure envelopes of unidirectional carbon fiber reinforced polymer composites. The proposed failure criteria have shown significant improvement according to the computational and experimental results obtained compared to classical failure criteria models.

Because of high tensile strength and modulus, lightweight, good compressive strength, temperature, and chemical resistance, excellent electrical and thermal conductivity, polyacrylonitrile-based carbon fibers are used as essential constructions in advanced composites [4-6]. In [7] polyacrylonitrile (PAN)-based carbon fibers (CFs) were electrochemically deposited and graphene oxide (GO) and three different kinds of comonomers, including diacetone acrylamide (DAAM), acrylic acid (AA) and phenol (Pheno), were electrografted onto CF surfaces. It was found a tensile

IOP Conf. Series: Materials Science and Engineering **568** (2019) 012041 doi:10.1088/1757-899X/568/1/012041

strength increase of 5, 22.6 and 17% for GO-DAAM, GO-AA, and GO-Pheno compared to untreated carbon fibers.

In this paper, we study the vibration behavior for some composite sandwich bars with the core made from polypropylene honeycomb reinforced with carbon fiber. We clamp the bar at one end, and we apply a force at the other end, which is free. At the free end, we put an accelerometer to record the free vibrations. From the free vibrations, we determine the damping factors per unit mass and length, the eigenfrequency, the dynamic Young modulus, the dynamic stiffness, and the loss factor.

2. Free vibrations of finite length bars. Analytical background

The differential equation of transversal free vibrations has the form [8]:

$$c^2 (\partial^4 w/\partial x^4) + (\partial^2 w/\partial t^2) = 0 \tag{1}$$

parameter c being determined with (2).

$$c = [1/(2\pi)]\cdot[EI/(\rho A)]^{0,5} \tag{2}$$

In (1) and (2) we have marked with: w the transversal movement of bar medium fiber; ρA the specific mass; ρ material density; A the bar transversal section; E the Young modulus; I the section moment of inertia.

By solving the differential equation of the transversal free vibrations (1), applying the Laplace transform and taking into account the Krîlov functions like in [9], we determine the eigenfrequency of the first eigenmode (3). In (3) we have marked with l the bars length, and β is determined from the bars boundary conditions.

$$\nu = (\beta/l)^2\cdot[1/(2\pi)]\cdot[EI/(\rho A)]^{0,5} \tag{3}$$

The free vibrations experimental recording gives the possibility of damping calculus in this way:

- there are determined the values where the displacement is zero;
- there is determined the cancellation movement period (more precisely T is the double time gap between two consecutive cancellations);
- the frequency ν and the pulsation ω are determined with (4)[10];
- the damping factor per unit mass is determined with (5)[10].

$$\nu = T^{-1};\ \omega = 2\cdot\pi\cdot T^{-1} \tag{4}$$

$$\mu = T^{-1}\cdot\ln(\Delta_j/\Delta_{j+1}) \tag{5}$$

- the damping factor per unit length is determined with (6)[18]:

$$C = 2\cdot\mu\cdot(\rho A) \tag{6}$$

In (15) we have marked with Δ_j, Δ_{j+1} the maximums separated by periods. The loss factor can be determined with (7) according to [10].

$$\eta = \mu\cdot(\pi\cdot\nu_\xi)^{-1} \tag{7}$$

In the relationship (7) we have marked with: ζ- critical damping; ω_ξ- eigen pulsation; μ- damping factor per unit mass; η- loss factor; ν_ξ- the eigen frequency; t- time; ξ is the number of the eigenmode.

From (3) we can establish the dynamic stiffness EI and Young modulus E values (8).

$$EI = (l/\beta)^4\cdot[\nu\cdot\pi\cdot2\cdot(\rho A)^{0,5}]^2,\ EI = (l/\beta)^4\cdot[\nu\cdot\pi\cdot2\cdot(\rho A)^{0,5}/I^{0,5})]^2 \tag{8}$$

IOP Conf. Series: Materials Science and Engineering **568** (2019) 012041 doi:10.1088/1757-899X/568/1/012041

3. Experimental and dynamic parameters determination

We have built some sandwich bars with polypropylene honeycomb core reinforced with carbon fiber in the upper and lower layers. We have obtained the next samples sets: **set 1**: thickness 11 mm, mass 55g, width 60 mm, 2 carbon fiber layers; **set 2**: thickness 11,4 mm , mass 78 g, width 60 mm, 4 carbon fiber layers; **set 3**: thickness 21 mm, mass 75g, width 60 mm, 2 carbon fiber layers; **set 4**: thickness 21,4 mm , mass 99 g, width 60 mm, 4 carbon fiber layers; **set 5**: thickness 16 mm, mass 59g, width 60 mm, 2 carbon fiber layers; **set 6**: thickness 16,4 mm, mass 90g, width 60 mm, 4 carbon fiber layers.

Figure 1. The built samples

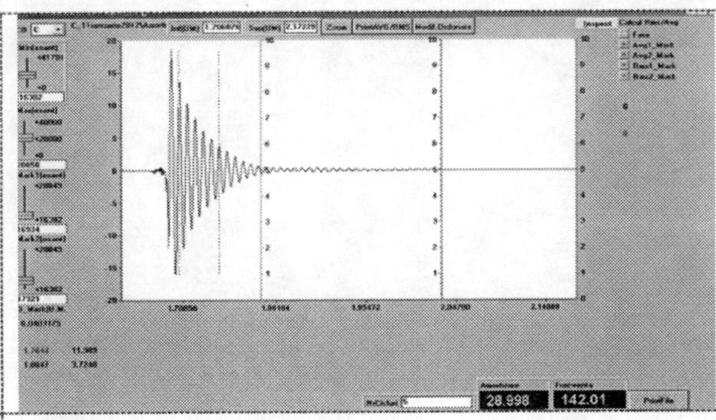

Figure 2. The eigenfrequency and damping factor per unit mass determination (set 5, 23 cm free length)

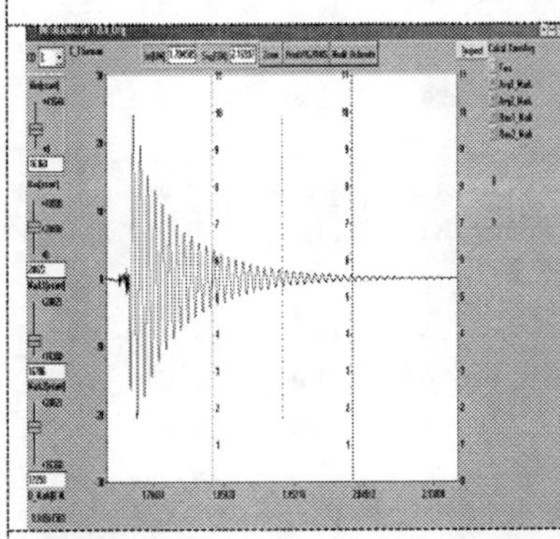

Figure 3. The vibration recording (set 1, 23 cm free length)

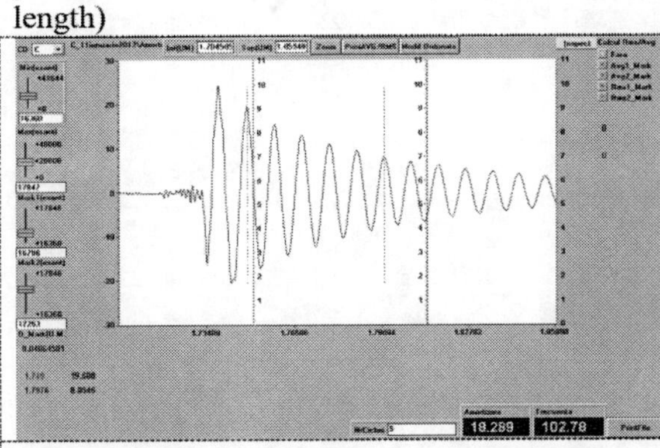

Figure 4. The eigenfrequency and damping factor per unit mass determination (set 1, 23 cm free length)

A general view of the samples is presented in figure 1. We have clamped the bars at one end, and we left them to vibrate freely. The vibrations were recorded with a Briel&Kjaer accelerometer placed at the free end. From the free vibrations, we have determined the damping factors per unit mass and length, loss factor, dynamic Young modulus, dynamic stiffness, and the frequency of the first eigenmode. We have considered the next free lengths for the bars: 20 cm, 23 cm, 26 cm, 29 cm, 32 cm and 35 cm. In figure 2, we have presented the eigenfrequency and damping factor determination for 5 cycles, set 5 - 23 cm free bar length. In figure 3, we have the vibration recordings for the set 1, the free length of 23 cm. In figure 4, we have presented the eigenfrequency and damping factor determination for 5 cycles, set 1 23 cm free bar length. All the dynamic parameters are written in table 1. The eigenfrequency and damping factor per unit mass were determined experimentally; the damping factor per unit length was determined with (6), the loss factor with (7), the dynamic stiffness and Young modulus with (8). If we make the arithmetic mean of all parameters, we obtain the dynamic mechanical characteristics written in table 2.

IOP Conf. Series: Materials Science and Engineering **568** (2019) 012041 doi:10.1088/1757-899X/568/1/012041

Table 1. Geometrical characteristics and dynamic parameters

set	width (mm)	Thickness (mm)	carbon fiber layers	mass (g)	free length (mm)	Eigenfrequency (1/s)	damping factor per unit mass (Ns/m)/kg	damping factor per unit length (Ns/m)/m	dynamic stiffness (Nm^2)	dynamic Young modulus (MPa)	Density (kg/m^3)	Specific mass (kg/m)	Loss factor
1	60	11	2	55	350	43.55	8.59	2.371	12.545	1878	208.33	0.138	0.061
1	60	11	2	55	320	53.87	10.33	2.851	13.413	2008	208.33	0.138	0.061
1	60	11	2	55	290	65.66	12.26	3.384	13.441	2012	208.33	0.138	0.059
1	60	11	2	55	260	82.33	15.75	4.347	13.65	2044	208.33	0.138	0.061
1	60	11	2	55	230	102.78	18.28	5.045	13.031	1950	208.33	0.138	0.057
1	60	11	2	55	200	122.6	23.38	6.453	10.258	1535	208.33	0.138	0.062
2	60	11.4	4	78	350	48.5	10.14	3.95	21.98	3303	295.455	0.195	0.067
2	60	11.4	4	78	320	61.22	12.88	5.023	24.478	3677	295.455	0.195	0.067
2	60	11.4	4	78	290	70.7	14.4	5.616	22.02	3308	295.455	0.195	0.065
2	60	11.4	4	78	260	85.72	17.24	6.724	20.914	3142	295.455	0.195	0.064
2	60	11.4	4	78	230	106	21.66	8.447	19.584	2942	295.455	0.195	0.065
2	60	11.4	4	78	200	130.1	26.25	10.238	16.868	2534	295.455	0.195	0.064
3	60	21	2	75	350	74.2	14.45	5.433	49.6	1068	148.81	0.188	0.062
3	60	21	2	75	320	93.6	19.14	7.197	55.165	1188	148.81	0.188	0.065
3	60	21	2	75	290	109.6	24	9.024	51.018	1099	148.81	0.188	0.07
3	60	21	2	75	260	126.71	28.31	10.645	44.058	948	148.81	0.188	0.071
3	60	21	2	75	230	168.09	36.21	13.615	47.48	1022	148.81	0.188	0.069
3	60	21	2	75	200	215.86	44.17	16.608	44.77	964	148.81	0.188	0.065
4	60	21.4	4	99	350	74.4	17.13	8.496	65.8	1340	192.757	0.248	0.073
4	60	21.4	4	99	320	93.96	20.1	9.97	73.33	1493	192.757	0.248	0.068
4	60	21.4	4	99	290	110.2	27.7	13.739	68.04	1385	192.757	0.248	0.08
4	60	21.4	4	99	260	136.99	36.3	16.021	67.932	1383	192.757	0.248	0.075
4	60	21.4	4	99	230	170.21	40.263	19.97	64.223	1308	192.757	0.248	0.075
4	60	21.4	4	99	200	225.66	49.6	24.602	64.541	1314	192.757	0.248	0.07
5	60	16	2	59	350	58.68	13.2	3.881	24.262	1188	153.646	0.147	0.072
5	60	16	2	59	320	75.7	15.63	4.595	28.214	1382	153.646	0.147	0.066
5	60	16	2	59	290	88.9	18.76	5.515	26.246	1286	153.646	0.147	0.067
5	60	16	2	59	260	108.8	22.77	6.694	25.399	1244	153.646	0.147	0.067
5	60	16	2	59	230	142.01	28.99	8.523	26.498	1298	153.646	0.147	0.065
5	60	16	2	59	200	183.4	36.11	10.616	25.269	1238	153.646	0.147	0.063
6	60	16.4	4	90	350	59.8	13.3	5.985	38.567	1748	228.659	0.225	0.071
6	60	16.4	4	90	320	75.9	17.4	7.83	43.413	1968	228.659	0.225	0.073
6	60	16.4	4	90	290	92.1	19.8	8.91	43.117	1955	228.659	0.225	0.068
6	60	16.4	4	90	260	108	24.9	11.205	38.307	1736	228.659	0.225	0.073
6	60	16.4	4	90	230	144.1	31.9	14.355	41.762	1893	228.659	0.225	0.07
6	60	16.4	4	90	200	186.9	39.4	17.73	40.168	1821	228.659	0.225	0.067

Table 2. Dynamic mechanical characteristics from the arithmetic mean values

Set	Stiffness	Young modulus	Loss factor
1	12.723	1904.5	0.06
2	20.974	3151	0.065
3	48.681	1048,17	0.067
4	67.31	1370.5	0.073
5	25.981	1272.67	0.067
6	40.89	1853.5	0.07

IOP Conf. Series: Materials Science and Engineering **568** (2019) 012041 doi:10.1088/1757-899X/568/1/012041

4. Conclusions

The damping factors analysis show that these factors must be experimentally determined for each type of material and sample, is difficult to deduce a quantitative correspondence with the parameters which influence the damping directly or indirectly.

The values of damping factors may depend on several features such as sample dimensions, specific mass, or the quantity of material from sample, elastic and damping properties of component materials. The sample width can influence the damping coefficient, by the fact that it determines the surface in which the air friction acts on the sample.

In addition to these general conclusions, by studying the tables 1 and 2, we can add new particular ones: the damping factor per unit mass and length increases with the thickness of the bar; the damping factor per unit mass of per unit length decrease exponentially with the free length of the platband; the eigenfrequency increases with the bars thickness increase; the dynamic stiffness increases with the thickness of the ba, but the Young modulus decreases with the bars thickness increase; there were obtained higher dynamic mechanical characteristics for the samples with four layers of carbon fiber compared to the ones with 2 layer - this fact can be explained by the addition of one carbon fiber layer in the upper and lower side of the sandwich bar; the loss factor tends to increase with the thickness of the bar

We consider that, the added values of this study can be: building some new composite platbands made by classical materials (such as carbon fiber, polypropylene honeycomb) combined in an original way; the experimental setup: the platbands are free at one end and clamped at the other where it is measured the vibratory response applied with a known initial force; the values of the damping factor (per unit mass and per unit length) for the built composite platbands; the frequency determination for the first eigenmode; the stiffness determination; the dynamic elasticity modulus calculus; the loss factor calculus.

References

[1] Q., Zhenchao, Z., Nanxi, L., Yong, C., Wenliang, 2012, *Prediction of mechanical properties of carbon fiber based on cross-scale FEM and machine learning*, Composite Structures, 212, p. 199-206

[2] J., Hu, K., Zhang, Q., Yang, et al., 2018, *An experimental study on the mechanical response of single-lap bolted CFRP composites interference fit joints*, Compos. Struct., 196, p. 76-88.

[3] Q., Sun, el. al., 2019, *Failure criteria of unidirectional carbon fiber reinforced polymer composites informed by a computational micromechanics model*, Composites Dcience and Technology, 172, p. 81-95.

[4] M., Andideh, M., Esfandeh, 2017, *Effect of surface modification of electrochemically oxidized carbon fibers by grafting hydroxyl and amine functionalized hyperbranched polyurethanes on interlaminar shear strength of epoxy composites*, Carbon, 123, p. 133-242.

[5] X., Qian, J.H., Zhi, L.Q., Chen, et al., 2018, *Evolution of microstructure and electrical property in the conversion of high-strength carbon fiber to high modulus and ultra-high modulus carbon fiber*, Compos. An ApplSci Manuf, 112, p. 111-118.

[6] X., Qian, R., Zou, Q., Ouyang, et al., 2015, Surface structural evolvement in the conversion of polyacrylonitrile precursors to carbon fibers, Appl Surf Sci, 327, p. 246-252.

[7] Z., Wen, X., Qian, et al., 2019, *Electrochemical polymerization of carbon fibers and its effect on the interfacial properties of carbon reinforced epoxy resin composites*, Composites Part A, 119, 21-29.

[8] M., M., StĂnescu, et al., 2009, *Experimental Researches Concerning the Properties of Composite Materials with Random Distribution of Reinforcement*, Materiale Plastice, 46, p. 73-78.

[9] W., Nowacki, 1969, *Dynamics of flexible systems*, Technical Publishing House, Bucharest.

[10] C.O., Burada, C.M., Miriţoiu, et al., 2015, *The vibration behavior of composite platbands reinforced with glass fiber*, Romanian Journal of Materials, 45, p. 244-254.

IOP Conf. Series: Materials Science and Engineering **568** (2019) 012042 doi:10.1088/1757-899X/568/1/012042

Considerations on the road safety situation in Sibiu county, between 2010 and 2019

A Gligor[1]

[1] Lucian Blaga University of Sibiu, Romania, Department of Industrial Engineering and Management, Emil Cioran str., 4, Sibiu, Romania

alina.gligor@ulbsibiu.ro

Abstract. Road transportation has become a significant part of human society and interactions, especially from an economic and social point of view. Road accidents represent major disruptions in this regard, and therefore, nowadays, road safety management is a critical topic. Sibiu county is located in a critical position in Romania, at the crossing of major North-South and East-West roads. Starting from these facts, the current paper aims to present an analysis of the road accidents that occurred in Sibiu county from January 1st, 2010 to December 31st, 2018, focusing on the number of victims but also on the causes generating or favoring the generation of these accidents.

1. Introduction

Modern civilization cannot be conceived without roads and cars. Unfortunately, in the conditions of an ever-increasing population, roads and cars to are increasing in numbers, and this, in turn, means that problems such as pollution or road accidents are also on the rise. Just as pollution is fought with the help of new, innovative, less polluting, renewable fuel sources [1, 2], road accidents can be fought by adopting specific road safety measures. These measures can be legal provisions, or they can be organizational or physical elements such as new motorways or lane-separating barriers. However, the essential precondition for efficient road safety management is to know the number of accidents, to know the evolution of this number over time and to know the causes of the accidents.

The member states of the European Union are among the best in terms of road safety. Still, in 2017, a total of 25260 persons lost their lives in road accidents across the European Union (while 1.45 million people have sustained severe or light injuries) [3, 4, 5]. While this represents a significant reduction compared to 54900 fatalities in 2001, it is nonetheless only a slight improvement in comparison to the figure of 31 500 fatalities in road accidents in 2010, especially considering the fact that the EU had set in 2010 a target of halving the number of road deaths until the year 2020 and that the ambitious “Vision Zero” goal seeks to eliminate fatalities and serious injuries stemming from accidents on European roads by the year 2050 [3].

In Romania, the government has adopted in 2016 the National Road Safety Strategy 2016-2020 [6], that reaffirms Romania’s commitment to the goal of a 50% reduction of the number of road deaths between 2010 and 2020 and this involves several measures in all aspects related to road safety, in all administrative units of the country [7, 8].

IOP Conf. Series: Materials Science and Engineering **568** (2019) 012042 doi:10.1088/1757-899X/568/1/012042

Still, the number of accidents recorded at the national level is relatively high, and there are required more efforts and solutions to achieve the goal, considering that in 2017, for example, there were recorded 1899 road fatalities compared to 2377 in 2010 [9, 10]

Sibiu county is crossed by several vital roads that usually see a high volume of road traffic. Therefore, road safety is paramount in this area, and the analysis of the number of road accidents and of the conditions in which they occur is a significant step in applying the right road safety measures.

In the current paper, the author analyses the number of accidents produced in Sibiu county between 2010 and 2018, the causes and the consequences of these accidents.

2. Analyzed area and methodology

Sibiu county is located in the center of Romania, and this location has allowed it to contain several important roads and road crossings (figure 1 [11]);

- DN1 crosses the county from east to west, coming from Braşov, passing through the city of Sibiu and going towards Sebeş and Alba Iulia;
- DN7 enters the county from the south, through the Olt River Gorge, enters the city of Sibiu and continues westwards towards Sebeş; from Sibiu to the border between Sibiu county and Alba county DN1 and DN7 share the same road within ;
- The A1 motorway segment from Sibiu to the border between Sibiu county and Alba county doubles the DN1/DN7 segment in that area.
- DN14 links the city of Sibiu with the city of Mediaş and the city of Sghişoara (the latter in Mureş county).

Figure 1. The leading road network of Sibiu county [11].

The statistical indicators for the number and causes of road accidents that occurred in Sibiu county between January 1st, 2010 and December 31st, 2018 have been interpreted using quantitative analysis with exploratory character, using mainly three sources:

- the National Road Safety Strategy 2016-2020 [6];
- the EAC (Registry of Road Accidents) application of the General Inspectorate of the Romanian Police [9];
- statistics published by the General Inspectorate of the Romanian Police [12].

IOP Conf. Series: Materials Science and Engineering **568** (2019) 012042 doi:10.1088/1757-899X/568/1/012042

It should also be mentioned that the numbers included in this analysis do not include road accidents from the analyzed period and the analyzed county (Sibiu County) that resulted only in material damage, without casualties.

Following analysis methods have been used:

- trends analysis;
- analysis of the intensity of an indicator in a specific time interval;
- comparative data analysis.

3. Results and discussion

The number of road accidents that occurred in Sibiu county between 2010 and 2018 and involved is presented in table 1.

Table 1. Number and consequences of road accidents that occurred in Sibiu county between 2010 and 2018.

Types of accidents and of consequences	2010	2011	2012	2013	2014	2015	2016	2017	2018
Light accidents	232	206	259	270	262	301	352	330	346
Serious accidents	202	217	204	196	208	200	187	184	196
Fatalities	47	44	43	31	46	29	33	31	34
Serious injuries	229	228	209	204	217	229	213	187	193
Light injuries	422	363	414	439	447	447	583	540	544

The leading causes of these accidents and the number of accidents attributed to each are listed in table 2. It can be noticed that the hierarchy of importance of these causes differs according to the category of accidents (light accidents or serious accidents).

Table 2. Leading causes of road accidents produced in Sibiu county between 2010 and 2018.

Main cause	Serious accidents	Light accidents	Total
Speed not adapted to road conditions	422	430	852
Failure to grant the right of way to vehicles	188	387	575
Illegal road crossing by pedestrians	227	229	456
Failure to grant the right of way to pedestrians	206	246	452
Failure to maintain a safe distance between vehicles	66	268	334
Illegal or unsafe actions by bicycle riders	112	143	255
Illegal overtaking maneuvers	100	96	196

The first four causes mentioned were responsible for 52% of all fatalities, severe injuries, and light injuries resulted from road accidents produced in Sibiu county between 2010 and 2018.

The first cause, speed not adapted to road conditions led to the death of 109 persons, to severe injuries in 506 persons and light injuries in 946 persons.

During the analyzed period, the second cause indicated in table 2, the failure to grant the right of way to vehicles, led to the deaths of 12 persons, the serious injuring of 214 persons and light injuring of 655 persons within the limits of Sibiu county. The highest number of victims from accidents having this cause was recorded on city/village streets, with a total of 595 casualties of various types. However, it should also be noted that this cause led to a much more significant proportion of light accidents than serious accidents (a trend that is also preserved in the case of the failure to maintain a safe distance as a cause).

The illegal crossing of roads was the cause of 46 fatalities, 184 severe injuries, and 237 light injuries. In this case, too, the highest number of severe consequences was recorded on city/village streets (346).

The failure to grant the right of way to pedestrians led to 19 fatalities, 193 severe injuries, and 270 light injuries. Again, most (448) victims were recorded on city/village streets.

However, it should be pointed out that the importance of the causes of road accidents varied on the various roads function of the intensity of the road traffic and the specific road conditions. As an example, table 3 shows the leading causes of accidents on the Olt River Gorge sector of DN7 [13]

Table 3. Leading causes of road accidents produced on the Olt River Gorge between 2010 and 2018 [13].

Curr. No.	Causes	Accidents
1	Speed not adapted to road conditions / excessive speed	93
2	Illegal overtaking maneuvres	31
3	Falling asleep while behind the wheel	25
4	Driving on the wrong side of the road	15
5	Failure to grant the right of way to pedestrians	14
6	Failure to maintain a safe distance between vehicles	16
7	Pedestrians indiscipline	14
8	Driving under the influence of alcohol	6
9	Failure to grant the right of way to vehicles	6
10	Driver distracted by other preoccupations	3
11	Medical conditions	2

For a more detailed analysis, table 4 shows the evolution of the consequences of road accidents by type of road.

Table 4. Numbers of casualties on the various types of roads in Sibiu county between 2010 and 2018.

Casualties by type of road	2010	2011	2012	2013	2014	2015	2016	2017	2018	Total
National roads										
Fatalities	24	31	23	13	34	25	20	14	17	199
Severe injuries	82	91	65	59	77	88	103	68	76	709
Light injuries	145	136	129	136	175	166	227	172	180	1466
County roads										
Fatalities	5	6	9	8	3	3	4	11	9	58
Severe injuries	39	32	31	34	37	22	37	38	31	301
Light injuries	44	43	66	73	64	64	82	117	87	640
Communal roads										
Fatalities	0	0	0	0	0	0	0	0	0	0
Severe injuries	2	3	2	2	1	3	0	3	1	17
Light injuries	4	3	1	9	5	8	3	5	8	46
City streets										
Fatalities	18	7	11	10	9	1	9	6	8	79
Severe injuries	106	102	110	107	102	116	73	78	82	876
Light injuries	229	180	217	218	203	209	271	246	263	2036
Other types of roads										
Fatalities	0	0	0	0	0	0	0	0	0	0
Severe injuries	0	0	1	2	0	0	0	0	3	6
Light injuries	0	1	1	3	0	0	0	0	6	11

IOP Conf. Series: Materials Science and Engineering **568** (2019) 012042 doi:10.1088/1757-899X/568/1/012042

It can be seen that most (83%) severe injuries and fatalities occurred on national roads or city/village streets.

Most accidents occurred on national roads (852), followed by county roads (247) and city/village streets (191). It is noteworthy that on public roads and other types of roads (forestry roads etc.) there were no fatalities and only 6 seriously injured persons.

When analysing the situation of the road accidents occurring on the streets of the city of Sibiu, for example, it can be seen that the first two streets in terms of the number of accidents (Alba Iulia street with a total of 128 accidents in the analyzed period and Dumbravii street with a total of 95 accidents in the analyzed period) recorded a higher number of accidents than the following four streets combined.

An essential element to note in table 4 is also the fact that in the last two years of the analysis, 2017 and 2018, the number of fatalities and serious injuries has decreased compared to the previous years. This is a consequence of the implementation of several measures, targeting mainly the road infrastructure, taken in order to increase the road safety, such as the installation of direction-separating barriers (New Jersey barriers), the improvement of the road markings, the introduction of appropriate speed limitations and not least the (re-)opening of the motorway segment from Sibiu towards the border between Sibiu county and Alba county.

4. Conclusions

When analysing the number of road accidents, their causes, the road conditions, the vehicles and persons (drivers, passengers, pedestrians) involved, it can be concluded that the majority of accidents that occurred on the roads of Sibiu county in the period 2010-2018 had as main favoring factor the human factor, which is the most important, but also the most difficult to manage.

This is especially significant when because 98% of the road accidents occurred on flat road segments and 77% of the road accidents occurred on dry roads.

The author intends to expand her studies about road safety to include the situations in other/neighboring counties and more influencing factors and causes.

References

[1] Dumitru M, Stanciu M, and Bibu M 2013 *Lucr. St. Manag. Agr.* **15** 156-159
[2] Bibu M and Dumitru M 2014 *Sci. Pap. Manag. Ec. Eng. Agri. Rural Dev.* **14** 35-38
[3] European Commission 2011 *Roadmap to a single European transport area – towards a competitive and resource efficient transport system* (Brussels: European Commission)
[4] ETSC 2018 *Ranking EU progress on road safety. 12th road safety performance index report* (Brussels: European Transport Safety Council)
[5] ERSO 2018 *Annual accident report 2018* (Brussels, European Commission)
[6] Government of Romania 2016 *National road safety strategy for the period 2016-2020* (Bucharest: Monitorul Oficial al Romaniei)
[7] Government of Romania 2012 *Law 265/2008 regarding the management of the safety of traffic on the road infrastructure, republished* (Bucharest: Monitorul Oficial al României)
[8] Government of Romania and AECOM 2015 *Final revised variant of the report on the master plan on short, medium and long term* (Bucharest: AECOM)
[9] Inspectorate-General of the Romanian Police 2019 *EAC Registry of road accidents* (Bucharest: Romanian Police)
[10] Tarnu L and Deac C 2018 *Matec Web Conf.* **184** 01012
[11] Google 2018 *Map of Sibiu county* (Mountainview, California: Google)
[12] Inspectorate-General of the Romanian Police 2019 *Statistics* (Bucharest: Romanian Police)
[13] Tarnu L and Deac C 2018 *Matec Web Conf.* **184** 01011

IOP Conf. Series: Materials Science and Engineering **568** (2019) 012043 doi:10.1088/1757-899X/568/1/012043

Aspects regarding corrosion and mechanical tests for metallic parts obtained by Powder Metallurgy

A Didu[1], I Pascu[1] ,S Gheorghe[1] and C Rotea[1]

[1] Automotive, Transport and Industrial Engineering Department, Faculty of Mechanics, University of Craiova, Calea Bucuresti Str., no.107, Craiova, 200512, Romania

ankadidu11@gmail.com

Abstract. The researches for this paper were made for knowing the corrosion behavior and mechanical resistance of some metallic powder's specimens obtained by powder metallurgy. This technology was chosen because of the technical and economic advantages, and the obtained parts can be used in all domains. The parts were obtained from iron powders by pressing at different pressures and sintered in an inert atmosphere. The innovation was to obtain metallic composites, having as a matrix an iron powder type Ancorsteel 1000B and as a reinforcement a spring steel wire. The specimens were tested in different solutions and environments and also two mechanical tests. This paper presents a part of tests that were the corrosion in salt spray environment and bending tests, aiming to compare the results for the types of proposed specimens. For the morphological characterization (size, shape, surface) and chemical composition of the material used, SEM electronic microscopy was used using an FEI Inspect S microscope. The sintering was made in a neutral controlled atmosphere (argon) furnace.

1. Introduction

Powder metallurgy is one of the top industrial branches, and it can be said that there is currently no industrial branch where it is not used in advantageous technical - economic conditions powder parts. Thus, in addition to the particle-specific characteristics of the powders, they must be more mechanically and chemically resistant. In this direction, it is necessary to know the resistance to the various types of corrosion of the parts, as they are used in polluting environments.

Due to these obvious technical and economic advantages, powder metallurgy offers multiple possibilities for many types of new materials and products, with special features required by modern technology, especially by the cutting-edge branches. Advantages of Powder Metallurgy have led to the use of this technology in large-scale and mass production, with MP products applying to the automotive and aerospace industries, the medical industry, the manufacture of self-lubricating bearings, the strength parts of the construction industry machines, power contacts, etc.[1-3].

In the contemporary era of intense industrial development, when the consumption of metals and machinery is continuously growing, it seems particularly important, especially for ferrous ones, the concern to protect them against corrosion damage. Costs due to corrosion from raw material and energy consumption are amplified by secondary damage due to activity interruptions, the loss of expensive products (oil, fuel oil), the need for oversize elements, etc.

IOP Conf. Series: Materials Science and Engineering **568** (2019) 012043 doi:10.1088/1757-899X/568/1/012043

2. Experiments and results

2.1. Obtaining iron powder parts used in researches. Characterization and stages

To obtain both unreinforced and reinforced specimens, with high alloyed steel wires used in the experiments, it was followed the specific stages of powder metallurgy, pressing and sintering. For researches, atomized Fe powder was used, the ANCORSTEEL 1000B sort. The medium size of the particles is 0,125 mm.[8]

For the morphological characterization (size, shape, surface) and chemical composition of the material used, SEM electronic microscopy was used using an FEI Inspect S microscope. For this purpose, a microscope was used to determine the shape of the powder granules with an analyzer - SiLi EDAX Inc detector with a wavelength of ± 0.00000014 mm.

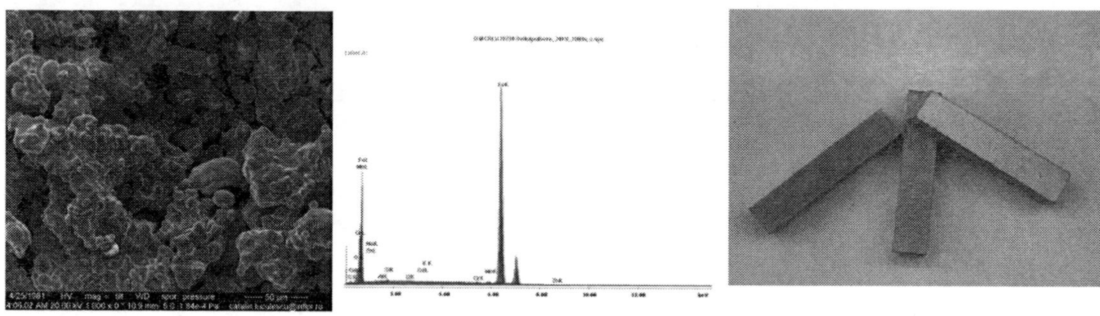

a) b) c)

Figure 1. a) Size, shape, dimensions of powder granules Ancorsteel 1000B; b) Chemical composition;c) The aspect of parallelepipedic samples.

Figure 1 shows the morphological aspects of the powder used: the chemical composition, respectively, the size, shape, and size of the granules. From the diagram of figure 1b, the chemical composition of the powder used is the same as that presented by the company in the product catalog.

Figure 1a also shows that the shape of the powder granules is very different, with rounded, irregular shapes, with a significant grain size of around 100μ. It was used1% of zinc stearate as a lubricant, blended with 99% iron powder, 1% other elements.The aspect of parallelepiped specimens can be seen in figure 1c. For reinforcement, spring steel wires with a tear resistance of 2300-2500 N / mm2 with diameter Φ 0.5 mm were used. [5], [6]

Porosity. Porosity was calculated after sintering, and the following values were obtained: In figure 2 can be observed that the porosity decreases with increasing compacting pressure.

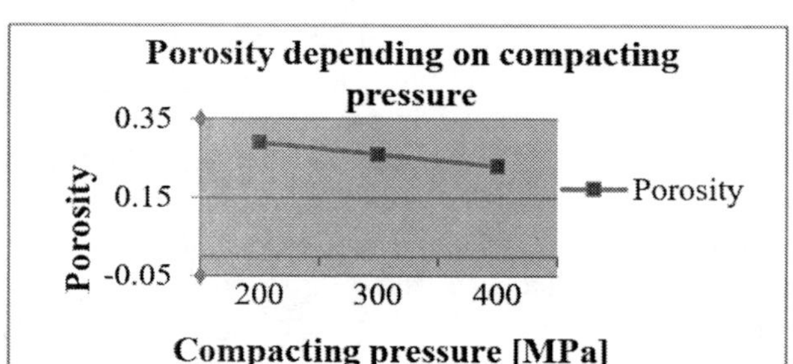

Figure 2. The porosity of specimens after sintering.

Pressing. The sampled were pressed unilaterally with a surface S = 5.5 cm^2 and was performed on a 100-tonne OMCN press. The specimens pressed with compacting pressure of 200 MPa, 300 MPa and 400 MPa were weighed, and their sizes were measured before and after sintering. From the SEM image analysis at 1000X and 5000X magnification of figure 3, it is observed that in the surface area, the iron granules are flattened due to contact with the die or punch.

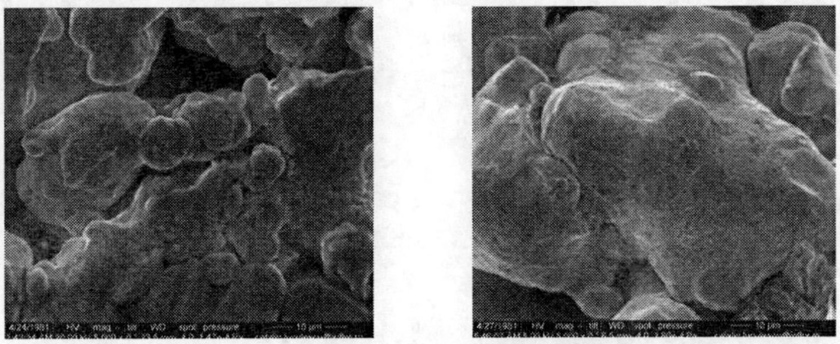

Figure 3. SEM images of pressed specimens, before sintering at magnification: 1000X and 5000X

Sintering. It was made in a neutral controlled atmosphere (argon) furnace. The sintering of the compacted samples was performed in an LL207 / 220 V-type resistive laboratory furnace using a specially designed enclosure to maintain an inert atmosphere to prevent the oxidation of the specimens as inert medium, argon was used, providing a flow rate of 2.5 l / hr. The sintering was made at 1150 ° C, step by step.

By analyzing the SEM images, it is observed that in the surface area, the iron powder granules are flattened due to contact with the mold or the punch (figure 4a) and are not flattened in the rupture zone. Spectral analysis showed a chemical composition similar to that before sintering(figure 4b).

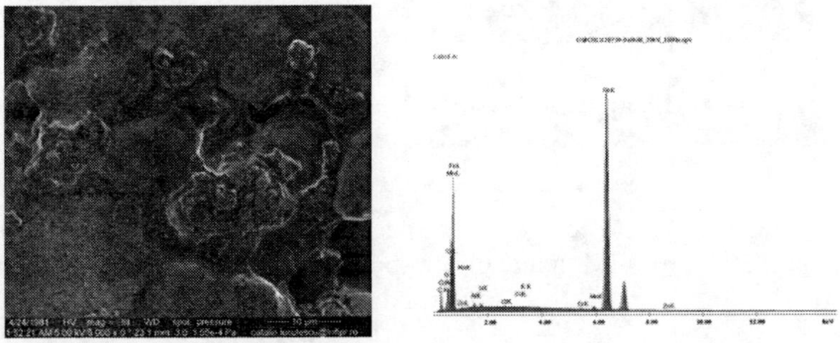

a) b)

Figure 4. SEM images of pressed and sintered samples a) surface) EDS spectrum of the pressed sample, b) after sintering.

2.2. *Salt spray corrosion*

The experimental tests were performed using a 5% NaCl solution (according to the salt spray corrosion standard SR EN ISO 9227/2007), [4] the temperature was maintained at about 28 ° C for the first set of specimens made of iron powders. The specimens were introduced into the salt spray chamber. The first check was made after approximately 3 hours, and the first corrosion points were observed on the uncovered specimens. After approximately 45 hours for standard samples, the corrosion process is becoming stronger, as seen in Figure 5.

IOP Conf. Series: Materials Science and Engineering **568** (2019) 012043 doi:10.1088/1757-899X/568/1/012043

The below images show that the iron powder, pressed and sintered pieces exhibit relative resistance in saline medium, with changing the appearance over the whole surface, due to the formation of layers composed of corrosion compounds of iron. Due to the low corrosion resistance of the used material, in the following researches methods of corrosion protection are searched and tested [7].

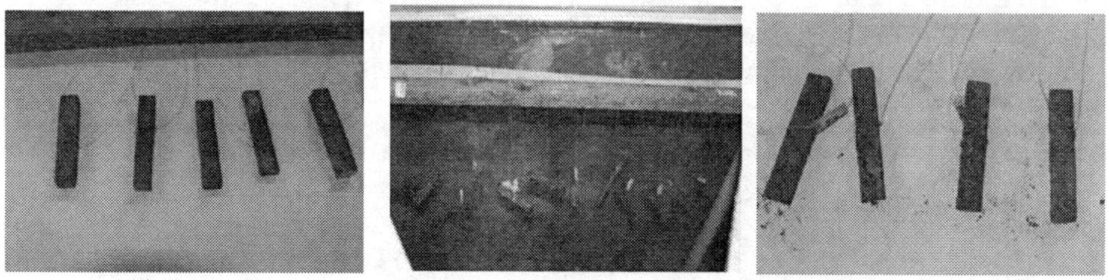

a) b) c

Figure 5. a)Pictures after 1 verification - 3 hours; b) 6th verification- after 45 hours; c) the last verification - after 90 hours.

a) b)

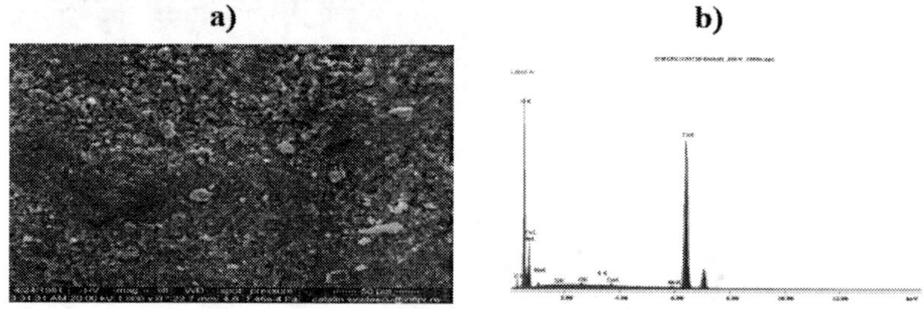

c) d)

Figure 6. a) SEM image of the specimen; b) EDS spectrum before corrosion; c) SEM image of the specimen; d) EDS spectrum after corrosion.

For the morphology of the sample surface at a compaction pressure of 400 MPa, a relative uniformity and relative porosity are observed. The EDS spectrum (Figure 6b) refers to a high proportion of iron and other elements as impurities. The difference between the non-corroded sample (Figure 6a) and the corroded sample (Figure 6c) is highlighted by the compact appearance and uniformity of the surface layer due to more or fewer iron compounds that can form from the oxidation processes.

2.3. Bending test

The bending test is more important in the case of cast iron and sintered powders which are sometimes required to determine the capacity to take up this type of request.

From the analysis of the force-deflection diagrams obtained, it was observed that the bending breaking strength is directly proportional to the density of the parts (implicitly with the compaction pressure) [4,5]. It is also clear that bending strength is higher for reinforced parts than for unreinforced ones for standard specimens.

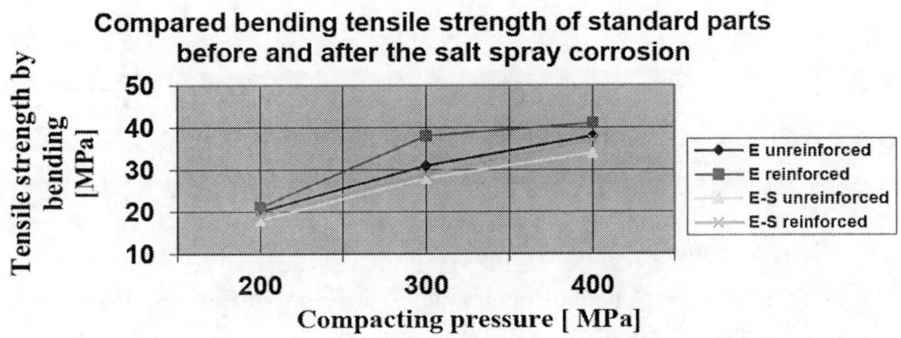

Figure 7. Relative bending tensile strength before and after salt spray attack of the samples.

From curve analysis in figure 7, it can be seen that the bending resistance decreases for salt spray corroded parts subjected to stress. It is also noticed that the reinforced pieces have bending strength higher than those unreinforced with yarns.

3. Conclusions

The salt spray corrosion resistance of metallic powder parts is relatively low; metallic coating is needed to improve the properties.

The corrosion process affects both the surface and the interior of the specimens, as well as the reinforcement yarn; the highest corrosion resistance in both salt spray and other studied environments corresponds to pressed pieces at 400 MPa.

After corrosion in salt spray chamber, SEM images and EDS spectra showed that at a compacting pressure of 400 MPa, a relative uniformity and porosity were observed.

Tensile strength by bending is directly proportional to the density of the parts (implicit with compacting pressure).

The tensile strength at bending is higher for the reinforced pieces than for the unreinforced ones.

The tensile bending strength of standard (not covered) corroded parts in the salt spray is lower than those not corroded.

References

[1] Aderhold J., Davydov V. Yu., Fedler F., Klausing H., Mistele D., Rotter T., Semchinova O., Stemmer J., and Graul J. 2001 *J. Cryst. Growth of* **222**.

[2] Bautista A., Velasco F., Guzman S., De la Fuente D., Cayuela F., and Morcillo M., *Rev. Metal Madrid* 2006 **42** 175

[3] Shankar J., Upadhyaya A. and Balasubramaniam R., *Corros. Sci.*2004**46** 487

[4] Didu A., Dumitru V., Stanescu G., and Ploscaru C, *Appl. Mech.s and Mater.* 2017 **880** 291-296

[5] Didu A., Studies on corrosion behavior in a pollutant environment of iron-based composites reinforced with high resistant steel yarns - Ph.D. thesis, 2013

[6] Vlasa A., Varvara S., Pop A., Bulea C., and Muresan M., *J. Appl. Electrochem.* 2010 **40** 1519-1527

[7] Didu M., Gruionu L, Radu S., and Ghermec C. *Mater. Sci. Forum.*, 2011 **672** 327-330

[8] Mureşan R., Powder Metallurgy (*Metalurgia pulberilor)*, Ed. U.T. PRES 2005 Cluj-Napoca

IOP Conf. Series: Materials Science and Engineering **568** (2019) 012044 doi:10.1088/1757-899X/568/1/012044

Introduction to the research of uncertainty in logistics management

Ł Marzantowicz[1] and I Dembińska[2]

[1] SGH Warsaw School of Economics, Collegium of Business Administration, Department of Logistics, Poland

[2] University of Szczecin, Faculty of Management and Economics of Services, Department of Logistics, Poland

E-mail: lukasz.marzantowicz@sgh.waw.pl; izabela.dembińska@wzieu.pl,

Abstract. The article aims to present the assumptions and results of the first part of the initial research of logistic management in conditions of uncertainty. As a result of the research, research gaps were identified which determine further directions of research in the problem above. The presented research results were analyzed using the statistical method of Spearman and Guilford, which allowed verifying the assumptions and determining the working version of the hypothesis and the partial purpose of further research. The evaluation of the survey results confirmed the need for conducting in-depth research in recognition of uncertainty conditions at the strategic management level.

1. Introduction

The critical challenge for many companies is to achieve the most effective adjustment of their volatility to the variability of the environment, to "eliminate the unacceptable level of variation in volatility", which then becomes a real problem for the company. However, one can not overlook the fact that there is a general pressure to improve all processes (including logistic) continuously. However, in every process (as well as in the management of it) events occur marked by randomness. This is mainly because economic instability, which enterprises face, is not characterized by a simple causal relationship and can not be directly identified with the phenomenon of volatility.

When making logistic decisions, managers must always consider issues of existing uncertainty and possible risk. Therefore, in terms of uncertainty, decision-making models of a deterministic nature (where the result verifies the correctness of the decision) or probabilistic (in which, apart from the result, the risk level is also taken into account) is often used. The essential feature characterizing uncertainty is the inability to predict the effects and consequences of events. Therefore, there is certainty about the occurrence of an event, but the set of effects is characterized by a limited possibility of forecasting and estimating possible consequences. This means that from the efficiency of logistics processes, it is impossible to determine the degree of their efficiency. Considering this category from mainstream economics (and management), uncertainty should be considered a situation that is in its essence an insoluble decision problem. One in which different decisions will have different possible consequences. This means that even if a certain event and the probability distribution of a given state of affairs has been known, it can result in many different values.

IOP Conf. Series: Materials Science and Engineering **568** (2019) 012044 doi:10.1088/1757-899X/568/1/012044

The problem raised in the article is a contribution to researching the field of uncertainty management in the company from the perspective of logistics management. The article aims to present the assumptions of the study of uncertainty in logistics management from the perspective of the company. The article has a research and demonstration nature, based on the results of the preliminary study and constitutes a voice in the ongoing discussion.

2. Determining the state of knowledge in the problem

The approach to understanding and defining logistic management is common and widely described in the available literature. Only the internet shows about 19,000 results corresponding to the logistic management tag. From the discourse pursued in the article, the approach to understanding uncertainty, and the occurrence of conditions of uncertainty is much more critical. As A.W.Willett [1] wrote at the beginning of the 20th century: "the risk is an objectified uncertainty as to the occurrence of an undesirable event. This risk depends on uncertainty and not on the degree of probability "[1].

It is worth noting that the definition of uncertainty also depends on the area of science and knowledge in which different authors move. However, it is noticeable that the uncertainty is related to risk and its scientific aspect (i.e., measurable but unpredictable effects) and the personal aspect (understood as a relation to risk) [2]. Notably, the risk concept also includes the expected high volatility [3].

There is always the possibility of an event occurring, but it may not be possible to assess the degree of probability. It can be concluded that the occurrence of various unpredictable factors, coming from various sources, and only partially measurable, forms the so-called conditions of inactivity. Uncertainty concerns changes and events are challenging to estimate. However, there are approaches to measuring uncertainty. One of the most popular approaches to measuring uncertainty is to say that uncertainty can be measured using statistical methods (if it is considered sufficient) – what again is the starting point for the considerations in the article, although the uncertainty in its pure form is here not accepted because the uncertainty that can not be measured is the uncertainty "in the strict sense"[4]. At the same time, it is not possible to predict the effects of actions in the long term, especially in the field of business conditions, technological progress, competition development, etc. [5], which results in the extreme claim that uncertainty cannot be measured in any way - there is no possibility to use any measurement tool [6].

The divergence in the approach to the possibility of measuring uncertainty in the light of management raises the question of whether there is a possibility of alleviating the level of uncertainty. The answer should be affirmative because the anticipation of the future should take into account at least the partially quantifiable variability. Thus, the volatility of the degree of uncertainty depends on the pace of identification of factors that should be taken into account in the forward-looking analysis [7]. Nevertheless, for the article, it can be assumed that uncertainty occurs when apart from unpredictable factors (usually the more significant part of the set) there are factors that are quantifiable. Not only the set of factors is essential, but also a set of effects that can create many options that make up the portfolio of results [8]. This suggests that there is the possibility of not only examining the level of uncertainty but also its impact on logistics management, which is the basis for the formulation of research tools in this regard.

3. Methodology and main assumptions of the first stage of the preliminary research

In January 2019, the methodology for qualitative research using the focus method was developed (FGI), aimed at finding answers to the following questions:

1. Do logisticians in the enterprise consider uncertainty in the management process?
2. Who takes responsibility for logistic management in conditions of uncertainty?
3. Should the impact of uncertainty conditions be measured on logistics management and how?

The answers to such questions were to facilitate the decision regarding the need to research logistic management in conditions of uncertainty, to build the concept of the pilot and proper research, including the model of logistic management in conditions of uncertainty and the possibility of measuring the impact of uncertainty on the management process.

The described stage is an original study that allows determining the main assumptions and premises for further research. Two focus sessions were conducted, attended by 19 representatives of enterprises employed at various organizational levels. So-called target selection was used (in the focus study the targeted sample is determined by the selection criteria), whose necessary qualifying criterion was to have in the employment and organizational structure the person (s) directly responsible for logistics management (e.g. logistics managers).

The participants of the research were presented with problems, and an external moderator, which ensures objectivity in formulating further conclusions, conducted the research.

In the analysis of the responses of the focus group participants, the methods of statistical analysis in the scope of question 2 and 3 were used, which aimed to give value to the answers enabling the determination of the highest value constituting the highest rank. The first question was developed only in terms of quantitative responses (not using statistical analysis).

The study uses the Spearman rank correlation coefficient. It has been defined as the Pearson correlation coefficient calculated for the variable ranks [9] (the rank is the number that corresponds to the place in the order of each feature).

When interpreting the results, the classification was based on J. Gilville, where:

$|r|=0$ - lack of correlation
$0,0<|r|\leq 0,1$ - slender correlation
$0,1<|r|\leq 0,3$ - weak correlation
$0,3<|r|\leq 0,5$ - average correlation
$0,5<|r|\leq 0,7$ - high correlation
$0,7<|r|\leq 0,9$ - very high correlation
$0,9<|r|<1,0$ - almost full correlation
$|r|=1$ - full correlation

4. Research results

1) In the scope of question 1. - Do logisticians in the enterprise consider uncertainty in the management process?
2)

Table 1. Percentage of responses to question 1 set by study participants

Established answers	Number of answers	Percentage
Factors shaping uncertainty are not taken into account. Management concerns the only risk.	4	21,05%
Uncertainty is taken into account, but only in terms of leveling the effects of its occurrence.	4	21,05%
Uncertainty in logistics management is taken into account. However, selected management methods are not used	9	47,37%
Yes, uncertainty is considered as a set of variable factors, including uncertainty using forecasting	2	10,53%

Source: Own study based on the focus research (FGI)

It should be indicated that the answers presented in the table come directly from the participants of the study. As can be seen, in the majority (47.37%) of enterprises - and in particular those responsible for logistics management - take into account the conditions of uncertainty in logistics management. However, uncertainty factors are included only in an intuitive way - depending on the competences (and qualifications) of logistics managers. There is no indication of the use of management tools about the occurrence of uncertainty conditions. Only two of the surveyed managers confirmed the development of their management tools taking into account the possibility of uncertainty conditions as well as their

IOP Conf. Series: Materials Science and Engineering **568** (2019) 012044 doi:10.1088/1757-899X/568/1/012044

application at the stage of forecasting the quality of the management process and estimating the effects of uncertainty.

3) In the scope of question 2. - Who takes responsibility for logistic management in conditions of uncertainty?

Table 2. The strength of correlation in question 2

Answers in the field of taking responsibility - the decision-making level	Correlation coefficient	Strength of correlation
Each level of management	0,5012	high
Strategic management level	0,571	high
The tactical level of management	0,2568	weak
Operational management level	0,3382	average

Source: Own study based on the focus research (FGI)

According to the correlation analysis, the answer with the highest impact (the strategic level of management) means that decisions regarding the uncertainty should be made at the strategic management level. Interestingly, a similarly high correlation coefficient received an answer stating that logistics management in the conditions of uncertainty requires management decisions at every level.

4) In the scope of question 3. - Should the impact of uncertainty conditions be measured on logistic management and how?

The difficulty in analyzing the participants' answers to such a question is dictated by earlier responses that speak about the lack of management tools in terms of uncertainty in logistics management. For this reason, it is concluded that among the surveyed companies, the tools for measuring possible conditions of uncertainty are also not used. Nevertheless, with the help of deduction, the participants of the study gave general answers shaping the cognitive value towards the need to measure the influence of uncertainty factors on logistic management in the way as in Table 3.

Table 3. Strength of correlation in the scope of question 3

Established answers	Correlation coefficient	Strength of correlation
measurement is not possible	0,0206	slender
measurement is possible using the available statistics and management tools	0,2568	weak
measurement is possible using a "sewn-in" tool - e.g., a measurement survey, checklist	0,6354	high

Source: Own study based on the focus research (FGI)

Answers whose correlation with question 3 is described as weak does not mean that this answer can not be answered in reality. It only means a reduced probability of the possibility of such a solution. However, it should be considered that measurement is possible, and the strength of the impact of uncertainty on logistic management is quantifiable at least partially - which confirms the assumptions of the study that some of the factors that shape uncertainty are subject to the possibility of partial estimation.

5. Discussion and Conclusions

The research being carried out is the first part of the preliminary study, allowing to put a working hypothesis in the following wording:

The uncertainty conditions belong to the set of inherent factors of the logistic management strategy and have an impact on its effectiveness.

IOP Conf. Series: Materials Science and Engineering **568** (2019) 012044 doi:10.1088/1757-899X/568/1/012044

The preliminary results of this part of the study are not sufficient to confirm the working hypothesis. However, in its scope, you can verify the assumptions of the study (questions posed in the focus session). As the study showed, the factors shaping the uncertainty identified by enterprises result instead of the perception of the phenomenon of uncertainty as to the basis for risk management. Hence the partial, though the strong, the tendency of managers to claim that management uncertainty is at least partially taken into account. This is a confirmation of this part of the working hypothesis that the conditions of inactivity have an impact on management, and a way to measure this influence should be developed.

The results of the survey also point to the fact that uncertainty is perceived by managers as a phenomenon requiring far-reaching actions in the field of its management. For this reason, decisions regarding the management of uncertainty in logistics should be made at the strategic level. This means that despite the universality of the uncertainty related to management at all levels, the conditions of uncertainty should, however, be part of the management strategy. Conditions of uncertainty should, therefore, be included in the process of strategic planning and implementation of long-term goals. The above conclusions partially confirm the hypothesis regarding the location of responsibility for managing uncertainty conditions among top management.

Most reservations at this stage of the study and at the same time, the most difficulties in interpretation leave the aspect of measuring the impact strength. At such an early stage of the study, when the qualitative research process failed to construct a measurement tool subjected to possible evaluation by the participants of the study, it is not possible to adequately assess the level of impact of uncertainty conditions on logistics management. Nevertheless, it is worth noting that the correlation of the answer with the question shows that among managers there is a need to find a way (including tools) to measure the impact of uncertainty factors on the way, efficiency and effectiveness of logistics management.

The preliminary study carried out at the same time allows us to determine the partial purpose of further research, which is to determine the typology of uncertainty about sources of variable factors and to determine how to react to the occurrence of uncertainty. The above discussion allows us to conclude a general nature, that the cognitive and scientific space in the field of logistics management in the conditions of uncertainty carries many challenges and is still open.

References

[1] Willett A.H. (1951), *The Economic Theory of Risk and Insurance*, University of Pennsylvania Press, Philadelphia.

[2] Williams Jr C.A., Heines R.M. (1989), *Risk Management and Insurance*, ed. 3. McGraw-Hill Publishing Co., New York.

[3] Jajuga K. (1999), *Nowe tendencje w zarządzaiu ryzykiem finansowym*, „Rynek Terminowy", nr 3/5.

[4] Knight F.H. (1993), *Risk*, Uncertainty, and Profit, London.

[5] Sudoł S. (2006), *Przedsiębiorstwo. Podstawy nauki o przedsiębiorstwie. Zarządzanie przedsiębiorstwem*, wyd. III, PWE, Warszawa.

[6] Janikowski, R. (2014), *Nieprzewidywalność w zarządzaniu przedsiębiorstwem*. „Modern Management Review", 19/21 (4).

[7] Bolesta-Kukułka K. (2003), *Decyzje menedżerskie*, PWE, Warszawa.

[8] Marzantowicz, Ł. (2017). *Niepewność i nieprzewidywalność w łańcuchu dostaw–rozważania teoretyczne*. *Nauki o Zarzadzaniu*,*31*(2), 62-70.

[9] Spearman Ch., *The proof and measurement of association between two things*, [in:] The American Journal of Psychology Vol. 15 No. 1 (Jan. 1904), p. 73.

IOP Conf. Series: Materials Science and Engineering **568** (2019) 012045 doi:10.1088/1757-899X/568/1/012045

Key assumptions of energy-consumption analysis for the functioning of a logistics center

I Dembińska[1] and Ł Marzantowicz[2]

[1] Department of Logistics, University of Szczecin, Poland
[2] Department of Logistics, SGH Warsaw School of Economics, Poland

izabela.dembinska@wzieu.pl

Abstract. The study aims to present the key assumptions of the energy-consumption analysis for the functioning of a logistics center. The leading energy carriers used in the logistics center were identified, and factors determining the level and structure of its consumption were indicated. Using the indicator method, a catalog of indicators for the energy consumption analysis for the functioning of a logistics center was created.

Introduction

The world is using more and more energy, and this trend is permanent. There is no possibility to balance the demand for energy both on the scale of our globe or continents, as well as individual countries without measures aimed at rational use of energy. Rational use of energy is most often more economical than the construction of new generation capacities based on still imperfect technologies. This fact determined that in terms of energy, the modern economy puts a strong emphasis on rationalizing energy consumption while maximizing the global economic effect. Increasing energy efficiency by 20% in 2020 is the goal set by the European Union.

In general, energy can be characterized in terms of sources and forms. The following energy sources are distinguished: energy of fossil fuels, solar energy, wind energy, water energy, geothermal energy, see tides energy, ocean heat energy. Because these sources come from nature, they constitute a group of primary energy carriers. In addition to primary energy, secondary energy (derivative) is also specified, which comes from energy derivatives, i.e., those obtained in the process of energy conversion. Among the secondary energy carriers, electricity, heat and gaseous fuels are most often mentioned in technological processes [3]. By exchanging these examples, one can see that the energy takes various forms, with each form being changed into another. The most important forms of energy include electricity, heat energy, chemical energy, nuclear energy, radiant energy. The first three forms of these are the most important for logistics infrastructure [2].

Currently, in the study of energy consumption has become an important control energy indicators. Methods using energy indicators are widely used for many tasks. They are a common approach in statistical analyzes regarding the energy condition of an object [4, 5, 6]. They are used for comparisons of benchmarking processes, as well as for monitoring the effectiveness of energy-saving implementations or forecasting the demand for energy carriers in the aspect of ecological effects. [7, 8, 9].

The study aims to present the critical assumptions of the energy-consumption analysis for the functioning of a logistics center. The logistics center is understood here as a warehouse building, i.e., a profiled logistics center, as well as a comprehensive center (logistic park, seaport). Due to the conceptual

IOP Conf. Series: Materials Science and Engineering **568** (2019) 012045 doi:10.1088/1757-899X/568/1/012045

(pragmatic) character of the study, the traditional literature review was abandoned. Thus the discussion was chosen as a form of reflection.

Discussion

When analyzing the energy consumption of storage infrastructure in the use phase, one should ask the fundamental question of whether to take into account the building itself or should its equipment be included as well? Because the analysis concerns the functioning of the warehouse, it is understandable that the analysis should cover not only the building but also all the devices that are in it. This approach is additionally confirmed by the energy performance certificate, in which the description is subjected to the energy needed for the operation of the warehouse, i.e., energy consumed for the implementation of its utility functions.

The underlying carriers of direct energy consumed during the use of the warehouse are: electricity for lighting, electricity for the drive of storage equipment, electricity for driving internal transport, electricity for powering communication devices, electricity for powering computer equipment, electricity for air conditioning, electricity for power chillers, hot water, heat for heating and ventilation. As can be inferred, the analysis of energy consumption by the warehouse in the phase of use performed on an ongoing basis boils down to three elementary issues:

- What is the structure of energy consumption in a warehouse at a given time (year, quarter, month)?
- What is the amount of energy consumption for a given medium at a given time?
- What is the share of individual energy carriers in total energy consumption in a given time?

Because the construction of warehouse facilities is carried out based on individual projects, the equipment is selected for the specific needs of the owner. This makes it difficult to give one formula for all energy carriers. Therefore, one can propose the following formula:

$$E_{cm} = E_{m1} + E_{m2} + E_{mi} \tag{1}$$

where:

E_{cm} - total energy consumed in the warehouse,

$\sum E_m$ - the sum of energy consumed by the i-th energy carrier at a given time during the warehouse operation phase,

i – the number of energy carriers.

In addition to this fundamental analysis - similarly to the analysis of cumulated energy consumption at the construction stage - the approach to analyzing cumulated energy consumption can also be used for the operation phase. Two streams of cumulative energy consumption can be considered. The first one expresses the cumulated energy consumption used to acquire and supply direct energy carriers consumed in the warehouse during use in the form of fuels and energy. The second stream relates to the cumulative energy consumption used for the production and delivery of direct energy carriers in the form of warehouse equipment.

If the analysis of energy consumption at the stage of use of the warehouse is to be complete, the performed operations should be taken into account as well, and more strictly - the service, renovation, and modernization procedures. Then, the energy needed to perform operations, manufacture of consumables and transport, or storage of these materials will be taken into consideration. The formulas for determining the level of cumulative energy consumption in these three cases will be analogous to the formulas presented for the cumulative energy consumption at the construction stage. To sum up, at the stage of warehouse operation, in the analysis of energy consumption, it is possible to consider the consumption of direct energy; energy consumed during the performance of operations and cumulated energy consumption.

From the economic point of view, the analysis of the energy efficiency of energy carriers is also essential at the stage of storage use. Then, the ratio of energy inputs to the effects of the warehouse should be taken into account, and in more detailed terms - the effects of devices involved in the implementation of the storage process. For example, one can analyze the ratio of the energy consumed within one hour of the forklift truck or the work of one refrigerating unit, energy consumption per 1 m^2

IOP Conf. Series: Materials Science and Engineering **568** (2019) 012045 doi:10.1088/1757-899X/568/1/012045

of warehouse space. Table 1 presents the most important indicators of the energy efficiency analysis of energy carriers used in the use of the warehouse.

Table 1. Fundamental indicators of the energy efficiency analysis of energy carriers used in the use of the warehouse

Indicator	Formula	Description
The basic index of the warehouse energy efficiency	1) $WEE_{Mbasic} = \frac{E_{cm}}{P}$ 2) $WEE_{Mbasic} = \frac{E_{cm}}{P_z}$ 3) $WEE_{Mbasic} = \frac{E_{cm}}{K}$ 4) $WEE_{Mbasic} = \frac{E_{cm}}{K_z}$	1) 1) expresses the amount of total energy consumed [MWh] about the total warehouse area [m^2] for the unit period 2) 2) expresses the amount of total energy consumed [MWh] about the occupied warehouse area [m^2] for the unit period 3) 3) expresses the amount of the total energy consumed [MWh] about the total cubic capacity of the storage [m^3] for the unit period 4) 4) expresses the amount of the total energy consumed [MWh] about the occupied cubic capacity of the storage [m^3] for the unit period
Energy efficiency indicator for lighting	1) $WEE_{light} = \frac{E_{light}}{P}$ 2) $WEE_{light} = \frac{E_{light}}{P_z}$ 3) $WEE_{light} = \frac{E_{light}}{K}$ 4) $WEE_{light} = \frac{E_{light}}{K_z}$ 5) $WEE_{light} = \frac{M_{light}}{P}$	1) 1) expresses the amount of energy consumed [MWh] for lighting about the total warehouse area [m^2] for the unit period 2) 2) expresses the amount of the total energy consumed [MWh] for lighting in relation to the occupied warehouse area [m^2] for the unit period 3) 3) expresses the amount of the total energy consumed [MWh] in lighting to the total cubic capacity of the warehouse [m^3] for the unit period 4) 4) expresses the amount of the total energy consumed [MWh] for lighting in relation to the occupied cubic capacity of the warehouse [m^3] for the unit period 5) 5) expresses the lighting power [W] in relation to the warehouse area [m^3] for the unit period
Indicator of heating energy efficiency	1) $WEE_{heat} = \frac{E_{heat}}{P}$ 2) $WEE_{heat} = \frac{E_{heat}}{P_z}$ 3) $WEE_{heat} = \frac{heat}{K}$ 4) $WEE_{heat} = \frac{heat}{K_z}$	1) 1) expresses the amount of energy consumed [MWh] for heating in relation to the total warehouse area [m^2] for the unit period 2) 2) expresses the amount of the total energy consumed [MWh] for heating in relation to the occupied warehouse area [m^2] for the unit period 3) 3) expresses the amount of the total energy consumed [MWh] in heating relative to the total cubic capacity of the storage [m^3] for the unit period 4) 4) expresses the amount of the total energy consumed [MWh] for heating in relation to the occupied cubic capacity of the warehouse [m^3] for the unit period
The efficiency of devices in heating and cooling mode	1) $EER = \frac{P_{cool}}{P_e}$ 2) $COP = \frac{heat}{P_e}$	1) 1) EER (Energy Efficiency Ratio) - the ratio of the supplied cooling power to the electric power consumed. If EER = 3, i.e., using 1 kWh of electricity, 3 kWh of cold is supplied to the room (or 3 kWh of heat removed). 2) 2) COP (Coefficient Of Performance) - a factor that determines the efficiency in heating mode. It is the ratio of the heating power supplied to the electric power consumed. If COP = 3, it means that by using 1 kWh of electricity, it is delivered to the room of 3 kWh of heat.

Source: Dembińska I 2018 *Infrastruktura logistyczna gospodarki w ujęciu środowiskowych uwarunkowań zrównoważonego rozwoju* (Szczecin: Wydawnictwo Naukowe Uniwersytetu Szczecińskiego p. 290-291)

It should be noted that considering the warehouse building itself in the analysis of energy demand is an incomplete approach. Warehouse buildings, whether for component or distribution purposes, are

IOP Conf. Series: Materials Science and Engineering **568** (2019) 012045 doi:10.1088/1757-899X/568/1/012045

always part of the facility, where apart from small buildings, there are also storage yards and maneuvering areas. Their functioning also involves energy consumption. It is mainly used for lighting. Therefore, the analysis of energy consumption, in addition to the analysis related to a warehouse building, should apply to the entire logistics facility, including energy consumed outside the warehouse building, throughout the entire logistics facility, but not other buildings with different functions, but with storage yards. Maneuvering and so-called green areas are belonging to the logistics facility. Speaking of other buildings than the warehouse, an office building should be considered, which can be an integrated part with a storage building or be a separate building. It should also be borne in mind that in the case of centers adapted for comprehensive service, in addition to warehouses, there are also buildings for other purposes, such as car workshops, car washes, production plants, etc. In the area of the transport center, due to defined functions, warehouse buildings it is not there. Energy consumption occurs as a result of the operation of mainly transshipment yards, maneuvering areas, although workshops and car washes may also be found here. Taking into account these remarks, in the analysis of energy consumption by a logistic object in the form of a logistic center, it is possible to propose using the formula:

$$E_{ccl} = \sum E_{em} + \sum E_{ea-b} + \sum E_{et} + \sum E_{ep} + \sum E_{ed} + \sum E_{ez} \tag{2}$$

where:
$\sum E_{ccl}$ – total energy consumed in the area of the center / logistic park,
$\sum E_{em}$ - total energy consumed in warehouse buildings,
$\sum E_{ea-b}$ – total energy consumed in administrative and office buildings,
$\sum E_{et}$ – total energy consumed in technical service buildings,
$\sum E_{ep}$ - total energy consumed on maneuvering yards, transhipment sites, container terminals, etc.,
$\sum E_{ed}$ - total energy consumed in connection with the functioning of additional components of the logistics facility, e.g. ra,ilway infrastructure, parking lots, hotel and catering facilities,
$\sum E_{ez}$ – total energy used for the so-called green areas.

In the energy-consumption analysis of a building, one can meet the concept of energy quality of a building. It is defined by a set of building properties, on which depends on the size of the annual energy demand related to its use. The energy quality of the building is determined by the thermal protection features of the building and the features of heating, ventilation and hot water supply systems. The lighting system is not considered here. Therefore, the energy consumption of a warehouse building is influenced by:

- the architecture of the building: the location of the building relative to the sides of the world, the layout of the rooms, the geometry of the building,
- warehouse location,
- constructional solutions of the building,
- type of material from which it is made,
- warehouse area,
- building insulation,
- degree of mechanization, warehouse automation,
- number of devices and means of internal transport powered by energy,
- type and number of lighting points,
- type and efficiency of the heating system,
- type and efficiency of the cooling system,
- type and efficiency of ventilation,
- the occurrence of refrigeration units,
- use of alternative energy sources,
- requirements for stored stocks related to energy consumption,
- organization of the warehouse work,
- warehouse management system, including the inclusion of energy-efficient policy.

The energy consumption of a logistics center is determined by the following factors: location, number and purpose of buildings, architecture, construction solutions, building insulation, type of material from which buildings are made, number and purpose of squares, surfaces of buildings and squares, energy consumption of components of the logistics center, organization of the logistics center work, level of mechanization, automation of the logistics center.

The impact of these factors is noticeable. It seems that you do not need these relationships here. One can only refer to the location factor because this issue was not discussed earlier. Speaking of location, in reality, it is about climatic zones, and more specifically - by the number of degree days. It should be indicated in which climate zone and the meteorological station there is a building. This is important both at the design stage and the current operation of the building. It can also be briefly mentioned that the degree of shading affects the building's energy consumption. Warehouses, centers or logistic parks are usually located in open space. This has a positive effect on shaping the energy-consumption factor.

Conclusion

In summary, the cumulative energy analysis of logistic facilities can be used for the following purposes:

- to compare energy expenditures used for constructing or decommissioning a logistics facility using various technologies,
- to study the variability and dynamics of energy consumption during the operation of logistic facilities under the influence of various factors,
- to study changes in the national energy demand in the construction of a specific unit of a logistics facility (1 km of road, 1 m^2 of warehouse space),
- to assess the impact of a specific substitution of energy carriers on primary energy consumption,
- to study the share (structure of consumption) of primary energy carriers in individual phases of the life cycle of a logistics facility,
- To study the share of renewable energy carriers in individual phases of the life cycle of a logistics facility.
- Besides, studies on the energy consumption of logistic facilities may concern:
- a single building object, i.e., a road, warehouse, etc.,
- groups of objects in a given area, e.g., a center or a logistics park, a network of highways,
- construction facilities throughout the country.

They may refer to the whole life cycle of a logistics facility, as well as may refer to a selected stage. They serve both analytical purposes and may form the basis for comparative analyses.

References

[1] Dembińska I 2018 *Infrastruktura logistyczna gospodarki w ujęciu środowiskowych uwarunkowań zrównoważonego rozwoju* (Szczecin: Wydawnictwo Naukowe Uniwersytetu Szczecińskiego).
[2] *Energochłonność skumulowana* 1983 ed. Z. Bibrowski (Warszawa PWN) p 38.
[3] Ustawa z dnia 15 kwietnia 2011 roku o efektywności energetycznej (Dziennik Ustaw nr 94 poz. 551).
[4] PN-EN ISO 13790 Energetyczne właściwości użytkowe budynków.
[5] PN-ISO 9836:1997 Właściwości użytkowe w budownictwie. Określanie i obliczanie wskaźników powierzchniowych i kubaturowych.
[6] Rozporządzenie Ministra Infrastruktury i Rozwoju z dnia 27 lutego 2015 r. w sprawie w sprawie metodologii wyznaczania charakterystyki energetycznej budynku lub części budynku oraz świadectw charakterystyki energetycznej (Dziennik Ustaw poz. 376).
[7] Amasyali K and El-Gohary N M 2016, Energy-related values and satisfaction levels of residential and office building occupants. *Build. Environ.* 95, 251–263.
[8] Dembińska I 2016 Energy-intensity's analysis of road infrastructure - methodology key assumptions (Annals of the Oradea University. Fascicle of Management and Technological Engineering 25, 2 Oradea University) p. 111-114
[9] Dembińska I 2016 Pomiar energochłonności infrastruktury drogowej: koncepcja i założenia metodyczne (Problemy Transportu i Logistyki 1, Szczecin: Wydawnictwo Naukowe Uniwersytetu Szczecińskiego) p. 107-116.

IOP Conf. Series: Materials Science and Engineering **568** (2019) 012046 doi:10.1088/1757-899X/568/1/012046

Numerical analysis of the SOC factor variations' influence on the autonomy of an electric vehicle

L Scurtu, B O Varga, F Mariasiu, T Buidin, Adela Borzan, D Moldovanu*

Technical University of Cluj-Napoca, Automotive Engineering and Transports Department, Bd. Muncii 103-105, Cluj-Napoca, Romania

dan.moldovanu@auto.utcluj.ro

Abstract. Electric vehicles are considered to be the immediate solution to drastically reduce the pollutant emissions from the road transport sector. However, one of the barriers of the massive penetration of the automotive market by electric vehicles is related to consumer/driver anxiety linked to the autonomy of electric vehicles. Autonomy depends directly on the battery technology that equips the electric vehicle, and from this point of view, the technology based on Li-Ion electrochemistry is the most accepted. This article presents a study by numerical simulation methods on the functional performance (autonomy) of an electric vehicle equipped with three different types of batteries (LiCoO2, LiFePO4, and a classic Acid-lead), as they have a low battery energetic charge. The obtained results showed that there are differences between the various technologies considered, the most efficient one being the LiCoO2 battery type.

1. Introduction

It is widely accepted that the development and use of electric vehicles (EV) is the long-term solution for reducing today's transport pollution. EVs are already produced as passenger cars and buses for passenger transport, and shortly an electric propulsion group solution will be developed for freight transport vehicles.

However, the widespread integration of EV into transport systems involves overcoming current market barriers generally defined as [1-3]: Range satisfaction (autonomy); Sustainable technology; Vehicle's purchase price; Reduced greenhouse gas emissions; Costs of exploitation and maintenance.

It is noticed that one of the critical barriers in accepting EV by consumers is the one related to the autonomy of an EV. The autonomy of an EV depends on three main factors of influence: Vehicle design, Driver, and Exploitation environment [4]. In turn, these factors may be direct or indirect and are characterized by parameters such as: the vehicle's geometric dimensions, passenger capacity, body type, rolling train and tires, specific front area, tires, type of heating, ventilation and air conditioning systems, the behavior of the driver, climate conditions, traffic, route, road infrastructure, state of charge (SOC), state of health (SOH), battery type and energy capacity etc.

The most crucial role in the autonomy (range) of an EV, the largest is the amount of energy stored in the energy source (battery) estimated by the SOC parameter and evaluated by different researchers as being 50-60% from total [4].

The SOC parameter represents/defines the amount of instantaneous energy in the battery, available to be supplied to the electric propulsion group, being a parameter characteristic for each electrochemistry used in the construction of the batteries' cells.

The battery's SOC is calculated using Peukert's Law as [5]:

$$SOC_t = SOC_0 - 100 \cdot \int_0^t \frac{1}{C_{nom}} \cdot I \left(\frac{I}{I_{nom}}\right)^{k-1} dt \qquad (1)$$

where: SOC_0 is the battery's initial state of charge, C_{nom} is the nominal capacity of the battery measured for the nominal electric current I_{nom}, I the electric current, k is Peukert's constant and t is time. Because Peukert's constant has a specific value for each type of battery cell (electrochemistry), the SOC value/variation will be different for different electrochemistry of cells. At present there are several types of batteries used as energy sources for EV (Li-ion, NiMH, Zebra, Li-Air etc.), each of which has different features in terms of energy performance, but Li-Ion technology has proven to be the fully-accepted solution for EV batteries, due to the immediate advantages it offers: high life (8 to 15 years), life cycles (2000 ... 5000) and high operating temperature range (-20 ... + 60oC) [6-8].

Lithium based battery technology also offers many functional advantages as the high potential of an electrode, high operating voltage, high discharge rate, the possibility of fast charging, low self-discharge rate, and no memory effect. Table 1 presents several features of the different technologies used in the construction of Li-ion batteries for EV, and Figure 1 is presented the performance of Li-ion batteries compared to other battery technologies [9].

Table 1. Characteristics of different Li-ion battery technologies (source [6-8])

Type	Energetic mass density	Energetic volumetric density	Charge/ discharge cycles	Price	Electric power	The maximal temperature of functioning	Cell's electric potential	The optimal range of operating temperature
	(Wh/kg)	(Wh/l)	-	(USD /Wh)	(C)	(°C)	(V)	(°C)
$LiCoO_2$	170-185	450-900	500	0.31-0.46	1C	170	3.6	-20...+60
$LiFePO_4$	90 -125	130-300	2000	0.30-0.60	5-10C	270	3.2	-20...+60
NMC	150-190	270-330	1500	0.50-0.90	2-5C	215	3.7	-20...+60
Li-Ti	65 -100	118-200	12000	1.00-1.70	10-20C	-	2.5	-50...+75
Li-Mn	90 -110	280	>1000	0.45-0.55	3-5C	255	3.8	-20...+50

Based on the above considerations, the paper aims to identify the effect of different SOC variation curves (specific to each type of battery considered) on the functional performance of an EV (autonomy). Both the functional parameters and the constructive parameters (Wh/kg) for each battery considered in the study were taken into account. SOC variation curves were experimentally determined within the experimental activities of the URBIVEL project.

IOP Conf. Series: Materials Science and Engineering **568** (2019) 012046 doi:10.1088/1757-899X/568/1/012046

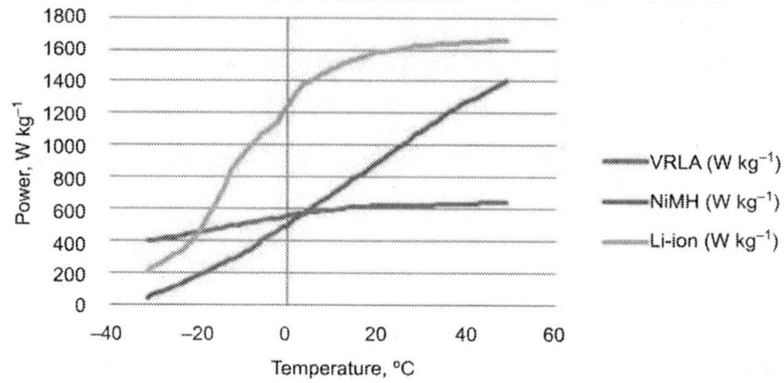

Figure 1. Energy performance of different battery types (technologies) depending on the operating temperature (VRLA - acid-lead accumulator) [9]

2. Methods

Numerical simulation methods (IPG CarMaker software) have been used, and the typical EV model considered was the Tesla Model S.

The Tesla Model S was chosen because it is a validated model in IPG CarMaker software, and the WLTC (Worldwide harmonized Light vehicles Test Cycle) was implemented, and the driver has to follow the velocity variation. The driver behavior was selected as a "normal" driver, meaning that the driver can go with a maximum velocity of 150 km/h, maximum longitudinal acceleration 3 m/s^2, maximum longitudinal deceleration -4 m/s^2, maximum lateral acceleration 4 m/s^2. The road was a straight line with no elevation, as required for the WLTC test cycle. The maneuvers were implemented as velocity requirements for the driver to follow.

The battery types that were considered in the present study were $LiCoO_2$, $LiFePO_4$, and a classic Acid-lead battery (APb). The significant difference between the chosen batteries is the SOC influence factor to idle voltage variation specific for each battery's chemistry, as shown in figure 2. For each battery type simulations were done with the initial charge of the battery at 10%, 20%, 40%, 60, 80% and 100% respectively. The most important output factors that were followed were the vehicle's autonomy (range), the vehicle velocity, and the variation of SOC of each considered battery.

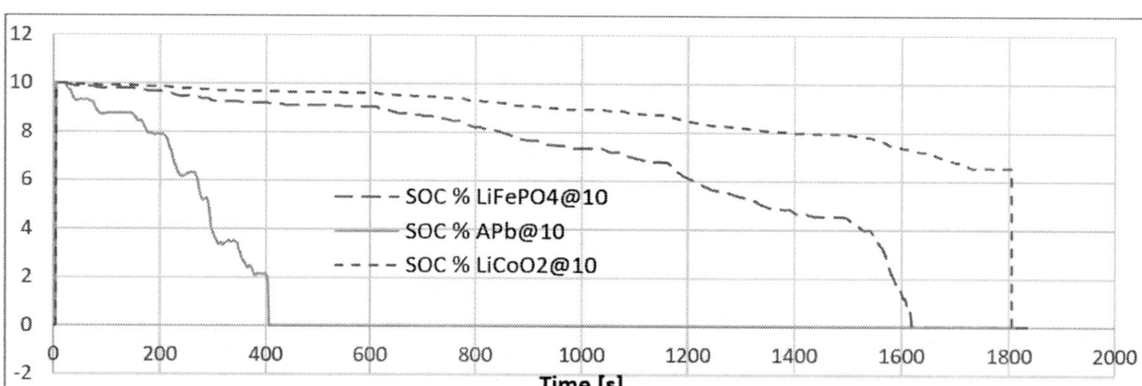

Figure 2. SOC for batteries starting with 10% initial charge

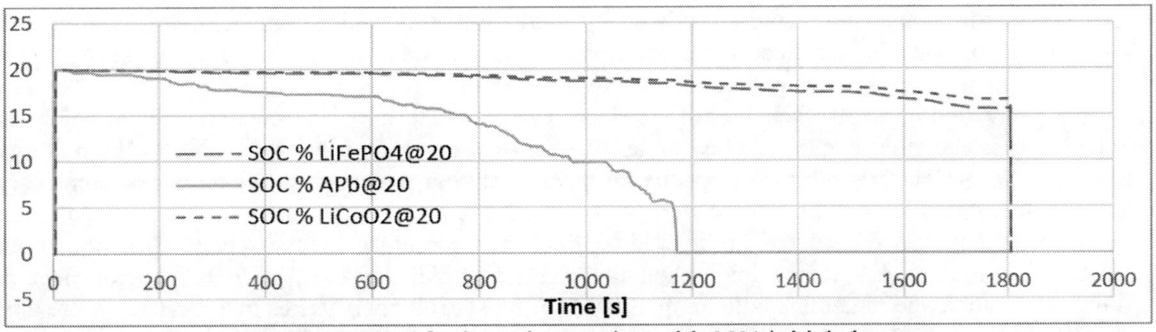

Figure 3. SOC for batteries starting with 20% initial charge

3. Results

The results of the simulation were extracted from IPG Control Data Window and exported to MS Excel to be further analyzed and interpreted. Figure 2 presents the variation of SOC parameter for the $LiCoO_2$, $LiFePO_4$, and Acid-lead battery, in the case when the vehicle starts on the WLTC test cycle with 10% of battery charge. Since the differences were significant, figure 3 underlines the SOC for all the batteries at an initial charge of 20%. Since the consumption of the batteries was small on the WLTC, the results for the 60%, 80%, and 100% battery charge are very similar and therefore not shown.

The typical evolution of the car distance when using the $LiCoO_2$ battery and the small difference when using the $LiFePO_4$ battery, with a considerable reduction in autonomy when using Acid-lead batteries which are not an energetic efficient solution for EVs, especially for low charging conditions (3127 m compared with 24078 m for $LiCoO_2$ and 19788 m for $LiFePO_4$).

4. Discussions

The analysis of the influence of the different types of batteries that can equip an electric vehicle is necessary because one of the major barriers that impede the massive penetration of the car market by the electric vehicle is the consumer's (the driver's) anxiety of being left with no energy. From this point of view, when designing and building an electric vehicle, it is necessary to take into account the battery's ability to provide the required energy for as long as possible.

From the numerical analysis, we can see that there are major differences regarding the autonomy that can be achieved both between the types of batteries used and the initial loading condition (considered as the initial moment of simulations). If the $LiCoO_2$ battery has reached the distance of 24078 m for an initial charge of 10%, the $LiFePO_4$ battery has been reached 19788 m and for the Acid-lead technology the result was only 3127 m. In the case of Acid-lead batteries, it can be argued that obtaining such a reduced distance was also due to the heavy weight of the battery pack, the weight that was taken into account in the numerical analysis performed. Therefore, it can be said that the best results (for each case considered) are the battery based on $LiCoO_2$ technology, the battery considered to be the second generation of Li-Ion batteries. The $LiFePO_4$ battery has lower performance than the $LiCoO_2$ battery but offers the advantage of having a much lower (approx. 15%) production price, which justifies the presence of this type of battery in the equipment of many electric vehicles (both passenger cars and buses). The $LiCoO_2$ battery has higher energy efficiency in that the maximum speed achieved is 134 km / h compared to 120 km / h for the $LiFePO_4$ battery.

All of these data lead to the primary conclusion that the $LiCoO_2$ battery offers the best performance in terms of efficient operation of an electric vehicle, especially when the battery has a low energetic potential.

5. Conclusions

The paper presents a comparative study of the energy potential of different technologies used in the construction of batteries for electric vehicles. Numerical analysis methods (IPG CarMaker software) were applied on a Tesla S electric vehicle, where three types of batteries were used.

The study highlighted the behavior of the electric vehicle in terms of its functional performance (the autonomy/distance that can be reached under different battery charging conditions), as the battery is relatively low at 10, 20 and 40% respectively from total charge capacity. For higher battery charge values ranging from 60 to 100%, there were no major differences about the achieved distance/autonomy between the batteries considered in the study. This difference arose for the special conditions under which batteries are discharged and when the electric vehicle systems need to provide accurate information about the remaining autonomy of EV. The most efficient batteries in this respect were those using $LiCoO_2$ technology, the worst results being the Acid-lead batteries.

It should be noted that the future hardware and software solutions developed for the SOC estimation must also take into account the electrochemistry of the cells from which the battery that equips the electric vehicles is formed.

Acknowledgments
This research was funded by the Program "POC-A1-A1.2.3-G-2-15 Partnerships for Knowledge Transfer, no. financing contract: 11 / 01.09.2016, project title: Intelligent Urban Intelligent Vehicle Technologies URBIVEL, ID: P_40_333, MySMIS: 105565, Subsidiary contract nr.6/16.01.2018".

References

[1] Park, E.; Lim, J.; Cho, Y. 2018 Understanding the Emergence and Social Acceptance of Electric Vehicles as Next-Generation Models for the Automobile Industry. Sustainability **10**, 662-675.
[2] Franke, T.; Günther, M.; Trantow, M.; Krems, J.F. 2017 Does this range suit me? Range satisfaction of battery electric vehicle users. Appl Ergon, **65**, 191-199.
[3] Labeye, E.; Hugot, M.; Brusque, C.; Regan, 2016 M.A. The electric vehicle: a new driving experience involving specific skills and rules. Transp. Res. Part F Traffic Psychol. Behav., **37**, 27-40.
[4] Varga B.O, Sagoian A., Mariasiu F. 2019 Prediction of Electric Vehicle Range: A Comprehensive Review of Current Issues and Challenges, Energies, **12**, 946-965.
[5] Vatanparvar K., Al Faruque M.A. 2015 Battery Lifetime-Aware Automotive Climate Control for Electric Vehicles, DAC '15 Conference, June 07 - 11 San Francisco, CA, USA.
[6] Jossen A. 2007 Overview on Current Status of Lithium-ion Batteries, 2nd International Renewable Energy Storage Conference (IRES II), Bonn, Germany, 19th-21st November.
[7] Buckley P. 2014 Samsung SDI Batteries to Drive Future BMW EV's, EE Times Europe
[8] Garnier C. Renault ZE 2013: Path Toward EV Battery Production, CAPIRE Workshop, Brussels
[9] Weighall M. J. 2003 Test requirements for 42 V battery systems, J. Power Sources, , **116**, 151-159.

IOP Conf. Series: Materials Science and Engineering **568** (2019) 012047 doi:10.1088/1757-899X/568/1/012047

Numerical study on the effects of using different concentrations of FAME blends on the spray structure of the injection process

M S Motogna[1], D Moldovanu[1,*] and F Mariasiu[1]

[1] Technical University of Cluj-Napoca, Automotive Engineering and Transport Dept., Cluj-Napoca, Romania

dan.moldovanu@auto.utcluj.ro

Abstract. The present paper takes a closer look at the injection process characteristics for different conditions when there are used different FAME blends. Using a numerical approach, this work highlights the main implications and differences in terms of geometry, structure, and physical parameters of the injected fuel spray, when alternating the used fuels. In this attempt, AVL FIRE was used as simulation software solution giving the chance to alternate the used fuels and furthermore to create a direct link between the fuel properties, internal flowing, and jet spray characteristics. Because of the environmental and economic reasons, there are a growing tendency of using biofuel blends in increasing concentrations, as ample consumption of commercial fuels. As the physical properties of the Fatty Acid Methyl Ester fuels are very different compared with conventional diesel fuel, we expect that the fuel blends between them will show different behaviors in terms of internal nozzle flowing and thus it will generate different injection characteristics, being closely related with the biofuel concentration. The main objective of this paper is to set a dependency between the used fuels, nozzle geometry, and injection parameters that should be adapted, to meet the emissions standards and to obtain a properly burning.

1. Introduction

In our days, the usage of biofuels is considered to be a potential solution for green gas produced by worldwide transportation systems. It also seems to offer an alternative for the dependency on fossil fuels. Because of this, biofuels gained fast popularity, and in almost every region, the law requires the commercialization of fuel blends instead of pure diesel fuel. Having a similar density but higher viscosity and surface tension, compared with the conventional diesel, the behavior of biofuels into the fuel system and specially into injection processes are closely studied by the researchers.

In one of the studies, Seoksu Moon[1] aims to investigate the spray dynamic spray structure, using biodiesel fuel. For a better understanding, he decided to highlight the results by making a comparison between the dynamic structure of the biodiesel and conventional fuel sprays obtained from a single and multi-hole diesel injectors and using a synchrotron X-ray velocimetry technique. The results showed that the high viscosity and density of biodiesel decreased the injection velocity compared to conventional fuels, but this behavior was less significant for the multi-hole injector. For the case of single-hole injector, the usage of biodiesel led to a slowed flow breakup and increased the intact core area, because of the lower velocity decay rate. This behavior was not recognised at the multi-hole injector, where the

IOP Conf. Series: Materials Science and Engineering **568** (2019) 012047 doi:10.1088/1757-899X/568/1/012047

flow breakup and the velocity decay rate along the spray core appeared unchanged, regardless of the used fuel.

Focused on internal flow characteristics of the fuel, the paper of Moon et al. compared in nozzle needle movement of biodiesel and diesel injected fuels from a single hole diesel injector.[2] Their observation includes that when the biofuels are used, because of their higher viscosity, the needle movement and the outlet nozzle flow velocity is slowed down, modifying the process timing and structure of the spray. Kuti et al.[3] put their focus on the droplet size distribution, entertained air mass and macroscopic spray structure of the jet spray, obtained by alternating the conventional diesel fuel, with biodiesel. They observe that, for the cases when were used biodiesel fuels, the resulted jet spray had a narrower spread angle and was characterised by larger droplet size, longer penetration distance and was capable of entertaining less air. Those results are assumed to be the direct consequences of higher viscosity and surface tension of biodiesel fuel.

In their paper, F.J. Salvador et al.[4] defined a numerical approach for spray characteristics determination. To highlight the effect of biofuel usage on the internal flow and further on the jet spray, they decided to simulate the injection process using the same parameters but switching between conventional diesel fuel with soybean oil. The parameters variation allowed to investigate two different situations in terms of inner flow: two-phase flow (cavitation presence) and one phase flow (no cavitation). The comparison between injection processes, in each situation, was made with the propose of following main parameters, as mass flow, momentum flux, sufficient velocity at the outlet, and cavitation/cavitation intensity. The pressure drop was defined as the difference between the injection pressure and back pressure. When the injected mass flow in relation with pressure drop was determined for each fuel, they conclude that due to its higher density, biofuel injects more fuel at the same time. Big differences between fuels were also found at reaching the critical cavitation conditions (CCC), and it was observed that the cavitation phenomena developed more quickly and were more intense at the diesel fuel. It is believed that, because of the increased viscosity and surface tension, the biodiesel tends to inhibit cavitation phenomena. It's well known that the air-fuel mixing process is better when the sufficient injection velocity is higher and also when there is a wide cone spread angle. However, when the injection process is obtained using soybean oil, both of those factors are smaller, generating a more compact spray jet capable of longer penetration distance and the air-fuel mixing is worst.

2. Method and simulation

As was already presented, changing the conventional diesel fuel with biofuels have a significant impact on the injection process. To obtain good burning performance, the conclusions of the studies shown that a couple of adjustments of the conventional injection system needs to be implemented for the cases of usage biofuels blends. Those adjustments are close related to the study cases and tend to be different at every change of used fuel, the thing that generates a real effort and represents a big challenge for developers.

However, modern computing techniques gives a potential helping hand in many study cases [5]. Using the numerical methods, it's possible to predict very close the behavior of the injection process and further the burning characteristics. One of the best software solutions used on a large scale in the automotive industry is represented by CFD Software AVL FIRE. AVL ESE DIESEL module allows computing the flow and jet spray using different types of fuel blends in different conditions. In our case of study, the conventional Diesel and B7 (Diesel and Biodiesel-7% blend) was taken into consideration, and several numerical simulations were done alternating those fuels, to monitories few parameters, as output velocity, dissipation rate, turbulence velocity, etc. Also, in this attempt, the effects of temperature variations were taken into consideration, at following values: 0, 40, and 80^{0}C.

The precise geometry of the injector nozzle was needed and used as a boundary condition. The injector that was taken as a reference is BL 76 S2 R, and the specifications were provided by the manufacturer. Both Diesel and B7 are widely used commercial fuels, and their properties were toked from official data of marketed fuels. The numerical simulation for injection event time is set at 0.005 sec. The cylinder inner diameter was set at 70 mm, and the mesh for this volume was generated automatically.

For time and calculation capabilities reasons and because the injector has one single hole, the computational domain was split and considered one-quarter of the hole domain.

Table 1 The boundary conditions

Injector	Type	Sac
	Hole number	1
	Hole diameter	0.17 mm
	Hole length	1.1 mm
	k- factor	0
Diesel	Density (300 K)	0.78 kg/l
	Liquid viscosity (300 K)	0.002 N-s/m^2
	Surface tension (300 K)	0.02 N/m
	Heat of vaporization (300 K)	385 kJ/kg
B7	Density (300 K)	0.86 kg/l
	Liquid viscosity (300 K)	0.0065
	Surface tension (300 K)	0.025 N/m
	Heat of vaporization(300 K)	360 kJ/kg
Environmental parameters	Environmental gas	air
	Environmental pressure	atmospheric

3. Results and discussions

One of the very first parameters that can be monitored at the injected spray and it's capableof characterizing the inner fuel flow is the particle distribution and particle size. To have more complex behavior, it was chosen to implement also a temperature variation, for both fuels. In the following images will be presented the simulation results in terms of those aspects:

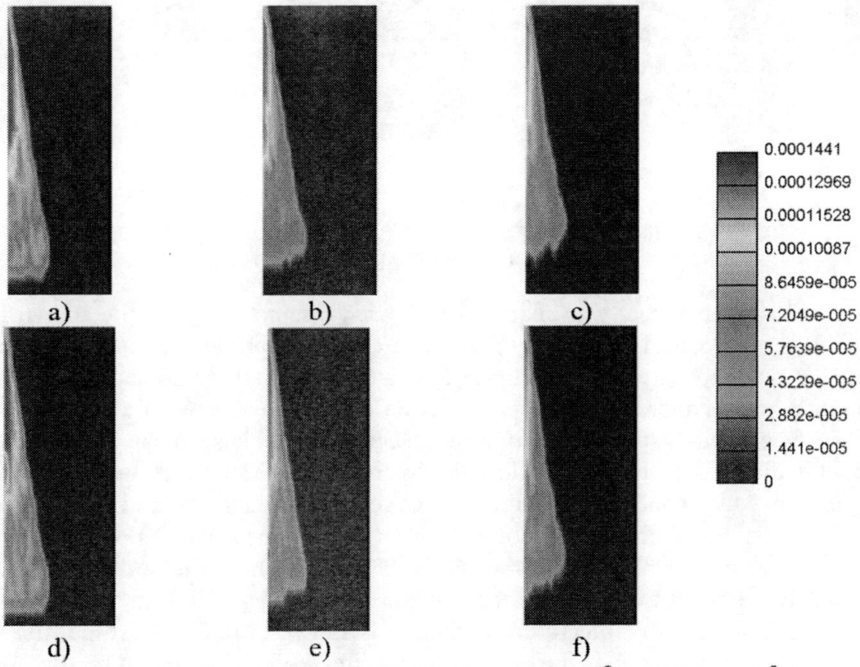

Figure 1. Sauter Mean Diameter (SMD) for a) B7 at 0^{0}C, b) B7 at 40^{0}C, c) B7 at 80^{0}C, d) D at 0^{0}C, e) D at 40^{0}C, f) D at 80^{0}C

It's not hard to observe that, once with the increasing of the fuel temperature, the particle medium diameter is drastically decreasing. Once with the decreasing dimensions of the particles, their kinetic

IOP Conf. Series: Materials Science and Engineering **568** (2019) 012047 doi:10.1088/1757-899X/568/1/012047

energy is also declining and because of this, they are not capable of penetrating such long distances and in the end leading to a smaller spray tip penetration. However, this observation is valid for both fuels, but not in the same manner. Once with the increasing of the temperature, the density, and also the viscosity of the fuel is decreasing. This effect enhances the particles breaking and thus generating a more fragmented spray. When it's looking at fuels, it can be observed that at small temperatures, the fuel which has higher surface tension and viscosity (B7) has also at the core of the jet, bigger particles.

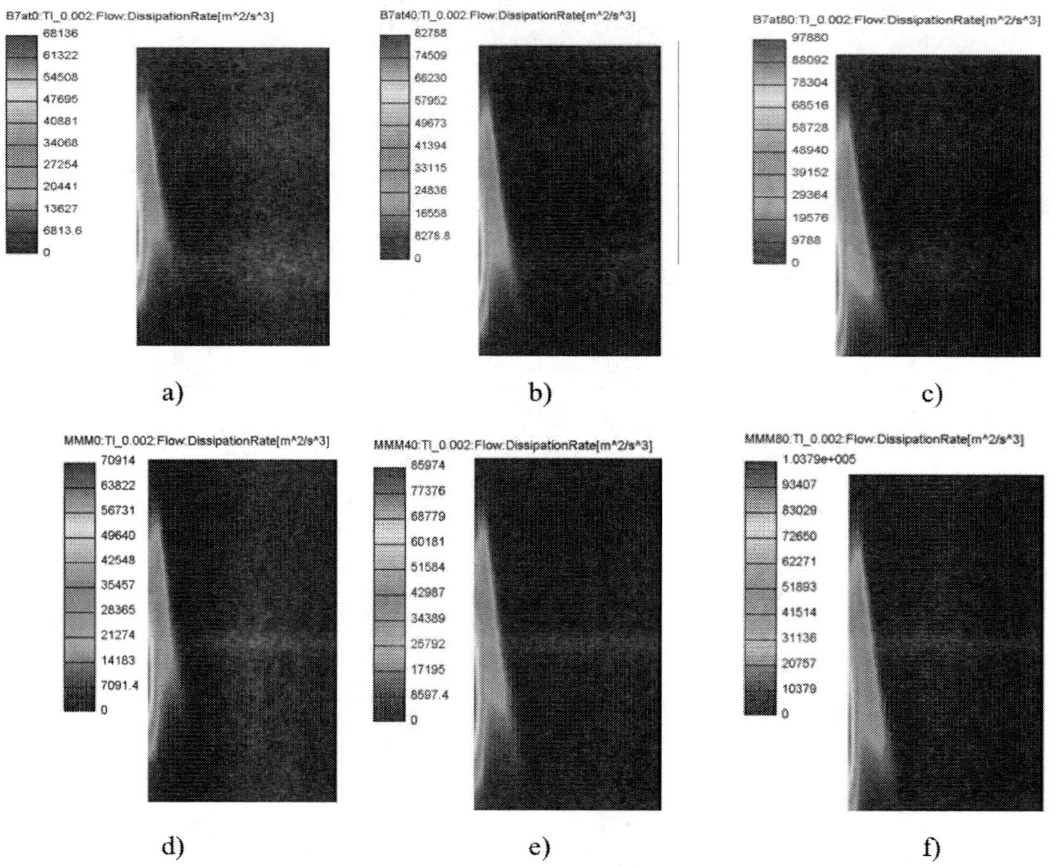

Figure 2. Dissipation rate for a) B7 at 0^{0}C, b) B7 at 40^{0}C, c) B7 at 80^{0}C, d) D at 0^{0}C, e) D at 40^{0}C, f) D at 80^{0}C

The second parameter that was analyzed is flowed dissipation rate. The shown results are easy to predict, taking into consideration the previous results of particle mean diameter. Once with the temperature is rising, because of the properties changes, the particle breakdown is amplified, and thus, the dissipation rate is increasing. Of course, this effect is more visible at fuels which have lower viscosity, surface tensio, and density properties. Because in our case, b7 and diesel fuels have very close properties, the differences in terms of flow dissipation rate are small. However, even those differences are enough to give us a conclusion, as follows: once with the increasing of the density and viscosity of the fuel (equivalent to an increase of biofuel concentration), the flow dissipation rate is decreasing, and furthe, generate a spray more compact which is capable to entrain less air. To be more obvious, the differences between those two fuels are represented also in the graph from figure 3. The output velocity of the fuels, in relation with temperature variation, is shown in figure 4. At the first sight, what can be conclude is that the velocity is increasing approximately linear and directly proportional with a temperature rising. Also can be mentioned that at every temperature value, the output velocity of the B7 fuel is lower than the diesel fuel. The explanation for this behavior is leading to the internal flow of each of the fuels. Because of the lower viscosity and surface tension, the critical cavitation conditions are reached more easily and this phenomenon is developing in more pronounced forms, for diesel fuel.

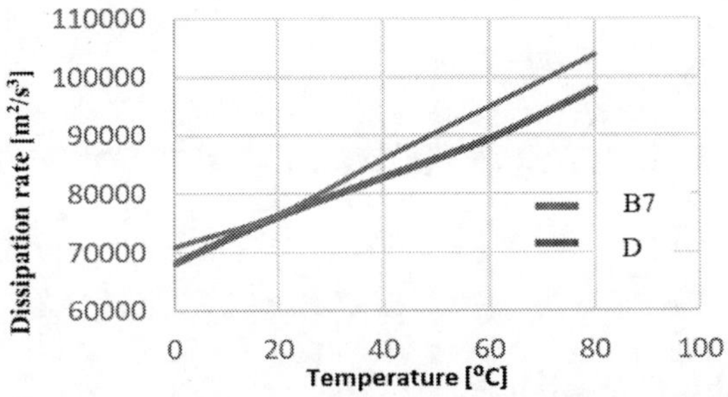

Figure 3. Dissipation rate of B7 and Diesel fuel, at different temperature values

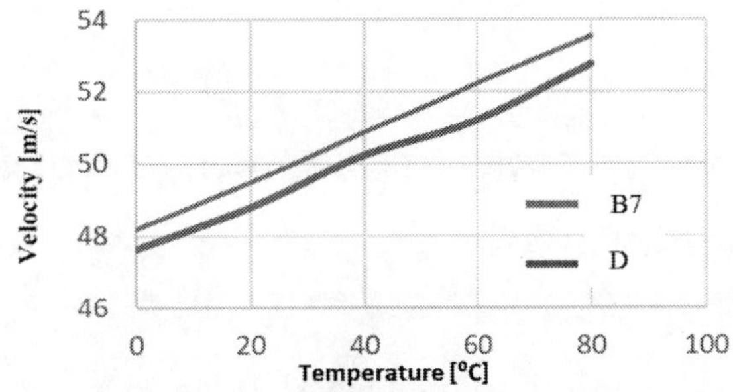

Figure 4. Output velocity of B7 and Diesel fuel, at different temperature values

4. Conclusions

Because there are many differences in terms of physical properties between the Diesel fuel and biofuel blends, several modifications should be implemented at injection systems in case of the fuel switch. For a better understanding and more accurate determination of the implications, the numerical approach remains one of the most convenient methods.

Conventional Diesel and B7 (Diesel+7%biofuel) were taken into consideration, at different temperature values: 0,40 and 80^{0}C. The fuel temperature increase leads to a decreasing mean particle diameter of the jet spray. It has also been observed that the penetration of the jet spray is decreasing once with the temperature of the fuels are rising, and this fact is believed that it's happening because smaller droplets lose their kinetic energy quicker and so, their travel distance is shorter.

Regarding the dissipation rate, this factor is more accentuate at conventional Diesel, because of the lower viscosity and surface tension properties. With the higher temperature, the viscosity of the fuels is decreasing, and particle breakdown phenomena are amplified, increasing the dissipation rate. Of course, the apparition and development of cavitation phenomena inside the nozzle have a significant effect on this factor. The outer fuel velocity was also measured, and the results showed that this parameter is increasing approximately linear and directly proportional with a temperature rising. At every temperature value, the output velocity of the B7 fuel was lower than the output velocity of Diesel, and this fact leads us to the cavitation phenomena which, mainly because it's physical properties, appears more often and develop itself in more intense phases, for this second-mentioned fuel.

References

[1] Seoksu Moon Research Institute for Energy Conservation, National Institute of Advanced Industrial Science and Technology, 1-2-1 Namiki, Tsukuba, Ibaraki, 305-8564 Japan

[2] Moon S, Tsujimura T, Gao Y, Park S, Wang J, Kurimoto N, et al. Biodiesel effects on transient needle motion and near-exit flow characteristics of a high pressure diesel injector. Int J Engine Res 2014;15:504e18.

[3] Kuti OA, Nishida K, Zhu J. Experimental studies on spray and gas entrainment characteristics of biodiesel fuel: implications of gas entrained and fuel oxygen content on soot formation. Energy 2013;57:434e42.

[4] Influence of biofuels on the internal flow in diesel injector nozzles F.J. Salvadora,*, J. Martínez-Lópeza, J.-V. Romerob, M.-D. Rosellób a CMT- Motores Térmicos, Universidad Politécnica de Valencia, Camino de Vera s/n, 46022, Valencia, Spain b Instituto de Matemática Multidisciplinar, Universidad Politécnica de Valencia, Camino de Vera s/n, 46022, Valencia, Spain.

[5] Chioreanu N, Mitran T, Rus A and Beles H 2014 Numerical simulation of thermal-hydraulic generators running in a single regime. Central European Journal of Engineering, Volume 4, Issue 2, June 2014, Pages: 119-124. Publisher: Springer-Verlag Wien, DOI: 10.2478/s13531-013-0145-2, ISSN: 18961541

IOP Conf. Series: Materials Science and Engineering **568** (2019) 012048 doi:10.1088/1757-899X/568/1/012048

Intake and exhaust pipe optimization for an internal combustion engine

V B Magdas[1], D Moldovanu[1,*] and D C Mastan[1]

[1] Automotive Engineering Department, Automotive Engineering, Mechatronics and Mechanics Faculty, Technical University of Cluj-Napoca, Cluj-Napoca, Romania

dan.moldovanu@auto.utcluj.ro

Abstract. Nowadays the stringent exhaust emission limitations of the internal combustion engines force the researchers toward an optimization process of new/prototype engines; but for limiting the overall emissions of an entire fleet of vehicles, an optimization of existing engines is also a right research direction. Solutions for this optimization process may come from using biofuels made from different plants that produce oxygen via photosynthesis during their vegetative state and therefore compensate the CO2 in the internal combustion engine exhaust. Other solutions include better exhaust treatment materials and devices. The current paper proposes a study on optimizing the intake pipe and exhaust pipe of a given internal combustion engine (Opel Astra F, code C18NZ, the production year 2000) using Lotus Engine Simulation software. First the model of the engine was implemented and validated with the full load power and torque variation. For the intake and exhaust the geometrical dimensions were then varied and an optimum was obtained.

1. Introduction

The four-stroke engine cycle begins with the intake stroke. The engine power is highly influenced by this stroke because in this process the air, which is used to burn the fuel, is entering the cylinder. So, the more air enters the cylinder, the more fuel can be burned, in order to produce power.

The parameter which measures how much air enters the cylinder compared with the maximum quantity of air that can be entering in the cylinder is the volumetric efficiency [1]. This parameter is reduced by every element that composes the intake system. In order to increase the quantity of air which enters the cylinder, we must build the intake elements with the optimum dimensions.

Regarding the exhaust phase, this is done in three ways: natural exhaust, forced exhaust and mixed exhaust. The main parameter that can influence the efficiency of the exhaust process is the residual gases coefficient, which represents the number of burnt gases that remains in the cylinder after the exhaust valve closes [2]. This coefficient is highly influenced by the interference of the exhaust gases with the exhaust manifold, which can make those gases turn back into the exhaust ports.

In this paper the intake and exhaust processes were analyzed, the geometry and construction of the intake and exhaust manifold were simulated to underline the importance of the construction parameters of the manifolds influence on the intake and exhaust processes. In the end the optimum construction of the manifolds regarding the best performances of the engine was chosen. These analyses were done using the Lotus Engine Simulation software.

IOP Conf. Series: Materials Science and Engineering **568** (2019) 012048 doi:10.1088/1757-899X/568/1/012048

2. Simulation of the engine

For the simulations, a Vauxhall engine was used which equipped the Astra MK3 model from the same constructor. This engine is naturally aspirated, equipped with a single point fuel system. In table 1, the main characteristics of the engine are shown.

Table 1. Engine technical data [3]

Engine code	C18NZ
Engine construction type	OHC (4 cylinders)
Displacement [cm^3]	1796
Max.Power [kW]/max. power RPM [rot/min]	66/5400
Max Torque [Nm]/max. torque RPM [rot/min]/	145/3000
Bore x Stroke [mm]	84.8 x 79.5
Compression ratio	9.2
Fuel/CO	Gasoline/95

2.1. Modeling the engine

The main components of the engine that need to be modeled are: the engine block (bore, stroke, con-rod length, pin-offset, compression ratio), the cylinder head (number of valves, valve open angle, valve close angle, maximum lift of the valves, the diameter of the valves), the intake system (air filter, ducts, intake manifold and the injection module with all the geometry and dimensions) and the exhaust system (the exhaust manifold, catalytic converter and noise silencers with all the geometry and dimensions).

For the engine block only the parameters need to be introduced, without any kind of geometry. These were taken out from the engine specifications. The most important objective in the modeling of the engine was to measure all the intake and exhaust elements, determine the exact shape of these components and then create the model using the measured data.

2.2. Modeling the intake system

The intake system is an assembly made of the injection module (throttle body, injector, idle air controller), the intake manifold, the EGR, the hose for the brake assistance and the fixing screws.

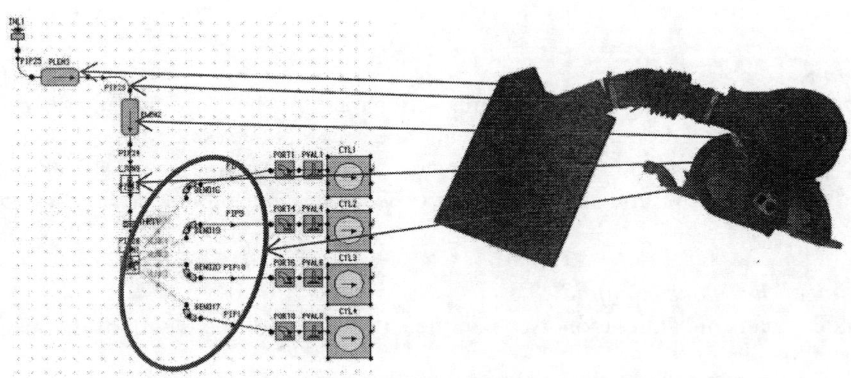

Figure 1. The intake system model

2.3. Modeling the exhaust system

The exhaust system is composed of the exhaust manifold, the connection piping, the catalytic converter and the noise silencers. In order to create this model, the following parameters need to be introduced: geometry and dimension of the exhaust manifold, dimensions of the piping, the type and dimensions of the catalytic converter and the type and dimensions of both noise silencers.

The exhaust manifold of the C18NZ engine combines the first and the fourth cylinder pipes in one outlet and the second and third cylinder pipes in the second outlet. This kind of construction decreases the migration of the burnt gases from the exhaust phase cylinder to the intake phase cylinder, when the valve openings are overlapped. This parameter will then be the input for the correct mixture of fuel and air, in order to obtain the full oxidation of the fuel mixture, thus the exhaust gases will be less pollutant. The construction of the exhaust manifold can be seen in the next figure:

The final model of the engine can be seen in the following figure

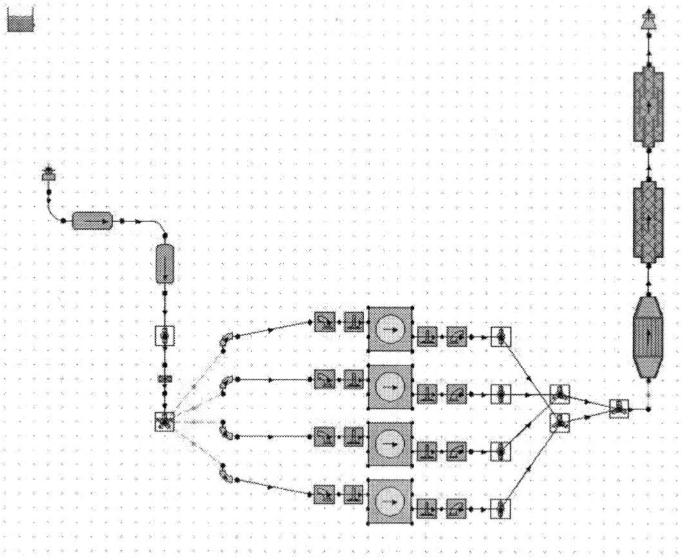

Figure 2. Final model of the C18NZ engine

The model is composed of the air inlet, air filter, throttle body, injection module, intake manifold, intake valve ports, intake valves, 4 cylinders, exhaust valves, exhaust valve ports, the exhaust manifold, the connection pipes between the exhaust manifold and the catalytic converter and two noise silencers.

3. Simulations

For the simulations the manifolds were optimized separately. There are two phases of the simulation: the first simulation was on the intake manifold and the second one was on the exhaust manifold.

The base parameters for both simulations were: minimum engine speed of 800 RPM, maximum engine speed 6400 RPM, the ambient air pressure of 1 bar and the ambient air temperature of 20 Celsius degrees.

3.1. Simulation of the intake manifold

On the intake manifold simulation, two parameters were used: the length of the manifold pipes and the diameter of the pipes.

The first simulation was used to check the model, so with the factory manifold the results should be the same as in the technical documentation of the engine. For this simulation the length of the pipes is 240mm and the diameter is 30mm.

After the simulation the following results were obtained: power – 65kW at 5000 RPM and torque – 140 Nm at 3800 RPM.

IOP Conf. Series: Materials Science and Engineering **568** (2019) 012048 doi:10.1088/1757-899X/568/1/012048

In the next step the parametric simulation tool was run. First the length of the intake pipes was used as a parameter and it varied from 50mm to 500mm with the step of 50mm. It can be seen that as the length increases, the volume efficiency increases too, but it drops after 4000 RPM abruptly.

The best choice is the 300mm length because it has a higher value and it decreases with the rpm typically, almost identical to the factory dimensions.

Regarding the power, for each variation of pipe length, the 300mm intake piping was chosen because the highest power is reached at lower rpm.

The last simulation was for the pipes length of 240mm and the variation of the diameter of the pipe from 10mm to 50mm with a step of 10mm.

In this case, the highest value of the volume efficiency is reached for the 20mm diameter, but at high rpm, it drops under 75% abruptly, so the best choice is the 30mm diameter of the intake pipes, as the factory diameters.

For the variation of power and rpm considering the variation of the intake pipes diameter we have the following results: if the 20mm pipe diameter is considered as the best choice for getting the highest volume efficiency, in this case, the 20mm diameter is not the best choice because it generates less power than the 30mm pipe diameter.

3.2 Simulation of the exhaust manifold

For the exhaust manifold simulation, three constructive variants of the manifold were taken into consideration: the manifold constructed over the interference principle (factory manifold), the manifold with short ramifications and the manifold with individual ramification.

In the first simulation, the base factory dimensions of the manifold were introduced: the pipe length for the first and fourth cylinders is 270mm, the length of the pipes for the second and third cylinders is 250mm, and the diameter of the pipes is 35mm. The results for the simulation are shown in figure 3:

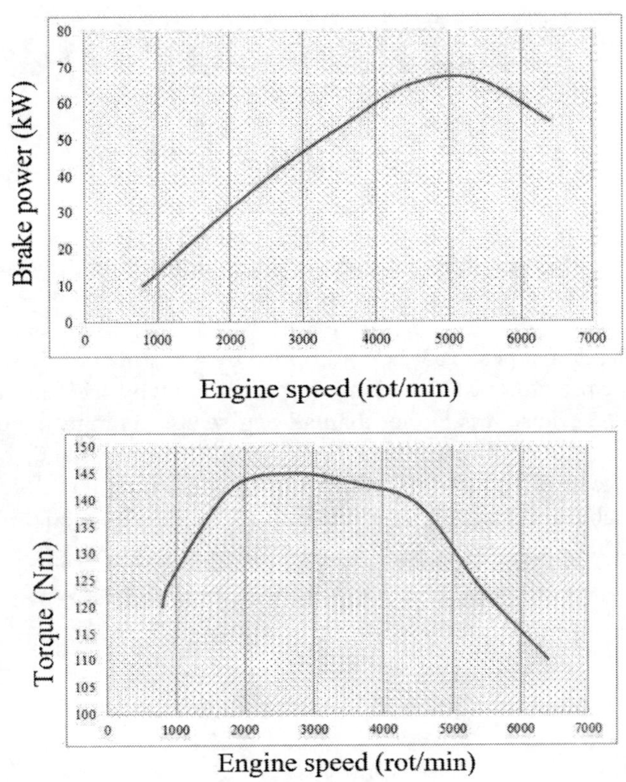

Figure 3. Simulation results of the exhaust manifold with the factory dimensions

IOP Conf. Series: Materials Science and Engineering **568** (2019) 012048 doi:10.1088/1757-899X/568/1/012048

One of the requirements of the exhaust manifold is to favor a proper evacuation of the exhaust gases out from the cylinder. This can be highlighted by the residual gas coefficient, which needs to have the lowest value. In the simulations this coefficient varies in the range of 3 – 13.2%. Because of the real functioning conditions, the results in the engine speed range of 2000 -3000 RPM (cruise conditions) must be analyzed.

For the second simulation the model was updated, thus the exhaust manifold was updated to a short ramification piping.

The results of this simulation can be seen in figure 4.

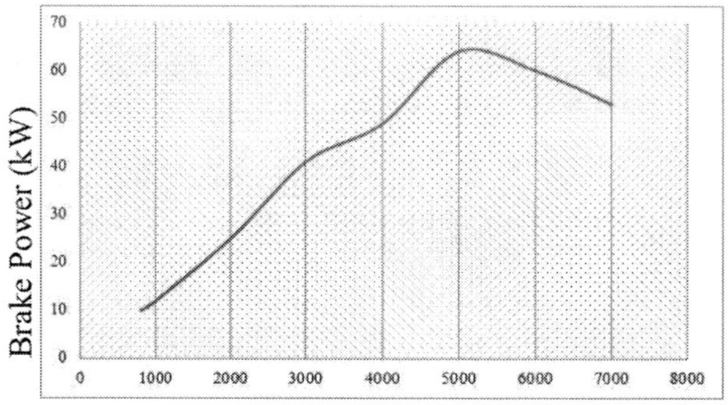

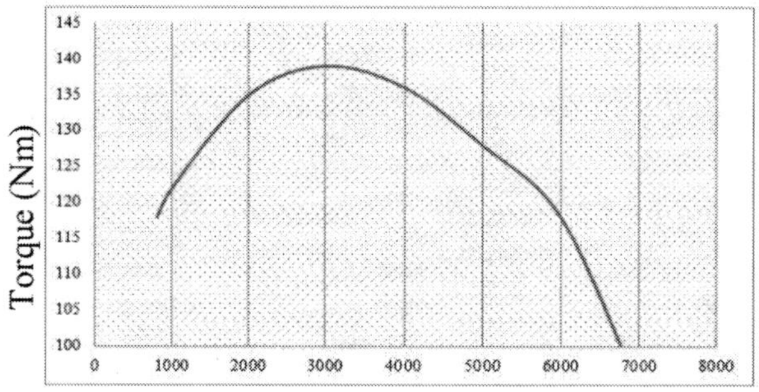

Figure 4. Results of the short ramification exhaust manifold

In this simulation, the power peak (63.2kW) dropped by 4.8% in comparison to the standard factory exhaust manifold and the peak torque (139.9 Nm) dropped as well by 4.01%. The residual gas coefficient has the worst behaviour, because it is higher than all engine speed ranges by over 4.5%. This behavior could be explained by the fact that there is a more significant flow resistance for the exhaust gases to exit the cylinders. For the next simulation the individual ramification manifold was used. The length of the ramification is the same and varies from 800mm to 1500mm. The model is shown in figure 5.

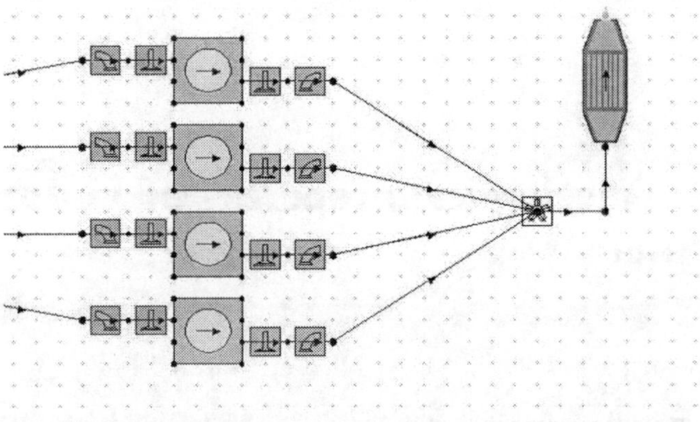

Figure 5. Individual ramification exhaust manifold

For power the peak is 72.2 kW at an engine speed of 5300 RPM. The increase is about 8.7% higher than the standard manifold. These results are for the 1300mm pipe length. The torque increased as well to a peak of 155.2Nm, at an engine speed of 2200 RPM, for the same 1300mm pipe length. The increase is 7.3% higher than the standard manifold.

4. Conclusions

The simulations proved that the intake and exhaust manifold has a significant influence over the performances of the engine.

In the intake manifold case, as seen in the simulation results, the best choice for this type of engine is the length of 30mm and the diameter of 30mm of the intake pipes. However, if the two parameters are combined, maybe there is a possibility to discover better combinations between the length and the diameter of the intake piping.

For the exhaust manifold, the best results were reached with the individual ramification manifold and the length of 1300mm. In this case, the counter-pressure was decreased and a noticeable increase in power and torque was underlined. The problem with this kind of manifold is the cost of production, because in series production casting the entire manifold is cheaper. Usually the individual ramification manifolds are used in high-performance engines.

References

[1] Abăităncei, D., Bobescu, Gh., „Automotive engines" („Motoare pentru automobile"), Editura Didactică şi Pedagogică, Bucureşti, 1975
[2] Grunwald, B., "The theory, calculus and automotive engine construction" ("Teoria, calculul şi construcţia motoarelor pentru autovehicule"), Editura Didactică şi Pedagogică, Bucureşti, 1980
[3] *** https://www.auto-data.net/ro/opel-astra-f-1.8i-90hp-2507
[4] *** https://www.megaparts.eu/index.php/en/product/418115036

IOP Conf. Series: Materials Science and Engineering **568** (2019) 012049 doi:10.1088/1757-899X/568/1/012049

Structural and mechanical properties of TiMnSnAlZrC alloy sintered in different ways

S Gheorghe[1,3], I Pascu[1], C Nicolicescu[2] and A Didu[1]

[1] Automotive, Transport and Industrial Engineering Department, Faculty of Mechanics, University of Craiova, Calea Bucuresti Str., no.107, Craiova, 200512, Romania

[2] University of Craiova, Faculty of Mechanics, Department of Engineering and Management of Technological Systems, Drobeta Turnu-Severin (IMST)

[3]stgheo@yahoo.com

Abstract. Powder metallurgy is a complex technology that allows obtaining various alloys with special properties which successfully replace alloys achieved by classical technologies. The ability to easily modify classic chemical compositions makes this technology an instrument in the hands of researchers that enables them to always improve the properties of sintered alloys. In addition, the technology itself is very permissive, and new technical solutions can always be found in the technological process of pieces manufacturing. Among the materials very used in the technique, which is very well suited to this technology is titanium. Titanium in combination with other chemical elements greatly improves its operating behavior of obtained material. Therefore, this study proposes a combination of titanium with tin and graphite, recommended for improving wear resistance. The other elements used in the mixture, Al, Zr, Mn help to improve the mechanical properties. The paper intends to present a study on the wear resistance of this Ti alloy, sintered in different ways. SEM and spectrometry analyzes for sintered parts are also presented.

1. Introduction

The sintered Titanium alloys are widely used in the technique due to their exceptional physic-chemical, mechanical and technological properties despite the fact that sometimes the cost price is high [1,2]. This is one of the reasons why the manufacturing of alloys with a lower cost price but similar properties has developed. One of the methods used in powder metallurgy is Two-Steps Sintering (TSS) technique which allows for the obtaining of compact components with special properties [3,4].

It is known that all titanium alloys generally have low wear resistance, the improvement of this behavior being always current. Many titanium alloys under light load conditions and relatively low movement can suffer rapid wear which affects the contact surfaces. The surfaces of titanium alloys that are subject to friction have a rapid wear, which leads to their deterioration [5, 6].

Relatively recently, a new technique has emerged that uses the vacuum sintering of titanium hydride powders (TiH2) instead of Ti metal powder. During vacuum or inert gas sintering, TiH2 will dehydrogenize at temperatures below the sintering temperature of Ti alloys, which temperature is generally above 1000^0C.

The titanium hydride has good chemical resistance and forms a stable oxide coating on its surface which makes it unreactive in water. Tin has a very good antifriction property, too. It is used both as a primary element in Sn-Sb type anti-friction alloys and as a secondary element in Cu-Sn or Fe-Sn alloys, for examples, which are recognized as having very good wear resistance.

Ti-Sn alloys have been studied and it has been found that an increase in the concentration of Sn up to 30% leads to the increase of alloy hardness. It has also been found that tin helps titanium in the grinding process, making from these alloys manufactured by mechanical alloying a good alternative [7].

Zr It is very corrosion resistant and is utilized extensively by the chemical industry where corrosive agents are used. Zirconium is similar to titanium, it is compatible with it and a wide range of combinations with other elements such as Al, Sn, C, Nb, are made. Due the properties like ductility and malleability, high melting point, the same atomic structure as titanium, with unlimited solubility, zirconium is an excellent alloying element of titanium, and its alloys are widely used in automotive and space industries [8-13].

Manganese is a hard metal, usually used like an alloying element and in many applications has the same behavior as iron. Manganese is used in steels and cast iron, and its concentration increases when special wear conditions are imposed; manganese is also added to certain types of bronze as well as certain nickel and aluminum alloys. Manganese forms a layer of protective oxide in contact with air, which in the friction conditions improves wear behavior [14,15].

It is also known that carbon in its allotropic graphite form has a very good wear resistance, mainly due to Van der Waals forces. Graphite is the most stable allotrope of carbon and the powder of graphite is an excellent dry lubricant, and its utility in a lot of wear-resistant alloys is so explained [16].

Due to the fact that aluminum, as well the titanium, has a low density but is highly appreciated in various combinations, Ti-Al alloys with high applications in automotive and aerospace technologies are properly studied [17]. The aluminum powder used in these experiments has Si in its component which leads to the increase of refractoriness of the titanium sintered alloy.

2. Experiments

The alloy obtained by powder metallurgy technology has the following mass distribution of the components used: 80%wt. TiH_2, 8%wt. Mn, 6%wt. Alumix321, 3%wt. Sn, 2%wt. Zr, 1%wt. graphite. This combination takes into account the fact that:

-aluminum, oxygen, carbon are the α stabilizing elements and they increase the allotropic transformation temperature and dissolve in Tiα;

- manganese and silicon are the β stabilizing elements and they decrease the allotropic transformation temperature, dissolve in Ti β or form eutectoids;

- tin, zirconium are elements with low influence on the transformation temperature.

Titanium hydride, (TiH_2), produced by Chemetall GmbH, used as precursor material, is the main powder of the mixture and is characterized by a grain-size distribution less than 100 µm. The chemical composition of titanium hydride powder, produced by the water-atomizing process, is: titanium min 95%, hydrogen min. 3.8%, nitrogen max.0.3%, Si max.015%, Al max.0.15% and Fe, Ni, Cl, Mg, C, Ni each below 0.1%.

Manganese powder, type MN006020, Goodfellow provenience, has maximum particle size: 45micron and Purity 99.5%. The alumix powder 321 is ECKA's provenance and contains Al, Mg, Si, Cu, Fe and lubricant. Zr powder Goodfellow provenance, has a maximum particle size 45micron and purity 99.2. Tin powder, SN006020 Goodfellow provenance has maximum particle size 45micron and purity 99.9%.

The Graphite Powder, is 2N5 American Elements type, with low friction coefficient and 99.5% purity.

The powder used in the mixture analyzed in this paper is presented in the table 1.

Table 1 The powders used

Elements	TiH_2	Mn	Sn	Alumix 321 [%]						Zr	Graphite
				Al	Mg	Si	Cu	Fe	lubricant		
properties				balance	0,95	0,49	0,21	0,07	1,50		

Mass [%]	80	8	6	3	2	1
Density[g/cm^3]	3,76	7,43	7,29	2,686	6,506	1,8

Planetary Mono Mill PULVERISETTE 6 classic line from Fritsch was used for obtaining a homogenized mixture. The technical conditions were: 250 ml grinding bowl filed with argon atmosphere, 200 rpm for main disk , ball: powders ratio 1.1, balls material stainless steel – 1.3541 ISO/EN/DIN codeX47Cr14, B50 and 60 min for mixing time.

A 90Plus particle size analyzer, Brookhaven Instruments Corporation, USA, with 35 mW solid state laser and 660 nm wavelength was used for established the particle size distributions. Grinding in the ball mill only followed a higher homogenization of the mixture and although the working time is relatively low, 60 min, it is found that the resulting particle size is nanometric. The grinding and homogenization operation in the planetary mill with balls results in a granulometric distribution which reveals that the powder obtained has a maximum size of 1,7293 μm while the minimum value is 0,1887 μm. The highest number of particles (28% from the total number) have the dimension equal to 0,2252 μm. The percent of the particles with higher size is lower than 1.5%. The mean hydrodynamic diameter is 0,2911 μm. The percentage of particles, depending on the granulometric distribution, is shown in figure 1.

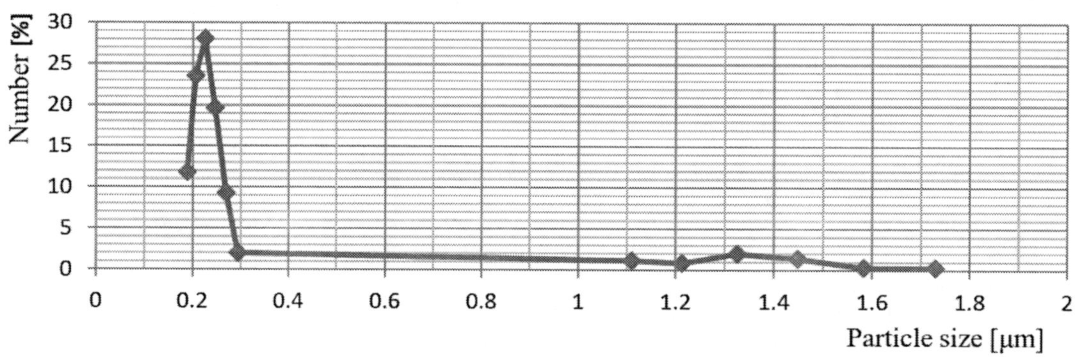

Figure 1. Relationship between percentage of particle and granulometric distribution

Selection of sinter mode for titanium alloys is influenced by a number of factors such as the purpose of the alloy, the mechanical and technological properties to be obtained, the alloying elements used, the behavior during heating and, last but not least, the operating costs [18,19,20,22].

Sintering was done in a neutral argon atmosphere and four cycles of work were chosen:

Cycle 1: heating to 1050^{0}C and hold for 15 minutes, return to temperature 1000^{0}C and maintain for 75 minutes; Cycle 2: heating to 1050^{0}C and holding for 90 minutes; Cycle 3: heating to 1050^{0}C and hold for 15 minutes, return to temperature 950^{0}C and maintain for 75 minutes; Cycle 4: heating to 1050^{0}C and hold for 15 minutes, return to temperature 1000^{0}C and maintain for 20 minutes, return to temperature 900^{0}C and maintain for 55 minutes.

In all four sintering cycles, the maximum temperature was 1050^{0}C and the total sintering time was 90 min. The cylindrical samples were analyzed microscopically by an electron microscope based on the 5th generation Phenom desktop SEM platform. Also a spectrometry elemental analysis for all samples was made [16]. These analyses are shown in figures 2-5.

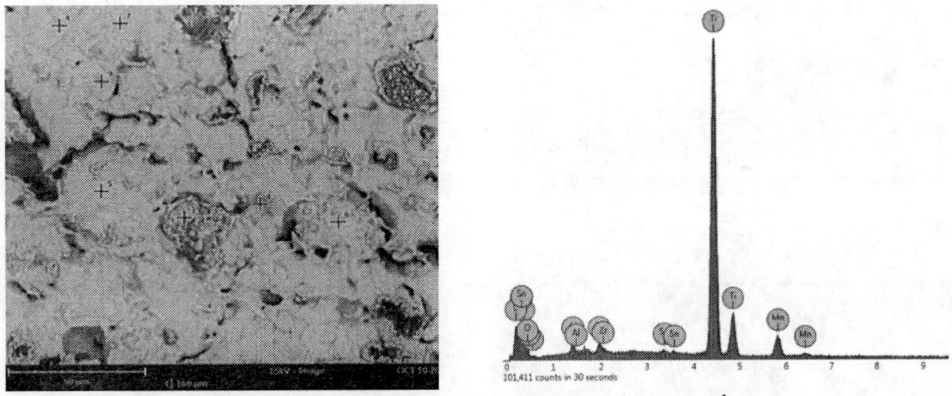

a b

Figure 2. a-Scanning electron microscopy (SEM) of sample for Cycle 1; b-Spectrometry elemental analysis for SEM cycle 1 spot 6

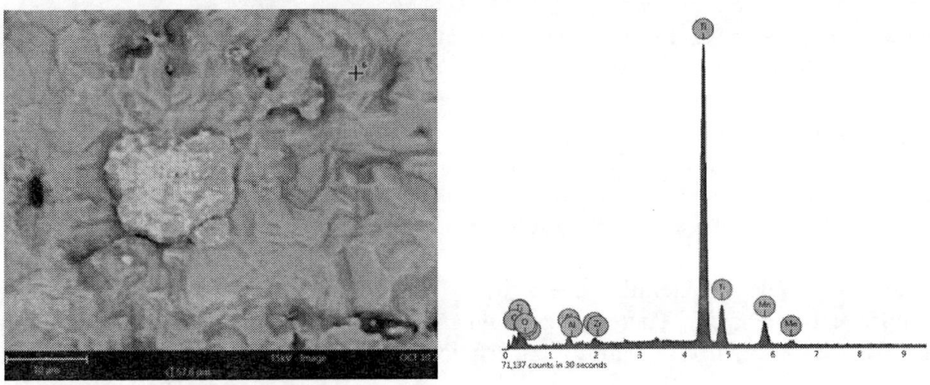

a b

Figure 3. a- Scanning electron microscopy (SEM) of sample for Cycle 2; b-Spectrometry elemental analysis for SEM cycle 2 spot 6

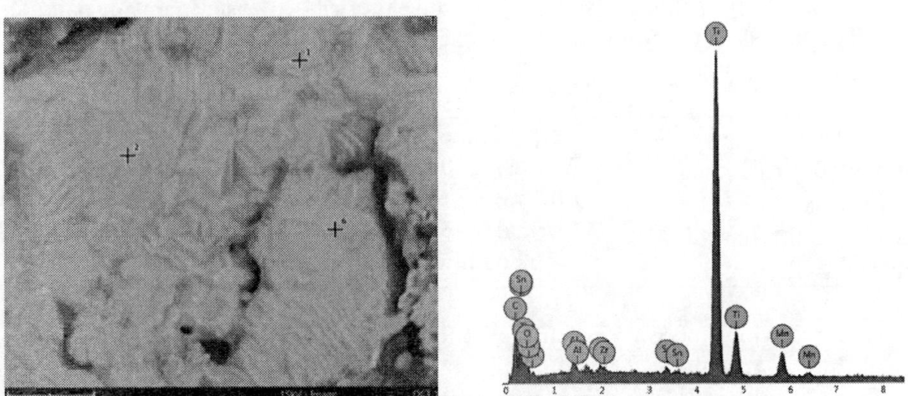

a b

Figure 4. a- Scanning electron microscopy (SEM) of sample for Cycle 3; b-Spectrometry elemental analysis for SEM cycle 3 spot 6

IOP Conf. Series: Materials Science and Engineering **568** (2019) 012049 doi:10.1088/1757-899X/568/1/012049

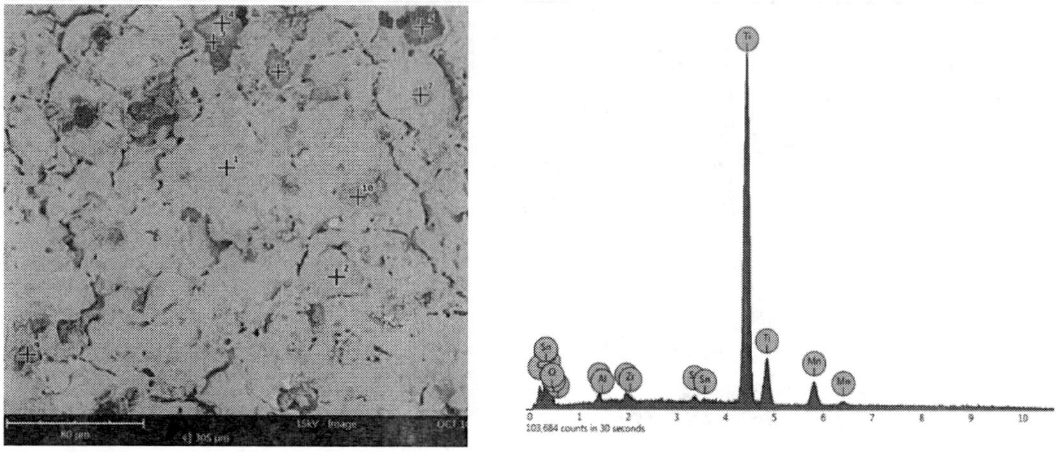

a b

Figure 5. a- Scanning electron microscopy (SEM) of sample for Cycle 4;
b-Spectrometry elemental analysis for SEM cycle 4 spot 1

Structural analyzes are chemically unattached and highlight a uniform and compact structure of the microstructure of titanium alloy heated in inert gas at 1050 ° C and slowly cooled, shows polyhedral grains. The sintered titanium alloy has α + β structure and this is influenced by the presence of α stabilizing elements (aluminum, oxygen, carbon), β stabilizers elements (Mn, Si), or low-impact elements (Sn, Zr). The presence of aluminum in these alloys consists in limiting the range of the solid solution β, increasing the allotropic transformation temperature, increasing the solubility of the isomorphic stabilizing elements β. Spectral analysis shows the presence of all structural elements and also oxygen traces [21].

Titanium and its alloys generally have a low wear resistance, which is why constantly looking for ways to improve this behavior. There are cases where, in addition to changing the chemical composition by adding elements recognized as improving friction behavior, superficial layers are proposed for this purpose.

Wearing attempts have been made in dry friction conditions, as these alloys have elements such as tin or graphite that can work properly in this type of application. For wear testing, sintered cylindrical specimens of titanium alloy 12 mm in diameter were used and the ball was made of steel 100Cr6. A TRB 01-2541CSM Instruments is used to determine the tribo measurement of samples and the test method was ball on disk in the following technical terms, table 2.

Table 2 Technical conditions for wear test

Static partner	**Acquisition**	**Environment**
Substrate: 100Cr6	Radius: 3.00 [mm]	Temperature: 25.00 [°C]
Supplier: CSM	Lin. speed: 6.00 [cm/s]	Atmosphere: Air
Dimension: 6.00 [mm]	Normal load: 2.00 [N]	Humidity: 30.00 [%]
Geometry: Ball	Stop condition: 2000 [lap]	

Surtronic S25 Taylor Hobson Precision profilometer that completes the tribometer product TRB-0254 was used to analyze the wear profile surface. The tribological magnitude curves during the test and the analysis of the profiles of these wear surfaces give us information about the size of the used surface and the magnitude of the resulting depth, are shown in figures 6-9.

For a force of 2 N for 1000 seconds, the experimental results show that the lowest average coefficient of friction, 0.146, was obtained for the fourth cycle. For cycles 1, 2 and 3 the values of the average friction coefficients were propelled by value, namely 0.467, 0.442, 0.431, figures 6-9.a. These values determined for the friction coefficient are considered as average values for this type of alloy. As far as surface profile analysis is concerned, max depth occurs in Cycle Number 4, 38.1μm^2, while the lowest depth of the used profile, 19.2 μm^2, is encountered in Cycle Number 3, figures 6-9.b. The increased presence of oxygen in the structure of alloys manufactured thru cycle number 4 can explain the lower value of friction coefficient and frictional force determined in the contact area.

The wear rate is determined by the amount of material removed, the pressing force and the length of the friction track, following a formula:

$$K=V/F\text{x}S \tag{1}$$

where K is the wear rate in [mm^3/Nm], F is the force of pressing on the specimens in [N] and S is the length of the friction track, in [m].

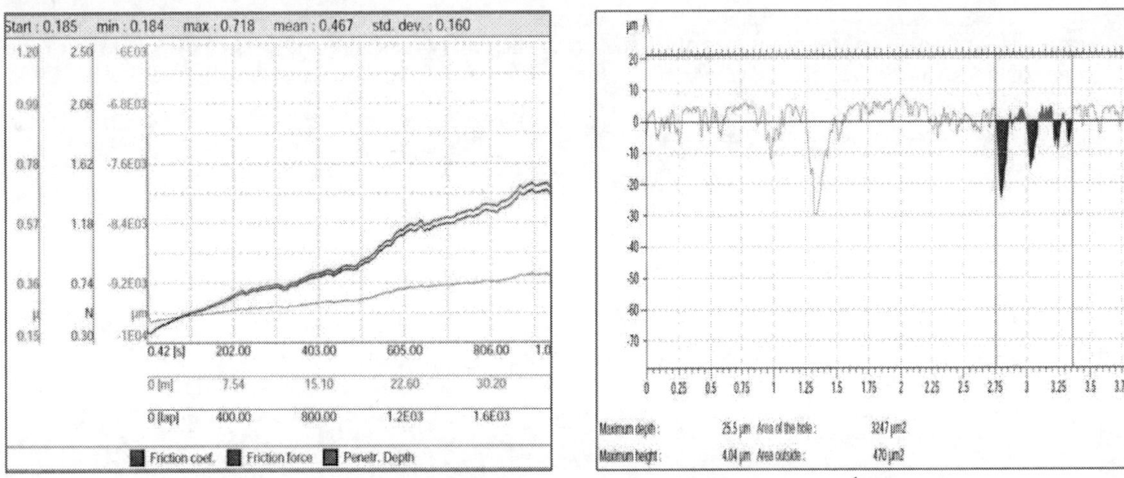

a b

Figure 6. Curves of tribological magnitudes during the test (a) and the analysis of the profiles of these wear surfaces for cycle 1 (b)

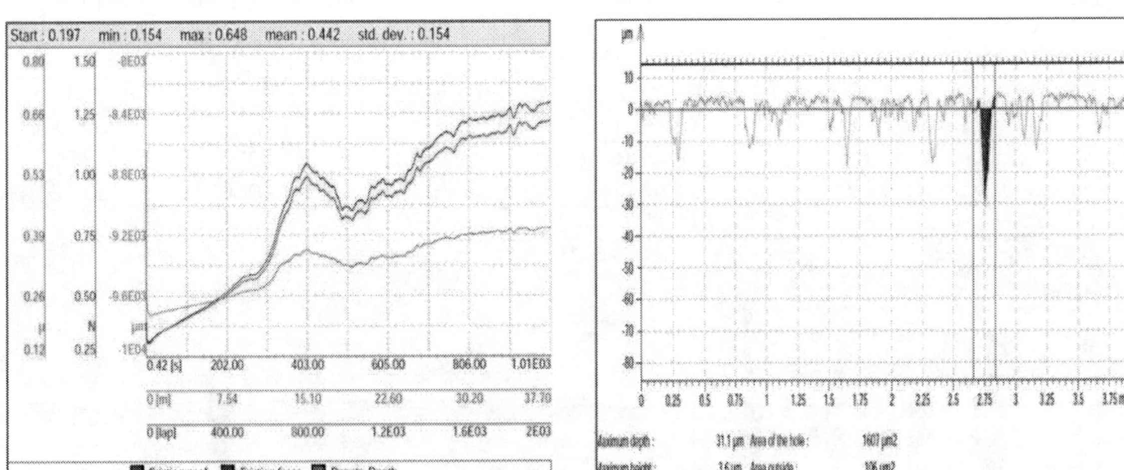

a b

Figure 7. Curves of tribological magnitudes during the test (a) and the analysis of the profiles of these wear surfaces for cycle 2 (b)

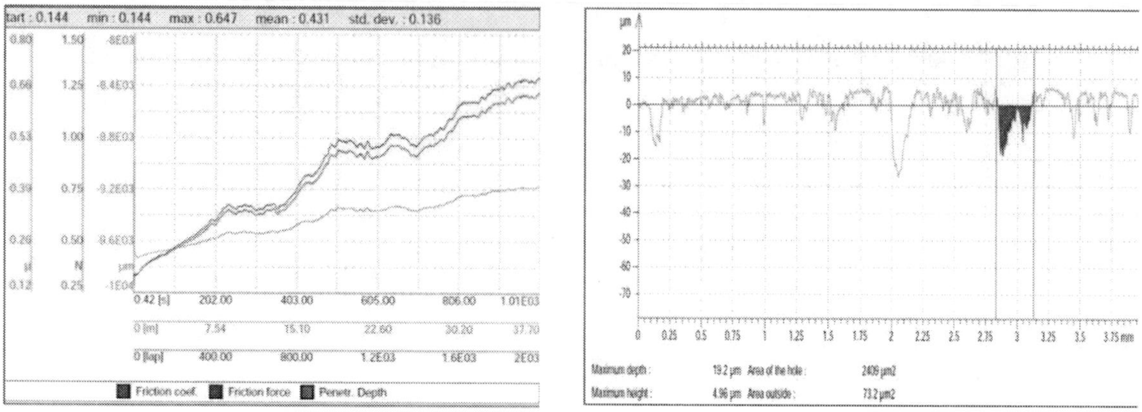

a b

Figure 8. Curves of tribological magnitudes during the test (a) and the analysis of the profiles of these wear surfaces for cycle 3 (b)

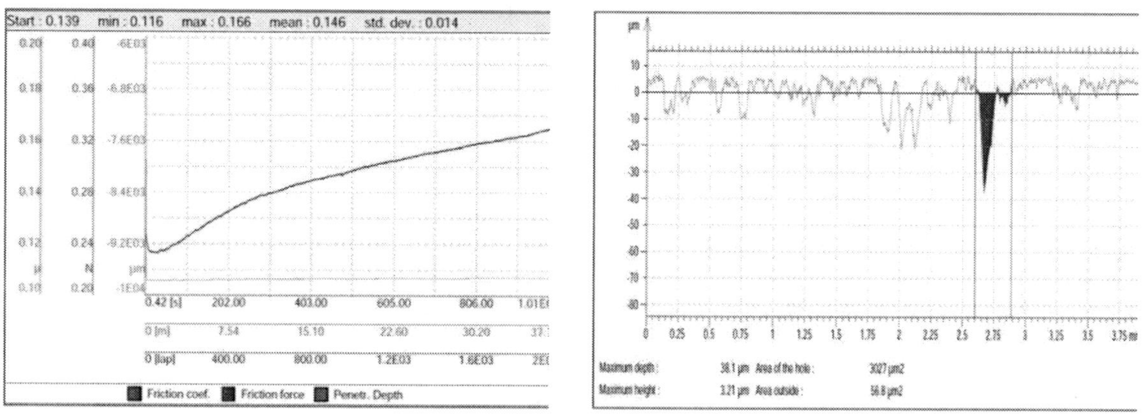

a b

Figure 9. Curves of tribological magnitudes during the test (a) and the analysis of the profiles of these wear surfaces for cycle 4 (b)

Table 3 Wear characteristics

	Cycle 1	Cycle 2	Cycle 3	Cycle 4
Worn track section, [μm²]	3247.0	1607.0	2409.0	4386.0
Sample wear rate, [mm³/Nm]	0.0008118	0.0004018	0.0006023	0.001096

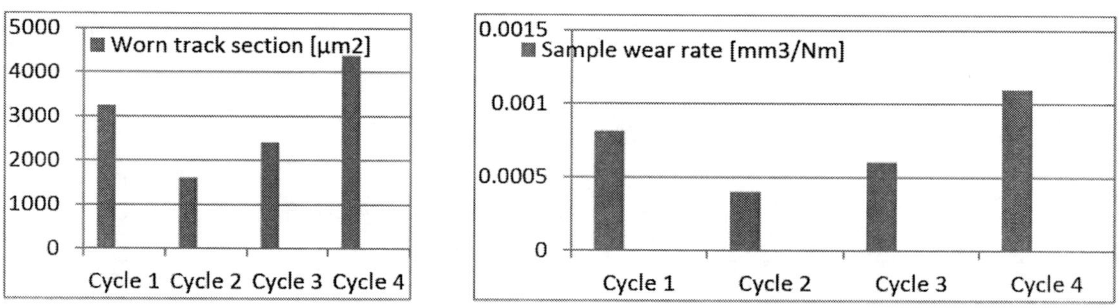

a b

Figure10. a - Worn track section; b - wear rate

The analysis of the experimental results shows that the best wear rates we encounter in the sintering cycle number 2, 0.0004018 mm^3/Nm, while the lowest value, 0.001096 mm^3/Nm, is obtained in the sintering cycle number 4, despite the fact that in this case the value of the coefficient of friction has the best value, 0.146, compared to 0.442 in Cycle 2, table 3 and figure 10. These results are also strengthened by the size of the wear sections, which are proportional to wear rates.

Conclusions

The grinding in planetary ball mills for one hour allows for getting a compact alloy and the particle size drops a lot, reaching nanoscale dimensions.

Multi-step sintering is a variant for titanium-based alloys and this technology may suffer improvements in the future.

The sintering temperature of 1050° C followed by a drop in range of 900° C is the lower limit value that can be used in this type of application.

The wear pattern of the specimens is one that respects the evolution of the phenomenon over time, the graph showing a stabilization of the phenomenon at average values.

The best results of the wear behavior are obtained in the case of sintering at 1050° C with a holding time of 90 minutes, but also the other sintering cycles have comparable values.

The presence of oxygen in the structure of alloys can increase the wear resistance explain the lower value of friction coefficient and frictional force determined in the contact area.

References

[1] Boyer R B 2010 *JOM* **62** 35
[2] Qian M, Schaffer G B, Bettles C J 2010 Sintering of titanium and its alloys *Woodhead Publishing Series in Metals and Surface Engineering,* pp 324-355
[3] Bartosz W, Waldemar P 2011 *PTCer* **63** 814
[4] Maca K, Pouch V, Shen Z, 2008 *Integr Ferroelectr* **99** 114
[5] Goutam D R, Raviraj S, Shrikantha S R, Vinayak N G 2016 *JMR&T* **6** 13
[6] Budinski KG. 1991 *Wear* **151** 203
[7] Hsueh-Chuan Hsu, Hsi-Chen Lin, Shih-Ching Wu, Yu-Sheng Hong, Wen-Fu Ho 2010 *J Mater Sci* **45** 1830
[8] Soyama J, Oehring M, Limberg W, Pyczak F 2015 *Mater Des* **84** 87
[9] Gabrisch H, Krekeler T, Lorenz U, Rackel M, Stark A 2018 *Mater Sci Forum* **941** 741
[10] Baker, H.: *Alloy Phase Diagrams* 2006 ASM Handbook, ASM International, Ohio Vol. 3
[11] Mohammed M T 2017 *Karbala International Journal of Modern Science* **3** 224
[12] Hsua H C, Wu S C, Sung Y C, Ho W F 2009 *J Alloy Compd* **488** 279
[13] Benic L, Strkalj A, Glavas Z 2019 *Eng Rev* **39** 115
[14] Gheorghe S, Teisanu C 2013 *Metalugia International* **6** 151
[15] Bolzoni L, Alshammari Y, Yang F 2018 *AMM* **884** 49
[16] Swee H T, Thampuran R, Goh J, Hong C 1995 Titanium-graphite sintered composite material with improved wear resistance and low coefficient of friction **DE19641023A1**
[17] Henriques V R et al., 2006 *Mater Sci Forum* **530** 10
[18] German R M 2014 *Sintering: from Empirical Observations to Scientific Principles* ed. Butterworth-Heinemann (Elsevier USA) pp 471-512
[19] Yuanshen Qi, Contreras K G, Jung H D, Kim H E, Lapovok R, Estrin Y 2016 *Mater Sci Eng* **59** 754
[20] Qian Ma, Yang YaF, Shudong D Luo 2015 *Science, Technology and Applications* ed. Butterworth-Heinemann (Amazon USA) pp 201-218
[21] Pascu C I, Gheorghe St, Nicolicescu C, Tarata D 2018 *Mater Res Proc* **8** 201
[22] Feng Q, Zhang Y, Dong Z, Zhang P, Wang J, Wang D, Gao Qi, Peng X, Wang F 2018 *Advances in Materials Processing, (*Springer China*)* **10.1007** 978

IOP Conf. Series: Materials Science and Engineering **568** (2019) 012050 doi:10.1088/1757-899X/568/1/012050

Flow studies on increasing the efficiency of the inlet manifold

D Moldovanu[1*], A Csato[1], N Bagameri[1]

[1]Technical University of Cluj-Napoca, Automotive Engineering and Transports Department, Bd. Muncii 103-105, Cluj-Napoca, Romania

dan.moldovanu@auto.utcluj.ro

Abstract. The current study focuses on improving the flow of air through the inlet manifold. With using CFD software, the phenomena occurring within the pipeline are studied. Modification of the intake results in an increase in the air mass flow, and thus in the increase of the engine performance. The air flow rate allowed at the boundary layer between the duct and the air tends to be zero. The high-rugged surface increases the flow velocity in the center of the tube. In this paper, the effect of a rough surface created by dimples in the intake pipe is studied. By applying small dimples to the surface of the piping, the velocity distribution becomes more uniform within the tube. Dimples cause lowering pressure at the start, thus increasing the air flow rate. The limit layer that has low flow rates relative to the speed at the center of the tube becomes thinner. By applying this method, the air flow rate in the intake manifold can be improved.

1. Introduction

In the automobile industry continuous research on the flows around bluff bodies and "immersed bodies. With a better understanding of the fluid dynamics can lead to an improvement in the design. Examining solutions from the past can provide new ideas, like the solution used in case of a golf ball where a small change in the surface geometry results in a significant change in drag. The dimples lead to a turbulent flow which is similar to adding roughness to the surface. Studying dimples may help to improve the air intake system on the internal combustion engine.

2. Inlet manifold

In the following paragraph short description of the flow in the intake manifold is presented, it is one of the central parts of the engine. The role of the intake manifold is to provide a uniformly distributed air flow to the cylinders,rational design can reduce the inlet pressure losses also increasing the air volume introduced in engine. Increasing the inlet manifold efficiency is a major challenge to increase the engine overall efficiency, in this way the emissions can be reduced. Volumetric efficiency of the engine is a measure of the effectiveness of the air intake system composed by intake manifold, intake port and cylinder [1]. This means that the velocity of air in the inlet manifold is increasing more air the intake system can deliver to the engine. This will increase the volumetric efficiency, this effect can increase the torque and the power of the engine [2]. Many studies are related to the shape and length of the inlet manifold [3]. On the effect of the boundary layer which is being generated by air flow through the inlet manifold there are only a few studies [4]. Computational Fluid Dynamics (CFD) is a helpful tool for predicting the flow characteristics in the inlet manifold.

Using dimples on the surface: Adding dimples to a surface is similar to adding roughness to the surface, the effect of the boundary layer has to be similar to the flow over a rough surface. On the boundary layer in dimples are vortexes created like in Figure 1. Flow over the dimpled surface can generate stable vortexes.

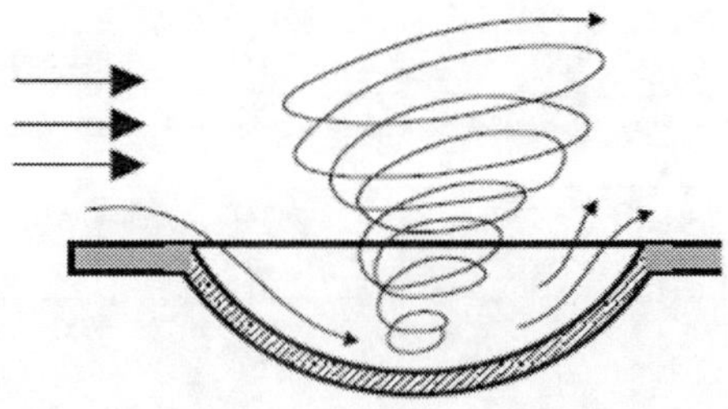

Figure 1. Mechanism of swirl motion in a dimple [5].

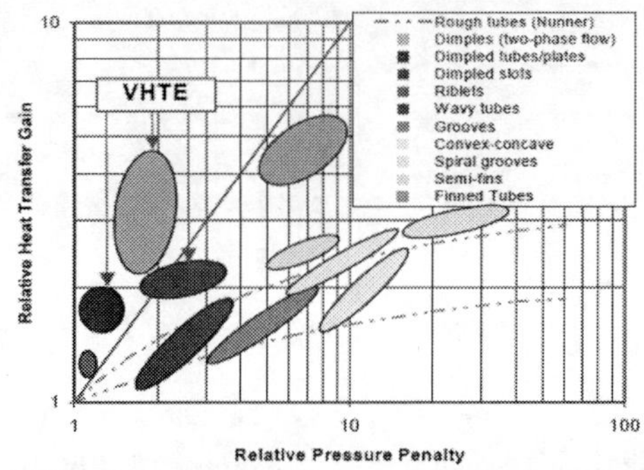

Figure 2. Effectiveness of different skin roughness technics [5].

The red line represents linear dependencies between heat transfer and pressure drop properties of surface roughness achieved with different techniques. Figure 2 shows that VHTE (vortex heat transfer enhancement) provide the best heat transfer with the least pressure drop then every other technique. Simulating the flow over golf ball shows that dimples create vortexes on the surface, these vortexes are increasing the momentum near the wall and serve to reattach the flow after the dimple[6]. Simulations prove that flow over a dimpled flat surface achieve higher velocity due to the fact they are creating a favorable pressure gradient [7]. This publication is focusing on using dimple technology in inlet manifold for achieving higher volumetric efficiency.

3. Simulation data

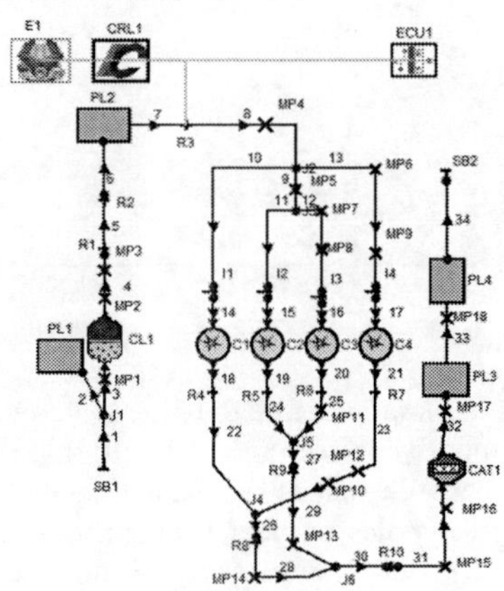

Figure 3. AVL Boost engine model for input data.

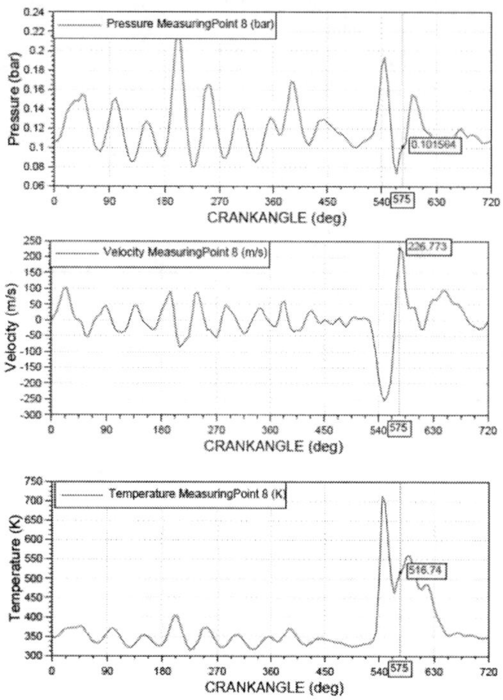

Figure 4. Results in inlet manifold measuring point at 2000 rpm.

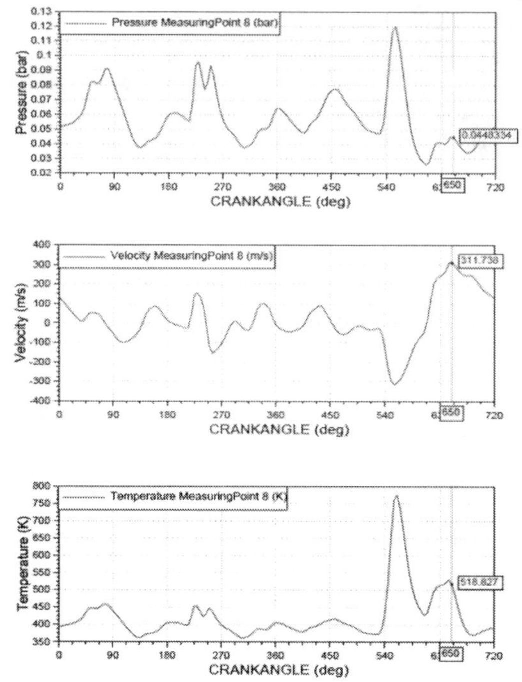

Figure 5. Results in inlet manifold measuring point at 6000 rpm.

Correct input data for CFD simulation is one of the most important requirement to have valuable results, just like seen in otherresearchers [10]. For having input data for simulation of inlet manifold pipes a one cylinder engine model (Figure 3) is used, the simulations are performed in AVL Boost. The load of the engine is controlled by the ECU via R3 flow restriction on the inlet system. Two different cases is taken into consideration one on 2000 rpm and one on 6000 rpm. The engine parameters are shown in Table 1 as input data for CFD simulation the air velocity and temperature in MP8 measuring point are taken in count.

Table 1. Simulated engine characteristics.

	Data	Unit
Bore	86	[mm]
Stroke	86	[mm]
Nr. of cylinders	4	[-]
Injection type	Indirect	
Control	Intake flap	

4. Simulation scheme

For accurate results the inlet will be in plenum before the pipe system to cover the turbulence generation then the flow is introduced in the smooth surface pipe, the entry pipe length is for developing the flow to have a fully developed velocity profile that is not changing with the length of the pipe. For the calculation of the entry pipe length Reynolds number is taken into the consideration and the equation **Error! Reference source not found.**) is used, where l_l is the entry length, D is the diameter of the pipe and R_e is the Reynolds number. The equation **Error! Reference source not found.**) is only available for turbulent flow:

$$\frac{l_l}{D} = 4.4R_e^{1/6} \tag{1}$$

Since velocities higher than 120m/s were used, the Reynolds number is over 4000 so the flow is turbulent[8].

After the entry pipe a 200 mm length dimpled pipe with the same diameter is used. Before the outlet boundary condition another smooth surface pipe with the same length as the first one is used to getting a fully developed flow at the outlet. The simulated system scheme is in Figure 6.

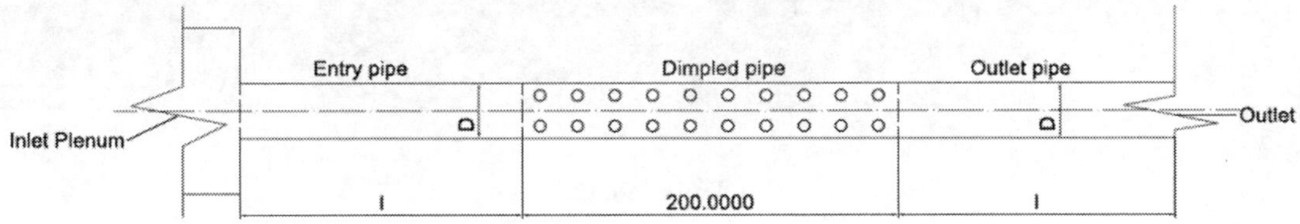

Figure 6. The analyzed pipe system scheme.

Dimples analysis: in this section different forms of dimples are studied. The flow over dimples is governed by the vortexes created in dimples. As the pressure is minimum in the center of the vortex [9] this phenomenon creates a reattaching flow after the dimple. Either laminar or turbulent flow they are affected by boundary layers. The shear stress created by the boundary layer has a different magnitude for laminar and turbulent flows. Due the upward eddy motion of the fluid particles has a negative effect on the velocity of the flow, this momentum transfer from the lower part of the flow to the upper part reduces the velocity of the lower part. The force is acting on the particle that is moving through differential area dA**Error! Reference source not found.**it depends on the horizontal (u'),vertical(v') velocity and also on density(r). Creating vortexes with dimples with reduces the vertical velocity of the particle, thereby the negative effect of this phenomenon on the velocity is decreased.

In this section five different shapes for dimples will be studied. For having accurate results for different dimples forms only one dimple is simulated with a simulation scheme in Figure 7. The best shape of the dimple is chosen the pressure distribution in the upstream of the dimple and also the flow reattaching length. Having lower pressure inside the vortexes created by dimples and smaller reattaching length makes the dimple shape more effective.

Figure 7. Studied dimples shapes.

Based on the results shape nr1 which is a half globe creates the smallest flow reattaching length and creates the highest vortex, therefore it was considered for further investigations/simulations. The inlet was populated with shape 1 dimples so that the air flow can be analyzed.

5. Simulation with shape nr. 1

Shape nr. 1 has the smallest reattaching length and maximal kinetic energy. Shape nr. 1 is a half globe which is a perfect shape for creating vortexes.

IOP Conf. Series: Materials Science and Engineering **568** (2019) 012050 doi:10.1088/1757-899X/568/1/012050

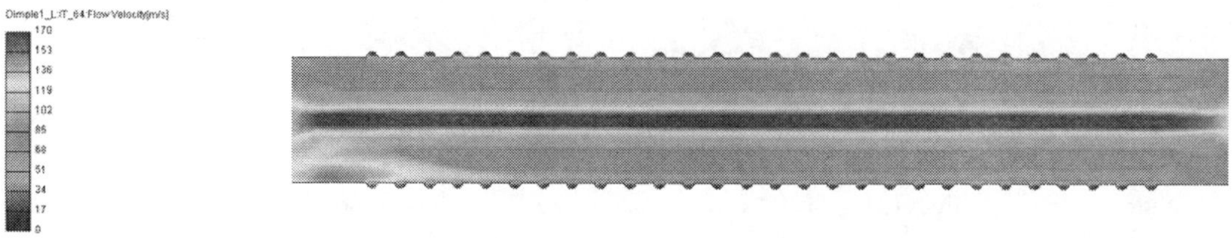

Figure 8. Air velocity with nr. 1 shape [m/s].

Figure 9. Turbulence velocity [m/s].

6. Conclusions

Figure 8 represents the air flow velocity on a dimpled intake tube with the length of 200 mm, after the flow is entering in the dimpled is shaped into a straight vane having a stable high-speed flow in the middle of the tube. The flow in the center of the tube has a 39.5% higher speed than at the inlet speed. The stabilization effect of the dimples creates a constant turbulence velocity (Figure 9) in the dimpled section. The density of the dimples and placement will be further analyzed including the simulations for bent intake pipes.

References

[1] Nicolao, G. De, Scattolini, R., Siviero, C., 'Modelling the volumetric efficiency of ic engines: Parametric, non-parametric and neural techniques', *Control Eng. Pract.*, vol. 4, no. 10, pp. 1405–1415, 1996.

[2] Ganesan, V., *Internal Combustion Engines*, 2nd edn. Singap: Tata McGraw-Hill Publishing Company Limited, 2004.

[3] Bayas, J., Wankar, A., Jadhav, N., 'A REVIEW PAPER ON EFFECT OF INTAKE MANIFOLD GEOMETRY ON PERFORMANCE OF IC ENGINE', 2016.

[4] Pahmi, M. A. A. H., Che Mat, S., Nasruddin, A. N., Ismail, M. F., Yusof, M. N., 'The Analysis of Intake Manifold Air Stream Velocity on Different Material Roughness', in *Applied Solutions of Engineering Science*, 2014, vol. 661, pp. 143–147.

[5] Chudnovsky, Y., Kozlov, A., 'Development and Field Trial of Dimpled-Tube Technology for Chemical Industry Process Heaters', 2006.

[6] Smith, C. E., Beratlis, N., Balaras, E., Squires, K., Tsunoda, M., 'Numerical investigation of the flow over a golf ball in the subcritical and supercritical regimes', *Int. J. Heat Fluid Flow*, vol. 31, no. 3, pp. 262–273, 2010.

[7] Michael Mode, J., 'Simulation of the Flow Over a Flat Dimpled Plate', 2010.

[8] Schobeiri, M., *Fluid mechanics for engineers. A graduate textbook.* 2010.

[9] Jeong, J., Hussain, F., 'Hussain, F.: On the identification of a vortex. JFM 285, 69-94', *J. Fluid Mech.*, vol. 285, pp. 69–94, 1995.

[10] Chioreanu N, Mitran T, Rus A and Beles H 2014 *Numerical simulation of thermal-hydraulic generators running in a single regime.* Central European Journal of Engineering, Volume 4, Issue 2, June 2014, Pages: 119-124. Publisher: Springer-Verlag Wien, DOI: 10.2478/s13531-013-0145-2, ISSN: 18961541t

IOP Conf. Series: Materials Science and Engineering **568** (2019) 012051 doi:10.1088/1757-899X/568/1/012051

Refueling strategies of autonomous hybrid vehicles

M Zöldy[1*], S Szalay[1], Z Szalay[1] and I Zsombók[1]

[1]Budapest University of Technology and Economics, Department of Automotive Technologies, 6 Sztoczek street, Building J, Floor V, 1111 Budapest, Hungary

mate.zoldy@auto.bme.hu

Abstract. Fuel consumption prediction and recharging strategies are crucial fields of autonomous hybrid vehicles. In our paper based on literature review the currents trends of hybrid vehicles are compared and working principals of hybrid strategies created for different types of hybrid drivetrains were described. The fuel management, basics of the range estimation in hybrid vehicles are displayed. Analyzations of the factors affecting the fuel consumption and energy management were made, along with an estimation of the order of magnitude of these factors, based on experiments done by EPA. Main fuel consumption influencers are listed and a real word test is carried out in a highway measurement. It was focusing on the affection on the fuel consumption, caused by the additional consumers of the vehicle has been carried out. Evaluation of this measurement was made and conclusions were drawn about the efficiency changes of the vehicle.

1. Brief history and relevance of the hybrid vehicles

Hybrid vehicles were present in automotive history from the beginning. At the last years of the XIX century, electric drive systems took a significant role. At this time, more electric vehicles were manufactured than the ones with internal combustion engines, so the opportunity was given to combining the two powertrains. In the first two decades of the XX century, the development of petrol-powered cars sky-rocketed, halting the development of the hybrid automobiles until the 1960s [1], [2].

The growing fuel prices during the oil crisis of 1973 and reduction plans of the harmful exhaust gas emissions during the 1980s meant a new start for the hybrid car development [1]. In the last few years the market-leading manufacturers started revealing plans of the full electrification and hybridization of their model lineup. The hybrid drive means a powertrain, where the energy required to propel the car is ensured by two or more power sources with different operating principles. The powertrain solutions can be seen today mostly use a combined system of an internal combustion engine and an electric motor [3].

The ICE-powered vehicles provide outstanding performance and long range, using the excellent energy density of their fuel. On the negative side, they aren't fuel efficient and are polluting. Electric cars have their advantages as well, such as better energy efficiency and extremely low harmful emissions while running, however their range per charge ratio is weak, because of the lower energy density of their batteries.

2. Categories of the hybrid systems

The power output of the electric drive, is the basis of categorization of hybrid systems. Hybrid systems with the smallest power output are micro-hybrids. In this case the typical performance of the electric motor is 2-3 kW. The electric motor, functioning as a starter and generator is connected to the internal combustion engine mediately, by a V-belt drive. Beneficially the ICE and the powertrain remains basically unchanged, so the system can be easily implemented even into a production-ready drivetrain.

Next step is the mild hybrid system, where the electric motor is directly connected to the powertrain. The ICE and the e-motor rotate together, as the two units are disconnectable, their torques add up. The electric machine can provide up to 10-15 kW of power [1]. Systems above 15 kW e-motor power output are full-hybrid, these can run without local emission. The electric drivetrain system operates on a voltage of 200-350 V. Plug-in hybrids are full-hybrids, chargeable not only by internal regeneration, but from an external power source [4].

Basic variations of the hybrid powertrains are series and parallel systems. Both have their advantages and disadvantages, which are combined with Series-Parallel hybrid systems. The series hybrid is a system, which uses two power sources to drive an electric motor which propels the vehicle. As its most significant benefit, the operation of the internal combustion engine is completely independent of the drive requirement of the vehicle, so it can be used in any range of its revolution-torque characteristics so it can be held at the range of its top efficiency. This drivetrain is the most variable and has the best opportunities to reduce emissions. On the downside, the energy given by the ICE is transformed at least twice, the efficiency is multiplied, so the total drivetrain loss can be significant [4].

The combined system architecture is basically the same as the series one, but the ICE can directly power the vehicle by using a clutch. In case of a power-split system, the internal combustion engine performance is split in a planetary gear set, one half powers the vehicle, the other half charges the battery or powers the car through the electric motor [5].

3. Parts of a hybrid drivetrain

As an internal combustion engine, modern petrol engines and turbocharged direct injection diesel engines can be used. After all, using an ICE of a traditional powertrain unchanged is not enough to operate a hybrid drivetrain efficiently. The ICE of a hybrid can be more straightforward and smaller, as it does not have to operate economically in the whole range of operation. However, the average load of the ICE is higher than an ordinary vehicle, needing optimization for the higher load. This will decrease the efficiency in case of a part load state but with using a suitable strategy, the scale of these states can be minimized [1], [5]. As a transmission, the shiftless (CVT) transmission is considered the best for the optimal efficiency. In hybrids using power split system, planetary transmissions are usually used.

For hybrid powertrains, proper energy storage is a key, the used solution depends on the features of the hybrid system. The possibly smallest and lightest battery must be able to handle the quickly changing charging and discharging processes. The dynamic features of the vehicle depend on the power density, the emission-free range depends on the energy density of the battery [1].

4. Hybrid drive modes

Depending on the use and torque demand the ICE and the electric motor usage may vary. The hybrid system ECU determines the torque distribution between the two power sources, considering the throttle pedal angle, the battery state of charge and data provided by the other systems of the vehicle.

The hybrid strategy determines the sequence of the drive modes. The purpose of the hybrid strategy is to operate the vehicle using the optimal drive combination. This could mean minimized fuel consumption or the best dynamics. The strategy is aligned to the operating states of the internal combustion engine, for example evading the less economical part load state. Depending on the system layout, the power output of the ICE can be different or the same as the momentary power need of the vehicle [1]. In the age of advanced driver assistance systems, GPS-technology and neural systems, development of predictive hybrid strategies are a possibility. These strategies are not only able to adapt to the driver, but they can also generate an optimal hybrid strategy by prediction using the data from the GPS and ADAS. Predictive strategies can reduce the fuel consumption and the swing of the battery state of charge, causing the potential reduction of the battery pack size [6].

5. Factors affecting fuel consumption

In case of an everyday car only the 12-30 % of the energy of the fuel is utilized to move the car, depending on the driving cycle. This proportion is 21-40 % in a hybrid vehicle, based on the city (FTP-75) and highway (HWFET) cycles used by EPA [7]. Overall energy losses are around 65-69 %. In Otto-engines, more than 62 % of the energy of the fuel is lost. Internal combustion engines have an extremely low efficiency altering the chemical energy of the fuel to mechanical energy, caused by friction, heat and pumping losses [8], [9].

Parasitic losses are around 4-6 %. It covers the energy required to power the water pump, fuel pump, oil pump, ignition, ECU, etc. Idle losses are negligible. While driving in a city a tremendous amount of energy is lost in idle state. This loss can be reduced by using a Start-Stop system. Auxiliary losses - air conditioning, power steering, headlights, etc. also use the energy produced by the internal combustion engine - are in the range of 0-3 %. The transmission and the other parts of the drivetrain cause losses as well. The magnitude of them is around 3-5 % Modern solutions such as continuously variable transmissions and double clutch gearboxes can adequately reduce these losses.

Aerodynamic drag is altogether 10-15 %. A rolling vehicle uses energy to conquer the resistance surrounding the medium. The energy used to that increases progressively with the speed growth. The aerodynamic drag is closely connected to the shape of the car, the shrinkage of the end-surface and the shape factor causes smaller energy consumption [10]. Rolling resistance is a force, which is needed to roll the wheels, and is directly proportional to the wheel load. It is approximately 10 %. Optimized tire tread area and use of materials can reduce rolling resistance effectively. The magnitude of overcoming inertia and brake loss could reach even 8-12 %.Energy recoverable by regenerative braking is around 5-9 %. Regarding energy efficiency, the amount of energy retrievable from the acceleration and moving is an important question [11, 12, 17, 18, 19]. Energy can be regenerated via braking, this time the electric motor used to drive works as a generator, charging the batteries and its torque can be used to slow the vehicle down [7, 8, 13, 14, 15]

6. Description of the measurement

The test vehicle uses the Honda IMA (Integrated Motor Assist), which is a parallel, mild hybrid system. Main parameters of the vehicle are presented in Table 1. The internal combustion engine is a 1,5 l, 4-cylinder power equipment, with variable valve timing. As a transmission, serves a 6-gear manual gearbox. Between these two a brushless DC motor is placed, which also serves as a starter, rated at 10 kW. The system output is 90 kW. The battery is a Li-ion pack, with a capacity of 0,6 kWh, consisting of 84 cells, with a voltage of 12V individually [11], [12], [13].

Table 1 Technical specifications of the measured vehicle [16]

Engine	1.5 i-VTEC	
Displacement (cm^3)	1499	
Bore (mm)	73	
Stroke (mm)	89.4	
Maximum Power (kW (LE)) /revolutions (1/min)	83(114) /6100	
Maximum Torque (Nm)/revolutions(1/min)	145/4800	
Compression ratio	10.4:1	
Maximum speed(km/h)	200	
Acceleration (0-100 km/h)	9.9	
Empty weight (kg)	1147	
Fuel consumption (l/100km)	city	4.4
	highway	6.1
	mixed	5.0

The way of the measurement was the following: the vehicle's cruise control was set at 110km/h speed, and we tested it in ECON and Normal modes. Measured on-board consumers were air condition, headlights and audio system.

IOP Conf. Series: Materials Science and Engineering **568** (2019) 012051 doi:10.1088/1757-899X/568/1/012051

The decrease of the range estimated by the vehicle on a 2 km long highway distance has been investigated. In every mode, we carried out 2 measurements with every consumer, and we measured 2 reference values with every consumer turned off. This data and the fuel consumption estimated by the vehicle gives the chance to calculate the added energy need of each consumer. When determining the performance of the environmental forces, I choose to consider the highway terrain flat, because determining the exact elevation angles would have gone far beyond this dissertation.

Analyzing the measurement, the fuel consumption increase caused by the consumers is not high for each but when summed, it is considerable. The effect of the air conditioning considering the inaccuracy of the own sensors of the vehicle is numerous, 0,2-0,3 l/100 km. Heating generated only 0,05 l more consumption than prior. The fuel consumption raising the effect of the radio and the lights are marginal according to the results. The environmental effects are severally significant [6, 18]. As a result of the distortions caused by the terrain, in Normal mode while all the consumers were turned off, the fuel consumption raised to 6,4 l/100 km. The fuel consumption on the plain road was 4,5-5 l/100 km, compared to this, the terrain can cause up to 30 % difference.

Table 2 Fuel consumption differences caused by the additional consumers

consumer	fuel consumption difference - ECON (l/100km)	fuel consumption difference - Normal (l/100km)
reference value	0	0
AC	0.3	-0.2
heating	0.35	-0.1
lights	-0.1	-0.8
audio center	-0.35	-1.05
all consumers	1	-0.7

7. Evaluation of the measurement

The fact that the board computer rounds the data to whole caused inaccuracy in the measurement of the range decrease. The range estimation of the own system of the car was accurate in half of the cases. When no consumers or low-powered consumers were turned on, the range decrease was only 2 km during the 4 km long measurement, and only twice did the range decrease by 5 km, so it can be said, that the board computer of the car miscalculates to the safety side.

The effective efficiency results were lower than expected, the performance of the resistance forces was 21 % of the engine performance. The calculated productive efficiency is in the range of 17-30 %, but if the measurement is carried out on a longer distance, it will be more representative, because the distortion caused by the environmental conditions will be negligible. The efficiencies for confident consumers were in the range of 22-27 %. In ECON mode the vehicle operated with 3 % better efficiency than in normal mode.

At ECON mode the use of the air conditioning worsened the efficiency by 1,6 %, so did the heating. Regarding the audio system and the lights, the efficiency was roughly the same. When turning on all consumers, the efficiency decreased by 5 %. At Normal mode the measurement when all consumers were turned off was distorted by an ascent. The use of air conditioning caused a 2,9-4,2 % decay in efficiency compared to the headlights and audio system.

8. Conclusions

Predictability of fuel and energy consumption effect of vehicle onboard consumers is the basis of the hybrid strategies of autonomous vehicles. In our work the necessary tests were done to measure different consumers effect on fuel consumption. The results showed that the use/non-use of these devices has a remarkable effect on vehicle fuel consumption. From the investigated equipment's the highest influence had the air conditioning on the fuel consumption of the hybrid test vehicle. The magnitude of the onboard consumer's influence on energy consumptions is a potential tool for autonomous vehicles to influence the driving distance and the used energy mix.

Acknowledgments
The research was supported by the Hungarian Government and co-financed by the European Social Fund through the project "Talent management in autonomous vehicle control technologies" (EFOP-3.6.3-VEKOP-16-2017-00001).

References

[1] Emőd I, Tölgyesi Z and Zöldy M 2006 *Alternative Vehicle Propulsion systems* (in Hungarian: Alternatív Járműhajtások), Maróti Plc, Budapest
[2] Berdman 2011 *History of Hybrid Cars*, https://www.hybridcars.com/history-of-hybrid-vehicles/
[3] Varga Z and Szauter F 2014 *Alternative und future drivetrains* (In Hungarian: Jövőbeli és Alternatív Járműhajtások), TÁMOP 2.1.4, Győr
[4] Hofmann P 2014 *Hybridfahrzeuge*, Springer, ISBN 978-3-7091-1780-4
[5] Polak F 2018 *E-REV's Hybrid Vehicle Range Modeling*, Journal of KONES Powertrain and Transport, Vol. 25, No. 2 pp. 281-286 DOI: https://doi.org/10.5604/01.3001.0012.2814
[6] Szalay Z, Tettamanti T, Esztergár-Kiss D, Varga I and Bartolini C 2018 *Development of a Test Track for Driverless Cars: Vehicle Design, Track Configuration, and Liability Considerations*, Per. Pol. Transportation Engineering, 46(1), pp. 29-35. doi: https://doi.org/10.3311/PPtr.10753
[7] Rhodes K, Kok D, Sohoni P, Perry E 2017 *Estimation of the Effects of Auxiliary Electrical Loads on Hybrid Electric Vehicle Fuel Economy*. SAE Technical Paper 2017-01-1155, https://doi.org/doi:10.4271/2017-01-1155.
[8] Široký J, Schroder S and Gašparík J 2017 *Comparison of operational and economic aspects of direct road transport and continental combined transport*. Communications - Scientific Letters of the University of Zilina. 19. 109-115.
[9] Zöldy M and Zsombók I 2019 *Influence of External Environmental Factors on Range Estimation of Autonomous Hybrid Vehicles*. System Safety: Human - Technical Facility - Environment, 1(1), pp. 472-480. Retrieved 6 Apr. 2019, doi: https://doi.org/10.2478/czoto-2019-0061
[10] Zöldy M and Zsombók I 2018 *Modelling fuel consumption and refueling of autonomous vehicles*, MATEC Web Conf. 235 00037 DOI: https://doi.org/10.1051/matecconf/201823500037
[11] Mitran T, Chioreanu N, Chioreanu C, Rus Al, Beles H 2017 *A Propulsion System for Means of Transport with Non-autonomous Electrical Traction* CONAT 2016: International Congress of Automotive and Transport Engineering, Pages: 491-496, DOI: 10.1007/978-3-319-45447-4_54, Published: 2017. Publisher: Springer Int Publishing AG, Switzerland, ISBN: 978-3-319-45447-4
[12] Thomas J 2014 *Drive Cycle Powertrain Efficiencies and Trends Derived from EPA Vehicle Dynamometer Results*. SAE Int. J. Passeng. Cars - Mech. Syst. 7(4):2014, doi: https://doi.org/10.4271/2014-01-2562.
[13] Carlson R, Wishart J and Stutenberg K 2016 *On-Road and Dynamometer Evaluation of Vehicle Auxiliary Loads*. SAE Int. J. Fuels Lubr. 9(1):2016, doi:https://doi.org/10.4271/2016-01-0901
[14] Chioreanu N, Mitran T, Rus Al, Beles H 2014 *Numerical Simulation of Thermal-Hydraulic Generators Running in a Single Regime*. Central European Journal of Engineering, Volume 4, Issue 2, June 2014, Pages: 119-124. Publisher: Springer-Verlag Wien, DOI: 10.2478/s13531-013-0145-2, ISSN: 18961541
[15] Baalen J V 2005 *Optimal Energy Management Strategy for the Honda Civic IMA*, Master's Thesis, University of Eindhoven, DCT 2005.148
[16] Honda 2011 CR-Z Powertrain, Honda North America, 1919 Torrance Boulevard, Torrance, CA 90501 http://www.hondanews.com/releases/2011-honda-cr-z-powertrain?l=en-US
[17] https://en.wikipedia.org/wiki/Honda_CR-Z
[18] Shah F 2013 *Delving into the details* http://paultan.org/2013/02/22/delving-into-the-details-interview-with-norio-tomobe-large-project-leader-of-the-new-facelifted-honda-cr-z/
[19] Török Ad, Török Ar and Heinitz F 2014 *Usage of Production Functions in the Comparative Analysis of Transport Related Fuel Consumption*. Transport and Telecommunication Journal. 15(4), pp. 292-298. https://doi.org/10.2478/ttj-2014-0025

IOP Conf. Series: Materials Science and Engineering **568** (2019) 012052 doi:10.1088/1757-899X/568/1/012052

Study regarding the influence of passive safety systems on the occupant in the case of Automatic Emergency Braking System activation

A I Radu, D D Trusca, G R Toganel and B Benea
University "TRANSILVANIA" of Brasov, Department of Automotive and Transport, Brasov, str. Politehnicii 1, Romania

Alexandru.radu@unitbv.ro

Abstract. The aim of this paper was to analyze the injury level of the occupants of a vehicle in the case using the AEB system to reduce the velocity of a front-back collision by utilizing virtual simulations in PC-Crash software. This was achieved by conducting simulation at different velocities and 3 passive safety systems configurations. The velocities tested were 70, 75 and 80 km/h to observe the speed reduction of the vehicle in a collision by the AEB system. Also, the occupant was configured in 3 types of scenarios, when the occupant does not have the seatbelt or airbag, when only the seatbelt is used and when both seatbelt and airbag are activated. The results revealed a great speed reduction by the AEB system, reducing the impact velocity and also the occupant presented a reduced injury risk when both passive safety systems are active.

1. Introduction

The AEB (Autonomous Emergency Braking) is a system introduced in 2011 by Volvo that utilises an array of sensors, radars and cameras in order to avoid a collision by braking the vehicle automatically [1]. After the system was introduced, more and more automakers install these type of systems on their vehicles in order to actively assist drivers in avoiding hazardous traffic situations [2]. The first 2 generations of the system introduced by Volvo were able to detect pedestrians and partially brake the vehicle with a deceleration of 5 m/s2 [3]. Studies show that the AEB system can reduce the speed before impact with 10% and 30% the risk of fatal injuries [4]. The AEB system is very useful and will be used in future fully autonomous vehicles where there will not be needed a human driver [5]. A statistic study shows that front-back collisions count as 63% of the total accidents on the road [6]. The AEB can be inserted into the ACC (Adaptive Cruise Control system). This system is also successful in assisting the driver in avoiding collisions on the highway by maintaining a reasonable distance between the vehicle and the vehicle in front at velocity [7]. The main core of a system designed of mitigating a collision is a situational awareness and risk assessment in case of an obstacle and predict a minimal distance necessary for stopping/or avoiding the collision [8, 9, 10]. This paper wants to analyze the functionality of the AEB system using the PC-Crash module software by simulating a vehicle to vehicle collision and also a MADYMO occupant model to simulate the movement inside the vehicle compartment.

In order to assess the injury level of the occupant, the head injury criteria were used. This is calculated using the following formula [11]:

$$HIC = \left\{ \left[\frac{1}{t_2 - t_1} \int_{t1}^{t2} a \cdot dt \right]^{2.5} \cdot (t_2 - t_1) \right\}_{MAX} \qquad (1)$$

IOP Conf. Series: Materials Science and Engineering **568** (2019) 012052 doi:10.1088/1757-899X/568/1/012052

Where a is the head acceleration, measured in m/s2 and t1 and t¬2 , measured in seconds, represent the interval time in between the HIC formula is calculated and has the maximum value. Also it is to mention that HIC is dimensionless.
HIC values can be correlated with the AIS (Abbreviated Injury Scale) in order to assess the injury level of the occupant. The AIS scale is a system of assessing the injury potential of the occupant or pedestrian in the form of category level of injury, ranging from 1 to 6, where 1 represents minor injury and 6 represents death. The scale is represented in figure 1 [12].

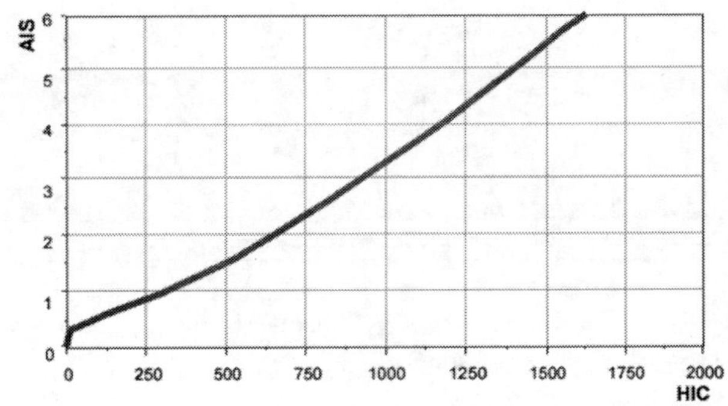

Figure 1. Correlation between HIC and the AIS scale.

2. Methods
The aim of the study was to analyze the active and passive systems of the vehicle to determine their influence on the injury level of the occupant by simulating a front-rear collision scenario at different speeds in which the autonomous braking system is activated using the PC Crash program and the MADYMO module. The study speeds were 70 km/h, 75 km/h and 80 km/h respectively. The chosen speeds are in line with the EuroNCAP for AEB tests, where the maximum speed is 80 km/h [13].
The test scenario consisted of the positioning of 2 vehicles at a distance of 20 m, one is stationary, and the second being imposed speed of movement. The scenario is shown in figure 2.

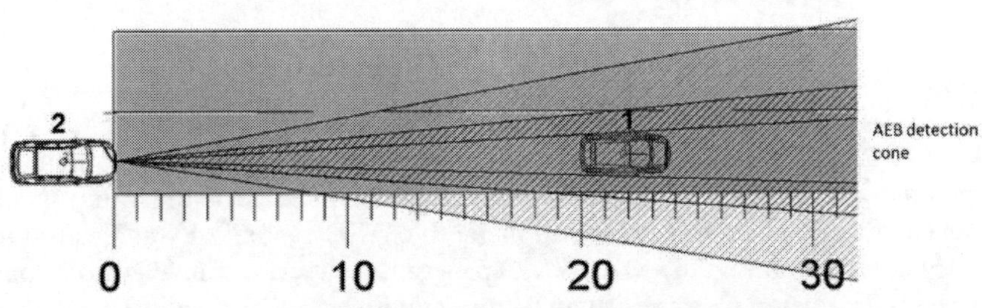

Figure 2. AEB simulation scenario in PC Crash.

In the simulation, vehicle 1 was stationary, positioned 20 m from the vehicle 2 which was in motion. The green detection cone, which is the basis for the operation of the AEB braking system, has been determined. The green cone is the hazard identification radar to inform the driver of imminent threats. The blue cone is the hazard detection and initial braking, and the red cone is secondary braking to avoid a collision. The AEB program uses a logical system scheme` to configure the braking system autonomy. Various parameters for modifying AEB system behavior can be configured in the operational schematic. The primary element of the system is the distance sensor based on simulation data (vehicle speed and distance from the front obstacle). Thus, using said data, the system decides the degree of braking of the vehicle by issuing the signal to the braking system. At the same time, it issues the audible warning signal to the driver to warn him of the imminent danger. In figure 4 the variation of the study speeds of the vehicle is presented under the conditions in which the AEB system is active.

3. Results

The initial result is the velocity variation for the tested scenarios as presented in the next figure.

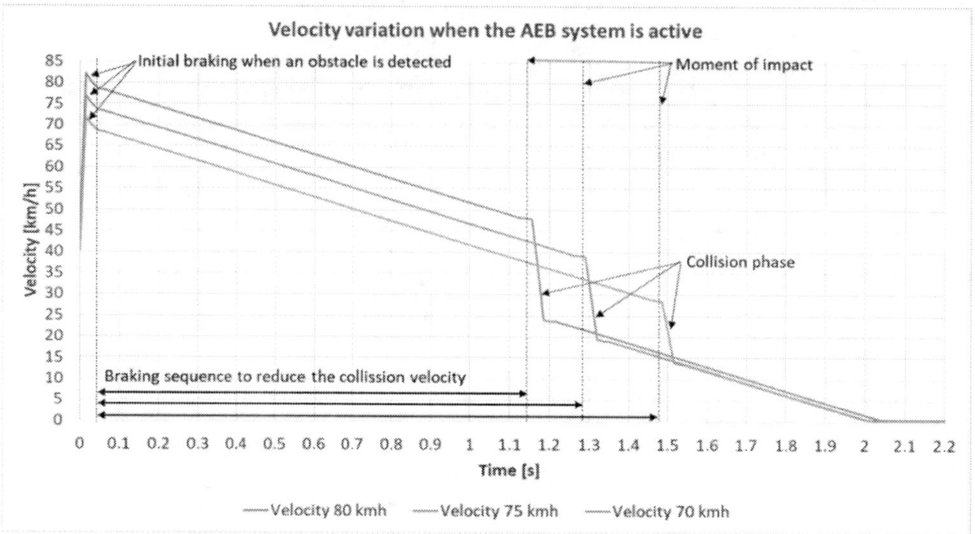

Figure 3. Variation of vehicle speed with active AEB system.

In the first phase, due to the small distance between the vehicles, the AEB system initiates a slight brake due to the detection of the first vehicle, followed by secondary braking to reduce the collision speed.

In table 1 the study rates and the percentage reduction of the collision velocities for the respective cases are presented

Table 1. Percentage reduction of collision rate depending on the speed of study.

Initial velocity [km/h]	Braking time [s]	Collision velocity [km/h]	Speed reduction [%]
70	0.5 – 1.48	28	60
75	0.5 – 1.3	40	47
80	0.5 – 1.15	50	38

The data obtained showed that the AEB system reduced the collision speed by up to 60%, from an initial speed of 70 km/h to 28 km/h. Following the trend in the operation of the AEB system, it was noted that the possibility of reducing the impact speed decreases with the increase in the initial speed of the vehicle. After obtaining the front-end impact data with the active AEB system, it was desired to determine the occupant's injury level by running the MADYMO program and performing 3 simulations for each study speed. In figure 5 the collision phases with an occupant without passive safety systems is presented.

Figure 4. Phase collisions of vehicles with MADYMO occupant without passive systems.

Phase 1 represents the time before the collision, where the occupant was in the normal seat position. Phase 2 is the collision phase where the occupant came in contact with the vehicle dashboard. In Phase 3, the vehicles were detached with the occupant's body rebound towards the seat.
In the next phase simulations were performed using the MADYMO module in which three situations were analyzed:

- The occupant is not retained in the seat;
- The occupant is retained by the seat belt with the pre-tensioning system;
- The occupant is retained by the seat belt and the airbag.

Data from the comparative analysis served to determine the degree of injury by calculating the head injury (HIC) criterion. Thus, in figure 6 the HIC36 diagram for the established situations is presented. In table 3 the maximum HIC criteria for the studied cases are presented.

Table 2. HIC for the tested velocities.

Velocity [km/h]	Passive safety systems (airbag and seatbelt)	Maximum HIC value
70		110
75	NO	150
80		215
70		9
75	YES	3
80		19

Based on the calculated maximum values of the HIC criterion, they correlated with the AIS injury scale. Since the HIC values are relatively small (less than 200), on the AIS scale it is considered minor injury (AIS-1). In figure 7 the probability of minor injury (AIS-1) was calculated.
From the diagram analysis we can see that the probability of injury increases with the speed of the vehicle in the absence of passive safety systems. In table 3 the probability injury is presented.

IOP Conf. Series: Materials Science and Engineering **568** (2019) 012052 doi:10.1088/1757-899X/568/1/012052

Table 3. Likelihood of occurrence of injury depending on the value of the HIC criterion.

Velocity [km/h]	Passive safety systems	Probability of minor injury (AIS -1) [%]	Reduction of the injury level by using passive safety systems [%]
70		37	0
75	NU	50	0
80		72	0
70		3	34
75	DA	1	49
80		6	66

The results of the study highlighted the effectiveness of safety systems in reducing head injury by up to 66% at 80 km/h.

4. Conclusions

The study of the influence of the autonomous braking system (AEB) has been concluded with positive results, so there was a significant reduction by up to 60% of the collision speed in the case of the front-and-rear impact. A tendency to increase the likelihood of injury of the AIS-1 occupant was observed, in the absence of passive safety systems, with increasing travel speed, from 37% to 72%. If passive safety systems (belt and airbag) are present, the probability of injury falls below 7%.

5. References

[1] Krafft M, Kullgren A, Lie A, Strandroth J and Tingvall C 2009 Conf. on Enhanced Safety of Vehicles
[2] Levinson J, Askeland J, Becker J, Dolson J, Held D, Kammel S, Kolter JZ, Langer D, Pink O, Pratt V and Sokolsky M 2011 IEEE Intelligent Vehicles Symposium (IV)
[3] Kaempchen N, Schiele B and Dietmayer K 2009 IEEE Transactions on Intelligent Transportation Systems
[4] Vahidi A, Eskandarian A. Research advances in intelligent collision avoidance and adaptive cruise control. IEEE transactions on intelligent transportation systems. 2003 Sep;4(3):143-53.
[5] Labayrade R, Royere C, Aubert D. Experimental assessment of the rescue collision-mitigation system. IEEE transactions on vehicular technology. 2007 Jan;56(1):89-102.
[6] Polychronopoulos A, Tsogas M, Amditis AJ, Andreone L. Sensor fusion for predicting vehicles' path for collision avoidance systems. IEEE Transactions on Intelligent Transportation Systems. 2007 Sep;8(3):549-62.
[7] McHenry BG. Head injury criterion and the ATB. ATB Users' group. 2004:5-8.
[8] Prasad P, Mertz HJ. The position of the United States delegation to the ISO Working Group 6 on the use of HIC in the automotive environment. SAE Transactions. 1985 Jan 1:106-16.
[9] Jörg, He. Objective Testing of Autonomous Emergency Braking Systems for the EuroNCAP AEB rating. VEHICO GmbH. 2016.
[10] Oana O, Nicolae D, Ilie D 2017 IOP Conference Series: Materials Science and Engineering Vol. 252, No. 1, p. 012017
[11] Popa V, Beles H, Tolea B, Dragomir G and Mitran T 2016 Applied Mechanics and Materials Vol. 822 Pages: 512-522
[12] Popa V, Beles H, Tolea B 2014 Annals of the University of Oradea Fascicle of Management and Technological Engineering Pages: 87-92
[13] EuroNCAP 2013

IOP Conf. Series: Materials Science and Engineering **568** (2019) 012053 doi:10.1088/1757-899X/568/1/012053

The study of the car's stability using a simplified model

L Simniceanu

Faculty of Mechanical Engineering, University of Craiova, Romania

lsimniceanu@yahoo.com

Abstract. This paper presents some research on vehicle stability, taking into account a simplified mechanical model of cars and analyzing the stability of vehicle movement using dynamic systems theory. The paper proposes a new mathematical model for the kinematic and dynamic study of the behavior of two-axle vehicles, a model using a nonlinear link between the lateral deflection of the tire and the external forces applied to it. The mechanical model is considered a model with two degrees of freedom, taking into account the following two variables: angular velocity and linear velocity in the transverse direction. To analyze the dynamic behavior of the vehicle, two displacement situations are considered: the first case in a case of displacement with a constant turning angle and the second considered case is the case with the sinusoidal variation of the wheel turning angle. Numerical solutions for these driving situations are determined and appreciations are made regarding the character of the movement in terms of stability.

1. Introduction

The automotive moving and its stability are significant problems for automotive constructors, whom permanent watching the customers interests. Those requirements are touching dynamic performances, design, mobile security, amenities. Various mechanical models for the study of dynamic behavior are proposed in the literature [1], [6], [7], [8], mechanical models with one to six degrees of freedom.

With this paper we purpose to realize a study over automotive stability using the dynamic systems theory [9], under a simplified form of plan automotive model.

2. The simplified mechanical and mathematical model for vehicles plane moving the study

The mathematical models used in the vehicle dynamics study contain a number of smaller or larger variables [2], [3], [4] in accordance with the relative magnitude in the dynamics analyzed. In this study the next degree of freedom is taking into account:

- x, define the displacement along longitudinal axle;
- y, define the displacement along transversal axle;
- φ, define the circular displacement of suspended mass around Oz axle;

Towards to obtain a simplified mechanical and mathematical model, who can be resolved both using a computer and analytical method, some simplified hypothesis may be considerate:

- The vertical forces reactions abide constant, this means that no redistribution of those are occurring by reason of rolling and pitching oscillations;
- The rolling and pitching displacement are neglected;
- No longitudinal forces are acting on tires (the free tire case);

2.1. Simplified plan model

A simplified plan model of the vehicle is presented in Figure 1, and the next notations are used:
a, b – are the coordinates of mass center point towards front axle, respectively rear axle;
F1, F2 – is the longitudinal forces of the contact area between tire and road;
Y1, Y2 – are the lateral forces of the contact area between tire and road;
δ_1 - is the cornering angle;
α_1, α_2 - are the sleep angles for the front axle, respectively rear axle (Figure 1).

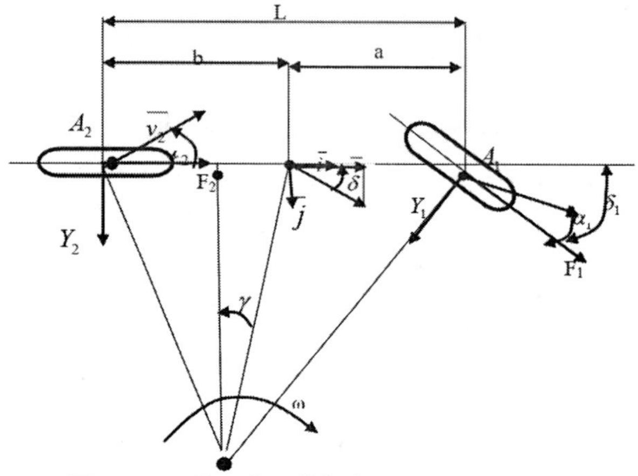

Figure1. The simplified model of vehicle

2.2. The mathematical model

The velocity of points A1, A2 are:

$$\vec{v}_{A_1} = \vec{v}_C + \vec{\omega} \times \overrightarrow{CA_1} = v_x \cdot \vec{i} + v_y \cdot \vec{j} + \begin{vmatrix} \vec{i} & \vec{j} & \vec{k} \\ 0 & 0 & \omega \\ a & 0 & 0 \end{vmatrix} \tag{1}$$

where ω is the yaw angular speed
v_x – is the longitudinal velocity of the vehicle
v_y – is the transverse velocity of the vehicle

$$\vec{v}_{A_2} = \vec{v}_C + \vec{\omega} \times \overrightarrow{CA_2} = v_x \cdot \vec{i} + v_y \cdot \vec{j} + \begin{vmatrix} \vec{i} & \vec{j} & \vec{k} \\ 0 & 0 & \omega \\ -b & 0 & 0 \end{vmatrix} \tag{2}$$

$$v_{1x} = v_{2x} = v \tag{3}$$

$$v_{1y} = v_y + a\omega \tag{4}$$

$$v_{2y} = v_y - b\omega \tag{5}$$

IOP Conf. Series: Materials Science and Engineering **568** (2019) 012053 doi:10.1088/1757-899X/568/1/012053

$$tg(\delta_1 - \alpha_1) = \frac{v_{1y}}{v_{1x}} = \frac{v_y + a\omega}{v} \tag{6}$$

$$\alpha_1 = \delta_1 - arctg\frac{v_y + a\omega}{v} \tag{7}$$

$$tg\alpha_2 = -\frac{v_{2y}}{v_{2x}} = -\frac{v_y - b\omega}{v} \tag{8}$$

$$\alpha_2 = -arctg(\frac{v_y}{v} - \frac{b\omega}{v}) \tag{9}$$

$$\vec{a} = \frac{d\vec{v}}{dt} + \vec{\omega} \times \vec{v}\ ; \quad a_x = \frac{dv_x}{dt} - v_y \cdot \omega\ ; \quad a_y = \frac{dv_y}{dt} + v \cdot \omega \tag{10}$$

The equations for car's moving are:

$$\begin{cases} m\left(\dot{v}_y + v\omega\right) = Y_1 \cos\delta_1 + Y_2 + F_1 \sin(\delta_1) \\ I_z \cdot \dot{\omega} = Y_1 \cos\delta_1 \cdot a - Y_2 \cdot b + F_1 \sin(\delta_1) \end{cases} \tag{11}$$

The lateral forces are given by the next relation:

$Y_{1,2} = k_{1,2} \cdot \alpha_{1,2}$, where $k_{1,2}$ – lateral stiffness coefficient of the tire;

$$\begin{cases} m\left(\dot{v}_y + v\omega\right) = k_1\left(\delta_1 - arctg\left(\frac{v_y + a\omega}{v}\right)\right)\cos\delta_1 + k_2 \cdot arctg\left(\frac{v_y - b\omega}{v}\right) + F_1 \sin(\delta_1) \\ I_z \cdot \dot{\omega} = k_1\left(\delta_1 - arctg\left(\frac{v_y + a\omega}{v}\right)\right)\cos\delta_1 \cdot a - k_2 \cdot arctg\left(\frac{v_y - b\omega}{v}\right) \cdot b + F_1 \sin(\delta_1) \cdot a \end{cases} \tag{12}$$

For small value of angular lateral deformation we can make the next approximation:

$$\begin{gathered} arctg\frac{v_y + a\omega}{v} \cong \frac{v_y}{v} + \frac{a}{v} \cdot \omega \\ arctg\frac{v_y - b\omega}{v} \cong \frac{v_y}{v} - \frac{b}{v} \cdot \omega \\ \cos\delta_1 \approx 1 \end{gathered} \tag{13}$$

The next system is obtained:

IOP Conf. Series: Materials Science and Engineering **568** (2019) 012053 doi:10.1088/1757-899X/568/1/012053

$$\begin{cases} \dot{v}_y = -\dfrac{k_1+k_2}{mv}\cdot v_y - \left(\dfrac{k_1a-k_2b}{mv^2}+1\right)\cdot\omega + \dfrac{k_1}{mv}\cdot\delta_1 \\ \dot{\omega} = -\dfrac{k_1a-k_2b}{I_z\cdot v}\cdot v_y - \dfrac{k_1a^2+k_2b^2}{I_z\cdot v}\cdot\omega + \dfrac{k_1a}{I_z}\cdot\delta_1 \end{cases} \tag{14}$$

In the case of constant velocity, a non-homogeneous system with constant coefficients is obtained. Using the matrices form, the system may be written:

$$\{\dot{x}\}+[A]\{x\}=\{B\}, \tag{15}$$

with: $\{x\}=\{v_y,\omega\}^T$ - is the velocity vector

$\{\dot{x}\}=\left\{\dot{v}_y,\dot{\omega}\right\}^T$ - is the acceleration vector

$$[A]=\begin{bmatrix} \dfrac{k_1+k_2}{mv} & \dfrac{k_1a-k_2b}{mv^2}+1 \\ \dfrac{k_1a-k_2b}{I_z} & \dfrac{k_1a^2+k_2b^2}{I_zv} \end{bmatrix} \qquad \{B\}=\begin{bmatrix} \dfrac{k_1}{mv}\cdot\delta_1 \\ \dfrac{k_1a}{I_z}\cdot\delta_1 \end{bmatrix} \tag{16}$$

If the variation of the cornering angle is known, the analytical solution may be founded. The next case may be taken into account:

_ $\delta_1(t)=\delta_{10}=const$ (this case corresponds with cornering with constant radius)

$$\begin{pmatrix}\delta(t)\\ \omega(t)\end{pmatrix}=\begin{pmatrix}\dfrac{k_1}{mv}\delta_{10}\cdot t\\ \dfrac{k_1a}{I_z}\delta_{10}\cdot t\end{pmatrix}-\begin{bmatrix} \dfrac{k_1+k_2}{mv} & \dfrac{k_1a-k_2b}{mv^2}+1 \\ \dfrac{k_1a-k_2b}{I_z} & \dfrac{k_1a^2+k_2b^2}{I_zv} \end{bmatrix}\cdot\begin{pmatrix}\dfrac{t^2}{2}\cdot\dfrac{k_1}{mv}\delta_{10}\\ \dfrac{t^2}{2}\cdot\dfrac{k_1a}{I_z}\delta_{10}\end{pmatrix}$$
$$+\frac{1}{2}\cdot\begin{bmatrix} \dfrac{k_1+k_2}{mv} & \dfrac{k_1a-k_2b}{mv^2}+1 \\ \dfrac{k_1a-k_2b}{I_z} & \dfrac{k_1a^2+k_2b^2}{I_zv} \end{bmatrix}^2\cdot\begin{pmatrix}\dfrac{t^3}{3}\cdot\dfrac{k_1}{mv}\delta_{10}\\ \dfrac{t^3}{3}\cdot\dfrac{k_1a}{I_z}\delta_{10}\end{pmatrix}+\ldots \tag{17}$$

The homogeneous system solution has the form:

$$\{x\}=e^{-[A]t}\cdot\{c\} \tag{18}$$

$\{c\}=\{c_1 \quad c_2\}^T$ – which depend on the initial conditions,

For the non-homogeneous system the solution is:

$$\{x\}=e^{-[A]t}\cdot\{c\}+\int_0^t e^{-[A](t-\tau)}\cdot B(\tau)d\tau \tag{19}$$

$$\begin{pmatrix}\delta\\ \omega\end{pmatrix}=e^{-A[t]}\cdot\begin{pmatrix}c_1\\ c_2\end{pmatrix}+\int_0^t e^{-[A](t-\tau)}\cdot\begin{pmatrix}\frac{k_1}{mv}\cdot\delta_1(\tau)\\ \frac{k_1 a}{I_z}\cdot\delta_1(\tau)\end{pmatrix}d\tau \tag{20}$$

For the initial conditions $\delta=0$ and $\omega=0$, $c_1=0$ and $c_2=0$ are obtained and using the serial extension of e^{-At} function, the system solution is obtained:

$$\begin{pmatrix}\delta\\ \omega\end{pmatrix}=\int_0^t\left\{\begin{bmatrix}1&0\\0&1\end{bmatrix}-\frac{1}{1!}A(t-\tau)+\frac{1}{2!}A^2(t-\tau)^2-\frac{1}{3!}A^3(t-\tau)^3+\dots\right\}\cdot\begin{pmatrix}\frac{k_1}{mv}\delta_1(\tau)\\ \frac{k_1 a}{I_z}\delta_1(\tau)\end{pmatrix}d\tau \tag{21}$$

The analytical solution is obtained by knowing the law of variation of the angle of rotation. It is considered the variation of the steering angle has $\delta_1(t)=A\sin(\omega\cdot t)$ form and the next solution is obtained:

$$\begin{pmatrix}\delta(t)\\ \omega(t)\end{pmatrix}=\begin{pmatrix}\frac{k_1\cdot A}{mv\omega}\cdot(-\cos(t\cdot\omega)+1)\\ \frac{k_1 a\cdot A}{I_z\cdot\omega}\cdot(-\cos(t\cdot\omega)+1)\end{pmatrix}-\begin{bmatrix}\frac{k_1+k_2}{mv} & \frac{k_1 a-k_2 b}{mv^2}+1\\ \frac{k_1 a-k_2 b}{I_z} & \frac{k_1 a^2+k_2 b^2}{I_z v}\end{bmatrix}\cdot\begin{pmatrix}\frac{k_1}{mv}\cdot\frac{A}{\omega}\cdot\left(\frac{-\sin(\omega t)}{\omega}+t\right)\\ \frac{k_1 a}{I_z}\cdot\frac{A}{\omega}\cdot\left(\frac{-\sin(\omega t)}{\omega}+t\right)\end{pmatrix}$$
$$+\frac{1}{2}\cdot\begin{bmatrix}\frac{k_1+k_2}{mv} & \frac{k_1 a-k_2 b}{mv^2}+1\\ \frac{k_1 a-k_2 b}{I_z} & \frac{k_1 a^2+k_2 b^2}{I_z v}\end{bmatrix}^2\cdot\begin{pmatrix}\frac{k_1}{mv}\cdot\frac{A}{\omega^3}\cdot\left(2\cos(\omega t)+\omega^2t^2-2\right)\\ \frac{k_1 a}{I_z}\cdot\frac{A}{\omega^3}\cdot\left(2\cos(\omega t)+\omega^2t^2-2\right)\end{pmatrix}+\dots \tag{22}$$

In Figure 2 are presented the numerical solutions in the case of initial velocity v=25 km/h and the sinusoidal trajectory with a variation of the angle having the next form $\theta(t)=15\cdot\sin(\frac{\pi}{2}\cdot t)$.

The numerical solutions are presented for the transverse velocity in Figure 2b and for angular velocity in Figure 2a.

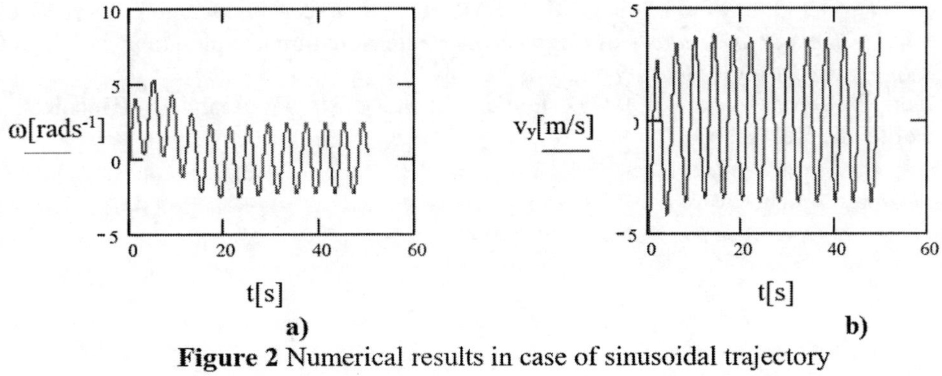

Figure 2 Numerical results in case of sinusoidal trajectory

The space of the phases is represented in the figure and shows the stable character of the motion of the analyzed system

IOP Conf. Series: Materials Science and Engineering **568** (2019) 012053 doi:10.1088/1757-899X/568/1/012053

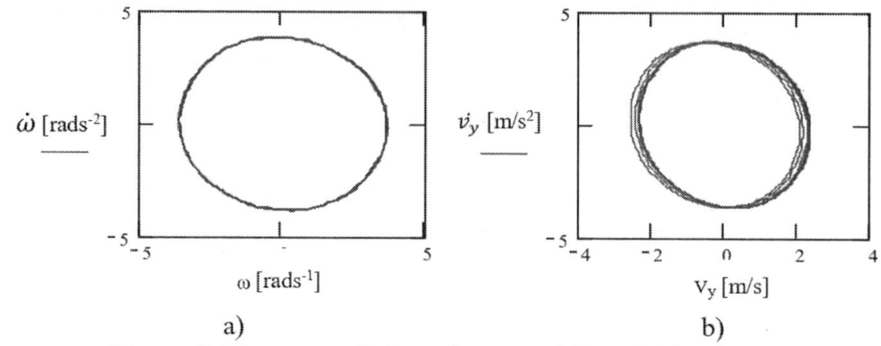

Figure 3. The space of phases in case of sinusoidal trajectory

3. CONCLUSION

This mathematical model is a simple but useful model for vehicle stability analysis, based on the theory of dynamic systems or the theory of classical mechanics. The values obtained are accurate values, compared to the measurements made on a real vehicle [2], having the same parameters as those used in the numerical simulation, so this can be considered a validation for this model.

References

[1] Oţăt V, Bolcu D, Simniceanu L and Thierheimer WW 2005 *Vehicle dynamics (Dinamica autovehiculelor)* (Craiova: Universitaria Publishing Craiova)

[2] Simniceanu L , Otat V, Trotea M and Bogdan M 2015 *Proc. of the European Automotive Congress EAEC-ESFA 2015* (Bucuresti) **vol 1** (Springer, Cham) p 741-47

[3] Otat V, Simniceanu L, Neagoe D, Bogdan M and Dincă L 2008 *Proc. of Sec. int. congress automotive, safety and environment SMAT* (Craiova) **Tome I** (SIAR Publisher) p 213-16

[4] Simniceanu L, Bogdan M, Oţăt V, Trotea M 2017 *Proc. of the 10th International Automotive Congress "Automotive and Environment" CAR 2011* (Pitesti) **Volume 252** (IOP Publishing, IOP Conference Series: Materials Science and Engineering) 012023

[5] Bolcu D and Oţăt V 2002 *The World Automotive Congress FISITA 2002* (Helsinki) paper code: FO2I240

[6] Neagoe D-tru and Bolcu D-tru 2008 *Stability and Manipulability Study of Motor Vehicles. The experimental and theoretical research* (Studiul stabilitatii si maniabilitatii autovehiculelor. Cercetari experimentale si teoretice) (Craiova: Universitaria Publishing, Craiova)

[7] Simniceanu L, Neagoe D-tru, Trotea M, Bogdan M and Constantinescu A 2016 *Applied Mechanics and Materials* **Vol. 822** ed. Trans Thec Publications pp. 54-59

[8] Simniceanu L, Otat V and Trotea M 2015 *Recent Advances in Mechanics, Mechatronics and Civil, Chemical and Industrial Engineering* (Determination to a plan model of the automotives moving used for the stability analysis) ed INASE pp 68 – 71

[9] Voinea P R and Stroe I 1994 *Dynamic systems* (Sisteme dinamice) (Bucuresti: Institutul Politehnic Bucuresti)

IOP Conf. Series: Materials Science and Engineering **568** (2019) 012054 doi:10.1088/1757-899X/568/1/012054

Numerical analysis of the influence of mechanical stress on the battery pack's housing of an electric vehicle

L Scurtu[1], I Szabo[1], F Mariasiu[1], D Moldovanu[1*], L Mihali[2] , Ancuta Jurco[1]

[1]Technical University of Cluj-Napoca, Automotive Engineering and Transports Department, Bd. Muncii 103-105, Cluj-Napoca, Romania
[2]Porsche Engineering Romania, Cluj-Napoca, Romania

dan.moldovanu@auto.utcluj.ro

Abstract. In this paper is presented a crash simulation analysis of the battery pack in order to increase the stiffness when the electric vehicle is involved in traffic accidents. In the first part is presented the state of the art, highlighting the advantages of using electric vehicles. The second part of the study presents the steps required to create the crash analysis of the battery pack. The CAD design of the assembly model is generated using advanced modelling techniques for two simulations battery pack models: the first model has a basic shape geometry and the second model has a shock absorber mounted on the external faces to reduce the crash impact. The crash analysis of the battery pack is determined for three velocity cases: 7 m/s, 14 m/s and 21 m/s. The final part of the paper presents the simulation results and different advantages of the battery pack geometry with the shock absorber.

1. Introduction

The current market for electric vehicles (EV) is no longer a question of speculation and analysis predictions. The introduction in the market in 2017 of the Chevy Bolt and Tesla Model 3 models (with an average range of over 250 km on a single charging of energy source) was the event that marked a new era for the automotive industry. The latest trends from the Scandinavian countries (Sweden, Norway and Denmark) is concerning the ban on the sale of vehicles equipped with internal combustion engines, underlining once again the use of EVs as a future sustainable means of transport.

Generally, if the battery (pack) is damaged during a road accident, it has the potential to cause a fire and/or explosion hazard. This immediate danger is amplified by the increasing capacity of energy storage for lithium-ion cells used in the construction of the battery pack of an EV, and thus the stored large energy can be released suddenly during a vehicle accident (as a thermal effect). Together, automotive manufacturers, battery/energetic sources producers, regulatory agencies and the car insurance domain should be prepared to cope with this growing problem resulting from the increased number of EVs in circulation. This implies the necessity of carrying out studies and researches aimed at identifying the constructive solutions (both at the micro-level of the electrochemical cell and at the macro level of the battery pack in its entirety) that provide the protection required for the battery pack (the energy source) in the case of impact or road accident. Experiments that were done included mechanical and structural stresses on the cell through penetration with a rigid rod, hemispherical head

IOP Conf. Series: Materials Science and Engineering **568** (2019) 012054 doi:10.1088/1757-899X/568/1/012054

penetration, three-point bending and axial and lateral compression, to determine the moment of the onset of the short circuit (detected by tracking cell voltage and temperature). The results of the tests showed that in the case of three-point bending the cell initially presented a linear elastic response (three distinct local peaks on the load-displacement curve were identified). The behavior of large NCO type Li-ion cylindrical cells (SOC = 0 %) was studied in quasi-static mode under three different loading scenarios: compression between two flat plates (lateral crushing), lateral penetration and three-point bending [1]. Observations on the cells after the three-point bending strength test case indicated that the short circuit was caused by a macroscopic fracture occurring in the cell's core due to mechanical stress. From the point of view of research on the effect of structural loads on the battery pack housing at the macro level, fewer studies were published because their geometric shape depends to a large extent on the geometric constraints given by the type of vehicle on which the battery pack will be mounted. A comprehensive analysis of the damage to a battery pack integrated into the vehicle body structure was done in order to develop a general methodology for impact situation [2]. The research results have shown that the right integration of the battery pack in vehicle structure is a key factor in controlling the damage severity of an electric car subjected to a ground impact. The car structural elements such as chassis, bottom protective plates and floor panel provide various levels of protection (function of design and materials used), but they can also endanger the safety level of the battery pack due to deformations caused by mechanical stress.

Full-vehicle crash analyses via finite element simulations were done for several battery pack configurations [3]. The research compared the multifunctional battery system to battery packs with batteries alone, battery packs and battery packs within housing where cellular solids are used as energy absorbers. The general conclusion stated the necessity to use a multifunctional battery system for EVs (damage tolerant and impact energy storage capable). It should be mentioned that there is a major restriction in the construction of the battery packs for EVs, namely a reference parameter in the EV design and use - the battery specific power (250…350 W/kg [4]) - a parameter that directly influences the autonomy of the electric vehicle (due to the total weight of EV). For this reason, materials used for the construction of the housing must provide a high degree of mechanical resistance in the case of structural loads for a minimum weight.

In the article, numerical simulation methods were used to present the effects of structural mechanical stresses (which can occur in the event of an electric vehicle accident) on the battery housing. Several scenarios / cases have been considered regarding the properties of the walls that form the battery housing by modifying the structure, material, and wall thickness.

2. Method

An important aspect of this study is to determine the optimal geometry of the battery pack in order to minimize the damage of the battery cells. In this paragraph are presented all the design and analysis processes used for the crash test of the battery module used for electric vehicles. The start point of the crash analysis is CAD design of the base battery module. The assembly model is loaded into Hypermesh software where applying an advanced technique it is meshed into finite elements. After the preprocessing is done the battery model is simulated using Radioss solver. The CAD model geometry was designed by using SolidWorks modelling technique. The base battery module assembly consists of eight parts for the base model and ten parts for the second model. In figure 1 the overall dimensions and the internal components of the battery's module are presented. The model of the battery assembly is composed from a component that ensures a proper function. For a better and safer assembly between the rows of the battery cell, the box module and the radiator, spacers are made of ABS material.

IOP Conf. Series: Materials Science and Engineering **568** (2019) 012054 doi:10.1088/1757-899X/568/1/012054

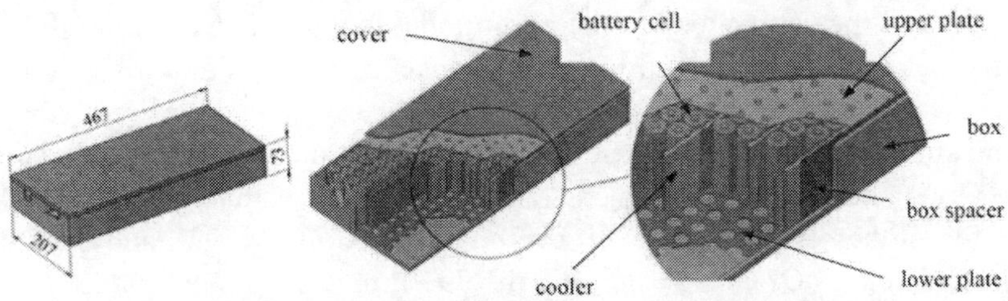

Figure 1. CAD model of battery pack

An important aspect of the electric battery module is for it to fit into a lower volume space and that it must have a lower weight. In the details of figure 1 it can be observed the fixing of the cell battery for a better distribution in the box space. The battery cell is modeled according to the dimensions of the real cell model 18650 type.

3. Finite element simulations

In this section are presented the steps that make possible the finite element simulation of the battery pack in order to view its behaviour in the crash events. Elaboration of a crash study requires three steps: in first step the CAD geometry is imported for cleanup and prepared to mesh into Hypermesh software, the second step consists of assignations of materials and creation of the load case, and in the last step the finite elements model is ran in Radioss solver and the results can be viewed in HyperView.

An important point of this research are the materials model of the battery pack and its components. The material type M2_PLAS_JOHNS_ZERIL is assigned to the components. In LAW 2 this material type is an isotropic elastic-plastic material, expressing the flow stress in a material as a function of strain, strain rate and temperature. For the spacer parts between batteries the isotropic elastic-plastic material type M36_PLAS_TAB is assigned. The stress-strain material curve is given by the value of the strain included in the range 0-16 and the value of the stress included in the range 1-17. After the cleanup geometry process the battery module assembly model has meshed into 374710 shell elements. During the crash and buckling simulation process a general contact interface type 7 is chosen. The shock absorber plate, the impact plate and the box are interconnected by rigid elements RBE2. The crash simulation scenarios for both batteries pack models were done. The initial velocity of the battery's module is set to 7 m/s in the first simulation case, 14 m/s in the second case and 21 m/s in the third case. For a better view of the results, the time of the simulations is set to 15 ms. During this time the battery pack assembly is crashed into a rigid pole and is rejected by the pole. Both simulation assemblies, the base model and the model with shock absorber, are impacted into a rigid pole haing a 100 mm diameter.

4. Simulations and results

In this section are presented the analysis results for all velocity cases considered.

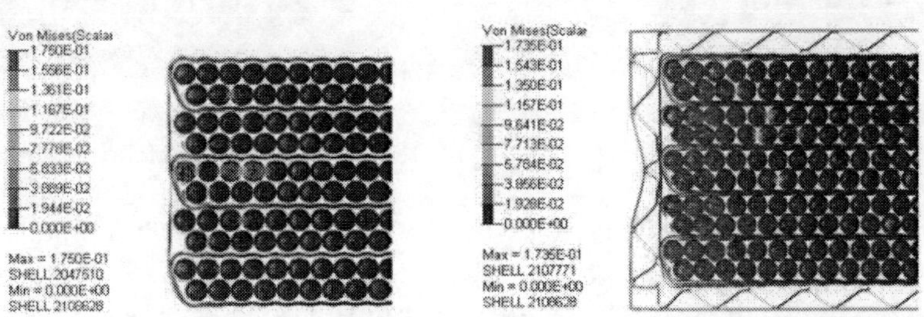

a) Von Mises stress results for the base model b) Von Mises stress results for the shock absorber model

Figure 2. Crash simulation of the first case (impact speed 7 m/s)

Distribution of the von Mises stress for the first simulation case are presented in figure 2, where it can be observed that the cell battery of the base model is not affected during the crash simulation. In figure 2b the shock absorber picks up the crash impact resulting into a lower value of the von Mises stress.

The second simulation case of the base model is presented in figure 3a. In this case the shape of the radiator is deformed due to the crash impact and the four cell batteries are affected. The displacement of the wall affected the cells batteries as it can be observed in figure 3, and the value of displacement for each indicated cell is: C1= 0.924 mm, C2= 1.797 mm, C3= 14.493 mm, C4= 1.697 mm and C5= 2.462 mm.

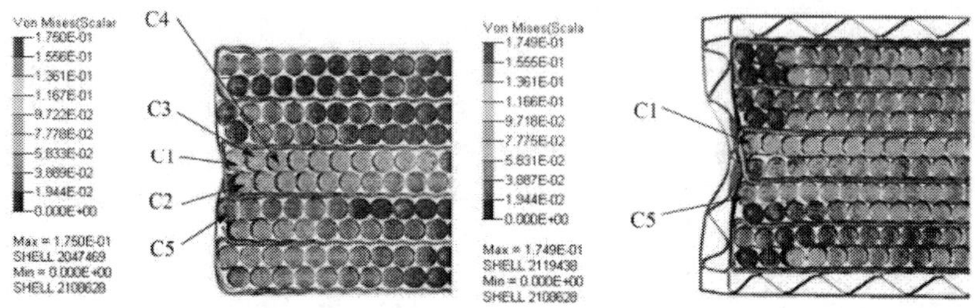

a) Von Mises stress results for the base model b) Von Mises stress results for the shock absorber model

Figure 3. Crash simulation of the second case (impact speed 14 m/s)

In the second case of simulation the impact energy is absorbed by the shock absorber mounted on the impact area of the battery pack. It can be observed that in the simulation of the model with the shock absorber the damage is lower, only two battery cells being affected. The value of the displacement of the wall affected battery is: for the first cell 6.775 mm and for the fifth cell 0.301. Due to the impact the cooler, contact plate and the shock absorber are bent, as it can be seen in figure 3b. In the third case the battery cells of the base model are the most affected as compared to the first two cases. The maximum displacement is located in the middle of the battery pack in the crash area. Displacement of the cell battery wall measured after crash impact is: C1=4.390 mm, C2=2.745 mm, C3= 4.094 mm, C4=2.56 mm, C5=4.588 mm, C6= 2.018 mm, C7= 1.808 mm, C8= 1.651 mm, C9=5.400 mm, C10=1.347 mm, C11=5.545 mm and C12=0.957 mm. The result of the simulation crash with the shock absorber can be observed in figure 4b. Due to the battery pack's shock absorber, the battery cells are less affected. The displacement values of the affected battery cells are (in mm): C1=5.859, C2=2.573, C3=1.966, C5=1.691, C8=0.92, C9=7.449 and C11=1.103.

a) Deformation of the base battery pack at 21 m/s b) Deformation of the base battery pack at 21 m/s with shock absorber

Figure 4. Results of the battery pack model simulation

In figure 4a are presented the deformation of the base battery pack model at the velocity of 21m/s. The kinetic energy curve starts from the initial value of 1410 joules, it remains constant for the first 0.55

milliseconds, after which the battery pack gets in contact with the rigid pole. The kinetic energy drops continuously until 2.7 milliseconds, being close to zero. The battery pack deformation stops when the kinetic energy of the model is absorbed in plastic deformation.

In the case study of the battery pack with shock absorber the crash results are presented in figure 4b, where the maxim deformation of the battery pack with shock absorber at the velocity 21m/s can be observed. The kinetic energy curve starts at 1607 joules, it remains constant for the first 0.55 milliseconds, after which the battery pack gets in contact with the rigid pole. After impact the kinetic energy drops continuously until 3.8 milliseconds, being close to zero. Deformation stops when the kinetic energy of the model is absorbed in plastic deformation, and it gets converted into internal energy.

5. Conclusions

An important advantage of using finite analysis in the design of the automotive industry is the possibility of using the models at many load cases and crash scenarios, without using the physical models. This way the material and human resources are reduced, and the final product satisfies the safety and comfort required conditions. According to the simulation results, it can be observed that the use of shock absorber in the construction of the battery pack improves the safety of the battery cells. In the first simulation case at a velocity of 7 m/s of the battery pack the battery cells are not affected by the crash impact in both simulations. In the second (14 m/s) and third (21 m/s) simulation cases the battery cells are affected by the crash impact but using a shock absorber the battery cells deformation is smaller compared to the simulation of the base model battery packs.

Acknowledgments

This research was funded by the Program "POC-A1-A1.2.3-G-2-15 Partnerships for Knowledge Transfer, no. financing contract: 11 / 01.09.2016, project title: Intelligent Urban Intelligent Vehicle Technologies URBIVEL, ID: P_40_333, MySMIS: 105565, Subsidiary contract nr.6/16.01.2018".

References

[1] Greve, L.; Fehrenbach, C. Mechanical testing and macro-mechanical finite element simulation of the deformation, fracture, and short circuit initiation of cylindrical Lithium-ion battery cells. *J. Power Sources* 2012, *214*, 377–385.

[2] Y. Xia,T. Wierzbicki, E. Sahraei, X. Zhang. Damage of cells and battery packs due to ground impact. Journal of Power Sources, 267, 2014, 78-97.

[3] J. Kukreja, T. Nguyen, T. Siegmund, W. Chen,W. Tsutsui, K. Balakrishnan, H. Liao, N. Parab. Crash analysis of a conceptual electric vehicle with a damage tolerant battery pack. Extreme Mechanics Letters, 9(3), 2016, 371-378.

[4] Hannan, M.A.; Lipu, M.S.H.; Hussain, A.; Mohamed, A. A review of lithium-ion battery state of charge estimation and management system in electric vehicles applications: Challenges and recommendations. Renew Sust Energ Rev, 2017, 78, 834-854.

IOP Conf. Series: Materials Science and Engineering **568** (2019) 012055 doi:10.1088/1757-899X/568/1/012055

Exploring the insights of a consignment stock program implementation in a leagile supply chain

M Faur and C Bungau

Engineering and Management Department of Doctoral School of Engineering Science, University of Oradea, 1 Universitatii, Romania

monica.faur@csud.uoradea.ro

Abstract. Supply chain performance, in terms of total cost, service level and delivery lead-time is an extremely important objective in an organization, being directly linked to inventory management. Therefore, well managed inventory is a main trigger to increasing profitability. In this regard, the present study attempts to explore the readiness of a leagile manufacturing company to embrace consignment stock model, in order to reduce inventory value for raw materials and packaging, as well as to avoid forecast accuracy problems and cash-flow issues. The methodology includes a brief analysis of different types of inventory models, focusing on consignment stock approach and proposes an exploratory case study, represented by a successful implementation of a consignment stock program in a multinational company, having already leagile prerequisites. Using as a basis the insights and learning from the real case, the aim of the study is to develop a business processes framework, accommodated with the IT support software, to serve as a guideline for future implementation of such programs.

1. Introduction

Tensions arising nowadays in most of the companies, regarding cost reductions, lead-times compression and enhanced responsiveness, created by the global competition, resulted in supply chains' efforts to adopt strategies or programs that reduce inventory cost and improve performance. When a company desires to be responsive, having a short lead-time and thereby the ability to rapidly react to market unpredictable requirements, it needs to maintain a high level of inventory [1]. In contrast, the same author argues that in order to operate effectively, inventory levels need to be continuously reduced [1]. Even though several authors agree about the view of a trade-off between responsiveness and efficiency in supply chains [1, 2, 3, 4 5], some of the researchers draw the attention on applying lean manufacturing on the parts of the supply chain that use forecasting to predict the demand and agile practices on the supply chain segments that deal with volatile demand or uncertainty. The use of leanness and agility in different areas of the supply chain is by Naylor et al. (1999) referred to as leagility [6].

The common goal of the lean and agile supply chain strategies is to meet the customers' demands at the lowest cost possible [7, 8]. Lean practices are mostly used to improve effectiveness, while agile tools enable the chain to become more flexible, increasing its adaptability, towards greater responsiveness. By these means, a company goes through a transformation process, being more suitable to further implementation of different programs or initiatives that address certain segments or processes of the value chain, improving overall performance.

IOP Conf. Series: Materials Science and Engineering **568** (2019) 012055 doi:10.1088/1757-899X/568/1/012055

Fisher (1997) lists some main supply chain priorities, regarding responsiveness and efficiency, which companies need to consider for continuous improvement. Among them, minimizing inventories throughout the chain, deploy buffer material stocks and adopt new approaches to reduce lead-time, are of the essence [9]. In manufacturing facilities, the levels of inventories at all stages indicate the company's competitive positioning, so, reducing inventory level and cost, especially for raw materials and packaging is an extremely targeted objective in business environments [10]. There are several options to be chosen in this view. Many firms adopted Just-in–Time (JIT) approach, or Vendor Managed Inventory (VMI) programs, while others implemented Consignment Stock (CS) policy, Among the above mentioned inventory management models, VMI has got the widest attention in the academic literature [11, 12], while quite little amount of literature was dedicated to CS [13, 14].

2. Research methodology

The present research methodology includes a brief analysis of different types of inventory models, focusing on consignment stock approach and proposes an exploratory case study, represented by a successful implementation of a consignment stock program in a multinational company, having already leagile prerequisites. Using as a basis the insights and learning from the real case, the aim of the study is to develop a business processes framework, accommodated with the IT support program, in order to help other companies to ease up the implementation of such models. The data collection in the field was carried out by the authors due to their involvement with the buying organization. The documentation consisted of material inventory levels and value, storage capacities evaluations, orders size and frequency, relationship with actual suppliers, employees' readiness, as well as process mapping procedures and the entire chain activities, in order to identify all the functions and processes that may be influenced by the implementation features of a CS program. Due to project's high complexity, this paper will only refer to the key business processes regarding the stock administration, that have to be conducted through an integrative software, namely the successive steps which are compulsory to follow and align with the supporting soft in case of such a program implementation.

3. Defining JIT, CS, VMI and comparing them with the traditional inventory model

JIT is an inventory management method whereby materials, goods, and labour are scheduled to arrive or be replenished exactly when needed in the production process [15]. Thus, in order to effectively employ JIT a company must accurately forecast demand. Operating in a JIT system stimulates suppliers to tighten inventory controls and ensure that demand is met on a timely basis [16].

CS is another approach in which the vendor removes his inventory and maintains a stock of materials at the buyer's plant. This model is being based on a close collaboration between the vendor and the buyer, aiming to create a win-win situation through revenue sharing [14]. The CS is legally owned by the vendor, but held by the buyer, meaning that the risk and rewards regarding the material stock remains with the first party. In accordance with a general CS policy, the vendor guarantees the buyer to continuously keep a stock between a minimum and a maximum level, which are agreed between the two parties [17]. The minimum stock level is termed as safety stock and the maximum level sets the limit the buyer can accept, concerning the storage space constraints. The entire process is governed by a consignment agreement, which represents the legal frame that enables both parties to perform the agreed activities. The buyer does not need to bare the financial cost associated with the inventory, as the materials are not bought until they are consumed in production [14].

VMI is a supply chain initiative where the vendor cooperates with its customers to reach a more efficient inventory management, the inventory being located to either the vendor or buyer, under the condition that the vendor manages the inventory [12]. This means that the vendor makes the inventory replenishment decisions on behalf of the buyer by monitoring inventory levels and product withdrawal, by deciding the appropriate inventory levels, as well as choosing the appropriate inventory policies to maintain the agreed inventory levels [12]. In VMI, the inventory could be owned by either the buyer, or by the vendor.

IOP Conf. Series: Materials Science and Engineering **568** (2019) 012055 doi:10.1088/1757-899X/568/1/012055

When referring to the similarities and differences between the inventory above management models, there are three main aspects to be taken into account: the inventory location, the ordering responsibility, and inventory ownership. Table 2 illustrates the differences and similarities of JIT, CS, VMI, and traditional inventory management models in regards to the three main points from the above description.

Table 1. Similarities and differences of JIT, CS, VMI, and traditional inventory model

	Inventory Models			
	JIT	CS	VMI	Traditional inventory Model
Inventory location	Vendor and Buyer	Buyer	Vendor or Buyer	Buyer
Ordering responsibility	Buyer	Buyer	Vendor	Buyer
Inventory ownership	Buyer	Vendor	Vendor or Buyer	Buyer

As shown in the table, whereas in traditional inventory management programs, the ordering, ownership, and physical control of the inventory stays with the buyer, in CS programs financial ownership of the material stock is with the vendor and shifts to the other party when the material is already consumed in production. Under VMI systems, the responsibility of ordering remains to the supplying company, while the ownership and the inventory location can be either at the buyer or the vendor. In the case of JIT delivery arrangement, the vendor holds the central part of the inventory at his facilities, but makes frequent deliveries to the buyers' location, based on pre-agreed schedules.

Based on the high specificity of each of the discussed inventory programs, it appears that the most suitable one to lower raw materials' inventory, as well as eliminating procurement lead-times is CS model. There are significant benefits for the focal firm, already mentioned in the industry reports or the research literature, as an outcome of CS implementation [17, 18]. Table 2 highlights the main benefits on the buyer's side.

Table 2. Buyer's benefits under CS

Buyer's Benefits under CS
- Reduced inventory costs; 'zero cost' for the consigned materials
- Improved cash flow; materials are invoiced and paid only after being consumed
- Procurement lead-times elliminated or drastically reduced; materials always on-hand
- Production continuity, with no disruptions from the perspective of the consigned materials
- Forecast accuracy problems and out-of stock instances are eliminated
- Extended payment terms, as the vendor is paid upon material consumption
- More flexible production mix and increased productivity
- Increased customer satisfaction, hence more responsiveness achieved

The next aspect of being investigated is at what extent is a company ready to adopt it, and how to proceed to achieve a successful implementation.

4. Case study insights and results

The focal company is an FMCG goods manufacturer, being since several years under leanness and agile strategies implementation, having already an integrated supply chain, with streamline processes, and clearly defined and allocated responsibilities. The software system that granted processes integration over the entire supply chain is represented by SAP (Systems, Applications, and Products), a German software product, which allows the business to track customers, suppliers and business functions interactions. SAP is especially well-known for its Enterprise Resource Planning (ERP) data management system and what is the most important, it is also designed to be adaptable for different business models, by specific input provided by the consultants, at request.

IOP Conf. Series: Materials Science and Engineering **568** (2019) 012055 doi:10.1088/1757-899X/568/1/012055

Pressures like unpredictable demands in the market, forecast accuracy issues, stock-outs, and cash flow, made the management team decide the adoption of CS policy. The proposal was to target the most suitable materials sourced by reliable suppliers, already engaged in a collaborative relationship with the company. All the involved suppliers have been assessed based on specific criteria [19]. An in-depth analysis has been performed on all the processes and resources involved in the transition to CS model, and the project team reached a final consensus regarding the actions to be followed to support the business case. The implementation was a success, being executed under the consistent guidance of the project team, even though it concerned only two types of materials, as a pilot start. An excellent communication among all the implied parties and stakeholders support helped to overcome the challenges arisen from the complexity of the process in certain areas, like increased reporting frequency and separate monitoring of the physical stock per each vendor.

Based on the insights of the case study, the action steps have been isolated by the research authors, structured and ordered in a process framework, along with the corresponding transactions' codes, roles and simulation outputs. The business process framework for CS implementation, accommodated with SAP software is presented in table 3.

Table 3. Business process framework for CS implementation accommodated with SAP

	Business Process Steps	Transaction Code	Role (execute this step)	Output Data / Results / Additional Information	Document Number out of the Simulation Program
1	**Create consignment Purchasing Info Record (PIR) with "zero" price**	ME1N	Master Data	Select consignment	PIR created under the number 5300274504
2	**Create yearly /monthly Purchase Order (PO) with item category, "K"-consignment assigned**	ME1N	Material Planner	Delivery date far away in the future in order not to interfere the APO planning	Standard PO created under the number 4500948441
3	**Create standard (real) PO against the vendor for the materials in scope**	ME21N	Material Planner	Communicate PO to vendor	Standard PO created under no. 4500948442
4	**Post Goods Receiving Note (GRN) against Consignment PO (in special stock "K") upon material receipt in the warehouse**	MIGO	Material Clerk	Movement type 101	Material document posted 5000003205
5	**Create physical inventory document for special stock "K" = consignment**	MIK1	Material Clerk		Physical inventory document created 100001221
6	**Transfer material to Manufacturing function (SL 2200)**				
6.1	Reverse quantity from initial GRN post against consignment PO	MIGO	Material Clerk	Movement type 102 (special stock "K")	Material document posted 5000003206
6.2	Perform GRN for the same quantity against the real PO	MIGO	Material Clerk	Movement type 101	Material document posted 5000003207
6.3	Stock transfer from SL 2000 to SL 2200	MB1B	Material Clerk	Movement type 311	Material document posted 4900019158
7	**Send to vendor weekly report with materials in scope, transferred to production, to be further invoiced**	MB51	Material Supervisor	Total quantity should be equal to weekly GRN posted in step 6.2	

5. Conclusions

Consignment stock program is an excellent option towards reducing inventory cost, as well as eliminating out-of-stock instances, forecast accuracy issues and bringing procurement lead-times to zero levels, all of these being targeted objectives in supply chain's management to achieve enhanced efficiency and responsiveness. Though, it is important for the focal company to be ready to adopt such initiatives to avoid implementation failure.

The case study demonstrates that a leagile company, already having in place streamlines processes, clearly defined and allocated responsibilities, along with a robust ERP software solution, represented by SAP software in the present case, is more prepared to adapt to this kind of changes and has a higher rate of a successful implementation than a typical firm. Furthermore, trust and commitment between supply chain members, as well as information sharing, are crucial for a positive outcome.

IOP Conf. Series: Materials Science and Engineering **568** (2019) 012055 doi:10.1088/1757-899X/568/1/012055

Based on the insights obtained through this case study, the key activities of the stock administration part of the project have been isolated by the authors and structured in a process framework, being accommodated with the SAP software program in terms of operating transactions and roles allocation. This framework could be used by other businesses in scope, who decide to adopt a CS program.

Both, theoretical and empirical research, contribute to the existing literature concerning inventory business solutions and represent a useful example for companies that aim to improve their overall supply chain performance by lowering inventory and shorten procurement lead-times, thereby building a competitive advance.

References

[1] Hugos M H 2003 *Essentials of supply chain management*, (New York: Wiley and Sons Inc) p12

[2] Abrudan I and Cândea D 2002 *Economic Engineering Textbook. Engineering and Management of Production Systems (Manual de Inginerie Economică. Ingineria şi Managementul Sistemelor de Productie, in Romanian*) Dacia (Cluj-Napoca) pp15-26

[3] Avasilcai S 2007 Performance evaluation in Romanian industrial organizations *Proc. Conf.: Intelligent Manufacturing & Automation, Annals of Daaam for 2007 The 18th Int.Daaam Symp* pp27-28

[4] Draghici A 2007 Adaptive technologies and business integration: social, managerial and organizational dimensions *IGI Global* 211-243

[5] Draghici A, Baban C F, Gogan M L and Ivascu L V 2015 A knowledge management approach for the university-industry collaboration in open innovation *Proceedings of the conference: Proc. Ec. Fin.* **23** 23-32

[6] Naylor B J, Naim M M and Berry D 1999 Leagility: Integrating the lean and agile manufacturing paradigms in the total supply chain *Int. J. Prod. Ec*, **62** 107-118

[7] Qi F, Xuejun X and Zhiyong G 2007 Research on lean, agile, and leagile supply chain *IEEE Int. Conf. on Wireless Communications, Networking & Mob. Computing* (Shangai) pp 4902–05

[8] Faur M and Bungau C 2019 Outsourcing towards greater agility by investigating decoupling points in leagile supply chains *Preprint* gr-qc/0401010

[9] Fisher M 1997 What is the right supply chain for your product? *Harvard Business Review* March-April 105-116

[10] Faur M and Bungau C 2018 Supply chain 'leagility' through adopting consignment stock strategy in manufacturing companies *Proceedings of the 6th RMEE Int. Mng. Conf.* (Cluj- Napoca) 623-632

[11] Kauremaa J, Småros J and Holmström J 2009 Patterns of vendor-managed inventory: findings from a multiple-case study *Int. J. Op. Prod. Mngm* **29** 11 1109-39

[12] Niranjan T T, Wagner S M and Nguyen SM 2012 Prerequisites to vendor-managed inventory *Int. J. Prod. Research*, **50** 4 939-951

[13] Gümüş M, Jewkes E M and Bookbindker JH 2008 Impact of consignment inventory and vendor-managed inventory for a two-party supply chain *Int. J. Prod. Ec.* **113** 502-17

[14] Battini D, Gunasekaran A, Faccio M, Persona A, and Sgarbossa F 2010 Consignment stock inventory model in an integrated supply chain, *Int. J. Prod. Research* **48** 477-500

[15] https://www.merriam-webster.com/dictionary/JIT accessed 20.03.2019

[16] Kraemer K L, Dedrick J, and Yamashiro S 2000 Refining and extending the business model with information technology: Dell computer corporation *The Information Society* **16** 1 5-21

[17] Valentini G and Zavanella L 2003 The consignment stock of inventories: industrial case in performance analysis *Int. J. Prod. Ec.* **81–82**: 215–224

[18] Xu K, Yin R, and Dong Y 2016 Stockout recovery under consignment: the role of inventory ownership in supply chains. *Decision Sci.* **47** 1 94-124

[19] Faur M, 2018 Perspectives in Developing the Suppliers' Strategy within Procurement Processes *Review Mngm. Ec. Engineering* **17** 1 99-111

IOP Conf. Series: Materials Science and Engineering **568** (2019) 012056 doi:10.1088/1757-899X/568/1/012056

Vehicle dynamics influence on traffic safety in case of a certain route deceleration

O V Oţăt[1], V Oţăt[1], L Racilă[1], I Geonea[1], D Tutunea[1]
[1]University of Craiova, Faculty of Mechanical Engineering, Craiova, Romania

otatoana@yahoo.com

Abstract. The present paper is aimed at determining the kinematic and dynamic parameters of a vehicle during braking on a road in alignment. Closely related to the determining the braking capability, we need to consider the key parameters that characterize the maximum braking capabilities of a vehicle, i.e. maximum deceleration, minimum braking space and minimum braking time. Hence, resorting to the ADAMS Car software platform, we carried out a case study to determine the kinematic and dynamic parameters of a vehicle in the braking situation on the road in alignment. The virtual modeling developed within our case study consisted of five full-vehicle analyses of the braking situation on a track in alignment. The initial speeds of the vehicle and the gearbox shift varied for each of five analyzed cases, while the rest of the input parameters did not change. The initial speeds considered were: 30 km/h, 50 km/h, 80 km/h, 100 km/h and 120 km/h. Following the numerical processing, we set out to determine the forces and moments at the wheel at the braking moment. The processing of the numerical results envisaged a comparative analysis of the vehicle kinematic and dynamic parameters for the five cases analyzed.

1. Introduction

The contemporary accelerated development of the automotive industry has opened the door for multiple research perspectives within the field of vehicle study, hence various research studies focused both on the increase of vehicles' safety and passengers' comfort, as well as on the achievement of higher dynamic performance.

Since vehicles are not subject to microscopic-level investigations, we need to analyze it within the road traffic system as a whole. Adopting this approach, various research papers such as [1] evaluate through a series of applied mathematical methods the effect of different lane grouping and critical lane group concept on the saturation flow rate.

Other research perspectives have focused on the study of the security systems. In [2] the authors address the topic of active security systems, and resorted to the "quarter" vehicle model to validate the ABS/ASR system working algorithm over the braking or starting processes. Passive safety and, in particular the influence of the geometric parameters of the frontal vehicle profile in case of car-to-pedestrian collisions are highlighted in [3].

Further in-depth studies on vehicle dynamics in terms of stability are approached in [4], as the authors put forward a mechanical model of the vehicles with two axles, taking into account the lateral deflection of the tire. Also, in [5] the speeds influence on vehicle stability when changing the lane is analyzed via experimental tests. Experimental determinations have been applied in [6], as well, in order to develop an analysis of vehicle behavior during a dynamometer performance.

IOP Conf. Series: Materials Science and Engineering **568** (2019) 012056 doi:10.1088/1757-899X/568/1/012056

Vehicle ride comfort has been studied by means of a half-car model in [7] that led to a comparison of the main semi-active suspension systems used in a passenger car. In the same line, in [8], further research studies have been carried out taking into account that the nonlinearity of the model is contributed by the damper models.

2. An analytical method to determine the braking capability

Among the vehicles' dynamic performances frequently analyzed, we highlight vehicles' braking capability. The parameters that feature the maximum braking capabilities of vehicles are: the maximum acceleration, the minimum braking distance and the minimum braking time. The implications of high efficient braking capability contribute to increasing road safety, hence reducing the number of traffic accidents.

2.1. Analytical determination of the braking time

The braking process starts from the moment when an actual need to stop arises, causing the vehicle to brake, and, ends when the vehicle stops. This process covers the following time stages: the time elapsed during the involuntary delays (T_{if}– the physiological delays, T_{im}– the mechanical delays) and the time elapsed between the start of the braking process and the moment the vehicle stops.

The general relation to determining the total stop time, if we know the initial speed, is [9]:

$$T_t = T_{if} + T_{im} + 0{,}5 \cdot t_3 + \frac{K_e v_a}{3{,}6\ g\ \varphi_{red}} \tag{1}$$

where:

t_3 - is the time elapsed between the start of the braking process and the lock of the wheels

φ_{red} – is the average value of the adhesion coefficient when braking

v_a[km/h] – is the vehicle speeds at the start of braking

The analytical determination of the braking time was achieved for five distinct cases, considering different speeds parameters, i.e. 30 km/h, 50 km/h, 80 km/h, 100 km/h and 120 km/h. The mathematical model was developed in a flexible structure, by means of the Maple software, a sequence that allows for rapid change of input parameters and new braking analysis. Table 1 below summarizes a part of the sequence developed:

Table 1. Maple sequence to determine the braking time and space

> $Tii := Tif + Tim + 0.5 \cdot t3;$

> $va_1 := 30;\ va_2 := 50;\ va_3 := 80;\ va_4 := 100;\ va_5 := 120;$

> **for** i **from** 1 **by** 1 **to** 5 **do** $Tf_i := \dfrac{(ke \cdot va_i)}{(3.6 \cdot g \cdot fred)}$ **end do;**

$Tf_1 := 1.061841658,\ Tf_2 := 1.769736097,\ Tf_3 := 2.831577756,\ Tf_4 := 3.539472195,\ Tf_5 := 4.247366634$

> **for** i **from** 1 **by** 1 **to** 5 **do** $Tt_i := Tii + Tf_i$ **end do;**

$Tt_1 := 2.061841658,\ Tt_2 := 2.769736097,\ Tt_3 := 3.831577756,\ Tt_4 := 4.539472195,\ Tt_5 := 5.247366634$

> **for** i **from** 1 **by** 1 **to** 5 **do** $Sii_i := Tii \cdot \dfrac{va_i}{3.6}$ **end do;**

$Sii_1 := 8.333333334,\ Sii_2 := 13.88888889,\ Sii_3 := 22.22222222,\ Sii_4 := 27.77777778,$
$Sii_5 := 33.33333334$

> **for** i **from** 1 **by** 1 **to** 5 **do** $Sf_i := \dfrac{ke \cdot va_i \cdot va_i}{26 \cdot g \cdot fred}$ **end do;**

$Sf_1 := 4.410726889,\ Sf_2 := 12.25201913,\ Sf_3 := 31.36516899,\ Sf_4 := 49.00807652,\ Sf_5 := 70.57163020$

> **for** i **from** 1 **by** 1 **to** 5 **do** $St_i := Sii_i + Sf_i$ **end do;**

$St_1 := 12.74406022,\ St_2 := 26.14090802,\ St_3 := 53.58739121,\ St_4 := 76.78585430,\ St_5 := 103.9049635$

IOP Conf. Series: Materials Science and Engineering **568** (2019) 012056 doi:10.1088/1757-899X/568/1/012056

2.2. Analytical determination of the braking space

Establishing the initial vehicle travel speeds [km/h], in accordance with the elements previously set, we can determine via computation the total braking space results from the sum of the space run during the duration of the involuntary delays and the actual braking space [9]:

$$S_T = \left(T_{if} + T_{im} + 0{,}5t_3\right)\frac{v_a}{3{,}6} + \frac{K_e v_a^2}{26 g f_{red}} \quad (2)$$

As with braking time, the total braking space was determined by means of the analytical method, considering the same input parameters, i.e. the speeds values.

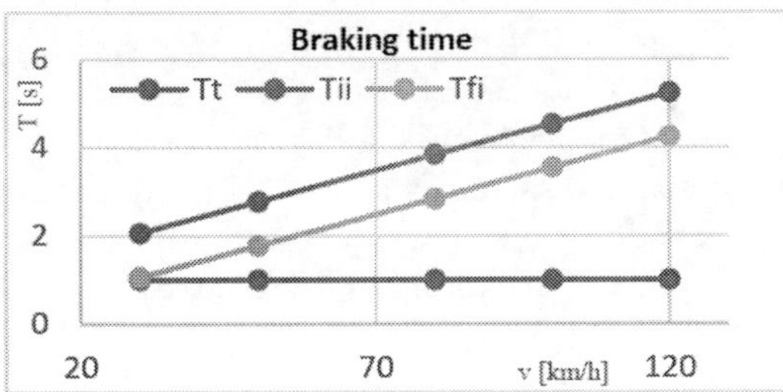

Figure 1. Analytical determination of the breaking time.

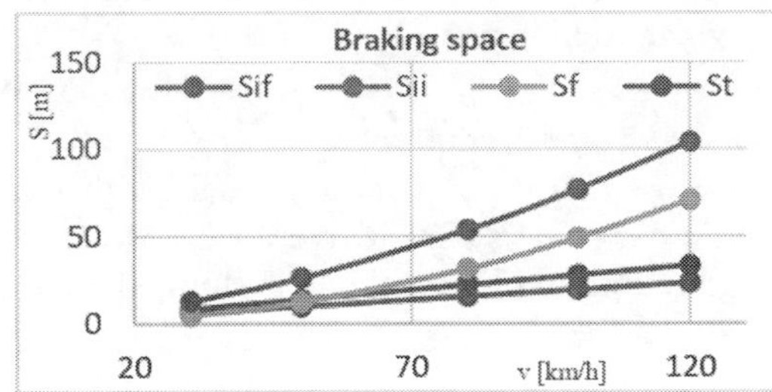

Figure 2. Analytical determination of the braking space.

3. Determining the kinematic and dynamic parameters via the Adams Car software

3.1. Input parameters and simulation development

The first step to develop our simulation was to create a new full vehicle assembly. For the purpose of defining a new assembly, we must select and input all the component subassemblies. The main subassemblies to be defined in the vehicle structure are: the front suspension system, the rear suspension system, the steering system, the front wheel drive, the rear running system, and the vehicle bodywork.

In order to analyze the vehicle kinematic and dynamic behavior in the event of braking, we need to take into the following components, as well: the braking system, the propulsion system of the vehicle, the traffic lane (road) on which the vehicle will travel.

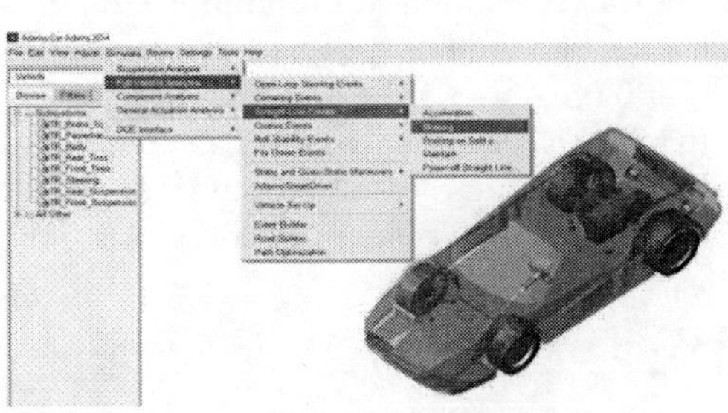

Figure 3. Selecting the analysis type.

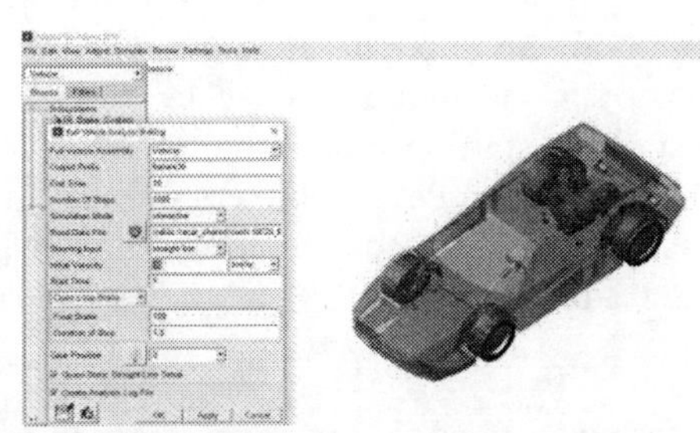

Figure 4. Defining the analysis parameters - braking at 30 km/h.

The case study carried out consisted of five full-vehicle (vehicle assembly) analyses for the braking situation on the road in alignment – see Figure 3. For the five analyzed cases, the initial speeds of the vehicle and the gearbox shift varied, while the rest of the input parameters remained the same. The initial speeds considered were: 30 km/h, 50 km/h, 80 km/h, 100 km /h and 120 km/h.

The input parameters for kinematic and dynamic vehicle analysis when braking on the road in alignment at an initial speed of 30 km/h are illustrated in Figure 4. Following the numerical processing,

IOP Conf. Series: Materials Science and Engineering **568** (2019) 012056 doi:10.1088/1757-899X/568/1/012056

as illustrated in Figures 5 and 6, we determined the forces and the moments on the wheel, at the simulation endpoint, for the speeds of 30 km/h, respectively of 120 km/h.

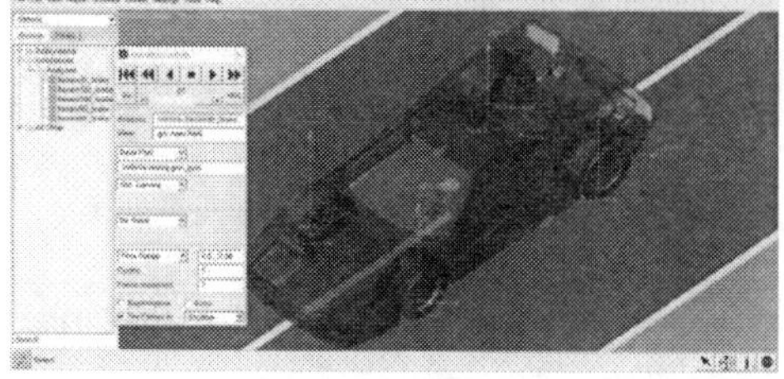

Figure 5. Forces and moments on the wheel when braking at 30 km/h.

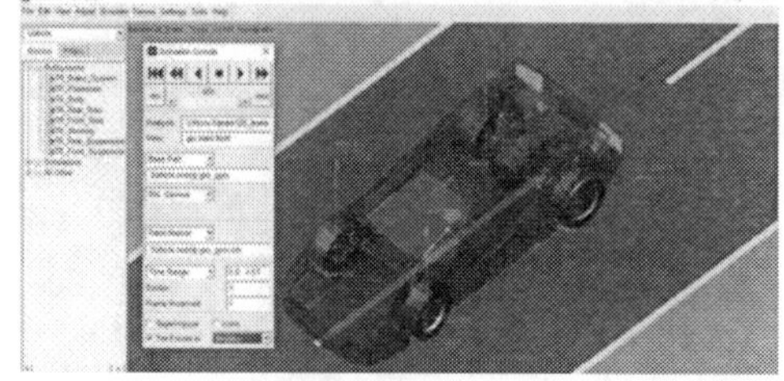

Figure 6. Forces and moments on the wheel when braking at 120 km/h.

3.2. Results interpretation

The processing of the numerical results aimed at a comparative analysis of the vehicle kinematic and dynamic parameters for all five analyzed cases. Figure 7 illustrates graphically the distance traveled during braking until the vehicle is stopped at an initial speed of 30 km/h, 50 km/h, 80 km/h, 100 km/h and 120 km/h.

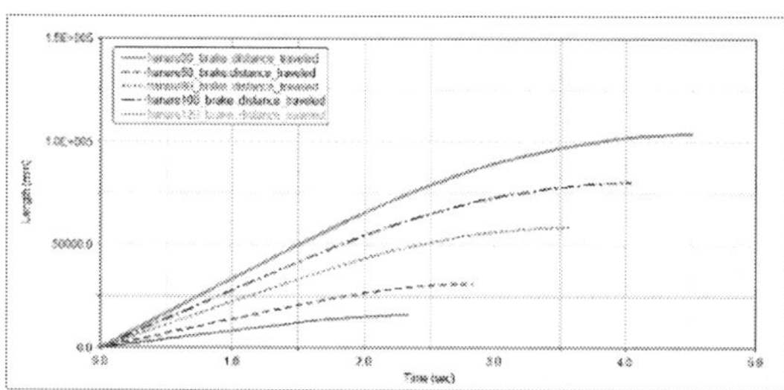

Figure 7. The distance traveled during braking.

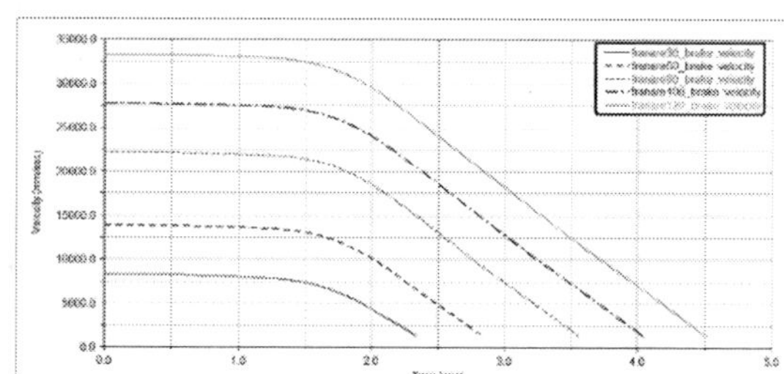

Figure 8. Speeds variation during braking.

The speed variation during the braking maneuver is indicated in Figure 8. According to the same figure, we can establish the total braking time for each analyzed situation.

The variation of the deceleration values during braking is illustrated in the Figure 9 diagram. The maximum longitudinal deceleration for the first case, i.e. the initial speed of 30 km/h, is 10000 mm/s^2≈1,01 g. The maximum deceleration reached is 11500 mm/ s^2≈1,17 g, and has been reached in all four situations, starting at an initial speed of 50 km/h, up to the last considered case, i.e. the initial speed of 120 km/h. The target parameters were applied to determine the tire forces for both front and rear axle tires. The components of the forces in the longitudinal direction for the front tires are graphically illustrated in Figure 10.

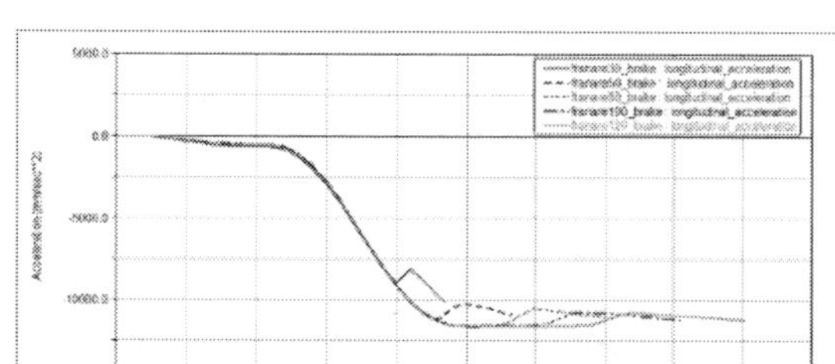

Figure 9. Longitudinal deceleration variation during braking.

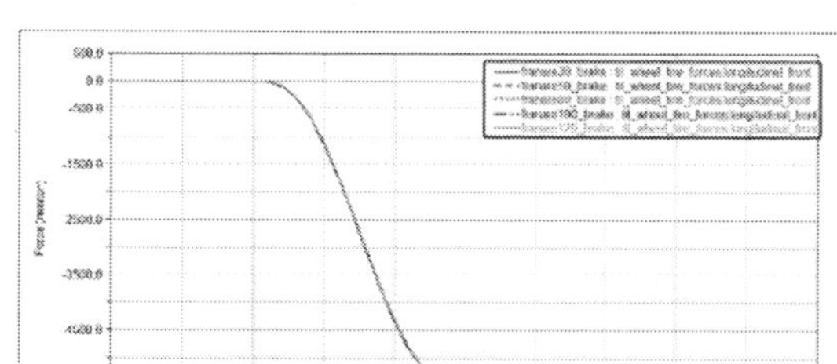

Figure 10. The force in the longitudinal direction of the tire - front.

IOP Conf. Series: Materials Science and Engineering **568** (2019) 012056 doi:10.1088/1757-899X/568/1/012056

4. Conclusion

In order to determine the vehicle kinematic and dynamic parameters during braking on the road in alignment, we carried out a case study while applying two distinct methods. First, we resorted to the Maple software platform to develop a flexible sequence for the analytical determination of time and braking space. The second method involved the realization of five virtual modeling cases via the ADAMS Car platform.

The case study carried out envisaged five full-vehicle (vehicle assembly) analyses for the braking situation on the road in alignment.

The processing of the numerical results aimed at a comparative analysis of the vehicle kinematic and dynamic parameters for the five analyzed situations. The minimum braking time is of about 2.1 s for an initial speed of 30 km/h, reaching about 5 s at an initial speed of 120 km/h.

The minimum braking distance is of 13 m (virtual model) and 12.75 m (analytical method) for a speed of 30 km/h; it increases to 27 m (virtual model) and 26.15 m (analytical) at a speed of 50 km/h; 54 m (virtual model) and 53.58 m (analytical) for an initial speed of 80 km/h, 77 m (virtual model) and 76.78 m (analytical) for 100 km h respectively, thus the maximum braking distance is obtained for the initial speed of 120 km/h and 104 m (virtual model), i.e. 103, 90 m (analytical). Another set of results focused on the determination of tire forces for both front and rear axle tires.

The values obtained as to determine the braking capability, at different speeds, indicate similar variations, both by applying the analytical method and the virtual simulation of a vehicle braking on the road in alignment, hence validating the model developed within our research study.

References

[1] Dumitru I, Nicolae D, Matei L and Racila L, 2016 Saturation flow mathematical model based on multiple combinations of lane groups *Efficient, Safe and Intelligent Transport, Transportation Research Procedia* **18** pp 399-404

[2] Beles H, Preda I, Dragomir G, Blaga V and Ardelean F 2008 Theoretical research regarding the active safety systems, *Annals of DAAAM and Proc.* 19th Int. Symp. of the Danube-Adria-Association-for-Automation-and-Manufacturing pp 87-88

[3] Tolea B, Radu A I, Beles H and Antonya C, 2018 Influence of the geometric parameters of the vehicle frontal profile on the pedestrian's head accelerations in case of accidents *Int. J of Automotive Technology* **19** pp 85-98

[4] Simniceanu L, Bogdan M, Otat V and Trotea M, 2017 Nonlinear differential system applied of a mechanical plan model of the automotive used for the nonlinear stability analysis *11th Int. Congress of Automotive and Transport Engineering - Mobility Engineering and Environment (CAR2017), IOP Conference Series-Materials Science and Engineering* **252**

[5] Heerwan P M., Asyraf S M., Efistein A N, Seah C H, Zikri J M and Syawahieda J N, 2017 Experimental study of the vehicle dynamics behavior during lane changing in different speedss *4th Int. Conf. on Mechanical Engineering Research (ICMER) IOP Conference Series-Materials Science and Engineering* **257**

[6] Dumitru I, Racila L, Tutunea D, Matei L, Dima A and Oprica A, 2017 Experimental Investigation of a Vehicle Behavior Using Different Complex Data Acquisition Systems *CONAT 2016: Int. Congress of Automotive and Transport Engineering* pp 136-143

[7] Koulocheris D, Papaioannou G and Chrysos E, 2017 A comparison of optimal semi-active suspension systems regarding vehicle ride confort *Int. Congress of Automotive and Transport Engineering - Mobility Engineering and Environment (CAR2017) IOP Conference Series-Materials Science and Engineering* **252**

[8] Darsivan FJ, 2016 The Investigation of a Ride Quality of Nonlinear Half-Car Model, 3rd Int. Conf. on Mechanical, Automotive and Aerospace Engineering *IOP Conference Series-Materials Science and Engineering* **184**

[9] Nistor N and Stoleru M, 1987 *Expertiza tehnică a accidentelor de circulaţie* (Editura Militară, Bucureşti)

IOP Conf. Series: Materials Science and Engineering **568** (2019) 012057 doi:10.1088/1757-899X/568/1/012057

Notes on the possibilities and limitations of using topological tools in mechanical modeling of media

G Lámer

University of Debrecen, Faculty of Engineering, Department of Engineering Management and Enterprise. 4028 Hungary, Debrecen, Ótemető utca 2-4.

glamer@eng.unideb.hu

Abstract. In the mechanical modeling of the media, the atomic-molecular structure of the material is usually neglected, furthermore, with regard to the characteristic distance of the atomic-molecular structure, the media is considered to be a continuous material distribution. Referring to the continuous distribution, we use the tools of differential and integral calculus for the mechanical-mathematical description of the media. The use of these mathematical tools has conditions and limitations: there must be a sort of topological order in the examined space. This order records the relative position of the points. The fixed position relative to each other makes it possible to describe only such mappings that retain the fixed position relative to one another. Thus, it is possible to distinguish between strain and rearrangement, and at the same time to define the boundaries of the possibilities offered by the applied mathematical tools: the continuum many points set provides the basis of the topological description for the classical continuum, a set of finite points in the grid network with its internal structural features, provides the basis of the topological description for the grid (that is generalized) continuum.

1. Introduction

When describing the behavior of elastic materials, the material is considered to be continuous. In the model, only the degree of displacement freedom is assigned to each point of the material considered as continuous (see e.g. [1,2]). The model can be expanded by assigning to each point of the material considered to be continuous, its rotational degree of freedom in addition to its displacement (see e.g. [3]), then we gain the micropolar theory of elasticity. The model is purely theoretical, the author [3] indicates (in fact complains) that no experimentally measured material constants can be found to the new theory. The expanded continuous media model can be further expanded to include pointers ("directors"), granules ("micro continuums"), and the rotation of directions, the strain states of granules will become the additional kinematic variables (see e.g. [4,5]).

The question arises as to whether the granules in the granular material can be attributed to a degree of distinct rotational degree of freedom that is independent of the material embedding the granule: in concrete, the gravel grain cannot rotate. It cannot rotate theoretically, because then a force independent of the cement stone has to act upon it to rotate, but we have no knowledge of such force. It cannot rotate practically because the fracture tests prove that concrete breaks in a way that gravel does not fall out of the cement stone but breaks.

The question arises as to what kind of continuous and discrete modeling is possible for materials with discrete, atomic-molecular, or larger scale internal structures [6,7,8,9]. In the depths of the research, the same question arose: in order to create a model of continuous media, the material must not only be

IOP Conf. Series: Materials Science and Engineering **568** (2019) 012057 doi:10.1088/1757-899X/568/1/012057

continuous but must also have an internal order. This order can be identified with the topological order [10]. In this paper, we review the conditions and consequences of using topological tools: continuous mapping, differential and integral calculus.

2. Description of the order. Consecutiveness of series and topology

The order in the body is interpreted as the atoms and/or molecules that make up the body adjacent to one another, making them belong together. The preservation of the order can be defined as the atoms and/or molecules adjacent to each other before a visible effect remains and belong together after the external effect.

In order to "describe" the order, the space occupied by the atoms and/or molecules constituting the body is considered fixed; the location of the atoms and/or molecules is given by the reference point of each atom and/or molecule. The point network thus obtained is considered a grid. (We take the grid of rock salt crystal as an example). In the grid the order is characterized by the fact that the order of the grid points in one line along each direction is unchanged, and that each grid point can be specified by a unique triple of numbers and this unique indicator does not change during an effect upon the body.

In addition to the triple, there is another possible way to specify the order: specify the set of all the points that belong to each other. This is the well-known neighborhood system from point topology [11]. In the topology, the order in the set of points, the "interconnection" between the points is interpreted with the neighborhoods and at the same time "interconnection" is fixed for all points of the point set once and for all. In the set of points, "interconnection" means that two or more points belong to the same neighborhood. Since we determine for each point which points belong to its neighborhood, the neighborhood system, i.e. the topological order, reflects the order (consecutiveness) of the points relative to one another (see e.g. [11]).

The topological order (and topological mapping) also allows two topological spaces to be mapped while preserving the internal structure of the topological space.

The concept of topology has been developed by describing the various accumulative and separability properties of the point sequences in the metric spaces. The topological space is a "generalization" of the metric space in which the topological properties are not expressed by the metric but more general, e.g. neighborhood system. It is noteworthy that there is no need to "construct" a "rich" topological space, because a set of real numbers (obtained with metrics from the Euclidean space) already has the topological qualities needed to perform differentiation and integration (e.g. existence of accumulation points, separation of neighborhoods, or in other words the spaces of the n-tuple numbers is Hausdorff space (see e.g. [11,12] for topological spaces and analysis).

Hereinafter, the topological space refers to Hausdorff space locally.

The consequences of the topological order in the topological space can be summarized as follows.

- The topological order makes it possible to interpret and apply the coordinate system: the straight line can be mapped to the topological space, the sequence of points is unchanged, and the coordinates of a point are clearly fixed even when the topological space (or a domain of it) is mapped to another topological space (or to its domain).
- The topological order makes it possible to interpret and take a limit, including differentiation and integration: completeness makes it possible for point sequences to have an accumulation point; the separability allows the uniqueness of a limit.

3. Conditions and limitations of using topological tools

In general, the condition of using any tool is also a limitation of using the tool.

3.1 Order, as a necessary condition

The prerequisite for using topological tools is that the atoms and/or molecules that form the examined body, and in general the "particles" that make up the internal structure of the material should be in a fixed "order" and remain so under (and after) an effect on the body. It is also the most restrictive condition for the use of topological tools: the behavior of a body in which there is no topological order,

IOP Conf. Series: Materials Science and Engineering **568** (2019) 012057 doi:10.1088/1757-899X/568/1/012057

or if a resting topological order does not remain under an effect on the body, cannot be examined with the tools of topology.

The basic idea of modeling: instead of the grid of atoms and/or molecules sitting in order in discrete points, we examine the points of a continuous point set, a domain of the Hausdorff space. The consequence of this is that the result obtained should not be referenced to any point of a domain of the Hausdorff space, but only to the reference point of each atom and/or molecule. In principle, this distinction is true, but it is irrelevant from a practical point of view: the human eye, the instruments used in everyday practice cannot distinguish an atom, but rather the location and position of a part of a material that can be well defined and incorporates a vast number of atoms and/or molecules; our examinations and measurements refer to such finite size "particles".

The "transition" from the discrete point set to the domain of a topological space is made possible by the fact that "there is order" in the material at the atomic-molecular level. This is considered a basic assumption. Consequently, we can only examine such systems with the tools of the topology for which the basic assumption is met, and in this case the obtained results produce correct findings for the particles sitting on the points of the grid, or if there is no basic assumption of modeling, i.e. there is no topological order, then the interpretation of the obtained results should take into account that the underlying basic assumption is not correct. In the latter case, the atoms and/or molecules mixing with each other lose their unique marker, so in the description obtained with the topological means, the "result" may not refer to the particle to which the mapping is to be applied, but to another particle. This does not always mean that modeling is not able to describe the phenomenon examined within measurement accuracy. Think about the plastic forming of metals in which the degree of plastic rearrangement refers to a small part of the body. Such is the production of cold bent sections. In the plastic zone, we may not be able to precisely predict the location of all the metal atoms, but it can be predicted, for example, that a flat plate becomes an angled bend plate. However, in cases where "disorder" extends to the body as a whole, we cannot even estimate the location of atoms and/or molecules. Examples include kneading dough or Brown-movement of tiny, solid particles in liquids.

The requirement for a topological order sharply limits the application of modeling: even in the case of gaseous and liquid bodies, there is no order at all. In these cases, we cannot consider the material itself to be a domain of the Hausdorff space, because the constant movement of atoms and/or molecules makes it impossible to create an unambiguous correspondence between the "resting" gaseous and liquid body particles and the points of the Hausdorff space. Instead, we use a different "approach": the embedding space is considered to be Hausdorff space, and we examine the mechanical characteristics of a particle of a matter flowing in space when it reaches the given point in space. That is, the topological order in the embedding space allows a continuous description of the flow. The flow of the body (granulated matter = powder, liquid, gas, the latter also requires thermodynamic formalism) is described in the embedded space as the topological space. The examined point is a fixed point of the embedding space, its place does not change; at this point in time different material particles pass through. We emphasize that we do not consider the material as a continuum, but consider the flow to be continuous, the flow itself occurs "*before*" a point set considered a continuum, as a coordinate system. That is to say that, the description model does not specify how a particle of a matter changes its location and velocity, and in fact does not specify the velocity of the particle arriving at a selected point in the space, but instead it determines the velocity of the particles arriving at a selected point in the space or in its smallneighborhood will be roughly the same as that calculated in the model; at the same time, minor or major deviations are possible. In principle, the model is not accurate either if we assume that the flow lines never intersect, and accordingly, this flow model does not give any indication of the flow occurring in the flow perpendicular to the flow lines (the transverse diffusion).

Topological tools are suitable for determining the continuous mapping of two topologically equivalent ranges. This means that a topological mapping of a cube on to a sphere can be made, and a topological mapping among a rod, a disc, and a cuboid can also be produced. We know that the atoms and/or molecules themselves do not significantly change their size during mechanical effects. So, if we require the volume of the cube and sphere, or the rod, the plate, and the cuboid to be the same, i.e., they consist of the same number of atoms (and/or molecules), then the atoms (and/or molecules) cannot be transferred from a spherical form to cube form, or among rod, disk, and cuboid forms by the topological mapping preserving the atomic-molecular structure. The explanation is trivial: finite-size atoms (and/or molecules) have to be rearranged, while in terms of points with no extension, two line segments of different lengths, two plane figures with different areas, and two bodies with different volumes (measured in some metrics) with the same (continuum) cardinality, can be mapped topologically, i.e. continuously and unique way one onto another. This means that during the shaping of the spherical dough into a cube, flattening it to a pie or rolling it into a roll, the alteration or rearrangement of the atomic-molecular order cannot be described by topological means (see Fig. 1.).

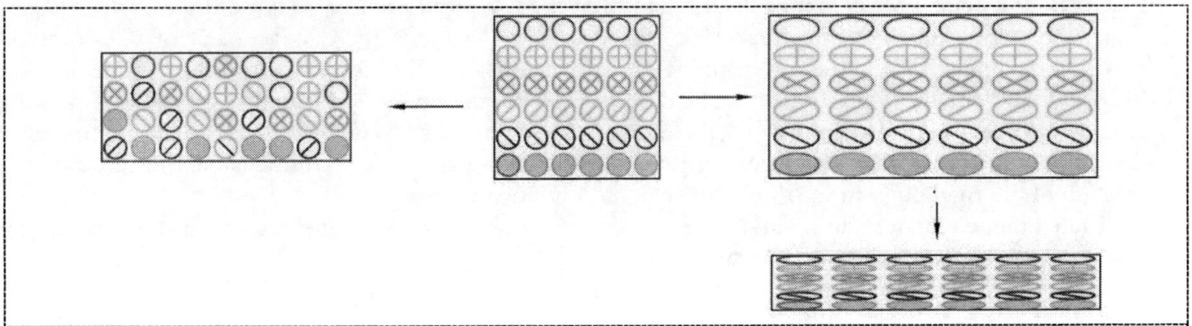

Figure 1. "Mixing", "rearranging" mapping of discrete ranges and topological mapping of point set

The two examples above show that the use of topological devices within the body must have a topological order that must be maintained during (and after the cessation of) the effect; atoms and/or molecules cannot exchange place, disappear from the body to "nothing", and cannot "join" to the body from "nothing". This means that the reversible, elastic change of the shape of the solid bodies is a phenomenon that can be described without limitation by the topological description.

3.2 Strain and rearrangement

The body changes its shape under effect. The change of shape, from a topological point of view, can only be of two types: the topological order is preserved or the topological order does not remain. In order to distinguish between the two phenomena, in the first case we refer to strain and in the second case to rearrangement. Accordingly, during the flow of liquid, the atoms and/or molecules that make up the liquid are rearranged, the particles of soil under the influence of the penetrating effect of a pile are rearranged around the pile, at the same time, steel is subjected to strain during tension, bending and twisting. In the plastic formation of the metals, atoms are rearranged in the plastic zone, and strain occurs in the elastic zone.

3.3 The scale of the order and its kinematic degree of freedom

The topological order is usually "imagined" on two scales. One is the atomic-molecular order, and the other is a much larger order of magnitude, the built order; for example, a spatial frame structure or a system of uniform particles periodically contained in a matrix. In the first case, it is natural to identify the particle with the atoms, which in atom → point modeling, we keep only three from the six degrees of kinematic freedom which three ones describe the displacement of the atom as a rigid body. In the second case, identifying the particle with a node of the frame structure, we feel that it is natural for a node in the frame → pointmodeling to keep all six kinematic degrees of freedom of a node of the frame as a rigid body, i.e., three displacements and three rotations. Also, note that the stubs of the node stretch and shorten (imagining three rods, intersecting at one point as a node, these are three degrees of freedom), and that the angles of each branch of the node change (also three degrees of freedom at the same node). Most of the time, we also think of the particles embedded in the matrix similarly. In other words, when we switch to e.g. from a finite number of nodes in a frame structure to a continuum point set as a mathematical model, then we want each point of the point set not to be three, but more, at least six, or possibly twelve kinematic degrees of freedom. This in itself is not special, think about the rigid body: the number of degree of kinematic freedom of it in the three-dimensional Euclidean space is six. At the same time, the space “including” of complete degrees of freedom, the “picked up” space for testing, is not physically the three-dimensional Euclidean space around us, but we constructed a space mathematically to perform the test. In the case of a rigid body, this is the six-dimensional phase space. The question is how many kinematic degrees of freedom are there for one (and so every) point in the set of points: equal to or more than the number of dimensions of the point set or possibly even more. A point in the point set has no size or an internal structure. Accordingly, neither rotation nor strain can be interpreted in relation to a point. Therefore, each point of a point set has exactly three degrees of kinematic freedom: three components of the displacement vector (assuming a three-dimensional space). Thus, the degree of kinematic freedom of the point set describes the displacement of the point set according to points, i.e. continuous mapping of a continuous manifold. Any other non-kinematic degrees of freedom other than displacement can be described by a function interpreted on the point set: e.g. body temperature, its electromagnetic state. To interpret the degrees of physical freedom, we do not increase the number of dimensions of space, nor do we consider the space as six-dimensional to describe the rotation of the rigid body in the three-dimensional Euclidean space.

The examined matter can be divided into matter containing structured parts (matrix) and particles; this way, the matter can be split into two parts.

Assuming that the topological order remains in both parts, there is the possibility of interpreting an embedded or generalized continuum. Take the case when the particles in the matrix are small. At this point, the location of the liftes parts of the matrix should be “removed”, and instead of a matrix we can consider a continuum. “Above” this is the system of lifted, periodically arranged particles. These particles are modelled with no-extension points. Only displacement will remain as the degree of kinematic freedom. Because there is a finite amount of them, the topology of the point set can be interpreted, but in this case a limit can not be taken. Additionally, to merge the two systems, the points must be “replanted” in the topological space modelling the matrix. However, there is no place there for another point; because we consider the topological space to be complete (otherwise it would not be possible to take a limit in a unique way). Consider the case when the particles in the matrix are large. At this point, the location of the lifted particles in the matrix can not be “eliminated”, and the matrix should be seen as a multiple connected continuum leaving the place of particles empty. “Above” this is the system of lifted, periodically arranged particles. These particles can be modelled as a rigid or deformable solid body. In this case, the degree of freedom is not only the displacement, but also the rotation and the internal deformation. In this case, a limit can be interpreted within a particle (the particles themselves can also be considered as a domain of a topological space), but because of the finite amount of them, the sequence passing through from one particle to the other cannot be interpreted and so there cannot be interpreted a limit in the case of particle set. (The topology of particle set can be interpreted in principle, see [13,14], but between disjunct domains, differential and integral calculations cannot). Furthermore, to unite the two systems, it is necessary to “replant” the particles in the topological space modeling the matrix. It is true that we left its place out, but we need to establish a connection between the lifted and replanted particles and the matrix that accommodates it (if you like, the neighborhoods that run on the two “boundary” surfaces must be merged into one), then the movement of particles independent of the matrix (revitalized by embedding) prevents it. (So, the lifted particles and the hole left in the matrix may be deformed differently, so the particles, which are deformed independently of the matrix, cannot be replanted into the matrix that deformed differently from particles. The condition of replantation is exactly this; an identical deformation of the mutual surfaces. In other words, they are non-autonomous continuums: the displacement vectors for the right and left sides of the contact surface are the same. This system can be considered as a continuum the material characteristics of which change with jump along one (or more) internal surfaces. This does not make the matrix and the granules separate into two individual continuums, or the two jointly into a generalized continuum.

IOP Conf. Series: Materials Science and Engineering **568** (2019) 012057 doi:10.1088/1757-899X/568/1/012057

3.4. Overview of topological characteristics of modeling matter

The following elements of the topological characteristics are highlighted. The number of "particles" that make up the model; its number can be finite or a continuum. Is there a topological order among the particles forming the model before deformation? Is there a topological order between the particles forming the model during deformation? Is the mapping of "particles" - their relative movements - continuous or discrete during deformation? Based on these topological features, the models can be divided into the following main classes.

Classical continuum. Continuum number of a set of points. There is a topological order before the deformation (mapping). The mapping is continuous. The topological order between the points remains during and after the mapping. The point making up the continuum, i.e. "particle" has no size, no internal structure. The degree of kinematic freedom is displacement. The functions used to describe the system are interpreted over the continuum as an interpretation domain [1,2]. Within the classical continuum as a model, the use of fracture conditions allows for the description of larger scale discontinuities, such as the determination of the surface of the plastic behavior area or of fracture surfaces in granular media. Within the classical continuum as a model, it is possible to describe rapid or stochastic variables by using averaging and homogenization methods [15,16,17,18].

Grid continuum (generalized continuum). Finite number of the set of points. There is a topological order before deformation. The mapping may also include continuous (collective, acoustic) and discrete (individual, optical) motion forms. The topological order between the points remains during and after mapping [19]. The point making up the grid continuum, that is, the "particle" has dimensions, has an internal structure. The degree of kinematic freedom may be rotation, elongation, angular variation, the second-order deformation tensor commonly used in the classical continuum, or the richer (multi-parameter) form that describes the shape change. Three (main) representations can be distinguished. One is to represent the "particles" in the grid point with directors. Then the relative change in the length of the directors and the change of angle between the directors give the degrees of kinematic freedom. The second is to classify the "particles" in discrete grid points into cells, and to group the individual degrees of freedom of displacement of the points in the cells into different types of strain, such as tension, shear, bend, or strain that are common in classical continuum mechanics. The third is that besides the functions describing the system, their gradients are also treated as unknown quantities. To describe the system, the grid continuum is embedded into an external space considered as the domain of functions describing the system, but not the entire set of values but only the values over grid points are applied to describe the grid continuum [20,4,5,19,21,22,23].

Granular medium. Finite number of the set of points. There is a topological order before deformation. The mapping can be continuous (the stress set of particles are in equilibrium due to the effects of external forces) and may be non-continuous (some parts of the particle set or the whole is rearranged for equilibrium). The particle forming the medium has dimensions and an internal structure. The degree of kinematic freedom is the displacement and the rotation, and in some cases the strain degree of freedom can be assigned to it. The functions used to describe the system are interpreted above the space enclosing the particles as the domain of interpretation. For a small number of particles, state equations for each particle are written. In the case of a large number of particles, the set of particles ismodeled as a medium, and the functions for each particle areas considered continuous functions in the embedding space over the domain of interpretation. Usually both the classical and the generalized continuum models are used. In these cases, the state functions are applied to a larger volume of particles (if the value is "averaged") [24,25,26].

Liquids and gases. The set of particles consists of a finite number of "particles". There is no topological order in liquids or gases. Statistical methods can be used primarily to describe the state. The continuous description should exclude that the particles are "disordered" relative to one another: this "chaotic" state is "smoothed", and it is considered that the particles forming the liquid and or the gas have a topological order relative to one another. With this condition, the mathematical apparatus offered by both the classical continuum and the grid continuum becomes applicable. At the same time, when interpreting the results obtained, one has to consider the fact that the solution was obtained by smoothing

IOP Conf. Series: Materials Science and Engineering **568** (2019) 012057 doi:10.1088/1757-899X/568/1/012057

the chaotic behavior, in reality there is chaotic movement, and even some flows would not be created without chaotic behavior.

4. Summary
In this paper, we have reviewed the possibilities and limitations of using topological tools.
Continuous mapping, differential and integral computation are referred to as topological tools.

The conditions for using topological tools are as follows:

- the particles that form the material to be modeled, let the particles be atoms and/or molecules, or larger-scale elements such as particles or cells of a constructed structure, must have a topological order,
- the material to be modeled must be replaced by either a topological space or the points representing one cell must be embedded in a topological space.

The consequences of using topological tools are as follows.

- Three types of models can be obtained
  - material flowing before the continuum (particle, liquid or gas),
  - classical continuum,
  - grid continuum or generalized continuum.
- By modeling the material flowing before the continuum, information about either the rearrangement of the flowing material or from the transverse flow perpendicular to the flow lines cannot be obtained.
- Classical continuum (continuum number of the set of points)
  - freezes internal degrees of kinematic freedom: only (continuous) displacement can be interpreted, independent rotational degree of freedom cannot,
  - the internal kinematic degrees of freedom can only be continuous: the displacement of two adjacent points is almost identical, only points between which there are several additional points can move face to face.
- The grid continuum or generalized continuum (finite number of the set of points, embedded into a point set of continuum cardinality)
  - preserves the degree of internal kinematic freedom: besides the displacement, the independent rotation can be interpreted, as well as any other degree of kinematic freedom in connection with the change of the shape of the solid body sitting in the node points of the grid continuum, or the cells form by the solid bodies sitting in the node points of the grid continuum,
  - the internal degree of kinematic freedom can be both continuous and discrete: the displacement, rotation, change of shape of two adjacent "points" can be almost the same, i.e. continuous, but two neighboring "points" can shift and rotate face to face, their strain states may be opposite,
  - continuous state functions only have mechanical content applicable to elements of the grid points in the node points of the grid continuum.

Findings analogous to kinematic degrees of freedom can also be made for dynamical degrees of freedom (stresses, internal forces).

References
[1] Love A E H 1927 *A Treatise on the Mathematical Theory of Elasticity* 4th ed. (Cambridge, At the University Press)
[2] Lurie A I 2005 *Theory of Elasticity* (Berlin: Springer-Verlag)
[3] Nowacky W 1970 *Theory of Micropolar Elasticity* (Wien: Springer-Verlag)
[4] Eringen A C 1970 *Foundations of Micropolar Thermoelasticity* (Wien: Springer-Verlag)
[5] Eringen A C 1998 *Microcontinuum Field Theories* (New York: Springer-Verlag)
[6] Lámer G 1984 Contradictions in the theory of micropolar elasticity and their causes *Newsletter* (Techn, Univ. of Budapest) vol II (1) pp. 12-16
[7] Noll W Lectures on the foundation of continuum mechanics and thermodynamics *Archives for*

Rational Mechanics and Analysis vol 52 pp. 62-92
[8] Lámer G 1984 Differential geometry, as the mathematical background of continuum physics *Newsletter* (Techn. Univ. of Budapest) vol II (3) pp. 17-21
[9] Lámer G 2008 Dry, wet, bounded granules and continuous media, or from grain to continuum *Proc. Engineering-Rock Mechanics 2008.* (Budapest, 26. November 2008.) Eds.: Török Á and Vásárhelyi B (Budapest: P. Co. of BTU) pp. 271-286 (In Hungarian)
[10] Lámer G 2003 Solid and soft body with and without structure *Quasi-static deformations of particular materials Proc. of QuaDMP'03 Workshop* 25-28. August 2003. Ed.: Bagi K (Budapest P. Co. of BTU) pp. 159-166.
[11] Kelly J L 1957 *General Topology* (New Yersey: Van Nostrand Co. Inc. Princeton)
[12] Riesz F and Szőkefalvi-Nagy B 1988 *Funkcional Analysis* (Budapest: Tankönyvkiadó) (In Hungarian)
[13] Lámer G 2011 Geometrical and topological charecterisation of grains and heaped nody. *Poc. Technical Science in the North East Hungary Region 2011.* (Miskolc, 19. May 2011.): Ed.: Pokorádi L (Miskolc: Technical Expert Committee of the Debrecen Academy Committee) pp. 273-285 (In Hungarian)
[14] Lámer G. 2011 Some geometrical and topological questions of grains and heaped body *Multidisciplinary sciences, Publications of the University of Miskolc.* vol I (1). Ed.: Tisza M (Miskolc: Press at University of Miskolc) pp. 79-86 (In Hungarian)
[15] Jikov V V, Kozlov S M and Oleinik O A 1994 *Homogenization of Differential Operators and Integral Functionals* (Berlin: Springer-Verlag)
[16] Lomakin V A 1976 *Theory of Elasticity of Inhomogeneous Bodies* (Moscow: Press at Moscow University) (In Russian)
[17] Shermergor T D 1977 *Theory of Mikro-Inhomogeneous Bodies* (Moscow: Nauka) (In Russian)
[18] Pavliotis G A and Stuart A M 2008 *Multiscale Methods. Averaging and Homogenisation* (New York: Springer)
[19] Kunin I A 1982-83 *Elastic Media with Microstructure I. One-Dimensional Models II. Three-Dimensional Models* (Berlin: Springer-Verlag)
[20] Kröner E (Ed.) 1968 *Mechanics of Generalized Continua. Proc. of the IUTAM-Symposium on The Generalized Cosserat Continuum and the Continuum Theory of Dislocations with Applications*, Freudenstadt and Stuttgart (Germany) 1967. (Berlin: Springer-Verlag)
[21] Maugin G A and Metrikine A V (Eds.) 2010 *Mechanics of Generalized Continua. One Hundred Years After the Cosserats* (Berlin: Springer-Verlag)
[22] Altenbach H, Maugin G A and Erofeyev V (Eds.) 2011 *Mechanics of Generalized Continua* (Berlin: Springer-Verlag)
[23] Maugin G A 2013 *Continuum Mechanics Through the Twentieth Century: A Concise Historical Perspective* (Berlin: Springer-Verlag)
[24] Revuzenko A F 2006 *Mechanics of Granular Media* (Berlin: Springer-Verlag)
[25] Tejchman J 2010 *Shear Localization in Granular Bodies with Micro-Polar Hypoplaszticity* (Berlin: Springer-Verlag)
[26] Sadovszkaya O and Sadovszkii V 2012 *Mathematical Modelling in Mechanics of Granular Materials* (Berlin: Springer-Verlag)

IOP Conf. Series: Materials Science and Engineering **568** (2019) 012058 doi:10.1088/1757-899X/568/1/012058

Notes on the possibilities and limitations of using metrical tools in mechanical modeling of media

G Lámer

University of Debrecen, Faculty of Engineering, Department of Engineering Management and Enterprise. 4028 Hungary, Debrecen, Ótemető utca 2-4.

glamer@eng.unideb.hu

Abstract. Changes in the size and shape of a continuous body can be described by metrical tools. The change in size is experimentally associated with relative elongation, the change of shape with the (local) angle change. Within the body, the concept of size (line length) and angle (angle between two lines) is based on the scalar product; the metric tensor is also needed to describe these relationships. When determining the relationship between size and size change as well as angle and angle change, it should be taken into account that, although both length and angle are interpreted by scalar product of metric tensor components, but in the context of relative length change there are fractional functions of radical expressions, in the context of angular change, there are fractional functions of radical expressions and partly inverse angle functions. The tensor cannot be formed directly from these two functions, but only by series expansion, and then there is a limit to the theoretical and numerical application of the introduced strain tensor: that the whole function can be approached well with the first few members of series expansion. This allows only a small strain (small relative elongation) to be described. This limit and the limitation of the use of topological tools (that is, the significant change of size and shape is not a deformation but a rearrangement) are consistent with each with other.

1. Introduction

Generally, the symmetrical linear expression constructed from the gradient of the displacement vector of the body is used for the local description of continuous media, $\boldsymbol{\varepsilon}_L = (\nabla\mathbf{u} + (\nabla\mathbf{u})^* / 2,$ or an expression including first degree and quadratic elements, $\boldsymbol{\varepsilon}_{NL} = (\nabla\mathbf{u} + (\nabla\mathbf{u})^* + \nabla\mathbf{u}(\nabla\mathbf{u})^*)/2.$ Generally, the ε_L expression is associated with small strain, while the ε_{NL} expression is associated with large strain. It is known from the literature [1,2], that nonlinear relationships are necessary in case of movement without strain for the strains to "disappear". According to this, it is not the linear but the nonlinear expressions that belong not even the small, but rather to the zero strains. Based on the principle of continuity - the expression suitable for describing small strains is closer to zero than the expression describing large strain - this is incomprehensible.

It is also known from the literature (see, for example, the above references) that the relationships explained above from the metric tensor to describe the physical strains are not given "automatically", but can be interpreted mathematically in relation to relative elongations and relative angular changes such that the root function and the angle function are associated with the first one or two elements of its power series. This assumes that the absolute value of the expressions tested is small (negligible to compare with 1 (one)).

IOP Conf. Series: Materials Science and Engineering **568** (2019) 012058 doi:10.1088/1757-899X/568/1/012058

These examinations are fundamentally based on previous research [3,4,5].

2. Differential geometric description of the continuous medium

2.1. Differential geometric background

On the geometric background see e.g. [6,7], and [8]; theory of elasticity [9], and [10,2]. In this subsection, the overview is based on results from literature; our aim is to list relationships to support further arguments. In addition to the list, we also focus on presenting step by step how each geometric concept is built upon each other.

The coordinate system $\{q^i\}$ should be given in space.

Mark the position vector before the deformation with **r**, and **R** after the deformation. The following relationships are written only for the state before the deformation. They apply to analog relationships following the change in strain.

First we look at the quantities describing the geometry.

Basis vectors in position vector **r**: $\mathbf{r}_i = \dfrac{\partial \mathbf{r}}{\partial q^i}$.

Components of metric tensor: $g_{ij} = \langle \mathbf{r}_i, \mathbf{r}_j \rangle$.

Partial differentials of basis vectors: $\mathbf{r}_{ij} = \dfrac{\partial \mathbf{r}_i}{\partial q^j}\left(= \dfrac{\partial \mathbf{r}_j}{\partial q^i}\right) = \Gamma^k_{ij}\mathbf{r}_k \left(= \Gamma^k_{ji}\mathbf{r}_k\right)$.

The affine coefficients (Christoffel-type symbols): $\Gamma^k_{ij} = g^{ks}\langle \mathbf{r}_{ij}, \mathbf{r}_s \rangle = \dfrac{1}{2} g^{ks}\left(\dfrac{\partial g_{is}}{\partial q^j} + \dfrac{\partial g_{js}}{\partial q^i} - \dfrac{\partial g_{ij}}{\partial q^s}\right)$.

Curvature tensor describing the curvature of space: $r^i{}_{jkl} = \dfrac{\partial \Gamma^i_{jl}}{\partial q^k} - \dfrac{\partial \Gamma^i_{jk}}{\partial q^l} + \Gamma^i_{sk}\Gamma^s_{jl} - \Gamma^i_{sl}\Gamma^s_{jk}$.

Then we look at those geometric terms, which can be expressed with the amounts above.

Scalar product of vectors **a** and **b**: $\langle \mathbf{a}, \mathbf{b} \rangle = g_{ij} a^i b^j$.

The (cosine of) angle between vectors **a** and **b**: $\cos(\mathbf{a}, \mathbf{b}) = \dfrac{\langle \mathbf{a}, \mathbf{b} \rangle}{\sqrt{\langle \mathbf{a}, \mathbf{a} \rangle \langle \mathbf{b}, \mathbf{b} \rangle}}$.

The length of the regular arc between two points of the space: $s = \int_{t_0}^{t_1} ds = \int_{t_0}^{t_1} \sqrt{\left| g_{ij} \dfrac{dq^i}{dt}\dfrac{dq^j}{dt} \right|}\, dt.$

The absolute differentials (presented for one contravariant component): $\nabla_j a^i = \dfrac{da^i}{dq^j} + \Gamma^i_{kj} a^k$.

Parallel displacement and the difference resulting from interchanging absolute differentials can be described with the curvature tensor.

2.2. Relationships among the geometric objects before and after deformation

See relationships in general e.g. [9,10,2].

The difference of position vectors is the displacement vector: $\mathbf{u} = \mathbf{R} - \mathbf{r}$.

The difference of metric tensors in the measure tensor of strain: $\boldsymbol{\gamma} = \mathbf{G} - \mathbf{g}$; different measure tensors of strain are usually interpreted, generally half of those given here. We will come back to this later.

The change of affine connexion coefficient: $\lambda^k_{ij} = \bar{\Gamma}^k_{ij} - \Gamma^k_{ij}$. Generally, no geometric content is associated with it. In case of small strains *and* small displacements, this change is negligible, in case of small strains *and* large displacements cannot be neglected (see e.g. [4,5]).

Change of curvature tensor: $\boldsymbol{\rho} = \mathbf{R} - \mathbf{r}$, that is $\rho^i{}_{jkl} = R^i{}_{jkl} - r^i{}_{jkl}$; give the compatibility equations (if it is made to equal zero).

To construct the theory, it is necessary to express the change of different geometric objects with the gradient tensor of the displacement vector. Here, we only give the relationship for one geometric object, the measure tensor of strain: $\boldsymbol{\gamma} = \nabla\mathbf{u} + (\nabla\mathbf{u})^* + (\nabla\mathbf{u})(\nabla\mathbf{u})^*$. We do not use the other relationships in this study, for relationships see [3,4,5].

3. Relationship of displacement and rotation during the strain of continuous medium

The change of continuous medium due to external effect is hereinafter referred to as mapping. The mapping of the continuous medium locally is usually split into the "product" of rotation and strain with polar division (see e.g. [11,12,13], or [9,10,]).

The mapping of a continuous medium is given with the displacement vector: $\mathbf{u} = \mathbf{R} - \mathbf{r}$. The displacement vector is the kinematic variable of the continuous medium.

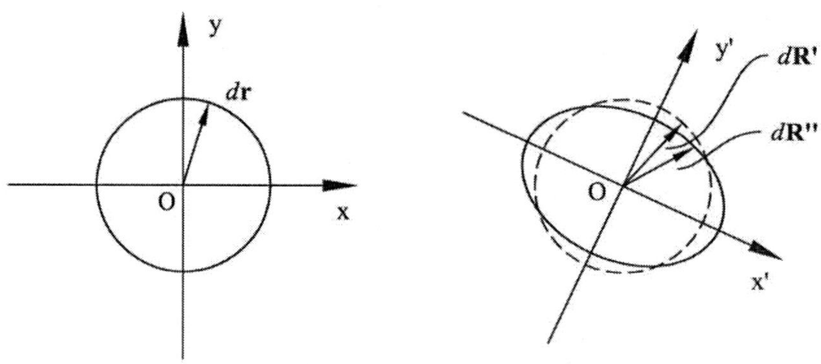

Figure 1. Rotation of a direction vector and the neighbourhood of a point:
$d\mathbf{R}' = \boldsymbol{\Omega}\cdot d\mathbf{r}$; $d\mathbf{R}'' = (\nabla\mathbf{R})^* d\mathbf{r}$

The tensor $\boldsymbol{\Omega}$ obtained by polar division ($\nabla\mathbf{R} = \boldsymbol{\Omega}\boldsymbol{\mathcal{A}}$) of gradient tensor $\nabla\mathbf{R}$ of mapping $\mathbf{u}$ gives the rotation of the neighborhood of the point, $d\mathbf{R}' = \boldsymbol{\Omega}\, d\mathbf{r}$ (see figure 1).

The gradient tensor $\nabla\mathbf{R}$ of mapping $\mathbf{u}$ gives the rotation of a direction vector marked with $d\mathbf{r}$ of position vector $\mathbf{r}$, and the same time the change in its length: $d\mathbf{R}'' = (\nabla\mathbf{R})^* d\mathbf{r}$ (see figure 1).

Based on the above, the following conclusions can be made. In the continuous medium we can interpret two rotations. One is the rotation of the direction, and the rotation of the neighborhood of a point. Both depend on the kinematic variable of the continuous medium, the displacement vector **u**.

The individual decomposition can be expressed by the gradient tensor of the displacement vector **u**.

In the linear case, for this reason, the subscript L is used, we get the following terms:

$$(\nabla\mathbf{R})^* = \mathbf{g} + (\nabla\mathbf{r})^* = \boldsymbol{\mathcal{A}}_L \circ \boldsymbol{\Omega}_L = \boldsymbol{\Omega}_L \circ \boldsymbol{\mathcal{A}}_L = \mathbf{g} + \boldsymbol{\varepsilon}_L + \boldsymbol{\omega}_L, \tag{1}$$

where

$$\boldsymbol{\mathcal{A}}_L = \mathbf{g} + \boldsymbol{\varepsilon}_L, \tag{2}$$

$$\boldsymbol{\Omega}_L = \mathbf{g} + \boldsymbol{\omega}_L, \tag{3}$$

$$\boldsymbol{\varepsilon}_L = \frac{1}{2}(\nabla\mathbf{u} + (\nabla\mathbf{u})^*), \tag{4}$$

$$\boldsymbol{\omega}_L = \frac{1}{2}((\nabla\mathbf{u})^* - \nabla\mathbf{u}). \tag{5}$$

IOP Conf. Series: Materials Science and Engineering **568** (2019) 012058 doi:10.1088/1757-899X/568/1/012058

In the linear case, both the right and left rotation tensors ($\mathbf{\Omega}_L$) as well as the right and left tensors connected with the strain ($\mathcal{A}_L$) coincide; the $\mathbf{\Omega}_L$ and $\mathcal{A}_L$ quantities can be interchanged with error negligible to compare with 1 (one).

Note. It is known from linear algebra that $\boldsymbol{\omega}_L = (\nabla\mathbf{u})^* - \nabla\mathbf{u})/2$ tensor quantity does not describe rotation but interprets vector product; the expression $\mathbf{\Omega}_L = \mathbf{g} + \boldsymbol{\omega}_L$ itself rotates, if $\boldsymbol{\omega}_L$ is small compare with g [7,14].

In the quadratic case, for this reason, the subscript *KV* is used, as well as subscript "*le*" refers to the left, subscript "*ri*" refers to the right decomposition, we get the following terms:

$$(\nabla\mathbf{R})^* = \mathcal{A}_{le\,KV} \circ \mathbf{\Omega}_{KV} = \mathbf{\Omega}_{KV} \circ \mathcal{A}_{ri\,KV} = \mathbf{g} + (\nabla\mathbf{r})^*, \tag{6}$$

where

$$\mathcal{A}_{le\,KV} = \mathbf{g} + \frac{1}{2}(\nabla\mathbf{u} + (\nabla\mathbf{u})^*) - \frac{1}{8}((\nabla\mathbf{u})(\nabla\mathbf{u}) + (\nabla\mathbf{u})^*(\nabla\mathbf{u})^*) - \frac{1}{8}(\nabla\mathbf{u})(\nabla\mathbf{u})^* + \frac{3}{8}(\nabla\mathbf{u})^*\nabla\mathbf{u}, \tag{7}$$

$$\mathcal{A}_{ri\,KV} = \mathbf{g} + \frac{1}{2}(\nabla\mathbf{u} + (\nabla\mathbf{u})^*) - \frac{1}{8}((\nabla\mathbf{u})(\nabla\mathbf{u}) + (\nabla\mathbf{u})^*(\nabla\mathbf{u})^*) + \frac{3}{8}(\nabla\mathbf{u})(\nabla\mathbf{u})^* - \frac{1}{8}(\nabla\mathbf{u})^*\nabla\mathbf{u}, \tag{8}$$

$$\mathbf{\Omega}_{KV} = \mathbf{g} + \frac{1}{2}((\nabla\mathbf{u})^* - \nabla\mathbf{u}) + \frac{3}{8}(\nabla\mathbf{u})(\nabla\mathbf{u}) - \frac{1}{8}(\nabla\mathbf{u})^*(\nabla\mathbf{u})^*) - \frac{1}{8}(\nabla\mathbf{u})(\nabla\mathbf{u})^* - \frac{1}{8}(\nabla\mathbf{u})^*\nabla\mathbf{u}. \tag{9}$$

The left and right decompositions do not match. The fact that the tensors connected with deformations in the left and right decomposition are not the same raises the possibility that nature chooses between the left and right decompositions: among the two possibilities of deformation, experimental results should be distinguished. The author is not aware of any theoretical foundations and experimental results. The components of the strain tensor will be determined using components of metrical tensors (in deformed and non-deformed states), so there is no need to use the tensors $\mathcal{A}_{le\,KV}, \mathcal{A}_{ri\,KV}$, see Section 4.

4. Interpretation of strain in a continuous medium

4.1. Experimental interpretation of strain

The strain was derived from experimental experience. Two types of strain are usually distinguished [13]: specific length change, independent of the length of the prismatic rod (specimen), and the specific rotation of each cross-section of the circular cylinder (specimen) independent of the length of the specimen.

The length of the prismatic specimen is marked with *l*, while the elongated (or shortened) length is marked with *L*. The relative elongation is interpreted with the relationship $\varepsilon_l = (L-l)/l$; (see figure 2). The thus interpreted specific elongation locally describes the strain of the body. This strain is also called linear strain. The linear strain can also be interpreted by comparing the elongation of the specimen with the length of the elongated (shortened) length, $\varepsilon_l = (L-l)/L$, rather than the length of the specimen. It can be shown that as long as with the ε_l is negligible to compare with 1 (one), then the two interpretations equal with errors negligible to compare with 1 (one). It also means that the above interpretation of linear strain has a limit: the linear strains cannot be large. This does not significantly affect the interpretation of strain, since the large-scale change of shape is not a strain but a rearrangement [15].

The length of the cylindrical specimen is marked with *l*, its radius with *r*, the rotation of two end cross-sections compared to each other, marked with Φ, the relative angular change is interpreted with the relationship: $\varepsilon_\varphi = (r\Phi)/l$ (see figure 2). The thus interpreted specific angular change locally describes the strain of the body. This strain is also called for change. The angle change strain can also

IOP Conf. Series: Materials Science and Engineering **568** (2019) 012058 doi:10.1088/1757-899X/568/1/012058

be interpreted as comparing the angle change of the specimen with the length of the elongated (shortened) length or the length of the twisted line, $\varepsilon_\varphi = (r\Phi)/L$. It can be shown that as long as with ε_φ is negligible to compare with 1 (one), then the two interpretations equal each to other with errors negligible to compare with 1 (one). It also means that the above interpretation of form change has a limit: the angular changes cannot be large. Regarding this, see the note made at the linear strain.

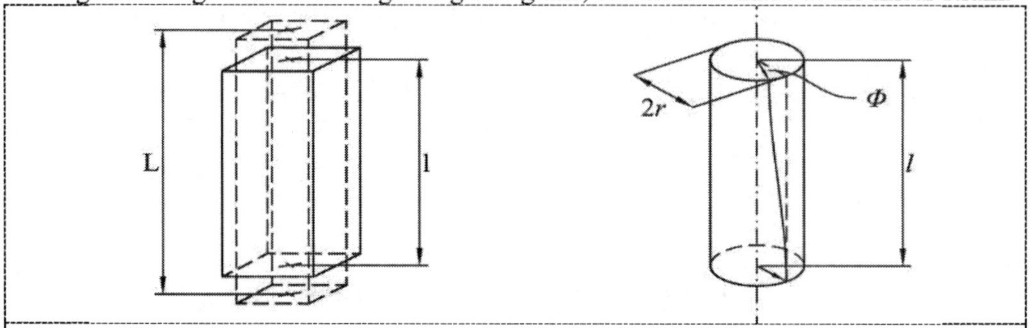

Figure 2 Experimental interpretation of strain: relative elongation and angle change

4.2. Mathematical interpretation of strain

The experimental interpretation of strain refers to a homogeneous strain state. This is not a general strain state. Therefore, in the case of arbitrary strain, it is not possible to characterize the strain of the body by the relative elongation of a particular selected curve. Analyzing the specific elongation of the specimen, we interpret the relative elongation of a (regular) arc (curve section) within a body: $\varepsilon_l = (L-l)/l$, where l is the length of the selected (regular) arc before the deformation, and L marks the length following the deformation. Then the strain of a body can be interpreted: the strain of the body is the set of the relatíve elongation of all regular arces (curve sections) in the volume of the examined body.

The strain is based on the relative elongation of the curve sections and does not formally say anything about the change of angle closed by the curves. In terms of a triangle, the relationship between elongation and angular change can be interpreted.

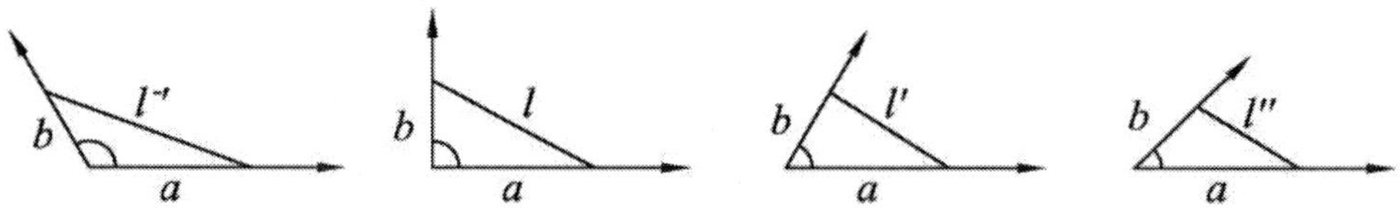

Figure 3 For the interpretation of the relationship between angular change and elongation

Consider the triangles shown in figure 3. The line segments a and b can lock different angles. Changes in angles are characterized by relative elongation applying to different lengths l. If the relative elongation of line segments a and b is negligible, i.e. negligible to compare with 1 (one), then the angle change is characterized by the relative elongation of the length l of the section stretched by the two line segments with errors negligible to compare with 1 (one).

4.3. Small and large strain

The interpretation of strain allows the strains to be grouped according to the scale of the strains.

The strain of a body is small if the relative elongation of any (regular) arc connecting any two points of the body within the body is sufficiently small, i.e. less than a negligible to compare with 1 (one) value.

The strain of a body is great if there is at least one regular arc that connects any two points of the body within the body with a relatively not small elongation.

5. Expressing strain with the change of the metric tensor

The strains described above characterize the strain of the body globally (in the integral form). For local characterization, arbitrary regular arcs should be located at every point of the body, with three fixed-length regular arcs – the base vectors as infinitely short arcs. In this case, it should be pointed out that the relative elongation in the three directions and relative angular changes of the three directions in one point already determine the relative elongation of the basic vector in any direction at the same point, and the relative angular change between any two basic vectors at the same point. If the relative elongations and relative angular change constitute a tensor quantity, they clearly determinate these strains.

5.1. Geometric content of metric tensor elements

The elements of the metric tensor are partly related to the relative elongation of the base vectors and partly to the angle between the base vectors (see e.g. [9,10,2]). Relative elongation of the base vectors can be expressed with the help of the main diagonal elements (there is no summary here and later according to the repeated indices)

$$\eta_{ii} = \frac{\sqrt{G_{ii}} - \sqrt{g_{ii}}}{\sqrt{g_{ii}}}, \tag{10}$$

and its angular change

$$\eta_{ij} = \alpha_{ij} - A_{ij}, \tag{11}$$

where

$$\cos\alpha_{ij} = \frac{g_{ij}}{\sqrt{g_{ii} g_{jj}}} \quad \text{and} \quad \cos A_{ij} = \frac{G_{ij}}{\sqrt{G_{ii} G_{jj}}}. \tag{12}$$

Note that indexing applying to the η quantity suggests that we are dealing with a secondary rank tensor; if it is true, it would have to prove it. The terms (10-12) above show that the nine η_{ij} amounts do not form a tensor.

5.2. Expressing small strain with metric tensor change

The explication below is based on our previous results [3,4,5].

Relationships (10-12) do not only mean that the nine η_{ij} quantities do not form a tensor, but it can also be seen that the η_{ij} quantities depend on components of the measure tensor of strain $\boldsymbol{\gamma} = \mathbf{G} - \mathbf{g}$ in a complex way.

Let's transform the (10) relationship!

$$\eta_{ii} = \frac{\sqrt{G_{ii}} - \sqrt{g_{ii}}}{\sqrt{g_{ii}}} = \sqrt{1 + \frac{\gamma_{ii}}{g_{ii}}} - 1. \tag{13}$$

We can have two assumptions. *The first assumption* is that the relative elongations, i.e., η_{ii} quantities, are considered to be small, that is negligible to compare with 1 (one). Then the square root expression can be expressed in a series and in it it is enough to keep some elements. At the same time, this also means that the main diagonal elements of the measure tensor of strain, γ_{ii}, are negligible in relation to the main diagonal elements of the metric tensor of the undeformed state, g_{ii}. *The second assumption* is to express the square root expression in a series, and some first elements approximate the square root with sufficient accuracy. This means, on the one hand, that the main diagonal elements of the measure tensor of strain, γ_{ii}, are negligible in relation to the main diagonal elements of the metric tensor of the undeformed position, g_{ii}, on the other hand, that the η_{ii} quantities are small and negligible to compare with 1 (one). The two assumptions are equivalent.

By series expansion of the radical expression and retained by the two elements (13) it can formulate as follows:

$$\eta_{ii} = \sqrt{1+\frac{\gamma_{ii}}{g_{ii}}} - 1 \cong \frac{\gamma_{ii}}{2g_{ii}} - \frac{1}{2}\left(\frac{\gamma_{ii}}{2g_{ii}}\right)^2 = \frac{\gamma_{ii}}{2g_{ii}}\left(1 - \frac{1}{4}\frac{\gamma_{ii}}{g_{ii}}\right). \tag{14}$$

Since the condition of series expansion is that the quotient of γ_{ii}/g_{ii} in other words, the relative elongation at 1 (one) is negligible, the relative elongation in the case of small relative elongations can be therefore described with the expression

$$\eta_{ii} = \frac{\gamma_{ii}}{2g_{ii}}\left(= \frac{1}{2}\frac{\gamma_{ii}}{\sqrt{g_{ii}}\sqrt{g_{ii}}}\right). \tag{15}$$

For the "explication" of the expression (11), consider the $A_{ij} = \alpha_{ij} - \varphi_{ij}$ expression, where φ_{ij} marks the angular change of the ith and the jth base vectors. Then the following relationship can be described:

$$\cos A_{ij} = \cos\left(a_{ij} - \varphi_{ij}\right) = \cos a_{ij}\cos\varphi_{ij} + \sin a_{ij}\sin\varphi_{ij}. \tag{16}$$

The relationship means that the correlation in the case of a rectangular coordinate system is greatly simplified.

The case of orthogonal coordinates

The relationship (16) is simplified to the relationship

$$\cos A_{ij} = \sin\varphi_{ij}. \tag{17}$$

From this, the function of the angle change φ_{ij} can be expressed as a first two member of polynomial series if the angle change itself is negligible at 90°, i.e. if the relative (and thus the absolute) angle change is small. You can then write instead (17):

$$\varphi_{ij} = \cos A_{ij} = \frac{G_{ij}}{\sqrt{G_{ii}G_{jj}}}. \tag{18}$$

Using repeatedly that the main diagonal elements of the measure tensor, γ_{ii}, are negligible to compare with main diagonal elements of the metric tensor of the undeformed state, g_{ii},, in the case of orthogonal coordinates, for the change of the angles of the base vectors we obtain the expression:

$$\varphi_{ij} = \frac{\gamma_{ij}}{\sqrt{g_{ii}g_{jj}}}, \quad (i \neq j). \tag{19}$$

Except for coefficient ½, this term is the same as that of the small relative elongation below (15).

The case of oblique coordinates

Assuming in expression (16), that the φ_{ij} function of the angle change is negligible to compare with 90°, and applying the series expansion for the cos and sin angle functions up to the quadratic elements ($\cos\varphi_{ij} \cong 1-\varphi_{ij}^2/2$ and $\sin\varphi_{ij} \cong \varphi_{ij}$) the (16) expression can be transformed into a quadratic function the roots of which:

$$\varphi_{ij} = \mathrm{tg}\,\alpha_{ij}\left(1 \pm \sqrt{1 - 2\frac{\cos\alpha_{ij}}{\sin\alpha_{ij}}\frac{\cos A_{ij} - \cos\alpha_{ij}}{\sin\alpha_{ij}}}\right), \quad (i \neq j). \tag{20}$$

Small angle change only results in a solution with a negative sign. We apply series expansion again, and the elements of the series expansion, up to the quadratic elements, give the following expression to describe the angle change:

$$\varphi_{ij} = \frac{\cos A_{ij} - \cos\alpha_{ij}}{\sin\alpha_{ij}} + \frac{1}{2}\frac{\cos\alpha_{ij}}{\sin\alpha_{ij}}\left(\frac{\cos A_{ij} - \cos\alpha_{ij}}{\sin\alpha_{ij}}\right)^2, \quad (i \neq j). \tag{21}$$

(To retain cubic or higher power elements, the term (20) cannot be used.)

To examine the term (21), first consider the second members. The member in front of the parentheses in the near-right coordinate system gives a value close to zero, and if the first member is negligible to compare with 1 (one), then the second member is negligible to compare with the first member.

Therefore, the angle change can be expressed in an oblique coordinate system as:

$$\varphi_{ij} = \frac{\cos A_{ij} - \cos\alpha_{ij}}{\sin\alpha_{ij}} = \frac{\dfrac{G_{ij}}{\sqrt{G_{ii}}\sqrt{G_{jj}}} - \dfrac{g_{ij}}{\sqrt{g_{ii}}\sqrt{g_{jj}}}}{\sqrt{1 - \dfrac{g_{ij}^2}{g_{ii}g_{jj}}}}, \quad (i \neq j). \tag{22}$$

Using again the main diagonal elements of the measure tensor of strain, γ_{ii}, are negligible to compare with the metric tensor elements of the undeformed state, g_{ii}, in (22) we get:

$$\varphi_{ij} = \frac{\gamma_{ij}}{\sqrt{g_{ii}g_{jj} - g_{ij}^2}}, \quad (i \neq j). \tag{23}$$

In an oblique coordinate system, the relationship (23) only equals with the (19) relationship with errors negligible to compare with 1 (one), if the coordinate system is close to orthogonal one, i.e. the base vectors close an angle between 85 to 95 degrees.

In the case of small strains, i.e. small relative elongations and small relative angular changes, and in a slightly oblique coordinate system, we found a relationship between the specific elongation, the specific angular change and the measure tensor of strain:

$$\eta_{ii} = \frac{\gamma_{ii}}{2\sqrt{g_{ii}}\sqrt{g_{ii}}}, \tag{24}$$

$$\eta_{ij} = \varphi_{ij} = \frac{\gamma_{ij}}{\sqrt{g_{ii}g_{jj} - g_{ij}^2}}, \quad (i \neq j). \tag{25}$$

In accordance with this, the tensor of small strains can be introduced.

We call to the tensor of *small* strains the tensor interpreted with physical components:

$$\tilde{\varepsilon}_{ij} = \frac{\gamma_{ij}}{2\sqrt{g_{ij}}\sqrt{g_{ij}}}. \tag{26}$$

The covariant components of small strains in the base $\{\mathbf{R}_k\}$ in the deformed state are:

$$\varepsilon_{ij} = \frac{\gamma_{ij}}{2}. \tag{27}$$

5.3. Conditions of introducing the tensor of strain and its consequences

We introduced the tensor of strain in such a way, that the tensor of strain let have a direct physical meaning: the main diagonal elements of strain tensor give the relative elongation of the tangent vectors of the coordinate lines, and the side diagonal elements of strain tensor gives the change of the angles between the coordinate lines (see e.g. [2]).

IOP Conf. Series: Materials Science and Engineering **568** (2019) 012058 doi:10.1088/1757-899X/568/1/012058

The conditions for introducing the strain tensor are as follows: relative elongation and relative angular changes are small; the coordinate system used to describe the problem is orthogonal at least nearly orthogonal.

The consequences of introducing the strain tensor can be summarized as follows. The deformations can only be small. During the derivation of formula, the small magnitude of relative elongation and relative angular change were used several times from the radical and trigonometric expressions. For this reason, it is difficult to assume that other strain tensors of large deformation could be derived by using a metric tensor-based expression of relative elongation and relative angular change from the above relationships. In pragmatic terms: there is *one strain tensor* and it only *applies to small strains*; *tensor of large strains cannot be derived*. This is not in contradiction with the physically experienced phenomena: the phenomenon considered to be a large strain is not deformation, but a rearrangement, for which the concept of the coordinate line cannot be interpreted, since the rearrangement of the particles excludes the topological order from being preserved [15]. *The strain tensor is symmetrical*. The side diagonal elements define the relative angle change between two vectors. Since the definition of the angle from one vector to the other or the angle from the second to the first is always the same, an asymmetric metric tensor cannot be formed. For the interpretation of an asymmetric metric tensor the interpretation of the metric tensor should be fundamentally changed. For this reason, the interpretation of an asymmetric strain tensor requires mathematical foundations other than the Riemann geometry, such as changing the definition of a scalar product, or defining an angle between two lines different from the scalar product. (Knowing the structure of geometry and differential geometry makes it difficult to imagine a different mathematical construction for these mathematical objects.)

Due to the interpretation of the tensors of small strains, it is non-linear: the strain tensor was not derived from the gradient tensor of the displacement, but from the measure tensor of the strain interpreted as the difference of the metric tensor in the deformed and undeformed state. However, this is by definition a quadratic function of the gradient tensor $\nabla \mathbf{u}$ of the displacement vector $\mathbf{u}$. Based on the above, the linearization of nonlinear relationships of the tensor of small strains is not related to the magnitude of the strain but to something else [3,4,5].

6. Summary

In this study we provided an overview of applying metric tools and their limitations.

By metric tools we mean the scalar product, the interpretation and measurement of distance and angle.

The conditions for applying metric tools are the following:

- we model the media to be examined as a domain of a topological space,
- a symmetric quadratic form, the metric tensor is given in the topological space,
- the length and angle enclosed by the two curves are given by a scalar product,
- relative strains are negligible to compare with 1 (one),
- the applied coordinate system is orthogonal at least almost orthogonal.

The consequences of applying metric tools are the following.

- In the classic continuum independent rotational degree of freedom cannot be interpreted:
  - the rotation of a point cannot be interpreted,
  - rotation can only be interpreted of a direction, or in a rigid body-like neighborhood of a point (three points, or a triple),
  - the displacement in the classic continuum clearly determines the rotation of any base vector as a direction, and the rotation of the neighborhood of every point,
  - as a result of it, an angle change which is independent of the displacement in any sense cannot be interpreted,
  - since the metric tensor is symmetric (the elements of the scalar product are interchangeable), therefore, an asymmetric strain tensor, and, consequently, a rotation field independent of the derivative of the displacement vector cannot be interpreted.
- In the classic continuum

- interpretable a strain tensor with mechanical content the main diagonal elements describe the relative elongation and the side diagonal elements describe the relative angular change,
- the strain tensor interpreted describes small strains,
- the strain tensor describing large strains cannot be interpreted.

– Tensor of small strains
- is half of the difference of the metric tensor of the deformed **G** and undeformed **g** state; and
- is expressed, because of its interpretation, by the quadratic function of the gradient tensor of the displacement vector **u**: $\boldsymbol{\varepsilon} = (\nabla\mathbf{u} + (\nabla\mathbf{u})^* + (\nabla\mathbf{u})(\nabla\mathbf{u})^*)/2$,
- is symmetrical, and
- linearization is independent of the magnitude of strain, more precisely, of that fact that the strains are small.

– The following conclusions can be regarding the rotations:
- the rotation of a direction vector is given by the $\nabla\mathbf{R}$ gradient tensor of position vector **R**,
- the rotation of the neighborhood of a point is given by the orthogonal tensor $\boldsymbol{\Omega}$ getting from the polar decomposition of the gradient tensor $\nabla\mathbf{R}$ of position vector **R** in the deformed state,
- the rotation of the neighborhood of a point is described in the first, linear approach by the tensor $\boldsymbol{\Omega}_L = \mathbf{g} + \boldsymbol{\omega}_L$,
- the tensor $\boldsymbol{\omega}_L = ((\nabla\mathbf{u})^* - \nabla\mathbf{u})/2$ does not describe the rotation, but instead interprets a vector product.

The tensor of small strains is consistent with the atomic-molecular structure and continuous mapping. The fact that a tensor of large strains cannot be created does not contradict the experience, since the significant change of shape is not a strain but a rearrangement.

References

[1] Goldenblat I I 1969 *Some Problems of Theory of Elasticity* (Moscow: Nauka) (In Russian)
[2] Lurie A I 2005 *Theory of Elasticity* (Berlin: Springer-Verlag)
[3] Lámer G 1985 Notes on the theory of large displacement with small strain *Periodica Politechnica* vol 29 (1-2) pp. 53-65
[4] Lámer G 1990 *Mathematical Foundation of the Theory of Perfectly Elastic Shells and Rods Underdoing Large Displacement with Small Strain* Thesis. (Budapest) (In Hungarian)
[5] Lámer G 1992-93 On the kinematics of continuum underdoing large displacement with small strain *Épités-, Építészettudomány* vol XXIII (1-2) pp. 35-59 (In Hungarian)
[6] Kobayashi S and Nomizu K 1963-69 *Foundations of Differential Geometry* vol I-II. (New York: Interscience Publisher)
[7] Korn G A and Korn T M 2000 Math*ematical Handbook for Scientists and Engineers. Definitions, Theorems, and Formulas for Reference and Review.* (New York: Dover Publications)
[8] Levi-Civita T 1926 *The Absolute Differential Calculus* (London: Blackie & Son Limited)
[9] Green A E and Zerna W S 1954 *Theory of Elasticity* (Oxford: The Clarendon Press)
[10] Love A E H 1927 *A Treatise on the Mathematical Theory of Elasticity* 4th ed. (Cambridge, At the University Press)
[11] Appell P 1928 *Traité de Mécanique Rationelle.* Tome Troisiéme. 3. Ed. (Paris: Gauther-Villars)
[12] Appell P 1926 *Traité de Mécanique Rationelle.* Tome Cinquieme. (Paris: Gauther-Villars)
[13] Budó Á 1972 *Mechanics.* 5th Ed. (Budapest: Tankönyvkiadó) (In Hungarian)
[14] Rózsa, P 1974 *Linear Algebra* (Budapest: Műszaki Könyvkiadó) (In Hungarian)
[15] Lámer G Notes on the possibilities and limitations of using topological tools in mechanical modeling of media (*In this Issue*)

IOP Conf. Series: Materials Science and Engineering **568** (2019) 012059 doi:10.1088/1757-899X/568/1/012059

Restoration and PLC control of a robot cell compatible MP-11 industrial manipulator platform

G Hegyi, T I Erdei, G Husi

Department of Mechatronics Engineering, University of Debrecen, Faculty of Engineering, Debrecen, Hungary

gabesz96@gmail.com

Abstract. The designing and usage of newer and newer tools and devices is an unavoidable part of industrial development. The robots employed this way either replace human work, or replace previous, obsolete robots. This advancement might cause otherwise functional robots to be discarded, or otherwise lost. Such robots might still be in usable condition after a suitable refurbishing, and their limitations can be alleviated by integrating these devices into a robot cell. In my project, I have set the goal to restore such a machine, which goal includes the full mechanical and electrical overview of the device and replacing its missing controller. The robot itself and other various resources necessary were provided by University of Debrecen, Faculty of Engineering, department of mechatronics.

1. Introduction

The robotics lab can be found University of Debrecen, Faculty of Engineering, department of mechatronics, where the subject of robot technology is taught to the students of the faculty. In this laboratory, numerous industrial devices can be found [1]. The goal of the laboratory is to have a collection of devices assembled and programmed in a way, that the students can find their way in any industrial environment where assembly lines are present. The MP-11 industrial manipulator can also be found in this room. I have chosen its restoration as my topic. During this work, the goal was the restoration and improvement of the manipulator, including its integration to the robot cell.

2. The robot's structure

The MP-11 manipulator, based on its first 3 axes of movement, can be categorised as an RTT work area manipulator, in other words, its work area is cylindrical. The work area itself is the curve of the cylinder, changing the curve's properties requires mechanical change. The pneumatic actuators' function is made possible by monostable, spring return, 3/2 electropneumatic valves. The manipulator is also equipped with hydraulic end-stop dampeners. The manipulator consists of three main parts: the first, is the body. This is where the valves are, and on the back panel of this part can the electrical and pneumatic connectors be found. The second, is the tower. This part is responsible for the translational, and rotational movements, meaning, that this part is able to rotate around, or move along its axes. This is where the movement limits can be set, by moving the end-stop bumpers, and the limit switches. The third part of the manipulator is where the two arms can be found. These are secured on top of the tower, and execute horizontal, translational movement. The limit switches are activated by magnetised bumpers found on the arms. After loosening the tightener screws, they are freely movable along the pistons' axes,

IOP Conf. Series: Materials Science and Engineering **568** (2019) 012059 doi:10.1088/1757-899X/568/1/012059

to the desired position. This first bump against the dampeners, and only later against the limit switches, so the dampeners' proper condition is crucial.On the second figure, the manipulator's kinematic structure can be observed. This specifies the amount of degrees of freedom. The first is a rotational direction of movement, around the "Z" axis. This is achieved by the tower's rotation. The second one is the translational (vertical) movement of the arms, along "Z" axis. The third movement is given by the movement of the arms themselves. The direction of these varies, as the angle between the two arms is adjustable. Another fact to note about the working of these axes, is that they always move in the same direction, as both cylinders and controller by the same valve. Both arms can be equipped with one gripper each, which have different properties. The first one is able to rotate in 180 degrees, along one axis, which axis is parallel with the axis of the arm's translational movement. This can be useful for rotating the workpiece by a desired amount, as the bumpers can be adjusted to adjust the rotation's amount.

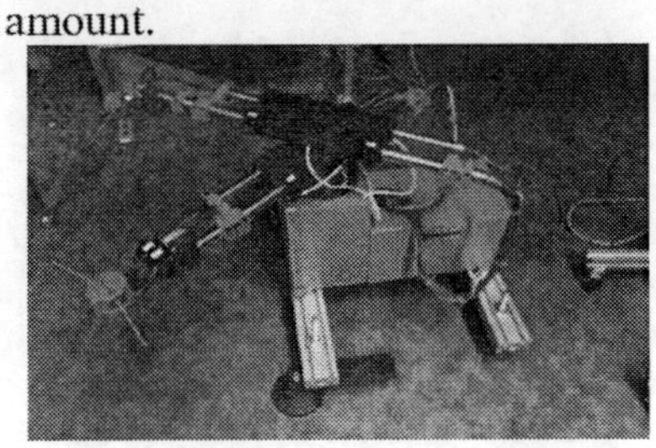

Figure 1. MP-11 manipulator

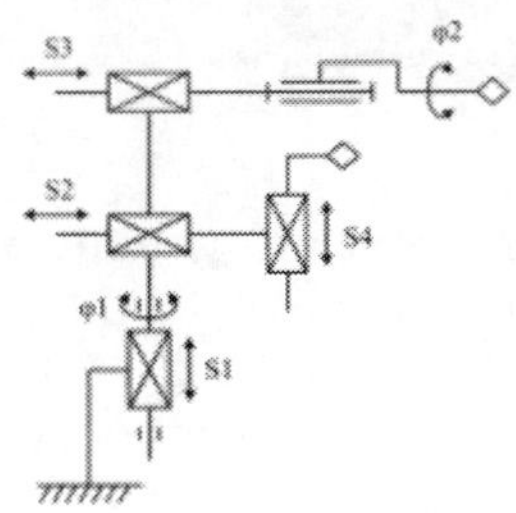

Figure 2. kinematic diagram [2]

The other gripper is able to do translational movement, perpendicular to its arm's axis. This can be used to lift out, or precisely move the workpiece. The magnitude of the movement can be similarly adjusted by moving bumpers. These grippers require jaws to grip the desired object. In this case, they are made from sheet metal.The heads are adjustable, depending on if we desire normally open, or normally closed functions. On the inside, a small piston can be found, which is returned to its starting position by a spring. As the head is entirely symmetrical, by rotating the piston and moving the spring to the other side, the gripper's resting state can be altered. These can be useful for lowering the load on the compressor, if multiple valves are operated at the same time.

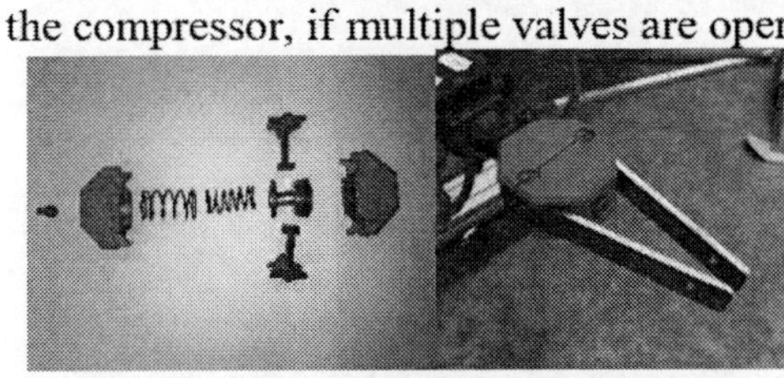

Figure 3. Normally closed gripper

Figure 4. Normally open gripper

Every actuator's end position, except the grippers, is sensed by reed-sensors. These sensors are placed in small, wooden sockets. In one half, the reed-relay can be found, while in the other, a spring-loaded bumper. When the bumper moves, the magnet of it approaches the reed sensor, closing the electromagnetic contacts, sensing the actuator's state. The advantage of the reed switch is its long lifetime, its insensibility to dirt and grime. Their maintenance is simpler than mechanical switches: in event of malfunction, the fix is to simply switch out the broken relay for a functional one.

3. Inspection and restoration

3.1. Maintenance

The first task was to inspect the manipulator's movements. After carefully moving the axes, it was found that not every actuator can reach its limits, as some cylinders were partially stuck. By thorough cleaning and lubrication this problem could be solved and has not returned later.The next task was to map the electronical connections. After removing the back panel, the 12 valves became visible, which control

IOP Conf. Series: Materials Science and Engineering **568** (2019) 012059 doi:10.1088/1757-899X/568/1/012059

the actuators. On them, it can be seen that they operate at a voltage level of 24 volts, however, no additional information was available concerning their power consumption. By using a bench power supply, by carefully and gradually increasing the current limit, I have found that the valves reliable open at 0,5 A.This was followed by the check-up of the limit switches, with the help of a digital multimeter's continuity mode. This process was greatly sped up by the fact, that near the connector, the same number is shown as near the limit switches. By the end of this test, every sensor seemed functional, except one, where the internal soldering was faulty. This was later fixed.

Figure 5. Valve

3.2. *Improvements*

The first upgrade was to improve the mobility of the manipulator. For this, four wheels (equipped with brakes) were placed at the bottom of the manipulator. As these wheels were not enough to provide necessary stability during operation, additional stands were added to the structure. The old connectors, which were used to connect the old controller, were damaged. As a replacement, three 9-pin D-SUB connectors were used. These connectors were fastened to plastic PCB panel material. These were then later cut to size and prepared in a way to allow the connector to fit through and be fastened by screws. The chosen PLC's output voltage is 24 V, however, it cannot operate two valves at the same time, as its total power is not high enough to do so. To overcome this problem, a device to switch logic power to a much higher power was created. In it, 12 TIP120G transistors were built in, using them as a switch. The PLC opens and closes these transistors, which are supplied by a power supply which is able to satisfy the power demands of multiple valves at the same time.

Figure 6. New connectors

In this device, galvanic separation was also added. The importance of this, is that the PLC and the manipulator are now completely separated, so in case of a valve's malfunction, the created power spike cannot damage the PLC. The device has a separate power supply.To connect the manipulator and the amplifier device, it was necessary to determine which connector is witch, and what purpose they fulfil. This is when the connectors were given the names "A", "B", and "C". These names were used to identify every pin, using the actual pin's number on the connector. For example, the first pin of connector "A" is called A/1. These connections were somewhat simplified within the amplifier device. Thanks to this simplification, the numbering of the PLC's outputs and inputs match the device's input and output numbering.

4. Control of the manipulator

4.1. *PLC*

To control the manipulator, a Schneider Modicon M340 modular PLC was chosen. The device was assembled on a BMXXBP0840 type rack. The power supply has its specific place on the rack, which is a BMXCPS2000. The processor must be placed at slot zero of the rack. The used processor is BMXP342030 [3], version 0.1. This processor is able to communicate via Ethernet, CanOpen, and USB networks. In the first and second position of the rack a BMXDDI1602, in third and fourth slots a BMXDDO1602 input and output modules were placed.

For PLC programming the Unity Pro S program was used. With this program, first the PLC itself must be assembled, using a graphical interface. The used modules must be chosen: the rack, the power supply, and the processor. After the processor is added, the rest of the modules can be freely added. Afterwards, the network itself must be created. The P342030 is able to communicate via Ethernet, as such, there's no need for a separate network module. After validating, the network is ready to use.

This is followed by the program's creation. The program was written in ladder diagram. During the addressing pf the discrete I/O modules, care must be taken that four pieces of information must be given. The first is defining if a given pin is an input, or output. In case of input, the letter "I2, in case of output, the letter "Q" is used. Afterwards, three numbers, separated by dots must be entered. The first number is only necessary if we use multiple PLCs within the same network, which was not the case in my project, so that number is 0. The second number is the module's slot on the rack. As the processor always goes to slot 0 in the first rack, the I/O modules are placed into slots 1-4. The final number is the number of the pin itself.

4.2. Human-machine interface

To faithfully represent an industrial environment, an interactive terminal (Schneider Magelis XBTGT7340) has also been used during a project. This is a touchscreen HMI, which can send to and receive data from the PLC. Vijeo Designer was used during its programming.

For the interactive terminal's use, it was necessary for it to be on the same network as the PLC. In Vijeo Designer, the HMI was given the IP address of the PLC. TCP/IP Ethernet network was used it this case as well. Registers had to be defined to be shared between the PLC and the HMI. The variables themselves may be named anything, but they must be of the correct format.

To control the PLC, the only required task is to use the created register variables on both the HMI, and the PLC. The HMI may function as an output (to show the PLC's states), or as an input (to command the PLC to do certain task, not unlike a pushbutton or switch). The created control panel has two uses. First, it's able to operate the valves separately, while also gives visual feedback, if they reached their end states. This can be useful during the creation of a new task. It might be time-consuming, or dangerous to run the entire program, so by using this method, the different states can be tested by the user beforehand. The other use is starting the automatic program.

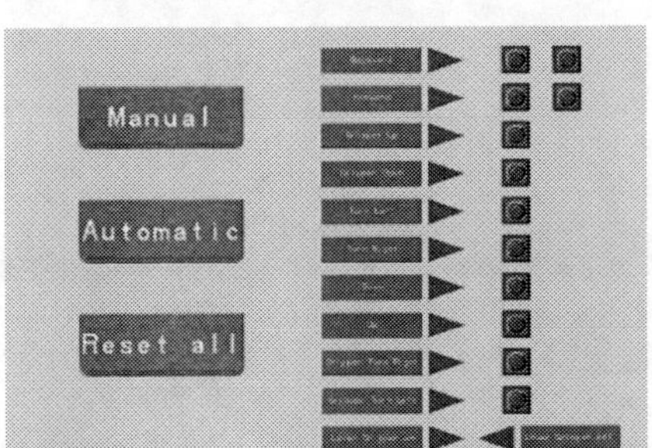

Figure 7. HMI

4.3. Automatic program

The manipulator's automatic control makes it possible for it to execute a task without further commands. The chosen task was the material handling of workpieces, modelled by steel cubes.

The program's "Start" function is activated when the HMI's "automatic" button is pressed. After that, the program continues, and sets the "HOME" position on the manipulator, by ordering three of its actuators to move. It moves upwards, left, and retracts the arms. This step has no additional conditions, as this position must be possible to reach from any other position. In every step, registers are used for the sequential control.

In the next step, the manipulator extends its arms. For this, the Y1 valve must be turned off, and Y2 valve must be turned on. In other words, set command is sent to output %Q0.3.2, and reset command is sent to output %Q0.3.1. The condition of this movement is that the manipulator must be in a lifted position, rotated to the left, and the R2 must be active. Every row also has an "A" register. This indicates the "automatic" switch's state on the HMI. If it's active, the program is running, if not, it stops. When later restarted, the PLC continues the program from the point where it earlier stopped. After the arm extends, the gripper grabs the cube. The gripper has no limit switch, so we cannot use this movement as condition in the next step. Because of this, a TON timer is used in the next step, which delays the execution of the following step, giving enough time for the gripper to successfully grip the handled material. Following the same logic, the manipulator retracts its arms, turns right, drops the cube, extends, grips another cube, retracts again, turns left, and drops the cube.

The program works cyclically, thus at this point, it jumps back to the beginning, after every register except R1 has been reset. The manipulator is in HOME position in the last step, so that particular step needn't be repeated. From this point, the program repeats until the "automatic" button is pressed on the HMI. For a completely clear state, the "rest all" button must be pressed. This resets every register and output to "low" state, and the system is no longer under load.

5. Conclusion

The restoration of the MP-11 manipulator was successful. The machine is mechanically intact. Its mobility was improved by wheels, but it's also possible to secure it with stands, during operation. Suitable modifications were made to make it compatible with modern PLCs. The new D-SUB connectors and the transistor device enable the system to use a wide variety of controllers. During testing, even an Ardunio Nano could drive the manipulator.It can be stated that the manipulator is suitable for educational purposes, because it's easy to follow the functioning of the manipulator, the programs are easy to upload using a PC via Ethernet, and they receive adequate feedback from the manipulator.As a potential upgrade, colour sensors or barcode scanners could be employed, which would enable the manipulator to sort products into two categories, based on this sensing method.

Acknowledgement

I'd like to express my thanks towards Erdei Timotei István for fulfilling his role as my consultant, and providing me with the necessary tools and information, which all helped during the realisation of this project, and to Dr. habil Husi Géza, who made this project possible.

References

[1] Erdei T I, Molnár Zs, Obinna N C and Husi G 2017 *Cyber physical systems in mechatronic research centre* Volume 126 MATEC Web Conf.

[2] Sabirov P A 2016 *Development and research of robotic assembly lines* (*Разработка и исследование роботизированной линии)*
Available: https://core.ac.uk/download/pdf/81248242.pdf [Accessed: 2018.12.04]

[3] *Schneider Electric, „BMXP342030,"* [Online]
Available:https://www.schneider-electric.co.in/en/product/BMXP342030/processor-module-m340---max-1024-discrete-%2B-256-analog-i-o---ethernet---canopen[Accessed: 2018.09.21]

IOP Conf. Series: Materials Science and Engineering **568** (2019) 012060 doi:10.1088/1757-899X/568/1/012060

Restoration and PLC control of a material handling MP-9S industrial manipulator platform

P P Horváth, T I Erdei, G Husi

Department of Mechatronics Engineering, University of Debrecen, Faculty of Engineering, Debrecen, Hungary

patrikgepe@hotmail.com

Abstract. In this project, a material handling MP-9S industrial manipulator was restored and put into operation. In its starting state, the only the manipulator itself was available, without a functioning controller. As a part of the restoration, the manipulator's mechanical structure was examined, the existing electrical connections and sensors were checked, its missing controller was replaced, and was later programmed to execute an example task. The control system was designed to flexible and able to accommodate a wide variety of controllers. In its currently realized state, the program can be observed and interacted with by the implementation of a Human-Machine Interface. These tasks and the necessary tests were completed at the robotics lab of University of Debrecen, Faculty of Engineering.

1. Introduction

At the „Cyber-Physical and Intelligent Robot Systems Laboratory" residing at University of Debrecen's building mechatronics research center, a laboratory showcasing new industrial solutions was created [1]. The new industrial robot cell makes it possible to emulate or simulate various industrial tasks [2]. Relevant data is available to those who participate in the research done in this laboratory, through its closed network [3]. Nowadays, mass production is considered an everyday phenomenon. This level of productivity can be most easily achieved by automated solutions. During modernizations, it's not uncommon to discard otherwise functional robots due to their obsolete controllers. As questions regarding sustainable development arise, with the desire of ever-higher profits still being present, reconstruction and modernization of an earlier discarded machine might be an option to consider. During this project, I have completed a similar restoration, during which my main goals were to restore the manipulator to its original (or near-original) state, and to replace the missing controller with a suitable, modern device. Keeping costs low was also an important consideration.The MP-9S manipulator making up a large part of this project was originally employed in a VAZ factory. Within the factory, they fulfilled material handling roles, more exactly, the handling of oriented half-done workpieces during hot-die manufacturing. Using these manipulators in pairs, the number of required workers reduced from six to just two [4].

2. Technical overview

2.1. *The physical and technical structure*

The manipulator's three degrees of freedom are given by one rotational axis (B), and two translational axes (A, C). Due to this, the manipulator is of TRT construction. The parameters defining the working

IOP Conf. Series: Materials Science and Engineering **568** (2019) 012060 doi:10.1088/1757-899X/568/1/012060

zone are the rotational angle, movement along X-axis, and movement along the Z axis. The manipulator has an open kinematic chain, which may be observed in figure 2. Apart from the three, degrees of freedom-giving movements, the manipulator can also move the gripper attached to it (D). The simplest possible gripper consists of two metal plates, which to be inserted into the gripper head's jaws. The actuators are all bistable construction, except the gripper, which is monostable (spring returned).

The movements of the manipulator happen from one end-state (limit) to the other, which are sensed by reed-sensors placed at every limit. The actuators have small permanent magnets built into them, which, after reaching the limit of movement, activate the sensors. The gripper does not have any sensors, as such they do not give feedback regarding their current position or state.

Figure 1. MP-9S manipulator

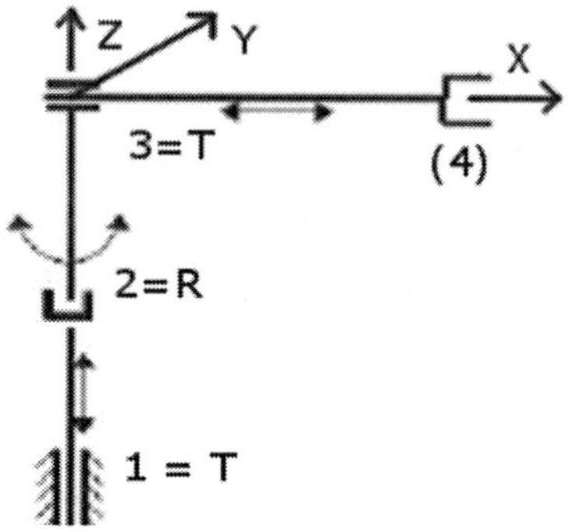

Figure 2. Kinematic chain based on [5].

The compressed air needed by the actuators is first moved though one of the previously mentioned, 3/2 monostable electro-pneumatic valves. The air supply must be connected to a connector found on the back panel of the manipulator. Their nominal voltage is 24[V], and their nominal pressure is 10 [bar]. The actuators require two valves each, except the gripper (which only needs one). As such, the manipulator's movements require a total of seven of these valves, leaving the eight one unused. The interior of the manipulator's housing can be accessed by lifting off the curved cover plate.

2.2. *Electrical structure*

Without a controller, the manipulator's electrical structure is limited to the wiring of the limit switches and electropneumatic valves. One of the connectors carry the signals send by the limit switches, why the other is responsible for controlling the valves. These valves, due to their electromagnets, generate a small backward current spike. To avoid damage from these spikes, diodes are connected parallel to them, on a separate component to which the valves' wires are connected to. In the manipulator, there are 17 separately numbered wires. Discovering the connections and purpose of these wires are one of the key tasks during the project.

3. Steps of restoration

3.1. *Preparations*

The restoration was preceded by preparations. First, the manipulator was cleaned, its outside surfaces received a new layer of paint. In addition, two aluminum profiles were fastened to the manipulator, which profiles were later equipped by wheels. This ensures the robot's increased mobility. Four threaded rods were added, serving as both a stand, and a way to move the manipulator along its Z axis, for finer positioning.

3.2. *Re-examination and restoration of electrical connections*

During this project, some of the main goals were to restore the equipment to a state where it may fulfill its original roles, and also achieve this by cost-effective methods. Due to this, keeping the existing wires and components (if they are safely functional) was justifiable. In the manipulator, most wires were already numbered, which numbering was extensively used during the restoration. Before reverse-

IOP Conf. Series: Materials Science and Engineering **568** (2019) 012060 doi:10.1088/1757-899X/568/1/012060

engineering said connections, the original connectors were removed: they were old, and partially damaged, finding an identical replacement was not feasible. Considering that only eight and nine pins were used per connector, it was more sensible to replace them with more modern parts. By using a continuity tester, tracing wires, and methodically activating limit-switches, the electrical connections of the manipulator were mapped. Later, DB-9 connectors were added as replacement connectors.

After finding the purpose of every wire, the valves' electromagnets had to be tested. Finding the value of electrical current required for the proper function of the electromagnets was found by using a bench power supply, by gradually increasing the current limit at 24 [V] voltage: the value was found to be between 500 and 600[mA].

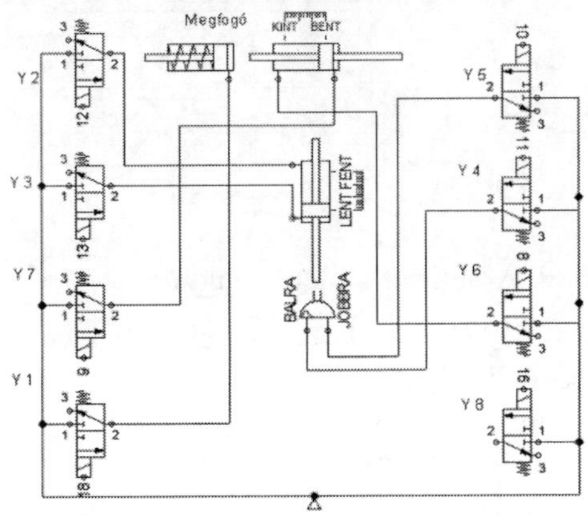

Figure 3. Pneumatic connections

3.3. Testing of pneumatics

Before the controllers' replacement, it was prudent to test the pneumatic components of the system. The manipulator was connected to a source of compressed air. The pressure value was chosen to be 2,5[bar], which is close to the minimum operational value. To complete the testing, the valves were tested using the bench power supply, one by one. The result of this test was satisfactory: all axes except axis "Z" functioned well. Initially, movement long axis "Z" could not reach its limit while lifting the arm. This was solved by finding the suitable pressure level, and repeated the movement of the mechanism.

4. Choosing the new controller

4.1. Parameters of the old control system

The ETsPU-6030 (ECPU-6030 / ЭЦПУ-6030) is a cyclical controller, designed especially to control machines consisting of two end-state actuators, and their technological processes [6]. This controller was able to operate machines using up to four axes, and six additional technological commands. Timers (delays) were also available, with time intervals of 0…0,7[s]. Each program could consist of a maximum of 30 steps, and four modes were available, which are as follows: manual (moving the actuators based on the moving axes), command (execution one part of the program), cycle (execution of one cycle), and automatic (the cycle repeats continuously).

While choosing the new controller, it was a critical condition, that the controller must have all original (and relevant) features, and if possible, it should be easy to use, and allow further improvements. These needs are all met by a modular PLC (with an amplifier stage). PLCs are wide-spread in industries, they are reliable, freely re-programmable, and physically meet all requirements set by an industrial environment.

4.2. Amplifier stage

Some controllers are unable to drive loads that they control, due to their construction. In such cases, a separate amplifier stage needs to be used, which can also fulfill other roles, such as separating incoming cables into wires within the device itself or provide additional protection by galvanic separation. During my project, such a device was at my disposal. Its main components are the 12 TIP120 NPN darlington transistors, which, using the output signal of the PLCs are able to provide the necessary performance to the electropneumatic valves. The device has its own separate power supply. The transistors are cooled by a 12 [V] fan, driven by a "7812" voltage regulator IC. The manipulator's cables can be connected to the device by the connectors placed on the front face of it.

Figure 4. Amplifier stage device

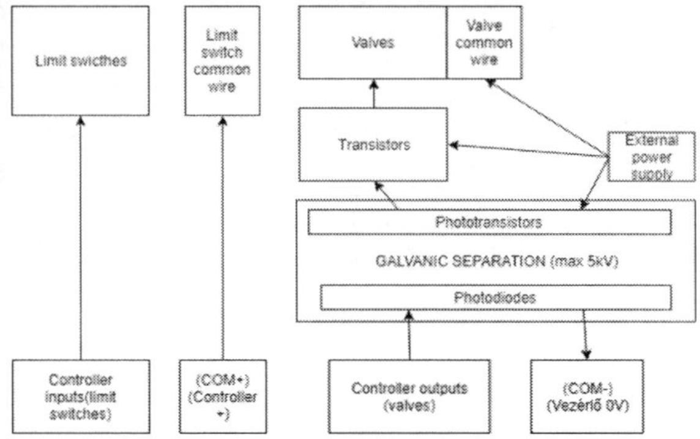

Figure 5. Amplifier's block diagram

The galvanic separation was later added in the form of 12 TLV-817 optocouplers, which offer complete galvanic separation up to voltages as high as 5[kV]. As the amplifier stage supports a wide range of controllers, initial tests were carried out using an Arduino Nano open source controller. Its low cost and ease of use make it ideal as an early testing device. The amplifier stage is usable both at the 5[V] output voltage of the Arduino, and the 24 [V] signals sent by the PLC. No other peripherals were used during this test, the manipulator executed the simple sequential program without error.

5. Final controller

The used device is a Schneider Electric Modicon M340. This modular PLC's 32 inputs and outputs (64 I/O total) can serve not only the following program's needs, but also enable us to connect additional devices, like pushbuttons or indicator lights, but may also pave the way to further improvements. The PLC has a separate power supply, connected via the rack.

The earlier mapped connections help during the programming. In the PLC program, the amplifier's inputs and outputs must be assigned to the variables found in the PLC program. In the program, sequential programming was used. During programming, it must be considered that at any time, only eight points from the manipulator's work area can be reached.

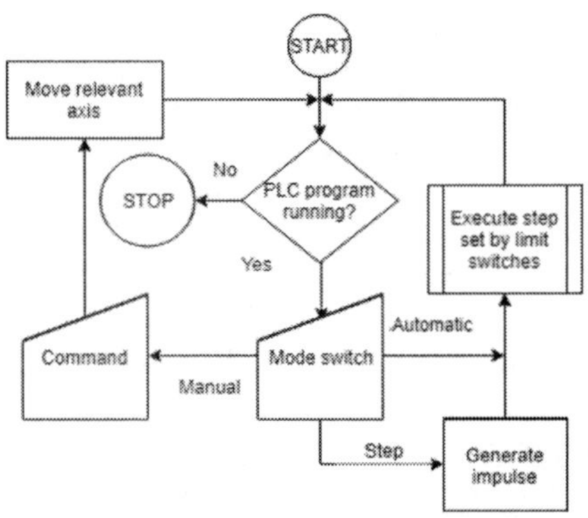

Figure 6. Simplified flowchart of the PLC program

The PLC program also contains the variables necessary for the different modes, and also sections responsible for communication with the HMI. After activating the main switch on the HMI, we can choose from three functions. In automatic, the movements repeat continuously, in manual-step mode, the manipulator only executes the next step of the program (the necessary impulse signal is created by a TP timer). These two modes can be switched between freely, the program remembers the last executed step. In manual mode, the actuators can be operated separately. Upon deactivating the main switch, all actuators and internal variables are reset.

6. HMI

A Schneider Electric Magelis touchscreen HMI was used, programmed with Vijeo Designer. The HMI's interface consists of various pushbuttons that move separate axes or select modes. In the upper right corner of the interface, a progress bar showing the current step can be seen, with a model right under it, that visualizes the manipulator's position in real time, serving as visual feedback. With the addition of the HMI, the control system becomes complete.

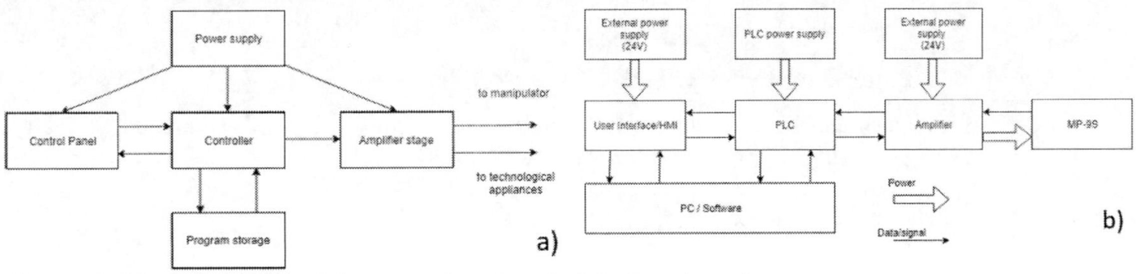

Figure 7. Block diagrams of the control system (original and new)

7. Conclusion

During this project, the restoration of the introduced MP-9S manipulator, and some steps ensuring its ease of use were completed. Its missing controller and user interface were replaced by up-to-date devices. During the project, most guidelines were successfully followed, the restored manipulator has retained all of its past functions, and even received new functionalities, thanks to its modern controller. To summarise, the restoration was a success, the previously non-functional device is now able to do material handling tasks and may also serve as a basis of the additional projects, such as a new gripper design.

Acknowledgment

I'd like to thank my consultant, Erdei Timotei István, for providing the necessary tools and conditions, and for his guidance, whenever I needed help, and to Dr. habil. Husi Géza, for making this project possible.

References

[1] Husi G, Molnár Zs, Obinna N C, and Erdei T I 2017 *A novel Design of an Augmented Reality Based Navigation System & it's Industrial Applications* 15th IMEKO TC10 – Technical Diagnostics in Cyber-Physical Era Budapest, Hungary

[2] Husi G, Erdei T I, Szemes P T, and Dávid E 2013 *Development of Simulation and Research Toolset for Domestic Energy Utilization* IYCE 2013 conference

[3] Husi G, Molnár Zs, Obinna N C and Erdei T I 2017 *LabVIEW Motion Planning and Tracking of an Industrial Robotic Manipulator (KUKA KR5 arc): Design, Modelling, and Simulating the Robot's Controller Unit* FMTÜ - XXII. International Scientific Conference of Young Engineers, Kolozsvár

[4] Akhmedova S N et. al 1985 *USSR Report - Machine tools and metalworking equipment* (Springfield, VA.: National Technical Information Service)

[5] Novgorod V 2005 *Programming of MP-9S industrial manipulator (Программирование промышленного робота МП-9С)* Moscow

[6] Popov A K 2005, *Structure and programming of MP-9S industrial robot (УСТРОЙСТВО И ПРОГРАММИРОВАНИЕ ПРОМЫШЛЕННОГО РОБОТА МП-9С)* Moscow

IOP Conf. Series: Materials Science and Engineering **568** (2019) 012061 doi:10.1088/1757-899X/568/1/012061

CodeSys modeling of an industrial cell, and linear drive control with PLC & ATmega2560 microcontrollers for testing and improvement purposes

T Pajkos, T I Erdei, G Husi

Department of Mechatronics Engineering, University of Debrecen, Faculty of Engineering, Debrecen, Hungary

tiborc97@gmail.com

Abstract. In this document, the upgrades of the robotics lab at the University of Debrecen, faculty of engineering will be detailed. As the first part of the task, the conversion and upgrade of the KUKA KR5's cube holder were done. As a result, the cube holder (feeder) can count the added and removed cubes, while also displays their current amount in the holder. The completion of the other task enriched the existing robot cell with the addition of a liner drive powered system, complete with a test panel, which can even be used for educational purposes in the future. And lastly, as a summary, the simulation of a manufacturing process in the robot cell was also completed.

1. Introduction

The upgrades and research tasks were completed in the Building-mechatronics Research Centre, robotics lab of University of Debrecen Faculty of Engineering [1]. The robotics lab, complete with an industrial robot cell, is able to satisfy all challenges set by the upcoming industry 4.0 [2].By the end of the 20th century, mass production became a wide-spread phenomenon. As the demand for mass production grew, so did the need for automation. As a result, in 1968 the first PLC was created. This new controller had to satisfy the following criteria [3]: to be simple and freely reprogrammable, to have galvanically isolated I/O modules, to be safer than electromechanical solutions, and lastly, to have a better price/value ratio than pre-wired logic control circuits. In the case of switching for the manufacturing of a new item, it's easier to reprogram them, than to create new machines for the changed workflow. I chose the enhancement of the robotics lab's robot cell as my topic (and the subject of the completed tasks), because I find automation an interesting field.

2. The cube counter

My first task was the upgrade of the KUKA KR-5's cube holder and feeder. I will start by showing the steps and methods used during the conversion and upgrading process.

2.1. Used methods and equipment

The original cube holder was only capable of storing the cubes, and, due to the fact that the construction mainly uses hollow sections, the user could not see how many items are currently available in the storage. To solve this issue, my proposal was to design a counting unit with a connected display

IOP Conf. Series: Materials Science and Engineering **568** (2019) 012061 doi:10.1088/1757-899X/568/1/012061

The first task was to realize the sensing of the cubes. For this, reflective object sensing was used, as only one side of the holder was usable for placing sensors, and this seemed the most suitable option. These sensing points were created using optical cable. To process the sensors' digital signals, an Arduino Nano was chosen as a controller. With this choice, it was possible to not only process the incoming signals, but to control the display as well. As a numeric display, an 8 by 8 LED matrix screen [4] was chosen, which is based on a MAX7219 controller. These appliances required both 5 volts and 24-volt power; to solve the issue of different voltage levels, a voltage divider was built (complete with pulldown resistors), making the 24-volt sensors compatible with the Arduino Nano.

2.2. *Assembly of the counter*

A solution was needed to fasten every component to the holder, using the least amount of modification. During assembly, two holes were drilled on the top and bottom parts of the holder, to place the optical sensors. On the side of the holder, a DIN rail was secured to provide mounting points for the equipment. Two power supplies, and a 5-connector terminal block was placed here. Since the 24 volt sensors and the 5 volts Arduino was used at the same time, the two power supplies were connected to a common ground. The box, which houses the display, the voltage divider, and the Arduino, has been placed at the upper part of the cube holder. The "clear" button has been placed on the side of said box.

2.3. *Creating the program*

Two libraries have been used in the program's creation (not counting the Arduino IDE's built-in functions): EEPROM.h, and LedControl.h. During programming, my main goals were to keep track of the added and removed cubes' number, and the display and storage (archival) of said number. The storage of the number of cubes was necessary, to ensure that the values are remembered after the system returns from a non-powered state. In addition, a "clear" button was included in the program.

In the first half of the program, the variables and their starting value are initialized, the inputs and outputs are defined, the used functions are called, and the hexadecimal forms of the shown images are declared, to be used by the screen. This is followed by the "setup", where the screen is set up, the number of cubes is loaded from the EEPROM, and the lower sensor's value is checked, whether there's a cube in front of it, or not. This is necessary, because in the program, the changes of the sensor's output are used, and if there's a cube in front of the sensor in the starting state, it must not be interpreted as a cube being taken out, to avoid erroneous down-counting.

```
if ((vs_1  == 1) && (folyamat_1 == false)) {
  counter++;
  folyamat_1 = true;
  EEPROM.update(1, counter);
  Serial.println( counter );
  update_Display();
  delay(200);
} else if ((vs_1  == 0) && (folyamat_1 == true)) {
  folyamat_1 = false;
}
```

Figure 1. The upper sensor's program block

```
void printByte(byte character [])
{
  int i = 0;
  for (i = 0; i < 8; i++)
  {
    lc.setRow(0, i, character[i]);
  }
}
```

Figure 2. Displaying characters

On the contrary, the "loop" part of the program repeats from beginning to end every time it's fully executed. This is where the current states of the sensors are read, and where the up and down counting happens, with the usage of the assigned variable. If the number of cubes changes, the screen and the value in the EEPROM are both updated. The "clear" button's state is also observed in this portion. Pressing this button causes the number of cubes to return to zero.

The pictures on the LED-matrix screen are achieved by multiplexing. During this process, every row and its matching columns are switched on only for a short time. The numbers used in this case were created in a program called "Pixel to Matrix"[5]. The process of this is as follows: first, the desired shapes are created in an eight by eight matrix, then, after pressing the "Generate" button, the program provides the necessary binary, and hexadecimal form of said characters. In this chase the hexadecimal form was chosen, because it takes up less space, and improves the code's readability. After creating the code to be displayed, a variable was created for it, to be able to point back to it every time there's a need to display it. In the program, a case structure was used to assign each number's picture to the number itself. After the code is assigned, the function draws it on the display, row by row.

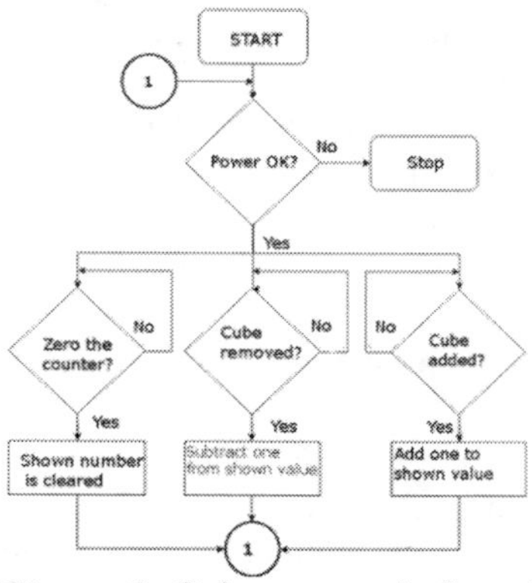

Figure 3. Cube counter's flowchart

Figure 4. The finished cube counter

3. Industrial linear drives

3.1. Designing the process

In my other task, I wished to use linear drives. The idea was to create a process using linear drives, which would cooperate with a KUKA KR-5 robot, by moving cubes in a specific way, after the industrial robot placed one on it. After working out a draft of the task its details had to be worked on. First, the KUKA KR-5 places a cube on one of the sliders residing on the linear drive, which will be sensed by a reflective optical sensor. Following that, the slider starts moving forward its other end position (limit), and after reaching it, another drive would activate, and push the cube into a box. This simulates the workpiece leaving the cell. During this task, I have used two ISEL LEZ-1 drive [6], which are driven by an MS045 HT motor [7]. The chosen optical sensors are Schneider Electric XUB0BKSNM12T [8].

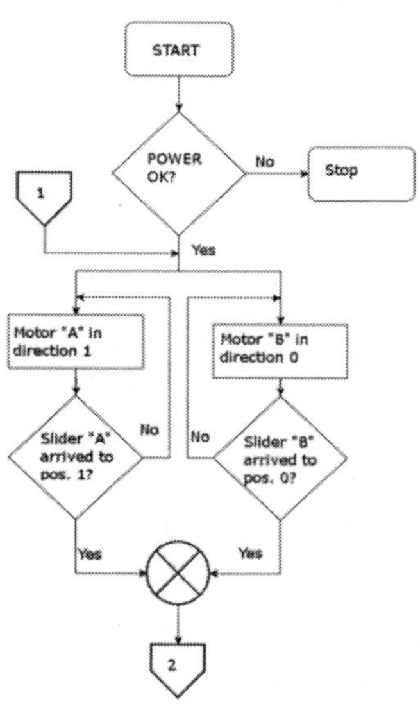

Figure 5. Linear drives' flowchart (1/2)

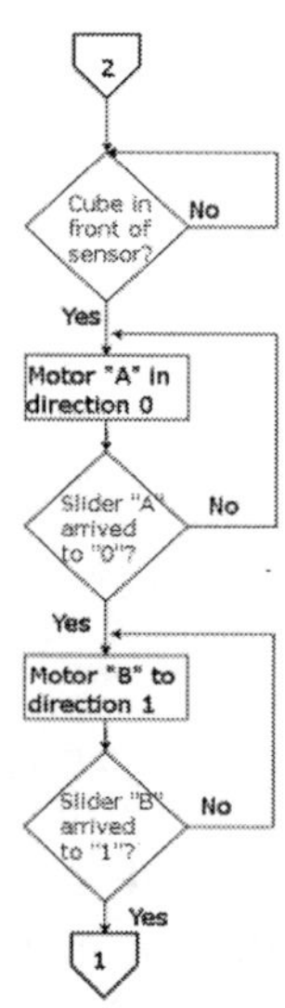

Figure 6. Linear drives' flowchart (2/2)

3.2. Programming the drives in CodeSys

As the next step of the task, I have created the model of the unit in CodeSys, and later simulated the entire process within CodeSys. First, options were created for both manual and automatic execution. A

TON timer was used in series with a TOF timer. This works in a way, that a button, which activates a bell, must be held for 3 seconds, and only after this can the program start. After releasing the "bell" button, seven seconds are given to start the process itself.

In automatic mode, the process was realized by using a form of the sequential method, in which a cycle consisting of steps was created, which are executed one after another. This solution is relatively easy to read, and making the modification of the existing program relatively simple. The content of the steps themselves also follows a modular, "5 ladder step" pattern, these 5 are: starting, safety, command, complete, and error. Commands are given in these steps, and their successful completion is later verified.

3.3. CodeSys simulation, the motor controller unit

As the first step of the modeling, the three-dimensional models of the drives had to be found, which were re-colored to improve their visibility, and later imported to CodeSys and assigned an ID in there, so that it's possible later to reference them during programming the animation. For the buttons, the build in CodeSys resources was used, or customized to suit the needs. To achieve the visual movements on the HMI, it was necessary to re-create the virtual linear drives, limit switches, and sensors. The position changes and their observation were achieved by a code segment, which detects which direction the motor gets the signal from at a given moment. When this detection and reading is complete, "Lin A pos" variable is checked and verified. If the limit is not yet reached, or in this particular case, the limit switches are not yet activated, then one unit is added, or deducted from the variable, depending on the movement's current direction. The limit switches function was realized by using a case structure.

```
IF Leptetes.Lin_A_0 THEN
    IF Lin_A_pos > 0 THEN
        Lin_A_pos := Lin_A_pos - 1;
    END_IF
END_IF

IF Leptetes.Lin_A_1 THEN
    IF Lin_A_pos < 1019 THEN
        Lin_A_pos := Lin_A_pos + 1;
    END_IF
END_IF

CASE Lin_A_pos OF
      0: GVL.Vegallas_Lin_A_0 := TRUE;
      1: GVL.Vegallas_Lin_A_0 := FALSE;
   1018: GVL.Vegallas_Lin_A_1 := FALSE;
   1019: GVL.Vegallas_Lin_A_1 := TRUE;
END_CASE
```

Figure 7. Programming of the virtual linear drive

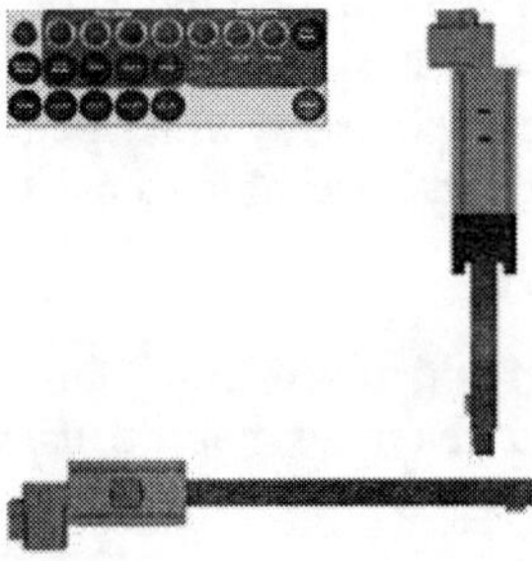

Figure 8. Test HMI

In the making of the motor controller unit, two Polulu A4988 bipolar stepper motor controllers were used. An Arduino Nano serves as a controller for these components. In addition, two capacitors were added to smoothen out the power spikes caused by the motors. For testing purposes, a test panel was created, which is compatible with an Arduino Mega, and with the help of an external program, can be programmed in ladder diagram. The test panel has eight inputs, and eight outputs, which are galvanically separated using optical couplers. To visualise the working channels, LEDs were added. Using the completed test panel, and the "PLC Ladder Simulator Pro" android application, the PLC program operating the linear drives was created. After uploading the necessary additional program, the assembled unit could be successfully tested.

Figure 9. The completed tool

Figure 10. Completed test panel

IOP Conf. Series: Materials Science and Engineering **568** (2019) 012061 doi:10.1088/1757-899X/568/1/012061

3.4. PLC and synchronized control of the linear drives

After making certain that that both the motor controller unit and the ladder diagram program work without error, it was deemed important to operate the linear drives using the PLC found in the robotics lab. The already designed and tested program was implemented in the Unity Pro XL software.

During operation, the system first returns to its starting position. Then, it checks the optical sensor's signal, and when it senses an object, the process starts. When that happens, the drive with the cube starts moving towards its other end position. When it reaches the limit switch, the other drive perpendicular to it starts moving, and eventually pushes off the cube, into the box. When this step is complete, the system returns to its starting position again. As the last task, the KUKA KR-5 industrial robot was integrated into this process, utilising the finished linear drives' program.

Figure 11. The completed robot cell

4. Conclusion

During the completion of these tasks, I have successfully applied my previously learned knowledge. As such, I can declare that not only I have acquired the necessary theoretical knowledge, but I only gained valuable practical skills as well. In my time spent at the robotics lab, I became acquainted with industrial devices and components, and I was able to use them in my project.

Acknowledgment

I would like to express my thanks to Erdei Timotei István, for his support during my work on the task. In addition, I'd also like to thank Dr. habil. Husi Géza, for making it possible for me to use the tools and devices found in the robotics lab.

References

[1] Erdei T I, Molnár Zs and Husi G 2016 *Robot visual and virtual control technology In industrial environment* WoS publication, International Symposium on Small-Scale Intelligent Manufacturing Systems, Narvik, NORWAY- IEEE

[2] Erdei T I, Molnár Zs, Obinna N C and Husi G 2017 *Cyber-physical systems in mechatronic research centre* Volume 126 MATEC Web Conf.

[3] Kulcsár B 2012 *Control and automation in material handling* (*Az anyagmozgatás irányítás- és automatizálástechnikája)* (Typotex Press)

[4] „*MAX7219*," [Online]. Available: https://datasheets.maximintegrated.com/en/ds/MAX7219-MAX7221.pdf [Accessed: 25. 11. 2018.]

[5] „*Pixel To Matrix*," [Online]. Available: http://educ8s.tv/arduino-8x8-led-matrix-tutorial/. [Accessed: 25. 11. 2018.]

[6] *ISEL* [Online]. Available: http://www.isel.hu/index.php/klub-7/piac/1886/mechanika/bordasszijas-linearis-egyseg/lez1-bordasszijas-linearis-egyseg.html [Accessed: 25. 11. 2018.]

[7] *ISEL* MS 045 HT [Online]. Available: https://www.isel.com/hu/dsh-s.html [Accessed: 25. 11. 2018.]

[8] *OsiSense* [Online]. Available: https://www.schneider-electric.com/en/product/XUB0BKSNM12T/photo-electric-sensor---xub---emitter---12..24vdc---m12/ [Accessed: 25. 11. 2018.]

IOP Conf. Series: Materials Science and Engineering **568** (2019) 012062 doi:10.1088/1757-899X/568/1/012062

CodeSys based HMI design and sensor upgrades for a PLC and Atmega2560 controlled industrial conveyor

R Varga, T I Erdei, G Husi

Department of Mechatronics Engineering, University of Debrecen, Faculty of Engineering, Debrecen, Hungary

vargarul94@gmail.com

Abstract. Engineers nowadays must be able to solve the challenges set by industry 4.0, in order to make existing devices viable for this new trend. This requires a complex, system-based way of thinking. Nowadays, in most cases, we usually find company-specific hardware and software in industrial automated production. However, in different industries, it is not unheard of that a given manufacturing cell, robot, or manipulator became unserviceable, because a manufacturer is no longer on the market, making it impossible to find replacement controllers, software, or updated firmware for our industrial robot, or manipulator. The following project aims to provide a solution to the problem detailed above, where an unserviceable Flexlink conveyor's controller is replaced with both an Arduino and a PLC, and before realisation of the new control method, simulations are possible before connecting the actual controller.

1. Introduction

In the building-mechatronics research centre of University of Debrecen, a new lab has been created with the name of „Cyber-Physical and Intelligent Robot Systems Laboratory". The goal of this laboratory is to design and apply new technologies in the field of robotics, which are later to be integrated into other fields as well. In our robotics lab, numerous upgrades and research are underway, which include the designing of a new conveyor belt system.

The functions of the KUKA KR5 found at the Department of mechatronics [1] have been extended during previous upgrades, namely the robot cell built around it. One of the key components of this cell is the conveyor belt. During its conversion, one of the main focus was to ensure simple connection and usage of various controllers, and to choose controllers which can be replaced in case of their malfunction, without having to modify the existing system. By this, we can avoid the situation where a manufacturer disappears, as the shown example in the abstract. Two particular controllers were chosen with specific properties: one of them is open-source, while the other conforms industrial needs and conditions. In both cases, the existence of a simulation environment is crucial.

2. ATMega2560 & 328 microcontroller

In order to ensure the open-source nature of the controller, and to guarantee its usability in an educational setting, the Ardunio product line was chosen, which also has extensive support.

The Ardunio Mega 2560 Rev3. has 54 digital I/O ports, 16 analogue inputs, 4 UART communications port, a 16MHz crystal oscillator, and an USB connector, power supply connector, an ISCP header, and a "Reset" button. This version of the Ardunio serves the role of a "pre-tester" device in the control of the conveyor belt.

In terms of programming, the Ardunio IDE development environment is available, which is a cross-platform, java-based tool. In terms of Ardunio, another model was also chosen, which is the Ardunio Nano. This device has been added to the system not with a pre-testing consideration, but as a device to connect the PLC to the conveyor belt, with the goal of stepper motor control. The panel's main component is the ATmega328 microcontroller, which is available with 32KB of internal flash memory, 14 digital I/O pins and 8 analogue inputs and works at a frequency of 16 MHz. Its operational voltage is 5V, although the input voltage may vary between 7 and 12V.

3. NEMA23 stepper motor and Pololu A4988

To drive the conveyor belt, a NEMA23 Nanotec Munich made ST5818L type stepper motor was used, with a step angle of 1,8 degrees, which can be operated in both bipolar, and unipolar modes [9].

For the control of the stepper motor, a stable, reliable motor controller was required. The chosen controller was the Pololu A4988, which sees wide-spread use in 3D printers, and CNC machines [10]. The five different stepping resolutions are set by MS1, MS2, MS3 inputs. The MS1 and MS3 have a pull-down resistor of 100kΩ, while the MS2 has a 50kΩ one [2].

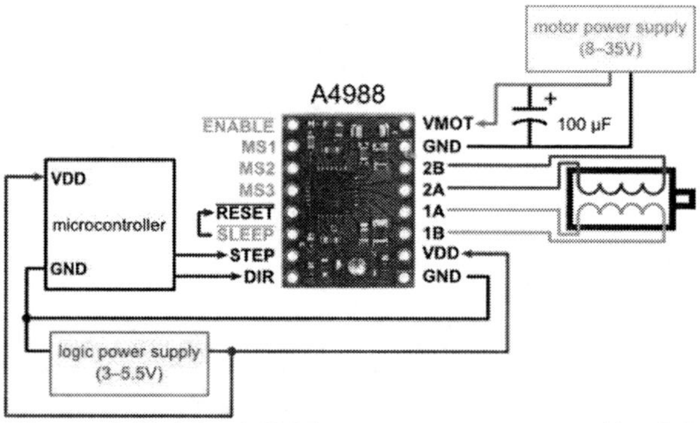

Figure 1. Pololu A4988 stepper motor controller [2]

4. PLC & conveyor belt

For the PLC itself, a modular Schneider Electric Modicon M340 [3] was chosen, as it's easily customisable. In the project, the to-be controlled conveyor belt is found in the robotics lab, which is part of the FlexLink XK product line used in car and food industries. This particular FlexLing conveyor belt was chosen, because there are no plans of moving workpieces during education, that are greater than 10[kg] in weight. The total maximum weight the conveyor can carry is 30[kg]. The plastic belt is 45[mm] wide, and 5200[mm] long [4]. Two XUB0BPSNM12 photoelectric sensors were secured on the conveyor, which are part of the OsiSense XU product line [5].

5. ATMega2560 pre-tester

Assembly of the conveyor belts and its stepper motor was done in an earlier project in the robotics lab. However, the original FlexLink controller is non-functional, and the original development environment is only compatible with the specific controller. As a result, without replacing the original hardware, the conveyor could not be operated.

IOP Conf. Series: Materials Science and Engineering **568** (2019) 012062 doi:10.1088/1757-899X/568/1/012062

Figure 2. Conveyor belt with sensors

In order to re-integrate the conveyor into future projects and education, the original hardware had to be replaced. The Ardunio Mega2560 was a precursor to the final controller. As it was shown earlier, two photoelectric sensors were fastened to the frame of the conveyor belt. Four terminal blocks were also secured on this frame, assigned to negative-positive, and A-B allocation.

The terminal blocks marked by "A" is connected to the sensor placed at the front part of the conveyor, while "B" is connected to the sensor on the other end. The usage of terminal blocks was warranted by the nature of PLC connections: the screw terminals of both the PLC and the terminal blocks ensure quick connection of wires to the desired devices.

The perforated sheet, serving as a mounting point for the "PLC controller wall" also holds the five button Schneider button box. A total of five inputs were connected to the PLC's input module. With these, "manual" control of the conveyor belt can be achieved (start/stop).

The "PLC controller wall" also contains another Schneider box, housing five red LEDs indicator lights, which are connected to the PLC's output module. These serve indication purposes and provide visible feedback about the program's state.

The Ardunio ATmega 2560 was also secured on this panel, including the breadboard housing the necessary electrical components. A 3D printed, DIN rail compatible adapter made of PRO PLA was used as mounting point. The Ardunio Nano is mounted in a similar fashion.

During pre-testing, an Ardunio ATmega2560 microcontroller and a suitable stepper motor controller, A4988 was used to drive the stepper motor itself, the Nanotec Munich ST5818L.

The external 12V power supply was an AX-3003D bench power supply [6]. This was required because the 3,3V/5V output level of the Ardunio was insufficient, and also because the rated maximum current output of the ports is also only 40[mA], which is not enough for the used NEMA23 stepper motors.

As the five pushbuttons and the two sensors both require 24 V voltage, a voltage divider was required to connect them to the Ardunio board.

After every device necessary for the electrical control was prepared, it was necessary to ensure their connection. This can be observed on the schematic drawing, seen on figure 8.

IOP Conf. Series: Materials Science and Engineering **568** (2019) 012062 doi:10.1088/1757-899X/568/1/012062

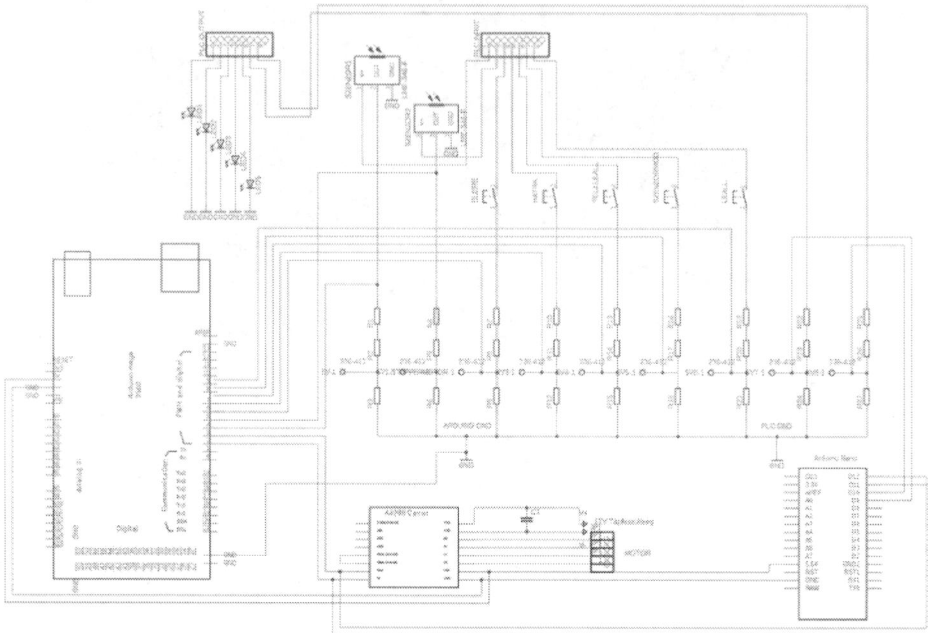

Figure 3. Circuit schematics

Every pushbutton received its own function. As such, pressing the button connected to pin 6, the conveyor belt automatically moves forward. Pressing the button connected to pin 8 stops the conveyor. Only after this can the button connected to pin 7 be activated, to start the conveyor to move in the opposite direction. This serves as a safety function, to avoid motor overload caused by the direction reversal. Sensor mode is activated, when the pushbutton connected to pin 9 is activated. This mode can be deactivated by pushing the button connected to pin 10, which deactivates the conveyor belt, and causes it to stop.

After finishing the program, it was verified by the Arduino IDE, and uploaded to the Ardunio Mega 2560 board. Afterwards, the setup was tested within the KUKA KR5's work zone.

6. CodeSys PLC and HMI

If we want to apply the previously written and tested program onto the final, industrial hardware, there's one serious limitation to consider.

The Arduino IDE provides means to check our written program, however, if we'd like to transfer this same program to PLC, we might not be given the same opportunity of troubleshooting.

One of the reasons for this, is the large number manufacturer-specific PLC programming environment, which often only allow simulation through their own, separate "plugin", to check the program's viability [7]. To combat this, a software called "CodeSys" was used, which is free, and its features include PLC programming, simulation, and HMI (human machine interface) design [8].

During the program transfer from Ardunio to PLC, first the previously detailed buttons were designed as part of the HMI. After designing the HMI, the Ardunio program was written in PLC Lad, and the HMI's buttons were assigned.

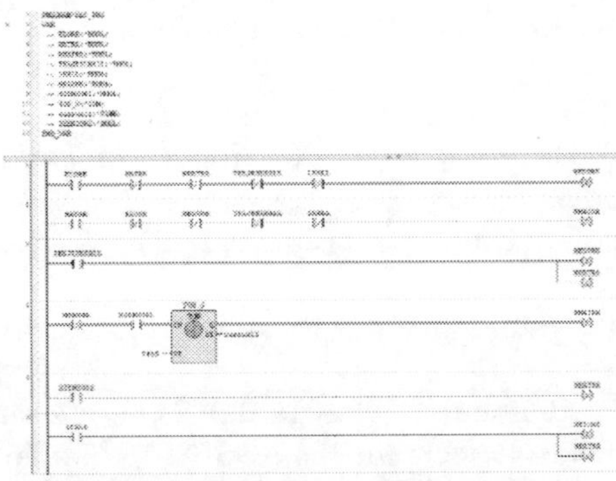

Figure 4. CodeSys – PLC program

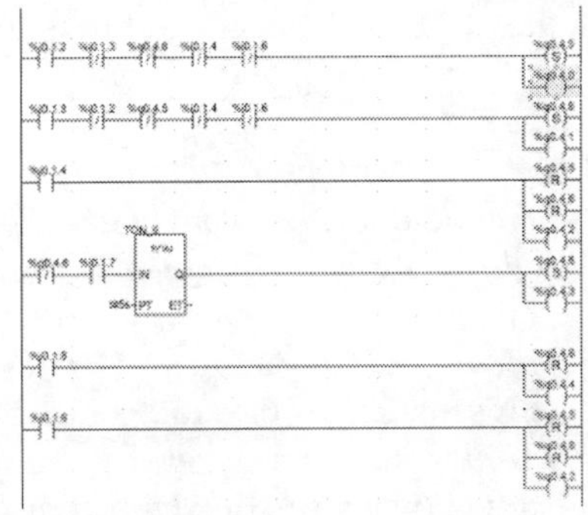

Figure 5. Unity - Ladder diagram program

Later the program written in CodeSys was ran, to ensure its error-free function. Another positive effect of this, is that a PLC based conveyor model was gained, which may be used in future controllers or HMI projects, as it is freely editable.

7. Industrial PLC finalising

After the previously written CodeSys program was checked, the already simulated PLC Lad program was copied to the Schneider Electric Unity development environment. The timer also only functions with a 5 seconds delay. The physical input addresses range form the second digital input to the eighth and are found on the first module. They are addressed in the following form: "%i.0.0.2". The physical outputs range from the zeroth digital output to the sixth and are located on the fourth module. They are addressed in a similar manner as inputs are (as an example: "%q0.4.0"). Timer values are also 5 seconds.

The sensors and pushbuttons were connected to the input module, while the indicators and the outputs necessary for the stepper motor's function were connected to the output module. In this program, lamps had to be set as normal coils (outputs), as they are responsible in providing feedback about the program's state. The connection between the computer and the PLC are establish via a UTP cable, and a router. Within Unity, the PLC's IP address must be defined, to guarantee stable communication with the Schneider Electric Modicon M340 PLC.

At the moment of controlling the conveyor belt, the PLC was equipped with BXM DDI 1602 input modules, and BXM DDO 1602 output modules. Because these modules are incapable of generating the high frequency signal necessary for controlling the stepper motors, the previously mentioned Arduni Nano board was chosen to generate signals to be used by the Pololu A4988 motor controller. The Ardunio, depending on its inputs, either sets the conveyor's rotation direction, or stops it. The angular speed of the NEMA23 is constant. As the Ardunio Nano operates on a 5V voltage level, the 24V signal coming from the PLC had to be connected to a voltage divider first.

8. Conclusion

The conveyor belt found in the „Cyber-Physical and Intelligent Robot Systems Laboratory" had its missing controller successfully replaced, which paves the way for the usage of new sensors, and controller shields.

In terms of hardware, two controllers are used: Ardunio fulfilling a "pre-tester" function, and the industrial PLC, serving as the final controller, using the results acquired by during pre-testing.

Simulation is possible in either cases, to check the created program's reliable operation, before testing with real hardware.

It can be said, that a cost. effective, flexible system was created, with easy-to-replace components. In addition, a previously inactive assembly line device was restored to working condition, which now works reliably.

Acknowledgement

Hereby I'd like to express my thanks towards Dr. habil Husi Géza and Erdei Timotei István, and also towards the „Cyber-Physical and Intelligent Robot Systems Laboratory" and its members.

References

[1] Erdei T I, Molnár Zs, Obinna N C and Husi G 2016 *Industrial KUKA Robot Manipulator in Simulation Environment and Position Read-back*
[2] *Pololu A4988 stepper motor controller* [Online]. (2019.01.12). Available: https://www.pololu.com/product /1182.
[3] *Modicon M340* [Online]. (2019.01.12). Available: https://www.se.com/hu/hu/product-range-presentation/1468-modicon-m340/
[4] *Flexlink Conveyor system XK* [Online]. (2019.01.12). Available: https://www.flexlink.com/zcms/zpublish/45/uploads/45/tech_lib/14170936898875465656_08XK.pdf
[5] *XUB0BPSNM* [Online]. (2019.01.12). Available: https://www.schneider-electric.com/en/product/XUB0BPSNM12/photo-electric-sensor---xub---multi---sn-0..20m---12..24vdc---m12/#search. Available: 2018.11.09.]
[6] *AX-3003D power supply* [Online]. (2019.01.05). Available: https://www.hestore.hu/prod_10025553.html
[7] T I Erdei and Zs Molnár 2018 *PLC – Applied Automation I.* (*PLC - Alkalmazott automatizálás I.*)
[8] *CodeSys* [Online]. (2019.01.04). Available: https://www.codesys.com/
[9] *Nanotec catalogue* [Online]. (2019.01.08). Available: https://en.nanotec.com/fileadmin/files/Kataloge_Flyer_Archiv/Nanotec_Catalog_2007.pdf
[10] Erdei T I, Molnár Zs and G Husi 2016 *Self-reproduction 3D printer in controlled environment* InterTalent UNIDEB - University of Debrecen International Conference

IOP Conf. Series: Materials Science and Engineering **568** (2019) 012063 doi:10.1088/1757-899X/568/1/012063

Some aspects of gas meter quality and analyzing the sustainability of a gas meter element

P Ilea[1], D Săvescu[2], A Stoica[3]

[1,2] Product Design, Mechatronics and Environment Department, Transilvania University of Braşov, B-dul Eroilor 29, 500036, Romania
[3] Engineering Department, Lucian Blaga University of Sibiu, Emil Cioran 4, Romania

paul.ilea@unitbv.ro;

Abstract. This paper refers to analyzing some components of a natural gas meter. There is a question of making the gas meter design and execution process more efficient. A description of the components of a gas meter is made. It is proposed to improve the natural gas meter. Designing is an important and vital stage in the product's lifetime, an analysis of the environmental impact and the durability of the product is done, so an analysis is made using the SolidWorks-CAD software. The object of the analysis is a toothed wheel, part of the mechanical bimetal temperature compensator. The case study presents a sustainability report showing a decrease in manufacturing cost, an improvement in the materials used, and a lower impact on the environment.

1. Introduction

In this paper, emphasis is placed on the design and execution quality of a natural gas meter in close connection with the production cost and its durability.

Due to the emergence of new raw materials and manufacturing industries in recent years, there have been a multitude of effects with negative impact on the environment. The negative effects of the manufacturing process are reflected on the environment through the occurrence of pollution and climate change. In this regard, the need to protect and preserve the environment has emerged immediately, such as environmental taxes, subsidies for companies that have preventive measures against pollutants, directives and laws for manufacturers, etc.[1].

Sustainable development / sustainability is defined by the World Commission for Environment and Development as a development that seeks to meet the needs of the present without compromising the ability of future generations to meet their own needs [2].

Sustainable development is an obligation that must be undertaken to ensure a future for future generations.

It is said that the absence of quality is expensive, so it is necessary to implement a quality control system.

For a successful project, we need to evaluate the product life cycle [3].

LCA (Life Cycle Assessment) is a management technique that involves assessing and analyzing the environmental consequences of the product, from product extraction and processing of raw material, design phases, manufacturing, use, repair, maintenance and recycling phase, going to product reintegration into the environment. Also, the environmental impact due to transport needs to be remembered [4].

IOP Conf. Series: Materials Science and Engineering **568** (2019) 012063 doi:10.1088/1757-899X/568/1/012063

A very important aspect is the clients trust, who have to occupy an important position in the organization activity. Improving product quality can be done either by improving existing features, either by increasing the number of product quality features. The American Quality Society (ASQ) defines quality as a continuous effort of organizations to achieve all the processes better and better, in other words to obtain goods and services to better meet the growing needs of consumers. Failure to appreciate all the features of a product or service may lead to a deacrease in quality.

To reach the quality level of a product, Deming recalls some principles [5]. which in his opinion are exclusively in the responsibility of the management and they can not be delegated: continuous improvement of product and service quality based on a plan; dropping the acceptable level of quality; introducing statistical quality verification methods; requesting proof of statistical evidence from suppliers; discover the problems; use modern staff training methods; ensure all employees the tools they need to conduct their activity properly; encourage communication and eliminate fear, so every employee can tell their opinions.

2. Components study

Gas meters with deformable walls are volumetric gas meters.

Gas volume measurement is carried out using deformable wall cameras, with or without temperature conversion devices.

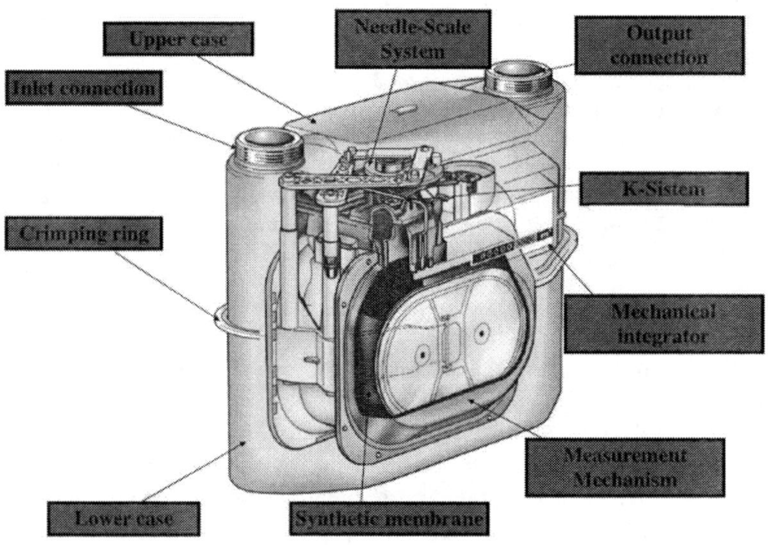

Figure 1. Gas meter [6]

Gas meters consist of a measuring mechanism, a housing assembly and a mechanical integrator for recording gas consumption. The measuring chambers are filled and housed periodically, and the membrane movement is transmitted via a gear to a crankshaft and it moves the gas intake control drawers. Through a toothed wheel system, the motion is transmitted by means of a magnetic coupling to the mechanical integrator of the gas meter. The gas inlet is possible via the inlet of the meter and the gas outlet through the outlet of the meter [6].

Gas meters with deformable walls must meet the following technical characteristics: the meter housing must be thick enough for the pressure range, and the joint between the upper and the lower case must be sealed, protected against corrosion. The housing should be made of aluminum or galvanized sheet; the meter components must be resistant to the corrosive substances that they come in contact with; to avoid fraud in the distribution system, the meter must be provided with a gas leakage locking system in the reverse direction; the meter's membrane must be made of synthetic, corrosion-resistant

IOP Conf. Series: Materials Science and Engineering **568** (2019) 012063 doi:10.1088/1757-899X/568/1/012063

material; the integrator must be protected by a transparent lid, light-proof and other environmental factors.

In the following papers it is proposed to develop a 3D model of a natural gas meter that is capable of preventing and discouraging all known unauthorized intervention methods [7].

The 3D model will incorporate the following safety features:

- the disappearance of the metrology marks, and the integrator will be integrated into the meter housing, so that everything is under pressure. Inscription on the transparent cover of the registration mechanism of the manufacturing year and the device series, plus safety belts;
- input / output connections will be provided with metalic filter;
- a flap, with the possibility of automatic gas closing;
- an automatic meter reading system based on radio frequencies and concentrators;
- an alarm system for removing the gas meter from the plant;
- temperature and pressure sensors [8].

3. Case study

For sustainability analysis, the SolidWorks CAD software was used, which incorporates a product life cycle assessment module (LCA). The subject of this case study is a toothed wheel, part of the bimetal mechanical temperature compensator (Figure 2).

For mechanical compensation with the gas temperature a bimetal is used.

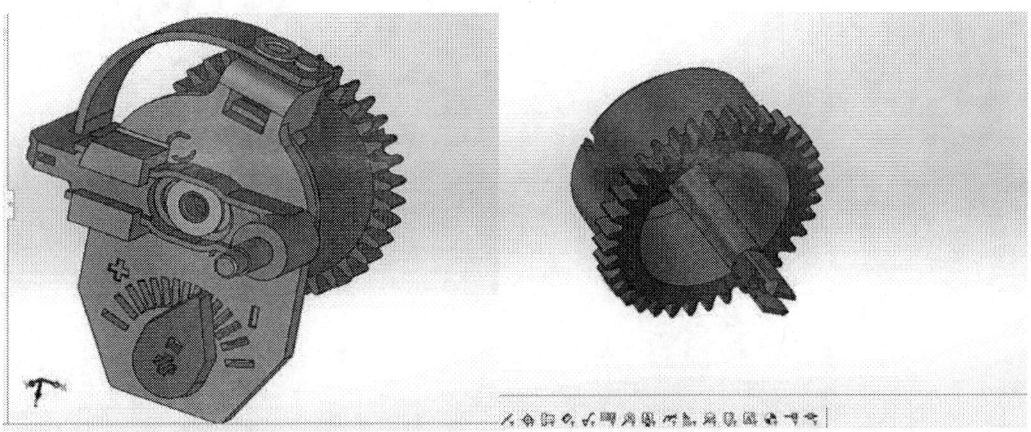

Figure 2. Mechanical compensator with bimetal/ cogwheel

A bimetallic element is fixed to the disc of the metering mechanism. This influences the movement of the measuring chambers so that the measuring accuracy within the stated temperature range remains within the permissible limits.

Bimetal is an assembly made of two metal welded elded blades, which under the influence of temperature can perform a limited movement, developing a certain force.

Given that a natural gas meter has plenty of plastic and metal components, a complete sustainability analysis should be done for each piece in order to reach a result that will have a major impact. In this case study we will only focus on a plastic gear wheel part of the bimetal mechanical temperature compensator.

In the first phase of this case study, 3D bimetallic temperature compensator assembly (Figure 3) was designed. After designing the assembly, the sustainability module was run, which is an integrated part of SolidWorks CAD software.

Subsequently, the type of toothed wheel manufacture was selected, injection molding and material type, POM Acetal Copolymer.

IOP Conf. Series: Materials Science and Engineering **568** (2019) 012063 doi:10.1088/1757-899X/568/1/012063

This module makes a comparison between the same product from an environmental point of view depending on what the material it is made of, the manufacturing mode, product duration usage, etc..

The module analyzes the input data and tracks the product from the manufacturing stage, the lifetime, the transport and removal of the product from use.

We are provided with information on the gas and electricity consumption consumed to produce/ pack the product, the manufacturing cost per unit. It can also be selected if the piece is painted or not (in this study the piece is unpainted), product duration usage and the length of time it is programmed to withstand the product.

It is possible to select the product manufacture place and the product place of use. Depending on this, the module makes a calculation that reflects transport costs (by truck, train, boat and airplane).

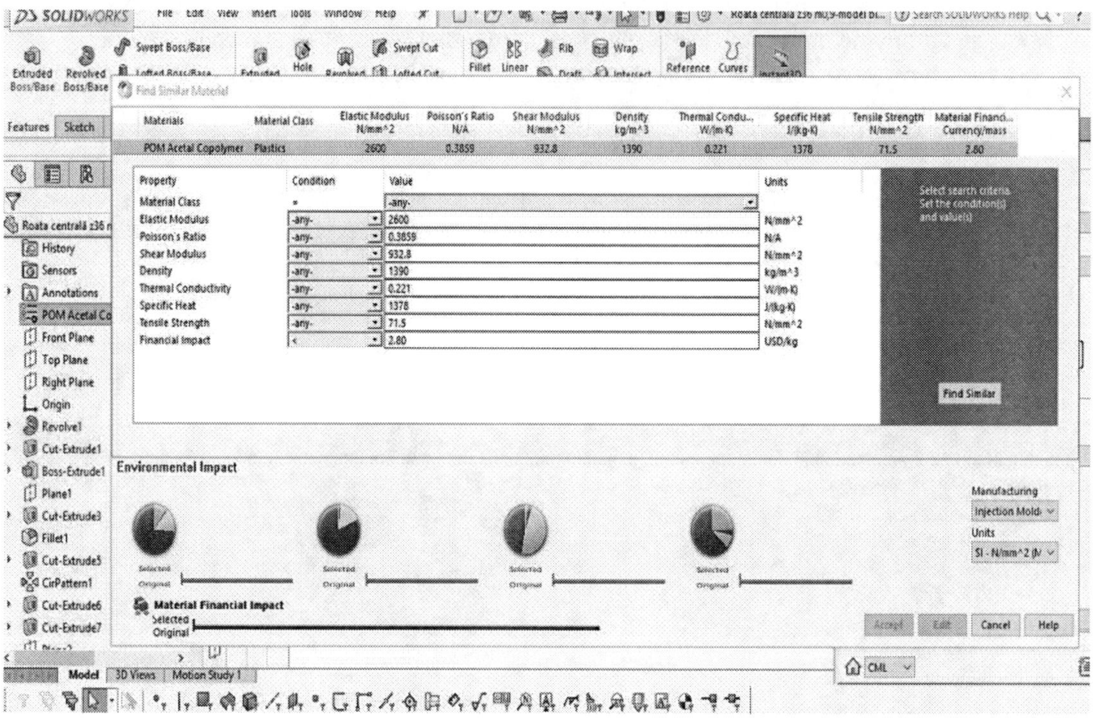

Figure 3. Material tab

Obviously, if the product is used at a great distance from the manufacturing site, the impact is greater on the environment due to transport.

In the sustainability report we also receive information about the product end of life, how much can we recycle, how much it is incinerated and how much it should be transported to a landfill.

In the next phase I changed the material from which the wheel is made of, from POM Acetal Copolymer in Medium / Low Density Polyethylene, and the manufacturing process remained the same, injection molded.

I have selected Europe as the manufacturing area and Australia as usage area. This shows us that the product has a negative impact on the environment, due to the transport cost from Europe to Australia (Figure 4).

IOP Conf. Series: Materials Science and Engineering **568** (2019) 012063 doi:10.1088/1757-899X/568/1/012063

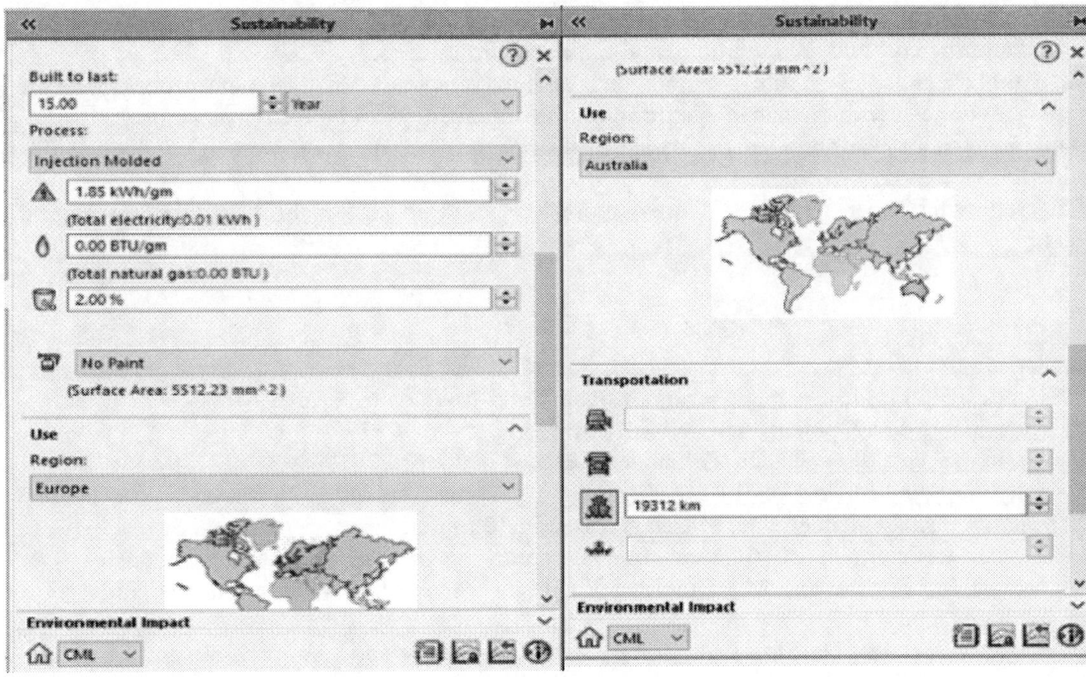

Figure 4. Input menu

By comparing the two types of materials we can see that medium / low density polyethylene has a much lower environmental impact, and the manufacturing cost is also lower.

The environmental impact for polyethylene is shown in (Figure 5).

So we have a 60% reduction in carbon footprint, a drop in total energy consumption by 45%, a decrease in air acidification by 44%, a decrease in water eutrophication by 65% and a decrease of material cost by 39%.

We have the opportunity to get a complete report using this module, which can help organizations.

Sessions can be organized within the organization and talk freely on the subject so that the best decision can be made.

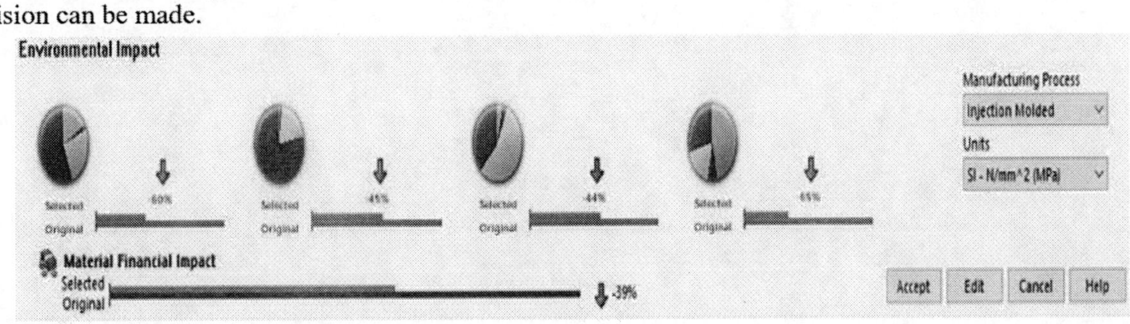

Figure 5. Results

4. Conclusion

As a result, we have to understand that the quality of a product is a necessity.

Gas metering devices must evolve in such a way that the safety in operation is greatest and the fraudulent part disappears.

IOP Conf. Series: Materials Science and Engineering **568** (2019) 012063 doi:10.1088/1757-899X/568/1/012063

Organizations must understand that if a product is not executed perfectly from the beginning, losses can be high or very high.

Following the sustainability analysis applied to only one piece of the gas meter, we can see a decrease in manufacturing cost and the environmental impact is diminished.

So I think companies should integrate such analyzes into their management process because the benefits can be seen with the naked eye, for both the organization and the environment.

This analysis can be done for the whole measuring device, and then we can compare the results obtained.

The following papers will also address more sophisticated sustainability studies for natural gas meters.

References

[1] Ecodesign in Industrial engineering, Mogan, 2005 Acessed in 5 April 2019
[2] Sustainable development, Acessed in 10 April
[3] Basic concepts of ecodesign, Fernandez Accesed in 2 April 2019
[4] Integrating the concept of "Life cycle analisys" , Toma Acessed in 30 March
[5] Quality management course, Mitran Acessed in 30 March
[6] Stoica, A., Feier, D., (2016). Measuring and control devices in the natural gas industry
[7] Neagu, I., Constantin, M., Ciocirlea-Vasilescu, A., (2007). Measurements and metrology legislation
[8] Ilea, P., Savescu, D., Stoica, A. Some aspects of quality and risk management for natural gas meters. PRASIC 2018

IOP Conf. Series: Materials Science and Engineering **568** (2019) 012064 doi:10.1088/1757-899X/568/1/012064

The theoretical bases and study parameters with FEA application approach on a bending tool design

D C Negrău[1,2], G Grebenişan[1] ,T Vesseleny[1]

[1] University of Oradea, Faculty of Managerial and Technological Engineering, Universităţii Street No. 1, Oradea, Romania

dan.negrau@yahoo.com

Abstract. This paper presents the theoretical basis of the approach to Finite Element Analysis (FEA) techniques and their application in the case of a bending request for a piece of aluminum sheet. The plate of a bending device deforms elasto-plastically under the action of the drive forces of the device, but also under the action of the reaction forces in the supports, more for a material such as aluminum and a few times less for a structural steel. If the forces acting on the device deform the carrier plate, which is in the operator's attention, in the elastic field, then there is no danger of permanent deformations occurring. Conversely, if deformation exceeds the threshold of elastic deformation, the material undergoes plastic deformation stresses, then the piece deforms plastic, with the risk of permanent deformations occurring. These deformations, in the first place, can generate production errors, or device failure, by modifying the execution rates and the games in the studied ensemble. This is precisely the purpose of this work: to prove, with the results of the FEA, that in the case of a certain material, with the same configuration of working conditions, the system can function without producing errors, obtaining qualitatively conformable parts.

1.Introduction

In the automotive industry, the future direction requires the continuous development of new concepts and the optimization of all business processes. Also, the growing globalization produces a growing distance from traditional business processes, both at the design, manufacturing, sales and logistics level, as well as convergence between information and communication technologies and process optimization:

- Market and customer requirements on each product are continually changing, which increases the complexity of production processes and costs together;
- Heterogeneous IT fleets and high maintenance costs for old systems generate unnecessary and growing IT costs;
- Strong and fast sales fluctuations require increasing flexibility in production;
- Planning, deploying, controlling and improving major production processes and resources requires the methodical implementation of a "digital factory", including IT-based equipment for a well-founded management decision;
- Comfort and safety are increasingly traded by automotive customers, and investment in electronics, and therefore in software solutions and applications, is needed.

The car exhaust manufacturers to request the construction of a manual assembly of the car exhaust protector parts. The device contain more much subassemblies, one of these subassemblies being designed to support a clamp . The clamp is attached to a support plate and at the acting o this clamp,I has been observed that when the device is actuated to fixing the subassemblies to the exhaust, the support

IOP Conf. Series: Materials Science and Engineering **568** (2019) 012064 doi:10.1088/1757-899X/568/1/012064

plate yield elastically [1,2]. The subassembly is presented in figure 1, elastically yields due to their support plate.

2. Support plate

2.1 General presentation

The support plate of a clamp is a component part of a subassembly (figure 1). When the clamp is acting, the support plate elastically (figure 1 A and B) deforms at the end of the work stroke by approximately 0.33 [mm], measured, as a result of the action of the force applied by the punch on the active plate, as shown in Figure 1 [4]. The elastic deflection of this component of the fixing assembly, generates incorrect assembly of the exhaust protect sheet.

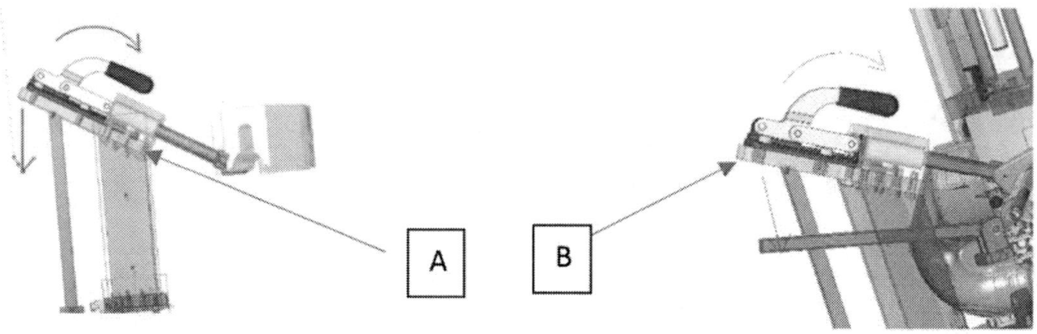

a) Subassembly acting clamp b) Assembly device

Figure 1. Clamp port-drive assembly

The support plate, as highlighted in figure 1, was built on the customer's technical prescriptions so as to have the dimensions present in figure 2 a). The material used is duraluminium, a material that meets the technological requirements, namely: it is easy to manufacturing, low specific weight, insignificant cutting tool wear compared to the machining of a harder material (soft steel, hard steel, stainless steel) and the manufacturing time is little.

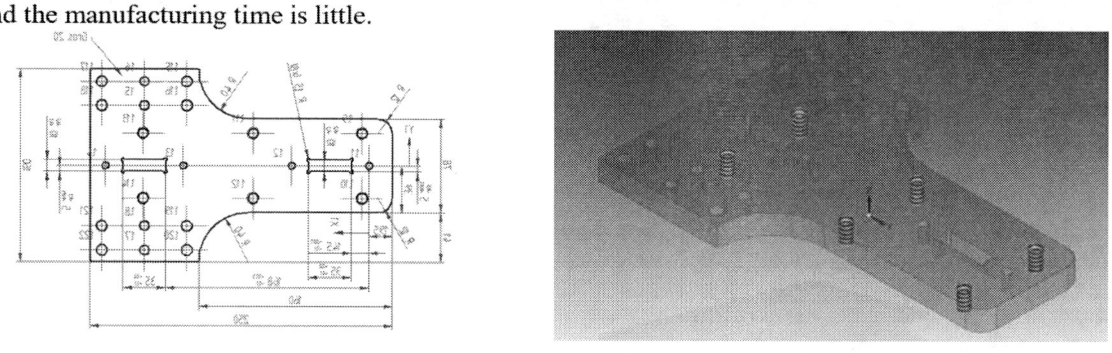

a) 2D model support plate b) 3D support plate

Figure 2 Geometrical properties

For the drawings has present in figure 2, was used the Cad software Solid Edge ST7, which offer facilities 3D drawing, and after on the base the 3D model results very easy 2D model [6]. Withal the 3D model is used at the simulation FEA.

2.2 Limit condition and mesh support plate – Ansys [5]

Computer Aided Engineering procedures allows the following types of analysis: Structural Analysis; Thermal analysis; Analysis of kinematics and dynamics of fluids; electromagnetism; mechanisms; Plastic deformation; Injection molding, etc .; analyzes that can be acquired helping of the following software: ANSYS; ABAQUS (simulated); ALGOR; COSMOS; I-DEAS NX; LS-DYNA; MSC.MARC; MSC.NASTRAN; …[3]. The analysis was solved by using the default finite elements

IOP Conf. Series: Materials Science and Engineering **568** (2019) 012064 doi:10.1088/1757-899X/568/1/012064

(tetrahedral and hexahedral elements type). The percentage of needed bending requires the load type, namely the acting force, loading scheme, and the force are perpendicular on the surface.

Few examples of engineering problems where one use the Computer Aided Engineering (CAE):

When it is possible to obtain an accurate solution, a set of equations provides details of the behavior of the modeling system under the given conditions:

$$EI\frac{d^2 \cdot Y}{dX^2} = \frac{PX(L-X)}{2} \tag{1}$$

Boundary conditions:
X=0, Y=0
X=L, Y=0
Beam deformation in the Y direction, as a function of X

$$Y = \frac{P}{24EI}(-X^4 + 2LX^3 - L^3X) \tag{2}$$

W(F)-Force [N/m^2]
E- Young Modulus $E = \frac{\sigma}{\varepsilon}$
I-Modulus of Inertia I=M·L^2 [kg·m^2]
L,a,b- characteristic dimensions
R_1, R_2 –Force reaction at supports

2.3 General Finite Element Analysis formulation, [6]

Generally, at any point in a mechanical structure, the stresses and strains are defined by the axes of a global coordinate system as six components:

- for stresses

$$\sigma_x, \sigma_y, \sigma_z, \tau_{xy}, \tau_{xz}, \tau_{yz}, \tag{3}$$

- for strains

$$\varepsilon_x, \varepsilon_y, \varepsilon_z, \gamma_{xy}, \gamma_{xz}, \gamma_{yz}, \tag{4}$$

The displacements field in a 3D element is defined using the N_i shape functions given by the relationships:

$$u = \sum_{i=1}^{N} N_i u_i \, ; v = \sum_{i=1}^{N} N_i v_i \, ; \; w = \sum_{i=1}^{N} N_i w_i \, ; \tag{5}$$

where u_i, v_i, w_i are node values of the displacement on the element, and N is the number of nodes of an element. Using these data, we can determine the stifness matrix with the formula:

$$k = \int_V B^T EBdV \tag{6}$$

in which B is the nodal displacement matrix, on the d direction d (ε = Bd). In coordinates (x, y, z) the displacements matrix is given as:

$$\varepsilon = \begin{Bmatrix} \varepsilon_x \\ \varepsilon_x \\ \varepsilon_x \\ \gamma_{xy} \\ \gamma_{yz} \\ \gamma_{zx} \end{Bmatrix} = \begin{Bmatrix} \frac{\partial u}{\partial x} \\ \frac{\partial v}{\partial y} \\ \frac{\partial w}{\partial z} \\ \frac{\partial v}{\partial x} + \frac{\partial u}{\partial y} \\ \frac{\partial w}{\partial y} + \frac{\partial v}{\partial z} \\ \frac{\partial u}{\partial z} + \frac{\partial w}{\partial x} \end{Bmatrix} = \mathrm{Bd} \tag{7}$$

The deformation energy is determined with the relationship:

$$U = \frac{1}{2}\int_V \sigma^T \varepsilon dV = \frac{1}{2}\int_V (E\varepsilon)^T \varepsilon dV = \int_V \varepsilon^T E \varepsilon dV = \frac{1}{2} d^T \left[\int_V B^T E B dV \right] d \quad (8)$$

In order to analyze a part using FEA in Ansys, it can be projected directly into ANSYS using drawing tools or as in the present case, the 3D model of the part to be analyzed will be imported from a system external design [7].

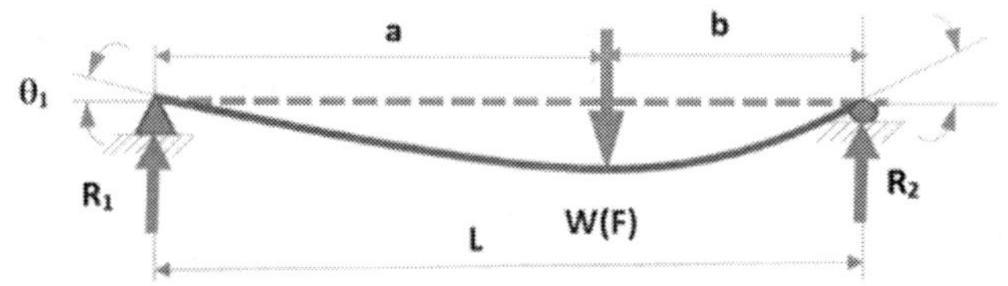

Figure 3. The generic scheme of solving the bending problem, [3]

The first step for FEA analysis, for the support plate, will be to define the boundary conditions. The boundary conditions represent the fixing elements (fixed support) and the force acting on the analyzed part, and at the same time their position [8].

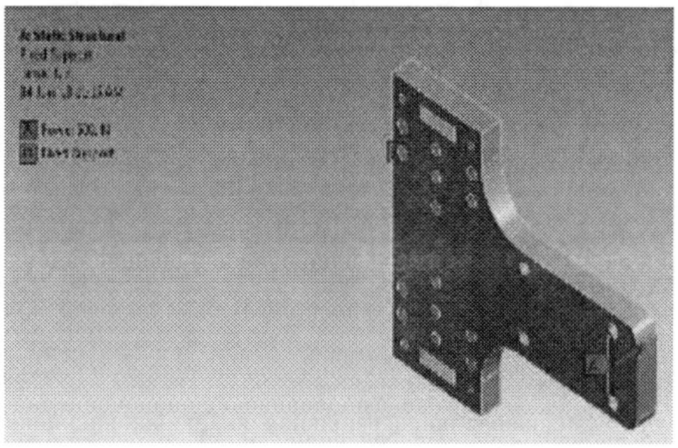

Figure 4 Boundary condition

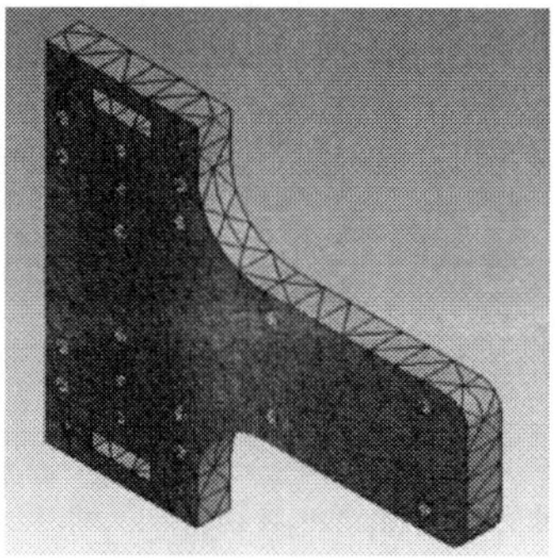

Figure 5 Mesh plate

In figure 4 the boundary condition, marking with A(Fixed Support) and B(Force=500 [N])is presented and in figure 5 the meshing of support plate is shown.

- Number of nodes:9994
- Number of finite elements:5403

After define the boundary condition, will be define the input parameters,variables of optimization from the set of constructive parameters of the support plate, parameters that will generate the output parameters of the best optimal plate, so that on the basis of the final results of the analysis we can manufacture a suitable support plate geometric, mechanical and technological.

2.4 Graphic representation results

Following the definition of boundary conditions, the software provides the 3D image of the support plate (Figure 5) bending request, which is submitted to the subassembly of which it is part.

The Total deformation value is equal to 0.17 mm, value that influences sub-assembly operation, and the Safety Factor 4.5, for aluminum, and 0.06 mm Total deformation, respectively 4.6 the Safety Factor, for Structural Steel, see figure 6.

3. Conclusion

In conclusion, today's graphical tools (3D design software and graphics simulation) can solve constructive and functional trouble of some components with some assemblies or assemblies without the need to build layouts that later or try tests and attempts to determine which is the best version.

IOP Conf. Series: Materials Science and Engineering **568** (2019) 012064 doi:10.1088/1757-899X/568/1/012064

To solve these issues, as in the present case, it is necessary to perform some steps as well as define the conditions at the limit, and once defined as input parameters, the software will provide details on the number of nodes and finite elements and at the same time of the deformation value of the support plate.

Aluminum | Structural Steel

Total deformation

Safety Factor

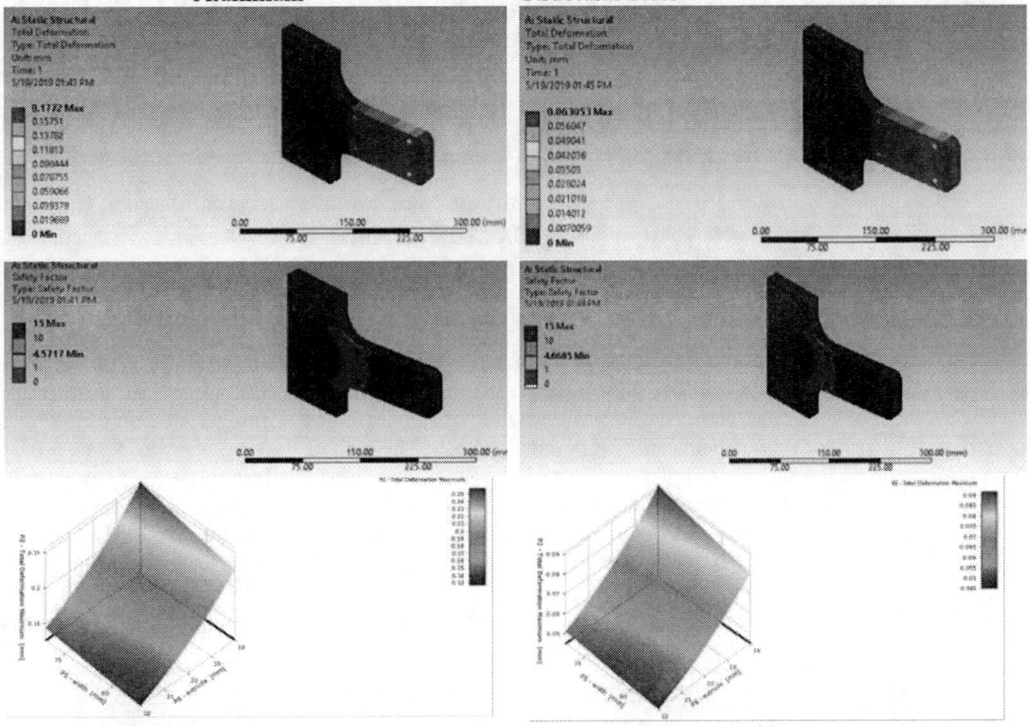

a) b)

Figure 6 Total deformation & Safety Factor
a)-for Aluminium; b)-for Structural Steel

Acknowledgments
This research was developed under financing of S.C. Isrom Impex S.R.L Oradea, The Agreement # 17250/2017. Thanks to advisors and all support of the company Manager Eng. Binselean Mihai

References

[1] Exhaustion https://www.autoexpert.ro/inlocuirea-tobelor-de-esapament/(accessed at 2019.03.04)
[2] Jianping L, Hongwei Z, Xiaoli H , Lin Z, 2016 Pengliang H , Cong L , Ning L , Yuexi Za , and Changyi L 2016, Ins. and Exp. Tech. Volume 59, Issue 5 P762–767
[3] Grebenişan G., Bogdan S. 2017 Parameterized Finite Element Analysis of a Superplastic Forming Process, Using ANSYS, MATEC Web of Conferences, **126**, 2017, https://doi.org/10.1051/matecconf/201712603001
[4] Negrău D C, *Finite Element Analysis and Solution Optimization for a bending device,* 2018, University of Oradea.
[5] Huei-H L, 2017 Finite Element Simulations with ANSYS Workbench 17- Theory, Applications, Case Studies, SDC Publications
[6] Xiaolin C, Yijun L 2018 Finite Element Modeling and Simulation with ANSYS Workbench, Second Edition CRC Press DOI: 10.1201/9781351045872
[7] ANSYS 19, Workbench User's Guide http://www.ansys.com (accessed on April 2019)
[8] Negrau D. C., Grebenisan G., Indre C. 2019 Experimental approach and finite element analysis of the behavior of a steel bending machine, IManE&E 2019

IOP Conf. Series: Materials Science and Engineering **568** (2019) 012065 doi:10.1088/1757-899X/568/1/012065

An optimization approach by Finite Element Analysis, using Design of Experiments and Response Surfaces- a survey

D C Negrău[1,2], G Grebenişan[1], C Gherghea[1]

[1] University of Oradea, Faculty of Managerial and Technological Engineering, Universităţii Street No. 1, Oradea, Romania

[2] dan.negrau@yahoo.com

Abstract. The deformation of an aluminum plate, called in this paper, a support plate, which is part of a bending plastic deformation device, is studied in this paper using the Finite Elements Method, respectively design points of the design by optimization using features such as Design of Experiments and Response Surfaces. The start of the analysis was determined by a functional requirement in the design and construction of a sophisticated bending device controlled by steel plates for the production of exhausts for cars. To optimize the project, it was necessary to analyze the behavior of the active material and components of the device. The analysis of material and piece behavior, in working conditions, has been developed by addressing the facilities offered by Ansys software. Following FEA analysis, optimizing the active components of the device has provided proven results in practice through the execution and operation of the device, according to the technology required by the product designer.

1. Introduction

In this research paper is presented the optimization of a support plate by modifying material and thickness, the plate which is part of a subassembly, the subassembly which is part of a sheet metal bending machine [1].

The support plate was built on the customer's technical prescriptions, so that it has the dimensions shown in figure 1 a). The material used is duraluminium, a material that meets the technological requirements, namely: it is easy to milling, low specific weight, insignificant cutting tool wear compared to the machining of a harder material (soft steel, stainless steel) and the production time is short [2].

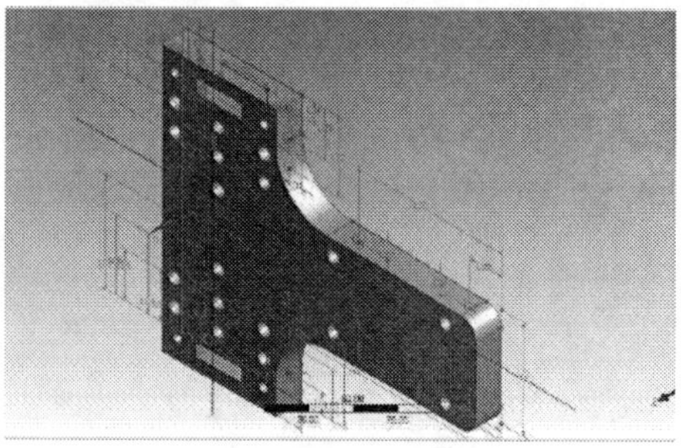

a) 2D support plate

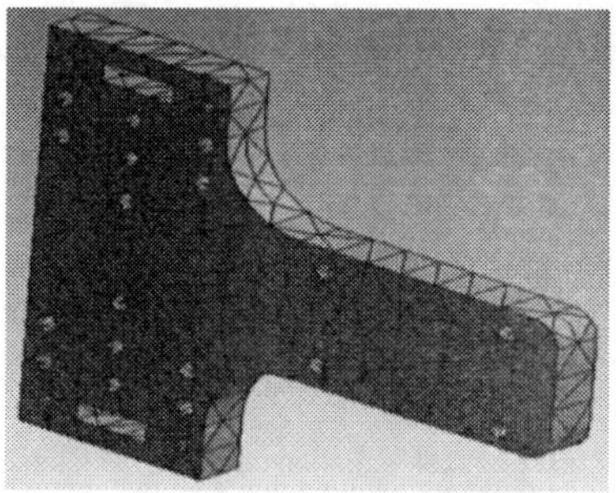

b) Mesh structure (9374 nodes, 5041 elements)

Figure 1. Geometric properties

IOP Conf. Series: Materials Science and Engineering **568** (2019) 012065 doi:10.1088/1757-899X/568/1/012065

2. Optimizing using Design Optimization Experiments (DOE)

The DesignXplorer module (from Design Optimization Exploration- DOE) proposes two optimization workflows: 1) - Response Area Optimization (RSO) based on DOE and Response Area; 2) Direct optimization. The behaviors of the optimization algorithm are identical for Response Area Optimization (RSO) and Direct Optimization (there are some exceptions). The default settings may differ for RSO or Direct Optimization. For RSO, output parameter outputs are retrieved in RS evaluations (very quickly). For Direct Optimization, the results of output parameters are found in the actual resolutions.

Work plan:

A. Establishing the optimization area (Limits and types of parameters (manufactured / discrete values));
B. Input Parameters Relationship (Eliminate Needlessness / Achievable Combinations of Input Values)
C. Objectives and constraints (minimizing or maximizing some limits);
D. Optimization methods (there are 6 optimization methods in DX:
1. Screening (Shifted Hammersley) [implicit mode]
2. MOGA (Multi-Objective Genetic Algorithm)
3. NLPQL (Non-linear Programming by Quadratic Lagrangian)
4. MISQP (Mixed-Integer Sequential Quadratic Programming Method)
5. Adaptive Single-Objective(just for direct optimization)
6. Adaptive Multiple-Objective (just for direct optimization)
E. Convergence and post-processing

For all types of optimization, the history of each parameter is reported by a curve, the convergence criteria are also represented graphically.

2.1 Optimization of design points (project) - proposed variants

Design points are generated according to the objectives and constraints of the system, depending on the user's settings. Out of the generated design points and for which the output parameter values were calculated, three points are generally selected that have the most convenient output parameter values, these points being candidate points. Among the candidate points, the user chooses the optimal option to continue the project. The design point data (figure 2), which has become the current point of the project, is inserted into the project input data, these being the key features on which the optimized project was created.

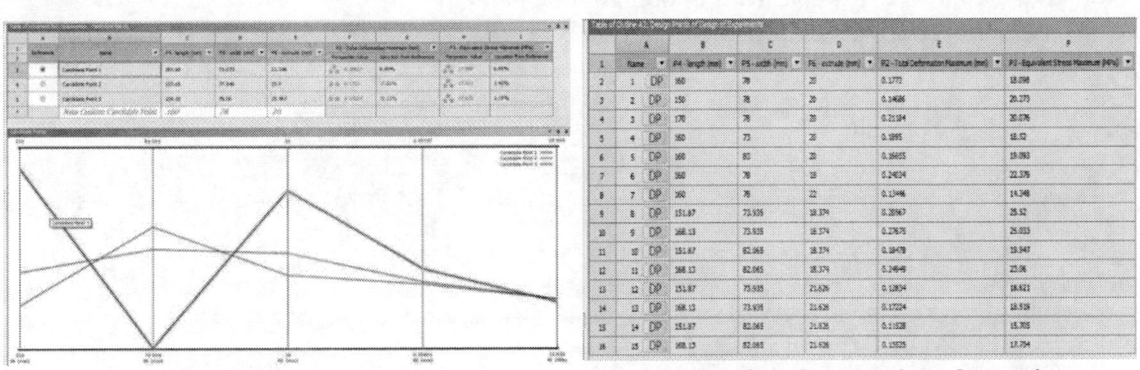

a)-Candidate Points b)-Design Points from Design of Experiments

Figure 2. The candidate points and design points –follow in choosing as a Design point

Design points, verification (for optimizing the response area). RSO design points (figure 3) are in fact approximations of the response surface. The algorithm verifies the design points by creating and updating the output parameters with a "real solution" using the input points of the design points [3].

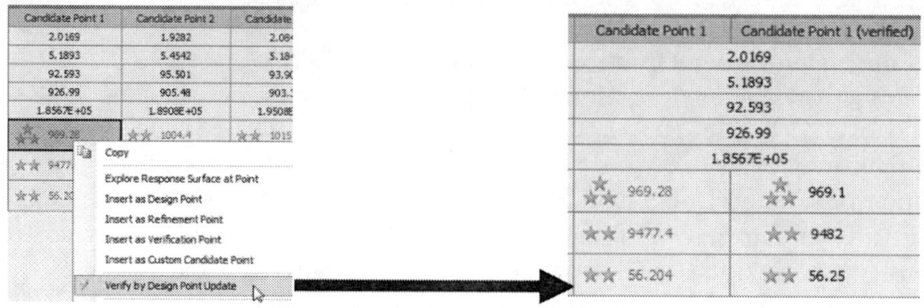

Figure 3. Design point in RSO

The output resolution parameter of the actual resolution is accordingly displayed below the response surface output values to allow easy comparison. If the results are not the same, it indicates that the response area is not entirely accurate in that area and refining or other adjustments may be necessary. It is possible to insert design points as well as a refinement point.

Optimization requirements:

- At least one of the output parameters should confirm an objective (Maximize, Minimize, Seek Target to run an optimization with MOGA, AMO, NLPQL, MISQP, and ASO methods.) If this is not done then the optimization problem is either undefined (without objective) or merely a problem of the satisfaction of the constraints (Objective set to> = Target or <= Target) When the problem is not defined, MOGA or NLPQL analyzes cannot run. AMO and MISQP.
- The Screening method does not depend on any parameter setting and can be used to perform preliminary design studies.
- External optimizations can be connected to DX via ACT Extensions. They can be written in IronPython, C ++ or C #.

For example, an extension is available in the Client Portal to run MATLAB Optimization algorithms in DX.

3. Parametric optimization procedure

After the analysis with finite structural elements has been carried out, for the conditions imposed by the client, a parametric optimization project is carried out, starting from the setting of the optimization input parameters: 1)- the thickness of the support plate (20 mm); 2)-The acting force (N). Then adjust the optimization output parameters:1)- strength to deformation; 2)- the equivalent deformation stress (von Mises), whose values are determined according to the results of the static structural analysis, as well as the limit values as objectives of optimization [4, 5, 6].

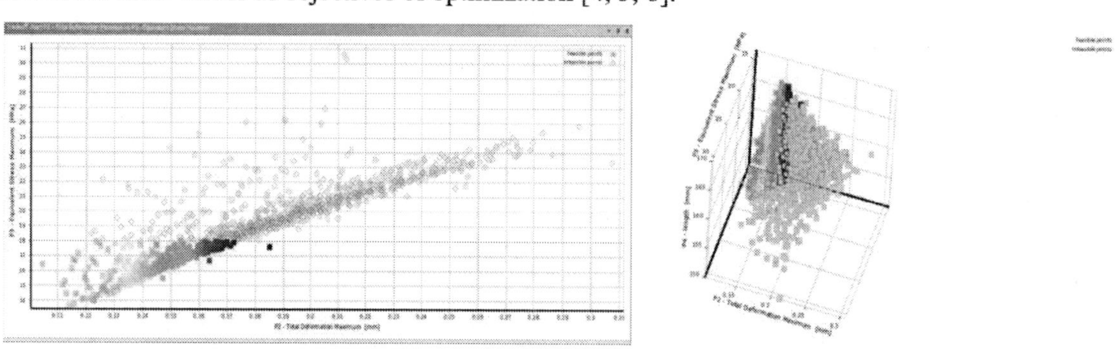

a)-2d graph b)-3D reprezentation

Figure 4. The Tradeoff Graph-contains the Feasible Points(colored points) and Infeasible Points (grey points)

Figure 4 shows the Equivalent Stress and Total deformation graph, several points, a graph which is determined based on the input parameters: thickness, material, arm force.

Based on the above input data, figure 4 shows the total deformation, equivalent stress for different material thicknesses chosen. By comparing the values of each plate thickness (20, 18, 16, 15, 14 [mm]), a graph provides information that the best thickness deformation values of 20 [mm] = 0.179 [mm], marked in the figure with the one star (*).

4. Support plate –steel

Since steel has better mechanical properties than aluminum, a more detailed analysis has been carried out on the steel plate. Figure 5 shows the influence of the change of thickness and material on Total Deformation, the best result is obtained in the case of a 16 mm thick plate.

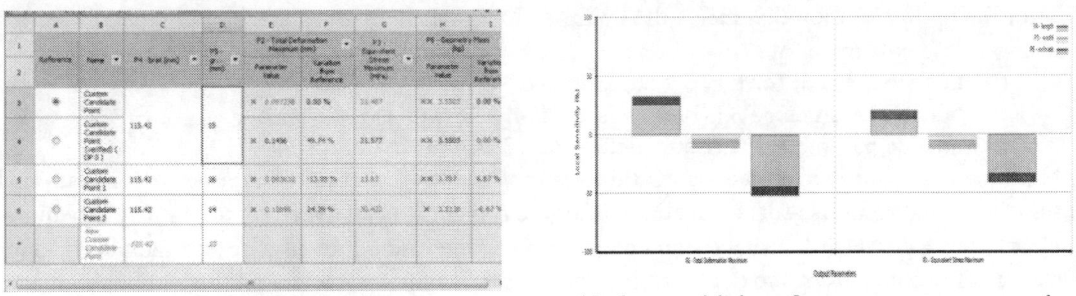

a)-Custom candidate points b)-the sensitivity of output parameters to the input parameters

Figure 5 Influence of the change of thickness an material

The total deformation in the case of the slab thickness = 16 [mm], the deformation which is= 0.083 [mm]. The value that was intended to be reached to meet the requirements of the work, which require that the deformation the maximum elasticity is 0.083 mm, which will make it possible to assemble the protective parts around the exhaust properly.

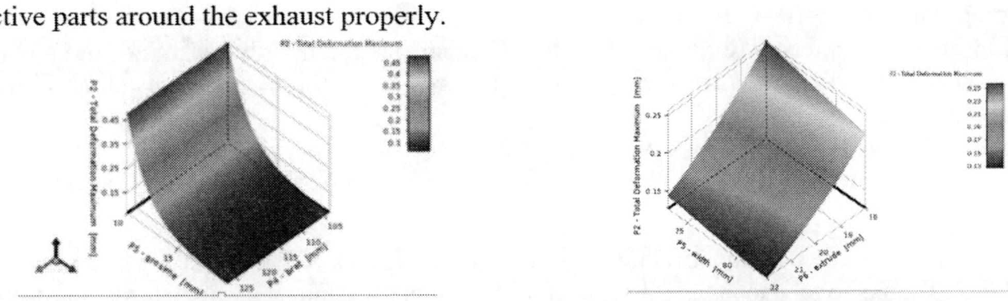

a)-thickness=16 [mm] b)-thickness=20 [mm]

Figure 6 Response surface,

Also, the safety factor is suitable = 4.4 (figure 7) so that the workpiece can be used without the problem that after several uses it will yield again elastically over the reachable limit = 0.1 mm.

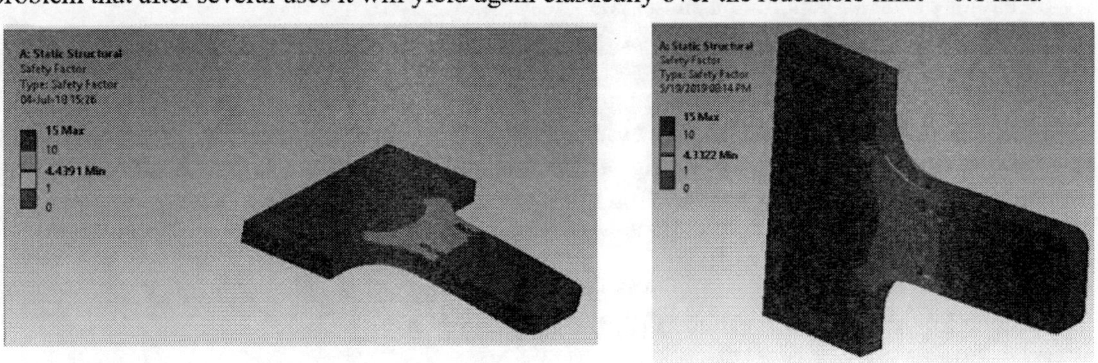

a)-thickness=16 [mm] & structural steel b)-thickness=20 [mm] & aluminium

Figure 7. Safety Factor

IOP Conf. Series: Materials Science and Engineering **568** (2019) 012065 doi:10.1088/1757-899X/568/1/012065

5. Conclusion

The safety factor of the assembly is given by the lowest safety factor determined for its components. From static structural analysis, FEA results in a minimum safety factor of 0.89 for the aluminum plate and 4.4 for the steel plate.

- have been achieved 1000 optimization templates for each input, output, and optimization objective was achieved, with a total of 5000 finite element analyzes to optimize the project.
- each analysis lasted approximately 2 seconds
- the three candidate points obtained have the following values(Figure 2 a))
- of those the 3 candidate point, Candidate Point 1 has the slightest variation from the reference (0.00%) and Candidate Point 2 has the worst variations compared to the reference
- for the optimal option that meets the optimization goals the values are:
    - low-cost material
    - the geometrical form is not complicated
    - outside parameter P3 (*Total Deformation Maximum*)-0.09 mm
    - outside parameter P4 (Equivalent Stress Maximum)- 21.48 MPa

Optimization was performed using three complementary methods: parameter correlation, linear optimization and response surface method, after the project was analyzed by the finite element method. Each of these phases of project implementation is determined by the results achieved. Thus, by correlating the parameters, the degree of influence of the input parameters on the output parameters was determined to decide which ones influence decisively or not the functioning and the degree of safety of the operation of the project. The degree of safety of the project is given by the degree of certainty regarding the quality of the product (the bent piece). This phase is viably linked to the finite element analysis of the perforated and bent punch. Thus, the lifetime and safety factor of the project are determining the quality of the obtained piece and the economic efficiency of the production.

Direct optimization takes into account the correlation of parameters, but does not use the results as a basis for optimization itself.

Optimization using response surfaces is a highly laborious method, but provides the most concise results, with optimization checkpoints performed on one or all three candidate points found in the optimization process.

Based on this analysis, the new device has been built, with a more robust construction that works mechanically at the desired parameters and a 40% lower cost of the semi-finished product.

Acknowledgments
This research was developed under financing of S.C. Isrom Impex S.R.L Oradea, The Agreement # 17250/2017. Thanks to advisors and all support of the company Manager Eng. Binselean Mihail,

References:

[1] D. C. Negrău, G. Grebenişan, C. Indre Experimental approach and finite element analysis of the behavior of a steel bending machine, 2019, IManE&E 2019
[2] D C Negrău, *Finite Element Analysis and Solution Optimization for a bending device,* 2018, University of Oradea.
[3] ANSYS 19, Workbench User's Guide http://www.ansys.com (accessed in April 2019)
[4] Grebenişan Gavril, Nazzal Salem-*The Multi-Objective Genetic Algorithm Optimization, of a Superplastic Forming Process, Using ANSYS*, MATEC Web of Conferences, **Volume** 126, 2017, https://doi.org/10.1051/matecconf/201712603003.
[5] Huei-H L, Finite Element Simulations with ANSYS Workbench 17- Theory, Applications, Case Studies, SDC Publications, 2017
[6] Xiaolin Chen, Yijun Liu 2018 Finite Element Modeling and Simulation with ANSYS Workbench, Second Edition CRC Press DOI: 10.1201/9781351045872

IOP Conf. Series: Materials Science and Engineering **568** (2019) 012066 doi:10.1088/1757-899X/568/1/012066

Research about the quality of the surface after turning out of duralumin alloy

I Pascu[1,3] , I Popescu[2], S Gheorghe[1], I. Geonea[2] and A Didu[1]

[1] Automotive, Transport and Industrial Engineering Department, Faculty of Mechanics, University of Craiova, Calea Bucuresti Str., no.107, Craiova, 200512, Romania
[2] Applied Mechanics and Civil Engineering Department, Faculty of Mechanics, University of Craiova, Calea Bucuresti Str., no.107, Craiova, 200512, Romania

i_pascu@yahoo.com

Abstract. This paper describes a study concerning the inner turning of a piece of duralumin, using 6 values of the feed. The shapes and micro-geometry of the resulting chips have been observed experimentally and these are explained by the large deformations in the cutting area. The processed surfaces, both by enlarging them to a microscope and by measuring the roughness have been experimentally studied. Also, the sections of chips are analysed, giving the theoretical and real dimension values, also interpreted on the basis of the deformations in the cutting area. Also, values of roughness parameters are presented.

1. Introduction

Duralumin has been used extensively in aircrafts construction [1] and aerospace applications [2,3], is a durable and lightweight material [4].

For a long time, only duralumin mats have been used, but other types of Al alloy bars have also been made over time, or have been machined [5]. Thus arose research on processing by turning and milling of this material [6-10]. Thus, in the [11] the process of turning the duralumin alloy 2017 is modeled. The stresses and temperature of the cutting area are studied for the purpose of processing machines with numerical controls. The roughness of the processed surfaces have been analyzed and photographs were taken at the microscope.

In [12] probabilistic methods for estimation the feed and cutting speed influence on the surface roughness of the finish machined at the turning for the following materials: brass, steel, alpax and duralumin are used. Numerous charts of roughness and estimated roughness are presented.

In [13] the milling of 5 types of aluminum alloys at cutting speeds from 60 to 250 m / min, using a 20 mm cylindrical-frontal cutter has been analyzed. They were studied: the processing precision, the cutting forces and roughness of the machined surface. Also, the roughness diagrams are given.

Machining of duralumin AL-2017-T4 alloy for aerospace applications is studied in [14].

The cutting force reduction and surface quality improvement have been investigated. In doctoral thesis [15] the milling processing on centers milling with head spherical having 8 mm radius of the sphere, using an experimental program based on 27 experiences is studied. The cutter speed range was between 1000 and 8000 rpm. The chips root and tool geometry, roughness and precision are modeled.

IOP Conf. Series: Materials Science and Engineering **568** (2019) 012066 doi:10.1088/1757-899X/568/1/012066

In [16] the turning and grinding of 4 cast aluminum alloys at different cutting regimes with different cutting fluids has been studied. The optimal geometry of the drawknife for each alloy are settled, the cutting forces, roughness and temperature in the cutting area have been measured. The forms of the resulting chips were studied, explained by the analysis of the phenomena in the cutting area have been established. Tools wear and determine the cutting speed relationships for each alloy have been determined.

In the [17] the almost pure aluminum turning with 99.39% aluminum with the hardness HB 24.2 2.5/31.2/30 are investigated. The shapes of chips resulting from different cutting regimes have been studied, the shape of the channel on the face of the groove has been optimized and they have been analyzed the depositions on edge, which are very pronounced.

In [18] turning of aluminum alloy ATSi7Mg, with HB 44 5 / 31.2/ 30 with σ_r = 14 daN/mm^2 was studied, analyzing in detail the forms of chips in various combinations feed - the depth of cut with different cutting speeds. The dimensions of the cuttings at different cutting speeds have been detailed.

Also, in [19-22] some research about duralumin turning and the effect of the process parameter on the surface quality are presented.

2. Initial Data

A duralumin Al Cu4 Mg1 Mn alloy tube was turned out inside with different feeds, without cutting fluid. After machining, the piece was cut in two for measuring the roughness and view the shaded surfaces (figure 1). It was run at a speed of 500 rot/min and a cutting depth of 0.5 mm.

Figure 1. Turned piece.

For machining, an inside boring tool from high-speed steel, , with the following geometry: a channel on the front face with the dimensions of 11.46 mm x 3.36 mm and depth of 3.5 mm, κ_r=110^0, κ_r'=8^0, α_n=18^{0,} γ_n=30^0.

3. Experimental Results

In figure 2 the forms of chips obtained a cutting feed of f=0.096 mm/rot. A bold used as a scale factor, with a length of 28 mm and a diameter of 0.76 mm, appears on the images.

Silver-white splinters of different shapes resulted. Some are in the form of propellers, not more than 90 mm long, 6 mm in diameter and 10.89 mm in pitch, others are in the form of a gem, starting from helical shapes but tangled after having a longer length 10 mm, reaching on random shapes with volumes entered in spheres with a maximum diameter of 45 mm. There are also comma-shaped chips, about half a spiral of an Archimedes spiral.

In figure 3 the chips micro-geometry on the surface that did not come into contact with the channel on the cutting tool front face is shown.

IOP Conf. Series: Materials Science and Engineering **568** (2019) 012066 doi:10.1088/1757-899X/568/1/012066

It can be observed many uneven scales, which demonstrates large deformation of the chip in the cutting area.

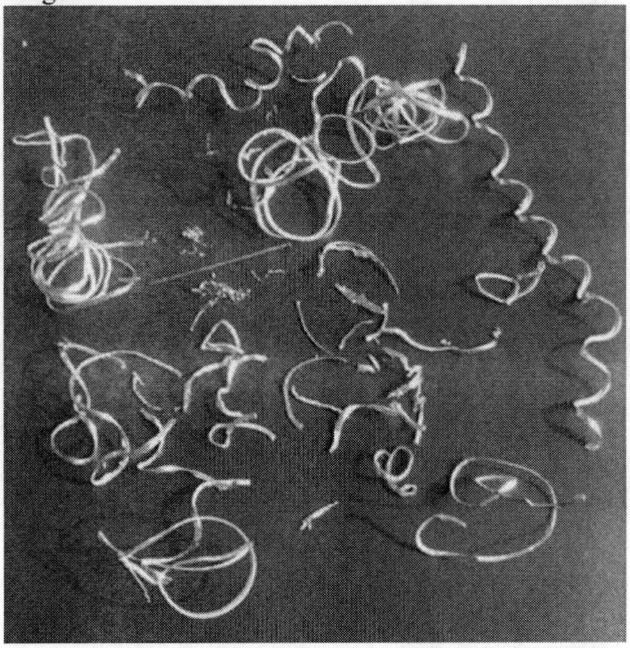

Figure 2. Resulted chips after turning with a feed f=0.096 mm/rot.

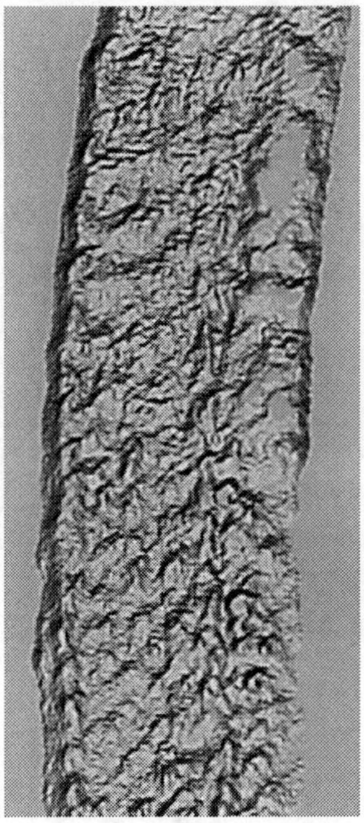

Figure 3. Chip micro-geometry for a feed of f=0.096 mm/rot

The surface roughness is given by the profile micro-unevenness presented in figure 4, measured with an electronic roughness tester Mitutoyo type, a SJ-201 P model. The device provided the image of the profile micro-unevenness and the following values of the roughness parameters (SR ISO 4287): Ra, defined as the arithmetic average deviation of the evaluated profile and represents the arithmetic mean of the absolute values of the deviations of the profile, within the limits of the basic length lp; Rz - the maximum profile height, defined as the sum of the largest of the projections of the profile, Zpmax, and the largest of the depths of the profile gaps, Zvmax, within the basic length limits; Rq - the square average deviation of the profile, as a quadratic mean of the values of the ordinates Z (x) within the basic length limit.

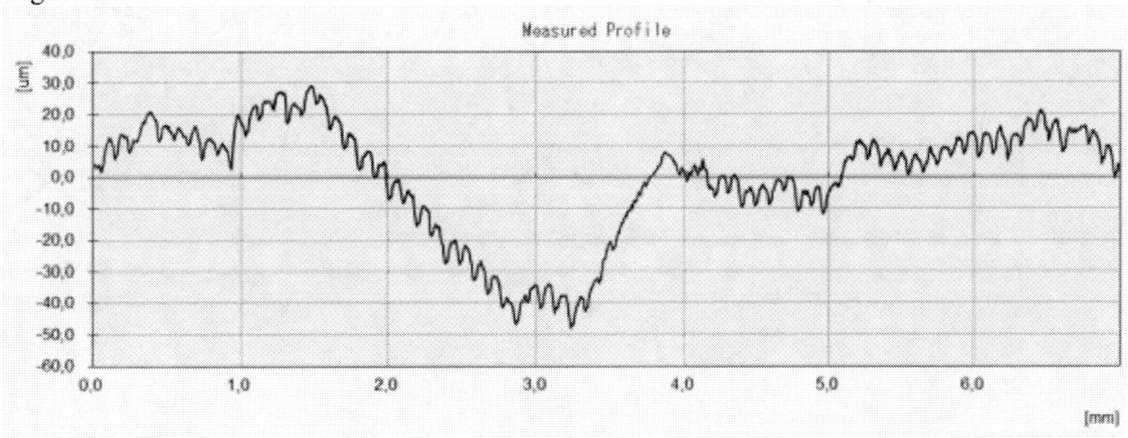

Figure 4. Surface roughness for bushing, after turning with f=0.096 mm/rot feed.

IOP Conf. Series: Materials Science and Engineering **568** (2019) 012066 doi:10.1088/1757-899X/568/1/012066

In figures 5 to 7 the roughness images for other feeds are presented.

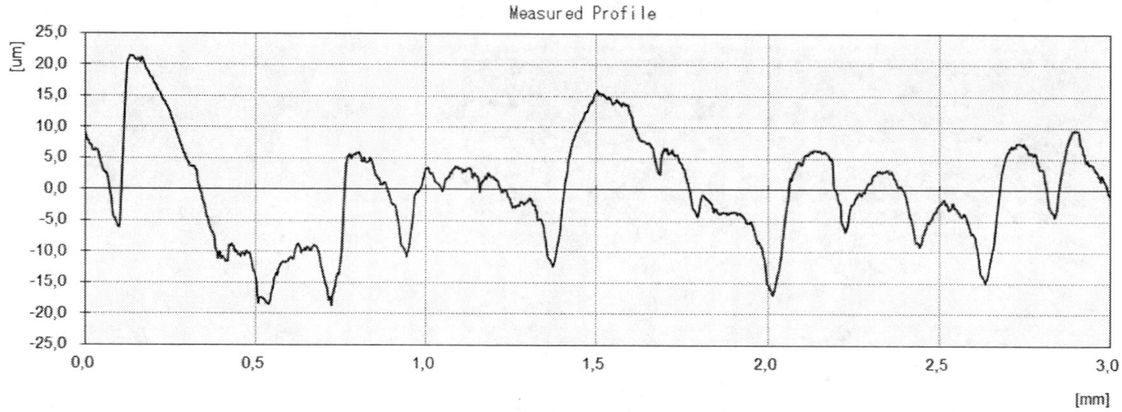

Figure 5. Surface roughness for bushing, after turning with f=0.208 mm/rot feed.

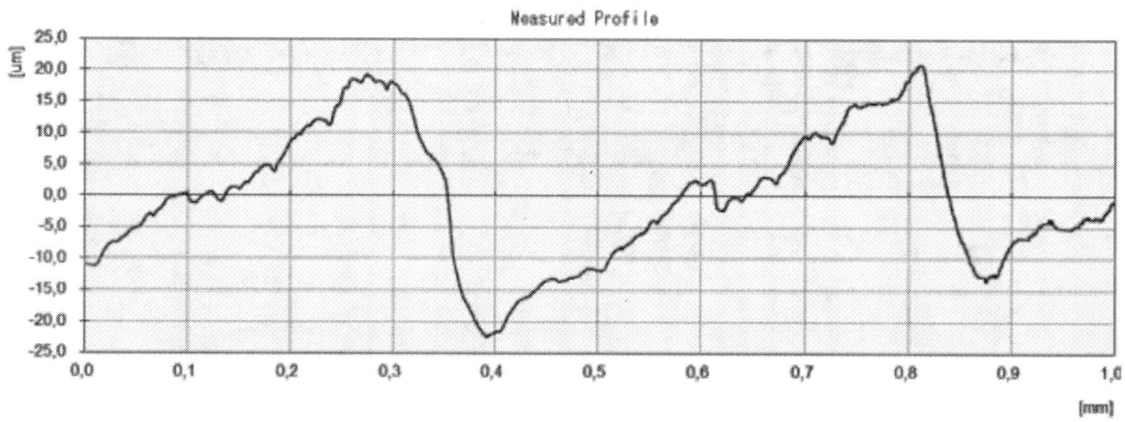

Figure 6. Surface roughness for bushing, after turning with f=0.5 mm/rot feed.

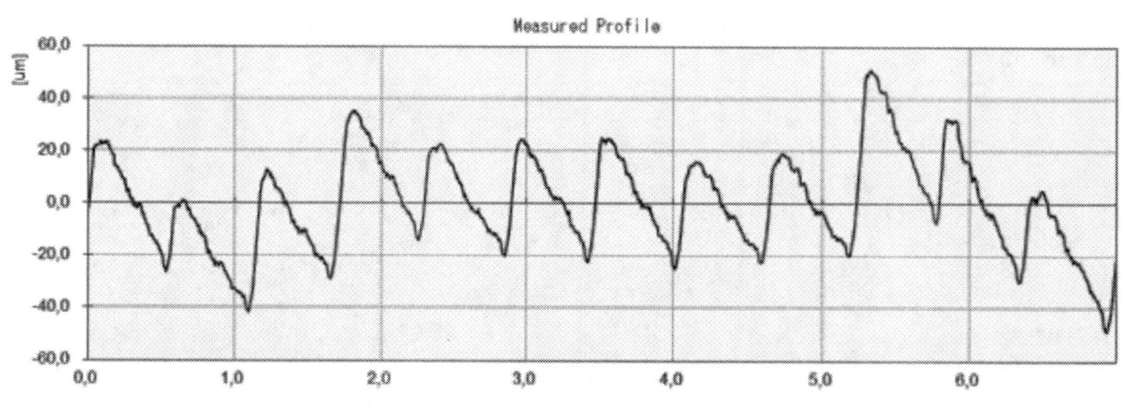

Figure 7. Surface roughness for bushing, after turning with f=0.584 mm/rot feed.

In the table 1 the values of roughness parameters are given.

IOP Conf. Series: Materials Science and Engineering **568** (2019) 012066 doi:10.1088/1757-899X/568/1/012066

Table 1. The values of roughness parameters.

f (mm/rot)	Ra (μm)	Rz (μm)	Rq (μm)
0.096	6.94	47.98	9.46
0.208	5.28	32.84	6.44
0.302	3.74	26.95	4.61
0.416	4.48	36.74	5.92
0.5	4.46	26.97	6.15
0.584	12.92	81.17	15.83

4. Conclusions

The turning out process for a duralumin alloy part with 6 different feeds, collecting the resulting chips, studying their shapes and micro-geometry have been studied.

The processed surfaces were evaluated by enlarging them to a microscope and measuring three roughness parameters.

All this evidence shows that the material is easy to process, but it does not produce deposits on the cutting edge to disturb the cutting process. However, the not too high hardness of the material causes large deformations in the cutting area, so that the process does not flow uniformly, the chip pieces overlap on each other forming the scales.

The overall conclusion is that this material is easy to process, but in the cutting area they appear large deformations of the chipping faces.

References

[1] Zronik J 2005 *Metals shaping our world.* (Crabtree, Canada)
[2] Sayuti M *et al* 2013 *Int. J. Adv. Manuf. Technol.* **65** 1493.
[3] Heinz A *et al* 2000 *Mater. Sci. Eng. A.* **280** 102.
[4] Yoshida H 2015 *J.Japan Inst. L. M.* **66** 97
[5] Cheremisinoff NP 1996 *Materials selection deskbook.* (Noyes, New Jersey)
[6] Benardos P G, Vosniakos G C 2002 *Robot Comput Integr Manuf* **18** 343.
[7] Lin W S, Lee B Y and Wu C L 2001 *J. Mater. Process. Technol.* **108** 286.
[8] Chang H-K *et al* 2007 *Int. J. Mach. Tools. Manuf.* **47** 1021.
[9] Singh D, Rao PV 2006 *Int. J. Adv. Manuf. Technol.* **32** 1115.
[10] Bogatu L[1] and Ionascu G 2015 *Int. J. Innov. Sci. Eng. Technol.* **3** 2348.
[11] Paszta P 2017 *Proc. MATEC Web of Conf. CoSME'16* **94** 02010 (Brasov Romania).
[12] Kryviy P, Dzyura V, Tymoshenko N 2018 *Metals* **8** 965.
[13] Mizugaki Y, Takafuji K, Kikkawa K 2009 *Key Eng Mater.* **408** 714.
[14] Maher I, Eltaib M E H, Sarhan A A D, El-Zahry R M 2014 *Int. J. Adv. Manuf. Technol.* **74** 531.
[15] Rusu O T 2015 *Contributions on the study of molding process milling, by processing centers*, PhD. Thesis (Iasi, Romania).
[16] Popescu I 1974 *Cutting aluminum alloys* (Technical Publishing House, Bucharest).
[17] Popescu I, Pascu C I 2011 *RECENT Ind. Eng. J.* **12** 384.
[18] Popescu I, Pascu CI 2012 *Annalls of the Oradea University*, **11** 4120.
[19] Dogra M *et al* 2010 *Int J of Precis. Eng. and Manuf.* **11** 341.
[20] Liu Y, Sun D, You H, Chung JS 2005 *Appl. Surf. Sci.* **246** 82.
[21] Sheasby PG, Pinner R, Wernick S 2001 *The Surface Treatment and Finishing of Aluminium and Its Alloys* (ASM International, Finishing Publications, Materials Park, USA).
[22] Kadlečková M et al 2018 *Materials* **12** 109.

IOP Conf. Series: Materials Science and Engineering **568** (2019) 012067 doi:10.1088/1757-899X/568/1/012067

Considerations on the influence of the moment of inertia on the movement of a vehicle

M Trotea, A Constantinescu, A E Romanescu

Faculty of Mechanical Engineering, University of Craiova, Romania

lsimniceanu@yahoo.com

Abstract. This paper presents a mathematical model for automotive stability study taking into account the influences of the inertias's moment. It is also presents an experimental method for determination of the inertias's moment for a vehicle related to the axis of a system which has the origin in its mass centre. The used procedure is based on the pendulous method which consists in the vehicle's postponement and the measurement of the oscillations period around the suspension axle. The inertia moments of the tested vehicle are determined experimentally for three cases of loading, taking into account the load of a passenger case, three passengers case and five passengers case. The numerical results are determined based on a simplified mechanical model and a nonlinear differential mathematical model. It is investigated several cases of vehicle movement with different load and thus is highlighted the influence of different moments of inertia over the dynamic behaviour of the considerate vehicle.

1. Introduction

The assessment of the dynamic and cinematic behavior of a car is the starting point in the conception phase of that. For this purpose, the mechanical models and mathematical models is designed according to sizes of interest. Mechanical models have different degrees of approximation of the real model [2], [5], [9] and lead to mathematical models with different levels of complexity [3], [4], [8].

2. Theoretical considerations on the method of determining moment of inertia

It is considered a rigid solid with a horizontal fixed axis, subjected only to the action of the gravitational field (Figure 1), relative to two reference systems, as follows:

-a fixed outer reference system (E) having the axes $Ox_1y_1z_1$, the axis Ox_1 being the vertical of the place;

-an own reference system (P) connected to a rigid system, with Oxyz axes, the X axis passing through the center of mass of the rigid body.

The equation of motion of the solid, in case of friction neglect, is:

$$J_o\ddot{\theta} + Gl\sin\theta = 0 \quad (1)$$

where: - J_0 is the moment of inertia of the solid to the oscillation axis of the suspension;

-θ is the angle between Ox_1 - Ox axes;

- G is the weight of the body;

-l is the distance from the suspension axis to the center of gravity of the solid;

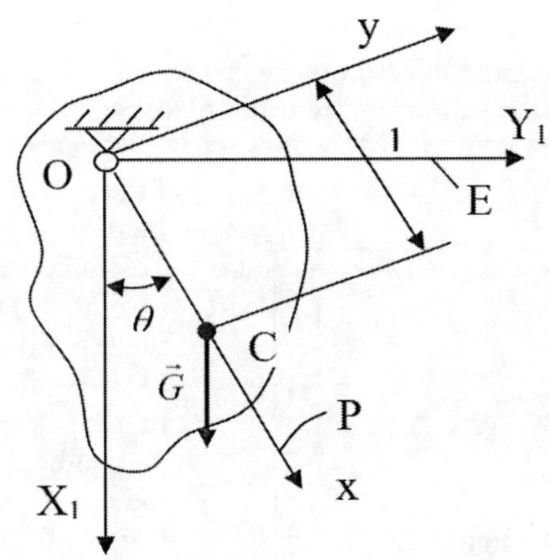

Figure 1. The mechanical model of the solid to determine the moment of inertia

In the case of small oscillations equation (1) takes the form:

$$J_0\ddot{\theta} + Gl\theta = 0 \tag{2}$$

The period of small oscillations is given by:

$$T = 2\pi \cdot \sqrt{\frac{J_0}{G \cdot l}} \tag{3}$$

Relationship (3) shows that small oscillations are autocrons, the oscillation period does not depend on amplitude.

In case of higher oscillations the period of movement is:

$$T = 2\pi \cdot \sqrt{\frac{J_0}{G \cdot l}\left(1 + \frac{\alpha^2}{16}\right)} \tag{4}$$

where: - is the angular oscillation of the oscillations.

Therefore, to determine the moment of inertia J0, the weight of the solid G, the position of the center of gravity (through the distance l from the center of the suspension) and the period of oscillations, T.

The moment of inertia of the solid to the axis of oscillation takes the form:

- for small oscillations:

$$J_0 = \frac{T^2 \cdot G \cdot l}{4\pi^2} \tag{5}$$

- for large amplitude oscillations:

$$J_0 = \frac{T^2 \cdot G \cdot l}{4\pi^2\left(1 + \frac{\alpha^2}{16}\right)} \tag{6}$$

The moment of inertia of the solid to an axis passing through the center of mass and parallel to the suspension axis is calculated using Steiner's theorem:

IOP Conf. Series: Materials Science and Engineering **568** (2019) 012067 doi:10.1088/1757-899X/568/1/012067

$$J_c = J_0 - \frac{Gl^2}{g} \tag{7}$$

3. Experimental results

The installation for determining the moment of inertia consists of a frame on which the vehicle is fixed, a frame which can be suspended by a beam, in order to allow it to oscillate (Figure 2).

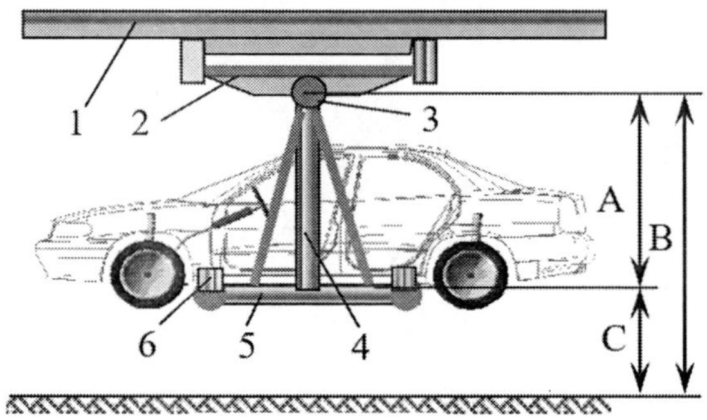

Figure 2 Stand for determining moment of inertia
1 - supporting beam; 2 - horizontal shaft assembly (having axis parallel to the Ox axis) made with hinges (radial ball bearings); 3 - horizontal shaft assembly (with axis parallel to the Oy axis) rigidly fixed by the perpendicular shaft 2, also made with ball radial ball bearings; 4 - vertical support; 5-car support frame; 6 - car fixation device.

Several test cycles were performed under the following conditions:
-the following angular amplitudes α and loads are intended:

α=10° – without load	α=10° –1 passenger load	α=10°–2 passengers load	α=10°–5 passengers
α=15° – without load	α=15° –1 passenger load	α=15°–2 passengers load	α=15°–5 passengers
α=20° – without load	α=20° –1 passenger load	α=20°–2 passengers load	α=20°–5 passengers

-at least 30 oscillations shall be made at each test.

The results of experimental research and their average are presented in the table below.

Moment of inertia \ Angle α		α			Average
		10°	15°	20°	
Without load	$J^{(0)}_{xx}$ [kg.m²]	262	525	525	394
	$J^{(0)}_{yy}$ [kg.m²]	1307	2614	2614	1961
	$J^{(0)}_{zz}$ [kg.m²]	1450	2901	2901	2176
1 passenger	$J^{(0)}_{xx}$ [kg.m²]	278	556	556	417
	$J^{(0)}_{yy}$ [kg.m²]	1384	2768	2768	2076
	$J^{(0)}_{zz}$ [kg.m²]	1536	3072	3072	2304
2 passengers	$J^{(0)}_{xx}$ [kg.m²]	293	586	586	440
	$J^{(0)}_{yy}$ [kg.m²]	1461	2922	2922	2192
	$J^{(0)}_{zz}$ [kg.m²]	1620	3242	3242	2432
5 passengers	$J^{(0)}_{xx}$ [kg.m²]	393	786	786	590
	$J^{(0)}_{yy}$ [kg.m²]	1689	3378	3378	2534
	$J^{(0)}_{zz}$ [kg.m²]	1874	3748	3748	2811

The values of these moments are used for the dynamic study of vehicles, namely for the analysis of their stability, and it is easy to see that their value is dependent on the load size and the maximum values are those based on the Oz axis.

IOP Conf. Series: Materials Science and Engineering **568** (2019) 012067 doi:10.1088/1757-899X/568/1/012067

4. Numerical results

For the displacement after a circular trajectory, the obtained solutions will be presented graphically considering for each of them two different values of the moment of inertia, corresponding to the case when the vehicle moves only with the load of the driver or with a load of five persons.

The following numeric values will be used:

- tire rigidity k1 = k2 = 250 N / grad
- wheelbase A = 2570 mm.

For the case of one passenger car the following parameters are used: distance from center of gravity to front axle a = 1113 mm, distance from centre of gravity to rear axle b = 1457 mm, mass m = 1358 kg, moment of inertia I = 2611 kgm^2;

For the case of circular trajectory [6] and one passenger load the next numerical results are obtained:

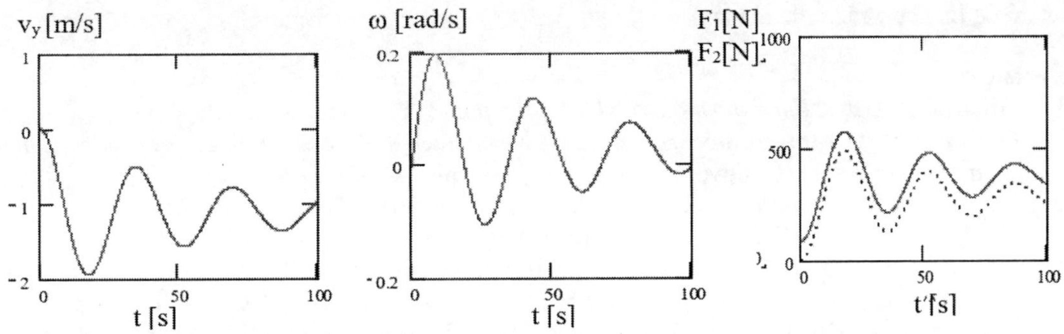

Figure 3. Numerical results in case of one passenger load

For the case of five passengers car the following parameters are used: distance from centre of gravity to front axle a = 1234 mm, distance from centre of gravity to rear axle b = 1336 mm, mass m = 1667 kg, moment of inertia I = 2802 kgm^2;

In this case the cars movement is stable, the amplitude of the angular velocity is 0.2 rad / sec and the wheel force amplitude is approximately 600.

For the case of circular trajectory and five passengers load the next numerical results are obtained:

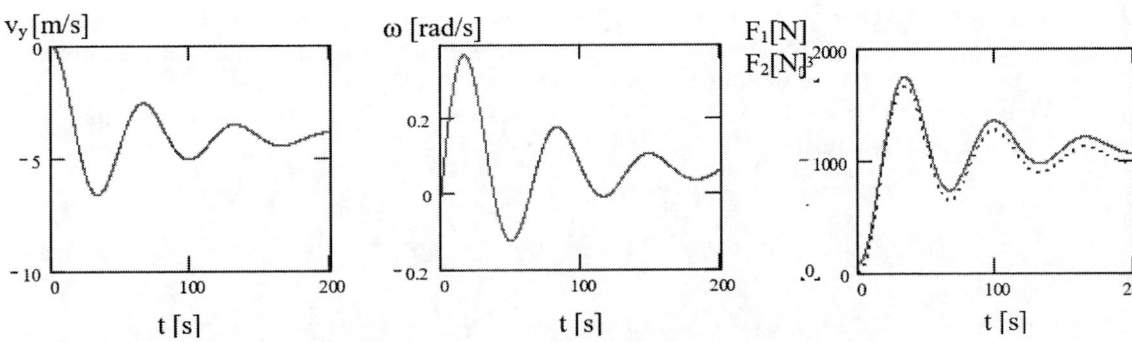

Figure 4. Numerical results in case of five passengers load

The movement has the same characteristics, it is stable, the magnitude of the rotation speed is 0.4radian/sec and the wheel force amplitude is approximately 1700 N.

5. CONCLUSION

This mathematical model is a simple but useful model for vehicle stability analysis, based on the theory of dynamic systems or the theory of classical mechanics. The values obtained are accurate values,

IOP Conf. Series: Materials Science and Engineering **568** (2019) 012067 doi:10.1088/1757-899X/568/1/012067

compared to the measurements made on a real vehicle [7], having the same parameters as those used in the numerical simulation, so this can be considered a validation for this model.

For the proposed model and for the types of motion analyzed, the variation of the moment of inertia (made by changing the car load) does not change the character of the movement (the movement remains stable in all the analyzed cases) but essentially alters the motion parameters both by the magnitude of the angular velocity and the lateral velocity, especially by the amplitude of the lateral wheel force.

A variation for inertia less than 10% leads to an increase for lateral forces to 280%, reaching values comparable to the grip force, resulting in vehicle skidding and loss of stability.

These conclusions refer to the simplified plan layout where the center of the table is placed in the plane of the tread and on the symmetry axis of the car. If calculation possibilities allow the analysis of some spatial models of the vehicle, including other physical and geometric parameters (unsuspended masses and suspended mass, rigidity of tires and suspension), then the moment of inertia taken into account for both suspended mass and non-suspended masses, can influence the character of the movement in a considerable manner.

References

[1] Gillespie Th 1992 *Fundamentals of vehicle dynamics*, SAE International, R-114

[2] Oţăt V, Bolcu D-tru, Simniceanu L and Thierheimer WW 2005 *Vehicle dynamics* (*Dinamica autovehiculelor)* (Craiova: Universitaria Publishing Craiova)

[3] Simniceanu L , Otat V, Trotea M and Bogdan M 2015 *Proc. of the European Automotive Congress EAEC-ESFA 2015* (Bucuresti) **vol 1** (Springer, Cham) p 741-47

[4] Otat V, Simniceanu L, Neagoe D, Bogdan M and Dincă L 2008 *Proc. of Sec. int. congress automotive, safety and environment SMAT* (Craiova) **Tome I** (SIAR Publisher) p 213-16

[5] Simniceanu L, Bogdan M, Oţăt V, Trotea M 2017 *Proc. of the 10th International Automotive Congress "Automotive and Environment" CAR 2011* (Pitesti) **Volume 252** (IOP Publishing, IOP Conference Series: Materials Science and Engineering) 012023

[6] Bolcu D and Oţăt V 2002 *The World Automotive Congress FISITA 2002* (Helsinki) paper code: FO2I240

[7] Neagoe D-tru and Bolcu D-tru 2008 *Stability and Manipultability Study of Motor Vehicles. Experimental and theoretical research* (Studiul stabilitatii si maniabilitatii autovehiculelor. Cercetari experimentale si teoretice) (Craiova: Universitaria Publishing, Craiova)

[8] Simniceanu L, Neagoe D-tru, Trotea M, Bogdan M and Constantinescu A 2016 *Applied Mechanics and Materials* **Vol. 822** ed. Trans Thec Publications pp. 54-59

[9] Simniceanu L, Otat V and Trotea M 2015 *Recent Advances in Mechanics, Mechatronics and Civil, Chemical and Industrial Engineering* (Determination to a plan model of the automotives moving used for the stability analysis) ed INASE pp 68 – 71

[10] Voinea P R and Stroe I 1994 *Dynamic systems* (Sisteme dinamice) (Bucuresti: Institutul Politehnic Bucuresti)

IOP Conf. Series: Materials Science and Engineering **568** (2019) 012068 doi:10.1088/1757-899X/568/1/012068

Developing a real-time accuracy monitoring system for industrial robot arms based on lean principles

Zs Molnár[1,3], G E Cascio[2], G Husi[1] and E G Szűcs[1]

[1] Department of Mechatronics, Faculty of Engineering, University of Debrecen, Ótemető st 2., Debrecen 4028, Hungary
[2] The George Washington University, Washington, DC 20052, 2121 I St NW, USA

[3] zsolt.molnar94@gmail.com

Abstract. In the Robot Laboratory of Mechatronics Department during practical classes of robot programming, the participants observed that the KUKA KR5 Arc robot arm's practicing tool does not accurately follow the programmed path in every case. In accordance to Lean approaches, the observations of the participants were taken into consideration and the possible sources of error were identified. One of them is the robot arm itself, which is responsible for the execution of the programmed movements. The goal of this research is to develop a system which makes it possible to exclude the robot arm as a potential source of error from the error analysis. For this purpose, a monitoring system was developed which supports the Lean production by implementing one of the main pillars of the Toyota Production System. Jidoka is one of the main pillars, which means that the production process is automatically stopped if an error condition arises. The method was first implemented by Sakichi Toyoda in 1896, who developed a simple device for automatic looms, which stopped the machine if the thread broke. Using this principle, the autonomation – automation with human touch – of the robot arm is achieved.

1. Introduction

The development presented in this article focuses on robot arm type robots with open kinematics chain and 6 degrees of freedom [1]. In the Robot Laboratory of the Mechatronics Department, a KUKA KR5 Arc robotic arm is available for students during robot programming classes and for research purposes. These type of robot arms are also used by the industry, due to their multifunctionality [2]. Several developments were realized having in focus the robot arm, including a custom designed 3 finger gripper for cylindrical object manipulation [3], which was created using 3D printing technology [4].

In the following sections the development of a monitoring system is presented, which is able to calculate the execution error of the robot arm in real-time. After executing different tests, if it is the case, the robot arm can be excluded as a possible source of error. Supporting Lean production was taken into consideration as a design guideline. This is implemented as Jidoka main pillar of the Toyota Production System (TPS) [5]. Jidoka contributes to the final product quality and cost.

The monitoring system will be capable to calculate the difference between the executed path and the preprogrammed one. Using these values, a threshold can be set and when the difference exceeds the value, a halt command will be sent to the robot arm. In this way, material loss and waste production can be reduced in a manufacturing process.

IOP Conf. Series: Materials Science and Engineering **568** (2019) 012068 doi:10.1088/1757-899X/568/1/012068

2. Design principles

2.1. Analysing the robot arm as a system

Before we start designing the monitoring system, we should analyze the robot arm in detail from mechanical, electrical and software aspects.

Mechanically, the robot arm consists of interconnecting rigid metal arms and revolute joints. In our case, we have six joints and every joint adds one degree of freedom to the system. This means that the robot arm Tool Center Point (TCP) can move along each axis in Cartesian space and can execute rotational movements around them. To control these movements, six servomotors are used, one for each joint. The rotational movement from the motor to the joint is transferred via different drivetrains.

Electrically, each servomotor it consists of three parts: the motor part, electrical brake and the digital resolver. The digital resolver gives feedback to the motor controller about the actual position of the motor. Using the provided information, the computer (PC and MFC) sends the proper signals to the motor controllers (KSD) to command the motors to move in the programmed position. In this way a closed loop control is realized. The electrical block diagram of the robot arm is shown in Figure 1.

From software perspective, the resolver data and the calculated TCP position from the robot arm's computer can be retrieved via Ethernet connection. Using the resolvers as sensors, the position of the robot arm can be monitored in real-time, without invasive measurement methods, which require external sensors and trackers mounted on the robot, which increase the cost of the system.

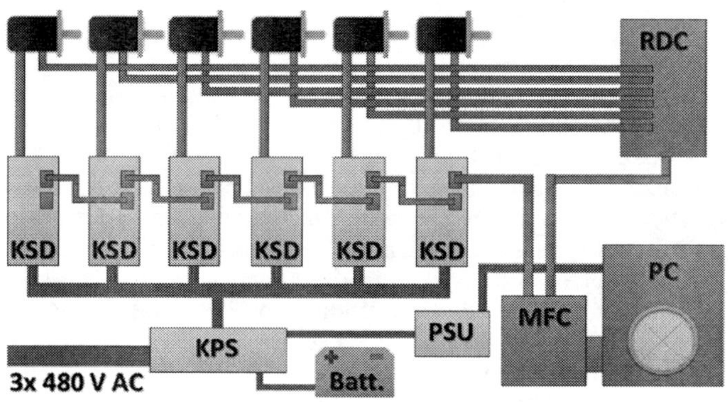

Figure 1. The electrical system of the robot arm.

2.2. The error calculation method

During the KUKA KR5 Arc robot arm programming process, the beginning and end points of a path section are stored by the robot's computer. These are fixed points and the path itself is calculated on-the-fly by the robot arm's computer and translated to servo motor rotations. Using an external computer and knowing the stored points of the path, a line can be described mathematically [6]. By retrieving the real-time position data of the TCP, the distance between the previously mentioned line and the TCP can be calculated. All of these points are in a three-dimensional space; therefore, every point has three coordinates. To calculate the distance (which is the execution error) between the mathematically described line and the actual TCP position, vector algebra is used [7]. Figure 2 illustrates how the execution error is calculated using vectors.

According to the Figure 2, consider the TP_1T' right triangle. By interpreting the sine function the equation (1) can be expressed. $|\vec{v}|$ denotes the length of the $\vec{v}$ vector.

$$\sin(\alpha) = d \cdot \left(\left|\overrightarrow{\mathrm{P_1T}}\right|\right)^{-1} \tag{1}$$

Reordering the equation and multiplying with $|\vec{v}|$, results the format presented by the equation (2).

$$|\vec{v}| \cdot d = |\vec{v}| \cdot \left|\overrightarrow{\mathrm{P_1T}}\right| \cdot \sin(\alpha) \tag{2}$$

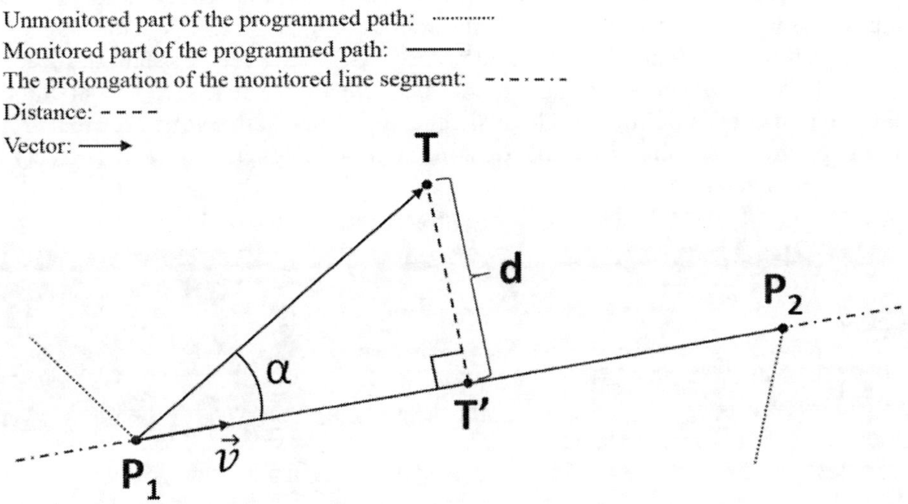

Figure 2. Distance calculation between the programmed path segment and TCP.

The right side of the equation (2) represents the length of a vector multiplication which can be written in equation (3) form:

$$|\vec{v}| \cdot d = |\overrightarrow{P_1T} \times \vec{v}| \tag{3}$$

By reordering the equation (3) we get equation (4) form:

$$d = |\overrightarrow{P_1T} \times \vec{v}| \cdot |\vec{v}|^{-1} \tag{4}$$

This can be used to calculate the distance between the TCP and the programmed path segment, which will be the execution error. Because the equation (4) represents the quotient of two lengths, this will be a fraction, which is $d \geq 0$.

Equation (4) cannot be used directly in C# programming language which is used for the development. To solve the issue, we have to transform equation (4) to a parametric form. This means that the input variables will be the three-dimensional coordinates (in millimeters) of each point, which represents the vectors. The result will be also in millimeters.

The parametric form of each point coordinate is presented in equation (5):

$$P_1(p_{1x};\ p_{1y};\ p_{1z}) \qquad P_2(p_{2x};\ p_{2y};\ p_{2z}) \qquad T(t_x;\ t_y;\ t_z) \tag{5}$$

The direction vector denoted by $\vec{v}$ can be written in equation (6) form:

$$\vec{v}(p_{2x} - p_{1x}; p_{2y} - p_{1y}; p_{2z} - p_{1z}) \tag{6}$$

With the parameters presented in equations (5), (6) we execute the vector operations (multiplication, length calculation). The final result is a clean equation, only containing coordinate values as parameters, as shown in equation (7):

$$\begin{aligned} d = \{&[(t_y - p_{1y}) \cdot (p_{2z} - p_{1z}) - (t_z - p_{1z}) \cdot (p_{2y} - p_{1y})]^2 + [(t_z - p_{1z}) \cdot (p_{2x} - p_{1x}) - (t_x - p_{1x}) \\ &\cdot (p_{2z} - p_{1z})]^2 + [(t_x - p_{1x}) \cdot (p_{2y} - p_{1y}) - (t_y - p_{1y}) \cdot (p_{2x} - p_{1x})]^2\}^{1/2} \\ &\cdot \{(p_{2x} - p_{1x})^2 + (p_{2y} - p_{1y})^2 + (p_{2z} - p_{1z})^2\}^{-1/2} \end{aligned} \tag{7}$$

Using the aforementioned form, the calculation method can be implemented in C# programming language and can be used for the development of the monitoring system.

3. Implementation of the monitoring principles

The monitoring system is a software, which retrieves data from the KUKA KR5 Arc robot arm and manages the necessary communication to halt the robotic arm if an error occurs. For the software to match further TPS principles, a graphical user interface (GUI) with a process status indicator is designed. The most suitable development environment for this project is Microsoft Visual Studio 2017 IDE (Integrated Development Environment). The main characteristics of this software are that it supports the C# programming language, it has a complex intuitive GUI designer. Lastly, it is freely available for research and educational purposes. The GUI of the presented version of the monitoring system integrates all the necessary control elements in a single window as can be seen on Figure 3.

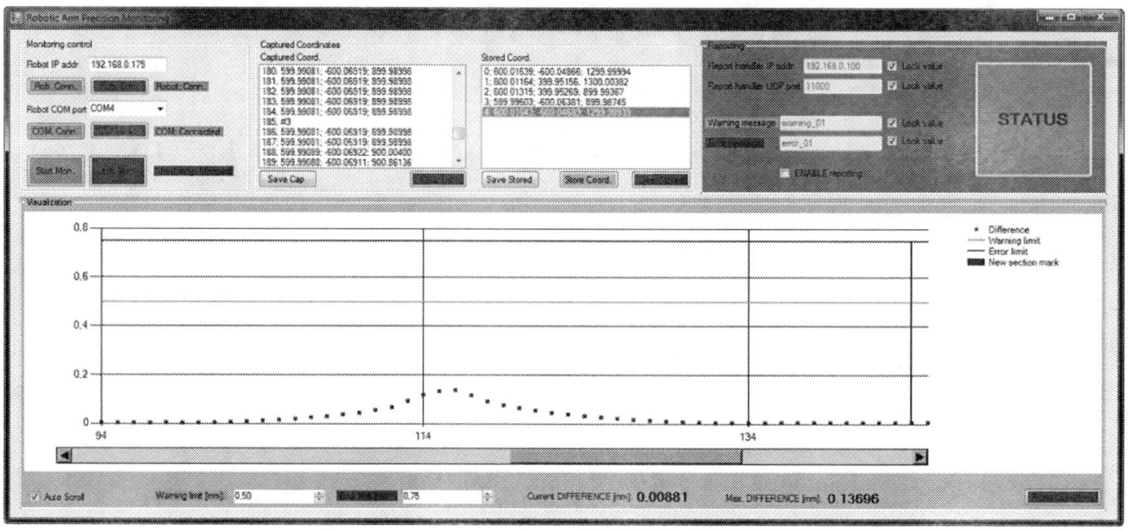

Figure 3. User interface of the monitoring system.

The GUI consists of four main modules. The first module, 'Monitoring control,' contains the controls necessary for the communication with the robot arm. Here, the IP address of the robot arm can be set and the monitoring process can be started or stopped. The 'Captured coordinates' contains the logging lists. This shows the retrieved TCP coordinates and the preprogrammed points.

The most important part of the user interface is the 'Reporting' module which is highlighted. This contains the necessary controls to connect to an external device which is able to receive User Datagram Protocol (UDP) packets. In case of an error, the system begins to send the preset warning or error message to the external device via UDP packets. By receiving the corresponding message, the external device, which can be an Ethernet connected Programmable Logic Controller (PLC), can initialize the shutdown process of a production line segment. In this way, the Jidoka principle of TPS is implemented. Along with the shutdown process, a visual indicator changes its normal color (green) to yellow in case of a warning or red in case of an error. In Lean manufacturing, this corresponds to an Andon signal, notifying the corresponding personnel about the abnormal operating conditions. The warning and error thresholds can be set in the 'Visualization' module of the GUI. Here a graphical representation of the execution error is shown, including the warning and error threshold lines.

4. Testing the monitoring system

To ascertain that the monitoring system works as expected, a series of testes must be conducted in a controller environment, where no material damage could occur in case of a malfunction. For this purpose, the KUKA KR5 Arc robot arm was used in the Robot Laboratory. The test consists of a predefined path in shape of a rectangle. The corner coordinates were stored by the robot arm control computer and, during the programming, they also were stored by the monitoring system. Then the programmed path was executed by the robot arm while the monitoring system measured the error in

IOP Conf. Series: Materials Science and Engineering **568** (2019) 012068 doi:10.1088/1757-899X/568/1/012068

real-time. During the test process the automatic halt function was disabled. The programmed path was executed multiple times with different speeds (10 times with 60 [mm/s], 200 [mm/s] and 600 [mm/s]). With each program cycle, the user interface displayed the maximum error during the whole path execution. This value was recorded in a spreadsheet at the end of each cycle. From these, the maximum values were selected for each execution speed category. The results presented in *Table 1* were obtained.

Table 1. Execution speed and the resulted errors.

Speed [mm/s]	60	200	600
Execution error [mm]	0.01059	0.01426	0.13820

5. Conclusion

We can conclude that the developed real-time accuracy monitoring system properly implements the automatic halt signal sending function and visual feedback, which supports the TPS and the Lean production. In present state the software can be deployed in a real-life scenario, on a production line for further testing. If the client requires, the user interface can be easily customized thanks to the Visual Studio IDE GUI editor. Considering the robot arm, the program execution accuracy greatly falls at 600 [mm/s] speed. Nevertheless this should not be considered as a disadvantage of the KUKA KR5 Arc, because it is intended for welding purposes. Theoretically the maximum obtainable welding speed by using a special MIG welding torch with the robot arm is 105.83 [mm/s] [8]. This value is half of the 200 [mm/s] where the robot arm showed a high execution precision and is 6 times lower than the 600 [mm/s], when the execution accuracy dropped significantly. Taking into account the obtained results, the Robot Laboratory's KUKA KR5 Arc robot arm is highly within its default parameters, because the repeatability specified by the manufacturer is ±0.04 [mm].

Acknowledgment
The development presented in this article was supported by University of Debrecen, Department of Mechatronics.

References

[1] Husi G 2015 Position Singularities and Ambiguities of the KUKA KR5 Robot *IJET.* vol **1**, no 1

[2] KUKA Roboter GmbH 2016 *KUKA KR 5 arc Specification* Version: Spez KR 5 arc V3. (Augsburg, Germany: KUKA Roboter GmbH)

[3] Varga V, Erdei T I and Husi G 2018 Redesigning of the gripping system on KUKA KR5 by utilizing 3 dimension printing technologies and Arduino *MATEC Web Conf., Annual Session of Scientific Papers IMT ORADEA* vol **184** (https://doi.org/10.1051/matecconf/201818402014)

[4] Erdei T I, Molnár Zs and Husi G 2016 Selecting Equipment and Supplies for Self–Replicating 3D Printer *Acta Technica Corviniensis - Bulletin of Engineering,* Tome IX, Fascicule 1, pp 59-62

[5] Liker J K 2003 *The Toyota Way* (McGraw-Hill Education)

[6] Strommer Gy 1988 *Geometria* (Budapest: Tankönyvkiadó)

[7] Boyd S 2018 *Introduction to Applied Linear Algebra* (Cambridge, UK: Cambridge University Press)

[8] Robotic Industries Association 2000 *Robots Facilitate High-speed Welding* [Online] (https://www.robotics.org/content-detail.cfm/Industrial-Robotics-Industry-Insights/Robots-Facilitate-High-speed-Welding/content_id/1186) [Accessed 03 02 2019]

IOP Conf. Series: Materials Science and Engineering **568** (2019) 012069 doi:10.1088/1757-899X/568/1/012069

The assessment of the financial performance based upon ratios. A comparative analysis

M B Tudose[1] and S Avasilcai[1]

[1] Department of Engineering and Management, FTPMI, Gheorghe Asachi Technical University, 29 Dimitrie Mangeron Bd., Iaşi, 700050, Romania

brindusa.tudose@gmail.com

Abstract. This study presents a comparative analysis model of the financial performance for two large companies operating in the field of production and trading of car parts. The model has as a starting point the assessment of the dynamics of liquidity, sales and profitability ratios (for a period of ten years). Based upon the preliminary results, the compound annual growth ratios were calculated and the trends (for the following seven years) were determined. All of these have made it possible to identify the performance profiles for the two companies. The study reveals that the financial performance was made more vulnerable by the insufficient cash availability (the cash reserves are not adequate for the current debt, assets and sales volume) and by the low levels of the recorded profit ratios (measured by the net profit margin, ROE and the return on shares). Although the cash and profit analysis reveals a cascade increase, the sales performance and return on equity ratios remain burdened by too high operating costs. To improve liquidity and profitability have been identified measures to meet the objective of ensuring the business sustainability.

1. Introduction

In a highly competitive business environment, managers need to monitor the performance and critical success factors of the companies they lead. The investors, the business partners, but also the competitors are directly interested in the financial performance of a company. Therefore, the performance analysis is pursued both inside and outside the business. This was the reason why the assessment of performance gained considerable attention in the research plan [1].

The objective of the research was to provide a comparative analysis model of the financial performance. To achieve this goal we structured the paper as follows: the first part presents the most important milestones of the research on performance assessment; the second part presents the methodology of research; in the third part the information describing the performance is presented and interpreted; then, the foreseeable performance profile has been identified and several measures have been highlighted, which can help to strengthen the financial performance; in the last part are presented the conclusions, limits and futures directions of the research.

2. Review of literature

During the last two decades, measuring the corporate performance has been the focus of attention not only in the academic field but also in the business field [2]. Since 1995, the performance assessment has been defined as a process of quantifying the efficiency and effectiveness of actions based upon a set of indicators [3]. More than 10 years later, Moullin [4] extends the scope of performance assessment and moved it to the area of organizational quality assessment and determination of the value for customers and other stakeholders

Subsequently, as performance was accepted as a barometer which presented "the current business situation and trends in its development, helping the company to decide where to go" [5], the performance assessment has incorporated into its sphere of interest and has identified the future business development trends. We subscribe to the opinion according to which "the company performance is the measurement of what had been achieved by a company which shows good conditions for certain periods of time" [6]. The ultimate goal of the performance assessment is to know the current state and to make predictions about business efficiency and effectiveness.

The theoretical and empirical research in the management area has facilitated the development of performance assessment methods. The first methods were based upon the prioritization of the financial-accounting information (such as ratio analysis, Du Pont analysis). Afterwards, the research focused upon methods based on [7]: accounting data; quality management; causal relations; assessment of business processes; system balancing. The latest concerns have led to the development of multi-criteria performance assessment methods.

The literature draws attention to the fact that the financial ratios are not absolute criteria; the financial ratios "are no substitute for a crystal ball. They are just a convenient way to summarize large quantities of financial data and to compare firms' performance" [8]. The significant financial ratios are utilized to reveal changes in the company financial position and performance and to illustrate the trends and nature of the changes.

3. Methodology

The assessment of the financial performance of the two companies was carried out using the information for the period 2008-2017, a period characterized by intense turmoil (crises, labour migration and political instability). Secondary sources of information (annual financial statements and audit reports) were accessed to build the database. In order to obtain significant conclusions (on the management efficiency, strengths and weaknesses of companies, the future evolution index, etc.), the trends for the analysed reports were determined (for a 7-year forecast horizon). Two performance profiles were drawn: one at the end of the 10 years analysed and one at the end of the 7-year forecast. Drawing on the identified situations for the two moments, measures have been identified to improve financial performance and ensure business sustainability.

The analysis model we proposed was simple in order to make it operational, not only by big listed companies (which perform computerized financial forecasts), but also by small and medium-sized enterprises. This was also the reason why the data processing was done using the analysis toolkit provided by Excel.

4. Data analysis and interpretation

The current ratio. The evolution of the ratio between the current assets and current liabilities for the period 2008-2017 has been analysed. The generic reference range for this indicator is [1-2]; if the indicator takes subunit values, the company does not have enough liquid assets to cover current liabilities. In practice, it is recognized that a current liquidity of less than 0.8 is a negative signal (especially for companies that are financed through operating credits). If the indicator exceeds the maximum limit of the specified range, the company has excess liquidity (for which it does not identify higher return alternatives). In this case also, the practice has revealed that sectors characterized by a higher asset rotation also need higher current liquidity to operate in good conditions.

From the perspective of current liquidity, the two analysed companies show relatively different situations (table 1). With the exception of 2008 (when the current liquidity was sub-unitary), company "A" recorded a good current liquidity (even if it presented a slight decrease during the last 3 years); the average of the indicator was 1.53. Company "B" had a supra-unitary and growing current ratio over the analysed period (indicator averaging 1.51). For company "A", the compound annual growth rate of the current assets is significantly higher than current liabilities (10.33% > 2.98%). In the case of the company "B", the compound annual growth rates are approximately equal (8.39% and 8.36% respectively). Therefore, there are significant differences in the annual growth rate of the current ratio (A: 7.13 > B: 0.03%).

Table 1. Current ratio, quick ratio, and cash ratio

	Current ratio		Quick ratio		Cash ratio	
Com-panies	Indicators (millions RON)	CAGR%	Indicators (millions RON)	CAGR%	Indicators (millions RON)	CAGR%
A	Current Assets	10.33%	Liquid Assets	10.29%	Cash	18.79%
	Current Liabilities	2.98%	Current Liabilities	2.98%	Current Liabilities	2.98%
	Current Ratio	**7.13%**	Quick Ratio	**7.10%**	Cash Ratio	**15.35%**
	Average = 1.53	-	Average = 0.93	-	Average = 0.04	-
B	current Assets	8.39%	Liquid Assets	13.18%	Cash	57.13%
	Current Liabilities	8.36%	Current Liabilities	8.36%	Current Liabilities	8.36%
	Current Ratio	**0.03%**	Quick Ratio	**4.45%**	Cash Ratio	**45.01%**
	Average = 1.51	-	Average = 0.84	-	Average = 0.23	-

Source: Own processing based on Annual Financial Statements (2008-2017)

The quick ratio. The ratio between liquid assets and current liabilities expresses the extent to which the company has sufficient liquid assets (receivables and cash availability) to discharge its short-term liabilities. The reference values for this indicator are [0.65-1]. In practice, lower values for this ratio are admitted only if the stocks have a large share in the total current assets.

The first table presents the liquidity situation without the influence of stocks (out of the total current assets the stocks have been eliminated). The first years of the analysis period are marked by the lack of liquid assets; this period is followed by a few years in which the *quick ratio* exceeds the maximum limit of the reference range. The explanation can be given by a too high volume of available funds or by a too high a volume of receivables; the clarification of this aspect will be achieved by analysing the cash ratio.

The differences between the two companies become more obvious when analysing the compound annual growth rate (GAGR). The gap between CAGR_liquid assets and CAGR_liquid liabilities is favourable and positive for both companies; although both are experiencing the *quick ratio* decline over the last four years of analysis (company "A" presents a better situation).

The cash ratio. The ratio between cash and current liabilities shows the ability of an enterprise to pay its short-term debt based upon the most liquid assets (cash). The reference values for this indicator are [0.35-0.65]. The data shows the most serious problem of the analysed companies: the lack of cash stocks. Although the compound annual growth rates of the cash was much higher than the compound annual growth rate of short-term debt, company "A" recorded a cash deficit throughout the analysed period.

The years that followed the onset of the crisis (2008–2012) were unfavourable (in terms of cash stock) for company "B" also. It has managed to recover, but for a short time (2013-2016). In this case, the analysis of compound annual growth rates reveals a large gap between cash availability and current liabilities; this aspect (normally favourable) should be interpreted with caution when the annual series start with extremely low values. The nominal increase of 46 times the stock availability (from 0.004 to 0.184 million RON) and a 45.01% compound annual growth rate has allowed only the temporary improvement of the cash ratio.

For company "A", the shortage of cash stocks is a problem that could not be fixed in 10 years (the data indicates only an improvement in the cash deficit). For company "B" the more intense concerns in this respect are obvious.

The cash to sale ratio. This ratio, which compares the cash balance at the end of the year with the revenues earned on sales indicates the effectiveness of current funding policies; a very low value signals not only the insufficiency of the cash stock but also the absence of buffer funds meant to ensure liquidity in case of late receipts. The data in Table 1 shows that the cash shortages affect not only the cash ratio but also the solvency (understood as the ability to deal with payments due).

Although the volume of sales was consistent and increased over the analysed period, the cash stocks of the two companies remained moderate. This is due to the fact that sales revenues are immediately directed to the payment of current liabilities (invoices, salaries, credit rates). From the very low cash-to-sales ratio, we can understand that the liquidity problem is associated with another problem, the business's profitability (the sustained sales growth is shaken by high operational costs, which makes the cash generated by sales to be immediately distributed for the payment of bills, wages, taxes, etc.). Compared to company „A", the data presented for company „B" reveals a wider concern to fix this problem.

The profit margin ratio. The ratio of net profit to total sales (known as *profit margin*) shows the efficiency with which a company uses (material, human, financial, informational and time) resources to execute and sell products in way profitably. In 2008, company "A" recorded 0.2 RON net profit for every 100 RON sales (which was very low); in 2017, for each 100 RON sales revenue, the company recorded 5.4 RON net profit. The situation has improved, but the profit margin remains at a sensitive level. Company "B" presents a more precarious position (table 2). The analysis from the perspective of the compound annual growth rate indicates a net profit growth (54.36% and 49.79%) faster than the sales growth (8.58% and 12.94% respectively). Even in this scenario, the situation of the two companies does not improve enough.

Table 2. Profit margin ratio, return on equity, and profit per share

	Profit margin ratio		**Return on equity (ROE)**		**Profit per share**	
Com-panies	Indicators (millions RON)	**CAGR%**	Indicators (millions RON)	**CAGR%**	Indicators (millions)	**CAGR%**
A	Net Profit	54.36%	Net profit (NP)	54.35%	Net profit (RON)	54.35%
	Total Sales	8.58%	Own Equity (OE)	4.43%	No. of shares	0.00%
	Net profit/total sales	**42.17%**	ROE = NP/OE	**47.81%**	Profit per share	**54.36%**
	Average = 0.04	-	Average = 0.07	-	Average = 0.10	-
B	Net Profit	49.79%	Net profit (NP)	49.79%	Net profit (RON)	49.79%
	Total Sales	12.94%	Own Equity (OE)	10.44%	No. of shares	4.65%
	Net profit/total sales	**32.63%**	ROE = NP/OE	**35.63%**	Profit per share	**43.13%**
	Average = 0.03	-	Average = 0.04	-	Average = 0.09	-

Source: Own processing based on Annual Financial Statements (2008-2017)

ROE (return on equity). Profitability shows the size of company financial performance during one period [9]. At the same time, the return on capital is an autonomous and complete criterion of economic efficiency. The ROE provides information on the business profitability and equity rotation. To assess the dynamics, the ROE can be compared to the average interest rate on the banking market (against which it should be higher). During the first two years of the analysed period, the ratios of return were extremely low (0.2%). Although annual compound growth rates of net profits are consistent (54.35% and 49.79% respectively), towards the end of the period the return on equity remains low (9.4% for company "A" and 5.8% for company "B"). Starting in 2015, company "A" manages to record a return on equity higher than the interest lending rate (on loans), but the gap is not enough to motivate to take the risk of a business (as opposed to assuming the risks of a credit relationship). The maximum gap was registered in 2016, when ROE was 14.2% and interest rate was 6.6%. For company "B", in the period 2008-2015, the return on equity remains lower than the interest rate on the banking market. In the period 2016-2017, it increased insignificantly above the interest rate (at the end of the period, return on equity was 5.8%, and interest rate was 5.5%).

Profit per share. This ratio (earning per share) reflects the effective return of a company's shares. The information in Table 2 shows that only one of the two companies maintained the same number of shares throughout the analysed period (this is the case of company "A", which has an annual compound earning growth rate of 54,35%). In 2010 company "B" has achieved an increase in the share capital. This operation did not materialize in the consolidation of the cash flow but only in the incorporation of the reserves (from re-evaluation) into the share capital. With a compound annual growth rate of 43.13%, the earning per share went from 0.004 to 0.145 RON per share. Given that the value of a share before the increase was 0.25 RON, and after the increase was 0.45 RON, the share return ratio (determined as the ratio between the profit per share and the nominal value of the share) decreased from 40% to 32%; this has occurred under the conditions the share capital has a 17.95% share in the total own equity.

At the beginning of the analysed period, company "A" recorded a return on 2% (profit per share/nominal value of share); at the end of the period it had a return ratio of 176%. This increase does not impress because the share capital (representing the shareholders' contributions) represented only 3.6% of the total equity (in 2017). By analysing the size of the profit on a share, we can notice that the performance balance is in favour of company "B" (although the data in table 3 indicates the opposite).

5. Findings and suggestions

Table 3 provides the information needed to build the financial performance profile from the perspective of *liquidity, cash and profitability positions*. Between 2008 and 2017, the current ratio and the quick ratio contributed to strengthening the financial performance. Instead, the trend line indicates values superior to those admitted; therefore, the attention of financial officers should focus upon monitoring the size of stocks and receivables. The measures may concern: a) the limitation of stocks only to those which ensure the continuity of production processes; b) adopting more restrictive policies related to customer payment facilities (their receiving at the time of the transaction). These measures will generate additional cash flows to improve the *cash ratio*. The financial managers should base appropriate cash flow management strategies, not just to ensure the compliance with financial rigor, but also to set up buffer stocks in order to minimize the liquidity risk. Although the data reveals that a company can survive without sufficient cash, we draw attention to the fact that the prospect of strengthening financial performance is badly affected. In order to avoid sub/super sizing of cash stocks, it is important to achieve and track receiving and payment budgets.

Table 3. Data for the "A" and "B" companies

Com-panies	Ratios	2008	2017	2008-2017 CAGR	2008-2017 Remarks	Predictions Trend line	Predictions 2024	Predictions Remarks
"A"	Current ratio	0.81	1.62	7.13%	Favourable	y = 0.1013x + 0.9717	2.69	Oversize
	Quick ratio	0.46	0.91	7.10%	Favourable	y = 0.0523x + 0.6438	1.53	Oversize
	Cash ratio	0.005	0.019	15.3%	Unfavourable	y = 0.0030x + 0.0273	0.08	Unfavourable
	Cash/ Total sales	0.002	0.004	9.4%	Unfavourable	y = 0.0005x + 0.0070	0.016	Unfavourable
	Net profit/Total sales	0.002	0.054	42.17%	Unfavourable	y = 0.0077x + 0.0012	0.129	Favourable
	ROE	0.002	0.094	47.81%	Acceptable	y = 0.0127x - 0.0017	0.215	Favourable
	Profit per share	0.002	0.176	54.36%	Irrelevant	y = 0.0227x - 0.0199	0.336	Favourable
"B"	Current ratio	1.69	1.70	0.03%	Favourable	y = 0.0328x + 1.3305	1.89	Favourable
	Quick ratio	0.65	0.59	4.45%	Favourable	y = 0.0766x + 0.4206	1.72	Oversize
	Cash ratio	0.004	0.184	45.01%	Unfavourable	y = 0.0611x - 0.1077	0.93	Oversize
	Cash/ Total sales	0.002	0.059	39.13%	Unfavourable	y = 0.0211x - 0.0342	0.324	Favourable
	Net profit/Total sales	0.002	0.037	32.63%	Unfavourable	y = 0.0054x - 0.0015	0.090	Favourable
	ROE	0.003	0.058	35.63%	Unfavourable	y = 0.0083x - 0.0018	0.139	Favourable
	Profit per share	0.008	0.145	43.13%	Unfavourable	y = 0.0193x - 0.0201	0.308	Favourable

Source: Own processing

In the emerging economies the crisis began with 2009. This scenario may explain the liquidity issues of the two companies which were analysed. The sudden shift from a normal state to a state of crisis has not provided companies with the time to take measures to strengthen liquidity (cash). During the crisis (2009-2011), the changes in financing habits (in the sense of diminishing the debt dependency) reduced the vulnerability towards financiers, but increased the risk of shortage of highly liquid assets (cash).

Although there were no acceptable levels in the 2007-2017 period, the last three indicators of the financial performance appreciation (*net profit per total sales, ROE, and profit per share)* have a

IOP Conf. Series: Materials Science and Engineering **568** (2019) 012069 doi:10.1088/1757-899X/568/1/012069

favourable trend line. In order to maintain this trend, the financial officers must achieve a balance between profitability and liquidity. The difficulty of this strategy is fuelled by the fact that the more liquid assets have a lower yield (and the less liquid assets have a higher return).

6. Conclusions

For managers and investors, the most accessible sources of obtaining data are the annual financial statements. However, these documents provide only a retrospective image of the financial position and operating results. Both managers and investors need forecasts that provide details of sales dynamics, management efficiency, business opportunities and threats, etc. In the support thereof, we proposed and exemplified a model of financial performance assessment, based, as a starting point, upon liquidity, sales and profitability ratios. The retrospective analysis of these ratios allowed us to identify a performance profile. At the end of the 7th-year forecast, the financial performance remains affected by the lack of cash stocks (for both companies). The efficient liquidity management is a prerequisite for increased profitability. Consequently, the managers must achieve a balance between profitability and liquidity.

During the three years after the onset of the crisis company „A" managed to raise the ratio of financial return. In the years that followed the profitability had a slightly upward trend. It can be inferred that its strategy has focused more upon profitability against the liquidity of the business. Throughout the analysed period, company "B" presented lower levels of profitability than company "A" (a part of the gap was taken over by the size differences between the two companies). In this case, the strategy aimed at controlling liquidity against profitability.

The limits and future lines of research. The study exploits only financial and accounting information and focuses only upon the financial performance research. The taking into account of the macroeconomic variables allowed only some logical deductions. For the research development, we are considering extending the analyses to other determinants of performance, especially of non-financial nature. The ultimate goal of the efforts is to perform a strategic performance measurement system.

References

[1] Salem M A, Hasnan N and Osman N H 2012 Balanced Scorecard: weaknesses, strengths as its ability as a performance management system versus other performance management systems *Journal of Environment and Earth Science* vol 2(9) 1-10

[2] Rajnoha R, Lesníková P and Korauš A 2016 From Financial Measures to Strategic Performance Measurement System and Corporate Sustainability: Empirical Evidence from Slovakia *Economics and Sociology* 9(4) 134-52

[3] Neely A, Gregory M and Platts K 1995 Performance measurement system design: A literature review and research agenda *International Journal of Operations & Production Management* vol 15(4) 80-116

[4] Moullin M 2007 Performance measurement definitions: Linking performance measurement and organisational excellence *International Journal of Health Care Quality Assurance* vol 20(3) 181-83

[5] Peleckis K, Krutinis M, Slavinskaitė N 2013 Daugiakriterinis alkoholio pramonės įmonių pagrindinės veiklos efektyvumo vertinimas *Verslo ir teisės aktualijos* vol 8 1-16

[6] Batchimeg B 2017 Financial Performance Determinants of Organizations: The Case of Mongolian Companies *Journal of Competitiveness* vol 9(3) 22-33

[7] Narkunienė J and Ulbinaitė A 2018 Comparative analysis of company performance evaluation methods *The International Journal Entrepreneurship and sustainability* vol 6(1) 125-38

[8] Brealey R A, Mayers S C and Allien F 2017 *Principles of corporate finance* (New York: McGraw-Hill Education)

[9] Simatupang H J, Purwanti L and Mardiati E 2019 Determinants of capital structures based on the Pecking Order Theory and Trade-off Theory *Jurnal Keuangan dan Perbankan* vol 23(1) 84-96

IOP Conf. Series: Materials Science and Engineering **568** (2019) 012070 doi:10.1088/1757-899X/568/1/012070

Digital twins of mechatronic machine tools for modern manufacturing

A A Kutin [1]*, V V Bushuev [1] and V V Molodtsov [1]

[1]Moscow State University of Technology «STANKIN», Vadkovskij per. 1, Moscow, Russian Federation, 127994

aa.kutin@stankin.ru

Abstract. The article is devoted to the problem of the development of “digital twins” of technological equipment — it’s specially organized and structured information models. The prospects for the use of digital twins for the organization of digital production and the problems associated with providing an adequate description of real mechatronic technological systems with their help are considered. It is shown that the creation of digital twins of technological equipment is a complex task associated with the integration of information and production technologies; therefore, its main solution is a systematic approach that defines the requirements for mathematical, informational and methodological support of digital productions. The advantages and disadvantages of various approaches to the formation of such models are considered. A combined technique for obtaining digital twins of mechatronic feed drives, forming a system for shaping CNC machines, which uses the results of finite element modelling and experimental data, is proposed. Using the example of drive identification, an experimental result shows the need to develop and implement scientifically based methods for constructing digital twins.

1. Introduction

Digital production accumulates advanced achievements in the field of manufacturing and information technologies and uses the scientific results of the modern interdisciplinary direction - mechatronics. Integration of new information and manufacturing technologies along with the creation of mechatronic technological systems mainly increases productivity and production efficiency.

Project “Industry 4.0” shows that to successfully solving of these problems in recent decades, the concept of a "digital factory" is being realized. In most cases, the term "digital factory" refers to the information model of high-tech production, covering the main directions of advanced manufacturing technologies, new materials and information and communication technologies [1]. This model (figure 1) includes information about all manufacturing processes, as well as the entire amount of information about the product, which can be distributed through the stages of its life cycle. Within the framework of the new paradigm, digital production is a system that supports the full life cycle of manufactured products and is characterized by a common information platform and the use of mechatronic equipment capable of generating information online and adapting to changing tasks. The implementation of the new paradigm requires the development of new methods of research and design of large manufacturing systems, focused primarily on the description of their production and technological state. The analysis of the structure of the CNC equipment should be carried out to determine the influence of physical and technological parameters on the possibilities of high-performance and reliable operation in conditions of "unmanned" production.

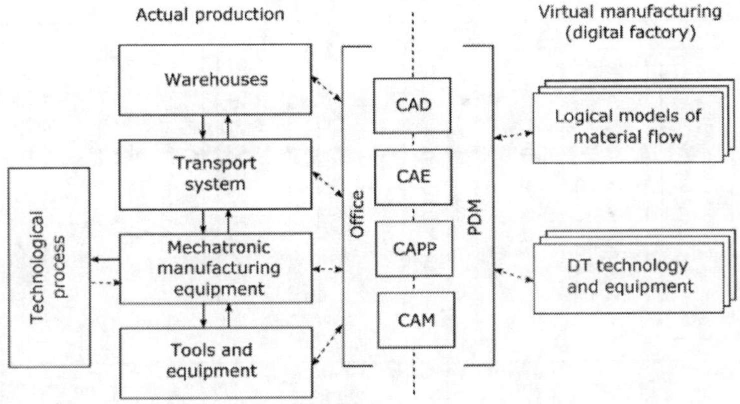

Figure 1. Modelling of technological processes of machining production

A fundamental problem is the lack of development of "digital twins" (DT) of CNC equipment – virtual model includes geometric model of the machine; a set of calculated data of parts and components, information about materials and technological processes of manufacturing. This model (figure 2) is the basis for informational support of the life cycle of technological systems in digital production, including the formulation of a technical task, the description of information and logistic links of a system operation, the quantification of performance indicators, the simulation of the system operation, processes of manufacturing products.

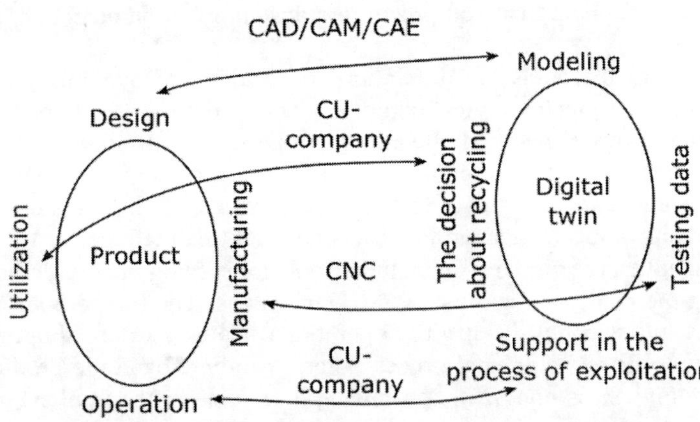

Figure 2. Information support of the product life cycle

Digital twins of mechatronic technological systems (MTS) will allow to solve a complex of problems, assessing the suitability of technological equipment for manufacturing parts of the required accuracy, preparing control programs that take into account the specific features of each machines, the formation of requirements for the designing of new equipment (figure 3). Such a variety of applications is explained by the fact that "digital twins" in structure are sufficiently adequate models of these objects, which makes it possible to obtain objective information about them and to form fairly complex control laws, in accordance with many input parameters that changing with time.

IOP Conf. Series: Materials Science and Engineering **568** (2019) 012070 doi:10.1088/1757-899X/568/1/012070

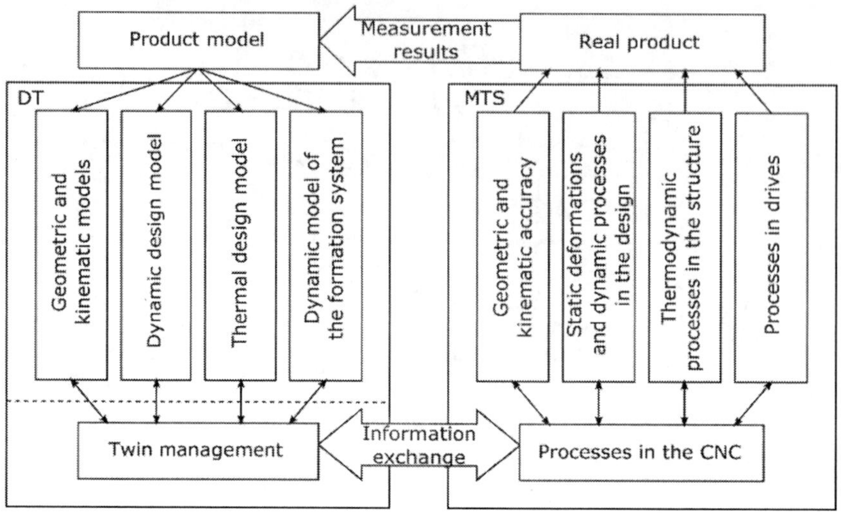

Figure 3. Digital twin mechatronic process system

2. Theory and Methods

The creation of digital twins of technological equipment is a complex task connected with the integration of information and manufacturing technologies, therefore the main approach is the systematic approach that determines the requirements for the mathematical, informational and methodical support of digital production. For their implementation it is necessary:

– to identify the relationship between the operational characteristics of the mechatronic system for the shaping of modern technological equipment, the designing parameters and control systems for the drives;

– to offer mathematical models for simulation and analysis of dynamic processes in mechatronic systems for shaping modern technological equipment and to develop methods for parametric synthesis and identification that provide specified operational properties of machine;

– develop the necessary software, create databases and provide recommendations that ensure the implementation of adequate reality sets of mathematical models of all stages of technological processes of forming the products of mechanical and physicochemical treatment.

It is very important that the digital twin includes tools for solving the problems of production process planning, including the control programs for CNC machines. The methods of CNC machine control, which allow to take into account the specific operational features of an equipment and adapt it to the requirements of a specific technological process, significantly expand the production possibilities and therefore their development is one of the promising directions in the implementation of new modern digital production technologies.

When machining precision parts with hard-profile surfaces, their contours are the result of axis movements of the tool and workpiece. Drives in mechatronic systems for forming CNC machines ensure the coordination of these movements. The task is transferred to them through the information network of the control device. The transition from mechanical to information connections made it possible to substantially increase the flexibility of the equipment and the different surfaces realized on the machine while simultaneously increasing the accuracy of machining, as the effect of the errors in the mechanical links of the kinematic chains ensuring the coordination of movements in CNC-free machines was ruled out. However, the dynamic processes in the drives, and the lag when operating the information network, have a direct impact on the output precision of the part.

The dynamic quality of the feed drives depends significantly on the elastic properties and distributed inertial characteristics of their mechanisms, since resonance phenomena appearing in the mechanical part of the drive prevent the increase in the speed and position contours to the required level. Despite the importance of the problem, it has not been studied enough, both theoretically and experimentally.

Resonance phenomena in drives with ball screw, which are the most popular type of traction devices for machine tools, are caused, as a rule, by axial and torsion oscillations in the design of these gears. The intrinsic frequencies of the axial oscillations of the moving assembly located in the band of the transmission of the position loop from (30-40 Hz to 200 Hz) or in the immediate vicinity of it have a direct influence on the quality of the machining. Resonance phenomena in torsion vibrations of rotating elements of the traction device structure (screw, coupling, engine rotor, etc.) arise at frequencies from several hundred to several thousand Hz. They practically do not appear on the moving node, but can lead to loss of stability of the speed loop and limit the possibility of its tuning [2].

Modern design and analysis tools, such as SolidWorks, MATLAB, ANSYS, and NX Motion Software, allow to simulate and analyze the design of modern machines at the designing stage. However, the finite element modeling built into modern computer engineering systems cannot be used directly in the modeling of the federate control loops, due to its cumbersomeness, redundancy and closeness [3, 4]. Dynamic models of mechatronic systems in MATLAB Simulink allow taking into account dynamic processes in the control circuits of feed drives, but obtaining adequate models of the mechanical part is a complex problem.

Models that allow describing the processes in the mechanical part of the drive with high accuracy, can be created on the basis of information about the natural frequencies and modes of the object's oscillation obtained from the experimental frequency characteristics or by calculation as a result of finite element modeling using means of modal analysis [5, 6]. The main drawback of these models is that they do not reflect the actual physical structure of the object, in fact representing a black box in which the connection between the input and the output is given by some abstract mathematical function - the modal series. For this reason, they are not used in the designing of feed drive mechanisms, although they can be effective enough when selecting and configuring its control loops.

A satisfactory description of the dynamic processes in the mechanical part of the drive with minimal resources is provided by models of the form

$$\left.\begin{aligned} J_1 s^2 \varphi_1(s) &= M(s) - k_{1,2}(s)[\varphi_1(s) - \varphi_2(s)] \\ &\vdots \\ J_i s^2 \varphi_i(s) &= k_{i-1,i}(s)[\varphi_{i-1}(s) - \varphi_i(s)] - k_{i,i+1}(s)[\varphi_i(s) - \varphi_{i+1}(s)] \\ &\vdots \\ J_n s^2 \varphi_n(s) &= k_{n-1,n}(s)[\varphi_{n-1}(s) - \varphi_n(s)] - M_c(s) \end{aligned}\right\}, \tag{1}$$

where Ji and $\varphi_i(s)$ – the moment of inertia and the angle of twisting of the i-th rotating mass, i = 1, ..., n; M(s) and Mc(s) – moments on the engine side and resistance forces; $k_{i,i+1}(s) = b_{i,i+1}s + c_{i,i+1}$ – a complex parameter that determines the elastic and dissipative coupling between adjacent masses; $c_{i,i+1}$ and $b_{i,i+1}$– stiffness and damping coefficient of the corresponding part of the structure. Determination of the parameters of such models on the basis of experimental data or finite-element modeling results is described in detail in [7].

The problem of determining the unknown values of the moments of inertia of the concentrated masses and the rigidity of the elastic constraints of such a system reduces to a joint solution of the inverse problems of eigenvalue problems for it and an analogous system with a rigidly fixed rotor of the type

$$([\tilde{\alpha}] - x_i[E])\{\Psi\}_i = \{0\}, \tag{2}$$

where $[\tilde{\alpha}] = [M]^{-1/2}[C][M]^{-1/2}$ – is the matrix of the squares of the partial frequencies, [M] and [C] – are the matrices of masses and rigidity, $\{\Psi\}_i = [M]^{-1/2}\{\Phi\}_i$, xi ≥ 0 and $\{\Phi\}_i = [\Phi_n \cdots \Phi_1]^T$– are the i eigenvalues and the vector.

The algorithm for determining the dimensions and parameters of the system includes the following basic steps:

1. Using finite element simulation or experimentally, nonzero eigenvalues of the rotating part of the drive with a free (pole) and fixed rotor (zeros) are determined.

2. For non-zero eigenvalues of poles and zeros, using the recurrence relations obtained from Newton's formulas for determining the power sums of the roots of a polynomial, the coefficients of the characteristic polynomials of matrices $[\tilde{\alpha}]$ for systems with a free and fixed rotor are determined.

3. From the coefficients of the characteristic polynomials of these matrices computed at the 2nd step, the values of the squares of the partial frequencies are found using the recursive computational procedure [8].

4. The unknown moments of inertia of concentrated masses and the rigidity of elastic bonds are determined on the basis of information about the total moment of inertia, the rotating elements of the mechanical part of the drive.

3. Results

As an example of the application of the developed method, a parametric identification of the model of the mechanism of the experimental stand presented in figure 4a was carried out using poles and zeros obtained experimentally.

To obtain the frequency characteristics of the drive, the model of its mechanical part (1) was used, which, in combination with a fairly simple model of the drive control loops described in [9], forms a mathematical model of the dynamic system of a multi-circuit electromechanical feed drive. The model of control loops takes into account delays in the transmission of information in current loops, speeds and positions that have a significant effect on the dynamic processes in them. figure 4b shows a comparison of simulation results in the MATLAB environment with the experimental frequency characteristics of the drive.

(a)

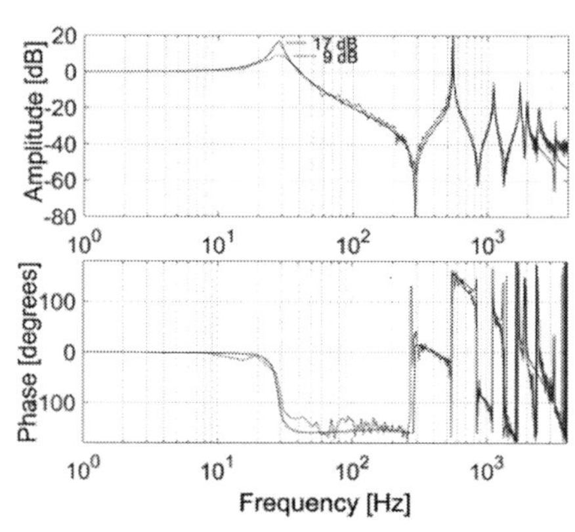

(b)

Figure 4. Stand design (**a**) and identification results (**b**)

4. Conclusions

It can be concluded that the development of methods and tools for creating digital twins is an important scientific and technical problem that is of decisive importance for engineering in general. The main difficulties are connected with the establishment of interrelations between the performance characteristics and design parameters of mechatronic systems made in the form of models and functional relations, and methods for their parametric and structural approximation.

Based on the research carried out by the authors, as well as the work of other researchers, we highly need for the development of "digital twins" of equipment for specific technological production processes, which allows us to conclude that the relevance, prospects and great practical significant are such works.

Acknowledgments
This work was carried out using equipment provided by the Center of Collective Use of MSUT "STANKIN". This research was made under financial support of Ministry of Science and Higher Education of the Russian Federation, project task N9.2731.2017/4.6

References
[1]. Kutin A, Dolgov V, Sedykh M 2016 *Information Links between Product Life Cycles and Production System Management in Designing of Digital Manufacturing*. (*Proceedings of the 48th CIRP Conference on Manufacturing System)* pp 423-426
[2]. Bushuev V V, Molodtsov V V 2016 *Influence of the elastic properties of supply drives on their dynamic characteristics* (*Russian Engineering Research)* vol 36 Issue 11 pp 940-947
[3]. Kim M S, Chung S C 2006 *Integrated Design Methodology of Ball-Screw Driven Servomechanisms with Discrete Controllers* (*Part I. - Modelling and Performance Analysis. Mechatronics)* vol 16 pp 491–502
[4]. Okwudire C, Altintas Y 2009 *Hybrid Modeling of Ball Screw Drives with Coupled Axial, Torsional and Lateral Dynamics* (*Transactions of ASME Journal of Machine Design)* vol 131(071002) pp 1-9
[5]. Reinhart G, Zaëh M F, Baudisch T 2004 *Simulation Environment for Designing the Dynamic Motion Behavior of the Mechatronic System Machine Tool* (*Journal of Production Engineering-Research and Development Jg)* vol XI Issue 1 pp 145-148
[6]. Weck M, Brecher C, Schulz A, Keiser R 2003 *Coupled Simulation of Control Loop and Structural Dynamics* (*Annals of the German Academic Society for Production Engineering)* vol X Issue 2 pp 105-110
[7]. Bratland M, Haugen B, Rølvåg T 2011 *Modal analysis of active flexible multibody systems* (*Computers&Structures)* vol 89 Issue 9-10 pp 750-761
[8]. Wang X, Mills J K 2004 *Experimental Modal Analysis of Flexible Linkages in a Smart Parallel Platform*. (*Proceedings of the 7-th Cansmart Meeting - International Workshop on Smart Materials and Structures, Montreal, Canada*) pp. 37-46
[9]. Bushuev V V; Evstafieva S V, Molodtsov V V 2016 *Control loops of a supply servo drive* (*Russian Engineering Research)* vol 36 Issue 9 pp 774-780

IOP Conf. Series: Materials Science and Engineering **568** (2019) 012071 doi:10.1088/1757-899X/568/1/012071

The structure of a sliding sensory system for a self-adaptive prehensor

I C Frincu[1] and I Stroe[2]

[1]Product Design, Mechatronics and environment Department, Transilvania University of, Brașov, Romania

cezarfrincu@yahoo.com

Abstract. The quality of the prehension is determined by the mobility state of the prehension object.The emphasis on sliding within the self-adaptive prehensor, proposed in the paper, is achieved by correlating the parameters generated by the rotation sensor (encoder), the contact sensor and the angular position of the stepper motor.The rotation of the encoder is influenced by the active and passive mechanical components in the prehensor jaws.In this way, it is obtained a 2D sliding sensory system. The sensory system is a complex mechatronic system that can adapt to most industrial robots of different sizes and types, including the cartesian type. The prehension of the objects is accomplished through a sensory technique which integrates all software and hardware operations to achieve the image of the two directions grip. The sensory system has a resistive position sensor in its structure which determines the angular position of the axis of rotation of the execution element. The inductive position sensor is the second sensor of the system. It highlights the mechanical tensionless contact on the prehensile object. The sliding of the object is emphasized by a resistive position sensor. An important role in programming the prehensor is the correlation between the software information from the three sensing elements with the stepper motor that achieves the force required for the prehension.

1. Introduction

In a prehensible system, the most important is the products safety system as they are moved from point A to point B. The safety of products is when the prehensible system handles a product and it does not leave prints on it, it does not plastically deform it and it is manipulated without the possibility of being dropped. The sliding sensory system does not allow the product to slip into the prehensor jaws, so it cannot escape during movement and knows at the same time how hard to squeeze it.

All parameters generated by the sensors are variable and can be changed during the movement, and in real time a correction is applied when the product is squeezed into the jaws. This sensory system can be adapted and mounted on classical prehensible systems, but the maximum yield is achieved on the self-adaptive prehensible system[1].

2. Presentation of the sensory system

The sensor system consists of a resistive encoder (1) having a spindle that exerts a rotational motion, generating values that are amplified by an electronic system for being readable.

IOP Conf. Series: Materials Science and Engineering **568** (2019) 012071 doi:10.1088/1757-899X/568/1/012071

The encoder is actuated by means of an elastic strap (6) which provides an amplified transmission between two drive wheels.The drive wheel (2) that transmits motion to the encoder also has the role of straining the drive belt.Each value generated by the encoder correlates with the position of the stepper motor spindle in real time to make a decision if it is necessary to exert a greater or lesser force on the product.The whole system is trained by a flat band that forms a lever with the first rotation system, while at the same time realizing a translational motion relative to the horizontal of the system.

The flat band makes a sensible translation movement, does not make any effort on the prehensible product, and once it is at the end of the stroke, the firm contact of the product is detected by an inductive sensor (3). The entire sensory system is activated with the contact confirmation of the inductive sensor'Figure 1'[2].

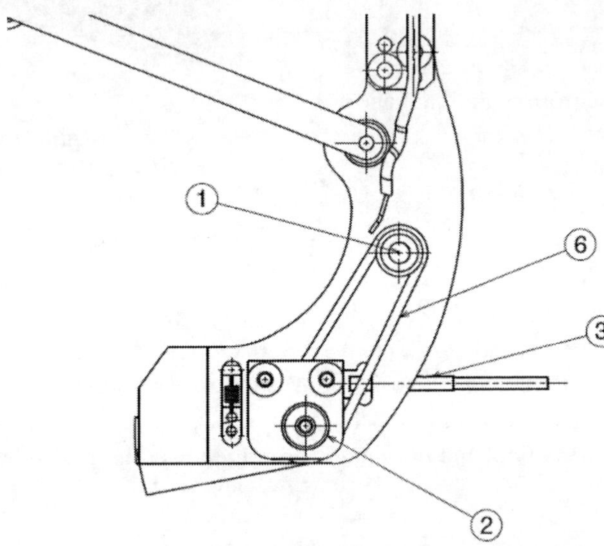

Figure 1. The self-adaptive prehensor jaw with the implemented sensory system.

2.1. Description of the movement made by the flat band

The flat band is an active part of a sensory system that plays an important role in detecting the movement of the prehensile object. The flat band is a stainless steel sheet of 1.5 mm that has adherent material on the contact surface with the object and at one end is fastened to a shaft through which it transmits the rotation movement.It has an atypical geometric shape with a length of approximately 40mm.After it performs the rotation motion, it returns to the initial position with a slightly tensioned extension spring.

The flat band together with the axis on which it is fastened and with the whole rotation movement system perform a 2 mm translation motion so that the prehensile object to reach the holding jaw and at the same time we can detect its presence.

The translation motion is made horizontally and returns to the initial position by means of a compression spring that is positioned around the inductive sensor and the flat band axle support bracket.

The axle on which the flat band is secured has a chamfer shape to provide its moment of rotation without allowing it to rotate about the axis 'Figure 3'[3].

The flat band has two holes by which is set the lever which makes the rotation and then we can generate the ideal position for the location of the extension spring.The position of the spindle is given by the

IOP Conf. Series: Materials Science and Engineering **568** (2019) 012071 doi:10.1088/1757-899X/568/1/012071

dimensions a + c and b, with these dimensions we can vary the transmission of the moment and the sensitivity of the transmission; therefore as the position of the spindle of rotation is further, the sensitivity decreases.During rotation, the flat band describes an Rx radius that is permanently in contact with the object, rolling on the prehensile object 'Figure 2'[4].

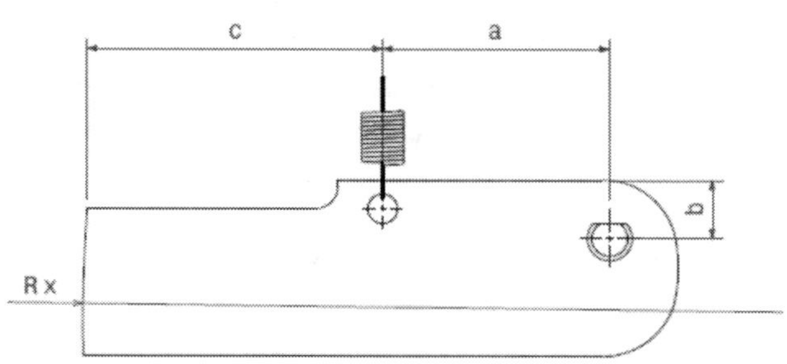

Figure 2. The geometric from of the flat band

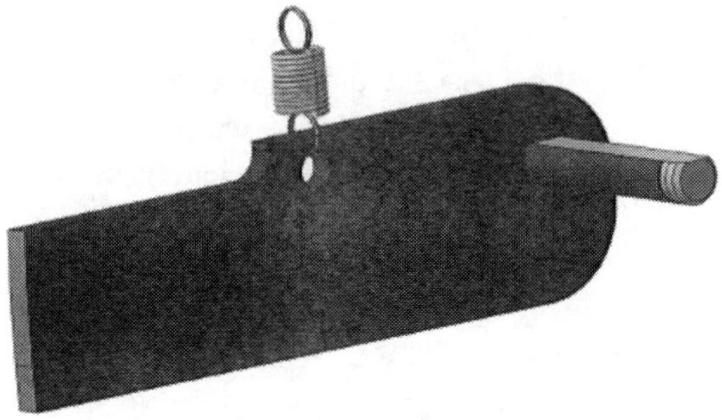

Figure 3. 3D view of the flat band with the spindle and the retraction spring

2.2. Flat band characteristics

The flat band is a piece of stainless steel sheet with a density of 7860 kg/m^3 and a weight of 5g which is cut into an irregular shape with a volume of 6.41835e-007 m^3.

In order to establish the position of the extension spring, we must find the center of gravity toward the flat band origin, and the coordinates are: Gx=0 mm, Gy=20.136 mm, Gz=5.543 mm.

For the flat band to return to its original position, the spring must overcome its weight, but it must also be sensitive. It should not exceed the force of 0.8N because it can prevent proper functioning of the sensory system. The entire weight of the flat band does not just lie in the spring, it also has a support point in the axis of rotation[5].

2.3. Gripping process description

The jaw of the self-adaptive prehensor with 5 arms in which the sliding sensory system is implemented is interchangeable.In the section represented it can be noted the rotation movement that it does after the contact with the object.

The stages of prehension with sensory sliding systems are:

- The stepper motor operates the prehensor arm until the flat band from the jaw reaches the product and confirms the inductive sensor;
- After the presence sensor confirms, then the motor tightens up to 5%;
- If the product moves in jaws due to inertia or weight greater than the force exerted upon it after it is gripped,then the flat band makes a downward rotation,and at the same time every 0.1°the stepper motor puts more pressure on the product until it does not move.The flat band performs the rotation movement from position (5) to position (4),thus achieving a maximum rotation of 11°,and then it returns to the initial position (5) by means of a traction spring'Figure 4'.

IOP Conf. Series: Materials Science and Engineering **568** (2019) 012071 doi:10.1088/1757-899X/568/1/012071

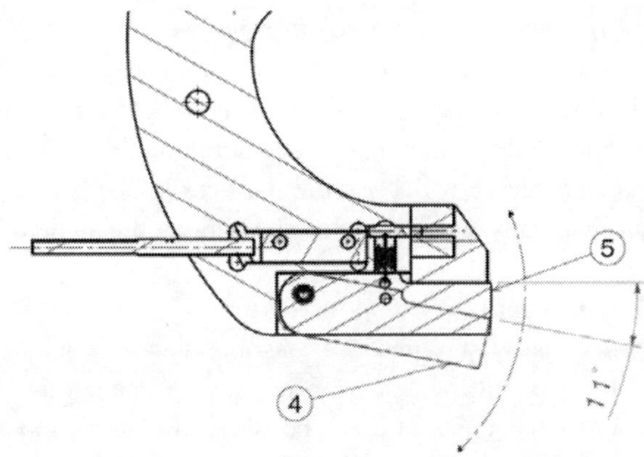

Figure 4. The section of the jaw where the sensory system is implemented.

3. Software design of the sensory system

The sensory system is a system where multiple electrical values generated by the sensors are cumulated to finally make a momentary decision.

3.1. Encoder structure and generated signals

The encoder is a sensory system that generates electrical signals or parameter variations according to the type chosen.In the self-adaptive prehensor sensory system is used a resistive encoder allowing a 360° rotation.It generates resistive variations to the programmer on its inputs.

Tabel 1. The values generated by the encoder and read by the controller.

Angular position of the flat band	The value generated by the encoder displacement	Stepper motor
0°	0 + (-0.03 to +0.03)V	0° = 0 steps
0.5°	0.05 + (-0.03 to +0.03)V	0.9° = 1 step
1°	0.07 + (-0.03 to +0.03)V	1.8° = 2 steps
1.5°	0.09 + (-0.03 to +0.03)V	2.7° = 3 steps
2°	0.11 + (-0.03 to +0.03)V	3.6° = 4 steps
---	---	---
6.5°	0.29 + (-0.03 to +0.03)V	11.7°=13 steps
7 °	0.31 + (-0.03 to +0.03)V	12.6°=14 steps
7.5°	0.32 + (-0.03 to +0.03)V	13.5°=15 steps
8°	0.35 + (-0.03 to +0.03)V	14.4°=16 steps
8.5°	0.37 + (-0.03 to +0.03)V	15.3°=17 steps
9°	0.39 + (-0.03 to +0.03)V	16.2°=18 steps
9.5°	0.41 + (-0.03 to +0.03)V	17.1°=19 steps
10°	0.43 + (-0.03 to +0.03)V	18°=20 steps
10.5°	0.45 + (-0.03 to +0.03)V	18.9°=21 steps
11°	0.47 + (-0.03 to +0.03)V	19.8°=21 steps

The signal read by the programmer is made through an analogue input with a read function, (val = analogRead(Ax)). Theresistive variations of the encoder make the output current to be variable with a linear feature but also

IOP Conf. Series: Materials Science and Engineering **568** (2019) 012071 doi:10.1088/1757-899X/568/1/012071

with a reading error.The reading error is declared with the function(if ((val „read value" > previous+3) || (val „read value"< previous-3))),where the declared error 3 is the sensitivity threshold of the encoder. This declared threshold is important because there are times when at a certain mechanical position the encoder can generate lower or higher values even if it is at rest.These error values are variable and do not have to further influence the mechanical system, and depending on the noise generated by the encoder at rest, the interval at which it can generate variable currents is established. The encoder performs the rotation motion until it is mechanically stopped by the flat band at the angle of 11° (Table 1).

3.2. Correlation parameters with stepper motor movement

In this case, the sensor system consists of a presence sensor and an encoder sensor. Values are read fromthe encoder as long as the proximity sensor confirms the presence of the object generating the signal (1 or true).In software, the proximity sensor of this type is declared as a simple contactor (int digitalread x) where x is the pin number chosen.

After declaring the connection between the sensor and the programmer, is declared the connection type (input).A very important parameter for stabilizing the system is the declared resolution in stepper motor programming.For a stepper motor with a 150 resolution declared as (steprezolution 150), it takes a step when the encoder error value is exceeded.To reposition the motor, the constant value generated by the encoder with the function (previous = val (current value)) is read in real time.

4. Conclusions

If the angular position of the flat band reaches 11°, then it means that the object is too heavy to achieve the prehension and it is not safe to move the object.

In order to achieve an even clearer prehension, it is possible to replace the stepper motors with servomotors.

5. Bibliography

[1] Frincu C and Stroe I 2018 *The state of the modular element structure of a prehensor through virtual prototyping* (Brasov: PRASIC)

[2] Dudita F and Staretu I 1988 *Structural synthesis and systematisation of flat elementary mechanisms with articulated bars usable as prehension robots(Sinteza si sistematizarea structurala a mecanismelor elementare plane cu bare articulate utilizabile ca robotomecanismelor de prehensiune, in Romanian)*(Cluj-napoca Buletinulcelui de-al VIII-lea Simpozion National de IR)

[3] Ghionea I 2009 *CATIA V5. Applications in mechanical engineering(Aplicatii in ingineria mecanica, in Romanian)* (Bucuresti: BREN)

[4] Staretu I 1996 *Systems of Prehension(Sisteme de Prehensiun, in Romanian)* (Brasov: Lux Libris)

[5] Stroe I 2000 *Sensors and transducers for industrial robots (Senzori si traductoare pentru roboti industriali, in Romanian)* (Brasov: ReprografiaUniversitatiiTransilvania din Brasov)

IOP Conf. Series: Materials Science and Engineering **568** (2019) 012072 doi:10.1088/1757-899X/568/1/012072

Analysis of tribological damages of plain bearings

S Sovilj-Nikić[1], B Sovilj[2], G Varga[3], R Antunović[4] and N Ungureanu[5]

[1] Iritel a.d. Beograd, 11 000 Belgrade, Serbia
[2] University of Novi Sad, Faculty of Technical Sciences, 21 000 Novi Sad, Serbia
[3] University of Miskolc, Faculty of Mechanical Engineering and Informatics, 3515Miskolc, Hungary
[4] University of Istočno Sarajevo, Faculty of Mechanical Engineering, Republic of Srpska
[5] North University of Baia Mare, Engineering Faculty, 430311 Baia Mare, Romania

E-mail: sandrasn@eunet.rs

Abstract. Tribological processes and failure of plain bearings significantly affect the availability, adaptability, reliability and productivity of modern technology, maintenance costs, energy costs and the costs of modern production. The research of tribological processes and failure of plain bearings should indicate the possible causes and the necessary measures to be applied in order to reduce the tribological processes and failure of plain bearings. In this paper, possible causes of tribological processes and failure of plain bearings were analyzed. Having in mind the most important causes of tribological processes and failure, a proposal for monitoring the condition is given.

1. Introduction

The development of modern mechanical engineering is significantly related to the development of plain bearings. The correctness and reliability of machines and devices depend on the correctness of the moving parts, of which the plain bearings are the most important elements [1,2,3,4]. Tribological processes and failure of these vital elements can cause long delays in the process-oriented complex production systems that usually result in large financial costs in the business of companies, and also include legal liability [3,5]. Today, in the world extraordinary attention is devoted to the improvement of procedures for the prediction of tribological processes and the failure of technical systems and mechanical elements.

Plain bearings have a significant application in heavy mechanical engineering in the production of mills, turbines, crushers, etc. They are applied where it is not possible to use roller bearings, such as inaccessible places where the bearings should be two-part. Also, plain bearings are used for sleeves below 15 mm and above 300 mm, and also in assemblies where roller bearings will create vibrations and noise. The advantage of plain bearings is reflected in their load and stability. Plain bearings have been greatly improved. New materials are used for the manufacture of plain bearings, lubrication and cooling systems have been improved, the load capacity has increased, etc. Research on the wear and failure of plain bearings are numerous and they are presented in literature through various classifications, research of causes, manifestations and corrective measures, lubricant impact studies, diagnostic and research procedures [2,3]. Tribological research and the application of knowledge from tribology [2,6,7,8,9,10] both in the construction process and in the production and exploitation of plain bearings

IOP Conf. Series: Materials Science and Engineering **568** (2019) 012072 doi:10.1088/1757-899X/568/1/012072

represent the necessity of the modern industry if its today's requirements are considered: economic use of materials, rational energy consumption and increased reliability of plain bearings.

2. Analysis of plain bearings wear and damage

Today there are clearly defined classifications of damage and failure for individual and the most important machine elements, and therefore for plain bearings. All documents and technical information are focused on identifying the modes of failure and determining the cause of the failure. The mode of failure is defined as the physical and/or chemical process that leads to the loss of a given function and the phenomena or symptoms that are caused by the failure process. Causes of failure are those conditions that lead to a registered failure.

In order to reduce the delay due to the failure of the plain bearings of the TE power plant Gacko, it was necessary to analyze tribological processes and damages. At TE power plant Gacko turbine K-330-240 and generator TVV 320/2 were installed. Figure 1 shows the disposition diagram of the turbogenerator. The TE power plant Gacko is disconnected from the drive after switching on the sub-frequency and grounded 100% protection of the generator stator, due to the disappearance of 400 kV and 110 kV voltages. The lubrication pumps of the turbogenerator of the DC voltage did not switch on. The sealing pumps of DC generators have been switched on by activating the automatic start of the reserve. Nine minutes after the disappearance 400 KV voltage comes and then the own power supply of power plant is turned on. Voltage 110 KV comes a minute later, after which the power supply state is normalized. During the stopping of the turbogenerator with the reduction of the vacuum, the flow of oil from gravity reservoirs to all plain bearings was activated. This paper presents a part of tribological research, damage and causes of plain bearing failure. Tribological processes and damage on a plain bearing number seven were analyzed. Figure 2 shows the vibration trends during normal stopping of the turbogenerator from 2950 rpm to 0 rpm. The display of the basic induction frequency (1x) on the plain bearing during interruption is given in figure 3. It shows a high vibration speed of 15 mm/s. Due to the touch of the rotor and the stator during the reduction of the rotation of the rotor, inadequate damping occurred.

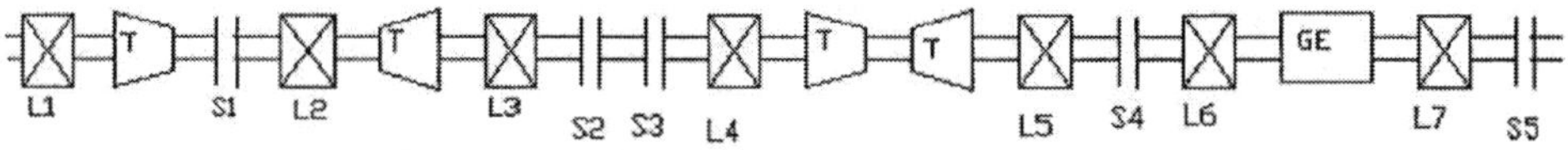

Figure 1. Disposition scheme of the plant.

During the stopping of the turbogenerator, the temperature of the plain bearing has an increasing trend, and after a few minutes the trend of decreasing and stabilizing. After establishing a regulatory lubrication system, the temperature of the plain bearing continues with the trend of increase. The absolute vibration of the turbogenerator of plain bearing has a leap after the unit is switched off to the resonant revolutions of turbine rotor and the rotor of the aggregate.

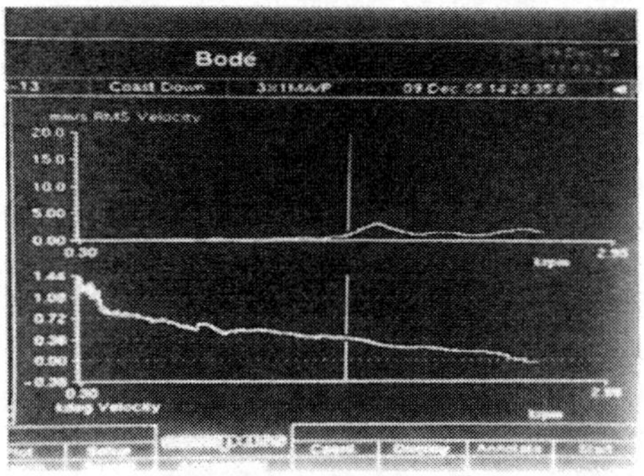

Figure 2. Diagram of the basic induction frequency of vibrations in the vertical direction on the plain bearing number 7 during the normal stop of the turbogenerator.

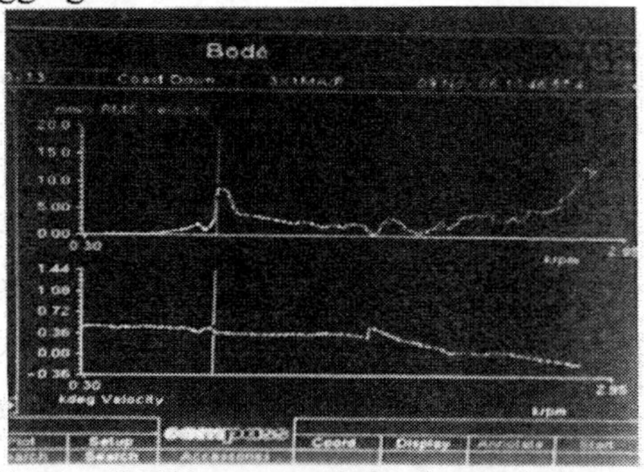

Figure 3. Diagram of the basic induction frequency of vibrations of zhe plain bearing number 7 during interruption.

IOP Conf. Series: Materials Science and Engineering **568** (2019) 012072 doi:10.1088/1757-899X/568/1/012072

On the plain bearing, the sensors of the relative rotor vibrations were not installed, so the rotation of the generator rotor during the emergency stop could not be monitored. The aggregate was stopped for 15 minutes, and after turning on the lubrication pumps of the turbogenerator plain bearings, the turning device was switched on to rotate the turbo generator with 3-5 rpm. After stopping the aggregate, cooling the turbine and switching off the lubrication system, the plain bearing and corresponding joint were opened. Damage of the joint is not registered. Greater damage was detected on the plain bearing. White metal was melted at bearings number 6, number 7 and number 8. On the journal of the generator rotor the rills were noticed which depths are about 2 mm in two rows with a total width of about 80 mm (figure 4). In the figure 4 traces of damage on the journal are visible, as well as belts of increased hardness (darkened areas). All damages are in the middle of the trigger journal. Circular surfaces are treated parts of the journal during the previous deviation at the places of residual cracks after the fine machining. The plain bearing journal was repaired and control of surface cracks was performed.

White metal deposits settled in the oil inlet on the right half (to the excitation), as can be seen in figure 5. Due to the closing of the part of the oil supply channel, partial interruption of the lubricating oil supply occurred and removal of the metal from the surface of the journal occurred in that part.

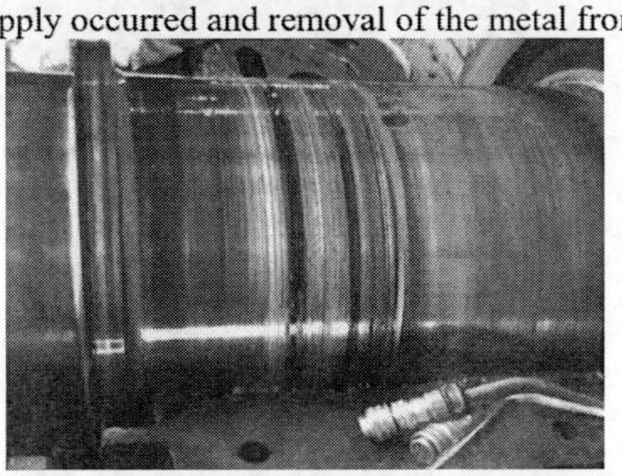

Figure 4. Damage of journal in the bearing zone.

Figure 5. Damage of bearing brass no. 7.

On the journal of the generator axle in the zone of segment bearing number 8 (journal diameter is Φ200 mm) the surface was damaged and the cracks were noticed as shown in figure 6. Due to the appearance of cracks, the continuation of the rotor is unusable and a new one was made of steel 34CrNiMo6 according to the existing documentation. The adaptation of bearing brass to the bearing journal number 7 is shown in figure 7.

Figure 6. Damage of axle extension of the generator.

Figure 7. Adaptation of bearing brass to the bearing journal number 7.

Vibration intensity of bearings number 1 and 3 of was directly dependent on the load on the block. This is clearly seen on the spectral overview of vibrations of bearing before the block power is reduced (figure 8). On the half-harmonic 1/2X vibration rate is 1 mm/s, and at 3/2X it is about 0.8mm/s. As the

IOP Conf. Series: Materials Science and Engineering **568** (2019) 012072 doi:10.1088/1757-899X/568/1/012072

power of the block was reduced vibrations decreased, at a speed of about 7-8 mm/s to a value of 6-7 mm / s. The component of the vibration on the half-harmonic 1/2X, 3/2X, 5/2X disappeared (figure 9).

In order to reduce the turbine's vibration level, a reduction in block power from 240-250 to 210-230 MW was carried out, i.e., the boiler was power from an electric power pump instead of turbocharging pump.

In an electric power pump mode with a load of 210-230 MW, the vibrations of the turbines are lowered to the B class for this type of device and, with additional control limits, the block operation is further extended.

Figure 8. Spectral diagram of vibration of bearing number 3 horizontally before reducing the block power.

Figure 9. Spectral diagram of vibrations of bearing number 3 horizontally after reducing the block power.

The block was stopped in order to determine the cause of the increased level of vibrations of the turbine bearings number 1 and 3. After cooling the turbine, the turbogenerator bearings were opened and an overview was performed. On the first bearing, there are visible traces of damage on the part of the white metal surface on the lower half of the insert (figure 10). The removal of white metal on the lower half of the surface of the plain bearing due to the failure of the rotor of 0.16 mm is also visible. According to the character of the damage, it is clear that it is an electro-erosion corrosion that occurs due to the breakdown of the vortex currents from the rotor to the stator part (contact surface of the bearing brass number 1). The rehabilitation consists of showering of the bearing surface, adjustments to the caliber and re-assembly.

In the figure 11 the trends of vibrations of the plain bearing at the time of the turbogenerator stopping from 2950 rpm to 0 rpm are shown. They were recorded by the Compass Schenk turbogenerator control system. The second speed peak of the total vibration level up to 10 mm/s is at 900 rpm, i.e. on the critical rotational speed of the generator rotor. Based on the vibrations during the turbogenerator stopping, it can be noticed that the aggregate on the plain bearing already at a speed of 2950 rpm., immediately after switching off the block, it had a contact of the rotor with the stator part.

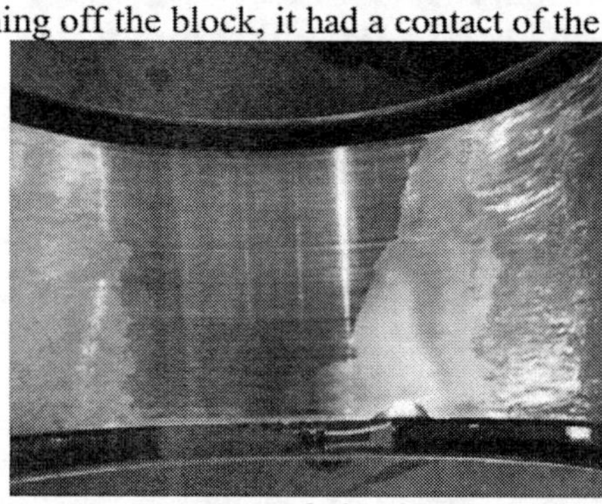

Figure 10. Damage of the lower half of the bearing brass due to electro-erosion corrosion.

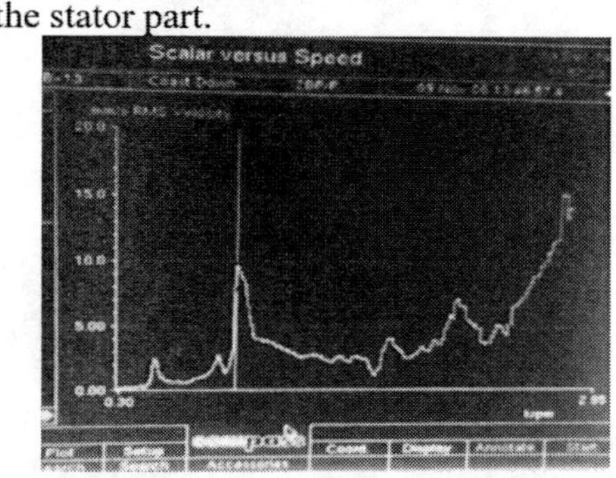

Figure 11. Diagram of the level of total vibrations on the plain bearing during stopping the turbogenerator.

The analyzed turbogenerator plain bearing has a reservoir for emergency lubrication, whose task is to prevent break-down of the journal of turbine rotor and generator. The bearing brass on the generator has been completely damaged. It can be noticed a significant damage of the journal.

Based on the analysis, it could be concluded that the resulting damages were the result of an irregular operation of the emergency oil reservoir. By analyzing the data from the figure of the control system, as well as the vibration monitoring system of the turbogenerator, it can be ascertained:

- the vibration of the plain bearing has increased sharply to 15 mm/s for 50 seconds after the unit has been switched off after the critical revolutions of the generator have passed
- the mentioned vibrations initiate on the contact of the rotor with the stator part of the plain bearing
- it is obvious that there are reasons that influence the shutdown regime of the generator rotor when problems arise due to the disappearance of the forced lubrication system.

3. Conclusion

Tribological processes are the main causes of damage and malfunction of plain bearings. Systematic study of the process of wear, standardization of tribometric methods and the development of real data on the tribological characteristics of the most important types of materials used today for plain bearings is significant field of research in the field of damage of plain bearings.

There are a number of causes that can lead to plain bearing failure. The most common cause of failure is inadequate and insufficient lubrication of the plain bearing.

Maintenance according to the condition, i.e. proactive maintenance of the entire system, including plain bearings, will provide significant technical and economic effects.

Acknoledgements

The research was funded by the Ministry of Education, Science and Technological Development of the Republic of Serbia within the project III 43008, and it is also the result of the cooperation within CEEPUS projects CIII- RO-0058-07-1819 and CIII-RS-0304-10-1819.

References

[1] Lamjua P 2006 *Tribological Behaviour Study of Sliding Bearing of Leaded Red Brass,* King Mongkut's Institute of Technology North Bangkok, Doctoral Thesis

[2] Tanasijević S 1989 The basis of trobology of machine elements (Osnovi tribologije mašinskih elemenata), Naučna knjiga, Beograd, Serbia

[3] Rac A 2003 Monitoring of the condition and diagnostics of failure of plain bearings (Praćenje stanja i dijagnostika otkaza kliznih ležaja), *Proceedings of the 8th International Tribology Conference*, 8-10 October 2003, Belgrade, Serbia, 271-275

[4] Ognjanović M 1999 *Machine elements (Mašinski elementi)*, Naučna knjiga, Beograd, Serbia

[5] Antunović R 2014 Defect factor of the journal bearings, *Proceedings of 3th Conference "Održavanje 2014"*, 11-13 Jun 2014, Zenica, Bosnia and Herzegovina, 97-101

[6] Ivković B and Rac A 1995 *Tribology (Tribologija)*, Jugoslovensko društvo za tribologiju, Kragujevac, Serbia

[7] Sovilj-Nikić I Modeling and optimization of hob milling process (Modelovanje i optimizacija procesa odvalnog glodanja), Fakultet tehničkih nauka, Novi Sad, Serbia, Doctoral Thesis (unpublished)

[8] Sovilj B, Sovilj-Nikić I and Ješić D 2012 The effect of specific relationship between material and coating on tribological and protective features of the product, *Metal.*, **51** (1), 21-24

[9] Sovilj B, Sovilj-Nikić I and Ješić D 2011 Measurement methodology of characteristics and election of materials of elements of tribomechanical systems, *Metal.*, **50** (2), 107-112

[10] Sovilj-Nikić I, Sovilj B, Kandeva M, Gajić V, Sovilj-Nikić S, Legutko S and Kovač P 2012 Tribological characteristics of hob milling tools from economic aspect, *J. Balk. Tribol. Assoc.* **18** (4), 577-585

IOP Conf. Series: Materials Science and Engineering **568** (2019) 012073 doi:10.1088/1757-899X/568/1/012073

The calculus of the temperatures in the characteristic points of the gasoline direct injection engine cycle

V Blaga, T Mitran, G Dragomir, D Făgădar

The University of Oradea, Universitatii st., no. 1, Oradea, Bihor County, Romania

tudor_mitran@yahoo.com

Abstract. Based on an own model, the authors have realized the analytical calculus of the temperatures in the characteristic points of the thermodynamic cycle for a GDI engine. These temperatures were calculated depending on the engine's rotational speed and on the environmental temperature (considered to be equal to the one in the intake valve port) and for an excess air factor λ=1. The first step was the development of the mathematical model of the engine's cycle. The model includes the presentation of the initial data for the thermal calculus and the correlation between the expressions of the parameters in order to develop the computing programs. The calculated data were compared with experimental data acquired at environmental temperature, at maximum power and at the rotational speed corresponding to the maximum power (full load). The results show that the proposed model is accurate enough to be a useful tool for the study of GDI engines.

1. Introduction

In fig. 1a is presented the theoretical cycle, with a constant volume burning, for the SI engines. In fig. 1b is presented the authors proposed cycle that includes: the constant volume burning, and the afterburning [1, 2, 3and 4]. Other authors used dedicated programs to develop models of the engine's cycle [5]. Other models don't include the reaction process but instead use suitable functions that describe the energy released from the combustion [6].

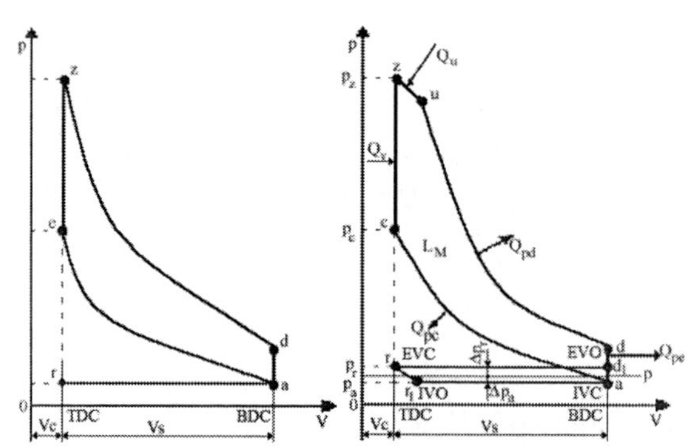

Figure 1. a) The cycle at constant volume burning; b) The cycle proposed by the authors.

The simplifying assumptions that were used to define this cycle are the following (see fig. 1b):

IOP Conf. Series: Materials Science and Engineering **568** (2019) 012073 doi:10.1088/1757-899X/568/1/012073

- the intake process r-a takes place at a constant pressure p_a, permanently smaller than the environmental one with the value of the pressure losses Δp_a, characteristic to the intake process;
- the exhaust process d-d_1-r takes place in two stages: the free exhaust of the gases at constant volume d-d_1, and the forced exhaust at constant pressure p_r, d_1-r;
- the link between the exhaust valve and the intake one is done through the isentropic expansion of the residual gases r-r_1;
- the compression a-c of the fresh charge is assimilated with a polytropic process with a constant exponent $n_c<k_c$;
- the burning process is schematized in two evolutions: the isochoric one c-z, in which the heat input is Q_v, and the polytropic evolution of exponent $n_u<1$, that defines the afterburning and in which the working agent receives the heat Q_u;
- the expansion process starts when the theoretical burning ends in the point u. This process is been assimilated with a polytropic one of exponent $n_{d1}>k_d$ that leads to the give up of the heat Q_{pd} to the cylinder's walls;
- the working fluid is considered to be a perfect gas with a specific heat depending on the temperature;

It must be mentioned that the assimilation of the afterburning process is based on experimental data [2] that show that the final burning temperature Tu is higher than the temperature at the end of the constant volume temperature Tz as a result of the heat release because of the fact that remaining fuel is burned during the phase z-u (fig. 1b). After that, the temperature drops because the increasing of the burning chamber volume and because no heat is released during the process u-d (fig. 1b).

2. The calculus of the temperatures in the characteristic points of the GDI engine cycle

The first step was to calculate the parameters of the theoretical cycle (fig. 1a).

The calculus of the temperature at the end of the intake stroke was presented in a previous paper [7].

Being given the temperature at the end of the intake process T_a=322 K, one can determine the temperature at the end of the compression process:

$$T_c = T_a \cdot \varepsilon^{n_c-1} \ [K]; \tag{1}$$

The polytropic exponent of the compression process n_c can be expressed depending on the adiabatic exponent k_c and on the coefficient that takes into consideration the heat ceded by the working agent to the cylinder's walls during this process:

$$n_c = 1 - \frac{k_c - 1}{\varphi_c}; \ \varphi_c = 1.06; \tag{2}$$

where: - φ_c is a coefficient that characterizes the deviation of the working process from the adiabatic compression process;

It's obvious the fact that, because $Q_{pc}<0$ and $\Delta U>0$, the coefficient $\varphi_c>1$, so that $n_c<k_c$. In the conditions in which the compression process takes place, the adiabatic exponent k_c is determined with the following relation:

$$k_c = 1.438 - 0.525 \cdot 10^{-4} \cdot (T_a + T_c); \tag{3}$$

where: - T_c [K] is the temperature of the fresh charge at the end of the compression process;

The pressure at the end of the compression process (point c in fig. 1b) is:

$$p_c = p_a \cdot \varepsilon^{n_c} \ [Pa] \tag{4}$$

One can determine the temperature at the end of the main phase of the burning process (point z in fig. 1b), after the calculus of the pressure at the end of the main phase of the burning process p_z:

$$\frac{p_z}{p_c} = \frac{T_z}{T_c} = \alpha; \tag{5}$$

and further: $T_z = \alpha \cdot T_c$ [K] and $p_z = \alpha \cdot p_c$ [Pa];
where: - p_z [Pa] is the pressure at the end of the main burning phase;
The temperature at the end of the afterburning process (point u in fig. 1b) is:

$$T_u = T_z \delta^{1-nu} \quad ; \; n_u = 0{,}9 \quad [2] \tag{6}$$

In fig. 2 is presented the variation of the temperature at the end of the compression process T_c, depending on the environmental temperature and on the rotational speed for an air-excess factor $\lambda=1$.

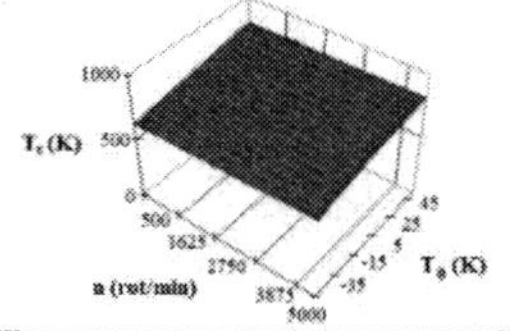

Figure 2. The variation of the temperature T_c.

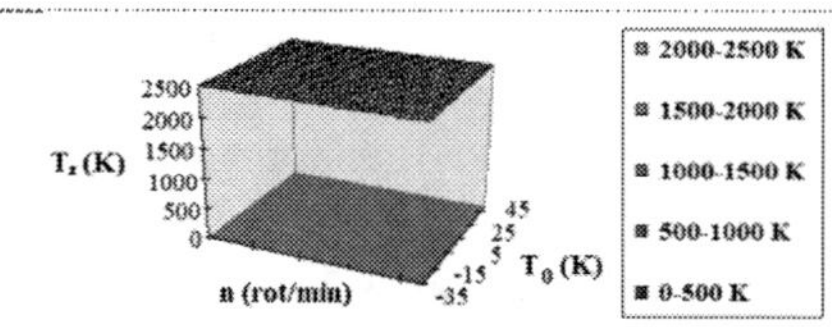

Figure 3. The variation of the temperature T_z.

In fig. 3 is presented the variation of the temperature at the end of the main phase of the burning process T_z, depending on the environmental temperature and on the rotational speed for an air-excess factor $\lambda=1$.

The temperature at the end of the burning process (the point u in fig. 1b) is [2]:

$$T_u = T_z \cdot \delta^{1-n_u} \; [K]; \tag{7}$$

In fig. 4 is presented the variation of the temperature at the end of the afterburning T_u depending on the rotational speed and on the environmental temperature.

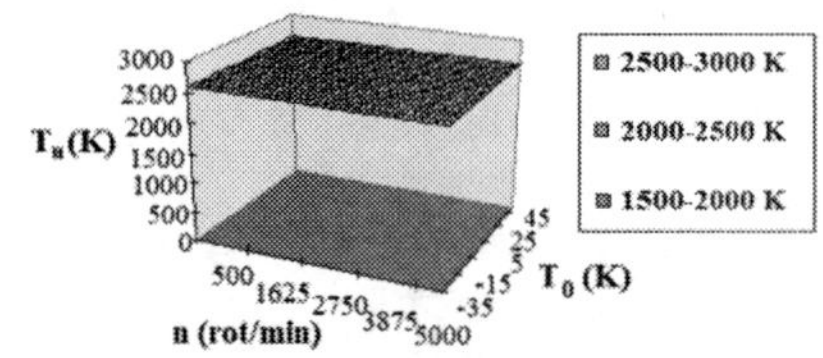

Figure 4. The variation of the temperature T_u.

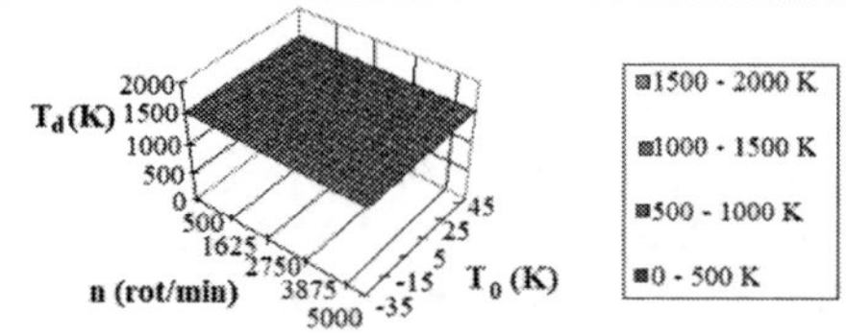

Figure 5. The variation of the temperature T_d.

The temperature at the end of the expansion process (point d in fig. 1b):

$$T_d = T_u \cdot \left(\frac{\delta}{\varepsilon}\right)^{n_d - 1} \; [K]; \tag{8}$$

In fig. 5 is presented the variation of the temperature at the end of the expansion process T_d depending on the rotational speed and on the environmental temperature.

The temperature at the end of the free exhaust phase (point d_1 in fig. 1b) is:

$$T_{d_1} = T_d \cdot \left[\frac{(\varepsilon / \delta)^{nd} - \delta^{n_u}}{\psi \cdot \alpha \cdot \varepsilon^{n_c}} \right]^{\frac{k_e - 1}{k_e}} [K]; \qquad (9)$$

In fig. 6 is presented the variation of the temperature at the end of the forced exhaust phase T_{d1} depending on the rotational speed and on the environmental temperature.

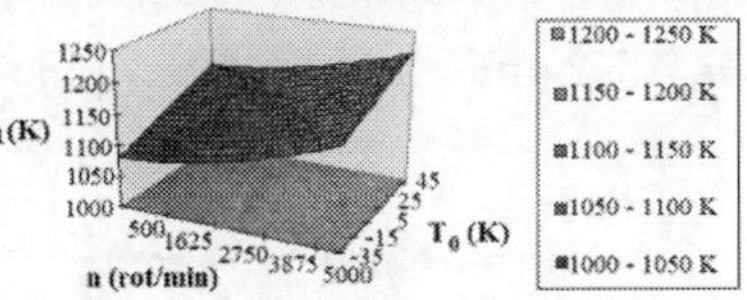

Figure 6. The variation of the temperature T_{d1}.

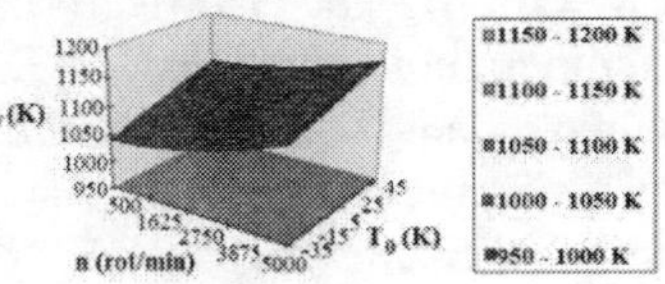

Figure 7. The variation of the temperature T_r.

The temperature at the end of the exhaust process (point r in fig.1.b) is:

$$T_r = T_{d_1} / \varphi_r \; [K]; \qquad (10)$$

In fig. 7 is presented the variation of the temperature at the end of the free exhaust phase T_r depending on the rotational speed and on the environmental temperature.

The temperature at the beginning of the intake process (the point r_1 in fig. 1b):

$$T_{r_1} = T_r \cdot \psi^{k_e - 1/k_e} \; [K]; \qquad (11)$$

In fig. 8 is presented the variation of the temperature at the beginning of the intake process T_{r1} depending on the rotational speed and on the environmental temperature.

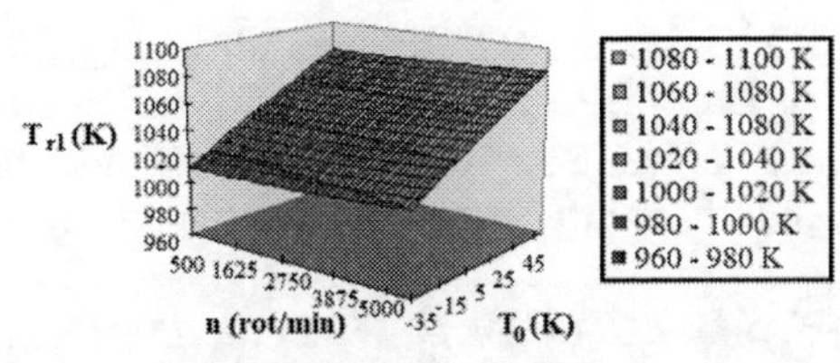

Figure 8. The variation of the temperature T_r

3. Conclusion

The computation results for the temperatures in the characteristic points of the engine's cycle were compared to those experimentally obtained. The experimental data were acquired at the environmental temperature T_0=293 K and at the maximum power rotational speed n=5250 rot/min.

The calculated temperature at the end of the intake stroke increases as the rotational speed rises. The calculated temperature is T_a=343.5 K, and the one determined from the experimental date is T_{am}=353 K.

The temperature calculated in the point c is T_c=762.5 K, and the determined from the experimental data is T_{cm}=783 K. The calculated temperature is with 2.7% smaller than the experimental one.

IOP Conf. Series: Materials Science and Engineering **568** (2019) 012073 doi:10.1088/1757-899X/568/1/012073

The temperature calculated in point z is T_z=2500 K, and the one determined from the experimental data is T_{zm}=2547. The difference is of 1.88% and appears as a result of the fact that the temperature in the point c was underestimated in the calculus.

The temperature calculated in the point u is T_u=2600 K, and the one determined from the experimental data is T_{um}=2680 K. The error is of 3.077% and appears as a result of the underestimation of the temperature in the point c.

The temperature calculated in the point d is T_d=1538 K, and the one determined from the experimental data is T_{dm}=1564 K. The error is of 1.7%.

The temperature calculated in the point d1 is T_{d1}=1193 K, and the one determined from the experimental data is T_{d1m}=1220 K. The error is of 2.26%.

The temperature calculated in this point is T_r=1147 K, and the one determined from the experimental data is T_{rm}=1140 K. The error is of 0.6%, that is insignificant.

The temperature calculated in the point r1 is T_{r1}=1147 K, and the one determined from the experimental data is T_{r1m}=1064 K. The error is of 1.5%.

One can observe that the errors are between 0.6 and 3%.

The greatest error is registered in the point u, at the end of the burning process (approx. 3%). That's an excellent value taken into consideration the fact the estimation of the heat exchange process inside the cylinder are very difficult to be estimated. This means that the model proposed by the authors is sufficiently accurate to estimate the temperatures in the characteristic points of the engine's cycle.

In conclusion, the model proposed by the authors, through small calibrations, can be a useful tool for the determination of the maximum pressure of the engine's cycle and can give information about the formation of the polluting emissions.

A good management of the thermal regime of the engine can reduce pollutant emissions [8,9] so, the model proposed in this paper can offer useful information about the thermal regime of the GDI engines.

References

[1] Bataga N., *Thermal engines (Motoare termice)*, Bucharest, Didactic and Pedagogical Publishing, 1979.

[2] Radcenko V., *Criteria for the optimization of the thermal processes (Criterii de optimizare a proceselor termice)*, Bucharest, Technical Publishing, 1977.

[3] Blaga V., *Motoare cu injectie de benzina (Engines with gasoline injection)*, Oradea, University of Oradea Publishing, 2013.

[4] Blaga V., *Processes and characteristics of the ICE (Procese si caracteristici ale motoarelor cu ardere interna)* Oradea, University of Oradea Publishing, 2013.

[5] Dan Moldovanu, Nicolae Burnete, Sanda Bodea; *Studies regarding model building for engine process simulation in AVL fire*, The 10th international automotive congress CAR 2011,

[6] Codan E., A program to simulate the thermal-dynamic working processes in Diesel Engines, MTZ (1996), 5

[7] Mitran T., Blaga V., Dragomir G., Fodor D., *The calculation algorithm for the determination of the temperature at the end of the intake stroke for GDI engines*, Conference of the 12th International Congress of Automotive and Transport Engineering (CONAT), Brasov, 2016, pp. 311-318, Published 2017.

[8] Leahu C I and Tarulescu S 2016 *The influence of thermal regime on gasoline direct injection engine performance and emissions,* ACME - The 7th International Conference on Advanced Concepts on Mechanical Engineering, 9-10 June 2016 Iaşi, IOP Conf. Series: Materials Science and Engineering, pg. 147.

[9] Leahu C I, Tarulescu S, Tarulescu R 2018 *The exhaust gas temperature control through an adequate thermal management of the engine,* ACME - The 8th International Conference on Advanced Concepts on Mechanical Engineering, 7-8 June 2018 Iaşi, IOP Conf. Series: Materials Science and Engineering, pp. 444.

IOP Conf. Series: Materials Science and Engineering **568** (2019) 012074 doi:10.1088/1757-899X/568/1/012074

Pedestrian visibility at night: the influence of the pedestrian clothing and the defective headlamp of the car

B C Benea[1], D D Trusca[1], G R Toganel[1] and A I Radu[1]

[1]Department of Automotive and Transportation, Mechanical Engineering Faculty, Transilvania University of Brasov, Brasov, Romania

b.benea@unitbv.ro

Abstract. Low night vision is a factor contributing to the production of traffic accidents. In this work, we investigated the distance from which pedestrians can be seen by night, by a driver driving a car with a single functional headlamp. Six pedestrian clothing conditions were tested (black, blue, red, green, yellow and reflective vest). Pedestrians wearing dark clothing (black, blue) were only visible from a distance of 10 meters; the best visibility was for the pedestrian that wearing the reflective vest. The results demonstrate the importance of pedestrian clothing during night time on visibility while traveling on the public road.

1. Introduction

Road accident is an unpredictable event occurring on a public roadway where at least one moving vehicle and other traffic participants are involved, resulting in material damage, injury or death of the person. The cause of most accidents is the low visibility of car participants at night [1]. This is primarily due to the difficulty in reading the traffic information, the colour differences of other traffic cars, the ability to see the edges of the low road, and the effort to the human eye [2].

In the case of drivers, their reduced night-time visibility is due to several factors:

- Fatigue is a factor that influences visibility. This leads to a decrease in driving capacity. Typically, in the time range of 03:00 - 07:00 this tends to the maximum, as in the period when he normally sleeps [3].
- Alcohol is the next major contributor to road accidents. This, along with night driving, is the leading cause of death among young drivers [3,4].
- Age, being a major influence of night vision. The decrease in visual acuity begins at the age of 20 years. Over time, the amount of light that goes through the cornea decreases, so somewhere around the age of 60 it is needed 3 times more light than at the age of 20 to see the same details [5].

In addition to driving factors, there are external factors that influence their visibility, such as:

- Lighting with headlights, where the visible road sector is much lower than during the day. Darkness allows perceiving distances with great difficulty, so that headlights, especially poorly tuned, can create a blinding effect [6,7].
- Unlit or poorly illuminated roads, which affect the visibility of the drivers, becoming much lower and pedestrian sensing, will be much slower [7,8].
- Weather conditions (fog, rain, snow, etc.) are an unpredictable external factor and cause night vision [9].

IOP Conf. Series: Materials Science and Engineering **568** (2019) 012074 doi:10.1088/1757-899X/568/1/012074

Pedestrians are the most vulnerable traffickers due to lack of protection during a collision with a car. A proper visibility of the pedestrian can reduce the consequences of an accident. During night driving, there is the possibility that the objects placed in the vehicle paths are not detected in time [8,9].

The objective of the work is the influence of the colour of pedestrian clothes on the visibility in the night conditions, in the case of vehicles running with defects of the lighting system.

2. Method

To perform this experiment, the measurements were made on the test track, with a length of 50 m, without street lighting. On this test track marks were made at intervals of 10 meters, so the test track was divided into 5 intervals.

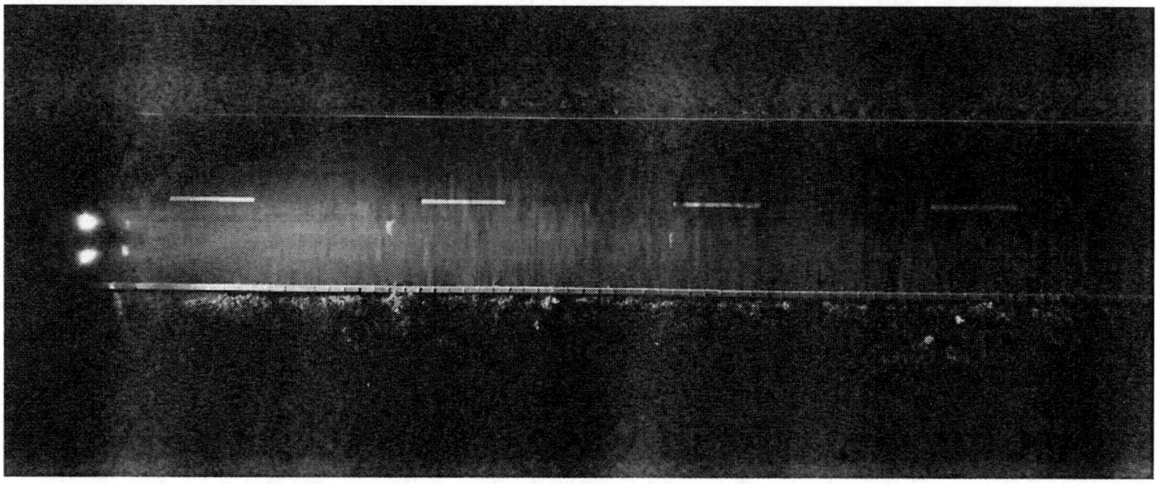

Figure 1. Test track.

Six pedestrians equipped with different colours clothes and a Dacia Logan were used to achieve the objectives. The pedestrian was located at a distance of 0.5 m from the right merge of the road, and the car was positioned at a distance of 10 to 50 m from the pedestrian. The tests were carried out using a single car headlight (simulating the failure of the other headlamp).

Pedestrians wore a black, blue, red, green, yellow blouses and black blouse with reflective vest.

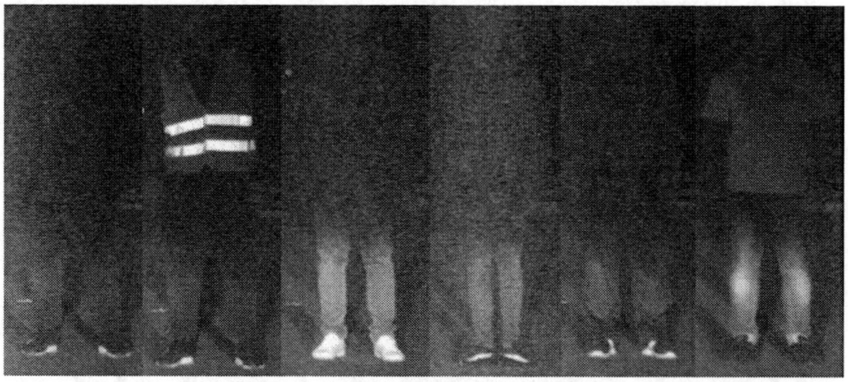

Figure 2. Pedestrians clothes.

A Fujifilm X-T1 camera was used to capture photos from the car. This camera allowed capturing photos at night without changing the focus.

IOP Conf. Series: Materials Science and Engineering **568** (2019) 012074 doi:10.1088/1757-899X/568/1/012074

For image processing was used the PC-RECT 4.2 software, developed by DSD Austria, specializing in photogrammetry. This software determines the brightness on different areas of the image and can calculate the contrast between the two analysed areas, so the software gives the visibility of the pedestrian by comparing the two luminance values and the contrast determination. The program uses Adrian Werner's method, which is the method of stability of the visibility of the object under different light conditions.
The analysed area of the pedestrian was the chest area, the wearing area of the reflective vest.
The program determines the value of the luminance of the object in the area of the pedestrian's chest (1) and compares it with the luminance of the environment (2), thus obtaining the contrast value.

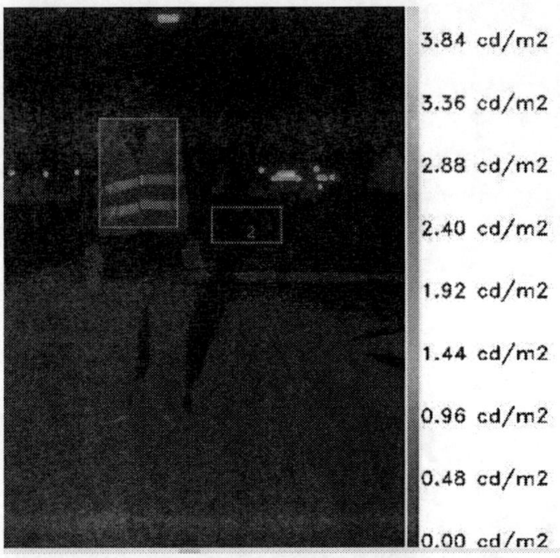

Figure 3. Photo processing to determine the visibility of the pedestrian.

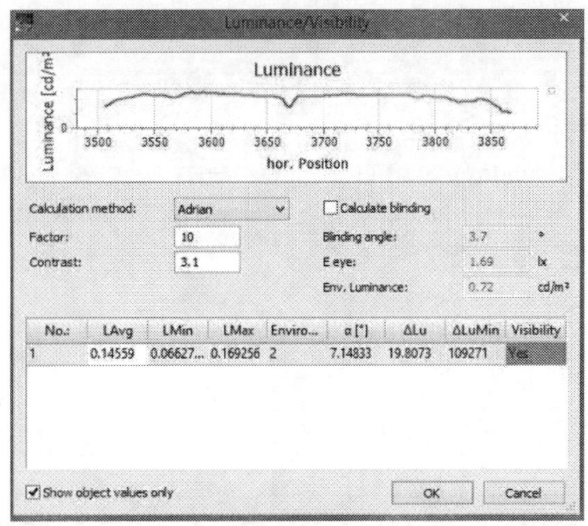

Figure 4. Luminance determination window from PC-RECT.

The software extracts all the data in a table showing the object's visibility based on the ΔLu parameter, which represents the difference between the two selected areas.
L_{avg} – average luminance; L_{Min}, L_{Max} – shows the minimum and maximum luminance of the respective area; α- object size; Environment – assigned background area; ΔL_u – luminance difference, it is the difference between LAvg of the defined area and LAvg of the background area;
ΔL – luminance difference threshold.

3. Results

The pedestrians wore a black (Bk), blue (B), red (R), green (G), yellow (Y) blouses and black blouse with reflective vest (BkRV).

In figure 5 are presented the results for the distance of 10 meters between the car and the pedestrian. The greatest contrast is for the pedestrian that wears the reflective vest, regardless of the functional headlamp.
The small contrast is found for the pedestrian that wears black clothing when the right headlamp works.
For the distance of 20 meters between the car and the pedestrian (figure 6), the biggest contrast is for the pedestrian that wear the reflective vest. At this distance, the pedestrians that wear black or blue clothing are not visible (they are confused with the environment). The pedestrians that wear green or red clothing are more visible when the left-hand headlight of the car works.

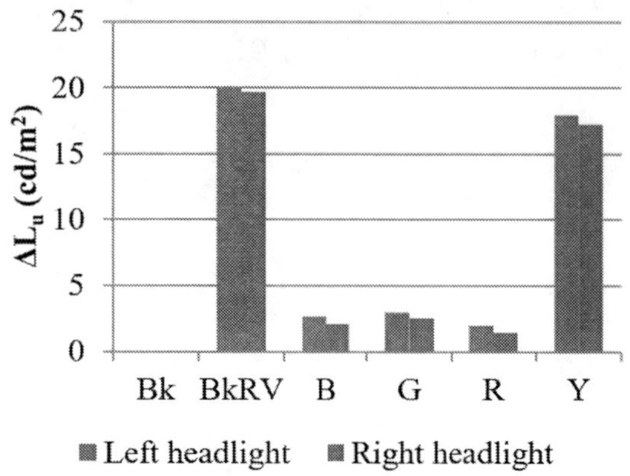

Figure 5. Contrast at 10 m distance.

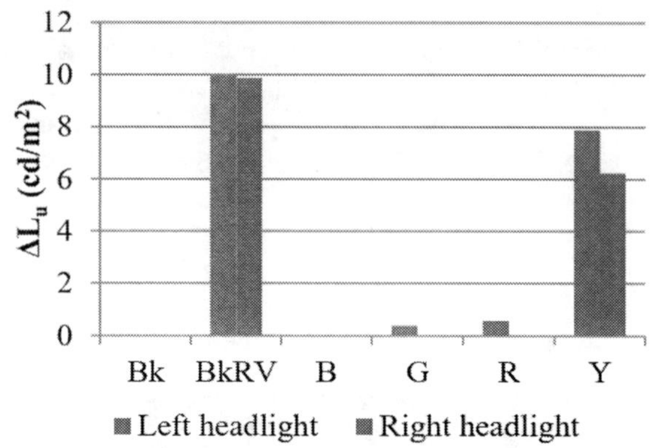

Figure 6. Contrast at 20m distance.

At a distance of 30 meters between the car and the pedestrian (figure 7), the pedestrian that wears yellow clothing or a reflective vest are visible, regardless of the headlight that works.

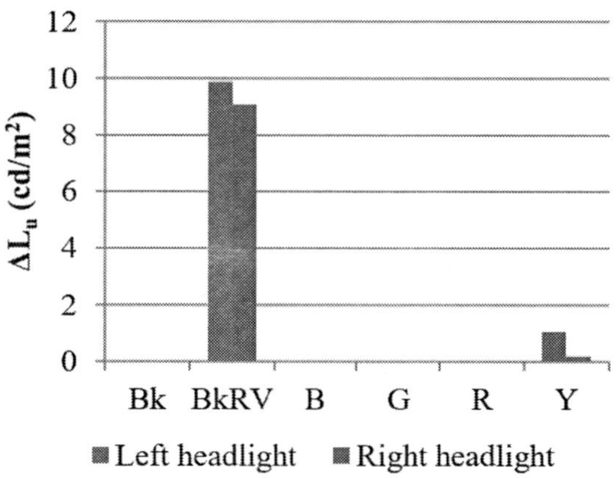

Figure 7. Contrast at 30m distance.

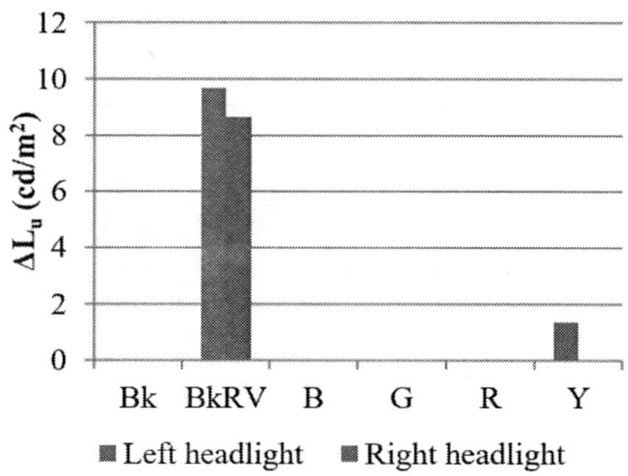

Figure 8. Contrast at 40m distance.

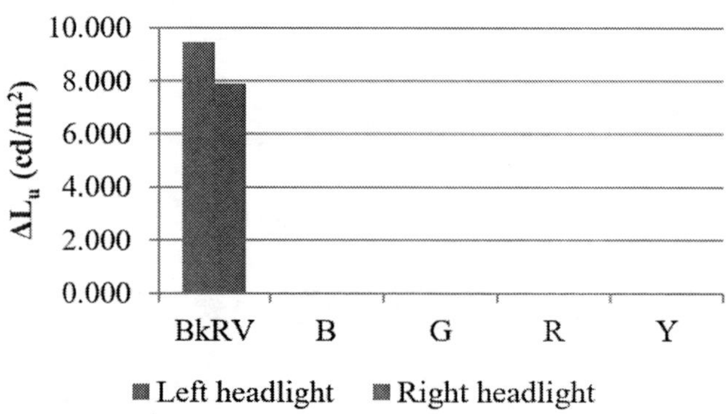

Figure 9. Contrast at 50m distance.

At a distance of 40 meters (figure 8), pedestrian with yellow clothing are not visible if the car has the left headlamp defective. At a distance of 50 meters (figure 9), only pedestrians with a reflective vest are visible.

IOP Conf. Series: Materials Science and Engineering **568** (2019) 012074 doi:10.1088/1757-899X/568/1/012074

4. Conclusions

The present experiment sought to determine the visibility of pedestrians depending on their worn clothing, in the event of damage to the vehicle lighting system (a defective light bulb). The data demonstrate that pedestrian clothing and functional headlamp significantly affect both the probability that a driver will recognize the presence of a pedestrian on the road and the distance to which pedestrians are recognized in the night.

Pedestrian clothing had a powerful effect on pedestrian visibility. When the pedestrian wore the black coat he was hardly sizable of it at a distance of 10 m. For the other distances tested, the pedestrian in black was confused with the background (it was invisible). The moment of visibility of the pedestrian wearing black clothes occurs at a point that may not allow drivers from stopping in time to avoid a collision.

The visibility of pedestrians is improved by wearing light-coloured clothes (greater contrast with the environment). In the case of green or red blouses, the pedestrians could be seen from a distance of 20 m. Pedestrians who wear yellow blouses can be seen from a distance of 40 m. Pedestrians who wear a reflective vest can be seen from a distance of 50 m.

The functional headlamp had a significant effect on pedestrians' recognition. In the case of the operation of the left-hand headlamp, pedestrians are more easily recognized.

5. References

[1] Rumar K 1990 The basic driver error: late detection. *Ergonomics*, 32, **1281-90**

[2] Sullivan J M and Flannagan M J 2002 The role of ambient light level in fatal crashes: inferences from daylight saving time transitions. *Accident Analysis and Prevention*, 34, **487-498**

[3] Summala H and Mikkola T 1994 Fatal accidents among car and truck drivers: effects of fatigue, age, and alcohol consumption. *Hum. Factors* 36 (2), **315–326**

[4] Owens D A and Sivak M 1996 Differentiation of visibility and alcohol as contributors to twilight road fatalities. *Human Factors*, 38, **680-689**

[5] Chrysler S T, Danielson S M and Kirby VM 1996 Age differences in visual abilities in nighttime driving field conditions. *Proc Human Factors Ergonomics Soc* 40th Meeting, **923-927**

[6] Tolea B, Antonya C and Beles H 2015 Assessment of the injury severity of the pedestrian lower limbs at the collision with a vehicle. *Proc of the Annual Session of Scientific papers "IMT-Oradea 2015"* XIV, **189-192**

[7] Brooks J O, Tyrell R A and Olson H M 2001 Pedestrian visibility at night: evaluating a method of teaching young drivers to be safer pedestrians. *80th Annual Meeting of the Transportation Research Board, Washington*

[8] Gegenfurtner K R, Mayser H and Sharpe L T 1999 Seeing movement in the dark. *Nature* 398, **475-476**

[9] Mannering F L and Bhat C R 2014 Analytic methods in accident research: methodological frontier and future directions. *Anal. Methods Accid. Res.* 1, **1-227**

IOP Conf. Series: Materials Science and Engineering **568** (2019) 012075 doi:10.1088/1757-899X/568/1/012075

Application of thermogravimetric analysis for thermal characterization of walnut oil and biodiesel

D Tutunea[1], D Ilie[1], L Racila[1], I Geonea[1], O Otat[1] and L Matei[1]

[1]Vehicles, Transport and Industrial Engineering, Faculty of Mechanics, Craiova, Romania

dragostutunea@yahoo.com

Abstract. The biodiesel is being promoted worldwide as an important alternative to petrodiesel fuel due mainly to availability, low emissions and simple industrial production from renewable sources. Analysis of thermal properties is important showing the distillation and decomposition of the fuel function of temperature providing the temperature range of the combustion processes. In this paper physical and thermal properties of walnut oil were analyzed using Thermogravimetry (TGA/DTG). According to the thermograms the walnut oil presents the highest thermal stability in comparison with diesel fuel and walnut oil biodiesel.

1. Introduction

The rising cost of petroleum derived fuels associated with climate change, global warming and energy security issues have caused the promotion and development of biofuels [1, 2]. Nowadays, biodiesel is considered as a solution for different applications like, transportation sectors, industrial, locomotives and marines [3]. 80% of the world's energy needs is assured by fossil fuels and these fuels are depleting rapidly all over the world causing an energy crises [4]. Beside the maintenance of the carbon cycle, biodiesel can reduce the emission of pollutant emissions which cause the damage of ozone layer and acid rains [5]. Biodiesel can be obtained by vegetable oils and animal fats forming mono-alkyl esters of long chain fatty acids with properties closer to diesel fuel. The key advantages of using biodiesel are: renewability, non-toxicity, thermal stability during handling due to its higher flash point, lubricity and easy biodegradability [6]. The main drawbacks in commercialization of biodiesel is the price of vegetable oil that cost much more than petro-diesel (80% of the total cost of biodiesel production is the raw material used). The solution is the use on non-edible oils [7]. The thermogravimetric analysis (TGA) is a quick and inexpensive technique to measure the properties of a fuel as a function of temperature or time. The thermal decomposition is influenced by chain molecules, the longer is the chain the higher is the boiling point of the fuel. The controlled atmosphere (nitrogen, oxygen, argon, and air etc.) combined with the increases of the temperature determines the weight loose due to the decomposition process. Dantas et al. [8] studied the thermal stability of corn biodiesel obtained by methanol and ethanol routes using a DTA/TG analyzer (SDT 2960, TA Instruments) in a temperature range of 30–600 °C in air and nitrogen and found that: in air atmosphere the corn oil was thermally stable at 225 °C, methanol biodiesel 139°C and ethanol biodiesel 159 °C; in nitrogen atmosphere the corn oil is stable at 336 °C, methanol biodiesel 145°C and ethanol biodiesel 169 °C. Chien et al. [9] investigated soybean biodiesel and found that is thermally stable at 119 °C and has a range of thermal decomposition from 119 to 237 °C. Patil et al. [9] studied the thermal stability of waste cooking biodiesel in a temperature range of 25–600 °C and found two mass loss steps, first between 125–130 °C attributed to the vaporization of biodiesel and second at 232 °C attributed to the non-transesterification of some waste cooking oil. Kok et al. [10] investigated the thermal behavior of canola oil biodiesel and reported a single reaction region from TG/ DTG curves of the combustion process using various heating rates (with a peak between 242–260 °C). Santos et al. [11] investigate the thermal stability of palm, sunflower and cotton in a temperature range of 30–600 °C to helium atmosphere with a heating rate of 10 °C min-1. With the exception of palm oil which had two steps of mass loss, the TG curve showed a single mass loss for sunflower and cotton oil biodiesel. Rodrıguez et al. [12] investigated thermal behavior of higuereta and soybean oil biodiesel at various heating rates (5, 10, 20 and 30 °C min-1) in air atmosphere at a temperature range between 25 and 450 °C and found that in biodiesel the mass loss occurs between 185 to 300 °C and in diesel fuel between 50 and 250°C. The objective of this study was to evaluate two different biodiesel samples by thermogravimetry (TG) and derivative thermogravimetry (DTG), and to compare with the thermogravimetric profile of petrodiesel fuel.

2. Material and methods

The walnut nuts were obtained from 2018 harvest in Dolj County. They were used as feedstock for biodiesel production using the methylic route. Diesel fuel was purchased from an OMV station in Craiova on 10 October 2018. The chemicals used for the transesterification reaction were purchased from Laborex Romania. The raw oil was obtained by an extraction process using a press PU-200 and then was filtered to remove impurities. Walnut biodiesel (WME) was obtained by a process called transesterification in order to modify some key properties of oil (viscosity, density) to make him closer to parameters of petrodiesel fuel. An alkaline-catalyzed transesterification was used keeping 6:1 molar ratio of methyl alcohol and walnut vegetable oil, using KOH as catalyst (1%) in a reactor of 30L per batch at a constant temperature of 65°C. After breakage of triglyceride molecules, a mixture of methyl esters appear corresponding to the fatty acids and a co-product called glycerin. After 24 h the glycerin layer was collected remaining only biodiesel. The product was washed with warm distilled water to remove impurities and then heated at 100°C for 2 hours. After the process is finished the physicochemical characterization of the diesel fuel, walnut oil and walnut oil biodiesel was performed according to the international standards. Thermogravimetric curves were obtained in a Diamond TG/DTG Analyzer from PerkinElmer Instruments. The applied heating rates were 5, 10, 15, 20 and 30°C min^{-1}, under air and nitrogen atmospheres, with a flow rate of 50 mL min^{-1} in the temperature range until 700°C, using an alumina crucible and a sample weight 10 mg. The differential thermogravimetric (DTG) curves were calculated by numerical derivation of the TG curves. All thermal conversion data were recorded in a computer coupled to the thermo analytical equipment and evaluated by Pyris software.

3. Results and discussion

The major properties including viscosity, density, flash point, calorific value, cetane number, cloud and pour point were determined using several methods according to ASTM and EN standards (Table 1). Kinematic viscosity is an indicator of the efficiency of the transesterification process and is related to the resistance of fluids to flow. The viscosity of WME is higher than diesel fuel and may cause problems by reducing the quality of the fuel, poor atomization and tubing blockage especially at low temperature. The density of the biodiesel is higher than diesel fuel which will led to the increase in fuel consumption and consequently of NOx emissions. The low calorific value of biodiesel is attributed of the higher oxygen content, however this presence of oxygen promote combustion in diesel engine. The higher flash point of biodiesel has advantages (safe to handle and storage) and disadvantages (low volatility properties). Cold flow properties of biodiesel indicate the biodiesel cannot be use safe in winter conditions.

Table 1. Pysical-chemical properties of fuels

Properties	Units	Method	Diesel	WME (B100)

IOP Conf. Series: Materials Science and Engineering **568** (2019) 012075 doi:10.1088/1757-899X/568/1/012075

Kinematic viscosity at 40 °C	mm^2s^{-1}	ASTM D445	3.078	3.88
Cloud Point	°C	ASTM D2500	-18	-6.1
Pour Point	°C	ASTM D97	-32	-10.0
Calorific value	MJ/kg	ASTM D4809	44	41.18
Flash point	°C	ASTM D93	68	170
Density at 40 °C	kg/m^3	EN ISO 3675	816	864
Cetane number	-	ASTM D613	49	54

For the thermal analyses there samples were used diesel fuel, vegetable oil (walnut oil) and walnut biodiesel. The TG and DTG curves indicate the behavior and thermal degradation during the process of pyrolysis or oxidation. From the pyrolysis thermograms (Figure 1) the vaporization process occurs in the case of diesel between 40 and 292 °C, biodiesel between 152 and 393 °C and oil between 320 and 493 °C. The intensity of the mass loss has a peak value at 257 °C (diesel fuel), 305 °C (biodiesel) and 424 °C (walnut oil).

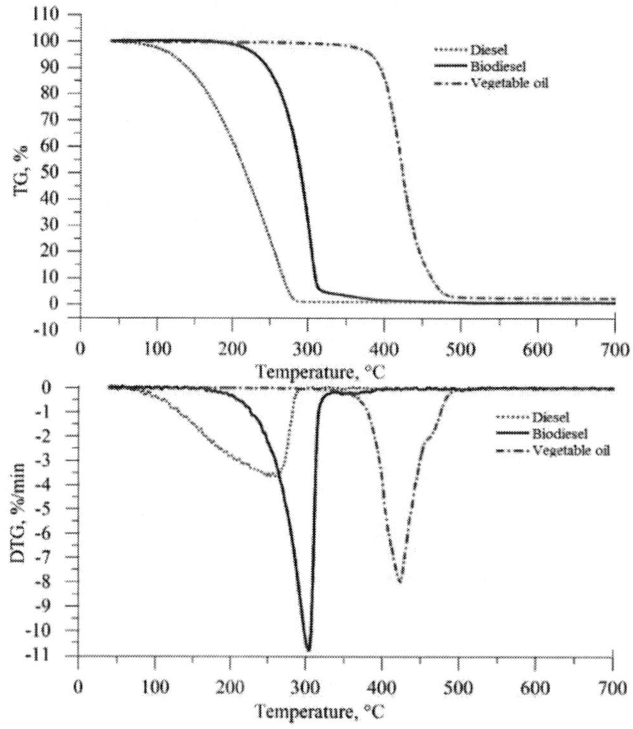

Figure 1. Mass loss and thermal behaviour of fuels during the process of pyrolysis

In the oxidation process the mass exchange processes and heat are more intense with changes in peaks of the DTG curve 252 °C (diesel fuel), 294 °C (biodiesel) and 418 °C (walnut oil) (Figure 2). The combustion temperature for the oil and biodiesel is higher due to the complex structure of methyl esters of fatty acids ((with long chains and oxygen bonds) in comparison with a simple structure of hydrocarbons of petrodiesel fuel. In the case of diesel fuel the combustion reaction starts at a low temperature burning and dissipating the energy to surrounding atmosphere. The walnut oil composed by triglyceride molecules has poor ignition properties since the combustion of fatty acids take place at higher temperatures.

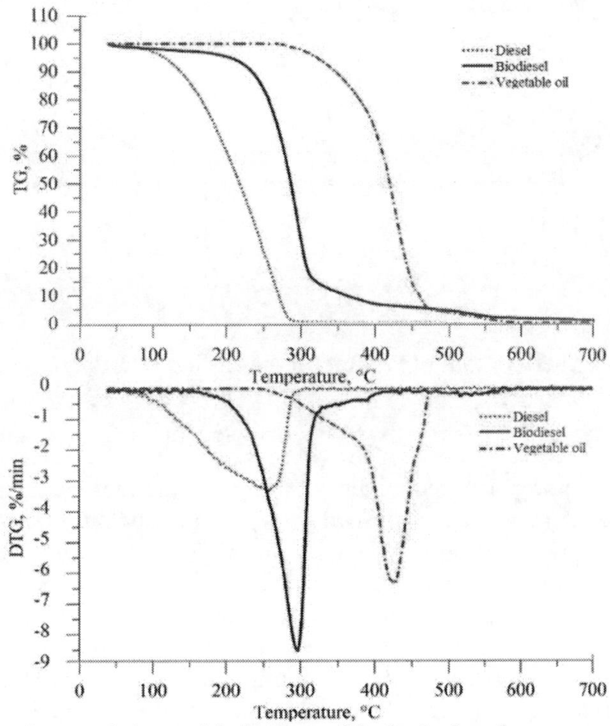

Figure 2. Mass loss and thermal behavior of fuels during the process of combustion

The thermal profile (TG curves) presented in all cases one mass loss step. In the case of diesel fuel the evaporation process take place in a range between 40 and 280 °C. Diesel fuel is thermally stable in nitrogen up to 40 °C. The smoothness of the curve suggests a single step reaction and the same behavior were observed to all heating rates with the exception of a slight displacement of the graph to higher temperatures (Figure 3).

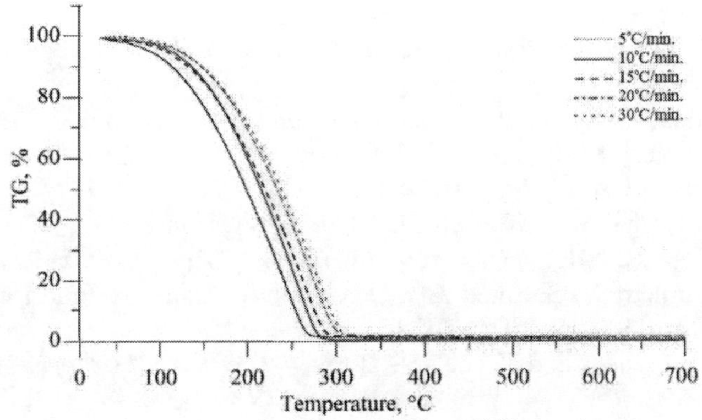

Figure 3. Thermogram of diesel fuel at different heating rates

The walnut oil biodiesel is more thermally stable than diesel starting the vaporization process between 150 and 400 °C. The thermal profile suggests two mass loss steps attributed to volatilization and/ or decomposition of methyl esters. With the increase of heating rate which generate a less uniform heat distribution and a higher temperature gradient, the TG curves shift to higher temperatures (Figure 4).

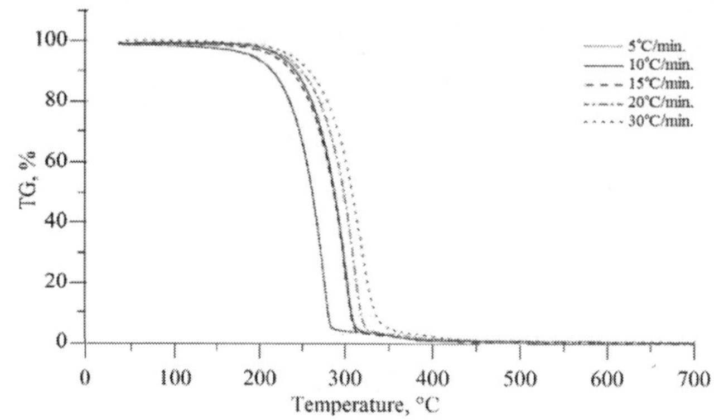

Figure 4. Thermogram of walnut biodiesel fuel at different heating rates

4. Conclusion

Thermogravimetry is an appropriate technique to study the thermal changes during combustion and pyrolysis process. An experiment was performed using a simultaneous TGA-DTG instrument for three samples of diesel fuel, walnut oil and walnut biodiesel. The stability of biodiesel depends on his chemical structure and fatty acid compositions (oils with fatty acids are more stable than the unsaturated). The physicochemical properties of walnut biodiesel are within the EN and ASTM specifications. Biodiesel has a higher viscosity than diesel fuel with problems at low temperature and a low calorific value with an increase in fuel consumption. As the heating rates increases in TG curves was noticed a shift to higher temperatures. The walnut oil was the most thermally stable.

References

[1] A Ganjehkaviri A, MohdJaafar MN, Hosseini SE, Musthafa AB, 2016, *Energies* 9.
[2] Siddharth J and Sharma M., 2012, *Fuel* 93, 252–257.
[3] Sayyed Siraj, Renuka Kale and Swapnil Deshmukh 2017 *American Journal of Mechanical and Industrial Engineering*, 2(1), pp 24-31.
[4] Dwivedi G and Sharma MP, 2016, *Egyptian Journal of Petroleum*, 25 (1), pp 33-38.
[5] Tathilene Bezerra Mota Gomes Arruda, Manoel Barbosa Dantas, Kelvin Costa de Araujo, et al., 2017, *Int J Energy Environ Eng*, 8, pp 135–141.
[6] Wang Y, Ou S, Liu P, Xue F, and Tang S, 2006, *Journal of Molecular Catalysis a: Chemical*, 252, pp 107-112.
[7] Demirbas A, 2009, *Energy Conversion and Management*, 50, pp 923-927.
[8] Dantas M., Conceicao MM, Fernandes VJ, Santos NA, Rosenhaim R, Marques ALB, et al., 2007, *J Therm Anal Calorim*, 87, pp 835–839.
[9] Chien YC, Lu M, Chai M and Boreo FJ, 2009, *Energy Fuels*, 23, pp 202–206.
[10] Kok MV and Topa E, 2015, *J Therm Anal Calorim*, 122(2), pp 955–961.
[11] Santos AGD, Caldeira VPS, Souza LD, Oliveira DS, Araujo AS and Luz GE Jr., 2016, *J Therm Anal Calorim*,123(3), pp 2021–2028.
[12] Rodrıguez RP, Sierens R and Verhelst S, 2009, *J Therm Anal Calorim*, 96(3), pp 897–901.

Experimental research regarding the influence of AdBlue solution temperature on internal combustion engines pollutants

D M S Capătă[1], M F Popa[1], N Burnete[2] and M L Ungureşan[3]

[1]Mechanical Engineering, Technical University of Cluj-Napoca, Romania
[2]Department of Automotive Engineering and Transports, Technical University of Cluj-Napoca, Romania
[3]Department of Physics and Chemistry, Technical University of Cluj-Napoca, Romania

dorin.capata@gmail.com

Abstract. The constraints related to the environment, the vehicle's pollution, as well as the increase in the number of the registered vehicles have to lead to the development and implementation of complex systems for the treatment of exhaust gases. The pursued purpose of the given regulations and environment norms has been the decrease of the main pollutants measured at the vehicle's exhaust pipe, that is: the nitrogen oxides, particles, carbon monoxide. This paper presents the experimental research and the results obtained on the influence of the solution AdBlue on the pollution performances – studies that have been achieved at Volvo Romania SRL. Thus, taking into consideration the evolution of the temperature of the reductant from -40°C to +40 °C, there are underlined the performances insured by SCR and the dependency of the pollutants on these temperatures, fulfilling the usual usage measures. The variations recorded for nitrogen oxides (after the catalyzer for catalytic reduction), the reductant consumption and the efficiency of SCRtransformation at different values of the AdBlue solution are being presented using graphic representations.

1. Introduction

The modern commercial vehicles benefit from complex and performant systems for treating the exhaust gases. Besides the optimization solutions of the ignition process, they also have oxidation catalyzers for the reduction of CO, particles filters for diminishing the PM evacuated in the atmosphere, catalytical reduction catalyzers (SCR) and EGR for diminishing NO_x, as well as catalyzers for the reduction of the ammonia excess that comes from the catalytic reduction reactions [1, 2, 3].The proposed experimental studies focus on the analyses of the efficiency of transforming the SCR catalyzer in the way of reducing NO_x by changing the AdBlue's temperature. The complexity of the paper is to maintain all the elements function within the standard usage limits, to ensure the flow of the exhaust gases constant throughout the tests and at the same value for each of the tests.

2. The experimental installation. Equipment description, and the evolution of the experiments

In order to underline the dependency of the efficiency of the transformation reaction of NO_x in the SCR catalyzer in conditions as real as possible, the tests have been achieved on a Volvo truck, model FH(4),

IOP Conf. Series: Materials Science and Engineering **568** (2019) 012076 doi:10.1088/1757-899X/568/1/012076

produced in 2018, equipped with a 13 liters engines (D13K of 500 HP), figure 1. The truck is Euro6 phase C, being equipped with a recirculation system for exhaust gases (EGR) as well as with a system for further treatment of the exhaust gases that acts mechanically, thermically and chemically on the exhaust pollutant products in the sense of diminishing and neutralizing them.

Figure 1. Preparing equipment of the experimental installation.

The experiments have involved the feeding of the SCR system with an AdBlue solution at different temperature, pursuing the performances of the selective reduction catalyzer and the consumption of reductant for each test. The reductant quantity for each test has been measured with a glass graded cylinder, using 200 ml of AdBlue solution. Two measurements have been achieved for each experiment, the first with the initial quantity of 200 ml and the second, after the test, with the remained quantity, in order to determine the consumption.

The quantity of the reductant has been warmed up and cooled down in order to reach the proposed temperature for the tests. For the conclusive results the starting temperature of the reductant was -10 ^{0}C, testing every 5 ^{0}C up to 50 ^{0}C. The chosen interval has been based on the freezing limit of the reductant, which is -11 ^{0}C, respectively 50 ^{0}C to avoid reaching the temperature where the risk of decomposing the AdBlue solution appears [4, 5].

Fueling the installation has been achieved from a thermos-isolated recipient in order to maintain a constant temperature. For each test there were two measurements, before and after, being accepted deviations of maximum 1^{0}C to the proposed value.

To reduce the alteration of the reductant's temperature, the heating system of the truck (SCR) has been eliminated by disconnecting the electrical resistances of the Ad-blue solution admission pipes and by the short of the heating serpentine within the pumping unit. Even more, water at the reductant's temperature has been circulated though the heating serpentine of the pumping unit, thus having a constant temperature of the reductant. For the negative temperatures of the ad-blue solution carbonic ice tampons have been used on the external surface of the pumping unit.

For determining the efficiency of the SCR transformation at different temperatures a test offered by the diagnosis computer Tech Tool has been used, the conversion test NO_x. The advantages of the conversion test NO_x are given by the fact that NO_x has a relative constant value throughout the injection of the reductant (1000-1300ppm), being able to follow up the way in which the dosage is being done, as well as the quantity of reductant and the time passed from the beginning of the injection up to the threshold of 75% for the transformation's efficiency.

For a correct determination of the quantity of consumed reductant for each test, before doing the next test, the SCR catalyser has been cooled up to the functioning temperature. Thus the reductant's injection

IOP Conf. Series: Materials Science and Engineering **568** (2019) 012076 doi:10.1088/1757-899X/568/1/012076

has been prevented at the start of the engine, before starting the actual transformation NO_x test. Pursuing the thermal state in the exhaustion pipe has been possible with the diagnosis computer TechTool by reading the values measured with the exhaust temperature sensors. The tests have been restarted only when the temperature in the SCR has dropped below 150^{0}C.

The validation of the measurements of reading NO_x at the exit from the catalyzer has been achieved by comparing the information offered by the NO_x sensor and the analyses apparatus for emissions Testo 350, figure 1. Installing the measurement probe has been done by adjusting a visitation window within the evacuation pipe of the truck's exhaust system, the flow of the evacuation gases being on the measurement probe's direction.

3. The analyses and interpretation on the results

The NO_x reduction process and the achievement of the efficiency target of conversion have been done by dosing an initial quantity of reductant, a dosage that has been adjusted according to the quantity of NO_x at the entrance in the catalyzer and the efficiency of transformation, figure 2.

As the quantity of the reductant increases, it can be noticed an acceleration in the transformation's efficiency. The NO_x upstream to the catalyzer is the basic element in choosing the quantity initially dosed, a smaller value of NO_x determining a smaller dosage of reductant and vice versa.

The period of time of the test has been determined by reaching the 75% for the efficiency of the conversion. It has been noticed experimentally that there is a direct link between the quantity of reductant and the period of the test, thus a smaller quantity makes a higher time.

For each of the tests the following have been determined: the reductant's temperature; the reductant's consumption (the difference between the quantity of 200 ml and the quantity measured after the test); the period of time for the test (directly determined from the test's graph); the quantity of treated NO_x (summing up the values given by the NO_x sensor at the entrance of the catalyzer with the time in which it was produced); the maximum limit of reductant for finalizing the transformation. Some NO_x transformation graphs at different temperatures are presented in figures 2 to 4.

If we refere to the maximum quantity of reductant in order to reach the target of transformation for $NO_{x,}$ where we can notice a decrease from 30 ^{0}C. Meanwhile, the time for dosing the maximum quantity of reductant decreases with the increase of its temperature.

The consumption of reductant used to neutralize NO_x decreases with the increase of its temperature. Figure 5, underlines the evolution of the reductant's consumption according to its temperature.

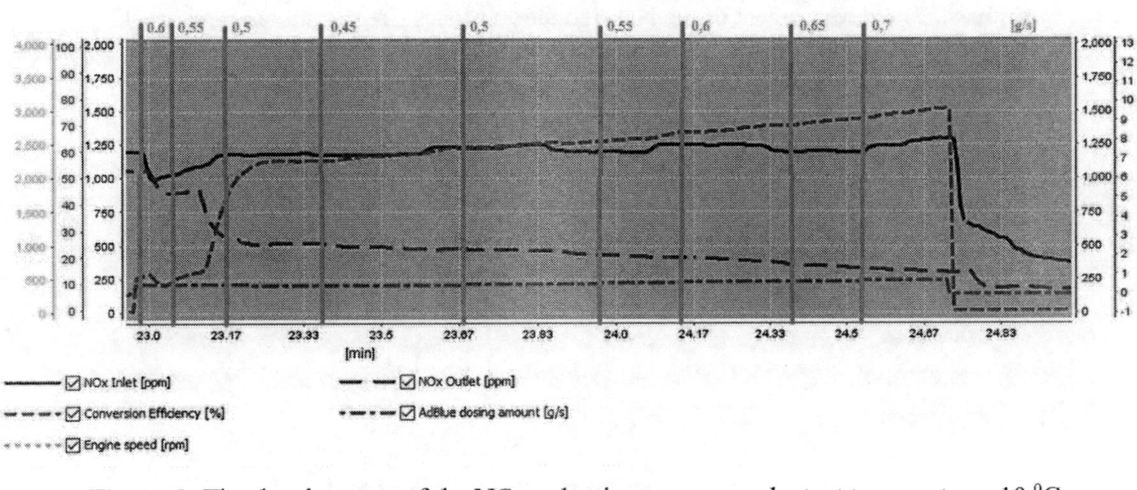

Figure 2. The development of the NO_x reduction process, reductant temperature -10 ^{0}C.

IOP Conf. Series: Materials Science and Engineering **568** (2019) 012076 doi:10.1088/1757-899X/568/1/012076

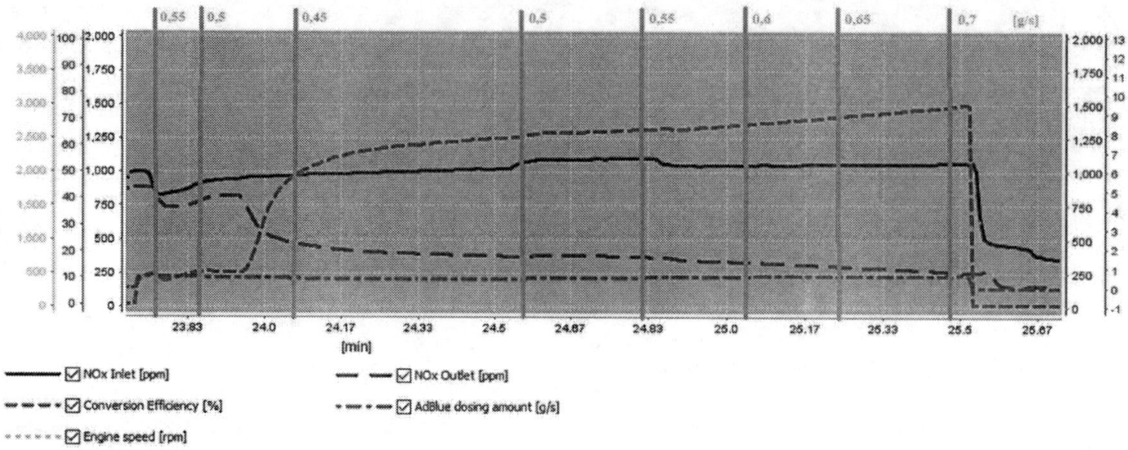

Figure 3. The development of the NO_x reduction process, reductant temperature 25 ^{0}C.

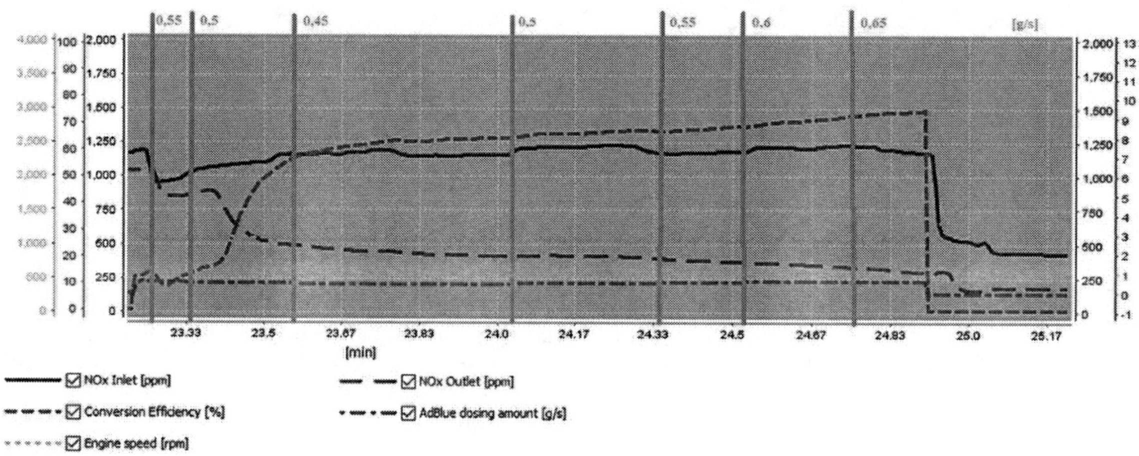

Figure 4. The development of the NO_x reduction process, reductant temperature 45 ^{0}C.

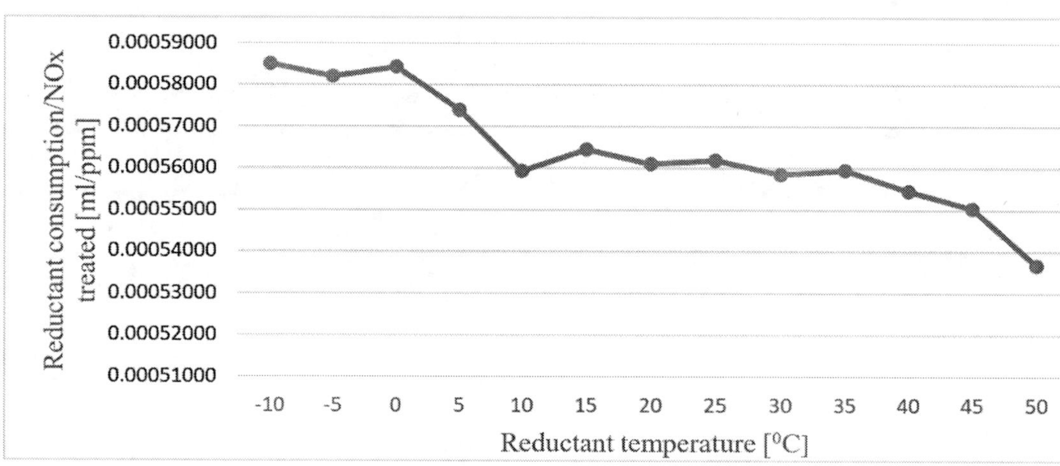

Figure 5. The variation of the reductant's consumption according to its temperature.

IOP Conf. Series: Materials Science and Engineering **568** (2019) 012076 doi:10.1088/1757-899X/568/1/012076

It can be noticed a significant decrease of the reductant's consumption between -10 ^{0}C and 10 ^{0}C, respectively 35 ^{0}C and 50 ^{0}C, intervals between which it maintains at a relatively constant value.

For negative temperatures, below 0 ^{0}C, it can be noticed an increase in density of the AdBlue solution, the appearance of solidification cristals being noticeable from -5 ^{0}C. The use of the reductant at these temepratures can deteriorate the SCR system by clogging the line from the tank through the pumping unit up to the injector.

The AdBlue solution, as it resulted from the laboratory determinations, starts to decompose at temperatures higher than 30 ^{0}C. The danger appears in the usage of the reductant at high temperatures if it decomposes before it is injected into the catalyser, thus shuterring its evacuation. The injector's warming up due to the increase of the temeprature of exhaust gases determines the overheating of the reductant, thus blogking it with decomposed solution

4. Conclusions regarding the experimental studies

The experiments have proved, at least for the SCR catalyser copper-zeolite, the existance of a link between the temperature of the reductant when injected into the SCR catalyser and the transformation performances for NO_x reduction. As underlined in figure 5, two lines of sustained reduction of AdBlue solution can be noticed: between -10 ^{0}C and 10 ^{0}C, respectively 35 ^{0}C and 50 ^{0}C. A level-headed reduction is uderlined outside these intervals.

It has been proven experimentally an increased efficiency with the increase of the reductant's temperature, but this is limited by its decomposition level. Even if the results of the transformation reach maximum limits at high temperatures of the reductant, the system's fiability and the protection of the components of the SCR system have to have the precedence.

When choosing the optimum temperature of the reductant, more studies are required related to the heating temperature of the AdBlue injector, respectively the possibility to cool it when overheated.

The minimum temperature at which the SCR copper-zeolite catalyser is able to facilitate the reaction of selective reduction is 190 ^{0}C [6]. The temperatures measured with the sensors from the truck's exhaustion have underlined temperatures between 230-260 ^{0}C at the entrance, after each test. Another interesting subject whould be to determine the optimum temperature of the SCR catalyser for a maximum efficiency of NO_x transformation.

By analyzing the results obtained for low temperatures of the reductant (-10 ^{0}C and 0 ^{0}C) we might say that this interval is also to be avoided, besided from the low performances related to the transformation of NO_x, the major danger being the appearance of solidification crystals and the increase in the solution's density. The sensitive elements of the pumping unit can be affected by this, the result being the blockage of the catalyzer's feeding with reductant and the alteration of NO_x transformation.

References

[1] Chiriac R.. 2012 *Internal Combustion Engines* (Bucharest: A.G.I.R. Press)

[2] Ivan, F. Liţă, D. and Buşoi, A. 2014 *Methods and resources for depoluttion of vehicles engines (Metode şi mijloace de depoluare a motoarelor pentru automobile)* (Bucharest: MatrixRom)

[3] Manea, Adriana T., Manea, L.C. and Sandu, V. 2004 *Thermal engines, pollution processes (Motoare termice, procese poluare)* (Bucharest: MatrixRom)

[4] BASF 2003 *Urea solution 32.5% AdBlue* (Technical Leaflet, BASF AG, Ludwigshafen, Germany)

[5] Fendri A., Kanoun O. and Buchholz A. 2015 *Temperature effect on the complex conductivity of Adblue*, (Mahdia: IEEE 12th International Multi-Conference on Systems, Signals & Devices (SSD15)) pp 1-4

[6] *** Volvo Trucks mechanical courses *The exhaust gas after-treatment systems, Euro 6* accesed April 2019

IOP Conf. Series: Materials Science and Engineering **568** (2019) 012077 doi:10.1088/1757-899X/568/1/012077

Dynamic Analysis of a Human Ankle Joint Prosthesis

C Copilusi[1], N Dumitru[1], A Margine[1] and I Geonea[1]

[1]Faculty of Mechanics, University of Craiova, Craiova, Romania.

cristache03@yahoo.co.uk

Abstract. This research addresses a human ankle joint prosthesis design based on a cam mechanism. Thus, a dynamic analysis of the entire human lower limb was performed in order to obtain the input data for numerical simulations of an ankle joint prosthesis. For this, experimental analysis was accomplished in order to obtain input and reference data for human ankle joint motion laws. These human ankle joint motion laws were considered from a healthy human subject and an amputee. A virtual model of the ankle joint prosthesis was designed with the aid of CATIA software and the working principle was based on a cam mechanism. This virtual model was simulated in dynamic conditions, similar to the imposed ones onto a mathematical model. Results represented through stress and deformations will validate the proposed ankle prosthesis prototype. Finally the proposed prosthesis prototype was manufactured and experimentally tested on the proposed amputee through a comparative analysis for validation.

1. Introduction

In the field of human lower limb prosthesis design it is well known the fact that prosthesis are in continuous growth and these should be resized or redesigned very often. Thus it can be remarked prosthetic systems with applications on human walking recovery especially designed on simple criteria like the ones from [1]. These can be purchased at low prices due to their simple functions and also it cannot contribute to a human natural motion achievement. For this inconvenient there were developed complex prostheses based on mechatronic systems like the ones from [2, 3, 4]. For these ones, in background, can be remarked complex analyses like kinematics and dynamics, by analyzing the whole human lower limb for certain activities like walking or running. Thus, by having in sight the state-of-the art from similar researches like [1, 5] it can be remarked the starting research point, which was to obtain input data for dynamic analysis of the proposed human subject. In this case, the input data are represented by the ankle joint motion law of a person with the same anthropometric data like the amputee, which is presented in second part of the paper. With the obtained data, a human lower limb dynamic model was elaborated and the obtained theoretical results are represented by ankle connection forces. The used method with the obtained results, which is applicable on a parameterized model, is presented in third part of this research. Theoretical results are important for virtual simulations for the proposed ankle prosthesis concept as input data. Thus, by using MSC Nastran, the proposed ankle prosthesis functionality was verified and approved by developing numerical simulations on dynamic modes similar with the real ones. These are described in the fourth section of this research and important results are synthesized on final conclusions.

2. Human lower limb experimental analysis

An experimental analysis was performed by taking into account a human subject with similar anthropometric data with the amputee (age, weight, height, gender). The amputee has the right lower limb amputated from above the ankle joint area. The analysis aim was to obtain specific human lower limb angular variations during walking for a single gait in case of hip, knee and ankle joints. These results represent the input data for a whole human lower limb dynamic analysis. On the other hand, a comparative analysis was performed in case of one analyzed amputee with a usual ankle joint prosthesis. This analysis was performed in specific clinical conditions by using high speed camera motion analysis equipment called CONTEMPLAS [6]. The workflow is schematized in figure 1.

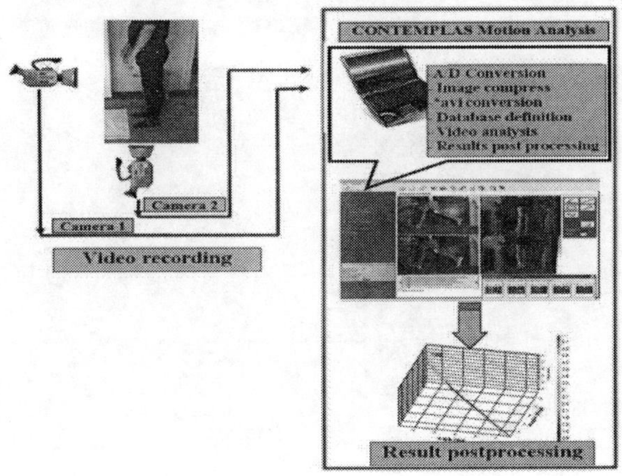

Figure 1. CONTEMPLAS Motion Analysis workflow [6].

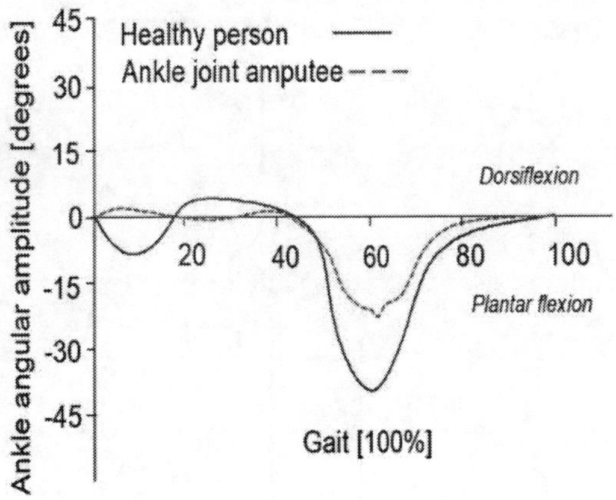

Figure 2. Knee angular amplitude for a complete gait of the analyzed human subjects.

The analyzed human subjects perform several walking steps in the defined workspace and from these steps were extracted only a single gait. It is important to know that the time period is different from an analyzed human subject to the proposed amputee. Thus, in case of a human subject, a complete gait was performed in 2.68 (seconds), and in case of the considered amputee, this was done in 3.42 (seconds). These time intervals are important to retain for dynamic computations and also virtual simulations.

The obtained results are materialized through graphs and a comparative graph is represented in figure 2, were it can be remarked the difference between a healthy subject angular variation and the chosen amputee. These results are reported for a 100 percent full gait in both cases. The obtained results were validated by considering similar researches reported in [7, 8].

From figure 2 it can be remarked that in case of an amputee ankle angular amplitude has some distortion zones on the resulted curve. A maximum obtained value was equal with 2.962 (degrees) for dorsiflexion and a minimum of 21.423 (degrees) for plantar flexion. In case of the analyzed healthy human subject, the obtained curve is smooth and the maximum values were around 9.674 degrees for dorsiflexion and a minimum one of -34.496 (degrees) for plantar flexion. In this case total ankle angular amplitude will be around 45 (degrees). Thus it can be also observed that curvature path is almost the same in both cases.

3. Human lower limb experimental analysis

The proposed mathematical model was elaborated by having in sight similar researches reported in [8, 9]. We consider the foot ground contact as it shows in figure 3. The kinematic constrain equations are:

$$\phi(\mathrm{q}, \mathrm{t}) = 0 \tag{1}$$

Where: q- generalized coordinates vector; t- time. By differentiating equation (1) depending on time and it will be obtained:

$$J_q \cdot \dot{q} + \frac{\partial \Phi}{\partial t} = 0 \tag{2}$$

IOP Conf. Series: Materials Science and Engineering **568** (2019) 012077 doi:10.1088/1757-899X/568/1/012077

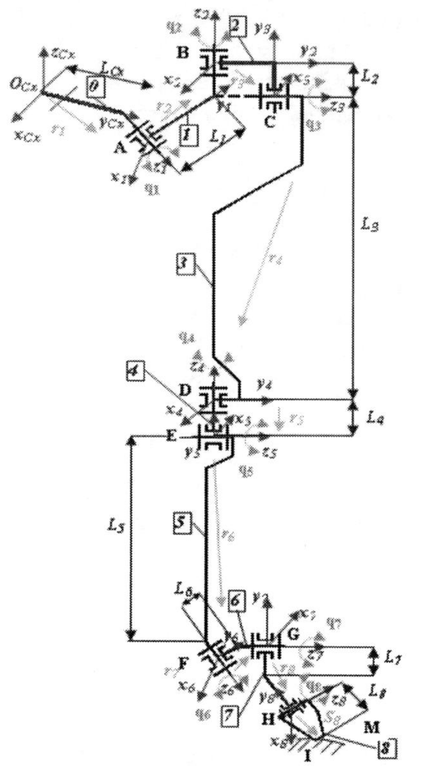

Figure 3. The human lower limb kinematic scheme.

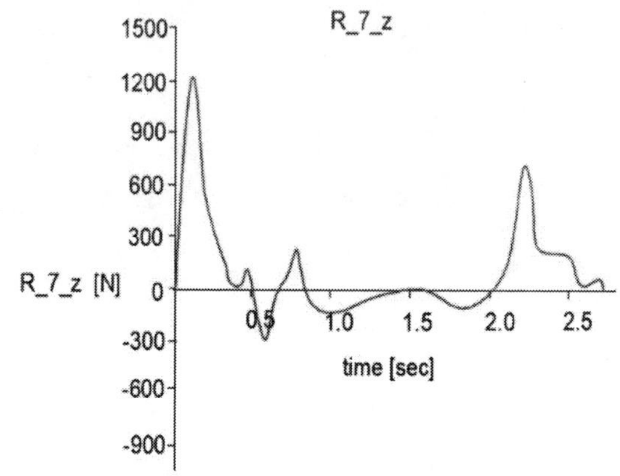

Figure 4. The variation of the connection force (Newton) for the ankle joint on z direction.

The motion equation will have the following form:

$$\begin{bmatrix} M & J_q^T \\ J_q & 0 \end{bmatrix} \cdot \begin{bmatrix} \ddot{q} \\ \lambda \end{bmatrix} = \begin{bmatrix} Q_a \\ a \end{bmatrix} \tag{3}$$

The mathematical equations, which can help for computing these connection forces, by considering the Lagrange multipliers, are:

$$\begin{cases} F_i^{nr(i,j)} = \left[R_{i,j"}\right]^T \cdot \left[A_{0i}\right]^T \cdot \left[\lambda\right]^{r(i,j)} \\ T_i^{nr(i,j)} = \left\{S_i^{iM}\right\}^T \cdot \left[P_{0i}\right]^T \cdot \left[I\right] - \left\{S_i^{iM}\right\}^T \cdot \left[P_{0i}\right]^T \cdot \left[\lambda\right]^{r(i,j)} \end{cases} \tag{4}$$

Motion laws for ankle joint will be considered as known, namely $q(t)$, $\dot{q}(t)$ and $\ddot{q}(t)$ from the experimental analysis, performed in dynamic modes. From equation (4) it can be determined Lagrange multipliers λ, with the aid of an algorithm performed in MAPLE program. After a numerical processing under MAPLE environment it will be obtained the connection force component variation for the ankle joint. This is shown in figure 4 and it will be further used on virtual simulations of the proposed ankle joint prosthesis as input data for motion variation during one gait. From figure 4 it can be remarked that the connection force reach a maximum value of 1223Newton when the gait was started. During one gait when the foot has lost the ground contact the force connection varies between 0 to -50Newton and it happens between time periods of 1 to 1.95seconds.

4. Ankle prosthesis numerical simulations

For designing an ankle joint prosthesis, this has to accomplish the following conditions: achieving an angular amplitude during gait almost identical with the one which was obtained on healthy human subject (45°). A suitable mechanism which can be implemented on the prosthesis structure was a cam mechanism. Thus, the cam component was fixed to the artificial foot, and the cam follower was fixed with the shock absorber's rod as it can be seen in figure 5. This was elaborated with the CATIA V5 R16 and the components are identified as follow: tibia component (1), which contain the shock absorber's body; shock absorber's support (2); shock absorber's rod (3); role of the cam follower (4) which slide over the cam profile; bolt (5) which glides over a imposed profile described by the motion law from a human subject without locomotion disabilities; foot (6). Also from figure 5 it can be remarked a cross sectional image for a better viewing of the cam mechanism.

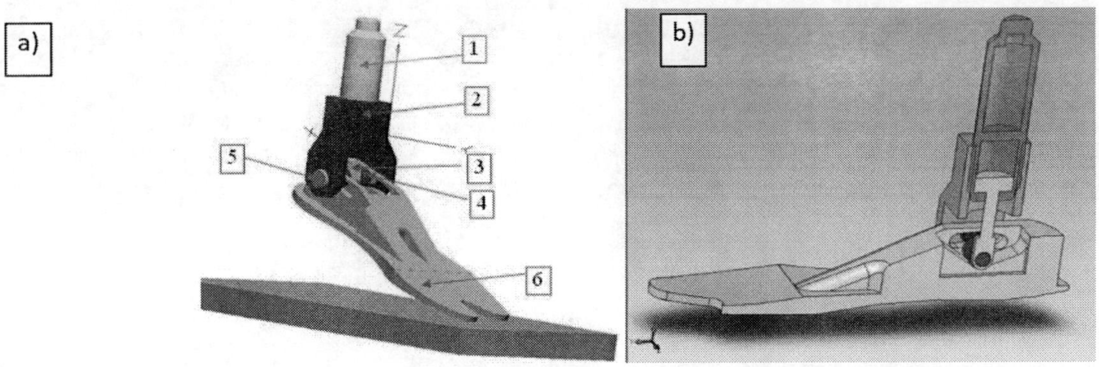

Figure 5. Ankle prosthesis virtual model: a) front view; b) section view.

The entire virtual model was imported under MSC Nastran software for virtual simulations in dynamic conditions. Material properties were defined such as rubber for the foot, aluminum alloys and steel characterized by density elasticity modulus and Poison ratio. Also friction was taken into account during virtual simulations. For meshing specific operations, there were individually meshed each component by choosing tetrahedral finite elements. Thus, at the level of shock absorber head it was applied the connection force variation mentioned in figure 4 in a polynomial form and the ankle prosthesis foot was considered as fixed one. According with the performed dynamic analysis, the computed sequence for virtual simulations was done for a time period of 2.68 (seconds) which corresponds for a complete gait in case of the analyzed healthy person. The obtained results after simulations can be seen in figure 6 and figure 7.

By analyzing the reported results from figure 6 and figure 7, the most significant ones are characterized by von Misses stress and displacements of the analyzed ankle prosthesis prototype. Thus, in Figure 7 it can be remarked a maximum value of von Misses stress equal with 57.8 (MPa), and this it was recorded in the area of cam mechanism. In particular, it can be remarked a smooth curve with large deflections during time period from 0 to 1 (second), which corresponds when the foot has lost ground contact and the same phenomena happens between time periods from 2 to 2.68 (seconds). Between time period of 1 to 2 (seconds) it can be remarked a smooth curve path and also a stress average of 15.378 (MPa), which is quite small.

IOP Conf. Series: Materials Science and Engineering **568** (2019) 012077 doi:10.1088/1757-899X/568/1/012077

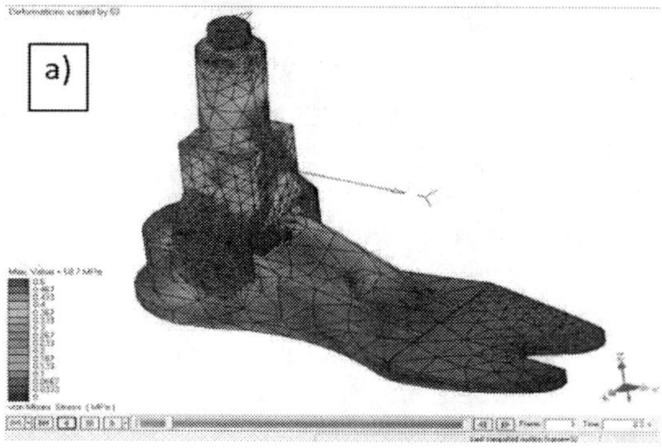

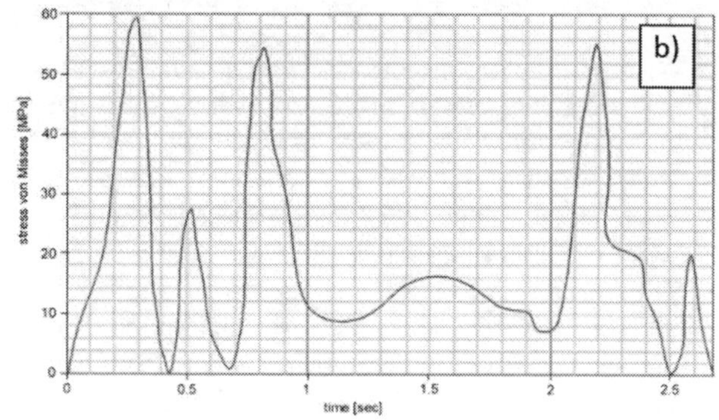

Figure 6. Ankle prosthesis Von Misses stress (MPa) vs. time (sec): a) stress map; b) graphic view.

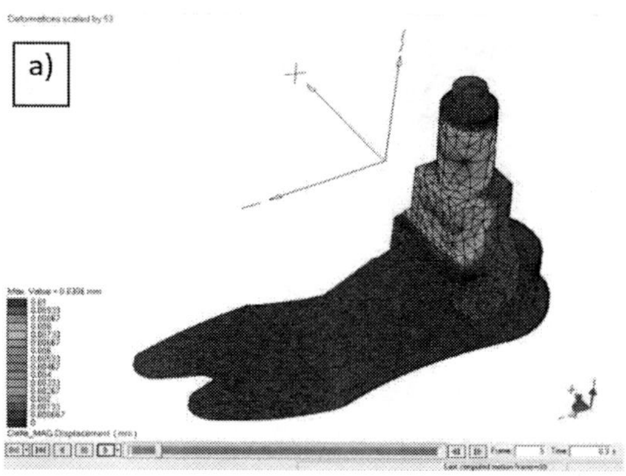

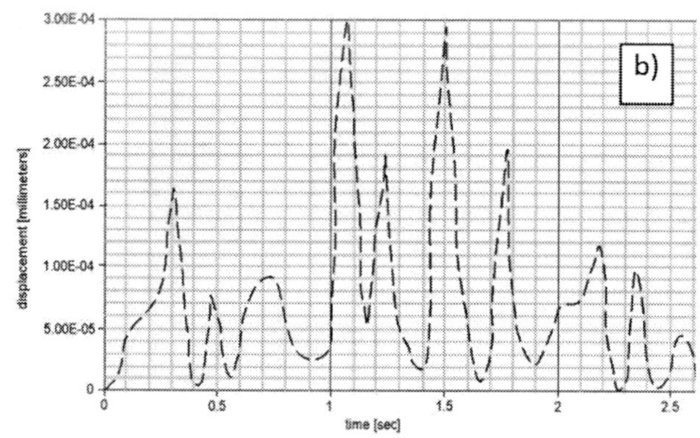

Figure 7. Ankle prosthesis Von Mises stress (MPa) vs. time (sec): a) displacement map; b) graphic view.

On the other hand, by considering the analyzed displacements it can be remarked a maximum value of 0.0306 (millimeters) which was obtained when the foot was left the ground contact and this value was obtained between time periods from 1 to 2 (seconds) according with the reported diagram from figure 7. In the other time periods like the one from 0 to 1 (second), the obtained displacements were not so high due to ankle prosthesis stiffness.

5. Ankle prosthesis prototype experimental analysis

By considering the design principles applied on the elaborated concept, a prototype was manufactured and it is shown in figure 8. This was created by modifying an old ankle prosthesis and the proposed shock absorber was acquired by FESTO Pneumatics and it is characterized by a maximum payload of 3500 (Newton). The elaborated prototype has 1.543 (kilograms) and this was resulted from FESTO shock absorber. According with the virtual simulations most of the ankle prosthesis components were manufactured from aluminum alloys. In order to validate the proposed ankle joint prosthesis prototype there were performed preliminary experimental tests with proper prosthesis adjustments and the amputee was let to wear this prototype for 10 days in order to accommodate with this. After 10 days, an experimental analysis was performed according with the presented protocol presented in previous section. This experimental analysis was done in his daily living activities accomplished by walking. Thus, a comparative experimental analysis was performed by having in sight the results reported in previous section, and this is shown in figure 9. By considering the obtained results reported in diagram from figure 9 it can be remarked the dotted curve which corresponds with the experimental tests of the amputee with ankle joint prosthesis prototype.

IOP Conf. Series: Materials Science and Engineering **568** (2019) 012077 doi:10.1088/1757-899X/568/1/012077

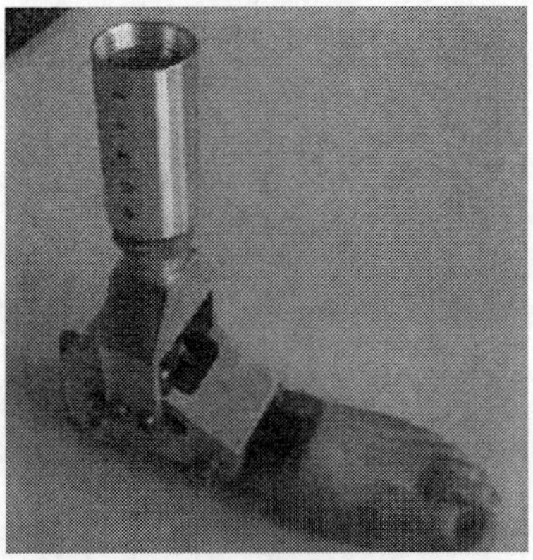

Figure 8. The new prosthesis used in human ankle disarticulations.

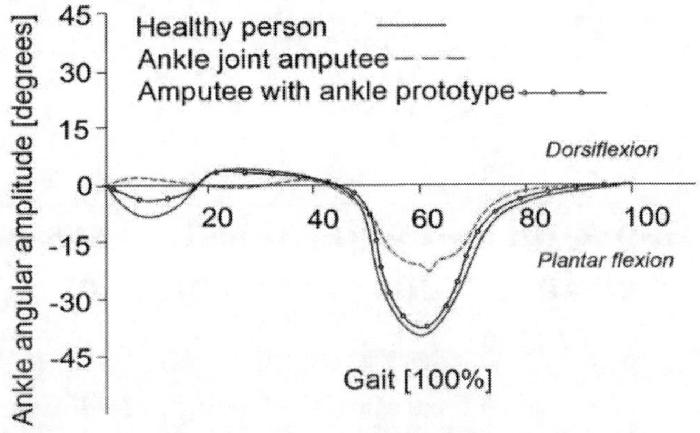

Figure 9. A comparative analysis after preliminary tests of the ankle prosthesis motion (deg.) vs. gait cycle (%).

It looks that this is very appropriate as pattern but also as values with the one of a healthy person during tests. Also at around 60% from a complete gait it can be seen that the obtained path is smooth. Thus, the maximum values were around 8.556 degrees for dorsiflexion and a minimum one of -32.223 (degrees) for plantar flexion. The maximum value in case of ankle angular amplitude during walking was around 40.7793 (degrees) and this curve was obtained in case of a complete gait in 3.16 (seconds).

6. Conclusions

A prototype of an ankle prosthesis was designed and elaborated. This will be used by persons with human lower limb amputation from above of ankle joints. Simulation results and numerical processing have outlined a good performance for suitable user-oriented walking specific operation and its design may require additional components in future developments. The concept is fairly simple wearable and lightweight and the novelty element is represented by the cam mechanism combined with a shock absorber. The obtained ankle prosthesis mechanism can be adapted for right or left human leg. This prosthesis can be personalized due to its parameterized form. In the future the mechanism principle will be extended for a human knee joint and also it will try to adapt this principle to children orthosis for knee and ankle joints.

References

[1] Copilusi C 2009 Research regarding applicable mechanical systems in medicine. *PhD Thesis*

[2] Sameer S 2004 Intelligent Robotics for Rehabilitation Robotic Lower Limb for Above Knee Prosthesis. *Grant type (83/ECE/2000) Department of Electronics and Communication Engineering* India

[3] Moreno J Meijer K Savelberg H and Pons J 2004 Characterization of an actuator system for a controllable knee ankle foot orthosis. *Proceedings of 8th International Conference ACTUATOR*, Bremen

[4] Hardt M Kreutz-Delgado K Helton J and Stryk O V 1999 Obtaining minimum energy biped walking gaits with symbolic models and optimal numerical control. *Workshop - Biomechanics Meets Robotics, Modelling and Simulation of Motion* Heidelberg, Germany pp. 1–19

[5] Wilson A B 1994 Limb Prosthetics. *Artificial Limbs J.* **14(1)** Spring Verlag. pp. 1-52

[6] Contemplas *User Manual*. Accessed in May 2014

[7] Williams M 1996 Biomechanics of human motion. *W.B. Saunders Co.* Philadelphia and London

[8] McGeer T 1990 Passive dynamic walking. *Int J of Robotics Research*, vol. **9**, no. **2**, pp. 62-82.

[9] Dumitru N Nanu Ghe Vintilă D 2008 Mechanisms and Mechanical Transmissions (Mecanisme şi transmisii mecanice, in Romanian) *Bucharest EDP Publishing House*

IOP Conf. Series: Materials Science and Engineering **568** (2019) 012078 doi:10.1088/1757-899X/568/1/012078

Experimental data collection system for reading pressure levels in a vacuum environment

D I Țarcă[1], L Csokmai[1], O Moldovan[1], R Ghincu[1], D Potroviță[1]

[1] University of Oradea, Faculty of Management and Technological Engineering, Oradea, Romania

dan@tarca.ro

Abstract. The University of Oradea is working on developing a research base in material sciences with capabilities of vacuum deposition. The laboratory will be equipped with two deposition system along with with other research equipment, mainly an atomic force microscope and a Nano indenter for material characterization. This paper presents the continuity of our efforts to install two vacuum deposition systems that will be part of the SMARTMAT Laboratory at the University of Oradea. As steps were made in cleaning the equipment, we now focus on obtaining high vacuum. In this process, a constant monitoring solution of the pressure inside the chamber is required, and the steps taken to realize it are presented in this paper. The main focus of the paper is the data acquisition system implemented for monitoring several parameters of which vacuum is the most important. The system will be based on two Pfeiffer Gauges for vacuum and a National Instruments PXI system.

1. The work so far

By now, we have successfully cleaned [1] two deposition systems (sputtering and evaporation). Tests were made by our team regarding the electrical/electronic parts [2]. Both pumps (primary plus turbomolecular and the other one that works by diffusion) were serviced externally.

As we are waiting for different parts to arrive, and the lack of research funding put us on a halt with the turbomolecular pump (the bearing needs replacement, and the pump needs service that can only be made by specialized personnel with the right equipment), we are not wasting time and decided to test the tightness of both vacuum chambers [3]. We primarily developed the system with the evaporation system [4], as the chamber is smaller, has fewer gaskets, and the vacuum system is 100% working. We also ran a few tests on the sputtering system (the one with the two pumps – primary and turbomolecular) but as the turbomolecular pump was making a whirring sound, we thought that the bearing is not in good condition, so we decided that the risk to ruin it entirely was not worth it to use it before we could replace the part.

2. Reading system architecture

The system was made with readily available equipment in the lab. For the hardware equipment, we used a PXI computer from National Instruments, a small router, and cables with standard Ethernet RJ45 connectors. On top of that, we already have the equipment up to the translator (two sensors connected to it). Surely any PC would have been great for this experiment, but as we would like to further automate the process and install new sensors on the vacuum systems, the PXI is an ideal solution as it has the

IOP Conf. Series: Materials Science and Engineering **568** (2019) 012078 doi:10.1088/1757-899X/568/1/012078

platform needed for the job. Our system is similar to the one presented in figure 1 and needs some additional interface panels

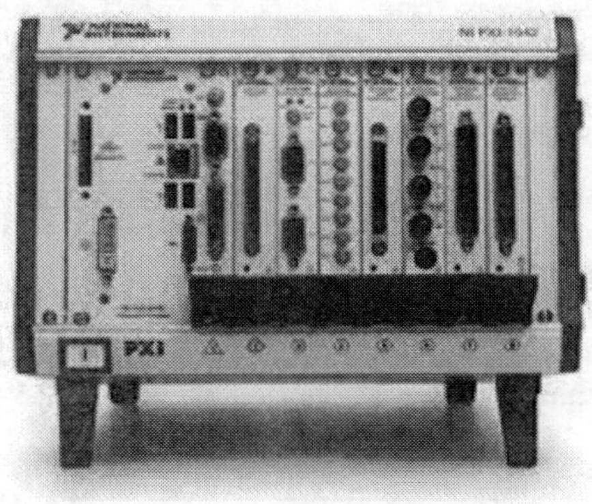

Figure 1. A PXI system same as our own, model number 1042, but with more interfaces

Figure 2. Pfeiffer Gauges translating sensor input into visual pressure output in mBar.

For the software part, all of the architecture was made inside the lab, and tuned specifically for the Pfeiffer DualGauge presented in figure 2. Unfortunately, the SingleGauge does not have an Ethernet port, so extra programming and connections are required to use it for history recording.

2.1. Hardware setup description

As mentioned, the physical setup consisted of the PXI, a router, cable, the Pfeiffer DoubleGauge, two sensors, and the actual system that was pumping out the air to make a vacuum.

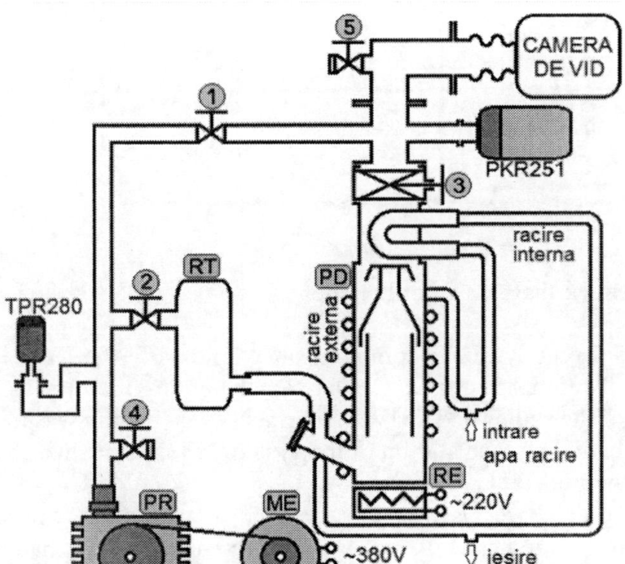

Figure 3. Vacuum system with the depiction of the position of sensors
ME – Electric Motor
PR – Rotating Pump
PD – Diffusion Pump
RT - Reservoir
RE – Electric Heater
TPR280 – pressure sensor
PKR251 – pressure sensor
Intrare apa racier – cold water input
Iesire apa racier – hot water output
Camera de vid – vacuum chamber
Racire externa – external cooling
1, 2, 3, 4, 5 - valves

IOP Conf. Series: Materials Science and Engineering **568** (2019) 012078 doi:10.1088/1757-899X/568/1/012078

The vacuum system of the evaporation deposition system, along with the two sensors mounted on it, is presented in figure 3. On channel one, we have the TPR280 sensor, and on channel two, we have the PKR251 sensor in our current configuration. The connection between the sensors and the Pfeiffer DualGauge is made with a special connector and cannot be directly connected to the PXI, as we need the translation for the signal coming out of the sensor. The gauge automatically recognizes the coupled sensor and reads its signal, outputting a value in mBar, as seen in figure 2. The -2 or E-3 represents the power of 10 with which you need to multiply the shown value indicated on display.

So, the DualGauge shows a value of 0.0326 mbar, and the SingleGauge shows a value of 0.0091 mBar. In this case, the differences observed are because one of the sensors is connected to the primary pump and the other one to the turbomolecular one. Usually there is a reading error between 0.3 to 0.5 display units on the DualGauge between the two sensors if they are connected to the same pump.

2.2. Software setup description

The software has 3 parts: the connection protocol that needed to be configured, the collection and storage of data in a database and the display module, as seen in figure 4.

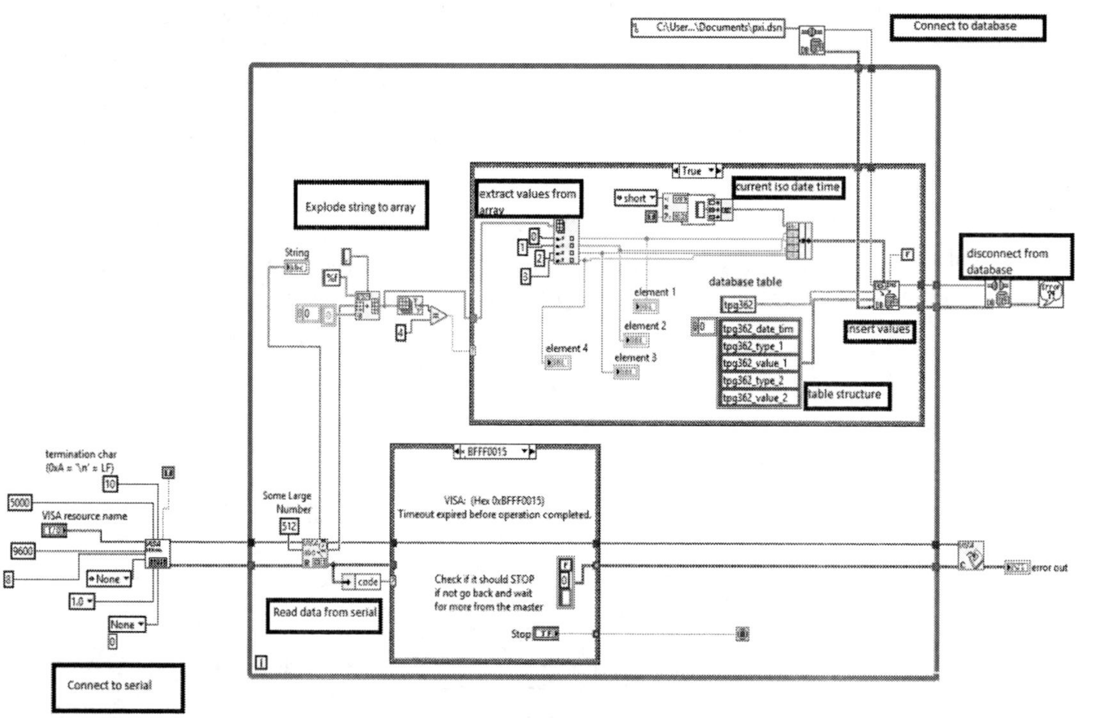

Figure 4. Block diagram from LabView

The connection protocol was configured for the system to communicate with the Pfeiffer DualGauge, with the help of proprietary software from National Instruments. Data for the configuration was taken from the technical book of the Gauge. All of the setups had a visual interface, so no coding was needed. It's still a manual connection for the time being, as each startup of the system requires manual input of the parameters to successfully connect to the Pfeiffer Gauge.

The collection and storage of data in a database was the most complicated part, as it needed a connection with a running database (made with phpMyAdmin) on the computer. This had to be connected to the program that processes the incoming data from the Pfeiffer and stores it for further use. Also, there was necessary preparation of the data to register it in a database, as the values came as a string and needed to be stored as an array. Date and time variables were introduced, which were read from the actual values of the OS. All the computing was made in LabView, which controlled all the processes – collection, reading, processing, adding time and date, insert value to DB

The last step was to read the data from the database and put it into a chart for real-time viewing. This was especially useful as we could see a history of about 1 minute on the graph and draw a fast conclusion if our efforts to seal the system are right or not. A little program was made in python, which collected data from the database and plotted it to a chart. The tricky part was to correctly adjust the viewing scale, as the data plotted was anywhere between 10^3 to 10^{-6}.

The export function of the database was beneficial, as we could export our data into an excel file, and then import it into MATLAB, comparing different setups.

3. Results and interpretation

Once the system was in place, the results were easy to extract and interpret, as seen in figure 5. The data collected gave us an idea of the approximate location of one leakage, and we managed to seal it by adding rubber seals and special vacuum Vaseline. Also, we could compare how fast it could achieve a high vacuum, as we tested in different experimental conditions (temperature and volumes).

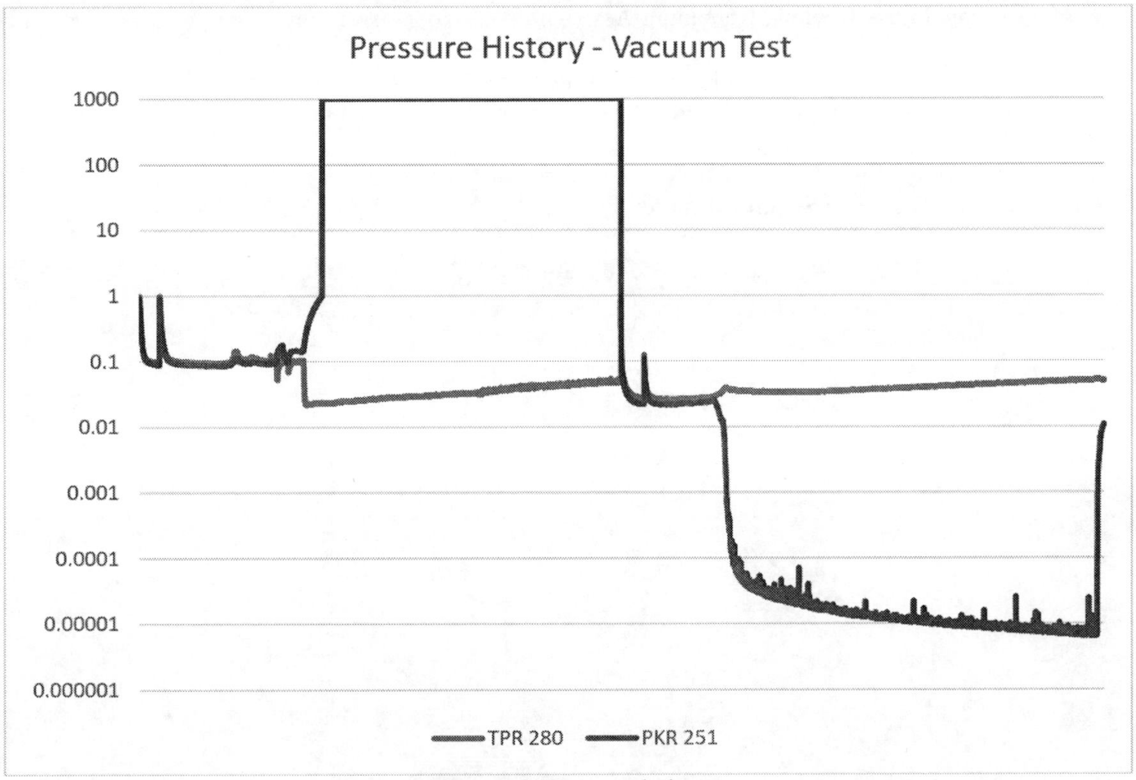

Figure 5. Exported chart

As valve 1, from figure 1 closes, and valve 3 opens, the PKR 251 sensor will measure vacuum generated by the diffusion pump. After sealing all leakages, we obtained a vacuum of E-6, which is recommended for the evaporation deposition system [5],[6].

Further tests will be made before starting the deposition process [7], as the fluctuations in pressure seen on the graphics need to be completely eliminated. By then, we estimate that a temperature sensor will be integrated into the vacuum chamber, electrical heater, and on the pump, to fully observe the temperature impact on the vacuum. The upward trend of the blue line is surely determined by the heating of the oil in the rotating pump, and the vacuum generated by the diffusion can also be improved by

IOP Conf. Series: Materials Science and Engineering **568** (2019) 012078 doi:10.1088/1757-899X/568/1/012078

tweaking the temperature with the help of the external cooling. Once we can gather this data, the next step is to automate the process, to obtain the best vacuum and maintain it relatively at a stable value.

4. Conclusion

This experimental setup helped us understand where leakages appeared in the system but also found out that the pumps (rotating and diffusion pump) have their limitations, and a temperature management system is critical in obtaining and maintaining a high and stable vacuum.

Our next steps regarding the experimental setup are to make it easier for any user to start it and export the registered values, to install some temperature sensors and find the right temperatures for obtaining a stable and high vacuum, and ultimately build a sustainable system, that can maintain a pre-set vacuum value through the help of all the installed sensors and a few actuators.

References

[1] Sasaki Y. T., A survey of vacuum material cleaning procedures: A subcommittee report of the American Vacuum Society Recommended Practices Committee, J. Vac. Sci. Technol. A, Vol 9, No. 3, May/Jun 1991

[2] Moga I., Ghincu R., Costea T., Ţarcă D. and Moldovan O. A., Some Experiments Regarding Magnetron Sputtering Deposition With SMALL CAPACITY DEVICES, Nonconventional Technologies Review, Romanian Association of Nonconventional Technologies, December, 2018

[3] Stuart R. V., Vacuum Technology, Thin Films, and Sputtering, Academic Press, 1993

[4] Sree H. K. S., Principles of Vapor Deposition of Thin Films, First Edition, 2006 Elsevier Ltd., ISBN 978-0-08-044699-8.

[5] Reid R, Ultra High Vacuum Guide, Daresbury Laboratory-issue-3, 1996

[6] Khaw J., "Technical Specification for Vacuum Systems.", SLAC, SLAC-TN-86-6, 1987.

[7] Krishna S., Handbook of Thin-Film Deposition Processes and Techniques, Principles, Methods, Equipment and Applications, Second Edition 2002, Noyes Publications, ISBN: 0-8155-1442-5

IOP Conf. Series: Materials Science and Engineering **568** (2019) 012079 doi:10.1088/1757-899X/568/1/012079

A neural networks approach of process fault diagnosis using time series collected data through oil condition monitoring

G Grebenişan[1,3], N Salem[2], S Bogdan[1], D C Negrău[1]

[1]University of Oradea, Romania
[2]Zarqa University, Jordan

[3]grebe@uoradea.ro

Abstract. In this paper was used the data set collected in a research project between private companies from Romania and Italy, for the development of a basic approach of artificial neural network techniques, as an application in Matlab, aiming to detect the degree of degradation of oil, an automated installation, measuring online the physicochemical properties of the oil. Physical-chemical parameters measured lead to the creation of generous time series, but accessible by numerical and statistical calculation, for the application of artificial intelligence techniques. Applying neural network techniques to parameters that measure oil degradation, oxidation and humidity have generated the results of this work. The main function of monitoring the state of operation of a mechanical system, machine, or plant is to provide the almost correct diagnosis of the machine's state and rate of change so that preventive measures can be taken at a given time .

1. Introduction

Condition monitoring and machine status classification are actions of great practical importance in the manufacturing industry as it provides online up-to-date state-of-the-art information, thus avoiding losses in production processes and minimizing chances of catastrophic machine failure [1].

Monitoring status, followed by the status classification, is an essential step in detecting failures in supervised machines, which is a model recognition problem as an approach. Currently, there are many techniques in this field, and the purpose of this paper is to investigate the various commonly used recognition techniques and to evaluate the results in correlation with the technical solutions given by the dynamic models and the transfer functions, respectively.

As has been mentioned here, [2] the literature abounds in surveys in which a wide variety of different neural network architectures [3] for the detection and diagnosis of malfunctions have been studied and experimented, [4], [5], [6]. The techniques for detecting and diagnosing malfunctions are typically divided into two categories: estimation methods and pattern recognition methods, respectively. Estimation methods are generally dependent on mathematical models, which must be as realistic as possible and not too complicated, with the cost of computing being excessive for complicated systems.

2. Some notions about simple Neural Network architectures

Error diagnosis literature, in the automation industry modern practice, focuses on linear or accurate methodologies, or with a high precision approximation rate. It is known that industrial processes are often difficult to model. These are complex and not always known precisely: induced noise and measurement uncertainty or manufacturing, error-prone sensors, corrupted measurements, vibrations,

IOP Conf. Series: Materials Science and Engineering **568** (2019) 012079 doi:10.1088/1757-899X/568/1/012079

damage to the state of lubricating oils, etc. Therefore, some researchers have perceived artificial neural networks as an alternative to representing knowledge about dysfunctions but also to find ways to solve technical problems with infinitesimal errors that are sometimes also influenced by an incorrect or overly relaxed approach, design errors or exploitation errors.

Neural networks can filter noise and disturbance; they can provide a stable diagnosis, failures without standard types of models, extremely sensitive, and economic efficiency due to insignificant computing and design efforts. Another feature accepted as defining neural networks is that exact patterns are not always needed to reach the decision stage.

Using a neural network with appropriate parameters and a logical structure and weights, any non-linear continuous relationship can be approximated with arbitrary accuracy, a significant feature that a relatively complicated but controllable algorithm as a mathematical process can be parametrically defined by MATLAB specific functions or applications, or dedicated software. Neural networks are composed of simple elements that work in parallel. Biological nerve systems inspire these elements. As in nature, the efficiency of a neural network is primarily determined by the connections between the elements. We can train a neural network to perform a specific function by weighing the weights.

Typically, conventional neural networks are adjusted or trained so that a special entry leads to specific target output. Such a situation is presented in Figure 1, [4]:

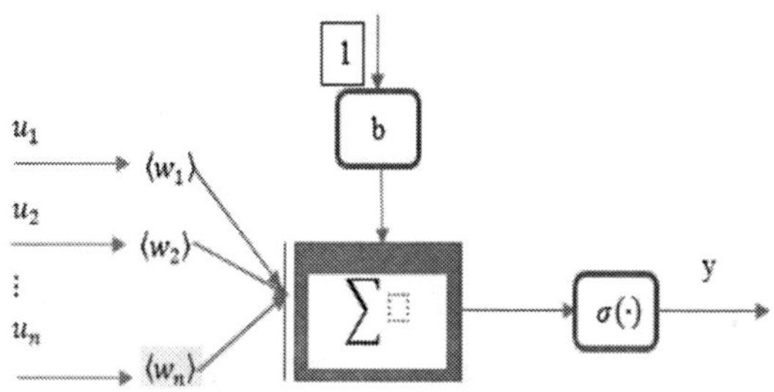

Figure 1. A simple Neuron scheme, with n inputs(u_i), w_i –weights, bias(b), $\sigma(\cdot)$ is the non-linear activation function and the only one output (y)

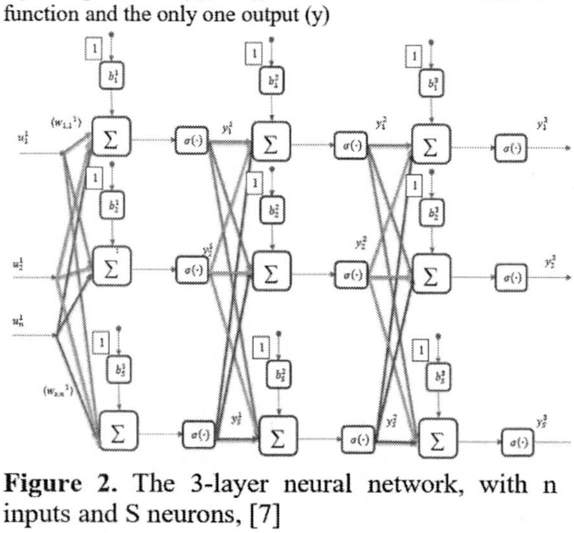

Figure 2. The 3-layer neural network, with n inputs and S neurons, [7]

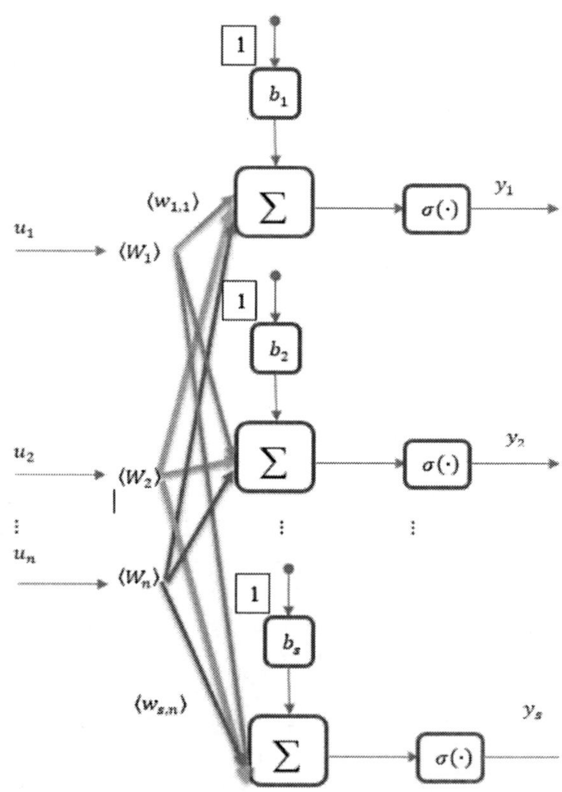

Figure 3. A single layer network, with n inputs and S neurons

Figure 2 shows a single-layer neural network with S neurons. Note that each of the network inputs is connected to each of the network neurons, for this reason the input matrix matrix now has S rows. Figure 3 presents a multilayer neural network (3 layers, here, for comprehensibility). This neural network, has an architecture defined as follow: each layer is characterized by its own weighting matrix, its own vector also biases the n inputs and an output vector, y.

IOP Conf. Series: Materials Science and Engineering **568** (2019) 012079 doi:10.1088/1757-899X/568/1/012079

3. Estimation methods [1]

3.1Fault Detection Based on State Variable Estimation,

The state variables are rarely all measurable, so unmetric variables have to be estimated (in this case, the dynamic model of the process linearises around the operating point). Estimation can be done using different methods, whether statistical or not, depends on how stochastic the system is. After estimation, an immediate assessment of the residuals is made as a difference between the estimated and actual measured values that were generated by the system, then the malfunction is detected, using statistical test and validation methods as well.

3.2 Fault Detection Based on Parameter Estimation

It is known that in many cases the process / system model parameters are in a complicated relationship with the physical coefficients of the process / system. Poor system operation generates effects that affect physical coefficients, the result of these influences being observable in process parameters. Since not all physical process parameters can be measured directly, the changes produced in the structure of some of the unmeasurable parameters must be calculated by the estimated values of process model parameters. Obviously, in this case, relationships between model parameters and physical coefficients must be known exactly.

3.3 Pattern recognition methods,

Used to detect and diagnose malfunctions, these methods do not strictly require mathematical process models. For these methods, the idea is that the operation of a process / system is ranked in full accord with the measured data. Formally this procedure is actually a mapping of the measurement space over the decision space. Generally, pattern recognition and classification methods can be structured into three stages of evolution: collecting measured data; feature extraction; and classification.

The mathematical model, quite comprehensive, of a process can be characterized by a equation, [1]:

$$Y = f\{U, N, \theta, X\} \tag{1}$$

in which Y represents outputs, measurable quantities; U represents the measurable input signals; N represents the non-measurable disruptive signals of the process, the measurement and control equipment; θ represent the immeasurable process parameters; X represents the state variables, partially measurable or partially nonmeasurable.

Depending on the category of processed signals, there are, therefore, four classes of fault detection methods:

(1) - measurable inputs U and Y outputs respectively - the input and output signals can be easily monitored to observe the accidental changes in the process based on the lower or upper limits of the output signal;

(2) -nonmeasurable state variables, X - this case requires the existence of a model to be defined on the basis of the measurable state variables, and on these grounds to estimate the values of the non-measurable variables, respectively the reconstruction of the data set characterizing this kind of variables. Defining models are defined by system types:

(2.a)- The static systems (the state variable, estimated, is $\hat{X} = f\{\bar{U}, \bar{Y}\}$);

(2.b)-Dynamic systems the state variable, estimated to be given by the state equation $X(t) = f\{U, Y, t\}$. By linearization we obtain:

$$\dot{x}(t) = Ax(t) + Bu(t) \tag{2}$$

$$y(t) = Cx(t) \tag{3}$$

where $y = \Delta Y, u = \Delta U$ $și$ $x = \Delta X$ are the deviations of **Y**, **U** and **X**.

IOP Conf. Series: Materials Science and Engineering **568** (2019) 012079 doi:10.1088/1757-899X/568/1/012079

(3)-nonmeasurable process parameters- the simple polynomial mathematical model is, in fact, a relation between the input parameters and the output parameters of the system, in which there are weighting coefficients of the various parameters or coefficients:

$$Y(U) = \beta_0 + \beta_1 U + \beta_2 U^2 + \ldots \quad (4)$$

In literature, [1] there are different procedures for addressing this class of study and analysis in the following stages:

I-establishing the mathematical model describing the dependence between measurable inputs and outputs:

$$Y(t) = f\{U(t), \theta\} \quad 5$$

II-determination of the relation between the parameters of the model θ_i and the coefficients of the physical process p_j: $\theta = f(p)$;

III-estimation of the parameters of the model θ_i as a result of the dependence of the inputs and outputs: U (t), Y (t);

IV-calculating process coefficients: $p = f^{-1}(\theta)$, and determining their deviation Δp_j;

(4) - non-measurable characteristics - these characteristics can be determined using the measurable state variables: $\eta = g\{U, Y\}$. Examples of unmeasurable quantities could be: process yield; fuel consumption; consumption of lubricating oil; wear of tools per unit of production, or per unit of time; wear residues deposited in oil tanks.

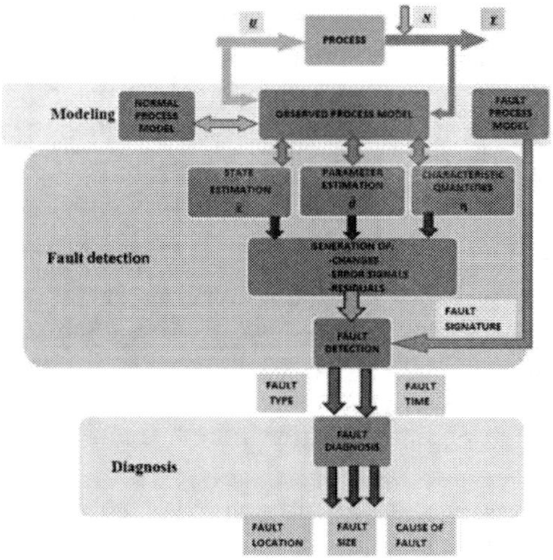

Figure 4- A general structure of process fault detection methods [1]

In figure 4 a structured general diagram is presented, [1], for the process fault detection and diagnosis of a process, based on the model for normal operation (which is supposed to be known), the fault model (which must contain, by means of definition, effects of faults on the analyzed quantities, effects concentrated in the fault signature), respectively the model for the observed process.

4. Conclusion and Results

The experiment was carried out in several stages, being part of a research program, supported by private companies in Romania and Italy, which proposed the creation of a data set allowing the development of a scientific opinion regarding the automatic monitoring system operation status, fault detection by determining the state of the lubricating oil [5], [8], [9], [10]. A first step in testing, an experimental physical model that took place at Mecoil (Italy) in the company's lab on a demo system was to test the operating parameters of the stand designed to simulate the operation of an automated mechanical system,

IOP Conf. Series: Materials Science and Engineering **568** (2019) 012079 doi:10.1088/1757-899X/568/1/012079

but also to determine by direct measurement and detection by specialized sensors the state of oil degradation in various positions: cooling with coolant, impregnation with debris or accidental increase in oil temperature, thus measuring its oxidation degree . The elements that were tested (physicochemical parameters) were evaluated in the first stage, on a single level. Data processing and computing have generated positive results in obtaining the necessary data points to advance in the actual monitoring of a physical system, [11], [12], [13], [14]. The overcoming of this early phase allowed the installation of a hydraulic system containing sensor sets and transducers designed and physically installed in industrial installations (pilot systems) at EMSIL Techtrans SA Oradea-Romania, as well as in four other locations in Italy, to Mecoil partners.

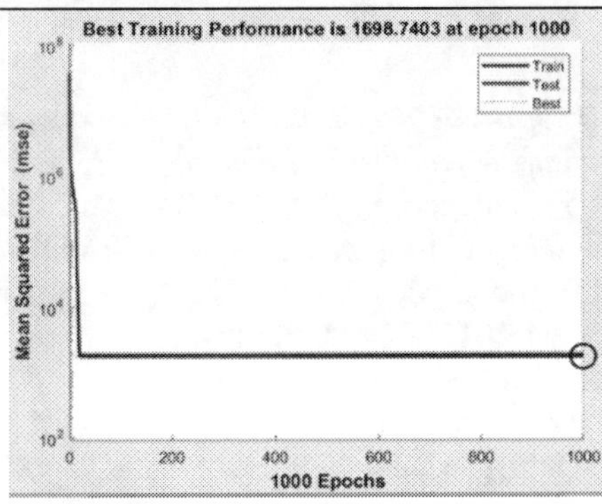

Figure 5. Performance

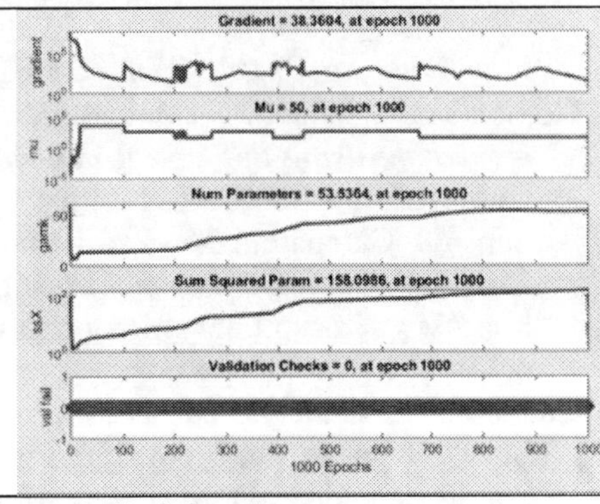

Figure 6. Training State

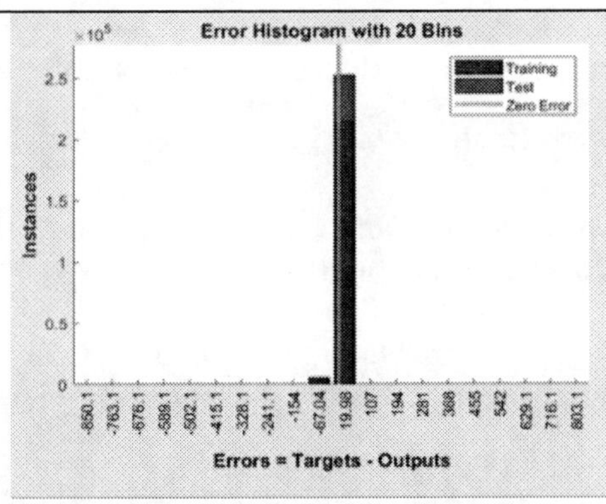

Figure 7. Error Histogram

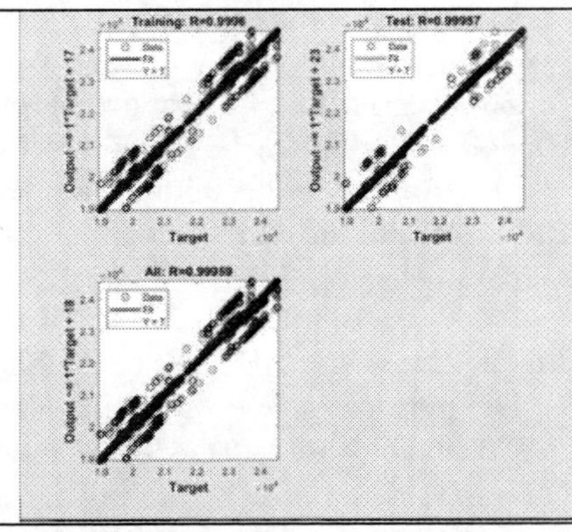

Figure 8. Plot regresion

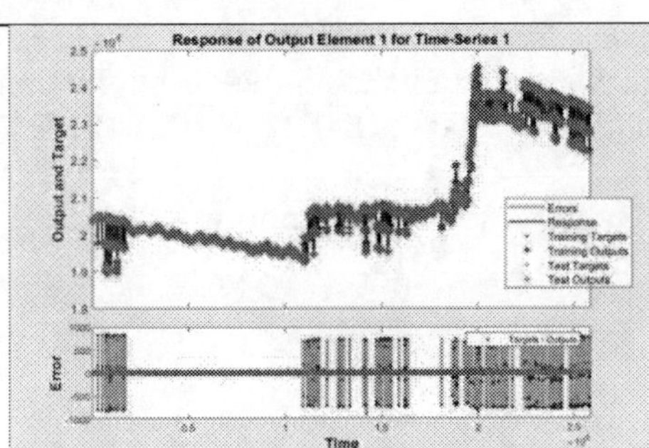

Figure 9. Time Series Response-Plotresponse

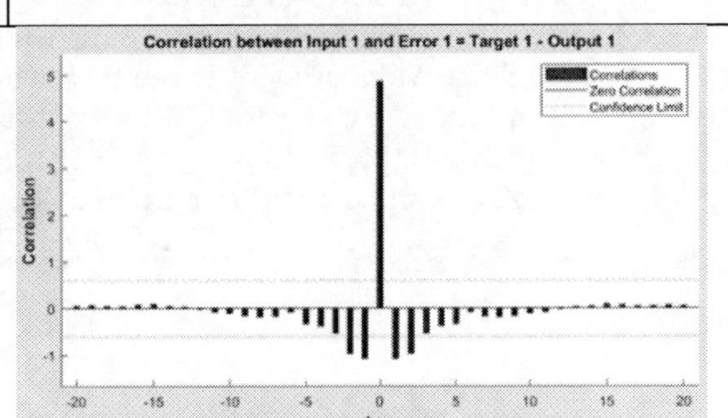

Figure 10. Input Error Cross Correlation

Acknowledgment

This work is supported by Machine Intelligence for Diagnosis Automation (MIDA), Research Program, National Program PN II, ERA MANUNET: NR 13081221/13.08.2013

References

[1] Isermann R. 1984 Process Fault Detection Based on Modeling and Estimation Methods A Survey, Published in Elsevier- Automatica, **Volume 20**, Issue 4, Pages 387-404, DOI:10.1016/0005-1098(84)90098-0

[2] Sorsa T., Koivo H. N., and Koivisto H., 1991 Neural Networks in Process Fault Diagnosis, IEEE Transactions On Systems, Man, And Cybernetics, **Vol. 21**, no. 4, , DOI: 10.1109/21.108299

[3] Grebenişan G., Salem N., and Bogdan S. 2017 *The lubricants parameters monitoring and data collecting,* MATEC Web of Conferences, **Volume** 126, 2017, https://doi.org/10.1051/matecconf/201818403008.

[4] Grebenişan G., Salem N., and Bogdan S. 2018 *An approach of classification and parameters estimation using the neural network, for lubricant degradation diagnosis*, MATEC Web of Conferences, **Volume** 184 https://doi.org/10.1051/matecconf/201818403009.

[5] Adriani G., Campatelli M., Paoli M., 2013 *Monitoil® - Online sensors for efficient and cost-saving oil conditions' monitoring*, Mecoil Diagnosi Meccaniche s.r.l., Florence, Italy, http://www.mecoil.net/wp-content/uploads/2013/04/20120914-Articolo-colonne_Monitoil_eng.pdf

[6] Chandola V., Banerjee A., and Kumar V. 2009 *Anomaly detection: A survey*. ACM Comput. Surv., 58 pages. http://doi.acm.org/10.1145/1541880.1541882

[7] B. Lei, D. de Ridder, D. M. J. Tax, F. van der Heijden, G. Xu, M. Feng, Y. Zou - *Classification, Parameter Estimation, and State Estimation An Engineering Approach Using MATLAB*, Second Edition, 2017 JohnWiley & Sons, Ltd.

[8] Gwidon W. Stachowiak, Andrew W. Batchelor, Engineering Tribology, Third Edition, 2005, Elsevier Butterworth-Heinemann

[9] Box G. E. P., Jenkins G. M., Reinsel G. C., Ljung G. M. 2015 *Time Series Analysis: Forecasting and Control*, Fifth Edition, John Wiley & Sons.

[10] B.C. Sharma and O.P. Gandhi, Performance evaluation and analysis of lubricating oil using parameter profile approach, Industrial Lubrication and Tribology 60(3):131-137, DOI: 10.1108/00368790810871057

[11] Junda Zhu, David He, Eric Bechhoefer, Survey of Lubrication Oil Condition Monitoring, Diagnostics, and Prognostics Techniques and Systems, Journal of Chemical Science and Technology July 2013, Vol. 2 Iss. 3, PP. 100-115

[12] N.K. Myshkin and L.V. Markova, On-line Condition Monitoring in Industrial Lubrication and Tribology, Applied Condition Monitoring 8, Springer International Publishing AG, 2018, DOI 10.1007/978-3-319-61134-1_1

[13] Qingfei He, Guiming Chen, Xiaohu Chen, Chunjiang Yao, Application of Oil Analysis to the Condition Monitoring of Large Engineering Machinery, Published in: 2009 8th International Conference on Reliability, Maintainability and Safety, IEEE *Xplore,* DOI: 10.1109/ICRMS.2009.5270033

[14] Junda Zhu, Online Industrial Lubrication Oil Health Condition Monitoring, Diagnosis and Prognostics, PhD Thesis, University of Illinois at Chicago, 2013.

IOP Conf. Series: Materials Science and Engineering **568** (2019) 012080 doi:10.1088/1757-899X/568/1/012080

Parameterized optimization analysis with finite elements, using ANSYS®, of a reference point under a torque action

F G Marian[1,2] and G Grebenişan[1]

[1]University of Oradea, Faculty of Managerial and Technological Engineering, Romania, str. Armatei Române nr. 5, 3700 Oradea, Bihor, România

marianflaviu77@yahoo.com

Abstract. To determine the elastic/plastic deformation value of the part, we used Ansys R19.0 software(CAE), to simulate a torque moment applied to the functional area part, which provided the information set about the part behavior. The mold of the part has 4 cavities, and the part material is a polyamide. Having a requirement imposed by the customer on the drawing, we considered this finite element analysis it is necessary before start manufacturing the molding and starting injection the parts to demonstrate that the customer requirement is achievable. The results obtained from the analysis with finite elements in ANSYS R19 showed that the constructional part could not fulfill the design requirement, namely to withstand a torque of 4500Nmm the part being fixed in the hole with a diameter of 2.5 mm. After analyzing the results, the client will have to modify the current drawing based on the results obtained by us, namely to reduce the torque value from 4500Nmm to 4000.3 Nmm.

1. Introduction

Injection of plastics has been a significant development, especially among the automotive industry. PLASTOR, one of the largest manufacturers of plastics and molds from Romania, has met along the years valuable clients with more prominent and drastic requirements for plastic injection from the automotive industry.

In this article will use the ANSYS software to analyze and optimized using as a variable value the torque moment, which is applied on the part in a specified area and compare the results, and then draw conclusions. Ansys software is used to design products and semiconductors, as well as to create simulations to test product durability, temperature distribution, fluid movements, and electromagnetic properties. Ansys develops and markets finite element analysis software used to simulate engineering problems (CAE-Computer-Aided Engineering), [1]. The software creates simulated computer models of structures, electronics or machine components to simulate strength, hardness, elasticity, temperature distribution, electromagnetism, fluid flow, and other attributes. Ansys is used to determine how a product will work with different specifications without building test products or performing impact tests. For example, Ansys software can simulate how a bridge will manage after years of trafficking, how to best process salmon at a factory to reduce the amount of waste, or how to design a slide that uses less material without sacrificing safety, [2], [3], [4].

Most Ansys simulations are performed using the Ansys Workbench software, one of the company's main products. Typically, Ansys users break larger structures into small components, each being individually modeled and tested. A user can start by defining the dimensions of an object and then adding weight, pressure, temperature, and other physical properties. Finally, Ansys software simulates and analyzes motion, fatigue, fractures, fluid flow, temperature distribution, electromagnetic efficiency, and other effects over time[5], [6].

IOP Conf. Series: Materials Science and Engineering **568** (2019) 012080 doi:10.1088/1757-899X/568/1/012080

2. Information about the part

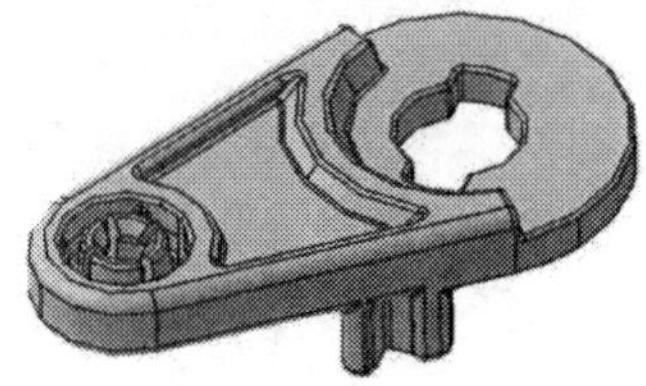

Figure 1. 3D model of the part

Table 1. Part geometry properties and material

Length	Value (mm)	Volume (mm³)	Material
X	41.05		
Y	22.00	2064.4	PC (polycarbonate)
Z	11.00		

3. Testing part requests

Besides dimensional control, the client has a particular requirement. This requirement is that the part has to withstand a torque T=minimum 4500 Nmm applied in the hole with diameter 2.5 mm +0.1 mm. The part must be blocked against rotation in functional area to sketch according to the drawing below

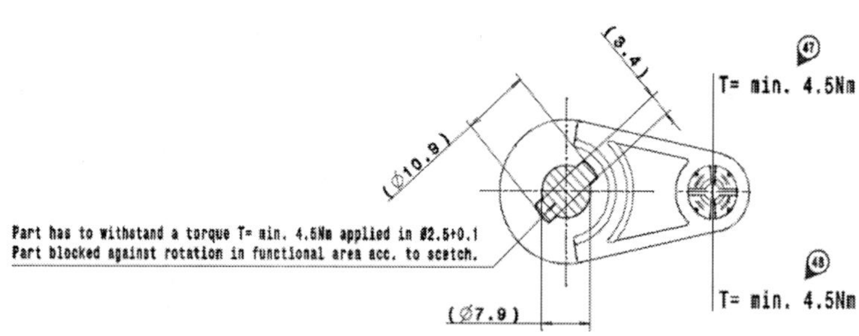

Figure 2. Part test requirement by the project

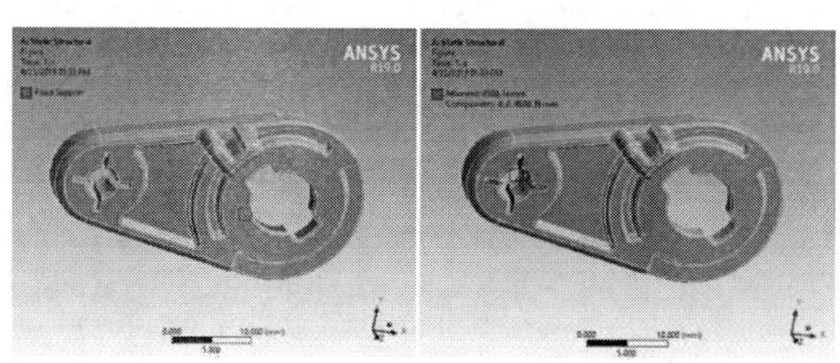

Figure 3. Part test requirement imposed and defined in ANSYS R19, [7]

IOP Conf. Series: Materials Science and Engineering **568** (2019) 012080 doi:10.1088/1757-899X/568/1/012080

4. Finite Element Analysis

Based on the values obtained, result that 4500 Nmm torque moment value it is to raised and the part can be damaged during his functionality.

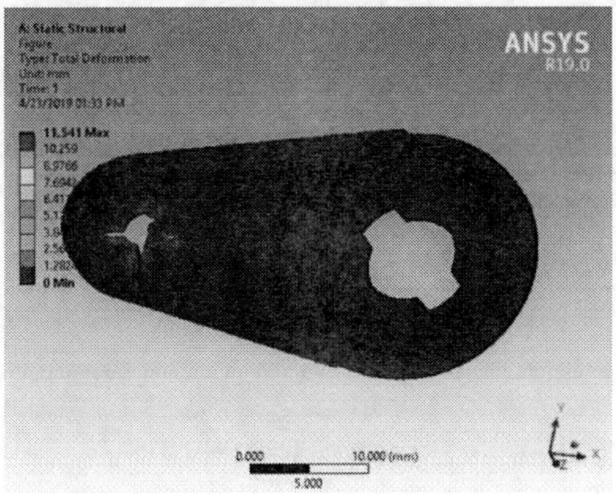

Figure 4. Total deformation (mm)

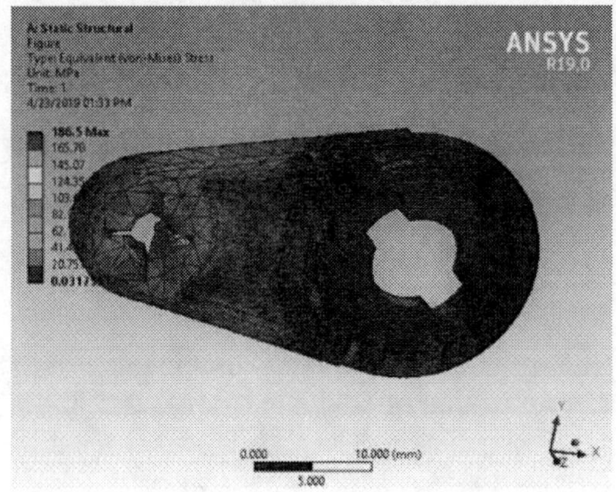

Figure 5. Maximum Equivalent stress (von Mises) (MPa)

The maximum value of the total deformation (plastic deformation), defined as displacement, is maximum 11.541 mm. The maximum value of the equivalent stress is 186.5 MPa.

To see the maximum torque moment value that can be applied on the part without to affect his functionality, an optimization procedure will be needed.

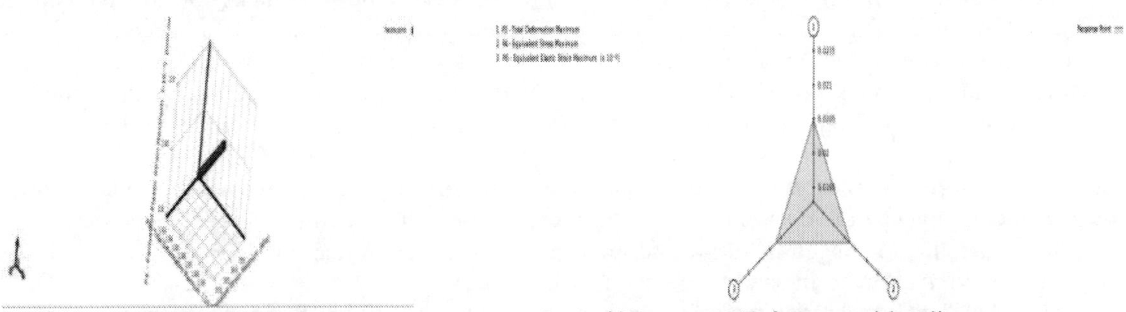

a)Response Points b)Response Points as spider diagram

Figure 6. Optimization Tradeoff and Spider sensitivity (Response Points)

5. Optimization using Design of Experiments, and results [7], [8]

Optimizing a project using the Design of Experiments technique involves performing a finite element analysis before allowing the setting of some convenient limits for the chosen input parameters, ie the achievement of objectives imposed by design, technological or design requirements operation. Do not aim here, minimize a cost function, or maximize efficiency, of any class. After the analysis of finite elements can be determined, through the algorithms characteristic of Design of Experiments techniques, the data set consisting of design points, generating new projects, similar to the one initially analyzed, but with results configured for the new input parameter values. Based on these design points, the behavior of the output parameters is intended, which in fact determines the behavior of the project itself. These design points, in relation to the results obtained (output parameter values), generate the behavioral diagrams of the project, called Response Surfaces. Response Surfaces provide the operator with interpolation results with the degree of influence indicated in the Local Sensitivity Diagrams. The new torque moment value is 4000.3 Nmm, which is the maximum value applied on the part without to affect

IOP Conf. Series: Materials Science and Engineering **568** (2019) 012080 doi:10.1088/1757-899X/568/1/012080

it. In this case, the client must change the draw and the value of the torque moment from 4500 Nmm in 4000.3 Nmm.

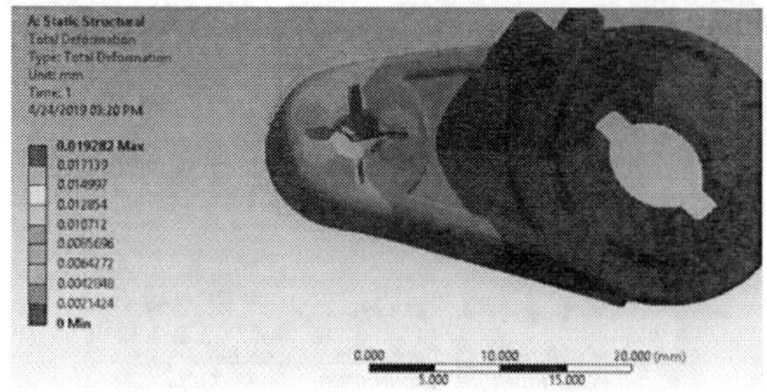

Figure 7. Total deformation (mm)

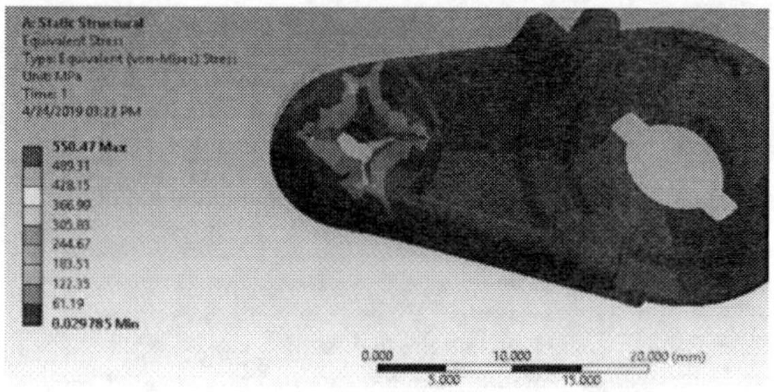

Figure 8. The maximum equivalent stress –von Mises, (MPa)

After optimization, the maximum value of the total deformation is 0.0192 mm. The maximum value after optimization of equivalent stress is 550.47 MPa.

6. Conclusion

The optimization was performed using several well-established, complementary methods: parameter correlation, linear optimization, and Response Surface's method after the project was analyzed by the finite element method. Each of these phases of project implementation is determined by the results obtained. Thus, by correlating the parameters, the degree of influence of the input parameters on the output parameters was determined to decide which of them decisively influences the functioning and the degree of safety of the operation of the project. The degree of safety of the project is given by the degree of certainty regarding the quality of the product, as well as the degree of satisfaction of the client, about the behavior of the finished product, which is also the high quality of the product. This phase is viably linked to the analysis of the finished part of the required piece at the torque imposed by the customer. Thus, the lifetime or safety factor of the project determines the quality of the obtained piece and the economic efficiency of the production. Direct optimization takes into account parameter correlation but does not use results as the basis for optimization itself. Optimization using Response Surfaces is an extremely laborious method but provides the most concise results with optimization checkpoints performed on one or all three candidate points found in the optimization process. The candidate points are a "statistical summary" of the best design points for which the algorithm provides the user with the optimization results but also the deviations from a statistical average deviation and the degree of compliance or convergence of the solution, the deviations being displayed in percentage.
Only after the client changed the draw and the torque value, the producer will proceed the project and begin the mold manufacturing and then injecting the parts.

Using ANSYS R19 before starting a project, everything that is on the drawing it is checked and analyzed reducing time, energy and material resources are efficient costs, the credibility and your professionalism to the client give you next opportunities for projects and the delivery term is respected.

This study we show as that sometimes even the client can make a mistake that`s why the automotive parts manufacturers need to be updated with everything means CAM, CAD, CAE. Following the analysis, the client sent the design with the modifications requested by us. The project will continue by designing the mold (CAD) and then manufacturing it (CAM).

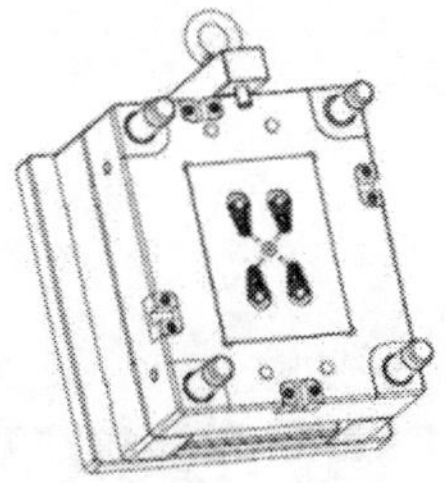

a)-Upper mold

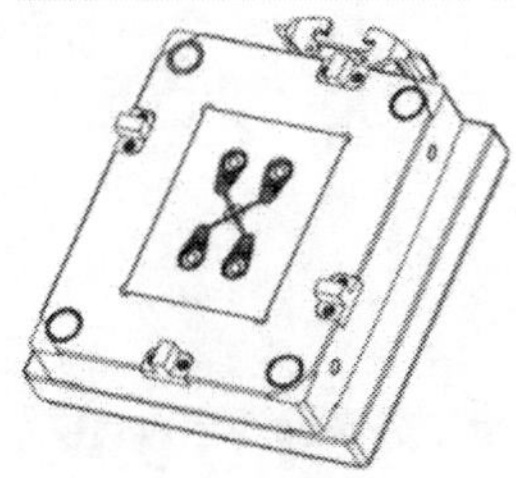

b)-Lower mold

Figure 9. The part mold

After the mold is received by the production department, the first injection trials will be made using the parameters agreed by the customer and then, after the CTC control give the feedback to the production department, the final parts will be injected and ready to be shipped to the customer.

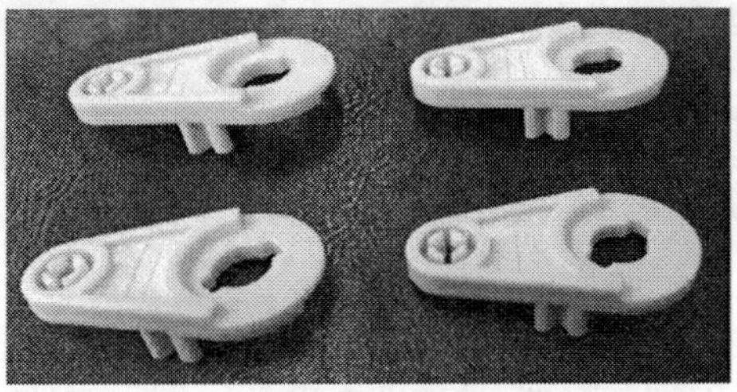

Figure 10. The injection parts (4 cavities)

Acknowledgments
Thanks to advisors for all support: Prof. dr. Eng Ion Sereş - General Manager, Eng. Dan Buzdugan – Quality Manager (PLASTOR SA Oradea)

References

[1] https://en.wikipedia.org/wiki/Ansys
[2] Jianping L, Hongwei Z, Xiaoli H, Lin Z, 2016 Pengliang H, Cong L, Ning L, Yuexi Za , and Changyi L 2016, Ins. And Exp. Tech. Volume 59, Issue 5 P762–767
[3] Grebenişan G., Bogdan S. 2017 Parameterized Finite Element Analysis of a Superplastic Forming Process, Using ANSYS, MATEC Web of Conferences, **126**, 2017, https://doi.org/10.1051/matecconf/201712603001
[4] Negrău D C, *Finite Element Analysis, and Solution Optimization for a bending device,* 2018, University of Oradea.
[5] Huei-H L, 2017 Finite Element Simulations with ANSYS Workbench 17- Theory, Applications, Case Studies, SDC Publications
[6] Xiaolin C, Yijun L 2018 Finite Element Modeling and Simulation with ANSYS Workbench, Second Edition CRC Press DOI: 10.1201/9781351045872
[7] ANSYS 19, Workbench User's Guide http://www.ansys.com (accessed in April 2019)
[8] Negrau D. C., Grebenisan G., Indre C. 2019 Experimental approach and finite element analysis of the behavior of a steel bending machine, IManE&E 2019

IOP Conf. Series: Materials Science and Engineering **568** (2019) 012081 doi:10.1088/1757-899X/568/1/012081

Designing and realization an ECG based the Arduino Mega 2560 development board

A I Tătaru[1], C N Drugă[1]

[1] Medical Engineering Department, Faculty of Design of Product and Environment, Transylvania University of Brasov, Romania

tataru_anca95@yahoo.com

Abstract. The paper presents several aspects concerning the design and construction of an ECG using an Arduino Mega 2560 development board and a click ECG module. In the first part of the paper are presented some notions regarding electrocardiography. Electrocardiography is used to diagnose cardiovascular disease and monitor heart health. This type of test is easy to use because electrodes are placed in certain regions on the chest, arms, and legs that are connected by wires to an ECG device. The second part of the paper presents the main components of the device: the Arduino platform, the ECG cables, the ECG click module, the Arduino Mega Click Shield, the WI-FI module and the ECG simulator. The recording system can be on thermosensitive paper, ink, cathode ray tube TFT, and LCD. The last part of the paper is dedicated to testing the device and explaining the way how to do a test.

1. Introduction to Electrocardiography

The device is required to receive the vital signs of the patient's heart rate (pulse and pressure) so that he or she can perform an electrocardiogram and a measurement of blood pressure and pulse to send them to the specialist doctor. In this way, even if the patient goes to the medical office less often, the doctor may notice improvement or worsening of the health condition during a recommended treatment to determine whether it is working or needs to be changed.

A standard ECG uses 10 cables to receive and interpret 12 representations of heart activity. These representations can be divided as follows: standard deviations (there are 3 standard deviations that are composed between the legs and hands resulting in the Einthoven triangle (figure 1)), unipolar deviations of limbs and thoracic deviation (precordial). Starting from this standard type with 10 cables, there may be other electrocardiographic variants such as:

- A 5-channel and 5-electrode device (4 limbs and one chest) used for continuous monitoring (during surgery or an ambulance);
- A 3-channel system and 4 electrodes placed on the limbs to monitor heart rate;
- Portable ECG monitoring systems;
- Wireless ECG systems. [1]

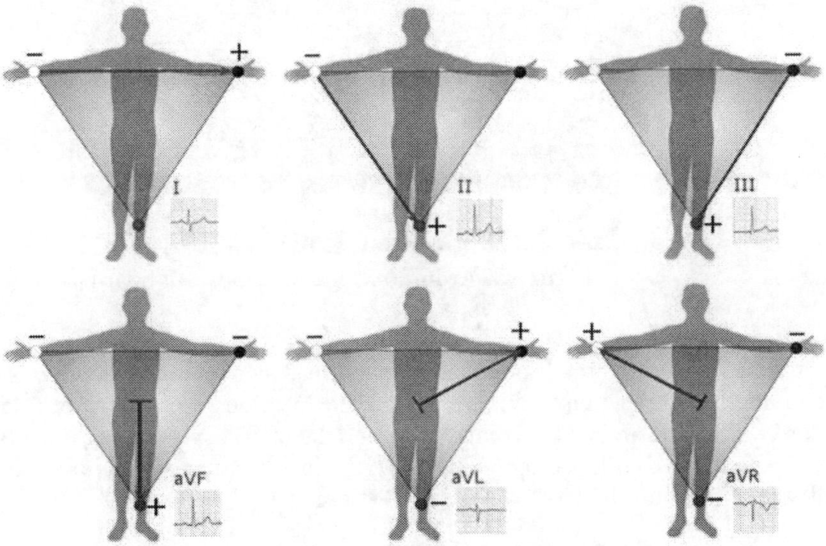

Figure 1 Graphical representation of Einthoven's triangle [2]

1.1. Analysis of the ECG route and identification of important parameters

The results of the ECG are represented on a special paper in the form of waves, which form a PQRSTU complex (figure 2). As exemplified above, the result of the EKG is represented as a wave complex (figure 2).

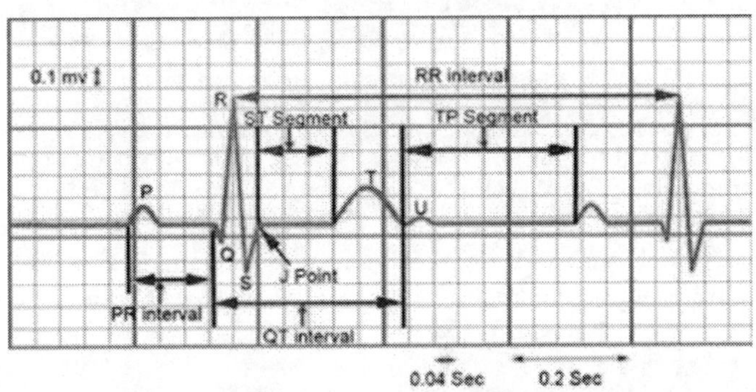

Figure 2 Example of electrocardiogram [3]

2. System components

An electrocardiograph must collect the signals from the activity of the heart to amplify them with a device and print them on a special paper. Signals can be interpreted as potential differences in heart activity and varying in time and space. The electrical activity of the heart is an analog signal that needs to be converted into a digital signal to be interpreted by the device that also contains an electronic amplification system.

2.1 Arduino Mega Development Board

The Arduino Mega (figure 3) is a microcontroller board based on the ATmega2560. It has 54 digital input/output pins, 16 analog inputs, 4 UARTs (hardware serial ports), a 16 MHz crystal oscillator, a USB connection, a power jack, an ICSP header, and a reset button. It contains everything needed to support the microcontroller; connect it to a computer with a USB cable or power it with an AC-to-DC adapter or battery to get started.

IOP Conf. Series: Materials Science and Engineering **568** (2019) 012081 doi:10.1088/1757-899X/568/1/012081

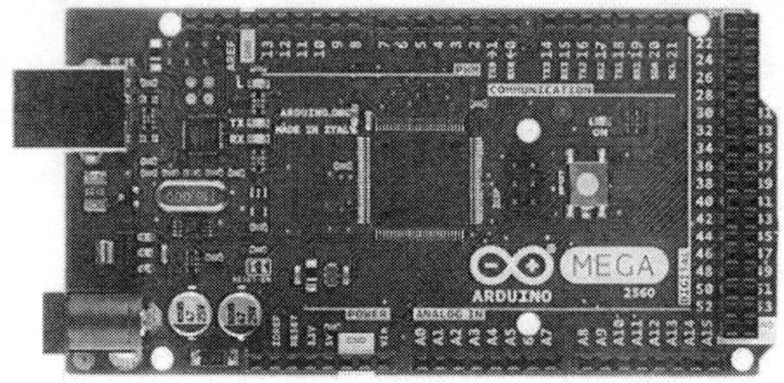

Figure 3. Arduino Mega development board [4]

2.2 Kit ECG click

The kit contains the ECG click bundle acquisition card, the 3 signal pickup electrodes and the disposable electrodes that stick to the patient's skin in 3 specific locations according to a scheme below. The acquisition card works with an input current of 5 V, and the interface is analog. It contains an operational amplifier MCP609 and is developed by MikroE. To achieve connectivity with the development board, use a special shield made for the type of plate used (figure 5).

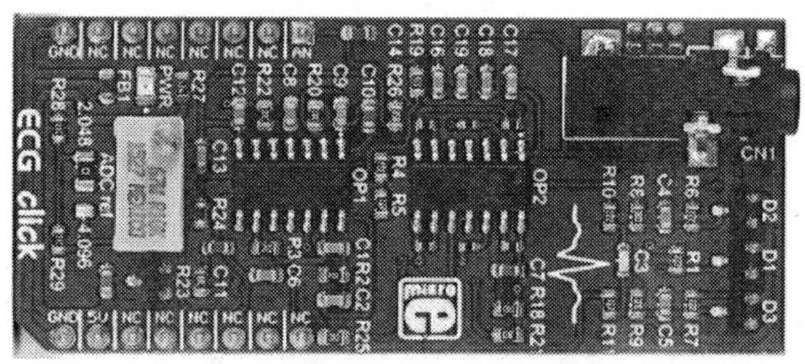

Figure 4. Acquisition board ECG; [5, 6]

This is an ECG (or EKG) machine on a click board. It measures the electrical activity of a beating heart through electrodes taped to the skin. The board requires little setup, and the final measurement results can be displayed as an Electrocardiogram using a free Windows app. With the ECG click bundle (click + cable + electrodes) you will be able to prototype your health-tracking, and fitness devices, quantified-self wearables or you can make an LED blink in sync with your heart [7].

2.3 Arduino Mega Click shield

Arduino Mega Click shield is an extension for Arduino Mega and any other boards that share its pinout. It's a simple shield with three mikroBUS™ host sockets that allow you to connect more than 150 different types of click boards to the Arduino Mega.[8]

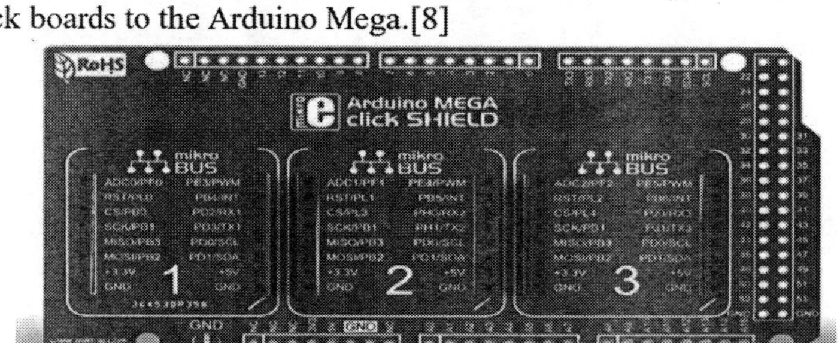

Figure 5. Arduino Mega Click shield [8]

2.4 ECG cables

Making connections with a 3-electrode cable makes it possible to use this electrocardiograph by people who do not have medical training. It is positioned, respecting the colors engraved on the cables

IOP Conf. Series: Materials Science and Engineering **568** (2019) 012081 doi:10.1088/1757-899X/568/1/012081

and connecting to the receiving device and displaying them as a graphic on a special paper or the screen. The cables and how to connect them can be seen in figure 6 a and b.

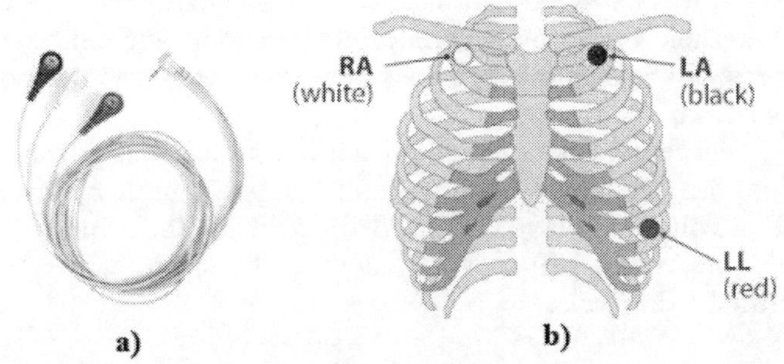

Figure 6. a) ECG cables [9]; b) Position of the electrodes [10]

2.5 Non-invasive blood pressure module

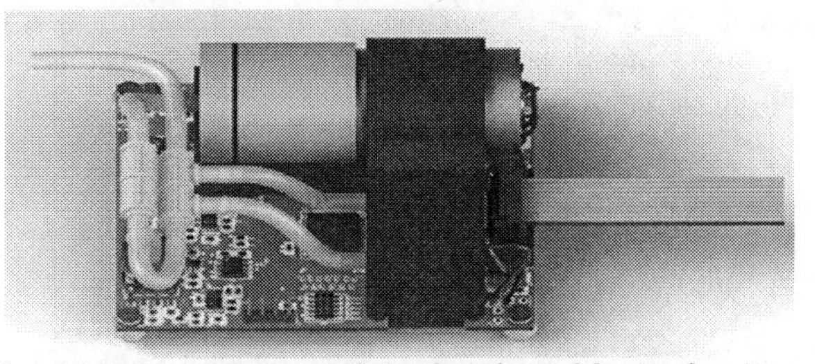

Figure 7. Non-invasive blood pressure module developed by Dulon Innovative Electronics [11]

The voltage module is used to measure blood pressure in the blood (figure 7). It is powered at 12V because it contains a motor and pump that introduce air into the blood pressure sleeve. The tension is divided into systolic, diastolic, and will also measure the pulse of the patient.

3. Programming your device

Example code for programming the ECG click card provided by the MIKROELEKTRONIKA plate manufacturer is shown below.

The code shown at their website initializes the ADC for measuring the ECG, and UART for talking to the PC. When the button is pressed, the timer interrupt will be initialized. An interrupt is generated every 3.9 ms, upon which the MCU gathers ADC and time data, and sends it to the PC over UART until the timer reaches 15 seconds. [12]

4. Device Testing

For testing this device, use an ECG simulator produced by SANATRONIC in Bucharest. This generates ECG signals similar to a real patient. With it, adjustments can be made before he can test patients. SimEKAR is an electrocardiogram simulator, with 12 derivations and arrhythmias, for periodic and post-repair checking of the technical and functional characteristics of electrocardiographs. This ECG simulator is endowed with the Medical Engineering Laboratory (Product Design, Mechatronics, and Environment Department).

5. How the ECG works

The operation of the electrocardiograph in this project is presented below. To begin with, there are several buttons that can be pressed to perform various functions. Pressing a button triggers an interruption in the program and performs the instructions for it. After the device is switched on, the necessary connections are made to record the signals from the simulator or the patient. These consist of connecting the 3 electrodes for the ECG according to the scheme in Figure 6.b and placing the sleeve on the patient arm. If only a test is desired, the other connection is dropped. Press the button specific to the desired function and wait.

To achieve the tension, it is expected that the sleeve swells to a pressure superior to the pressure it feels as systolic pressure. Then it will slowly deflate until you notice the last beat that represents diastolic pressure. These pressures will be displayed on the screen, and the sleeve will completely deflate and be pulled out of the patient's arm.

For the ECG the sensor captures signals with each interrupt coming from the program at 3.9 ms, enough time to re-record the electrocardiogram signal. The program then converts the signals from the sensor into a graphic that will be displayed on a specific grid and the printing paper (figure 2). Various parameters can be calculated based on the wavelength of the electrocardiogram. A representation of these places is exemplified in Figure 2.

6. Conclusions

This system is easy to use in family doctor's offices at home, but also in hospitals or clinics. The device can be improved by adding more sensors to measure several important parameters for health status measurement (eg: an expired air sensor - the introduction of a spirometer, a sensor for measuring muscle activity - useful when patients are at bed more time and need for monitoring and muscle activity to help them recover more efficiently, a pulse oximeter sensor, etc.). The device can be easily maintained and used, requiring minimal knowledge in this area.

Acknowledgments

Practical realization of this project was made possible thanks to funding received in the| ***My License Project-2019*** competition held in the Transylvania University of Brasov

References

[1] ***https://newsmed.ro/interpretare-ekg-normal-notiuni-de-fiziologie/, (accessed at 18.03.2019).
[2] ***https://en.wikipedia.org/wiki/Einthoven%27s_triangle (accessed at 27.03.2019).
[3] ***https://emedicine.medscape.com/article/1894014-overview, (accessed at 20.03.2019).
[4] ***https://www.sparkfun.com/products/11061, (accessed at 25.03.2019)
[5] ***https://www.tme.eu/ro/details/mikroe-2458/placi-de-extensie/mikroelektronika/ecg-click-bundle/, (accessed at 25.03.2019)
[6] ***https://www.tme.eu/ro/details/mikroe-1900/accesorii-pentru-kituri-de-dezvoltare/mikroelektronika/arduino-mega-click-shield/, (accessed at 25.03.2019)
[7] ***https://www.mikroe.com/ecg-click-bundle (accessed at 26.03.2019)
[8] ***https://www.mikroe.com/arduino-mega-click-shield (accessed at 26.03.2019)
[9] ***https://www.mindraynorthamerica.com/cmsAdmin/uploads/ecg-lead-placement-proc-7664reva.pdf , (accessed at 25.03.2019)
[10] ***https://www.mikroe.com/ecg-cable , (accessed at 25.03.2019)
[11] ***https://doulonelectronics.com/case-studies/non-invasive-blood-pressure-module/"case-studies/non-invasive-blood-pressure-module, (accessed at 25.03.2019)
[12] ***https://www.mikroe.com/ecg-click, (accessed at 28.03.2019)

Concept and premises of the logistic management model in conditions of uncertainty - research approach

Ł. Marzantowicz[1] I. Dembińska[2]

[1]SGH Warsaw School of Economics, Collegium of Business Administration, Department of Logistics, Poland
[2]University of Szczecin, Faculty of Management and Economics of Services, Department of Logistics, Poland

lukasz.marzantowicz@sgh.waw.pl

Abstract. The article aims to present the results of the preliminary study, defining the premises of the concept of creating a model of logistic management in conditions of uncertainty. The focus group interviews (FGI) determined the premises for building the concept of logistics management in conditions of uncertainty, and then a catalog of methods for building this model was selected. The results allowed to make general conclusions that the model of logistic management in uncertainty is possible to develop, which in turn allows setting the target for the proper research.

1. Introduction

Conditions of uncertainty become an integral part of the environment in which the company functions. Therefore, what is more, important is the approach to effective logistic management in the situation of unpredictable variable factors and, consequently, uncertainty. One of the ways to extend the management methodology to include the impact of uncertainty on the management process and activities is to build an effective logistic management model. The challenge is to find or create management methods that will ensure the stability of processes and at the same time allow them to increase their efficiency while increasing the efficiency of management decisions.

The aim of the article is to present the continuation of the preliminary research of logistic management in the conditions of uncertainty (described in the "Introduction to the research of uncertainty in logistics management"). A synthetic approach to the results of the research leads to discourse regarding the premises of modelling logistics management in conditions of uncertainty, as well as sets directions for the proper research.

The classical review of literature was abandoned due to the conceptual and research character of the article, transferring the theoretical point of view into the discussion contained in the article.

2. Methodology and scope of the preliminary research

In January 2019 the methodology for qualitative research using the focus method was developed (FGI - Focus Group Interview), leading to the designation of the main research objective in the field of logistic management modelling in conditions of uncertainty. The article discusses the continuation of the preliminary study (supported in the article "Introduction to the research of uncertainty in logistics management"). This part of the initial FGI study included questions related to establishing the prerequisites for the need to create a logistic management model in conditions of uncertainty, resulting

IOP Conf. Series: Materials Science and Engineering **568** (2019) 012082 doi:10.1088/1757-899X/568/1/012082

in an attempt to deletion of the concept of this model. In two focus sessions, 19 managers from Polish enterprises took part. The research sample was selected on the basis of the so-called purposeful selection, which basic qualifying criterion was to have in the employment and organizational structure the person (s) directly responsible for logistics management (eg logistics managers).

Participants in the preliminary research conducted a discussion on the following issues:

- What are the reasons for considering the conditions of uncertainty in the logistics management strategy?
- Is it possible to develop a logistic management model in conditions of uncertainty?

Treating the above issues as a starting point for specifying the proper research, it should be noted that the participants of the research were free to speak. Next, the answers were cataloged and the participants of the study were asked to give weight to the cataloged responses.

The results of this part of the preliminary research presented in the next part of the article have a synthetic character, which allowed conducting discourse and putting forward conclusions.

3. Research results

1) Regarding the first research problem regarding the establishment of reasons for considering uncertainty and modeling of logistic management in conditions of uncertainty, a synthetic statement of results together with the weighting is presented in Table 1.

Table 1. Premises for the logistic management model in conditions of uncertainty

No	Established premises	Rank 1-most important, 6-least important
1	There is a need to increase the level of management effectiveness by eliminating or reducing the impact of uncertainty factors	1
2	There is a need to measure the impact of uncertainty on the management process	2
3	There is a need to measure the effects of uncertainty	5
4	There is a need to create management mechanisms	3
5	There is a need to set basic criteria for making management decisions	6
6	There is a need to build tools that enable effective logistic management in conditions of uncertainty	4

Source: study based on FGI

The results indicating the premises determining, in a broad sense, the need to incorporate conditions of uncertainty into strategic management determine another thesis about the legitimate need to build a logistic management model in conditions of uncertainty. The participants of the study, giving the range to each of the premises, made a gradation in terms of setting priorities. It should be noted that giving the lowest importance does not mean that the premise is irrelevant, but it means giving it a rather ancillary character.

As it follows from this part of the preliminary research, the most important premise indicated by the survey participants is the need to increase the level of management effectiveness by eliminating or reducing the impact of uncertainty factors. It can be clearly stated that this is an obligatory condition, contributing directly to improving the efficiency and managerial effectiveness. This means that logistics management at a strategic decision-making level depends on the skills, qualifications and competences of managers and their ability to absorb new conditions for management. At the same time, the aforementioned condition determines, in the second place, an efficient model of logistic management in conditions of uncertainty.

High importance, according to the participants of the research, has a premise stating the need to measure the impact of non-compliance on the management process. The participants of the study determined that the proper measurement of the occurrence of uncertainty conditions themselves, their impact on management processes is an obligatory element in the effectiveness of strategic decision making. The measurement of both the impact and also the effects of uncertainty conditions determines long-term decision-making and is at the same time sine qua non for building a logistic management model.

It is important and interesting to note that the participants of the research considered the premise on the need to set decision criteria to be the least important, while on the other hand the premise of a slightly higher rank is finding or building tools enabling effective management.

Participants of the study also recognized that the list of premises remains open, because it depends on the type of business activity, the environment of its operation, resources and formal and legal conditions and geopolitical.

2) In the area of the second research problem, the participants of the study considered the problem of building a logistic management model in conditions of uncertainty. It should be noted here that the participants in two stages considered such a complex problem. First of all, it was decided about the flow and the possibilities of developing the model, and then proposed the types of possible models. The results of this part of the research are presented in Tables 2 and 3.

Table 2. A summary of the answers regarding the possibilities of developing a logistic management model in conditions of uncertainty and their rank

No	Established answers	Rank 1- most important, 4- least important
1	There is a possibility to develop a logistic management model in conditions of uncertainty. However, it should be constructed on the basis of knowledge and tools already available. Knowledge and skills are to allow for a response in a situation where there is no possibility of estimating the impact and the effects of uncertainty.	2
2	There is a possibility to develop a logistic management model in conditions of uncertainty. It is necessary to develop measuring tools and create your own - tailor-made catalog of decisions understood as prognostic decisions and responding to effects.	1
3	There is a possibility to develop a logistic management model in conditions of uncertainty. However, it is a model of decision making and management process after the occurrence of uncertainty variables.	3
4	It is difficult to say whether a logistic management model can be developed in conditions of uncertainty, due to the fact that the vast majority of variable factors are not measurable, it is difficult to estimate the effects. However, you can accept a response model.	4

Source: study based on FGI

As the study showed, there is a need to build logistic management models that take into account the conditions of uncertainty at every decision-making stage. The participants of the study recognized as the most important the possibility of developing methods and tools for measuring the impact of uncertainty on the management process and tools for measuring the effects that uncertainty causes. Therefore, it can be said that they should be solutions based on the needs and resources available to managers. Participants also expressed doubts in this regard, stating that due to the fact that uncertainty creates unpredictable variables that can not be estimated unambiguously, there is a risk that individual solutions can not be developed and that only partially can the existing solutions be used.

Based on the above results, the participants of the study attempted to determine the catalog of possible logistic management models in conditions of uncertainty. However, it should be added that this

IOP Conf. Series: Materials Science and Engineering **568** (2019) 012082 doi:10.1088/1757-899X/568/1/012082

part of the study concerned only the attempt to indicate the nature of the management model, but it did not concern the development of the models themselves (modeling in econometric meaning), their assumptions and the method of concept development, implementation and measurement. This part of the study is presented in Table 3.

Table 3. List of proposals for logistic management models in conditions of uncertainty and their rank.

No.	Established answers		Rank 1- most important, 3 - least important
1	Scenario 1.	Model-based on ex-post response. Taking management decisions are determined by the occurrence or non-occurrence of unpredictable and predictable phenomena and concerns the response to the effects of uncertainty.	2
2	Scenario 2.	Model-based on predictive response. The anticipation of the future and the prediction of events, as well as the estimation of the effects of uncertainty, determine undertaking management decisions. A catalog of uncertainty effects determined by the degree of impact of variable factors is required.	1
3	Scenario 3.	The model relies on the possibility of blocking the effects of the conditions of uncertainty. The level of hermetic management process determines taking management decisions. Preventive decisions are required.	3

Source: a study based on FGI

From enterprises, the management model based on forecasting methods and tools as well as future anticipation analysis is the most desirable - scenario 2. It should be pointed out that the other indicated scenarios are the answer to the premises declared by the participants (described in table 2). At the same time, it should be noted that each scenario is possible, although it will probably differ in the degree of effectiveness of decisions made about the level of effectiveness of the management process.

4. Discussion

In the current turbulent environment, the decisive factor in achieving the goal in the time provided is the effectiveness of making decisions. As the preliminary study showed, despite the uncertainty conditions being created by unpredictable factors, which are a larger part of the set of variable factors, the conditions of uncertainty should be included in the logistic management strategy. In addition, the assumption was made that it is possible to develop a model of logistic management in conditions of uncertainty based on the scenario. The results of the study confirm the goal set in this part of the study. Positive verification of the working hypothesis allows attempting to set the main goal of the research relevant to logistics management in conditions of uncertainty.

Making the right, effective management decisions requires the ability to assess the reliability of the source of variable factors, as well as knowledge about the possibilities of predicting future events and estimating the consequences. Measurement - because it is referred to in the context of building a logistic management model in conditions of uncertainty - according to the participants of the study is necessary but at the same time can be burdened with the problem of imprecision. There are many ways and approaches, but in each case it is necessary to parameterize variable factors [1]. There is a valid question about the precision of the measurement. It is impossible to completely avoid errors and mistakes in measurement [2]. It is worth noting that for the correct measurement it is necessary to reference the type of decision to sources of uncertainty [3], which determine the typological layout of uncertainty

IOP Conf. Series: Materials Science and Engineering **568** (2019) 012082 doi:10.1088/1757-899X/568/1/012082

conditions [3] [4]. Although building a logistic management model in conditions of uncertainty requires precision in the scope of measurement, it should be pointed out that the essence of logistic management from the point of view of effectiveness of management decisions is striving for stability of logistic processes and increasing their efficiency [5]. According to the principle that in every seemingly chaotic system of actions, there is some causal relationship [6] [7], it is a contribution to the statement that the conditions of uncertainty are subject to forecasting in terms of data analysis, statistics and econometrics.

5. Conclusions

Analysing the results of the preliminary research of logistics management in conditions of uncertainty, the following general conclusions were made:

- despite the fact that the so-called unpredictable factors creating uncertainty conditions there is a possibility of modelling logistic management using existing or tailored methods and tools,
- the scenario method is the most adequate decision-making method in logistic management, provided that it is possible to assess sources of variable factors
- the portfolio of the effects of uncertainty conditions can be effectively managed, provided that they have knowledge and skills in the creation and use of management tools
- the logistic management model in conditions of uncertainty is an integrated part of the business management strategy.

References

[1] Wdorozhovets, M., & Warsza, Z. L. (2007). Improving the methods of determining the uncertainty of measurement results in practice (Udoskonalenie metod wyznaczania niepewności wyników pomiaru w praktyce).*Organ*, *1*, 07.

[2] Kubisa, S., & Moskowicz, S. (2004). Measurement uncertainty. An attempt to systematize the concepts and methods of calculation (Niepewność pomiaru. Próba usystematyzowania pojęć i metod obliczeń). *Pomiary Automatyka Kontrola, 50.*

[3] Domurat, A., & Zieliński, T. (2013). Uncertainty and ambiguity as determinants of economic decisions (Niepewność i niejasność jako uwarunkowania decyzji ekonomicznych). *Decyzje*, (20), 21-47.

[4] Wawiernia, A. (2013). Taxonomy of uncertainty (Taksonomia niepewności). *Zarządzanie i Finanse*, *11*(1/3), 445-454.

[5] Marzantowicz, Ł. (2018). Stability of the supply chain - condition, levels and resource conditions (Stabilność łańcucha dostaw–stan, poziomy i zasobowe uwarunkowania). *Gospodarka Materiałowa i Logistyka*.

[6] Wyciślak, S. (2009). The use of chaos theory in enterprise management (Wykorzystanie teorii chaosu w zarządzaniu przedsiębiorstwem). *Organizacja i kierowanie. PAN-SGH Warszawa, č*, *1*, 37-49.

[7] Malawski, M., Wieczorek, A., & Sosnowska, H. (2012). Competition and cooperation: game theory in economics and social sciences (*Konkurencja i kooperacja: teoria gier w ekonomii i naukach społecznych)*. Państwowe Wydawnictwo Naukowe PWN.

IOP Conf. Series: Materials Science and Engineering **568** (2019) 012083 doi:10.1088/1757-899X/568/1/012083

Basic effects types of logistic infrastructural investments - theoretical approach

I Dembińska[1] and Ł Marzantowicz[2]

[1] Department of Logistics, University of Szczecin, Szczecin, Poland
[2] Department of Logistics, SGH Warsaw School of Economics, Warsaw, Poland

izabela.dembinska@wzieu.pl

Abstract. The study aims to determine the basic effects catalog of logistic infrastructural investments. The catalog proposes eight criteria for the division: the way the effects are impacted, the measurability of effects, the range of impact effects, the distribution of effects in time, types of effects, effects impacts, channels of impact effects and sustainability impacts. Each criterion was characterized. The result is a catalog of epistemological and pragmatic meaning.

1. Introduction

Planning the macro-logistics infrastructure is a complex, challenging and extremely responsible process. First of all, the type, size, and functional scope of the infrastructure investment should be determined. To do this well, you need to demonstrate the legitimacy of the investment and accurately assess what will be the demand for the planned infrastructure. For this purpose, it is necessary to analyze the dynamics of demand for both direct and indirect users of the future infrastructure. The basic decision at the planning stage of the logistics infrastructure is the selection of the location. Financial issues are also raised, mainly concerning sources of investment financing as well as the method and schedule of spending funds. Planning also includes the definition and analysis of effects that are considered in the category of benefits, as positive (desirable) effects, and as unfavorable (undesirable) effects. The effects should be presented in an evidential way, which means that they should be associated with an event occurring at a given time and in a given space.

The study aims to determine the basic effects catalog of logistic infrastructural investments. The catalog was built mainly based on a literature review but also includes a proprietary view. It has epistemological and pragmatic meaning.

2. Discussion

About the logistics infrastructure of the economy, the effects of investment can be considered in various classification sections. The most important classification criteria can be included (Table 1) [1] [2] [3] [4] [5]: the effect of effects, measurability of effects, range of impact effects, distribution of effects in time, areas where effects occur, who they influence effects, channels of impact effects, durability of effects.

The visible effects of logistic infrastructure investments are objectively stated and have an economic nature. In turn, visible and irrational effects are attributed to subjectivism and are usually classified in the group of socio-economic effects, and the effects of environmental protection and security referred to in the literature as external effects.

IOP Conf. Series: Materials Science and Engineering **568** (2019) 012083 doi:10.1088/1757-899X/568/1/012083

Table 1. Effects classification of logistic infrastructure investments

Classification criterion	The division results
The way the effects are affected	• Positive effects: benefits, savings • Negative effects: disadvantages, losses
Measurable effects	• Measurable effects • The effects are difficult to measure • Immeasurable effects
The range of impact effects	• The effects felt local, regionally, in the country on an international scale • The effects affecting the entire economy, the effects acting on selected sectors of the economy
Distribution of effects over time	• Short-term, medium-term and long-term effects
Types of effects	• Economic, social, environmental (external) effects • Demand and supply effects
Entities impact	• Effects felt by TSL enterprises, enterprises from other industries • External effects (felt by the environment)
Effects channels	• Direct effects • Indirect effects
Durability of effects	• Constant effects • Temporary effects

Source: own study based on [1] [2] [3] [4] [5].

In a large group of effects generated by the macro-logistic infrastructure, it seems to be superior to positive and negative ones. The positive effects of logistic infrastructure investments are manifested in the form of benefits and savings, although sometimes the indication of a clear border between these forms is challenging. The benefits of the logistics infrastructure of the economy are related to its impact on the functioning and development of the economy, recognized as both a structural and functional whole, as well as a division structure. This means that the benefits created by logistic infrastructure may arise at the micro, meso and macroeconomic levels and are a function of its impact on revenue growth, employment growth, investment growth, increase in commodity exchange, etc.

A well-developed macro-logistics infrastructure does not necessarily have to bring only benefits. One should also identify negative effects, which mainly include effects of the impact on the natural environment. Because this issue is the subject of further chapters, hence the detailed characteristics of the negative impact of construction, development, modernization, and operation of the logistics infrastructure on the natural environment will find its place there.

Benefits derived from logistic infrastructure investments can be quantifiable (measurable) or hard to quantify. Rational benefits include, above all:

a) regarding road infrastructure:
- an increase in the speed of travel (among others due to better quality and technical parameters of the road), leading to indirect benefits in the form of shortening the delivery time, reducing the value of inventory in transit, increasing the efficiency of transport means;
- reduction of effective operational costs of transport - among others reducing fuel consumption, reducing repair costs;
- increased transport accessibility, resulting in a shorter route, shorter transport times, lower transport costs;
- increased traffic safety translating into lower accident rates, and indirectly - reduction of losses and damage to cargo, means of transport as well as infrastructure itself;
- reduction or elimination of congestion, which is reflected in the time and cost of transport;
- reduction of transport costs in the total cost of logistics processes;

b) regarding logistic centers:
- reduction of service time thanks to modern technical and organizational solutions;

IOP Conf. Series: Materials Science and Engineering **568** (2019) 012083 doi:10.1088/1757-899X/568/1/012083

- reduction of employment due to mechanization and automation of warehouse and handling processes;
- reduction of service costs (including transport costs), which results from dependencies defined as cost concessions;
- reduction in the level of inventories, and what is related to this, and inventory costs, in distribution channels by reducing or eliminating intermediate links and introducing deconsolidation and consolidation reloadings, i.e., cross-docking;
- increase in the outsourcing of warehouse functions leading to a reduction in storage costs in the total cost accounting of logistics processes;
- centralization of logistics services provided by logistics service centers, resulting in lower total costs of logistics processes.

One should remember about the synergistic effects [6] resulting not only from the complementary relations of individual elements of the logistics infrastructure of the economy but also from the recognition of them as an integrated structural and functional coplanar of the implementation of logistics processes (network). The synergy will, therefore, result in the benefits gained as a result of the development of the logistic infrastructure multiplied. It can also trigger completely new effects that will not occur in the case of investments of individual elements of the logistics infrastructure carried out independently of each other. One can, in particular, mean the coherence of a logistics system, e.g., a supply chain, which is one of the basic conditions of its efficiency.

When it comes to the hard-to-come benefits of infrastructure investments, their precise definition causes some problems, and therefore, they are usually established based on estimates. As an example: social savings resulting from the reduction of the number of accident victims on the used transport infrastructure, reducing the level of external costs through a smaller impact on the natural environment, increasing the added value in goods and services as a result of extending the network of cooperative links between business entities, increased revenues as a result of expanding sales markets, improving the flow of goods between regions, increased revenues of transport companies due to better development opportunities for transit transport.

In the literature mentioning measurable and hard-to-measurable effects of transport infrastructural investments, the non-measurable effects are omitted. Undoubtedly, they should be included, because the development of logistics infrastructure brings not only effects that can be expressed using numbers, but also those that can not be measured or counted. Such effects include aesthetic feelings caused by changes in the environment as a result of the construction or expansion of the logistics infrastructure. More specifically, the aesthetic evaluation may concern elements of logistic infrastructure as well as the landscape in which particular infrastructure elements are integrated. In the first case, one may ask whether users, in addition to functionality, expect from some roads or logistic centers also some aesthetic impressions and what would have to consist of them because they are the largest group of evaluators. In the second case, the group of potential assessors consists primarily of members of the local community and tourists who want to use the aesthetic qualities of a given landscape.

The issues of shaping and protecting the aesthetic values of the logistics space of infrastructural investments are still insufficiently appreciated in Poland. It is also visible in the technical field because few authors deal with this subject [7] [8]. In the case of logistic infrastructure investments, however, the highest priority is given to technical, legal, functional, or ecological issues. The reasons for this state should be found in the legal system regulating spatial development issues because these problems are not identified in Polish legislation. There is the problem of methods for assessing aesthetic impressions, even for the simple reason that their reception is always subjective and depends on many factors, such as education, sex, age, whether or mood of the recipient. It is also known that human perception is inherently multisensory, although in a particular situation it can be compelled to be dominated by one sense, e.g., in the situation of excessive noise. In the case of landscape, beside the dominant visual stimuli, sound and smell stimuli play an especially important role in perception. Beautiful patterns regarding, for example, engineering structures or storage centers prevailing in the eighties of the twentieth century are difficult to compare to current standards.

The significant logistical effects of infrastructure investments can also include satisfaction when using the logistic infrastructure obtained due to the quality of the road surface, road network density, convenient connections in the road network, spatial availability of logistics centers, logistics centers, etc. Analyzing the phenomenon of satisfaction, usually, the states are accented positive, such as, for example, joy, a feeling of pleasure, but you can also be interested in negative states such as stress, effort or tension. However, in both cases, there are no clearly defined measures to measure these states. Another substantial effect of the development of logistics infrastructure is the improvement of the region's image, which is manifested by the increase of investment attractiveness.

The effects of logistic infrastructural investments can be felt locally, regionally, nationally or internationally. The range of radiation effects depends primarily on the size and function of the logistics infrastructure. Logistics service centers located in seaports will generate effects on an international scale, while the effects of the cross-docking center's functioning can be felt within the limits of the regions served. It is important that some effects may occur in parallel on several scales. It is difficult to delimit local and regional effects, and if we take into account Euroregions separated from abroad (e.g. Euroregion Pomerania, Euroregion Meuse-Rhine, Euroregion Szeszupa), then it is challenging to distinguish regional effects from international ones..

The time-based criterion allows distinguishing the effects of logistics infrastructure development on short, medium, and long-term. In terms of short-term impact, the effects of logistic infrastructural investments are related to the period of investment implementation and result from the involvement of production factors (labor resources, raw materials, machinery, etc.) in the construction process. The medium and long-term effects that appear after the end of the construction phase of infrastructure facilities are of key importance and are associated with the greater attractiveness of the area for functioning enterprises and potential investors. The most important medium- and long-term effects, involving business entities operating in an environment with favorable infrastructure potential, include the increase in productivity of resources used by the company, streamlining logistics processes related to supply, production and customer service, reducing the company's costs.

The effects of logistic infrastructure investments can be analyzed as demand and supply [9]. Demand effects result from short-term development impulses, directly related to the implementation of the investment. They concern creating demand for work and materials (resources).

The effects generated by the logistics infrastructure are the participation of various entities. In the simplest terms, they will be operators from the TFL sector and cargo holders that are entities using logistic services offered by the TFL sector and entities that independently implement logistics processes, have their fleet or own warehouse. Looking more broadly, the entire national economy will contribute to the emergence of the functioning of the logistics infrastructure. Therefore, the effects of logistic infrastructure investments can be seen in the category of impacts of logistics infrastructure on management processes implemented on the micro-, meso- and macroeconomic scale.

J.G. Heggie [10] divided the effects of transport investments on direct and indirect. One can use this division and refer it to logistic infrastructure investments, seeing as direct costs of capital and periodic costs of investment maintenance, value of travel time or storage of cargo, accidents costs, noise costs, and impact on real estate or land value, increased demand for transport or storage, impact on the environment as indirect effects.

In the works of such authors as F. Rietveld [11] [12], F. Bruinsma [12], T.R. Lakshmanan, P. Nijkamp, E.T. Verhoef [13] effects of transport infrastructural investments are characterized in terms of infrastructure projects appearing at various stages of the cycle, simultaneously taking into account the duration of impacts. Based on these criteria, they have distinguished temporary and permanent effects. About logistic infrastructure investments, temporary effects will be generated at the construction stage, and they will be demanding effects of a direct and indirect nature. In addition, there will be temporary external effects, such as noise or other environmental impacts when carrying out construction works. In turn, the permanent effects of logistic infrastructural investments are visible in the operation phase and occur for as long as the use of the logistic facility continues. It is therefore obvious that the permanent effects of logistics infrastructure development may imply logistic solutions based on the more intensive use of transport, such as on-time delivery, various small quantities of goods, cross-docking, warehouse

IOP Conf. Series: Materials Science and Engineering **568** (2019) 012083 doi:10.1088/1757-899X/568/1/012083

function outsourcing as well as multi-variant reconfigurations supply chains made at various geographical levels..

The permanent effects of the development of logistic infrastructural investments identified at the operational stage are also noticeable on the macroeconomic scale. For example, in the long-term perspective, assuming the slow pace of changes, permanent effects will be visible in the size of employment or GDP. During the operation of the logistics infrastructure, external effects will also arise. However, noise, which will be the source of infrastructure users, accidents, or air pollutants, will constantly be, but with varying intensity. The original landscape aesthetics will be permanently disturbed.

3. Conclusions

Considerations on the effects of logistic infrastructural investments lead to the belief that all empirical analyses should be made in a multifaceted way, using the matrix relations that occur between special effects. Also, the effects of logistic infrastructure investments can be subject to ex-ante and ex-post analysis. In the first case, the anticipated effects serve as a justification for the commencement of an infrastructural undertaking. On the other hand, ex-post analyses - apart from the cognitive value - show which effects should be intensified and which should be limited or eliminated.

References

[1] Grzywacz W 1982 Transport infrastructure (Infrastruktura transportu) (Warszawa: WKiŁ) p. 77-79.

[2] Kamińska T 1999 Macroeconomic assessment of the effectiveness of infrastructure investments on the example of transport (Makroekonomiczna ocena efektywności inwestycji infrastrukturalnych na przykładzie transportu) Gdańsk: Wydawnictwo Uniwersytetu Gdańskiego

[3] Rolbiecki R 2009 Transport infrastructure as a factor shaping the conditions for the development of enterprises in the socio-economic environment (Infrastruktura transportu jako czynnik kształtujący warunki rozwoju przedsiębiorstw w otoczeniu społeczno-gospodarczym) (Gdańsk Wydawnictwo Uniwersytetu Gdańskiego) p. 61-67.

[4] Rosik P and Szuster M Expansion of transport infrastructure and economy of regions (Rozbudowa infrastruktury transportowej a gospodarka regionów) (Poznań: Wydawnictwo Politechniki Poznańskiej) p. 18-24.

[5] Dembińska I 2018 Macrologistics infrastructure in terms of environmental determinants of sustainable development (Infrastruktura logistyczna gospodarki w ujęciu środowiskowych uwarunkowań zrównoważonego rozwoju) (Szczecin: Wydawnictwo Naukowe Uniwersytetu Szczecińskiego), p. 112.

[6] Mańkowski C 2009 Synergy in logistics (Synergia w logistyce) (Gdańsk: Wydawnictwo Uniwersytetu Gdańskiego) p. 115.

[7] Forczek-Brataniec U 2008 View from the road. Landscape in dynamic perception (Widok z drogi. Krajobraz w percepcji dynamicznej) (Katowice: Wydawnictwo ELAMED)

[8] Jędraszko A 2005 Spatial development in Poland - roads and off-road legal regulations (Zagospodarowanie przestrzenne w Polsce - drogi i bezdroża regulacji prawnych). Unia Metropolii Polskich (Warszawa: Wyd. PLATAN)

[9] Domańska A 2004 Impact of transport infrastructure on regional development (Wpływ infrastruktury transportu na rozwój regionalny) *Problemy Ekonomiki Transportu* 2 p. 95-99.

[10] Hegge J G 1978 Economics of transport investments (Ekonomika inwestycji transportowych) (Warszawa: WKiŁ) p. 78.

[11] Rietveld P 1989 Employment Effects of Changes in Transportation Infrastructure: Methological Aspects of the Gravity Model *Papers of the Regional Science Association* Vol. 66, pp.19-30.

[12] Rietveld P and Bruinsma F Is Transport Infrastructure Effective? Transport Infrastructure and Accessibility: Impacts on the Space Economy (Berlin: Springer Verlag) p. 46.

[13] Lakshmanan T R, Nijkamp P, Rietveld P and Verhoef E T 2001 Benefits and costs of transport. Classification, methodologies and policies *Regional Science* Vol.80 pp. 139–164.

IOP Conf. Series: Materials Science and Engineering **568** (2019) 012084 doi:10.1088/1757-899X/568/1/012084

Approaching the labor market from a generational perspective

D Pop[1] and M T Pop[2]

[1] Department of Industrial Engineering, FIMT, University of Oradea, Romania
[2] Department of Engineering and Management, FIMT, University of Oradea, Romania

dpop@uoradea.ro, pomt@uoradea.ro

Abstract. Generation research has grown even broader and has deeply extended in the last 30 years. It can be said that the use of research results related to generations in the human resources management is a certainty in to use the best recruitment, selection, and integration methods of employees, to implement the best methods and techniques of stimulating the employee's creativity and innovation in organizations, communication, etc. An important phenomenon emerged as a result of the generational conflict, which has been highlighted in recent years, is that of the young people named "idlers", i.e., those inactive on the labor market, called NEET.

1.Introduction

The study of human evolution at the individual, community, or even global level, as human society has been a constant concern of researchers from all areas of its existence. This has led, in essence, to the current human development. The study of human generations is subsumed to this idea. Thus, the research works on human generations have become a constant concern of the past decades.

The study of generations was originally associated with the artistic field (painting, sculpture, literature, etc.), historiography, etc. Subsequently, generation research has grown even broader and has deeply extended in the last 30 years. However, there are also opinions that consider the generational theory as a superficial view regarding human evolution, sometimes even of no scientific substance.

With all the existing controversies, it can be said that the use of research results related to generations in marketing is a certainty - to study the consumers behavior, in the human resources management - to use the best recruitment, selection and integration methods of employees, to implement the best methods and techniques of stimulating the employees creativity and innovation in organizations, communication, etc.

2. Essential Aspects of the Generational Theory

William Strauss and Neil Howe (1991)[1], experts in public policy, are the researchers who, at the beginning of the 1980s, defined a theoretical cycle of human generations for the United States of America starting with the year 1584. They define a social generation as an ensemble of all people born during about twenty years, a period they consider as equivalent to the stages of human life: childhood and adolescence, young age, maturity, and advanced age. These generations belong to cycles longer than 80-100 years that they call "saeculum." The end of each saeculum is marked by a crisis that is then followed by a recovery period characterized by the fact that the community institutions and values become stronger. The next generations conflict with the previous ones, this fact generating a weakening of the institutions and values of the previous saeculum, which leads to a tumultuous political

IOP Conf. Series: Materials Science and Engineering **568** (2019) 012084 doi:10.1088/1757-899X/568/1/012084

environment creating the probability of another crisis occurrence, this phenomenon being cyclically resumed. The succession of cycles has also been demonstrated in the evolution of countries other than the United States of America, which has led to an extensive and thorough study of generations.

Subsequent research made by the two authors has led to refinements and thorough approaches of the original theory. Also, other researchers from different fields of activity have approached this topic in their studies. Among them, we can remember the following: David Kaiser, Martin Keller, David Riesman, Douglas Coupland, etc.

Karl Mannheim spoke long before Strauss and Howe about generations, even in the 1930s. There are even some voices who believe that Mannheim is the father of the theory of generations, but he tackles generations differently, considering them as a succession of stages in the historical evolution of society, as well as from the perspective of the relations between them. He considers the generations as specific social-historical groups with defined development intervals that have particularities related to geographical or even to different countries and which have generational characteristics adapted to the historical developments in the respective territories. Subsequent studies have deepened this theory, have outrun the stage of sociological analysis and have led to its use and expansion in areas such as human resources management, marketing, psychology, sociology, etc.

William J. Schroer [2] is one of the American researchers who deepened aspects of the theory of generations initiated by Strauss and Howe, analyzing the characteristics of generations from the beginning of the last century up to present. He delimitated the generations from the beginning of the 20th century by defining them as it can be seen in Table 1. It contains the years of birth and maturity, the current age, and the features of each generation briefly described.

Table 1. The generations from the beginning of the 20th century

Generation	The interval of birth years	The interval of maturity years	The interval of the current age	Characteristics
The Silent Generation	1925 - 1945	1943 - 1963	74 - 94	Born in a time marked by the Great Depression and World War II; They have a secure moral, are managing, responsible, hardworking, and love their family.
The Boomers	1946 - 1966	1964 – 1984	53 - 73	The optimistic generation that had significant economic opportunities specific to the post-World War II period; become individualists, oriented to their interests, and skeptical towards the media.
Generation X	1967 - 1976	1985 – 1994	43 – 52	It is the generation of children with divorced parents, raised by a single parent; it is considered the most educated generation ever so far, with the highest percentage of graduates; they are pragmatic but cautious; in Romania, it is sometimes called the generation of sacrifice, due to the fact that they have passed in there through the historical episode of 1989, a moment after which this generation found it very difficult to define itself, being formed and educated in communism, but living after 1990 in a totally different society.
Generation Y (Echo-	1977 - 1994	1994 - 2012	25 - 42	Technology-oriented generation; the individuals of this generation are creative, sophisticated in

IOP Conf. Series: Materials Science and Engineering **568** (2019) 012084 doi:10.1088/1757-899X/568/1/012084

Boomers, Millennials)				knowledge,skeptical of the traditional marketing elements, loyal to the brands they like; it is considered to be the generation that seeks mentoring, counseling, but also wishes respect, inclusively by the quality products and services it targets; it is the first generation of digital natives.
Generation Z	1995 - 2012	2013 - 2030	24 - 7	The individuals of this generation are growing up with the computer and mobile phone at hand; they have easy access to the internet; they are multi-tasking; they are getting bored quickly; in short, they are: digital, social, global, mobile, flexible, independent and visual; have entrepreneurship ability;
The Alpha Generation	2013 -	2013 -	6 -	It is a generation that is now born and the studies on it are still in progress; it is considered to be the most educated generation, willing to learn, best placed financially and the most longeval; they are oriented to the present, the past is less critical; are very social but on-line; it is difficult to communicate face to face with the individuals of this generation, even the communication with their parents being considered severe.

Regarding the data in Table 1, the following aspects should be specified:

a. The field literature defines the Boomers as having two stages: Boomers I or Baby Boomers (1946-1954) and Boomers II or Generation Jones (1955-1966); for the United States, these periods have different generational characteristics, but for Romania we believe that this is the generation born during communism, and from this perspective the two periods, Boomers I and Boomers II, have similar generational features; at the same time, we consider that, for Romania, the end of the Boomers generation is 1966 and not that mentioned by the American researchers (1965); therefore, we consider that 1967 is the first year of Generation X, this being the one in which children were born after the entry into force of the Decree no. 770 of 1966 for the regulation of the interruption of pregnancy, legal regulation which led to doubling the number of children born in 1967 as compared to 1966; these children have specific generational characteristics: they are always in competition (they started their life in the context of competition for places available in kindergartens, schools, lyceums, facultie, etc.), they planned the resources of any kind they had (the generation that best understood the phrase "make economies at ...") and have had a transitory period of childhood / adolescence / youths of significantly different historical stages (communism and democracy, the period of transition to capitalism) through the December Revolution from 1989; in Romania, this generation is called the Generation of Sacrifice as, after the historical moment 1989, they were forced to learn to live in a completely new society, thus requiring great efforts to adapt in economic, cultural, social, civic, emotional way, etc. [3].

b.Generation Z is a generation that has just entered the labor market, so their employee behavior is still under analysis; however, studies have shown that due to their "multi-tasking" skills they are willing to work more individually than in teams, thus being characterized by a developed entrepreneurial spirit.

c.Mark McCrindle[4] calls the Alpha Generation (this is the most common name) as Tech Generation or Net Generation; McCrindle considers this to be the most educated and dynamic generation of mankind; also, he estimates that another feature of this generation is longevity, and from the family's point of view, the individuals of this generation will form families late and will be very preoccupied with the study, they will be those who will experience the transition from auditory and visual to a kinesthetic process; specialists studying the generational features of those in the Alpha Generation believe that its members will be difficult to motivate and provoke; at the same time, the technology, whose evolution is explosive, will be the one that will create a gap between the generations Z and Alpha; many studies are in progress and disputes have arisen regarding the physiological availability of the human body to process the high level of information that members of this generation are accessing, the limitation of the physiological rest time, the time necessary to be spent in nature, without "intelligent devices", the detachment from the realities of everyday life because of the virtual world they are constantly accessing (in many cases even preferring it to the real one), the loss of orientation towards their own interests etc; we appreciate that there are still many uncertain things regarding the characteristics of the Alpha generation and many unknown factors, the studies that have been made so far have an estimated value; just pursuing and analyzing the evolution of this generation at different stages of life will define their indubitable features.

d.An important, topical, common element is the generational identity, with which all human generations are analyzed, described and characterized; although there are denials of it, "the specialized analyses identify three main elements that determine the" identity "of a generation, namely: (1) the socio-economic context in which the individuals were born and living, (2) the attitude towards family and parenting style, and (3) the relationship with technology."[5]

3.The approach of the labor market from a generational perspective

According to a ManpowerGroup [6] study conducted globally between February and April 2016, it emerged that, in 2020, the generation-based labor force will have the structure from Figure 1: 35% belong to X and Y Generations, 24% belong to Generation Z, and 6 % are from the Boomer generation.

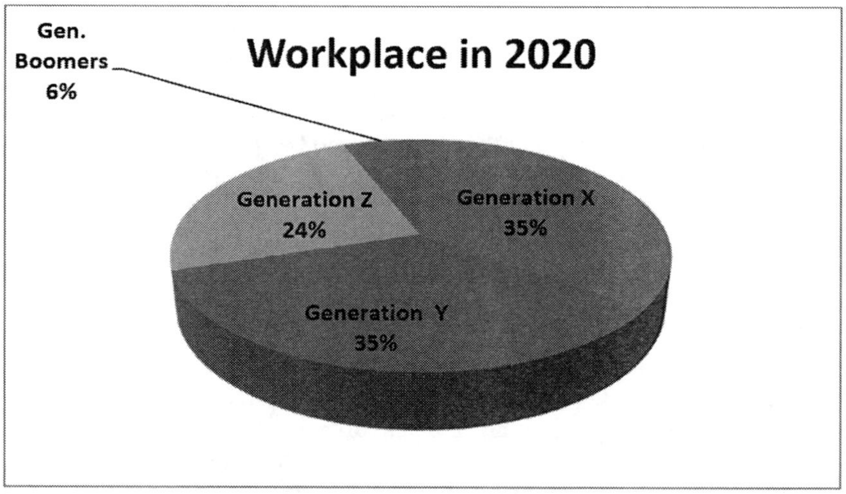

Figure 1 - Labor Market Structure on Generation in 2020

As a result of the above, the existence of four generations with different characteristics on the labor market, the joint work in the same organization of the exponents of these four generations is a real challenge. Beginning in 2025, the Boomer generation will naturally withdraw from the labor market and will remain the three generations that are now active: X, Y, and Z. Moreover, with 2029, they will slowly make their presence on the labor market the first representatives of the Alpha generation. In 2050, generation X will exit this market, being replaced by the Y generation and completed by generations Z and Alpha. As a result, the overall characteristics of the labor market are constantly changing according to the characteristics of the predominant generation (s), the supply of labor is influenced by the generational specificity.

A constant feature of the economic life is related to the fact that the labor market is permanently assured with a labor force of a specific structure of age, studies, environment, etc. but, at the same time,

IOP Conf. Series: Materials Science and Engineering **568** (2019) 012084 doi:10.1088/1757-899X/568/1/012084

the characteristics of the supply are also influenced by the types of generations that are predominant. This requires modeling of the behavior of the exponents of labor demand, of the organizations, that are forced to adapt and change aspects of relationships with employees and potential candidates for the available work. This can be already noticed in the following areas of human resources management of companies:

- the recruitment, selection, and integration of staff
- the management of employee rewards
- the way of organizing work and rest time
- in terms of working conditions, the endowment of workplaces, work environment, etc.
- the communication with employees
- their involvement in the decision-making system;
- the organization of work - individually or in teams (including team building), but also in the way of organizing the work itself, of managing the organizational structure of the firm;
- the defining of the type of work and the work place;
- training at the workplace and the support of training outside of it, gaining the necessary skills etc.

All these aspects will be totally differently accomplished compared to the way in which these organizations are doing these activities at present, and this can be already noticed: attractive and interactive employment announcements, social media recruitment channels, flexible work schedules and fractions norms, homework, virtual offices, virtual internal communication systems, lifelong learning facilities, the existence of an employer brand etc.

4. Particularities of the current labor market – the NEET category

An important phenomenon, emerged as a result of the generational conflict, which has been highlighted in recent years, is that of the young people named "idlers", i.e. those inactive on the labor market, called NEET. This notion defines a category of individuals but has expanded as the name of an indicator that evaluates the labor market. The term NEET ("Neither in Employment nor Education and Training" or "Not in Employment, Education or Training") is used to describe the category of young people aged 15-29 who are neither employed nor follows an education or training program. In this category, NEET includes the unemployed in the mentioned age group, but also the inactive persons who are not included in the form of education or training.

The NEET's young people are sometimes referred to as the "lost generation" or "social parasites". They are young people who, even if they have completed a form of education, do not engage (for various reasons), obtain their income either from undeclared or illegal or non-taxed sources, make successive internships without subsequently employing in any of the firms, prefer to stay with their parents and be supported by them etc. There is also the category of those who are the result of school dropout, who have not completed any form of education and thus significantly limits their access to a job that could provide them a decent life.

Regardless of the manifestations of the NEET phenomenon, we consider that among the causes of its appearance and proliferation, there are elements related to the specifics of generations that can be detailed as follows:

- the current education system is the result of the way of thinking and action of those in the Boomer and X generations and its adaptation to the demands of the generations Y and Z does not have the necessary pace for these last generations to consider the school as attractive and beneficial to their interests; this has led to the phenomenon of early school leaving, to the widely accepted idea of young people that information taught in school is useless for their active life at work, leading to frustrations and dissatisfaction, as well as to the attendance of the educational cycle with a minimum performance (far below the potential of their generation) which only provides them with a diploma (without correspondent in their knowledge); at the same time, the communication with teachers from Boomer and X generations, generations in the mature stage of human life, is deficient and ambiguous; because of this cleavage, there is even a generation-to-generation conflict; we believe that the Alpha generation is influenced by the current education system, but its effects on children can only be estimated, they will be evaluated at the time of their access to the labor market.

- similar to education, the systems and institutions in the economic, political, social, and cultural spheres created and developed by the older generations did not follow such rapid and profound development as compared to the particular development and characteristics of the current young generations: Y and Z;

- a characteristic of the Z generation, for example, is the orientation towards volunteering, which leads to their inclusion in the NEET category, but they are in fact useful for society, even if they are not active on the labor market; from this perspective, we believe that stimulating their entry into the labor market with part-time and flexible programs would provide the need and satisfaction of volunteering with integration into the active population, with the right benefits for the individual and society; in order to meet them, the employers can turn to partnerships with organizations where their employees can volunteer;

- the different types and ways of performing work offered by employers have led to a contradiction between their workforce needs and the expectations (wishes) of the younger generation; this has led to the following process: many young people are giving up or permanently looking for a job to meet their expectations; at the same time, their orientation towards individual work, on their own, without the constraints of an organizational system with strict procedures, with a manager to "give orders", with fixed work schedule etc., has turned many young people towards a freelancer type job that provides income from work (even if they are variable in time), but which is in the gray or black area in terms of declaring work and taxing it, which makes them appear in no official statistics.

The above elements are only a few aspects that prove the existence and causality of NEET. The phenomenon itself and its study are just at the beginning, but the economic effects of their existence in society have been continuously evaluated and monitored because they represent an essential resource for society. Thus, at the European level, for the year 2017, the average NEET share was 14.2% of the 18-24-year-olds (over 38 million people), is different in the European Union countries, as shown in Figure 2.

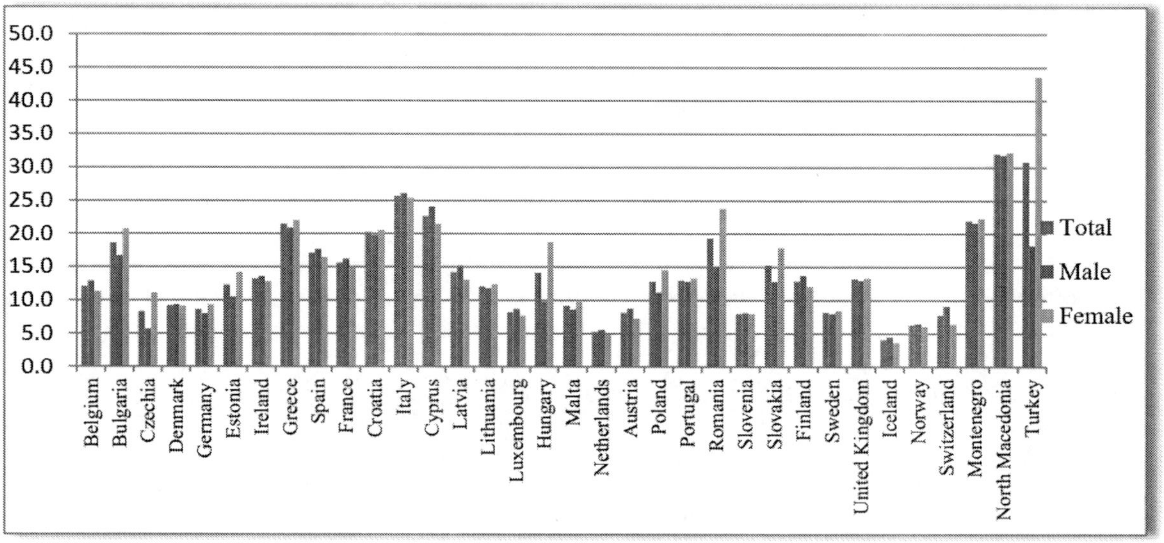

Figure 2 - The share of NEET in the total population of the EU 18-24 years (2017)[7]

For Romania, the situation is better than in countries such as Italy, Cyprus, Greece, and Croatia, this percentage being at the level of 19.3% (below 20%). However, it is alarming that almost a quarter (23.8%) of young women aged 18-24 are not active in the labor market.

All these young people, in addition to the economic, social, emotional issues they have at an individual level, generate, for society, a series of costs that are not, however, insignificant. Thus, Eurofound estimated, for 2011, the economic losses generated by the no-commitment of young people on the labor market, this being of 153 billion EUR, equivalent to 1.2% of European GDP, with variations per each member country. In 2015, estimates led to about 142 billion euros, representing benefits or profits and taxes lost by each country[8]. These figures demonstrate the significant impact on the EU economy that the maintaining of a part of NEET in the inactive area may provide.

5.Conclusions

Aspects related to the generational approach of the labor market are becoming more and more apparent, and the researchers in the field are increasingly concerned, as well as the governments of the world concerned with the needs and interests of the citizens of the countries they lead. We believe it is of interest to deepen this topic and to find the most appropriate solutions for all parties involved: individuals, employers, communities, states, etc.

References

[1] Strauss, William, Howe, Neil (1992). *Generations: The History of America's Future, 1584 to 2069*, William Morrow & Company, 1992, ISBN 0-688-11912-3.
[2] Schroer, William J. (2008) Generations X, Y, Z and the Others, The Portal, Household Goods Forwarders Association of America, Inc., vol. XL, pg. 9.
[3] IRES Survey– X Generation (2015), February
[4] McCrindle, Mark (2014) The ABC of XYZ: Understanding the Global Generations/ Mark Sydney: UNSW Press, pp. 218 – 228
[5] Pîslaru, Dragoş Nicolae (coordinator), Botezatu, Elena, Popescu, Andrei, Stăniloiu, Iolanda (2018) European Social Rights Pillar: Reducing inequalities and social disparities in Romania. Evolutions of European Youth Policy and Initiatives, European Institute of Romania, Policy Studies and Policies Studies SPOS 2018, Study no. 4, p.71;
[6] ManpowerGroup - Career and Y generation: Vision 2020. Information, data and practical advice from labor experts, 2016
[7] Source: Eurostat: http://appsso.eurostat.ec.europa.eu/nui/setupDownloads.do, ac. 27 Apr. 2019.
[8] https://www.eurofound.europa.eu/ro/topic/NEETs

IOP Conf. Series: Materials Science and Engineering **568** (2019) 012085 doi:10.1088/1757-899X/568/1/012085

Focusing Device Based on Overconstrained Mechanism

L D Racila[1], I Dumitru[1*], D Tutunea[1], L Matei[1], I Geonea[1], O Otat[1], M Racila[2]

[1] Faculty of Mechanics, University of Craiova, Romania
[2] Department of Applied Mathematics, University of Craiova, Romania

* corresponding author: dumitru_ilie@yahoo.com

Abstract. The paper presents an application based on an exceptional 6R overconstrained mechanism. This mechanism is started from 6R Wohlhart symmetric mechanism, not in its classical spatial position but a different one, with three non-adjacent joints constrained to remain in a fixed plane. This disposition allows the conception of an interesting device, a focusing table, capable of obtaining good accuracy.

1. Introduction

The paper analyzes a mechanism with six revolute joints in a closed loop, namely Bricard rectangular mechanism or Wohlhart symmetric mechanism. Overconstrained mechanisms with revolute joints are composed of four, five, or six joints. Sarrus [1] presents in 1853 a 6R closed-loop mechanism with six revolute joints, transforming a rotation movement in a translation one. Raoul Bricard described in his papers [2,3] six types of overconstrained linkages, three mobile octahedral mechanisms, and three symmetric mechanisms. J.E. Baker analyze in 1980 these mechanisms [4], giving the closure equations for all these mechanisms. Karl Wohlhart [5] presents in 1987 a new overconstrained mechanism with six revolute, considered as a generalization of Bricard rectangular (or threefold symmetric) mechanism. Other researchers have studied these types of mechanisms, like Bennett [6], Myard [7] or Goldberg [8], Waldron [9], Mavroidis and Roth [10] or Dietmaier [11].

Overconstrained mechanisms have been studied by researchers, initially for kinematical point of view and in the last years from their application point of view. Many authors, like Schatz [12], Gan and Pellegrino [13], Chen [14] and more recently Ding [15] or Song [16] have presented possible applications of these mechanisms.

A new possible application of overconstrained mechanisms have been presented by Racila [17, 18], based on 6R Wohlhart symmetric mechanism, obtaining a 6R translational device. A generalization of this translational device was made it by Zeng [19], with a new design based on general forms of spatial overconstrained linkages.

The 6R translational device have three non-adjacent joints in a fixed horizontal plane, while the other three joints will be forced by the new design to remain in a plane, parallel with the horizontal plane

2. The new spatial disposition

In its classical position (figure 2a), the Bricard rectangular mechanism is analyzed with the help of Denavit-Hartenberg parameters. In figure 1 is shown three revolute joints using Denavit and Hartenberg notations, where θ_i is the kinematical variable and a_i, α_i and d_i are geometrics parameters.

IOP Conf. Series: Materials Science and Engineering **568** (2019) 012085 doi:10.1088/1757-899X/568/1/012085

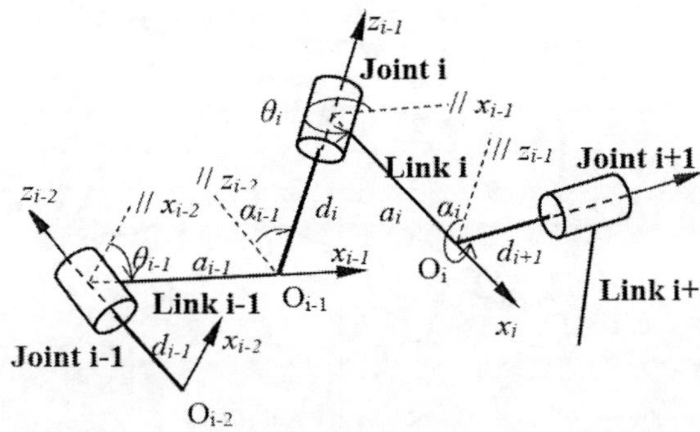

Figure 1. Denavit – Hartenberg parameters

The condition to obtain a closed loop mechanism is that the six transfer matrix product to be equal with the unity matrix:

$$^{6}Q_1 \cdot {^{1}Q_2} \cdot {^{2}Q_3} \cdot {^{3}Q_4} \cdot {^{4}Q_5} \cdot {^{5}Q_6} = I \tag{1}$$

Using D-H parameters, the input - output equation between the input angle θ and the output angle φ is [17]:

$$cos\,\theta \cdot cos\,\phi \cdot (1 + cos^2\,\alpha) + (cos\,\theta + cos\,\phi) \cdot sin^2\,\alpha - 2 \cdot sin\,\theta \cdot sin\,\phi \cdot cos\,\alpha + cos^2\,\alpha = 0 \tag{2}$$

In equation (2) α is the twist angle.

In the new spatial disposition (figure 2b), three non-adjacent joints are constrained to have movements in a horizontal plane to the centre of their circumscribed circle. The horizontal plane is supposed to be a fixed one. The others three joints will describe a plane, parallel with the horizontal one.

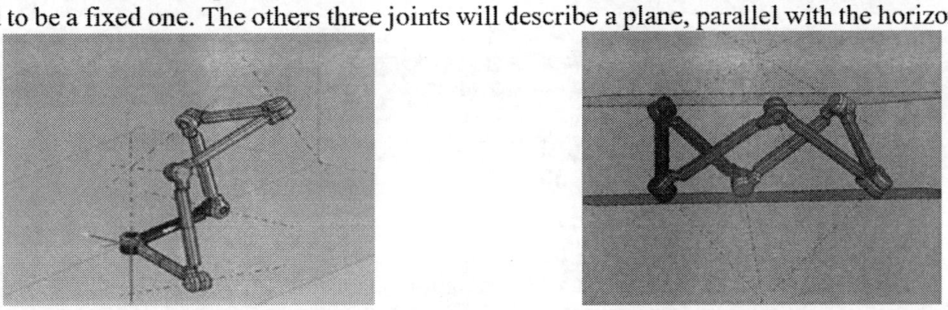

a) b)

Figure 2. The 6R mechanism in classical position (a) and the new spatial disposition (b)

The new design for the mechanism is considering (figure 3) and in the initial position all the six joints are in the horizontal plane and the input angle θ has zero degrees. The actuated joint is the revolute joint O_1, situated in the initial position in $O_1{}^1$. When O_1 is actuated, all the odd joints O_1, O_3 and O_5 make movements to the centre O of circumscribed circle of the triangle $O_1O_3O_5$. The even joints O_2, O_4 and O_6 will be situated in a parallel plane with the horizontal and fixed plane. The distance between the two planes is noted h and this distance depend to the input angle θ and the input distance b.

IOP Conf. Series: Materials Science and Engineering **568** (2019) 012085 doi:10.1088/1757-899X/568/1/012085

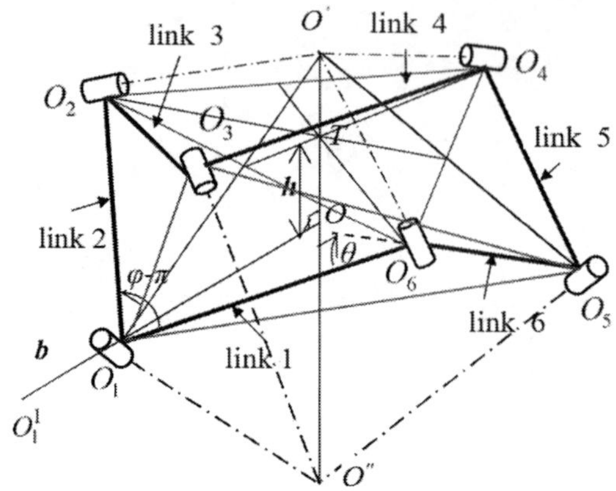

Figure 3. The the new spatial disposition in an intermediary position

This variation of the distance h between the horizontal plane (135) and the mobile plane (246) according to the input displacement b is [18]:

$$h = a \cdot \sqrt{\frac{4}{3} \cdot \left(1 - \frac{(2a - b\sqrt{3})^2}{4a^2}\right) \cdot \left(1 - \frac{a^2}{(2a - b\sqrt{3})^2}\right)} \tag{3}$$

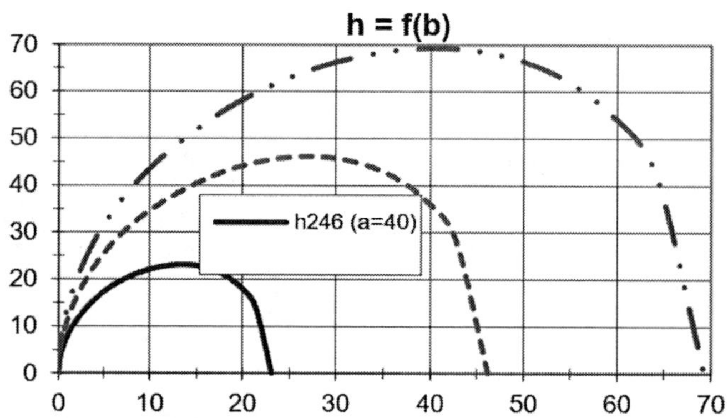

Figure 4. The variation of the height h according to the input distance b

This variation of the distance h between the horizontal plane and the mobile one, according to the input displacement b is given in figure 4 for three links length.

3. Focusing table

Starting from this new design, an interesting device can be obtained. The cross point of even joints axes noted O' and the cross point of odd joints axes, noted O", form a straight line O'O" always orthogonal to the two planes, the horizontal fixed plane and the mobile one.

For a twist angle $\alpha = \pi/2$, the initial position is defined by the all six links in the horizontal plane. In this position the point O' is to infinity because the axes of the odd joints are parallel. The O" point is situated in the same horizontal plane, with all six revolute joints. When the input parameter b increases the point O' will move to the horizontal plane and the point O" will move under the horizontal plane, arriving finally to infinity, when all six joints will be in the horizontal plane, but in a different position.

This property can be utilized in a opposite direction, when the input movement is the displacement of the cross point O" and the output parameter is the distance between the horizontal (135) and the mobile plane (246).

To obtain this dependence, we start from the input-output equation of Bricard rectangular mechanism, with a twist angle $\alpha=\pi/2$, a particular case of equation (2):

$$\cos\theta + \cos\phi + \cos\theta \cdot \cos\phi = 0 \tag{4}$$

We note by z the distance OO''. From figure 3 it can be written:

$$(OO'')^2 = z^2 = a^2 \cdot \cos^2\frac{\theta}{2} \cdot \left(\frac{1}{\sin^2\frac{\theta}{2}} - \frac{4}{3}\right)^2 \tag{5}$$

and finally:

$$\cos^2\frac{\theta}{2} = \frac{a^2-3\cdot z^2+\sqrt{(a^2-3\cdot z^2)^2+48\cdot a^2\cdot z^2}}{8\cdot a^2} \tag{6}$$

We note m(z) the expression:

$$m(z) = a^2 - 3\cdot z^2 + \sqrt{(a^2-3\cdot z^2)^2 + 48\cdot a^2 \cdot z^2} \tag{7}$$

The expression of the height *h* between the two planes according to the distance *z* is:

$$h = \sqrt{\frac{[m(z)-2\cdot a^2]\cdot[8\cdot a^2-m(z)]}{6\cdot m(z)}} \tag{8}$$

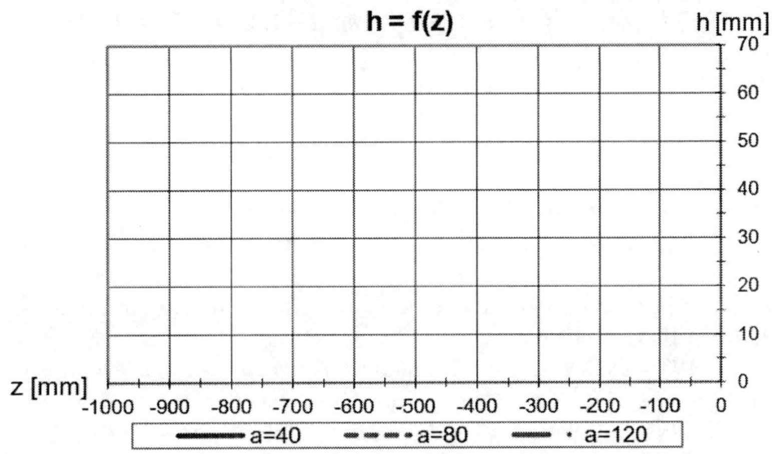

Figure 5. Variation of the height *h* according to *z* for three links length

In figure 5 is shown the variation of the distance OO", noted *h*, according to the input distance *z*, for three links length. A similar analyze can be obtained for different twist a variation twist angle, for example $\alpha=2\pi/3$ and $\alpha=5\pi/6$.

A virtual model of this device has been realized, confirming the accuracy of the analytical calculus. This accuracy can vary from 1:100 to 1:1000, depending of the links length and O"O variation domain.

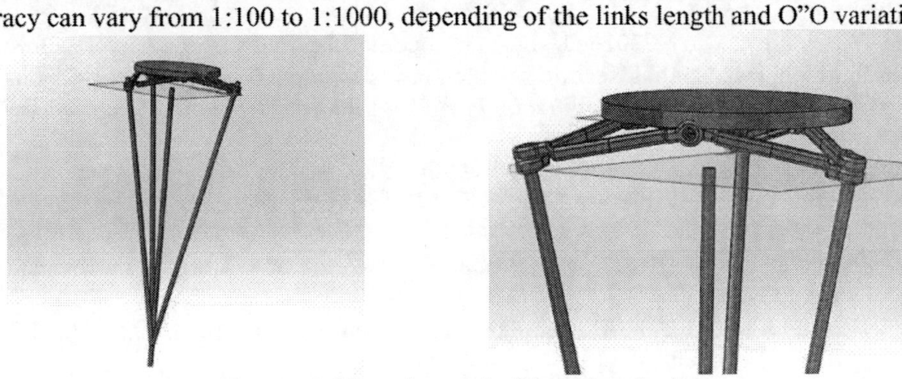

Figure 6. Virtual model of focusing device

IOP Conf. Series: Materials Science and Engineering **568** (2019) 012085 doi:10.1088/1757-899X/568/1/012085

4. Conclusion

The paper presents a 6R overconstrained closed-loop mechanism, not in its classical position, with one fixed element, but in a new spatial disposition, with no fixed elements. This special spatial position is used to obtain a new device, with three non adjacent joints imposed to make a particular movement in a fixed plane.

This device can be used to develop many applications, in the industrial area, or even in the entertainment domain. With the actuator placed in the cross point of the three nonadjacent joints of fixed plane, this device can be used, for example, to realize a focusing table for optical microscopes, or as plate table in rectifying machines, due to the its accuracy in this particular position. Some other devices are can be imagined, depending on the domain in which they are intended to be used.

References

[1] Sarrus P F 1853 Note sur la transformation des mouvements rectilignes alternatifs, en mouvements circulaires et réciproquement, *Comptes Rendus des Séances de l'Académie des Sciences*, vol 36, p 1036-1038

[2] Bricard R . 1897 Mémoire sur la théorie de l'octaèdre articulé, *Journal de Mathématiques pures et appliquées*, vol 3, p 113–148

[3] Bricard R. 1927 *Leçons de Cinématique, Tome II – Cinématique Appliquée*, ed Gauthier (Villars, Paris)

[4] Baker J E 1980 An Analysis of the Bricard Linkages, *Mechanism and Machine Theory*, vol 15, p. 267–286

[5] Wohlhart K 1987A New 6R Space Mechanism, *Proc. of the 7th IFToMM World Congress*, (Sevilla), p 193–198

[6] Bennett G T 1903 A new Mechanism, *Engineering* vol 76, p 777–778

[7] Myard F E 1931 Chaîne fermée à cinq couples rotoïdes, déformable au premier degré de liberté, *Comptes Rendus des Séances de l'Académie des Sciences*, vol 192, p 1352-1354

[8] Goldberg M 1943 New five–bar and six–bar linkages in three dimensions, *Transaction of the A.S.M.E.*, vol 65, p 649–663

[9] Waldron K J 1979 Overconstrained Linkages, *Environment and Planning B*, vol 6, p 393-402

[10] Mavroidis C and Roth B 1994 Analysis and Synthesis of Overconstrained Mechanisms, *Proc. of the ASME Design Technical Conferences*, vol. DE-70, p 115-133 (Minneapolis)

[11] Dietmaier P 1995 A new 6R space mechanism, *Proc of the 9th IFToMM World Congress*, (Milano) p. 52–56

[12] Schatz P 1975 *Rhytmusforschung und Technik* Ed Verlag Freies Geistesleben (Stuttgart)

[13] Gan W W and Pellegrino S 2003 Closed-loop deployable structures, *Proc. of 44th AIAA Structures, Structural Dynamics, and Materials Conference*, (Norfolk)

[14] Chen Y, You Z and Tarnai T 2005 Threefold-symmetric Bricard Linkages for Deployable Structures, *International Journal of Solids and Structures*, vol 42 p 2287-2301

[15] Ding W, Qiang R, Yao Y 2016 Design and Locomotion Analysis of a Novel Deformable Mobile Robot with Two Spatial Reconfigurable platforms and Three Kinematic Chains *Proc. of the Institution of Mechanical Engineers, Part C Journal of Mechanical Engineering Science*, vol 231 issue 8, p 1481-1499

[16] Song X, Deng Z, Guo H, Liu R, Li L and Liu R 2017 Networking of Bennett linkages and its application on deployable parabolic cylindrical antenna, *Mechanism and Machine Theory*, vol 109, p 95–125

[17] Racila L and Dahan M 2010 Spatial properties of Wohlhart symmetric mechanism, *Meccanica*, vol 45, p 153-165

[18] Racila L and Dahan M 2011 6R Parallel Translational Device *Proc. of the 13th IFToMM World Congress*, (Guanajuato)

[19] Zeng Q and Ehmann K F 2014 Design of parallel hybrid-loop manipulators with kinematotropic property and deployability, *Mechanism and Machine Theory*, vol. 71, p 1-26

IOP Conf. Series: Materials Science and Engineering **568** (2019) 012086 doi:10.1088/1757-899X/568/1/012086

Challenges of digitalization process in different industries. Before and after

V C Vartolomei[1] and S Avasilcai[1]

[1]"Gheorghe Asachi" Technical University of Iasi, Department of Engineering and Management, Mangeron Blvd. 28, TEX1, 700050, Romania

cristian.vartolomei@tuiasi.ro

Abstract. Since the appearance of the internet and new technologies, the business and socioeconomic sphere has taken a 360° turn. This gives us the opportunity to generate new business models, improve the analytical and predictive capacity both internally or externally and improve the company's processes in terms of efficiency and effectiveness. The digitalization of processes implies adapting to the new technological environment and represents an added advantage for all companies regardless of the sector they are engaged in or the projects they carry out. In this way, we investigate the challenges and needs of companies from different industries before and after the implementation of the digitalization process, which is defined as being a part of the Industry 4.0, also referred to as the Internet of Things (IoT). Also, this paper describes the benefits of a digitalized company and the impact of cybersecurity in the results along with the definitions of threats which can occur and the type of strategy which shall be adopted in order to avoid security risks.

1. Introduction

The last 10-15 years have seen a phenomenal development: the internet and later smart phone apps have changed almost every facet of our daily life. They have altered the way how we book travel, do our banking, watch TV, keep in contact with our friends, etc. [1] Nowadays, all these state of the art trends are possible due to the process of digitalization, which significantly replaced the way we spend our time, we live, we work and add value, and finally will update to a next level version the economy and the business environment.

The digitalization will influence or change every company in the world, regardless of the industry it belongs to. Adopting an appropriate strategy, along with the development of skills and competences that will meet adaptation to a new environment will be necessary to follow the use of the digitized context. For example, in the automotive industry, digitalization of approaches such as production, sales, distribution and, last but not least, suppliers, will significantly improve the quality of planning and management, along with streamlining processes by making more accurate decisions in a shorter time. Also, as a result of digitalization the companies will replace in a significant way their interaction with the clients, the management side will be improved and its efficiency will be increased, along with the quality, the speed and the relevance of procurement of the information. The changes that will occur due to these actions will also create new processes, which will be perceived at a different and increased level of understanding, due to the disappearance of technological barriers.

At this point, the financial organization will also have in its componence new roles, responsibilities and new methods of work, but it has to be prepared since the beginning of the digitalization process about the risks which can occur and the own future positioning.

2. Industry 4.0

Digitalization is a part of the Industry 4.0, which implies the promise of a new revolution that combines advanced production techniques and intelligent technology operations that will be integrated into organizations, people, and goods.

Industry 4.0 is a concept which has been developed in 2010 by the German government in order to describe a vision of the future of Internet of Things (IoT), having in its componence all the interconnected processes. It is also the current revolutionary industry that secures the digitization of industrial actions and processes through the implementation of the concept of artificial intelligence in production machines along with the optimization of existing resources used in the development of new, high-efficiency commercial methodologies. This industry interconnects in the digital environment different departments that aim at customer satisfaction and personalization of the services offered. One of the main objectives of this digitization concept is the combination of digital technologies that ensure the development of new industrial markets and the interaction between economic factors.

The results provided by Industry 4.0 along with the use of connected disciplines are:

- Permanent adaptability on demand;
- Offering products or services to the client in a more personalized and futuristic way;
- Providing a post-sales service to the customer in a more efficient way;
- Design, develop and sell your product or service in a shorter time;
- Developing shorter and more profitable production series;
- Use analysis information from various areas such as help desk, social networks, IoT for analysing and exploiting them in real time.

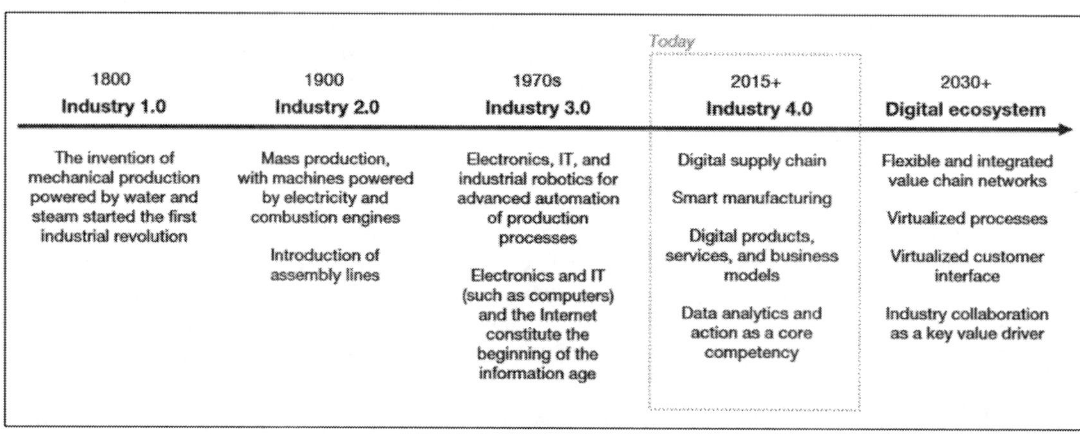

Figure 1. The long road to Industry 4.0 [2]

At the heart of all this activity sits the digital supply chain, and it is key to the operations of every company that manufactures or distributes anything. Indeed, for many companies the supply chain is the business. It extends the vertical integration of all corporate functions to the horizontal dimension, knitting together relevant players — the suppliers of raw materials and parts, the production process itself, warehouses and distributors of finished products, and finally the customer — through a network of sensors and social technologies, overseen via a central control hub, and managed through an overarching data analytics engine. [2]

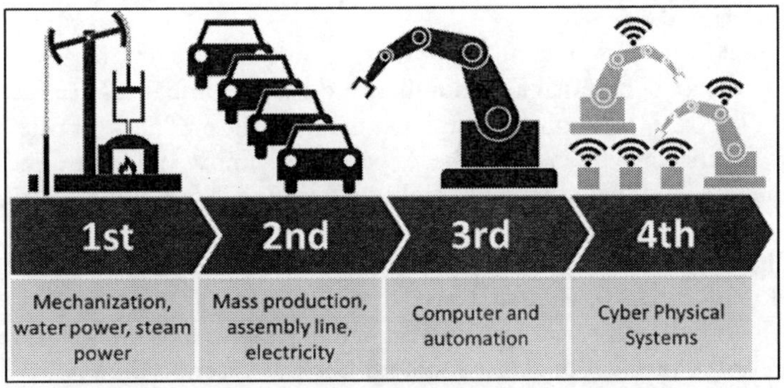

Figure 2. Industrial revolutions and future view [3]

The impact of the development and implementation of industry 4.0 can be felt at several levels, such as large ecosystems at the organizational and individual level (employees and customers). As far as ecosystems are concerned, the new industry affects all their agents, such as customers, suppliers, investors and regulatory considerations, facilitating interaction, communication and information sharing among them. Organizations can become more responsive, dynamic, and predictive to the extent that they have the capacity to implement, adapt and learn real-time new technologies. Also, for individuals, this industry can be perceived in a completely different way, such as updating the enterprise's way of doing business within the organization, while for customers it can mean a new way of personalizing and adapting the products and services purchased in meeting needs.

3. Challenges before digitalization

Information is created in various formats at an accelerating rate through various media and it is become increasingly complicated to remain abreast in this overflow of literature without the help of information technology. [4] For example, the challenges faced by a company which did not digitalized its workflow are the following [5]:

- Time and data intense business processes;
- Overwhelmed by information/data pools;
- Quality and integrity of the data is underwhelmed;
- Content or the documents are unmanageable;
- Missing out on relevant information or news;
- Important information about the company are missing;
- Product perception by the customers;
- Missing out business opportunities due to a lack of insight and connections;

These challenges and delays in advancing to digitization will lead to a low efficiency of the company, because of:

- Decreased productivity by spending valuable time to find a paper document; [6]
- Less space and uncomfortable workspace because of the paper documents which imply the need for physical storage;
- Data protection does not exist in a highly and valuable manner, because the paper support is vulnerable to various external factors that can damage the material. For example, the risk of moisture, or even its own footprint;
- Time management in data query due to the fact that a physical file may pose a risk of disorder or loss of documents;
- Missing of simplified searches for different products, values, papers, processes, etc., inside the company;
- Missing Corporate Social Responsibility. A company can demonstrate its social commitment by engaging in important causes such as environmental sustainability. Digitizing documents

IOP Conf. Series: Materials Science and Engineering **568** (2019) 012086 doi:10.1088/1757-899X/568/1/012086

promotes this economy in paper consumption. Also, if a copy of a document is required, it can always be printed.

Old systems and processes have to be rethought, and new technologies have to be put in place in order for business to stay competitive within their industry. Companies are faced with many challenges as they navigate their way through this digital transformation. These hurdles can range from limited budgets for new technology to lack of the expertise or know-how required to roll out new digital initiatives. But the one challenge that is crucial to overcome is creating an effective digital strategy. [7]

4. Benefits of digitalization

In order to develop an effective digital strategy, it is necessary for a company to be fully aware about the advantages which will overcome, in this way will be easy to exploit and develop the processes which will create value for the product or service that will be delivered to the client.

Some of the most important benefits due to digitalization are as the following:

- Reducing the use of paper and printed elements;
- Savings due to transportation costs;
- Reducing the space and costs of storing the documentation;
- Centralizing information by storing documents in digital formats that can later be accessed simultaneously by more users;
- Fighting the deterioration of documents with the passing of time through their storage in the virtual environment;
- Agility in searching for documents and their content;
- Safe and fast access and information structure allows for better decision making;
- Accurate copies of documents can be printed at any time, these copies serving as a backup copy of the documents;
- Digitized information may be modified and used for different purposes;
- The distribution of online documents is facilitated, they can be sent without waiting for confirmation of receipt;
- Contribute to maintaining and improving the environment by reducing the pollution generated by paper;
- Savings up to 30% on printing processes;
- Complexity, costs and risks are reduced due to work process analysis services, giving priority to the most efficient, safer and more optimized;
- Some tasks involving the intervention of the staff are simplified by means of automation;
- Permanent monitoring and control of document transfers between different systems;

Developing a digital substitute of rare, brittle or fragile, original documents can provide access to users while preventing the original from damage by handling or display. This is the motivation behind the digitization of many artefacts. [...] Apart from the general benefits that are discussed above, there are certain specific benefits of digitization that may help in making the society economically and ecologically sustained. [4]

5. The impact of cyber security in a digitalized company

The transformation from print to digital media for communication of information to the larger community is resulted from the growth of the Internet and now enables the tremendous amount of information accessible to everyone. [4] Along with this, as is described in [8] the cyber risk is on the rise in a hyper-connected world. The Internet of Things means that physical targets such as connected machinery will become of interest to hackers. For example, systems installed for remote access will become highly susceptible to attacks with the goal of, e.g., sabotaging infrastructure or industrial facilities. Companies need to protect themselves against attacks from state-or business-sponsored entities, politically motivated hackers, and organized crime. Cybersecurity possess significant risks and needs to be a top management issue.

What is the reason that cyber is so important nowadays? The hacker's motivations are generally financial, selling real opportunities that are hard to detect in the theft of information and financial resources through digitalization, either directly or indirectly, but there are other types of attacks, such as:

- **Threats to data** – The company must protect itself against any attempt to theft, loss or dissemination of its information (financial, technical, commercial) following the implementation of technological systems; Private data of the company can also be affected and at the same time have an impact on market reputation, jeopardizing the responsibility of current customers and bringing susceptibility to future customers;
- **Threats to information systems** – Information systems risks are considered important to the company as far as they relate to the viability of their business. The challenge lies in the company's ability to recover rapidly from a cyber-attack while operating in a degrading way.

Establishing a strategy that includes all aspects of IT security needs to be addressed from the identification phase of the technologies to be implemented as a result of the digitization process. Developing such a strategy has the role of generating trust both in a digital future and not only in the technology field, such as collaborators, information in company databases, processes used to perform activities, culture, and the physical environment.

Digitalized companies need to develop a strategy that covers the cyber security area, which involves a radical change in business models, the main idea being to determine the information security management in the company as a key factor for survival.

6. Conclusions

Technologies accelerate the digital transformation of all sectors, and digitalization affects all types of companies, without exception, although with different intensity.

During this work, we approached the digitization process which belongs to the Industry 4.0 concept, along with the benefits that come with it, and concluded that digitalization is necessary to increase the productivity, efficiency and competitiveness of all companies. Also, for many, it will be an opportunity to introduce new businesses or markets, to evolve products to services, or to cause disturbances in adjacent markets as it promotes dematerialisation and disintermediation.

Finally, we introduced the concept of cybersecurity, which is a result of digitalization, given the risks that companies expose when transferring from paper to computer, theft or loss of data, affecting the company's image on the market and in dealing with customers or partners.

References

[1] Isaksson A J, Harjunkoski I and Sand G 2017 *The impact of digitalization on the future control and operations* International Journal of Computer Applications in Chemical Engineering **1** p.2
[2] Scharuf S and Berttram P 2016 *How digitization makes the supply chain more efficient, agile and customer-focused* **1** pp.7-8
[3] Marr B 2016 *Why everyone must get ready for the 4th industrial revolution* Forbes Magazine
[4] Khan Sa, Khan Sh and Aftab M 2015 *Digitization and its impact on economy* International Journal of Digital Library Services **5** p.139
[5] Sorf M 2017 *Process of digitization and data management* Better data insights business Magazine p.3
[6] Corp A 2018 *10 Advantages of Digitization and Data Capture You Must Know* Latest Thinking (Aptara Digital Press)
[7] Wertz J 2018 *Digital Transformation Is Critical For Business Development* Forbes Magazine
[8] Bauer H 2016 *How to navigate digitization of the manufacturing sector* McKinsey Report p.46

IOP Conf. Series: Materials Science and Engineering **568** (2019) 012087 doi:10.1088/1757-899X/568/1/012087

The effect of changing training data on a fixed deep learning detection model

A N Nasser[1], A Alsabbagh[1] and H Géza[1]

[1]Mechatronics Department, Debrecen University, Ótemető utca 2–4, 4028, Debrecen, Hungary

aramoo4@gmail.com

Abstract. Within the lack of accurate data, for some computer vision applications, researchers usually use other pictures collected from different sources for the training. To know the effect of these added data, we compare the detection results of a customized dataset of objects, using the same detection model, while changing the training data fed into the network. For our work, we run the detection on images captured by the Microsoft Kinect sensor after training the network on different combinations of training data. The first part of the training data is captured by the Kinect itself, and the second is collected from several sources from the internet, referred to as collected images. We then change the distribution of these images between training and validation to feed them into the fixed training model. The results prove that this distribution of data can considerably affect training and detection results under the same model parameters. In addition, mixing the captured images with other collected ones can improve these results.

1. Introduction

The recent decade has witnessed a dramatic increase in the ability to classify, localize and detect objects in images. This success is not only the result of the advent of powerful General Purpose Unites (GPUs,) but also designing deep structures of convolutional neural networks, in addition to the availability of large datasets.

In our work, we focus on the detection problem, where the model has to decide what objects are in the image, as well as where do they appear. Our system depends on an object detection API [1], published by Google as an open source code for researchers, after making several changes to it.

The purpose of our work is to know how to tune the dataset to get better results under a fixed detection model. To make work easier, we use images of fruits as they are available and cheap. The detection runs on a picture captured by the Microsoft Kinect sensor. We first train and validate the network on images captured by the Kinect. Second, we replace the training set with collected images and move some of the previously captured images to the validation. Third, we enrich the original dataset of captured images by collected images. Different number of images is used to keep a balance among instances of each class in the three experiments.

2. Literature review

2.1. Datasets

IOP Conf. Series: Materials Science and Engineering **568** (2019) 012087 doi:10.1088/1757-899X/568/1/012087

Large datasets, such as Common Objects in COntexts (MS COCO) [2], ImageNet [3] and PASCAL Visual Object Classes (VOC) [4] have enabled significant advances in image classification, object detection, and object segmentation to be achievable. Each of these datasets varies significantly in size, list of labelled categories and types of images.

In this paper, we will focus on the MS COCO dataset, which is used in the pre-trained model. MS COCO is designed for the detection and segmentation of objects occurring in their natural context, with 91 categories, and a total of 2.5 million labelled instances in 328000 images. While MS COCO has fewer categories than ImageNet, it has more instances per category. In comparison to PASCAL VOC, MS COCO has both more categories and instances [2].

2.2. Classification Models

One of the most successful and remarkable algorithms that won the 2012 ImageNet Large Scale Visual Recognition Competition, ILSVRC, is the AlexNet [5]. This was the first successfully applied convolutional neural network at that time, which has changed the future of computer vision algorithms.

In 2014, there were two remarkable models, GoogleNet, also known as Inception V1[6], which won the competition, and VGGNet [7], which was ranked second in the contest but first in the localization task. VGGNet has 19 layers and 7.3% top-5 error rate in classification performance while GoogleNet has 22 layers with 6.7%. It is noteworthy to mention that the number of parameters used in GoogleNet, 5 million, is 12 times less than the ones used in AlexNet, 60 million, yet deeper and more efficient. It also introduced the concept of Inception model, that apply several filters of different sizes on the input from the previous layer producing not only deeper, yet wider layers without significant performance penalty [6]. To reduce dimensionality and remove computational bottlenecks, 1x1 convolutional layers followed typically by the rectified linear activation were added to the inception model.

In [8], the authors argue that adding the Batch Normalization to a state-of-the-art classification model can significantly increase its learning speed. Although, in some publications, this paper is referred to as Inception v2, such as in [10], it was not officially introduced as Inception v2.

The inception modules v2 and v3 were officially introduced in one paper [9]. In this paper, the authors added to [8] and replaced the 5x5 convolution by two 3x3 convolution operations after noticing that the 5x5 convolution is 2.78 times more expensive with the same number of filters. This architecture and the naïve Inception model are explained in figures 2 and 1 respectively.

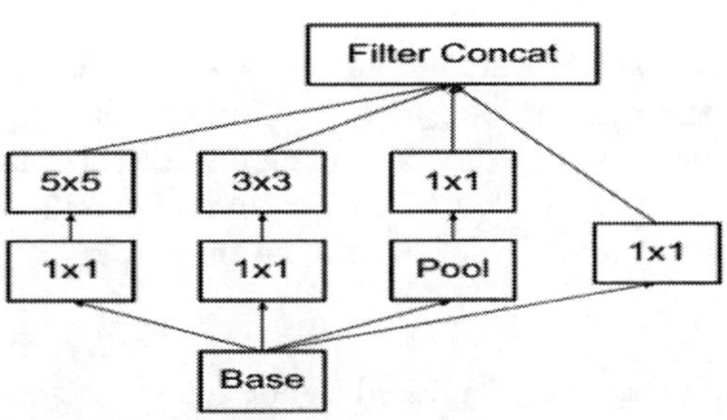

Figure 1. Original Inception module as described in [9]

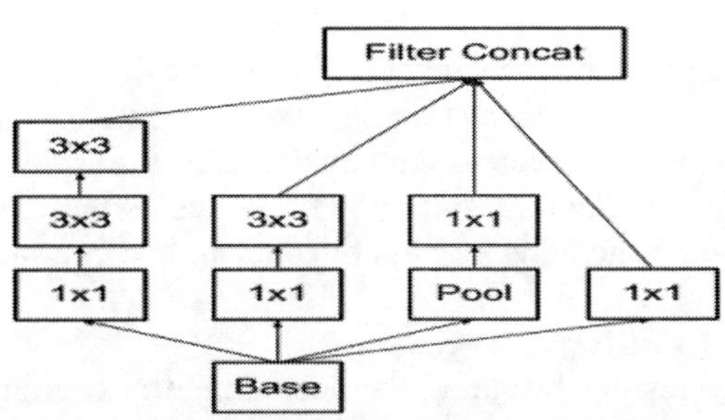

Figure 2. Inception modules where each 5 x 5 convolution is replaced by two 3 x 3 convolutions [9]

2.3. Detection models

For the detection problem, it is not enough to know whether an image contains an object or not. The detection model, however, has to know where the instances of the detected objects appear in the image.

Several detection models are existed in literature to solve this problem. Some of them use a sliding window, such as OverFeat [11]. Other models deal with the entire image during training and test time like the You Only Look Once (YOLO) [12] [13].

IOP Conf. Series: Materials Science and Engineering **568** (2019) 012087 doi:10.1088/1757-899X/568/1/012087

Other detection models run the classification on parts of an image proposed by a first stage region proposal process. The most common models that use this algorithm are the R-CNN [14] and Fast-RCNN [15], where the Selective Search algorithm [16] is used to predict region proposals. An updated version of [14] and [15] is the Faster-RCNN [17] had introduced a Region Proposal Network (PRN) that shares full-image convolutional features with the detection network, thus enabling nearly cost-free region proposals and improving the model's speed and accuracy. [17] also introduced anchor boxes the solve the multi-scale and size problem.

3. Detection methodology

3.1. Transfer Learning

Training models from scratch is time and resource consuming. Some models need days and maybe weeks to converge. Because of this, researchers usually benefit from others' previously trained models in order to initialize their models.

To explore the speed/accuracy trade-off of some of the modern detection systems, researchers at Google had published an object detection API, as an open source code [1]. In this module, the authors tried to apply the meta-architectures of object detection, described in [1], to conduct their experiments and comparisons. In our work, we depend and build on this API to train the detection model on our own dataset. The result of this training is a frozen inference graph, that can be used for the test purposes.

3.2. Used model's parameters

For the detection, we choose a detector Faster-RCNN with Inception v2 as its feature extractor. This model was pre-trained on COCO dataset, where it achieved 28 mean average precision (mAP) detector performance on a subset of the COCO validation set and a running time of 58 (milliseconds) per 600x600 image (including all pre and post-processing).

We choose this model because its accuracy and speed are suitable for our application. L2 regularizer is used in addition to the Truncated normal as the model's initializer. For optimization, we use the Momentum optimizer. We start the training with an initial learning rate of 0.0002, then we make it smaller to be 0.00002 after 5000 steps. The learning rate is set to 0.000002 if the learning exceeds 10000 steps.

3.3. Customized dataset

We created our own datasets to train a fixed model on. To run the first experiment, we captured 160 images using the Kinect and distribute them as 142 training and 18 validation images. Then, we collected 425 pictures from [18], as well as several websites, such as Adobe Stock and Shutter Stock, while keeping 72 images taken by the Kinect. For this experiment, we run the training on the collected images and validate the captured. For the last experiment, we used 385 collected images and 37 captured ones to train the network. 46 collected and 36 captured images were used for the validation process.

4. Results

The result of running the detection after training on images purely captured by the Kinect is shown in figure 3. As can be seen, there are three misclassified classes, in addition to the low probability for the detected objects. In addition, as we notice in figure 4, the training process takes a lot of time and the model is unstable. These results prove that the limited captured images are not enough for the training and more data need to be fed into the model.

IOP Conf. Series: Materials Science and Engineering **568** (2019) 012087 doi:10.1088/1757-899X/568/1/012087

Figure 3. Training and validating on images captured by the Kinect, (limited number of images)

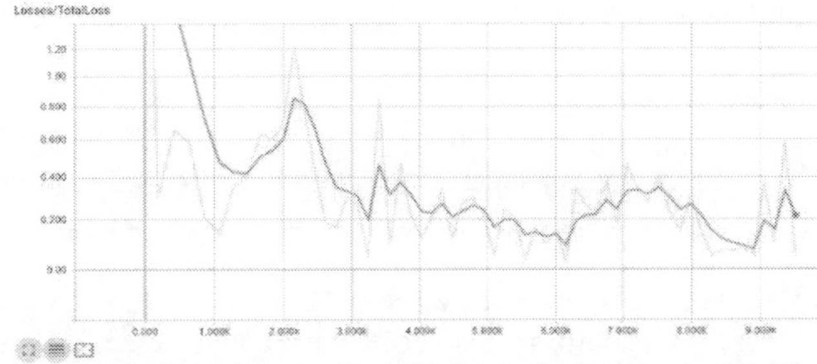

Figure 4. Training process when the model is trained and evaluated on captured.

In the second experiment, we change the training data with images collected from different resources as illustrated in 3.3. We can notice, in figure 5, that the detection results are improved, but still have two misclassified classes. Additionally, the training processed is improved but still showing instability between steps 6000 and 8000.

Figure 5. Training on collected images and validating on captured

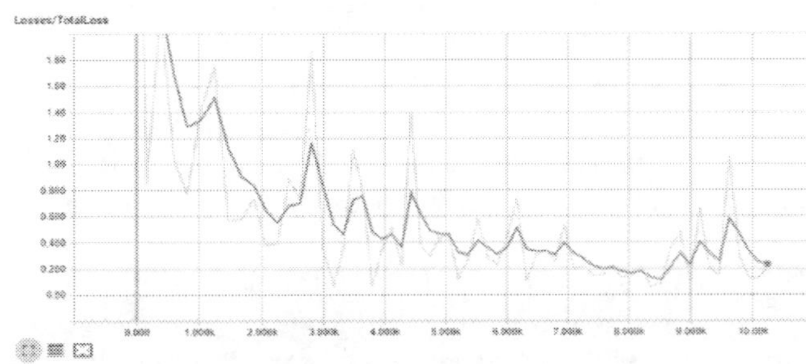

Figure 6. Training process when the model is trained on collected images and evaluated on captured

Finally, figure 7 shows more accurate detection results by only manipulating the dataset. It can be noticed in figure 8 that the training process becomes relatively stable after being unstable for the first 5000 steps.

Figure 7. Training and validating on mixed images.

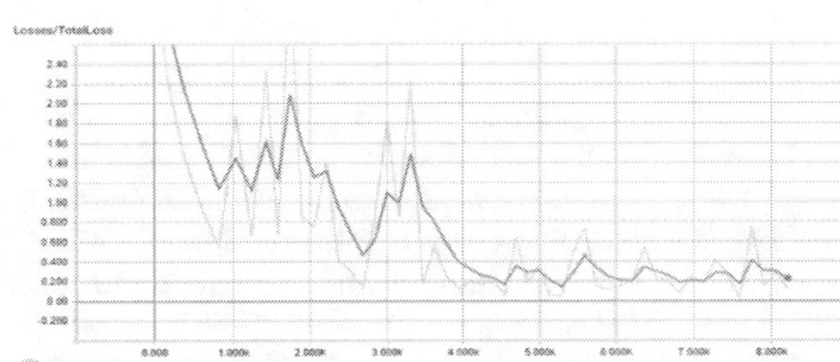

Figure 8. Training process when the model is trained and validated on mixed images.

5. Conclusion

In conclusion, this work shows the importance of choosing the right dataset for the training and validation processes under limited training data. We show the detection results of three different situations where only the dataset is changing while maintaining the model's parameters fixed. As a result, we see that when we have a lack of information, the best way is to use mixed images to train and validate the network.

Acknowledgment
The work/publication is supported by the EFOP-3.6.1-16-2016-00022 project. The project is co-financed by the European Union and the European Social Fund.

References

[1] Huang J, Rathod V, Sun C, Zhu M, Korattikara A, Fathi A, Fischer I, Wojna Z, Song Y, Guadarrama S and Murphy K 2017 Speed/accuracy trade-offs for modern convolutional object detectors *Proceedings of the IEEE conference on computer vision and pattern recognition* 7310-7311
[2] Lin T Y, Maire M, Belongie S, Hays J, Perona P, Ramanan D, Dollár P and Zitnick CL 2014 September. Microsoft coco: Common objects in context *Springer, Cham* (pp. 740-755)
[3] Russakovsky O, Deng J, Su H, Krause J, Satheesh S, Ma S, Huang Z, Karpathy A, Khosla A, Bernstein M and Berg A C 2015 Imagenet large scale visual recognition challenge *International journal of computer vision* **115(3)** 211-252
[4] Everingham M, Van Gool L, Williams C K, Winn J and Zisserman A 2010 The pascal visual object classes (voc) challenge *International journal of computer vision* **88(2)** 303-338
[5] Krizhevsky A, Sutskever I and Hinton G H 2012 Imagenet classification with deep convolutional neural networks *Advances in neural information processing systems* 1097-1105
[6] Szegedy C, Liu W, Jia Y, Sermanet P, Reed S, Anguelov D, Erhan D, Vanhoucke V and Rabinovich A 2015 Going deeper with convolutions *Proceedings of the IEEE conference on computer vision and pattern recognition* 1-9
[7] Simonyan K and Zisserman A 2014 Very deep convolutional networks for large-scale image recognition *preprint* arXiv/1409.1556
[8] Ioffe S and Szegedy C 2015.Batch normalization: Accelerating deep network training by reducing internal covariate shift *preprint* arXiv/1502.03167
[9] Szegedy, Christian, et al. "Rethinking the inception architecture for computer vision." Proceedings of the IEEE conference on computer vision and pattern recognition. 2016
[10] Szegedy C, Ioffe S, Vanhoucke V and Alemi A A 2017 Inception-v4, inception-resnet and the impact of residual connections on learning *Thirty-First AAAI Conference on Artificial Intelligence*
[11] Sermanet P, Eigen D, Zhang X, Mathieu M, Fergus R and LeCun Y 2013 Overfeat: Integrated recognition, localization and detection using convolutional networks *preprint* arXiv:1312.6229
[12] Redmon J, Divvala S, Girshick R and Farhadi A 2016 You only look once: Unified, real-time object detection *Proceedings of the IEEE conference on computer vision and pattern recognition* 779-788
[13] Redmon J and Farhadi A 2017 YOLO9000: better, faster, stronger *Proceedings of the IEEE conference on computer vision and pattern recognition* 7263-7271
[14] Girshick R. Donahue J, Darrell T and Malik J, 2014 Rich feature hierarchies for accurate object detection and semantic segmentation *Proceedings of the IEEE conference on computer vision and pattern recognition* 580-587
[15] Girshick R 2015 Fast R-CNN object detection with Caffe *Microsoft Research*
[16] Uijlings J R, Van De Sande K E, Gevers T and Smeulders AW 2013 Selective search for object recognition *International journal of computer vision* **104(2)** 154-171.
[17] Ren S, He K, Girshick R and Sun J 2015 Faster r-cnn: Towards real-time object detection with region proposal networks *Advances in neural information processing systems* 91-99
[18] Hussain I, He Q and Chen Z 2018 Automatic fruit recognition based on DCNN for commercial source trance system

IOP Conf. Series: Materials Science and Engineering **568** (2019) 012088 doi:10.1088/1757-899X/568/1/012088

The state of stress and deformation by the finite element method of the mechanical structure for a self-adaptive prehensor

I C Frincu[1] and I Stroe[1]

[1]Product Design, Mechatronics and environment Department, Transilvania University of, Braşov, Romania

cezarfrincu@yahoo.com

Abstract. The paper presents the state of stress and deformation of the component elements of the mechanical state of a self-adaptive prehensor. The theoretical and experimental model puts into evidence the links between the component elements and the forces acting on the two axes of symmetry of the physical model. The forces generated by the objects'prehension determine the state of stresses and deformations of the elements subjected to these forces. The 3D analysis of the prehensor structure is realized with the finite element method using the CATIA software. The structural elements of the prehensor are studied on each component by applying different forces in the three directions. The forces and moments that act upon the base elements of the prehensor's components are studied in the angular displacement generated by the drive motors. The composition of forces and moments for each component of the prehensor reveals the reaction forces appearing in the mechanical components of the proposed structure. It is taken into account that each module of the prehensor does not influence the state of tension and deformation of the prehensor assembly during the tightening of the objects.

1. Introduction

In a robotic system, the prehensor has the role of gripping objects of different shapes and sizes. The prehensor can be made of different mechanical structures and can be operated in different ways (electric, pneumatic, hydraulic) so that it can grip the desired object. Mechanical prehensors designed with 5 fingers are complex systems that imitate a human hand [1].

Generally, there are prehensile systems that generate a rotation movement with the "fingers" and, those are created in particular about gripping a single object. The deformation analysis is performed in the Catia program for the construction components of a self-adaptive prehensor module. The self-adaptive prehensor generates with its fingers a translation motion, and it can be used in many processes without it being mechanically modified by a particular object, as it adapts to the shape and weight of the object. The deformation analysis is performed only on components with higher stress with parallel and perpendicular forces, on advance and retreat [2].

IOP Conf. Series: Materials Science and Engineering **568** (2019) 012088 doi:10.1088/1757-899X/568/1/012088

2. The following steps were taken to analyze the mechanical components:

- Selecting the most wear-resistant components;
- Applying parallel and perpendicular forces for advance and retreat on selected mechanical components;
- 3D realization of deformations by finite element method;
- Generating some deformation parameters and calculation error;
- Geometric component modification based on results.

2.1.Selecting the most wear-resistant components

The prehensor module is driven by an electric stepper motor that generates a torque of 8.2 Kg*cm, which means it exerts a force on the components of 0.911 Nm'Figure 1' a).

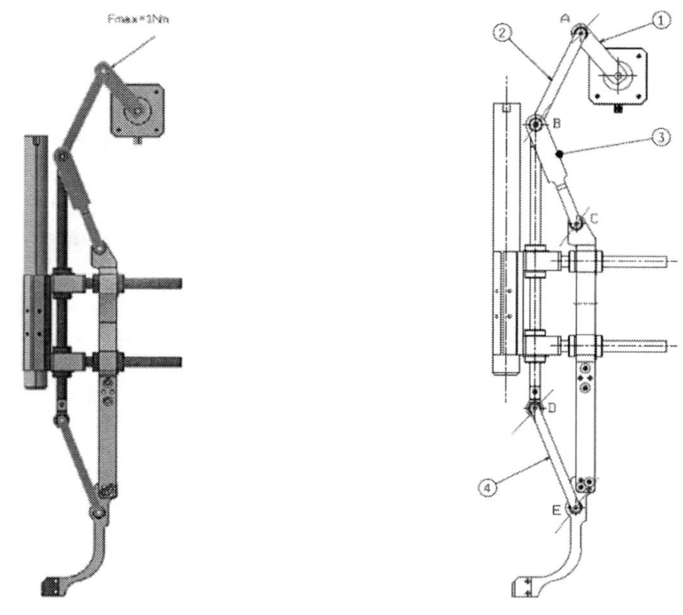

a) b)

Figure1. The representation of rotation joints and the selection of the most wear-resistant components.

The self-adaptive prehensor consists of 5 arms that are independently controlled both mechanically and sensory, and each arm is composed of several joints that have constant wear over time.The greatest stresses are found in the rotation joints, and they are marked with a red line passing through the center of the joint.In the 2D representation of the prehensor module are shown the rotation couplers (a red line in points A, B, C, D, E) where the components can be mechanically blocked and numbered.The mechanical component 1 is locked in joint A, the mechanical component 2 is locked in joint B, the mechanical

IOP Conf. Series: Materials Science and Engineering **568** (2019) 012088 doi:10.1088/1757-899X/568/1/012088

component 3 is locked in joint C, and the mechanical component 4 is locked in joint D or joint E depending on the advance or retreat motion'Figure 1' b) [3].

2.2. Applying parallel and perpendicular forces for advance and retreat on selected mechanical components
The components 1, 2, 3 and 4'Figure 1' (b)are removed separately and are applied the forces and constraints required to simulate locking in points A, B, C, D, and E.

Each component must be positioned, and the applied force must apply for in the sense generated by the whole module, where the Fc is the force comprised of the parallel force and the force perpendicular to the mechanical component analyzed.The application forces are equal and opposite, in our case called Fr (reaction force)'Figure 2'.

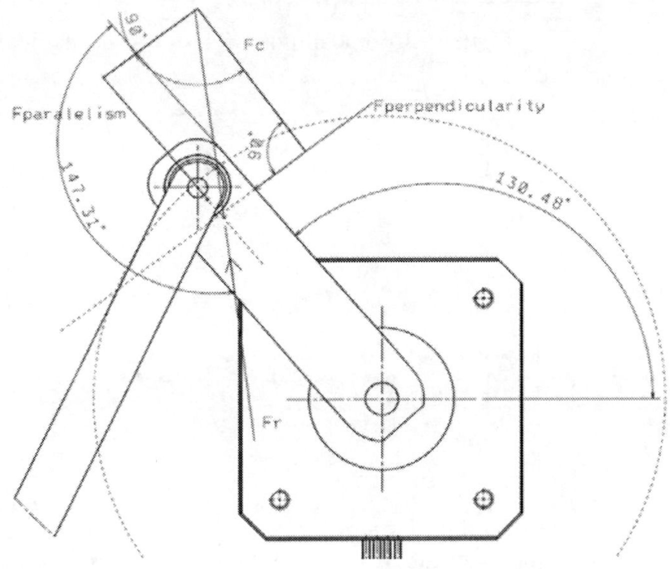

Figure 2. Example of composition of forces for mechanical component 1.

To generate the position of force Fr at an angle of approximately 147°, the values of forces Fx, Fy, Fz, Mx, My and Mz must be summed up. (Table 1.)

Tabel 1. Composition of forces for mechanical component 1

Force	Value of the applied force	Unit
Fx	3*817e-012	N
Fy	-1	N
Fz	-1	N
Mx	2*900e-002	N
My	-2*500e-003	N
Mz	2*500e-003	N

IOP Conf. Series: Materials Science and Engineering **568** (2019) 012088 doi:10.1088/1757-899X/568/1/012088

Respecting the conditions of the mechanical component 1, the reaction force Fr can be generated for the other mechanical components 2,3 and 4.

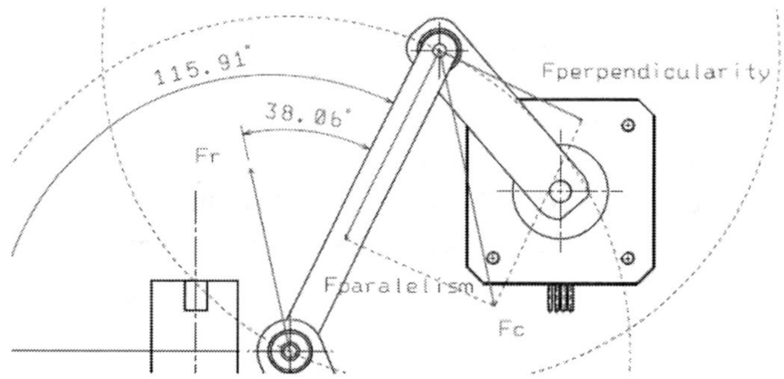

Figure 3. Composition of forces for mechanical component 2.

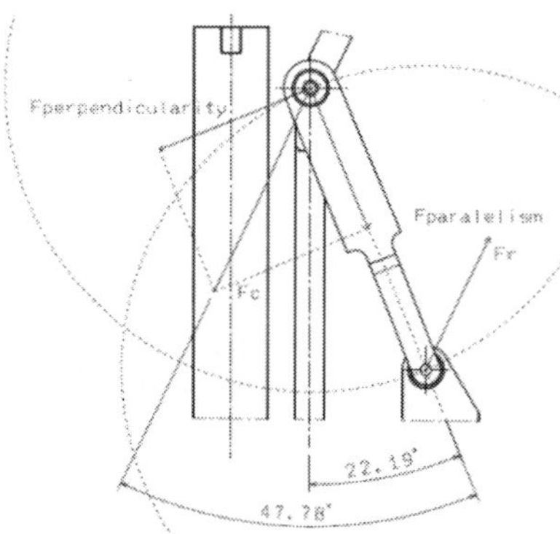

Figure 4. Composition of forces for mechanical component 3.

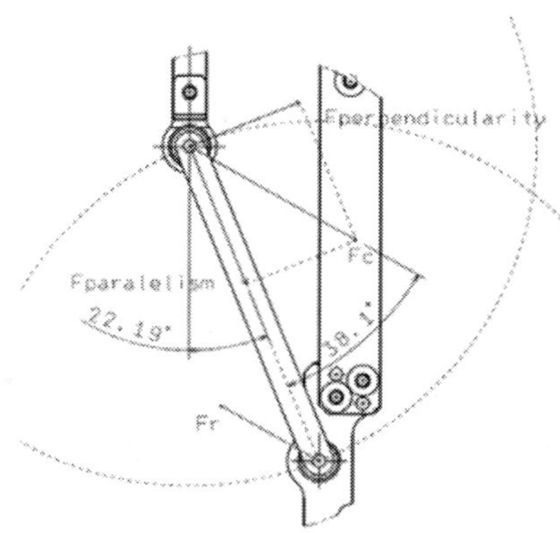

Figure 5. Composition of forces for mechanical component 4.

2.3. 3D realization of deformations by finite element method;

Summing up vectorially the forces Fx, Fy, Fz, Mx, My and Mz we determine the vectorial orientation of the reaction force Fr.

Once the forces on the mechanical component are used we can apply the finite element method where we can observe the plastic or elastic deformation a), b), c), with an amplification of 10^7'Figure 6'[4].

a) b) c)

Figure 6. Stages of amplified deformations.

2,4, Generating stress parameters of the mechanical component 1 and geometric modification if necessary;
After the analysis of stress type finite element, an automated report from the Catia program was generated where we can observe the affected areas and subjected to small deformations but also the areas where there are no deformations at all.

The report generated is in the form of a graduated column from red to blue, where red shows the largest deformation and the smallest blue'Figure 7'[5].

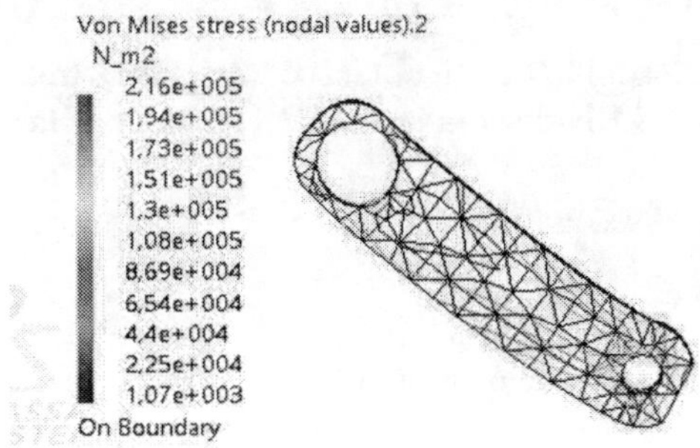

Figure 7. Finite element report.

3. Conclusions

The geometry of the mechanical components resists the maximum force exerted by the stepper motor, on advance and retreat motion.

The joint effort is taken over by the bearings and distributed throughout the system in case of locking.

Bearings depletion over time may be the cause of the grips in the mechanism.

4. Bibliography

[1] Staretu I 1996 *Systems of Prehension(Sisteme de Prehensiune, in Romanian)* (Brasov: Lux Libris)

[2] Dudita F, Staretu I 1988 *Structural synthesis and systematisation of flat elementary mechanisms with articulated bars usable as prehension robots (Sinteza si sistematizarea structurala a mecanismelor elementare plane cu bare articulate utilizabile ca robotomecanismelor de prehensiune, in Romanian)*Cluj-napoca: Buletinulcelui de-al VIII-lea Simpozion National de IR

[3] Frincu C, Stroe I, 2018 The state of the modular element structure of a prehensor through virtual prototyping, Brasov: PRASIC

[4] Ghionea I 2007 *Assisted design CATIA V5. Theoretical elements and applications (Proiectareasistata CATIA V5. Elemente teoretice si aplicatii, in Romanian)*(Bucuresti: BREN)

[5] Ghionea I 2009 *CATIA V5. Applications in mechanical engineering(CATIA V5.Aplicatii in ingineria mecanica in Romanian)* Bucuresti: BREN

IOP Conf. Series: Materials Science and Engineering **568** (2019) 012089 doi:10.1088/1757-899X/568/1/012089

Strategic Communication Process: Support for Performance Measurement and Popularization of SMEs Innovations – I

A HULUBEI (GEORGESCU)[1] and S AVASILCĂI[1]

[1] "Gheorghe Asachi" Technical University, Department of Engineering and Management, 28 Blvd. Mangeron, TEX1, 700050, Iasi, Romania

amalia.georgescu@tuiasi.ro

Abstract: To perform and be relevant, SMEs must continuously innovate, be in constant contact with the market, and communicate their news strategically. Entrepreneurs need to go through strategic management system training and organization development processes, understand the importance of internal and external communication processes, invest in technology and automated processes, optimize their innovation, consistently train their employees, develop new product and service flows on the market and keep constant contact with the stakeholders. This paper analyzes, in theory, the advantages of the strategic communication process, highlighting the Balanced Scorecard multidimensional analysis model and the impact of innovation in SMEs at all levels. Through this model, the company can predict and analyze the qualitative and quantitative factors needed in the process of innovation and strategic communication, becoming part of an ecosystem that can generate change and improve the standard of living. Through this process, the company becomes part of an ecosystem that helps the economy develops and innovates.

1. Introduction

All the technological changes and innovations in the last ten years have completely changed the way we develop businesses, communicate, and live our life today.

Romania is adapting more and more to these trends and gradually succeeds in assimilating new changes. However, at an entrepreneurial level, there is a considerable number of SMEs that do not have a strategic planning system, a well-defined program of innovation, a new product or service generation or an internal and external strategic communication plan. These small companies act by the immediate demands of the market and on the entrepreneur's flair, which often limits the development of effective management programs, so SMEs cannot survive major changes or unforeseen political and economic crises. These companies risk not being sustainable in the long run and can not generate a major impact in their field of activity and in the society in which they operate.

Entrepreneurs need to go through training and organization processes of strategic management systems, understand the importance of communication processes, invest in technology and process automation, optimize their innovation flows and co-operate with the marketplace in generating new products and services. By optimizing all of these processes, the company is becoming a unitary one that works better with all the resources at its disposal, anticipating market demands, becoming part of an ecosystem that can influence a certain area and consumer lifestyle. The company thus becomes relevant to the market and maintains its place in the top of consumer preferences wherever they are.

IOP Conf. Series: Materials Science and Engineering **568** (2019) 012089 doi:10.1088/1757-899X/568/1/012089

2. Theoretical and practical aspects

New concepts of new technologies, processes, and products, as key factors for economic success, make innovation communication an integral part of a company's strategic communication. Strategic communication explains the higher-lever concern behind the communicative efforts of an organization to advance and to fulfill its organizational mission [1]. Thus, this multidisciplinary and multi-departmental communication becomes the basis of the public relations program, and will generate enough information, will meet change and consider market feedback as an extremely valuable asset in methodological research to optimize future communication programs and innovation [2].

Recent concepts in innovation management (Block Chain, Industry 4.0, Internet of things) and measures taken to comprehensively help consumers to understand these innovations in day-to-day interactions, create new opportunities for communication and setting up contacts with them, becoming necessary for society's evolving needs. Thus, the communication process becomes vital for the strategic planning of the company's future, in close connection with active opening and listening to the target audience, always paying attention to its requirements, preferences and direction - essential elements that can determine the course and strategic development of the process long-term company innovation.

In the era of "open innovation" [3], the various internal or external stakeholders are systematically involved in the innovation process by integrating the crowdsourcing concept into the company. Open source software platforms and mobile communication technologies indicate that innovation networks between companies, customers and their partners are increasingly important. In this context, communication and systematic cooperation with the various actors involved is essential. Such bilateral relations with the public are controlled by means of communication in the context of specialized or informal events. It is in the analysis of stakeholders ("the co-interested groups" [4]) and in the collection of relevant information to be able to identify all the factors that can affect certain individuals and organizations in the process of strategic planning and innovation - a detailed SWOT analysis can be created, focusing on the "forces field" based on the theory of Kurt Lewin.

The answer to these issues can clarify the company's strategy planning process, and the way communication tests the fairness of the assumptions about who its audience is, what they know, what they think, how they are involved or affected, how innovation can influence them, what information they consider to be relevant, how they use it and where they get the information from.

In the modern context of communicating innovation, strategic communication facilitates processes on three levels: micro (internal communication), medium (relationship with relevant stakeholders) and macro (lobby, public debates). In the context of innovation management, all these levels generate different forms and types of ambassadors/spokespersons to promote, explain, communicate and facilitate information on the market through various channels and different forms of communication. There ambassadors know how to involve and convince people, know how to attract other supporters, potential investors or fundraising.

In the European Union, SMEs account ~99% of all registred companies, contributing about 60% to the value-added UE market. The large share of SMEs in Europe generates their particularly important role, but also the source of a number of specific issues. Their innovative strength is generally lower than large units with higher financial resources and, in particular, research capabilities [5]. The spread of innovation to SMEs can be done through market channels or outside the markets - from parent companies to SMEs. Many SMEs in Romania benefit from the spread of innovation from other countries where the head office of the company is located, which has contradictory effects. On one hand, a qualitative development of processes and products is achieved through the achievement of results obtained with great efforts and expenditures from other countries. On the other hand, the development process of research and development in Romania is very difficult, our country being dependent on other countries. Romania is a "modest" innovative country, taking the last place in 2017 according to EIS (European Innovation Scoreboard), with innovation performance falling by 14.1% compared to the EU as compared to 2010, and in the last places at other various categories of economic indicators.

Any attempt to estimate the contribution of SMEs to economic growth must start from the reality of a strong variation in the growth rate of enterprises. Most companies have low growth rates, innovation being absent or hardly assimilated in the company. At the same time, a small number of businesses are experiencing very rapid growth, being responsible for a considerable part of the new jobs created, the increase in sales and the added value created on the market, generating to a large extent the dynamism generally attributed to certain sectors.

These companies perform because they better capitalize existing opportunities and create competitive advantages either through the introduction of new technologies and radical innovation or throughout incremental innovation, stakeholder's engagement in various processes, cost reduction, increased quality standards, increased organizational flexibility and improved management processes [6].

The implication of this aspect draws the conclusions as follows: local economic growth conditions are supporting or not the important role of SMEs and not vice versa; you cannot blame a company for the fact that there is no developed entrepreneurial environment and creating value-added, but you have to take into account the entire economic ecosystem in which large firms and business have their very important role, providing the conditions for capitalizing on the potential of SMEs.

3. The strategic communication process, an important pillar in the development of the company

The Balanced Scorecard model (BSC) [7], one of the most advanced management systems and strategic planning tools, improves external and internal communication, and also monitors the performance of an organization, starting from its vision, strategic objectives, up to analyzing the impact of actions taken with resonance for the future.

Within the BSC, managers' attention is directed mainly to non-financial indicators to assess the satisfaction of clients, employees and partners, the length of processes and the quality of the results to be achieved. The mission statement, in the expected results, according to Kaplan and Norton, starts from the company's mission, which has to answer a simple question: "*Why do we exist?*" and then to the vision (what do we want to do), the values that the company believes in, to the strategy that proposes into an action plan, implementation process and innovation initiatives that can generate material and non-material strategic results.

Typical BSC objectives and performance indicators are grouped into four perspectives, linked by cause-effect relationships: Financial Perspective (How do shareholders perceive us?), Customer Perception (How do users perceive us?), The internal processes perspective (In what processes will we be successful?), Personal Perspective and Innovation (How do we improve our performance?).

The customers perception that can generate added value and it is relevant in the market, it can make the company visible and attractive to meet the requirements of its customers, but it can also generate a change in the field or the ecosystems the company belongs to. Recent management philosophy has shown an increasing focus on customer satisfaction in any organization. Thus, the process of strategic communication to the outside is proving extremely important [8]. According to the BSC, as shown in figure 1, the Customer & Stakeholders section emphasizes the value a company offers to interested groups and the importance of measuring their satisfaction, their loyalty, but also their involvement in the innovation process. As long as a company innovates and meets the requirements of its customers, it means that it wants to create a positive impact and generate added value in society. The company has to generate Customer Value Proposition (CVP) processes, which make a real contribution to the net benefits a consumer gets from interacting with the company.

Unlike usual public relations practices, strategic communication focuses on the main issues that can generate and control organizational success. Strategic communication also enlarges the area of traditional communication, to manage messages sent across all types of interactions with internal and external stakeholders and to highlight the impact the company may have in its field of work.

Communication will thus play a new role in promoting the innovation management that can be gained through understanding from a new perspective of crowdsoursing communication strategies. Theoretical concepts and experimental discoveries can be integrated into a circular innovative communication system that combines the social, cognitive, emotional and connotative dimensions of the new role of communication management in the innovation process. Therefore it can be placed the foundation of an

IOP Conf. Series: Materials Science and Engineering **568** (2019) 012089 doi:10.1088/1757-899X/568/1/012089

integrated, ever-evolving ecosystem where stakeholders and other organizations generate or support the main lines of innovation in the company.

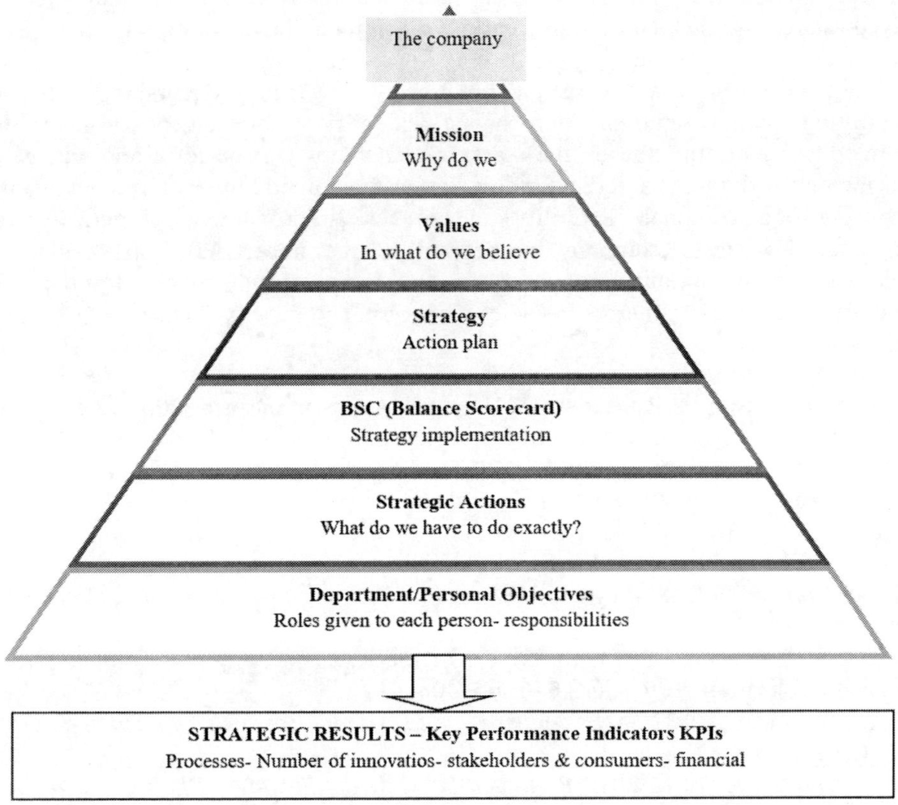

Figure 1. Strategic Management Model according to BALANCED SCORECARD [9]- personal adaptation.

4. The importance of crowdsourcing in the marketing and public relations strategy

International companies such as Uber, Google, Apple, Dreamtrips, Tesla or Romanian companies such as Mobexpert, Dedeman, Bit Defender have emerged from the desire to generate products, services and experiences that are different from what exists at some point in the market. They have identified the market niche, demand, preference or dissatisfaction with prospects and sought to innovate at the highest level. Today they are considered innovative companies with destructive marketing and thousands of users. These companies have generated a culture of innovation and are looking to surprise day by day with everything they market.

The event industry is an important channel in this area because it is among the few where the consumer can actually interact with the company's products, services, employees or partners. They can experience and consume their products and services in different ways, understand the company's philosophy, test and actively participate in the innovation process. When creating an event, the company accesses and works with lots of resources and channels, integrates technological, artistic and hospitality facilities. Efforts in this regard are important and must be optimally programmed and organized. Teams of specialists must constantly innovate and optimize channels for addressing and obtaining information from the public. The tools and materials that will result from this process of information and research

IOP Conf. Series: Materials Science and Engineering **568** (2019) 012089 doi:10.1088/1757-899X/568/1/012089

gathering, can be a powerful basis for predictable and future planning. They will also be at the bases of the future innovation and communication processes.

5. Conclusion

Entrepreneurs need to understand and accept the integration of the company into a wide network of relevant stakeholders, to cooperate interinstitutional in order to innovate and communicate the results obtained and to constantly use methods of information, planning and evaluation (Balanced Scorecard) or different type channels as crowdsourcing, in the era of open innovation.

Innovation is thus an essential and constant direction of new entrepreneurship, with companies focusing on identifying optimal communication channels, identifying relevant spokesmen to their field of activity, and facilitating information on the market. Companies will perform and will be oriented towards introducing new technologies, radical innovation or incremental innovation, increasing quality standards, enhancing organizational flexibility and process improvements. Practically, strategic communication, at any level in the company and beyond, reduces uncertainty in the decision-making and supports the creation of optimized methods for the management process, for the company's performance and the impact it can generate in society, triggering a new era of change.

References

[1] Hallahan K, Holtzhausen D, Ruler van B, Vercic D and Sriramesh K 2007 *Defining strategic communication* International Journal of Strategic Communication **1** pp .3–35

[2] Cutlip S, Center A, and Broom G 2010 *Effective Public Relations* **9** *(*Bucuresti: Comunicare.ro Press) pp 308-309

[3] Chesbrough H 2006 *Open Innovation: The New Imperative for Creating and Profiting from Technology (*Harvard Business Press)

[4] Grunig J and Fred C 1992 *Strategic Management, Publics, and Issues, Excellence in Public Relations* pp. 127-146

[5] Hoffman O and Glodeanu I 2011 *Green Carta and Innovative SMEs financing* (Sociology Romanian Publication, year XXII, **5–6**) pp. 403–419

[6] Paslaru D and Modreanu I 2012 *SMEs contribution to economic growth- present and perspectives,* (POSDCA study), p.13

[7] Stefanescu D and Silivestru M 2012 *Balanced Scorecard - Strategic Planning Tool*, Romanian Statistics Magazine **2**

[8] Argenti P A, Howell R A and Beck K A 2005 *The strategic communication imperative* MIT Sloan Management Review **46.3** pp. 83–89

[9] Kaplan R and Norton D 1996 *The Balanced Scorecard: translating strategy into action* (Boston: Harvard Business School Press)

IOP Conf. Series: Materials Science and Engineering **568** (2019) 012090 doi:10.1088/1757-899X/568/1/012090

Strategic Communication Process: Support for Performance Measurement and Popularization of SMEs Innovations – II

A HULUBEI (GEORGESCU)[1] and S AVASILCĂI[1]

[1] "Gheorghe Asachi" Technical University, Department of Engineering and Management, 28 Blvd. Mangeron, TEX1, 700050, Iasi, Romania

amalia.georgescu@tuiasi.ro

Abstract: This paper analyzes the advantages of the strategic communication process from an online and a digital point of view, as most companies make their changes from the off-line into the online market. Through this aspect, the company can quickly adapt to the fast changes of the industry or of the market requirements, to connect better with its stakeholders, to fulfill their requirements, and generate more revenues. Having in mind that the market is always changing, entrepreneurs must constantly reanalyze and rethink their actions and their communication strategies to remain relevant and desirable for their stakeholders. In the last part of the article, we have analysed the example of a traditional SME manufacturer company from Romania that had to make its way in the digital world and develop new markets, in order to strengthen its position in the many years of economic existence and to respond to a possible crisis that could have drastically affect its business.

1. Introduction

In the previous article, *Strategic Communication Process: Support for Performance Measurement and Popularization of SMEs Innovations - I (*Georgescu A, Avasilcai S, 2019), we have explained the importance of innovation as an essential and constant direction of new entrepreneurship, with companies focusing on identifying optimal communication channels, identifying relevant spokesmen to their field of activity, and facilitating information on the market. We have shown that strategic communication, at any level in the company and beyond, reduces uncertainty in the decision-making and supports the creation of optimized methods for the management process, for the company's performance and the impact it can generate in society, triggering a new era of change.

Continuing our research, the definition of strategic communication shows a communication aligned with the company's overall strategy, to enhance its strategic positioning [1].

Creating, developing, and implementing an effective business management strategy in a period of rapid change, evolving emerging technology, shifting customer tastes, regulatory changes, and intense competition is a real challenge. How to communicate it can be also challenging and here comes the importance of strategic planning of all the ways and channels the company will use to speak about its innovations, products and services, in order to make itself relevant on the market, heardable and desirable by the stakeholders.

The companies, especially the SMEs, have to have the ability to develop strategic narratives that enable innovation at all levels or departments, and to implement them strategicaly in order to become easily notable on the market. Entrepreuneurs and managers have to constantly "rethink the past and the present and reimagine the future" [2], in order to choose new strategic directs or adjust the ones already existing. This managerial behaviour can create momentum for innovative strategies and show that SMEs can be of great importance in the evolution of technology, the market or of the society.

2. What to do for strategically communicate on the market?

The transformation that a company will feel every year and the abundance of decisions that managers have to take daily shows that they have to be proactively open to face transformations, to understand their impact on the market and rapidly adjust at any important changes. They have to be able to incorporate these changes at certain levels in their companies in order to develop their markets, their products or services, to innovate, to train their people, to increase their profits and to truly impact the society. This "strategic renewal" becomes a market imperative and involves "making changes ahead of a crisis" and not be seduced by the success that you can have on the market at a certain time [3]. These market changes can have different forms: digital technology, new competitors, new customer needs and desires, emerging markets reshaping economy, new discoveries in medicine or political changes.

The change within is often driven from the outside changes of the industry, so the managers have to find new ways to make money, to remain relevant on the market and to straighten their core capabilities. They have to manage and to renew themselves "ahead of market disruption" and to enforce their main goals in order to emotionally touch their stakeholders and show the positive impact they can have at all levels [3]. At this level, the Balance Score Card (BSC) and a benchmark analysis have to be a routine in the eyes of the CEOs.

At these point, the importance of co-working, ecosystems, collaboration and crowdsourcing becomes of great interest, having the possibility to help the company develop a "leadership community in the work of renewal" [3] in order to see different aspects from a different point of view, to learn more from the needs and wishes of the co-interested groups, to explore new ideas and to develop new occasions to interact with the market through different types of events, marketing or public relations campaigns in the on-line or the off-line fields.

3. The strategic communication, part of the digital transformation process

Today's markets require changes everyday especially in the digital and the on-line field. Customers are increasingly demanding to interact with companies anytime and anywhere.

To create more value and enlarge their markets in order to access clients anytime and anywhere, every business has to have an on-line presence: a website page, social network accounts, e-mail or connect with their potential clients through chat messaging, on-line stores, blogs with high quality branded content, video content channels, newsletters, video-conferences, webinars or other type of multisided platforms. Multisided platforms (MSPs) are technologies, products or services that create value primarily by enabling direct interactions between two or more customer or participant groups in the on-line medium [4].

The benefits of creating these on-line platforms is that the company enters easily in contact with its direct and indirect public, can access new markets, can ask for feedback, reviews or involve the public as "outside innovators" or supplies of new ideas [5]. This gives the company the possibility to create a digital business model that could help build a strong on-line presence and gain new markets. For example, 72% of customers in an US survey said they would replace some traditional channels with mobile apps if the capability was available [6]. An approximate study made by Flurry Analysis in UE in 2016 showed that 69% of the Europeans customers use apps (especially UK, German and French customers) to connect and communicate with different companies [7]. So, companies as they grow, they should move more from the physical world of "place" to the digital world of "space" in order to strengthen their digital business models [8]. These digital opportunity gives the company the opportunity to innovate more and even create important content that could be shared and linked in the on-line, and also in the off-line field, through the usage of a various number of platforms- phone, PC, TV, CRM, e-mail or other interactive live or on-line events. This is an important part of the customer- driven innovation strategy model that helps the company develop its digital model (through quality content, customer experience and the creation of internal and external platforms).

If SMEs companies manage to quickly adjust their communication strategies and connect thought a relevant number of on-line and off-line channels, they can generate, in time, a bigger ecosystem and become important or even leading professional in their field of activity. Digital business models are transparent to all, give the chance to share experience and enhance "the age of customer voice" that all the companies experience more and more nowadays, where the users expect a brilliant digital experience in all of their interactions with the company [8].

Being in direct contact with the public could create higher consumer loyalty, a higher interest for the products or services, for the educational information or any other activity released on different channels that show to be effective and to be accessed by the co-intersected groups or the stakeholders. In the end, these range of activities can create multiple sources of revenues and profit for the company. At a periodical and straightforward cost-benefit analysis, the company will establish the best channels that gave best results in the communication or in developing a relationship with the stakeholders and will forward decide which of them to be developed more and need extra resources to invest in. But “future projections are intimately tied to interpretations of the past and the present” [2] and new shaped strategic choices can be made only if they are connected to the past and the present actions and results.

4. The creation of relevant strategy

Every opportunity for learning and reconfiguring the strategy have to be closely analyzed and search for new interpretations and novel possibilities in order to discover the sources of competitive advantages. The SMEs that manage to deal with the routine of the business and also to access conventional wisdom within or outside the company will manage to reinvent its strategy and develop new directions for achieving success. Each strategy should have in mind one of the three strategical directions: strategy of position, strategy of leverage and strategies of opportunity [9]. Each of these have to make the entrepreneurs understand the particular circumstances and forces that may shape the company’s competitive environment, its core values and its mission.

In order to choose the right strategy, choosing from different frameworks and archetypes can help the company to better identify its potential, its circumstances, available resources and how management should link those resources, as shown in table 1.

Analyzing this framework model, a company can easily identify its potential impact on the market and decide which strategy represents best their goals and core values.

5. Strategy adjustment example- BIOCOMP, a Romanian SME manufacturer that understood its potential and turned a crisis into an opportunity

For example, BIOCOMP, a SME Romanian company from Iasi, that produces bactericidal devices for disinfection based on ultraviolet light since 1991 (these devices have the ability to disinfect and destroy bacteria, viruses and molds up to 99,99% through the UV-C light, known the most efficient method of disinfection), has a set of core values and a strategy that helped the company make its market predictable until 2008.

During the economic crisis, it had to radically rethink its strategy, being strongly affected in losing almost 60% of the market. As a company that acted for years on a traditional and predictable market, it was concerned to keep its leader position, to deliver its products just in time and to enforce the relationships with its customers, adopting a position strategy.

But when the crises stroked during the period of 2009, the company had to quickly analyze its options, its past business model and look for a new one in order to survive and even gain new markets. The managers established that it is time to be bold and develop a new communication strategy on totally new markets.

Its first move was to focus on the areas where there is a high risk for diseases outside the medical and veterinarian areas, meaning the educational institutes (schools and kindergartens). The children that go to school or kindergarten are exposed to get all sorts of diseases and affect their immunity system. The company first developed a statistical research study, one-to-one interviews with parents and pediatric doctors and decided to develop a social responsibility campaign named “Free Illnesses Childhood”. Choosing a combination between leverage and opportunity strategy model, BIOCOMP developed specific events where the participants could understand better the impact of diseases in childhood and how to better prevent them in the educational communities, made videos and content that helped parents to understand better the impact on using the UV-C bactericidal devices in classrooms, developed a new website and organized dedicated conferences where specialists in different areas came to openly talk to parents and help them in different areas concerning health, parenting, hygiene or other subject of interest according to the parents' interests.

Table 1. Choosing the Right Strategy Model [9].

	Position Strategy	*Leverage Strategy*	*Opportunity Strategy*
Strategy	Build mutually reinforcing resource systems with many resources in an attractive strategic position. Deepen their links.	Build strategically important resources for current markets. Leverage them into attractive new products and new markets.	Pick a few strategic processes with deep and swift flows of opportunities. Learn simple rules to capture opportunities.
Circumstances best for	Stable environments	Moderately dynamic environments	Dynamic environments
Resources	Often mundane	Strategically important (i.e., valuable, rare, inimitable and non-substitutable)	Opportunity-rich strategic processes guided by simple rules
Relationships	Tightly interlocked resources	Moderately linked resources	Loosely linked resources
Basis of Competitive Advantage	A cost leadership or differentiated strategic position that is defensible	Ownership of specific strategically important resources that can be leveraged	Capture of attractive opportunities before rivals
Sustainability of Advantage	Long term	Medium term	Unpredictable
Inimitability of Advantage	Through causal ambiguity of tightly linked resources plus time to develop the resource system and path dependence	Through property rights, path dependence and time needed to develop the same resources	Through first-mover advantage and the challenge of inferring rules from partially improvised outcomes
Challenges	Adjusting system of tightly linked resources quickly enough and without producing negative synergy	Adjusting resource portfolio without being blocked by cognitive and political rigidities	Maintaining "edge of chaos" with the right number and types of rules. Timely pivoting to better strategic processes

The program was developed for two years and the educational field developed 23% new share market in the company's turnover. From this point, new markets have naturally developed: office buildings disinfection needs and health protection for the employers that work in the open spaces, the food industry, cosmetics, home use. The company understood that interdependent resources create complexity. In total, since 2009, BIOCOMP developed seven new markets and increased the range of products and services to over 100 types offered all over Romania and Eastern Europe.

The company understood its innovation potential and constantly collaborates with its clients in developing new products and applications, worked with professional resellers through on-line stores and specialized websites and developed a multisided platform to optimize its production flux. The company managers understood that they are getting from a relative stable to a dynamic industry once the new markets have started to be developed and it embraced the innovative strategy.

Nowadays the company provides one of the most effective environmental biosecurity concept worldwide, by disinfecting air and surfaces with bactericidal devices produced in Romania, for medical, food, educational, office, zootechnical, veterinary, cosmetic, hospitality and household applications. The bactericidal devices eliminate nosocomial infections in the environment, reduce disease transmission capacity, eliminate dangerous pathogens from people's work environment and allow healthier food production.

BIOCOMP is looking for internationalization opportunities of its products and services in UE and Asia, is analyzing the opportunity of coevolving their strategically important resources, developing new digital business models in order to easily connect with its stakeholders and create a constant flux of revenues from multiple markets.

6. Conclusion

The survival of a small medium company stays in its ability to constantly adapt at the changes of the market and, even better, to outcome them and generate the change before they come. The constant contact with the stakeholders is vital, today more then ever, no only in the off-line area, but in the on-line area especially. The on-line markets are always open, always in connection and can offer a constant source of relevant ideas of innovation that can help the SMEs companies to gain their place in the market. Managers that look for optimising their communication strategy and digital business model will remain on the market, will become noticeable, remarkable and will win in medium and long term form all point of view.

References

[1] Argenti P A, Howell R A, Beck, K A, 2005 *The Strategic Communication Imperative,* MIT Sloan Management Review, Massachusetts Institute of Technology

[2] Kaplan S, Orlikowski W, 2014 *Beyond Forecasting: Creating New Strategic Narratives,* MIT Sloan Management Review, Massachusetts Institute of Technology

[3] Binns A, Harreld B J, O' Reilly III C, Tushman M L, 2015 *The Art of Strategic renewal*, MIT Sloan Management Review, Massachusetts Institute of Technology

[4] Hagiu A, 2015 *Strategic Decisions for Multisided Platforms,* MIT Sloan Management Review, Massachusetts Institute of Technology

[5] Hagiu A, Wright J, 2011 *Multi-Sided Platforms*, Harvard Business School, Boston, working paper no. 12-024

[6] ClickFox, 2011 *Mobile Apps Consumer Survey,* (sample size=650). The survey also found that 78% of the sample reported using apps to interact with companies with which they do business, like banks and retailers

[7] Flurry Analysis, 2016 *European app trends in 2017: Business apps on the rise*, study interpretation published by Digital Strategy Consulting Company

[8] Weill P, Woerner S L, 2015 *Optimizing your Digital Business Model,* MIT Sloan Management Review, Massachusetts Institute of Technology

[9] Bigham, Christopher B, Eisenhardt, Kathleen M, Furr, Nathan R, 2011 *Which Strategy When?,* MIT Sloan Management Review, Massachusetts Institute of Technology

IOP Conf. Series: Materials Science and Engineering **568** (2019) 012091 doi:10.1088/1757-899X/568/1/012091

An overview of strategic issues in the evolution of its industry

N Bagiu[1], **S Avasilcăi**[1], **L Alexa**[1]

[1] Gheorghe Asachi Technical University of Iasi-Romania, Engineering and Management Department, Blvd. Dimitrie Mangeron, No. 28, 700050, Iasi, Romania

neculai.bagiu@tuiasi.ro

Abstract. The purpose of this paper is to find a balanced way to analyse IT (Information Technology) industry, from its value perspective, but first we need to know how to define the IT industry, what are its component and the dynamic on global industry. This is an empiric study due to an enormous information quantity – only a simple search on EBSCO platform, looking only for Academic journals, retrieve 183.508 titles where "Information Technology" keywords were used. For same keywords Google Academic retrieve approximately 6.000.000 results in its search. As part of the economy modernization trends, the software industry plays a natural engine role - most of the company's efforts being based on the IT, especially the software component. With the omnipresence of mobile phones and the customers demanding more and more personalized experiences, the software industry continues to stay at the forefront of the business agendas in most of the companies.

1. Introduction

The purpose of this paper is to find a balanced way to analyze IT (Information Technology) industry, from its value perspective, but first we need to know how to define the IT industry, what are its component and the dynamic on global industry.

This is an empiric study due to an enormous information quantity – only a simple search on EBSCO [1] platform, looking only for Academic journals, retrieve 183.508 titles where "Information Technology" keywords were used. For same keywords Google Academic [2] retrieve approximately 6.000.000 results in its search.

2. Some History

IT was use first time in the context of appliance of semiconductors, after more than 50 years of researches in this area, starting from first discovery of the transistor in 1947 that was built in Bell Labs. That was the very first step in modern information technology era, the transistor is a semiconductor device that encode information in 0 and 1, acting as an electrical switch that commute the position between on and off stages. Starting that moment, the construction of integrated circuits was possible, in 1958, memory circuits and logic chips in 1971 by Intel. In 1965 Gordon More [3] issued the Gordon's law captured the fact that successive generation of semiconductors are faster and better. The chips were used also to equip the computers but also the communication and peripherals, mainframes, storage devices, etc. Jorgenson [4] describe an accelerate drop cost of semiconductors and hardware equipment with the 1/5000 index from 1958 to 1994; the point here is that the decline of computer prices follows the semiconductor prices. The software component starts to become more important than hardware

IOP Conf. Series: Materials Science and Engineering **568** (2019) 012091 doi:10.1088/1757-899X/568/1/012091

starting with 2000 when Robert P. Parker and Grimm [5] describe the new estimates of investments in software. Another important moment was 1989 when www was launch and start the internet era.

3. IT profile industry

In 1999, Bureau of Economic Analysis (BEA) reclassified computer software as an investment and describe three type of software:

1. Prepackaged – sold or licensed on packages,
2. Custom – created for a specific need of a particular client (including all the services needed to be created beside coding),
3. Own-account – created for a specific application of the user.

This is first classification that was done on software area and is based on the economic value that an asset can have – and defining software as an asset was possible to attach an economic value. There were four direction of analysis made by Jorgenson [4] that capture the whole IT industry: 1. Computers; 2. Communications; 3. Software and 4. Services.

On the other hand, the very first moment of information technology development was the introduction of personal computers by IBM in 1981 [4]. That was the moment when Intel release 8086-8088 processors and the MS-DOS, the product of Microsoft Corporation founded by Bill Gates and Paul Allen, was licensed by IBM, the moment when hardware and software was clearly divided form the usability point of view, so the moment when BEA report the classification of software mark an turning point in investments showing the importance of investments in software area.

IT was defined as the outputs of computers, communication equipment and software [4]. In 2016 in the CompTIA's [6] report the IT is defined as the utilization of computing via hardware, software, services and infrastructure to create, store, exchange and leverage information in its various forms to accomplish any number of objectives. Harvard Business Review [7] define IT the use of any use of any computers, storage, networking and other physical devices, infrastructure and processes to create, process, store, secure and exchange all forms of electronic data. HBR suggest that an option to classify the IT industry could be: 1. Physical equipment's – all the hardware used: Virtualization and management or automation tools; Operating systems and application (software) used to perform essential functions; 2. Business application include: Databases – SQL Server; Transactional systems such as real-time order entry; Email servers like Exchange, Web servers, Apache; Customer relationship management; Enterprise resource planning.

Another classification provided by McAffe [8] divide IT industry intro three types:

- Function encompasses technologies – application that streamline individual tasks;
- Network IT – help people communicate each other;
- Enterprise IT – let companies re-create interactions between groups of workers or between business partners;

CompTIA's reports [6], [9], [10], [11], on the other hand, have proposed along all four reports since 2016 to 2019 different category.

In the *Table 1* we can see a dynamic in defining IT component due to a lack of clear criteria. From the details of the component we can see the difference in views from 2016 to 2019. Each of them suffered adjustments according with the new trends in every year.

Looking into these efforts to define the IT component we have saw two major constants across all the classification, and these are Hardware and Software. It's very hard to define a hardware component, in this IT area, without a small part of software, and vice versa. This could be a side effect of cost dropping in hardware area, where the pressure to build cheap product it's high, but adding a software component, could add much value to the final product that selling separately. Another explanation could be the higher degree of software re-usage in the case of generic and cheap hardware pieces that could bring. Based on this observation we build the illustration from ***Figure 1*** for IT industry.

Classifying the IT industry in this way, should be much easy to compare different type of products among similar ones, and also, would be much easy to identify the major players in the whole industry but also for specific products positioned into those four quadrants, compare the historical trends for hardware and software and apply the current technological trends should be much easy to identify further trends – knowing the evolution of hardware, so far, we can observe what kind of products are in the quadrants with high hardware component and predict an high dynamic among products and producers.

Table 1 The IT industry components 2016 - 2019, from CompTIA's perspective

IT category			
2016	2017	2018	2019
Software – apps for productivity, business, network, systems, security, mobile apps.	Software – applications, system infrastructure software	Software	Software
Services – deployment, integration, custom development, break/ fix, managed services	Services – planning & implementation, support services, operation management, training	Services	Services
Information – data, documents, voice, video, images, social streams	Telecom services – fixed voice, fixed data, wireless voice, wireless data	Telecom services	Telecom services
Hardware – computers, servers, storage, mobile devices, printers, network equipment, etc.	Hardware – servers, personal computers, storage, smartphones, tablets, network equipment, printers & other peripherals	Devices + infrastructure	Devices + infrastructure
Digital business – commerce, communication, collaboration, automation, governance		Other emerging tech	Other emerging tech – such as IoT SW/ HW/ connectivity, SaaS+ PaaS, robotics/ drones, AR/ VR, AI platforms, etc.

Software		Hardware: Low	Hardware: High
	High	II. education, e-commerce, administer data bases, fintech, social network, collaborative platforms, online markets/ marketplaces, augmented reality,	III. databases administrators, CAD project builders, Enterprises service providers, AI, Blockchain, Cryptocurrencies, Big data, Machine learning, Infrastructure providers for content producers and/or content administrators, Defense industry,
	Low	I. very dedicated devices, smart phones, smart devices, home appliance,	IV. robotics, e-heath, IoT, Drone 3D printing Space equipment,
		Low	High

Hardware

Figure 1 The IT industry classification based on HW/ SW components

The *Table 2* it's rather an overview of IT industry (or ITC – Information Technology and Communication like EU described) in both major area, US and Europe, based on NAICS and NACE

IOP Conf. Series: Materials Science and Engineering **568** (2019) 012091 doi:10.1088/1757-899X/568/1/012091

classification than a comparative description. In both cases the IT industry cover all areas described above.

Table 2 Codes for IT in Europe and US

US codes According to NAICS [12]	EU Codes According to NACE rev.2 [13]
Information technology (IT) services	ICT manufacturing industries (operational)
Computer systems, design & related services: 541511, 541512, 541513, 541519	Manufacturing: 261, 262, 263, 264
Computer training: 611420	ICT services industries: 58.2, 58.21, 58.29
Repair services: 811211, 811212, 811213	Telecommunications: 61.1, 61.2, 61.3, 61.9
Manufacturing: 3341, 3342, 3343, 3344, 3345, 3346	Computer programming, consultancy and related activities: 62.01, 62.02, 62.03, 62.09, 63.1, 63.11, 63.12
Software publishers: 511210	Repair of computers: 95.11; 95.12; 95.21
Telecommunications services: 517110; 517210; 517911; 517410; 517919	
Internet and data services: 518210, 519130, 519190	
Distribution and wholesale: 423420, 423430	

The IT industry is estimate by CompTIA [11] for 2019 at 5 trillion US dollars globally, where 31% represent US, 26% Asia-Pacific region, 19% Europe, 7% Far-East region, 6% South America, 5% Africa, 3% Central and Eastern Europe, 2% Canada. Form components point of view, the biggest percent is registered by Telecom service with 30%, followed by Devices and infrastructure with 22%, IT and business services 21%, Emerging tech 17% and Software with 11%. In the US the strongest component is IT and business services with 30%, and software have 18%, with one percent greater than Devices and infrastructure. The 2018 PREDICT report [14] analyze three dimensions of the European IT industry: Value added (VA) of the industry, employed numbers and business expenditure with R&D (BERD). At the EU level in 2015 the ICT sector report a VA to 581 billion euros, 5.8 million people employed and 30 billion euros BERD and those represent 3.9% of the EU VA, 2.5% employment and 15.7% of total BERD. In 2015 the ICT sector was more dynamic than the whole EU economy. The distribution by component of the VA is 8,6% for ICT manufacturing industries, 45,7% ICT services, 15,8% Telecommunication and 29,9% for Computer and related activities (software is included here). The Atradius's [15] report, citing sourcing like Macrobond and Oxford economics, present a 2019 forecast for some global key players in the ICT industry – Table 3

Table 3 The ICT sector growth

	China ICT	Germany ICT	Japan ICT	UK ICT	US ICT	Australia ICT	France ICT	India ICT	Italy ICT	The Netherlands	South Korea	United Arab Emirates
GDP growth %	6,0	1,8	0,9	1,7	2,4	2,6	1,9	7,0	1,1	1,7	2,6	3,8
Sector VA growth	6,1	3,4	1,0	3,4	3,0	4,3	3,4	8,7	3,4	2,8	3,0	4,5

Beside China and India, we can easily observe the growth of ICT industry in United Arab Emirates and Australia, countries where there is no tradition in developing technology.

4. Conclusions

The ITC evolution could not be ignored anymore. Follow the very first development steps, we could see the trends in whole area where the hardware component of the industry competes the software component in term of the value and importance. The rapid development and innovation input in the computer world, the society was transformed from producing culture to information exchange [16 - 18]. As part of the economy modernization trends, the software industry plays a natural engine role - most of the company's efforts being based on the IT, especially the software component. With the omnipresence of mobile phones and the customers demanding more and more personalized experiences, the software industry continues to stay at the forefront of the business agendas in most of the companies. As Steiniger [19] noted, this is not a new revolution - the IT started to change the business landscape 50 years ago. Furthermore, Steiniger notes that IT (and software in particular) is generating „waves of technological disruptions "in a periodical manner. The software industry is as prevalent as a growth engine, that one can barely open any business magazine without a reference to it - be it direct or indirect. Thus, both research and empirical evidences broadly agree that the software industry and its offerings are becoming more and more a daily part of the business operations in most sectors.

References

[1] EBSCO Apr 03 2019 EBSCO host Online retrieved from: http://web.b.ebscohost.com/ehost/resultsadvanced?vid=5&sid=df4b9631-c433-485f-adb5-0e5ee878e5c6%40sessionmgr101&bquery=information+technology&bdata=JmRiPWE5aCZkYj1seGgmZGI9OGdoJmRiPXRyaCZkYj1oZXYmdHlwZT0xJnNlYXJjaE1vZGU9U3RhbmRhcmQmc2l0ZT1laG9zdC1saXZl
[2] Google Academic Online 2019 retrieved from: https://scholar.google.ro/scholar?hl=ro&as_sdt=0%2C5&q=information+technology&btnG=&oq=information+
[3] Moore G E 1965 Cramming more components onto integrated circuits *Elect* **38(8)** 114-17
[4] Jorgenson D W 2001 *Information technology in the us economy* New Orelans LA American economic association
[5] Robert P Parker B and Grimm T 2000 *Recognition of business an government expenditures on software as investment: methodology and quantiative impacts* Washington DC: US BEA 1959
[6] CompTIA 2016 IT industry outlook 2016 s l: www.comptia.org
[7] Rouse M Feb 2019 www.techtarget.com Online Cited: Apr 4 2019 https://searchdatacenter.techtarget.com/definition/it
[8] McAffe A 2006 Mastering the three worlds of information technology *Harv Bus Rev* retrieved from: https://hbr.org/2006/11/mastering-the-three-worlds-of-information-technology
[9] CompTIA 2017 IT industry outlook 2017 retrieved from: https://www.comptia.org/resources/it-industry-trends-analysis-for-2017
[10] CompTIA 2018 IT industry outlook 2018 retrieved from: https://www.comptia.org/resources/it-industry-outlook-2018
[11] CompTIA 2019 IT industry outlook 2019 s l www.comptia.org
[12] NAICS 2017 North American Industry Classification System retrieved from: https://www.census.gov/eos/www/naics/
[13] Eurostat 2008 NACE Rev 2 Statistical classification of economic activities in the eu luxembourg: office for official publications of the european communities 1977-0375
[14] Mas M, Fernandez de Guevara J, Robledo JC, Cardona M, Lopez-Cobo M, Righi R and Samoili S 2018 The 2018 predict key facts report luxembourg: publications office of the EU 1831
[15] Atradius 2018 *Atradius market monitor* Amsterdam Atradius NV
[16] Kurzweil R 2005 *The singularity is near: when humans transcend biology* NY Penguin Books
[17] Zittrain J 2008 *The future of the internet: and how to stop it* New Haven Yale University Press
[18] Lee S M, Olson D L and Trimi S 2012 Co-innovation: convergenomics, collaboration, and co-creation for organizational values *Mng Dec* **50(5)** 817-831
[19] Steinger D M 2019 Linking information systems and entrepreneurship: A review and agenda for it-associated and digital entrepreneurship research *Inf Syst J* **29(2)** 363-407
[20] Khasawneh A M 2008 Concepts and measurements of innovativeness: the case of information and communication technologies *Int. J Arab Cult Manag Sustain Dev* **1(1)** 22-33

IOP Conf. Series: Materials Science and Engineering **568** (2019) 012092 doi:10.1088/1757-899X/568/1/012092

Study regarding the implementation of an Ackerman steering geometry in MATLAB

D Moldovanu[1*], A Csato[1], N Bagameri[1]

[1]Technical University of Cluj-Napoca, Automotive Engineering and Transports Department, Bd. Muncii 103-105, Cluj-Napoca, Romania

dan.moldovanu@auto.utcluj.ro

Abstract. In dynamic vehicle behavior the steering system has an important role. The vehicle stability, maneuverability depends on steering system characteristics. Majority of passenger vehicles uses the rack pinion steering linkage system. In this article, a simplified planar six-bar linkage model of the rack pinion steering system is presented. Kinematic relations are used to determine every linkage position in every possible situation. This simplified system contains a rack, tie rods, steering arms, and wheels. The six-bar linkage model is introduced in MATLAB SIMULINK. In the function of the rack displacement, the wheel angels are determined. The dimensions of this planar model can be estimated due to the minimum error between the calculated wheel angels for the rack and pinion system and an Ackerman principle based system with similar dimensions. For optimizing a system and for verifying the method numerical calculation are used.

1. Introduction

The rack-and-pinion steering mechanism is widely used in passenger cars, majorly cars with independent front suspension. The rack and pinion steering mechanism has several advantages: simple construction, easy and cheap to manufacture, compact, and easy to handle [1]. The rack and pinion steering system consists of the rack, two tie rods and two steering arms of the wheel knuckle. Steering mechanisms are complex spatial linkages. Restrictions regarding on the positioning of the steering mechanism in the engine compartment and any significant kinematic correlation to the suspension, such as the cross-linking effect of steering and suspension, may further increase the complexity of describing the steering mechanism. Disregarding the dynamic effect of elasticity and a lateral slip of tires, the turning characteristics of the vehicle are remaining unaffected. The inclination angle of kingpin only slightly influences the transmission function of the steering, and it is chosen by other consideration as a vehicle moves straight forward provides stability of steering, not letting the wheels to get out from straight direction and provides a self-adjusting effect when cornering[2]. As a result the rack and pinion steering mechanism which is a spatial mechanism can be modelled as a planar six bar linkage system for verifying the Ackermann condition. Therefore, in this study the kingpin inclination angle and the caster angle are not taken into consideration and a six bar planar linkage system is considered as steering mechanism. The six bar linkage system consist of six revolute joints and one prismatic joint. Such a steering system has been used by other researcher [3], [4].

Another important characteristic of the vehicle is steady-state cornering. Vehicle at constant speed and constant steering angel making a circular motion within a circle with constant radius, this motion is called steady-state cornering. Vehicle motion characteristic can be described by vehicle steady-state

IOP Conf. Series: Materials Science and Engineering **568** (2019) 012092 doi:10.1088/1757-899X/568/1/012092

cornering. For simplicity a single track bicycle model is used theoretical analysis of the vehicle cornering characteristics [5].

2. Objectives

In this research a rack and pinion steering mechanism with McPherson suspension is analysed. From the analysis of steering system of a projected six bar linkage mechanism with projected lengths are elaborated. By solving the nonlinear problem, a function is developed between the rack position and the wheel steering angle. A steering angel error is defined between the six bar linkage mechanism and the Ackerman condition. For minimizing the error a one parameter optimization problem is solved using the tie rod length as variable.

3. Ackerman condition

For the vehicle inner and outer wheel in steady-state cornering there is a kinematic relation called the Ackerman condition and it is defined by the equation (1) where δ_0 is the steer angle of the outer wheel, δ_i is the inner wheel steer angle, w is the track and l is the wheelbase of the vehicle. This condition means that the inner wheel has a bigger steer angle than the outer wheel, because is closer to the centre of rotation.

$$\cot\delta_0 - \cot\delta_i = \frac{w}{l} \quad (1)$$

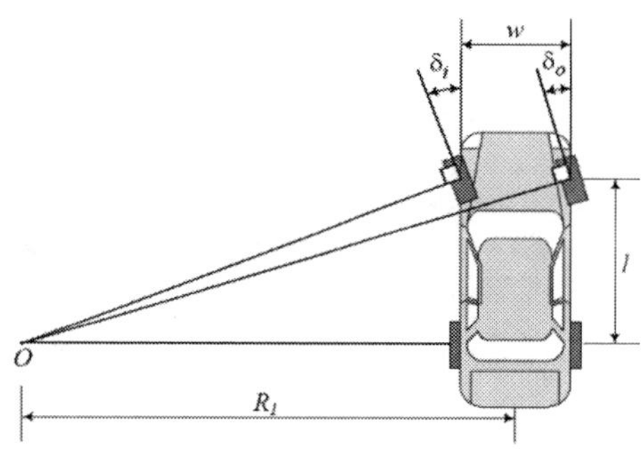

Figure 1. Ackerman condition for front wheel steering vehicle [6].

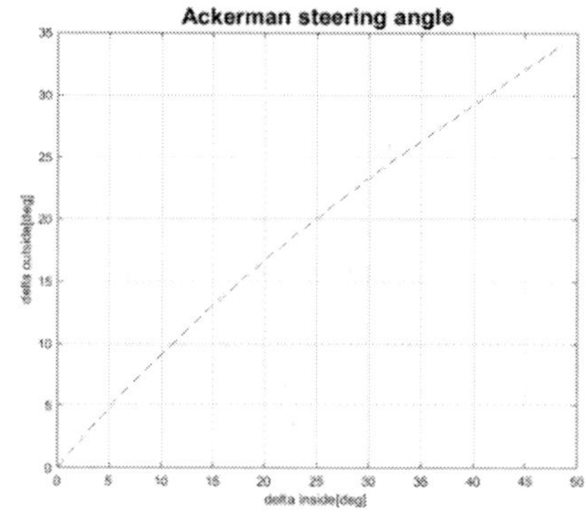

Figure 2. Relations between inner and outer wheel angle for Ackerman condition.

There are no four bar linkage system that can satisfy the Ackerman condition [6], but there are multiple linkage systems that are capable to work closely to the Ackerman condition.

4. The six bar linkage planar model

The planar model consist of the rack, two tie rods and two steering arms. The analytical method to describe the relation between the rack position and the steering angle of the wheels is based on the equation (2). Because the system is symmetrical in mechanical point of view the the nonlinear system can be solved on one half of the system (Figure 3). For eliminating the θ_2 member equation (2) is squared and the two equalities are summed and became equation (3). For simplifying the equation A, B, C is introduced. Knowing the equations (8), (9) equation (7) easily became a second order equation that can be solved with discriminant (12) and gave two solutions for wheel steering angle (13).

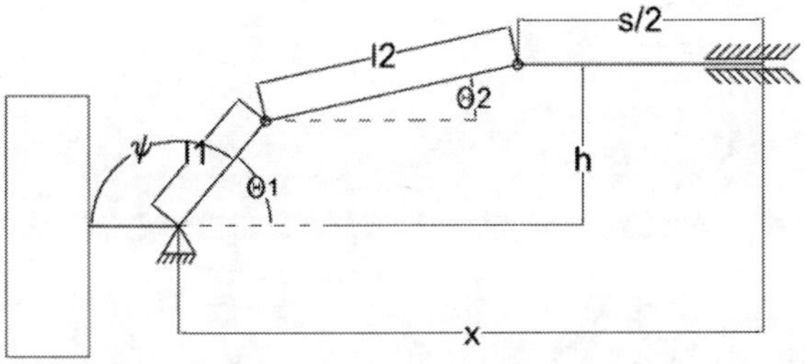

Figure 3. Half of the six bar linkage for solving the nonlinear equation.

$$\begin{cases} l_1 \sin\theta_1 + l_2 \sin\theta_2 - h = 0 \\ l_1 \cos\theta_1 + l_2 \cos\theta_2 + {}^{s}/_{2} - x = 0 \end{cases} \tag{2}$$

$$l_1^2 - 2l_1(\sin\theta_1 + \cos\theta_1) + h^2 + l_2^2 - x^2 \tag{3}$$

$$A = -2l_1 s \tag{4}$$

$$B = -2l_1 h \tag{5}$$

$$C = l_1^2 - l_2^2 + h^2 + s^2 \tag{6}$$

$$A\cos\theta_1 + B\sin\theta_1 + C = 0 \tag{7}$$

$$\cos\theta_1 = \frac{1-\tan^2\frac{\theta_1}{2}}{1+\tan^2\frac{\theta_1}{2}} \tag{8}$$

$$\sin\theta_1 = \frac{2\tan\frac{\theta_1}{1}}{1+\tan^2\frac{\theta_1}{2}} \tag{9}$$

$$\tan\frac{\theta_1}{2} = d \tag{10}$$

$$d^2(C-A) + 2Bd + C + A = 0 \tag{11}$$

$$\Delta = 4B^2 - 4(C-A)(C+A) \tag{12}$$

$$d_{1,2} = \frac{-2B \pm \sqrt{\Delta}}{2(C-A)} \tag{13}$$

5. Simulation model

For calculation of the wheel steering angels a MATLAB Simulink model was developed. The data for the projection lengths and angles of the different parameters of the six bar linkage mechanism are based on a VW Golf V short wheelbase model [7]. The parameters are presented in Table 1.

Equations from section 4 are implemented in Simulink model for each wheel separately. Inside and the outside wheel are masked in separate blocks to get accurate results, but the major difference between the wheels is only the input, the rack position. The rack position movement is simulated with a signal builder which is linearly changing values from 0 to 80.

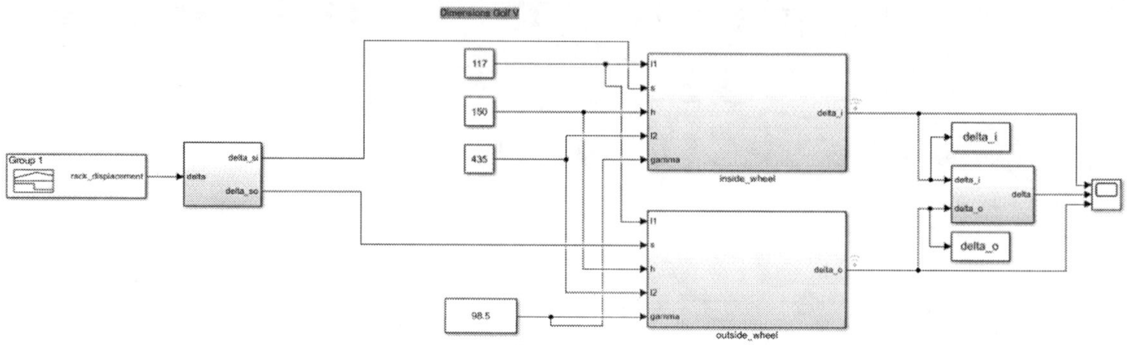

Figure 4. Simulink model of the six bar linkage.

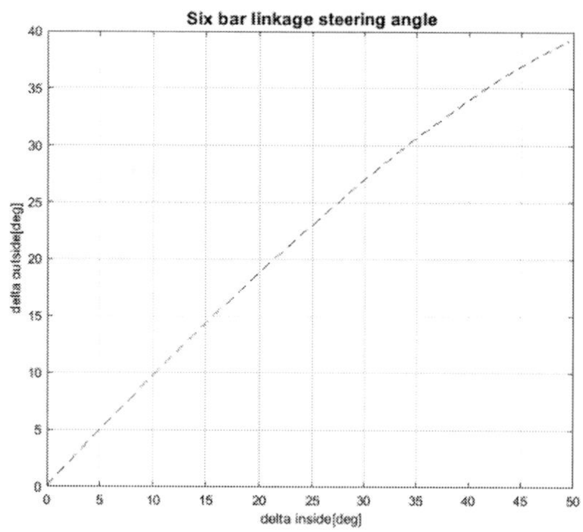

Figure 5. Relations between inner and outer wheel angle for six bar linkage.

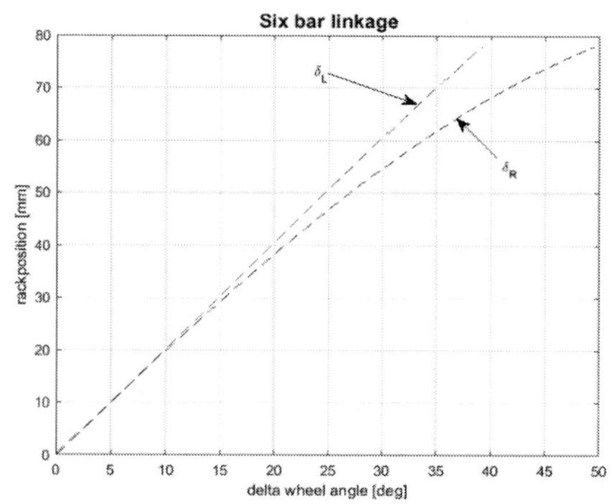

Figure 6. Relations between rack position and wheel steer angle, δ_L for left (outer side) δ_R for right (inner side)

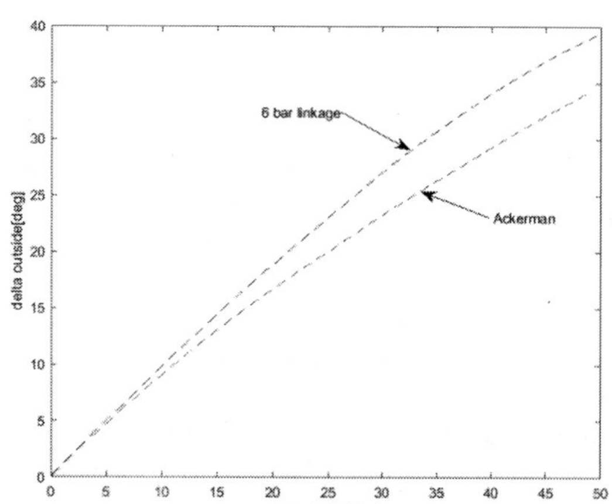

Figure 7. Relations between inner and outer wheel angle for Ackerman condition and for six bar linkage

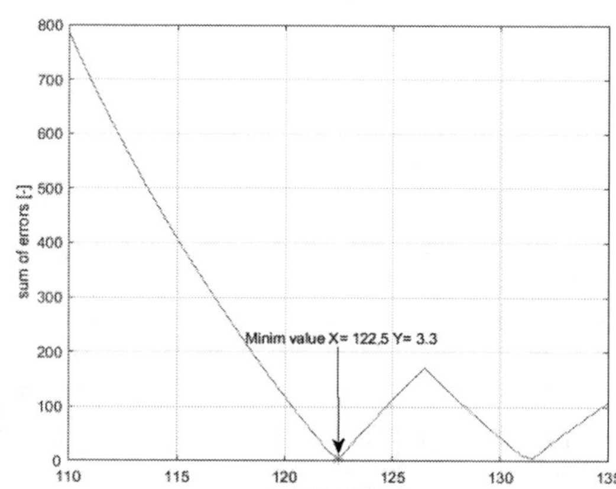

Figure 8. Sensitivity analysis for tie rod length.

IOP Conf. Series: Materials Science and Engineering **568** (2019) 012092 doi:10.1088/1757-899X/568/1/012092

6. Optimization

To have steering system working at very closely to the Ackerman condition the tie bar length is modified until de error between the two systems achieve the minimum. The error is calculated on each points of the two curves summing up this values in the sum of errors. For optimizing the six bar linkage a nonlinear program solver is used 'fmincon' which uses Hessian matrix for minimizing the cost function in this case the sum of errors.

After 18 iteration the algorithm has converge to the value of 122.5 mm of the tie rod and 3.3 for sum of errors (Figure 8). The optimized steering mechanism behaviour is represented in figure 9.

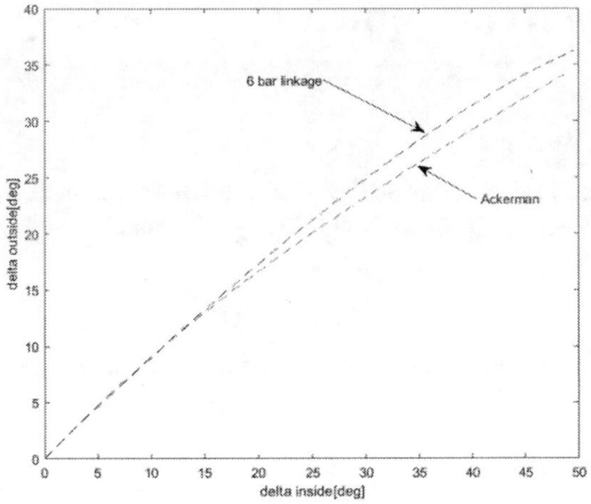

Figure 9. Relations between inner and outer wheel angle for Ackerman condition and for optimized six bar linkage.

7. Conclusion

The purpose of the paper was to show the modelling of the Ackerman condition as a simplified model with a six-bar linkage. By using MATLAB Simulink the calculations were made and optimized by using 'fmincon' which uses Hessian matrix for minimizing the cost function in this case the sum of errors, which has brought the relations between inner and outer wheel angle for Ackerman condition very close. Further simulations will include the control of the steering as a closed loop system.

References

[1] Wasiwitono, U., Sidarta, I., Sigit Pramono, A., Sutikno, S., Wikarta, A., 'Steering System Kinematic and Steady-State Cornering Analyses of the ITS Electric Car', *IPTEK J. Proc. Ser.*, 2014.

[2] Mazilu, A., Preda, I., 'Kinematic Optimization of the Rack and Pinion Steering-System of an Automobile: An Example', 2016.

[3] Habibi, H., Shirazi, K. H., Shishesaz, M., 'Roll steer minimization of McPherson-strut suspension system using genetic algorithm method', *Mech. Mach. Theory*, vol. 43, no. 1, pp. 57–67, 2008.

[4] Rahmani, A., Rao, P. V. M., Saha, S., 'Kinematic and sensitivity analysis and optimization of planar rack-and-pinion steering linkages', *Mech. Mach. Theory - MECH MACH THEOR*, vol. 44, pp. 42–56, 2009.

[5] Yang, S., Lu, Y., Li, S., 'An overview on vehicle dynamics', *Int. J. Dyn. Control*, vol. 1, no. 4, pp. 385–395, Dec. 2013.

[6] Jazar, R., *Vehicle Dynamics: Theory and Application*. 2008.

[7] Knapczyk, J., Kucybała, P., 'Simplified planar model of a car steering system with rack and pinion and McPherson suspension', *IOP Conf. Ser. Mater. Sci. Eng.*, vol. 148, p. 12011, 2016.

[8] Antonya Cs, Butnariu S and Beles H 2015 Geometric identification of a four-bar linkage from noisy tracking data. 2015 IFToMM World Congress Proceedings, IFToMM 2015, Pages: 217-222. 14th; Taipei; Taiwan; 25 - 30 October 2015; Code 127160. Publisher: National Taiwan University, DOI: 10.6567/IFToMM.14TH.WC.OS2.012.

IOP Conf. Series: Materials Science and Engineering **568** (2019) 012093 doi:10.1088/1757-899X/568/1/012093

Aspects of wear braking process on brake discs and brake pads in the construction of commercial vehicles

M F Popa[1], D M S Capătă[1] and N Burnete[1]

[1]Technical University of Cluj-Napoca, Department of Automotive Engineering and Transports, 103-105 B-dul Muncii Street, Cluj Napoca 400641, Romania

madalin.popa@live.com

Abstract. Economy and low pollution are essential requirements imposed on modern cars. The continuous increase in travel speeds, coupled with the massive increase in the number of vehicles worldwide, has led to the need for the development and use of braking systems to a very high standard. The exploitation of commercial vehicles has as a central factor the profit, which in turn is built on the highest availability of the vehicle and the lowest possible maintenance costs. Both aspects of profit are found in the process of wear of the elements in the braking system. For the owner of a truck, wear is measured in costs (parts and time for repairs). This study shows the result of analyzing multiple sets of data collected from trucks brake control unit in various stages of maintenance and highlights key factors influencing the braking process, also seeking solutions for increased braking performance, reduced wear, reduced maintenance costs for the braking system and increasing the availability of a truck. The processing of braking data and analysis of the braking process, based on the measured data, allowed the identification of some predominant influence factors in the wear of the disc brakes of the analyzed commercial vehicles.

1. Introduction

Since the invention of cars, they radically changed the world and people's way of life. A personal vehicle means independence from specific factors. Commercial vehicles are the ones that support an essential part of the economy, primarily through freight and passenger transport services, and secondly through the jobs in the many adjacent areas. Modern vehicles have evolved with technological advancement and have become complex structures capable of moving at higher speeds to the highest standards. At the same time, the steep increase in the number of vehicles has made it necessary to increase traffic safety in several ways, starting with the development of the central safety systems and continuing with the creation and development of driver assistance systems.

The braking system is one of the most critical safety systems on a vehicle, and it contributes significantly to avoiding accidents and increasing traffic safety. The complexity of current braking systems tends towards a technological peak, represented by their use in the construction of autonomous vehicles. Already on a series of vehicles, there are encountered braking assistance systems, the operation of which is based on analyzes, tests, and results from the actual operation of the vehicles [1].

2. Basics of the study. Elements of analysis

Elgharbawy, M. et al. in [1] points out that data collection and analysis can be used to improve performance in operation or to create programming algorithms for rolling and braking assistance systems.

IOP Conf. Series: Materials Science and Engineering **568** (2019) 012093 doi:10.1088/1757-899X/568/1/012093

Jingliang, D. et al. in [2], they are talking about introducing into the systems of the assist systems of predefined data sets, taken from the exploitation or experimental analysis, which underlie their operation. Thus, if the braking system has delays due to mechanical defects, it will not be possible to calculate and perform the brakes in the most efficient way, so in this case, in order to eliminate the errors and in order to be braked at the maximum parameters, it is necessary knowing the exact functioning of the mechanical systems and avoiding unforeseen failure.

Classical braking systems convert kinetic or potential energy into heat by friction [3] and as such, the braking process must be analyzed by the tribological aspect of friction, and the main problems being studied are related to the variation of some parameters during braking (distribution and variation temperature friction coefficient, heat flux variation, wear, noise and vibration, particle emissions, durability).

The interdependence between the friction coefficient and the temperature has a decisive role in influencing the braking performance. Alexey, V., et al. [3], analyze the heat flow distribution for plate and disc braking and find that on the model used the maximum temperature is recorded on the disc (490°C) and then the temperature drops to the hub area (100°C). Thus, almost 93% of the energy is taken up by the disk in the form of heat and only 6% in the brake pads. The temperature problem also has a significant influence on wear, emphasizing the process according to the working conditions, and in the case of thermal overloads, the lack of braking capacity can be achieved. An excellent heat transfer can mean faster heat dissipation and hence reduced wear.

Amirhossein H. et al. in [4], notes that the frictional force is transferred at disk trough brake pads. As a result of the friction, because the brake pads are made of a softer material, they will wear out more strongly, opening the way and studying the influence of the materials in the braking process. Also, in the same study [4], it is stressed that although friction is a great source of energy loss in mechanical systems, in automobiles, this is an essential factor in vehicle control. Among the factors of influence of wear, identified in the same study can be mentioned: brake pad geometry and the presence of channels in the middle of the plate for the disposal of waste material or impurities; the distance traveled between slip points, etc.

Most studies find that simulation of real conditions comes with several challenges and compromises [5], and the mathematical models that are created become very complex if too much data is introduced. To validate the results, both methods are used in parallel: finite element computerized simulation and experimental experimentation on the stand [6].

To find an optimal solution for improving braking performances, it is necessary to analyze several factors such as operating conditions and external factors of influence; properties of materials; wear process; variation of heat flux; noises and vibrations; system or component durability, availability and maintenance.

3. Experimental procedure

Commercial vehicles are often used in substantial, long-term and high-load conditions. This emphasizes the wear of the elements in the braking system and makes it necessary to continuously monitor the wear degree in order not to cause severe damage to the discs. In light commercial vehicles with a maximum authorized mass of fewer than 3.5 tons, the braking system is hydraulically operated and does not differ significantly from that of cars. Brake pads wear monitoring is carried out through a car dashboard that will alert the driver when one of the plates has exceeded the maximum wear limit. This is done by an electric circuit whose signal can have only two values.

For heavy trucks, due to the need for much higher braking forces, the brakes are pneumatic. The proper braking is done by actuating the brake piston mechanism (using levers) from the brake caliper, which in turn presses the plate from the inside towards the disc. At the same time, the brake caliper slides on its guide pins and allows the outer brake pad to contact with the disk.

This mechanism also contains a sensor that records the position of the pistons in the caliper, which it sends to a control unit in the form of an electrical signal. The signal sent by the position sensor is processed by the brake control unit (or the maintenance unit), and the information can be used to display

IOP Conf. Series: Materials Science and Engineering **568** (2019) 012093 doi:10.1088/1757-899X/568/1/012093

on board the degree of wear of the brake pads or to calculate the remaining time until full wear. There are cases in which braking will be differentiated on each wheel, depending on the degree of wear, as far as possible and without significantly affecting the braking properties or trajectory. This means that the brake control unit will try to brake more, the wheels at which the wear is lower and will brake with a smaller force, the wheels at which the wear is higher.

For this paper, most of the mileage data on which the service interventions took place were taken from the truck control units through the diagnostic socket (OBD) using diagnostic software. The data are in the form of tables and have been categorized into a database created explicitly for this purpose. Another part of the data was taken from the service booklet.

The study also contains some practical measurements to link the wear, expressed as a percentage, of the actual value of the used parts. Both used parts and new spare parts were also compared and analyzed. The paper also exemplifies some cases of specific wear on the brake pads, which were captured in pictures and represented cases with a reduced but essential frequency.

One of the objectives of the paper starts from the analysis of the values provided by the position sensor in the caliper, regarding the wear of the brake pads and makes a report of the wear on the number of kilometers traveled for different types of use. Another aspect studied in the paper relates to the monitoring and analysis of the operating conditions until the first intervention in service on the braking system. In this idea we have analyzed the number of kilometers traveled until the first replacement of the brake pads (front or rear axle); the dimensions of the brake pads and discs were measured and compared with the initial values.

Based on the analysis of these aspects and based on the stored data it was found that it is necessary to divide by exploitation categories because the values of the wear and the maintenance intervals vary significantly for the trucks running in Romania compared to the trucks frequently running on Europe routes. In this respect, three main factors have been identified which favor the increased use of brakes: the average road inclination in Romania, which favors the intense use of the braking system [8]; dust, mud, salts and non-slip material during the winter season - these cause deposits that most often block the stirrup mobility or favor abrasive wear; the human factor through the driving style becomes essential in the wear process of the braking system. On vehicles running on motorways, braking is slower and much lower in number; in this situatio, it is useful to use an auxiliary braking system, but this comes with its advantages and disadvantages.

4. Results and discussions

Heavy trucks are equipped with various types of auxiliary braking systems. The majority of trucks have an auxiliary braking system with engine gas discharge or compression release (known as "Jake Brake"). This type of compression-releasing auxiliary brake is very used and can generate a braking force of up to 400 KN (about 85% of the engine power [7]), but it has the main disadvantage of the braking noise. For higher braking forces, retarders are used (hydrodynamic deceleration). They can achieve braking forces of up to 4000 KN [7] but come with disadvantages such as mass and extra costs. The current study also took into account the retarder auxiliary braking system and in particular its influence on the wear of the main brake components. The results indicate a significant increase in the use of brake pads in retarder trucks, as shown in figure 1, where: *pf* and *pr* represent the replacement of the brake pads on the front or rear respectively; *with Ret* and *without Ret* - with or without auxiliary braking system (retarder); *Ro* and *EU* refer to the exploitation conditions (*Ro*- run in Romania and *Eu* - run on motorways in Europe).

It is noted that there are no significant differences in wear between a retarder truck running in Romania and one without a retarder running on highways in Europe. This aspect, together with the analysis of interventions on the braking system on trucks running in Romania, highlights the need to identify wear reduction solutions caused by external factors such as driving style, infrastructure and road conditions. On the other hand, the importance of the "Jake Brake" auxiliary braking system, which offers a reduced braking torque, covers a significant percentage of the braking demand. It should be noted that all non-retarder trucks in the study are equipped with an auxiliary braking system such as Jake Brake.

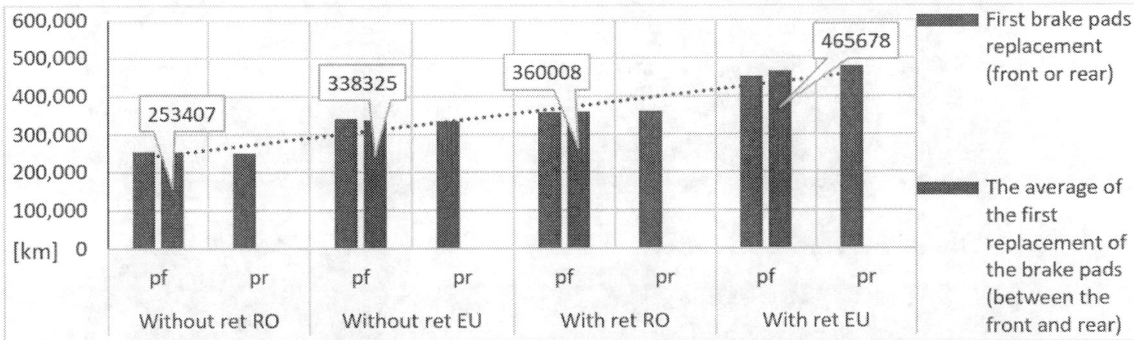

Figure 1. Average mileage at which the first brake pad replacement was performed

Retrieving data from the Brake Control Unit provides details of the replacement of the brake pads and the wear values at the time of reading. Based on these data, the frequency of replacement of the front and rear axle brake pads from the tractors were analyzed and the main trends were identified. Thus, for vehicles traveling in central and Western Europe, in about 40% of cases, the front axle pads are worn for the first time, and in 20% of cases, the first intervention is done at the rear axle. This can be explained by the analysis of the transported cargo and its distribution in the semi-trailer. In the rest of the cases, the brake pads were replaced by the same intervention on both the front and the rear. Some carriers prefer an advance replacement to just 80% of the brake pad thickness to reduce the number of service entries and to avoid excessive wear on the plates (which can also damage brake discs and increasing repair costs). In this respect, if the front side pads are worn 100% and 80% worn on the rear, it is possible to choose the option of total replacement (preventive), which will guarantee a longer service life without any intervention to the braking system.

During the study it was found that the wear values of vehicles with retarder compared to those with only the Jake Brake type system can`t be compared in the same graph without taking into account the route. It is necessary to divide into two categories: with or without a retarder and the type of the route. Under running conditions in Romania only a 50...70% wear may occur after a run of only 20,000...40,000 km although in highway conditions this wear corresponds to a distance of 150,000...200,000 km. This underlines once again the massive influence that this system has on the wear and tear and the fact that the analysis of the influence of infrastructure in developing countries has a particularly important role in determining the maintenance costs of the braking system.

4.1. Experimental determinations

In operation, the main problems that lead to excessive wear are caused by blocking the brake calipers due to dust and mud on the road; suppressing the movement of the brake pads; nonconforming parts; overstretching, uneven and unexpected wear, aggressive driving style. There are, however, situations where it has been found that the brake assembly (brake caliper, sliding mechanism, brake pads) operate in parameters, but the reported wear was powerful. Such a case was found in replacing the brake pads in a truck running in mixed conditions (Romania and Europe). At a distance of 225,538 km, the wear indicated by the control unit (after processing the signal received from the piston position sensor in the brake caliper) was 107% (more detailed, on the rear - left the wear was 100%, on the right 72% and the front left the wear was 64%, respectively 72% on the right). After disassembly and careful analysis of the braking system, it was found that on the rear left of the outer pad, the friction seals were worn out and the braking was done for a while with the metal plate support, as shown in figure 2. A mechanical fault could have caused a wear abnormality like this. It is known that there is a software in the brake computer that attempts to uniform wear on each wheel (by managing the braking force on each wheel).

As a result of the measurements on the worn surface of the inner brake pad holder, the profile of the wear can be contoured (in the radial direction from the hub to the outside). There is thus slower wear on the edges and very pronounced towards the inside, having two maximum zones, figure 4.

IOP Conf. Series: Materials Science and Engineering **568** (2019) 012093 doi:10.1088/1757-899X/568/1/012093

a)

b)

Figure 2. Excessive wear of the interior brake pad a)Top view; b)Side view

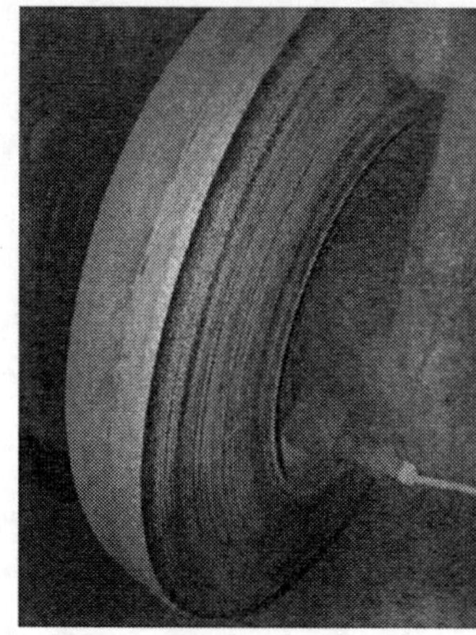

Figure 3. Wear on the brake disc and brake pads

The central area shows traces of thermal overload. In most cases of advanced wear, both the disc and the plates form a strip on the outer edge where wear is slower. A border is often formed at the edge of this area as shown in figure 3. This means that the edge of the disc remains a very narrow strip but much higher than the surface of the brake disk. This can be explained by the fact that the heat transfer is very active on edge, which also reduces wear. In the central area of the disc the temperatures are high, so the wear is high and decreases in the radial direction towards the hub.

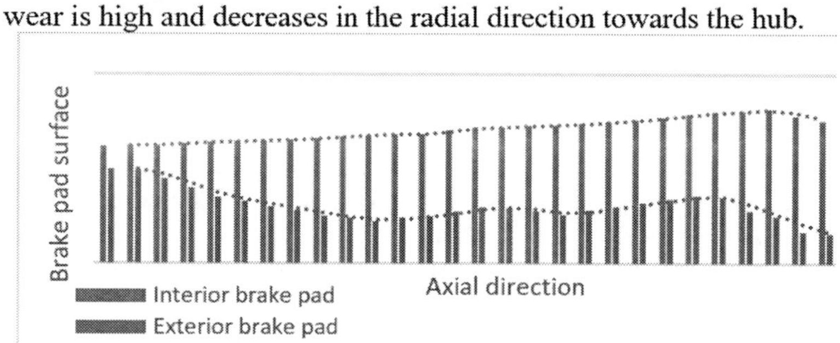

Figure 4. The wear profile of the left rear brake pads after a run of 225,538 km

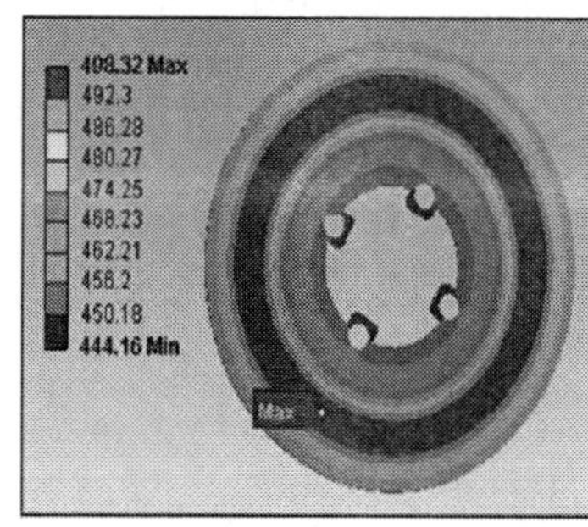

Figure 5. Temperature contours in °K [9]

A similar profile is also obtained in the measurement of the brake discs, estimated by measuring the wear profile in a radial direction from the hub to the outside, figure 6. The measurements were carried out on a truck run exclusively in Romania with 444,476 km. Again, there is more pronounced wear on the inner part, which decreases in intensity towards the extremities. This confirms that the heat flow propagates through thermal conductivity from the plates and disc, to the hub and then to the other parts, and by convection and radiation from the parts to the air. The wear profile also fits with the mathematical model in the study [9], the maximum wear being in the area of the maximum temperature indicated in the study as it shows in figure 5. In [9] maximum temperature zones, they correspond to the maximum pressure zones.

Increased wear of the inner part may be due to the fact that the inner brake pad is pressed directly by the piston but, to operate the outer brake pad, it is necessary to move the caliper on the guide bolts, which is not monitored by the sensors and which is often suppressed due to mechanical defects or the influence of external factors. Also, no wear on the disc is monitored by any sensor. This can pave the way for a system to monitor both the brake disk wear and the brake caliper mobility on its bolts. Identifying such solutions would significantly contribute to reducing the cost of premature wear of parts of the system.

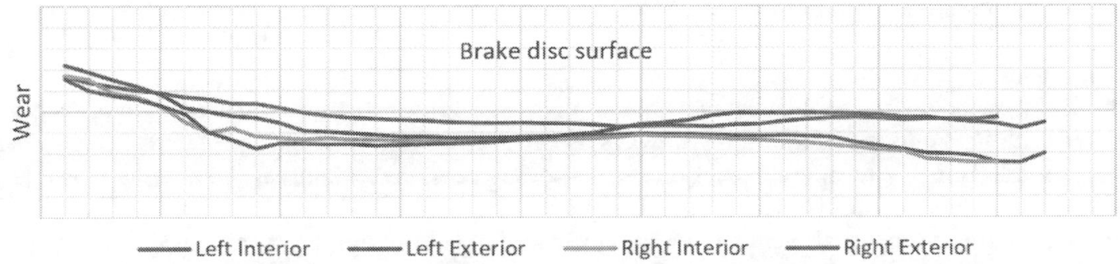

Figure 6. Wear profile on the surface of the brake disk

Another aspect of the data analyzed shows that around 60% of cases, brake pad wear is significantly higher on the right, figure 7. Data from 55 trucks sorted by different criteria was analyzed. This aspect of more pronounced wear on the right side could be explained by the fact that the road has a slight inclination to the right. In the literature of road construction, in alignments, the inclination of the transversal profile (road inclination and sloping) is right to the right direction and is currently 2% - 2.5% [8]. Road inclination must have the normative value because an insufficient incline leads to slow drainage of water from the road, which can lead to loss of adhesion at high speeds. In this case, when decomposition of the weight, on the right wheel comes to a further force of pressure, the effect of which is observed in the wear process.

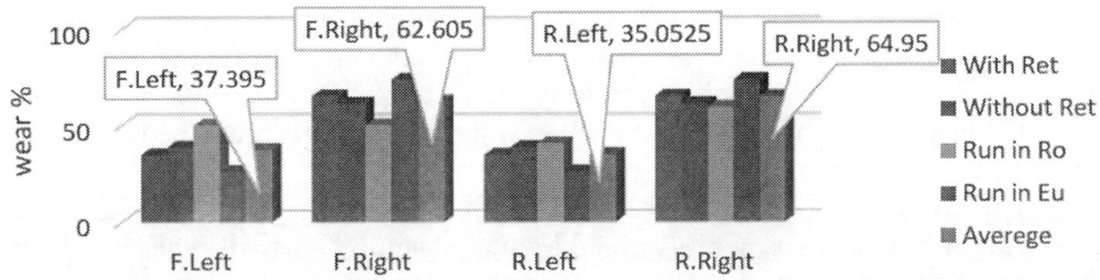

Figure 7. Situations in which the wear is higher on the right side [%]

Removing wear differences on each wheel is a challenge and is a necessary measure to reduce maintenance costs because when wear differences more significant than 10% have been reported, the life of the set on the deck decreases with at least 15%. Also, depending on how late the problem is detected, the brake discs can be affected in various forms (scratches, deformations, or cracks due to thermal overload).

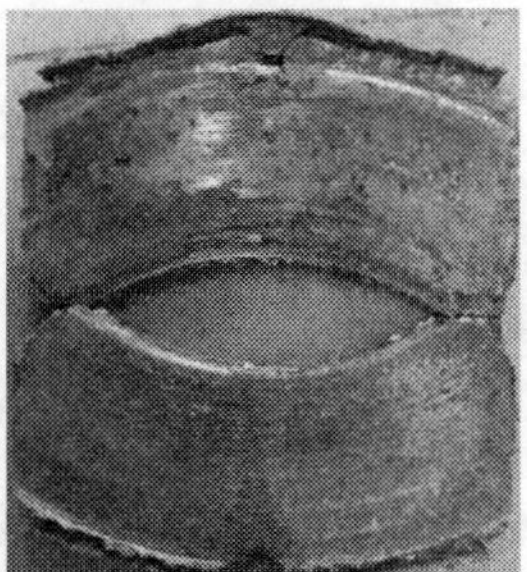

Figure 8. Abrasive and thermal overload (truck without retarder)

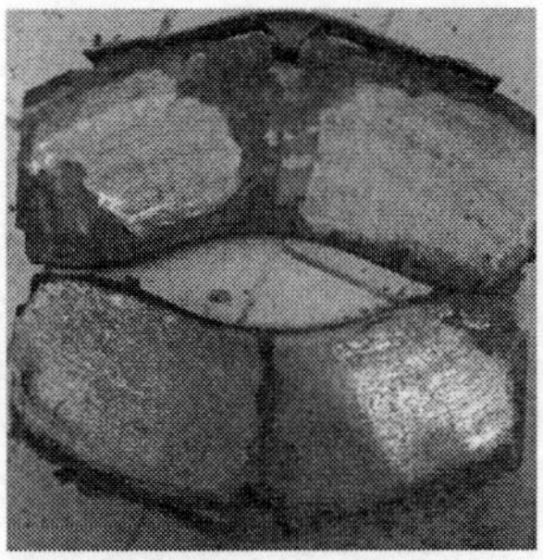

Figure 9. Corrosive wear 132,000 km / 2 years

Figure 10. Cracks caused by thermal overload

During the study, it was found that abrasive wear, corrosion wear and wear due to thermal overload figures 8 to 10 have a significant frequency in influencing the premature removal of brake pads and discs. In some cases, figure 9, worn plates at 80% of the capacity have been replaced due to the high

IOP Conf. Series: Materials Science and Engineering **568** (2019) 012093 doi:10.1088/1757-899X/568/1/012093

degree of decomposition due to superior corrosion. At this stage, the wear is exponential, and the traffic safety conditions are no longer met.

5. Conclusions

Commercial vehicles are sophisticated vehicles that, in addition to the increased gauge, frequently circulate with heavy loads on the same runways as cars. However, the braking systems fitted to trucks must respond promptly to operating requirements, be as economical and sustainable as possible and comply with, or even contribute to improved traffic safety.

Infrastructure and external factors have a significant impact on braking system usage and implicitly on maintenance costs. On the one hand, the average inclination of national roads, in the absence of highways, is much higher and leads to an increase in the number of brake applications and on the other hand these conditions make the driver a significant factor influencing the durability of the brake pads by the adopted driving style. Another aspect is corrosion and abrasion, favored by moisture, mud, and anti-skidding on the road. These types of wear have the highest frequency at the studied trucks.

The presence of an auxiliary braking system is all the more significant as the maximum braking force is more considerable. The importance of an auxiliary braking system is most noticeable for vehicles frequently traveling on highways, with the use of the same set of brake pads exceeding an average of 450,000 km. Also, the presence of an auxiliary retarder system almost eliminates overloading of the braking system and greatly reduces maintenance costs.

The faster wear of the brake pads on one of the truck's axles is conditioned by operating factors such as load and distribution of the load in the semi-trailer and road conditions. The different wear of the brake pads on the same axle has as its main causes, mechanical defects (primarily the brake caliper) most frequently caused by deposits and corrosion. More substantial wear on the right side of the brake pads (in a significant percentage) may be caused by the right-hand slope of the road (according to the design standard). It is also to be noted that the increased wear of a single brake pad or uneven wear leads to a drastic decrease in the durability of the whole system.

As a way to prevent local overloads caused by blocking the mobility of the brake caliper or pads, a monitoring system for both caliper mobility and disk status would be useful. This could increase traffic safety and reduce maintenance costs.

6. Acknowledgments

The practical part of the paper and the data acquisition took place at RMB Inter-Auto, dealer, and Mercedes-Benz partner.

References

[1] Elgharbawy M., et al., Adaptive functional testing for autonomous trucks, *International Journal of Transportation Science and Technology*, 2018.

[2] Duan J., et al., Driver braking behavior analysis to improve autonomous emergency braking systems in typical Chinese vehicle-bicycle conflicts, *Accident Analysis and Prevention* **108** 2017, 74–82.

[3] Vdovin A., et al., A coupled approach for vehicle brake cooling performance simulations, *International Journal of Thermal Sciences* **132**, 2018, 257–266.

[4] Hatam A., Khalkhali A. Simulation and sensitivity analysis of wear on the automotive brake pad, *Simulation Modelling Practice and Theory* **84**, 2018, 106–123.

[5] Blau P.J., Jolly B.C.: Wear of truck brake lining materials using three different test methods. *Wear* **259**, 2005, 1022–1030.

[6] Adamowicz A., Grzes P.: Influence of convective cooling on a disc brake temperature distribution during repetitive braking. *Therm. Eng.* **31**, 2011, 2177–2185.

[7] ***Daimler Media Website and Xentry portal, www.media.daimler.com, last accessed 09.09.2018.

[8] ***National company for road infrastructure administration, *Road technical bulletin*, no **12**/2011.

[9] Dhir D.K.,: Thermo-mechanical performance of automotive disc brakes. *Materials Today Proc*. **5**, 2018, 1864–1871.

IOP Conf. Series: Materials Science and Engineering **568** (2019) 012094 doi:10.1088/1757-899X/568/1/012094

Comparison of heating system alternatives for a rural house employing analytic hierarchy process

C Cristea[1], M Cristea[2] and R Tirnovan[3]

[1,2,3] Electrical Engineering Faculty, Technical University of Cluj-Napoca, 26-28 Baritiu Street, Cluj-Napoca, Romania

[1]ciprian.cristea@emd.utcluj.ro

Abstract. In this study were analyzed different heating systems to select the most appropriate type for a rural house. The performances of each alternative have been evaluated and compared using the analytic hierarchy process (AHP), a well-known and widely used multi-criteria method. There were considered five heating options: wood boiler, electric boiler, gas boiler, pellet boiler, and electric heating carpet, which were combined with two different heating terminals: radiant floor heating and the classic radiator. The alternatives were analyzed based on several criteria, which were defined considering economic, environmental, and social aspects. After the variants were ranked, and the most appropriate solution was determined, a sensitivity analysis was performed. The results indicated that the final ranking of the alternatives is robust, with slight variances, depending on the percentage with which the criteria weights were varied.

1. Introduction

There are different options that can be considered when choosing a heating system for a house and several factors that can influence the decision maker's opinion. The most important factor is to identify the heating sources that can be used to ensure thermal comfort. Considering the impact on the environment, the decision maker may choose between alternative heating sources (e.g., geothermal heat pumps, heating oil, solar heating) and traditional sources (e.g., wood, electricity, natural gas). Also, the heating sources may vary depending on the area where the house is located: if it is a mountain area or a rural zone wood is a reliable source, while electricity or natural gas is not that accessible; if the house is in a city, electricity or natural gas are the most obvious choices, because of the existing infrastructure, while wood may be considered expensive (because of the shipping and storage costs).

Different approaches were applied in previous related research papers to obtain the optimal solution for a heating system. In [1] the district heating planning tool (DHPT), developed considering the expected reduction in heat consumption, was used to evaluate five different district heating system development scenarios, to determine the most efficient variant (in Latvia). A similar study was conducted in [2], where six district heating systems from Upper Bavaria were monitored for a year and assessed by different Key Performance Indicators (KPIs) to identify the optimization potentials of each analyzed district heating system. In [3] were studied and evaluated different heating terminals (cast-iron radiator, radiant floor heating, bimetal radiator, and small temperature difference fan-coil unit) used in 'Coal-to-electricity' heat pump systems based on their heating performances. Also, the multi-criteria decision methods are frequently used to determine the most appropriate heating system for a given location. In [4] PROMETHEE (Preference Ranking Organization Method for Enrichment Evaluations) was used to evaluate four alternative energy heating systems for a single-family house in Kuujjuaq (Quebec, Canada), against twelve criteria, to identify the optimal solution. The Stochastic Multicriteria Acceptability Analysis (SMAA) was used in [5] to rank eleven heating systems alternatives for a

IOP Conf. Series: Materials Science and Engineering **568** (2019) 012094 doi:10.1088/1757-899X/568/1/012094

sustainable residential area in Loviisa (Finland). The AHP (Analytic Hierarchy Process) and FCE (Fuzzy Comprehensive Evaluation) were used in [6] to evaluate the performances of five heating systems for a rural house located in North of China to select the best performing system.

In this paper, the AHP method was used by the authors to select the most appropriate heating system for a rural house, located in Satu-Mare county, Romania.

2. Methodology

AHP is a well-known multi-criteria decision-making method developed by T.L.Saaty to identify the optimal solution from a set of alternatives, based on multiple criteria analysis. The method involves several steps: the first two steps are to identify the problem, define the goal of the analysis, determine the criteria and the alternatives used to achieve the goal and construct the hierarchical representation of the problem [7,8]. The judgment matrix is constructed based on the decision maker's preferences, assessed by Saaty's major rating scale and contains the priority weights obtained after the pair-wise comparisons of the criteria [8,9]. Saaty's major rating scale measures the intensity of the importance of the criteria, from 1 (equal importance) to 9 (extreme importance) [7].

The judgment matrix A (pair-wise comparison matrix) is given below in (1), for n elements at each hierarchical level (criteria and alternatives), where a_{ij} represents a pair-wise comparison and is defined as a division between the weight value of criterion C_i and weight value of criterion C_j $(i, j = 1, 2,..n)$ [8,10].

$$A = \begin{bmatrix} 1 & a_{12} & \cdots & a_{1n} \\ a_{21} & 1 & \cdots & a_{2n} \\ \vdots & \vdots & 1 & \vdots \\ a_{n1} & a_{n2} & \cdots & 1 \end{bmatrix} \tag{1}$$

The next step is to obtain the priority vector, which contains the relative weights used to form the ranking of importance for all decision elements, by normalizing the main eigenvector of judgment matrix A. This operation is repeated for each criterion and for all the alternatives to obtain the local and global priorities [10]. The advantage of the AHP method is that it allows to brief verification of the consistency of the judgments made by the decision maker. The pair-wise comparisons are considered consistent if the consistency ratio CR is less than 0.1. To calculate CR, the consistency index CI and the random index RI must be determined, using the largest eigenvalue of A (λ_{max}) [8,10,11]:

$$CI = \frac{\lambda_{\max} - n}{n-1} \tag{2}$$

$$RI = \frac{1.98 \cdot (n-2)}{n} \tag{3}$$

$$CR = \frac{CI}{RI} \tag{4}$$

The pair-wise comparisons matrix is considered consistent if $CR < 10\%$, otherwise the decision maker must revise the priorities established for each criterion and alternative, because the AHP method may not return the proper results. The final ranking matrix is obtained by synthesizing the local and global priorities for all the decision variables [10].

3. Case study

This case study was conducted based on the information obtained from the management of a company dealing with electrical installations work. The company has been contracted to install a new heating system for a house located in Satu-Mare county, Romania, having a constructed area of 120 square meters. Based on the preferences of the customer, nine alternatives were identified, considering five heating sources and two heating terminals : wood boiler combined with radiant floor heating (A1) or radiators (A2), electric boiler combined with radiant floor heating (A3) or radiators (A4), gas boiler

IOP Conf. Series: Materials Science and Engineering **568** (2019) 012094 doi:10.1088/1757-899X/568/1/012094

combined with radiant floor heating (A5) or radiators (A6), pellet boiler combined with radiant floor heating (A7) or radiators (A8) and electric heating carpet (A9).
The alternatives were evaluated against eleven criteria (C1-C11), which were defined considering economic, environmental and social aspects: raw material price ([RON]), boiler price ([RON]), installation price ([RON]), fuel price ([RON/year]), yield ([%]), boiler guaranty ([years]), system guaranty ([years]), system reliability ([years]), fuel availability ([%]), CO_2 emissions ([kg/year]), system autonomy ([points]).

4. Results

Satty's major rating scale was used to determine the judgment matrix, which was normalized to obtain the criteria weights. These first results are presented in table 1. The most important criterion is C4 (29.39 %), followed by C1 and C2 with the same percentage (15.93 %), and C3 with 11.29 %. Regarding the consistency of this judgment matrix (table 1), CR is less than 10% (3 %), which means the priorities established by the decision maker are consistent.

Table 1. Judgment matrix, criteria weights, and consistency results.

	C1	C2	C3	C4	C5	C6	C7	C8	C9	C10	C11	Criteria weights
C1	1	1	2	0.33	3	5	5	4	6	7	7	0.159
C2	1	1	2	0.33	3	5	5	4	6	7	7	0.159
C3	0.5	0.5	1	0.25	2	4	4	3	5	6	6	0.113
C4	3	3	4	1	5	7	7	6	8	9	9	0.294
C5	0.33	0.33	0.5	0.2	1	3	3	2	4	5	5	0.08
C6	0.2	0.2	0.25	0.14	0.33	1	1	0.5	2	3	3	0.38
C7	0.2	0.2	0.25	0.14	0.33	1	1	0.5	2	3	3	0.38
C8	0.25	0.25	0.33	0.17	0.5	2	2	1	3	4	4	0.056
C9	0.17	0.17	0.2	0.13	0.25	0.5	0.5	0.33	1	2	2	0.026
C10	0.14	0.14	0.17	0.11	0.2	0.33	0.33	0.25	0.5	1	1	0.018
C11	0.14	0.14	0.17	0.11	0.2	0.33	0.33	0.25	0.5	1	1	0.018
Consistency results				CI = 0.046				RI= 1.51				CR=0.0307

The same procedure was used to determine all the comparison matrices, and their local weights of the alternatives analyzed. In tables, 2-4 are presented the judgment matrix, local weights and consistency results for the three most important criteria: fuel price (C4), raw material price (C1) and boiler price (C2).

Table 2. The comparison matrix, criteria weights, and consistent results concerning fuel price.

	A1	A2	A3	A4	A5	A6	A7	A8	A9	Local weights
A1	1	1	6	6	2	2	3	3	9	0.225
A2	1	1	6	6	2	2	3	3	9	0.225
A3	0.17	0.17	1	1	0.2	0.2	0.25	0.25	4	0.034
A4	0.17	0.17	1	1	0.2	0.2	0.25	0.25	4	0.034
A5	0.5	0.5	5	5	1	1	2	2	8	0.142
A6	0.5	0.5	5	5	1	1	2	2	8	0.142
A7	0.33	0.33	4	4	0.5	0.5	1	1	7	0.092
A8	0.33	0.33	4	4	0.5	0.5	2	2	7	0.092
A9	0.11	0.11	0.25	0.25	0.13	0.13	0.14	0.14	1	0.016
Consistency results				CI = 0.038			RI= 1.45			CR=0.026

According to the results presented in table 2, the most appropriate solution concerning fuel price is the wood boiler (A1 and A2), followed by the gas boiler (A5 and A6). The alternative A9 is positioned at the bottom of the ranking, which means it has the highest fuel price from all alternatives.

IOP Conf. Series: Materials Science and Engineering **568** (2019) 012094 doi:10.1088/1757-899X/568/1/012094

Table 3. The comparison matrix, criteria weights, and consistent results concerning raw material price.

	A1	A2	A3	A4	A5	A6	A7	A8	A9	Local weights
A1	1	0.5	1	0.5	1	0.5	1	0.5	0.25	0.061
A2	2	1	2	1	2	1	2	1	0.33	0.116
A3	1	0.5	1	0.5	1	0.5	1	0.5	0.25	0.061
A4	2	1	2	1	2	1	2	1	0.33	0.116
A5	1	0.5	1	0.5	1	0.5	1	0.5	0.25	0.061
A6	2	1	2	1	2	1	2	1	0.33	0.116
A7	1	0.5	1	0.5	1	0.5	1	0.5	0.25	0.061
A8	2	1	2	1	2	1	2	1	0.33	0.116
A9	4	3	4	3	4	3	4	3	1	0.293
Consistency results				CI = 0.004			RI= 1.45			CR=0.003

Table 4. The comparison matrix, criteria weights, and consistency results concerning boiler price.

	A1	A2	A3	A4	A5	A6	A7	A8	A9	Local weights
A1	1	1	1	1	1	1	3	3	0.33	0.107
A2	1	1	1	1	1	1	3	3	0.33	0.107
A3	1	1	1	1	1	1	3	3	0.33	0.107
A4	1	1	1	1	1	1	3	3	0.33	0.107
A5	1	1	1	1	1	1	3	3	0.33	0.107
A6	1	1	1	1	1	1	3	3	0.33	0.107
A7	0.33	0.33	0.33	0.33	0.33	0.33	1	1	0.2	0.038
A8	0.33	0.33	0.33	0.33	0.33	0.33	1	1	0.2	0.038
A9	3	3	3	3	3	3	5	5	1	0.284
Consistency results				CI = 0.006			RI= 1.45			CR=0.004

It can be noticed in tables 3 and 4 that the alternative A9 has the highest local weights concerning raw material and boiler prices, which means is the most preferred solution considering these two criteria.

The local weights, obtained after performing the same procedure for the remained criteria, contributed to the construction of the final matrix, presented in table 5. The final ranking (Table 5), was obtained by multiplying the final matrix with the criteria weights and indicates the most appropriate heating system for a rural house, located in Satu-Mare county, Romania.

Table 5. Final results.

	C1	C2	C3	C4	C5	C6	C7	C8	C9	C10	C11	Final ranking
A1	0.061	0.107	0.086	0.225	0.058	0.112	0.204	0.176	0.168	0.209	0.029	0.138
A2	0.116	0.107	0.161	0.225	0.058	0.112	0.032	0.059	0.168	0.209	0.029	0.142
A3	0.061	0.107	0.086	0.034	0.185	0.058	0.204	0.176	0.096	0.025	0.163	0.087
A4	0.116	0.107	0.161	0.034	0.185	0.058	0.032	0.059	0.096	0.025	0.163	0.091
A5	0.061	0.107	0.086	0.142	0.106	0.112	0.204	0.176	0.021	0.05	0.163	0.113
A6	0.116	0.107	0.161	0.142	0.106	0.112	0.032	0.059	0.021	0.05	0.163	0.117
A7	0.061	0.038	0.086	0.092	0.058	0.112	0.204	0.176	0.168	0.209	0.063	0.088
A8	0.116	0.038	0.161	0.092	0.058	0.112	0.032	0.059	0.168	0.209	0.063	0.092
A9	0.293	0.284	0.014	0.016	0.185	0.211	0.053	0.059	0.096	0.018	0.163	0.132

It can be noticed that the optimal solution for the considered problem is A2, being the adequate alternative according to the analyzed criteria, followed closely by A1 and A9 with a tight score. Finally, the last places are occupied by A7 and A3 because they have the worst evolution.

A sensitivity analysis was performed to investigate the effect of varying the criteria weights of the four most important criteria, by seven percent. For the most part, the final ranking of the alternatives is robust, the alternatives A2, A1 and A9 change places between them if the criteria weights are varied only for the raw material price, results presented in figure 1. It can be noticed that the new ranking after the raw material price goes up is A9, A2, A1, A6, A5, A8, A4, A7, A3 (figure 1a), while the new ranking after the raw material price goes down is A1, A2, A9, A5, A6, A7, A8, A3, A4 (figure 1b).

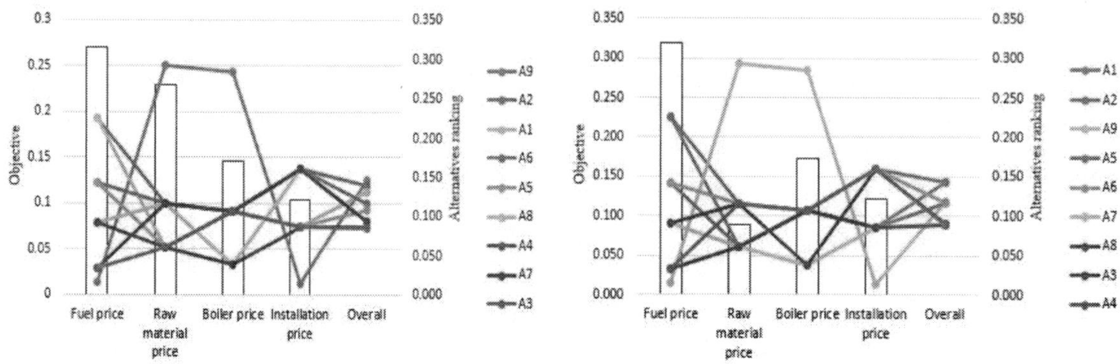

a. b.

Figure 1. Sensitivity analysis with regards to raw material price change: a - upward, b - downward

5. Conclusions

The selection process of the most appropriate heating system is complex and requires the evaluation of various alternatives against multiple criteria, considering the beneficiary's preferences. This paper proposed an application of the AHP method to identify the optimal solution of a heating system for a rural house, located in Satu-Mare county, Romania. The results obtained suggest that the AHP may be an important tool to support the electrical designer, or even the beneficiary, in resolving this type of decisional problems. The results obtained, after the sensitivity analysis was performed, indicate that the ranking of the heating systems is robust and represents a reliable tool for the given purpose.

6. References

[1] Zajacs A and Borodinecs A 2019 *Sustain. Cities Soc.* **48** 101540
[2] Bucker D, Jell P and Botsch R 2018 *Energy Procedia* **149** 5
[3] Hu B, Wang R Z, Xiao B, He L, Zhang W and Zhang S 2019 *Int. J. Refrig.* **98** 274
[4] Yan C, Rousse D and Glaus M 2019 *J. Clean. Prod.* **208** 1488
[5] Kontu K, Rinne S, Olkkonen V, Lahdelma R and Salminen P 2015 *Energy Build.* **93** 169
[6] Zhang X, Yang J and Zhao X 2018 *Energy* **150** 631
[7] Saaty T L 2008 *Int. J. Services Sciences* **1**
[8] Cristea C, Cristea M, Birou I, Onaca C and Serban F 2018 *Review of Management and Economic Engineering International Management Conference* (Cluj-Napoca: Technical University of Cluj-Napoca)
[9] Giamalaki M and Tsoutsos T 2019 *Renew. Energ.* **141** 64
[10] Kadavil R, Lurbe S, Suryanarayanan S, Aloise-Young P A, Isley S and Christensen D 2018 *Sustainable Energy, Grids and Networks* **16** 196
[11] Garcia-Cascales M S and Lamata M T 2009 *Comput. Ind. Eng.* **56** 1442

IOP Conf. Series: Materials Science and Engineering **568** (2019) 012095 doi:10.1088/1757-899X/568/1/012095

The assessment of pedestrian's head injury risk at the contact with the vehicle's hood

H Beles[1], B A Tolea[1], G F Crisan[2], C A Dogar[2], V V Ciotachiev[2]
[1]University of Oradea, Department of Mechanical Engineering and Automotive, Oradea, 1 Universității street, Romania
[2]University of Oradea, Faculty of Managerial and Technological Engineering, Oradea, 1 Universității street, Romania, Student at Systems and Advanced Technologies in Automotive Engineering Master

horia.beles@yahoo.com

Abstract. The main paper objective was to determine pedestrian's head injury risk, at the impact with the vehicle's bonnet. To achieve this objective, experimental tests on a test bench were performed. It is intended to determine the head injury risk, in different areas of the bonnet, to determine hood's "friendliness".

1. Introduction

Pedestrians represent the most significant fraction from overall road users. Simultaneously, they are considered "vulnerable road users".

Romania represents the country with the highest rate of deceased pedestrians per million inhabitants from the EU, with a 33% higher rate than the EU average [1].

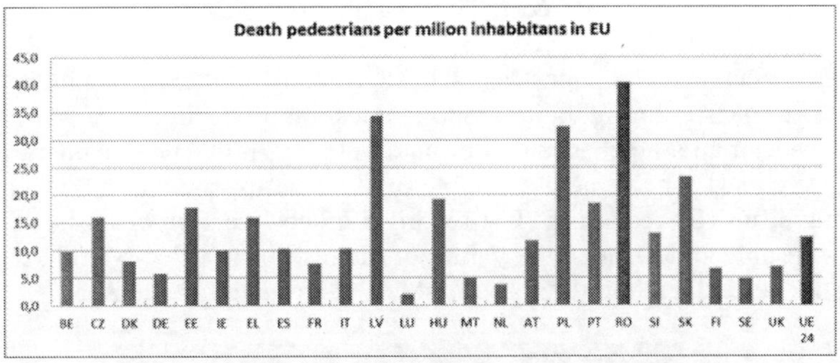

Figure 1. Pedestrian fatalities per million inhabitants in the function of EU countries [1]

Otte reported that the most common injured part is the pedestrian's head [2]. Usually, the contact between the pedestrian's head and the bonnet occurs in the final phases of the primary impact [3].

The most recent pedestrian passive safety devices are the pop-up bonnet, respectively, the pedestrian's head airbag.

The pop-up bonnet is classified according to its activation time as follows [4]:

- Passive hood – the bumper enters in contacts with the pedestrian's lower limbs
- Active hood – usually is activated before the first contact takes place

IOP Conf. Series: Materials Science and Engineering **568** (2019) 012095 doi:10.1088/1757-899X/568/1/012095

In 2012, Volvo introduced the first pedestrian airbag, consisting of the following elements: the airbag control unit, the bonnet lift cap, the bonnet hatch release mechanism, the bonnet hinges and the sensors [5].

EEVC (European Experimental Vehicles Committee) established several types of experimental tests to assess the pedestrian's safety. Thus, pedestrian safety tests are so-called component tests, avoiding full-scale tests. One of the tests consists of using a head impactor that is accelerated to hit the vehicle's bonnet, respectively windshield, in specific areas. This type of test, assess the pedestrian head injury risk, at the impact with the vehicle [6], [7].

2. Methodology

To assess the pedestrian's head injury risk, in functions of the bonnet area, three experimental tests were performed. In the first test, the area of interest was the bonnet's end part, at the connection with the windshield, for the second test, the impact took place in the center of the bonnet, while for the third test, the impact took place on the bonnet's lateral upper end.

The bonnet was mounted and fixed on the test bench, being positioned in the vertical plane. On the top side of the bench, a pendulum with the head-neck assembly was mounted. The designed test bench can be used in different scenarios: impact between the pedestrian's head with the bonnet; impact between the pedestrian's head with a pop-up bonnet; impact between the pedestrian's head with the windshield.

In figure 2 is presented the designed test bench, together with the head-neck assembly pendulum.

Figure 2. The designed test bench

In all three test scenarios, the pedestrian's head velocity was approximately 40 km/h, the same velocity used by Euro NCAP during their component testing procedures.

To determine pedestrian's head acceleration and velocity, high-speed camera recordings available on smartphones were used. The camera available on the smartphone was able to record at 480 frames per second, having a full HD recording. The recorded videos were processed in the Tracker software, from which we were able to determine the values of interest (head velocity and acceleration).

The pedestrian's head kinematics after the first test is presented in figure 3.

Figure 3. Pedestrian's head kinematics in the first bench-tests scenario

IOP Conf. Series: Materials Science and Engineering **568** (2019) 012095 doi:10.1088/1757-899X/568/1/012095

In the second bench-test scenario, the head impacted the center of the hood, and its kinematics is presented in figure 4.

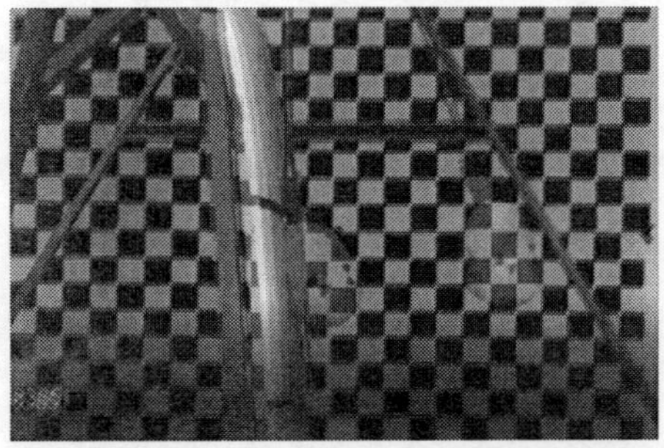

Figure 4. Pedestrian's head kinematics in the second bench-tests scenario

In the third bench-test scenario, the head impacted the left lateral edge of the hood, and its kinematics is presented in figure 5.

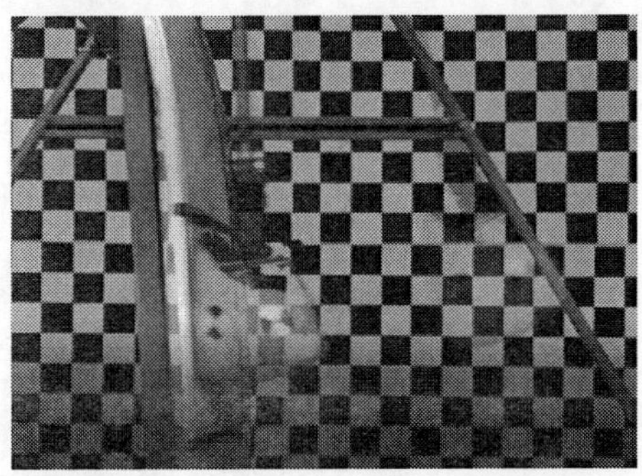

Figure 5. Pedestrian's head kinematics in the third bench-tests scenario

3. Results

The acceleration diagram resulted from the performed bench-tests, are presented in the figures below (figure 6).

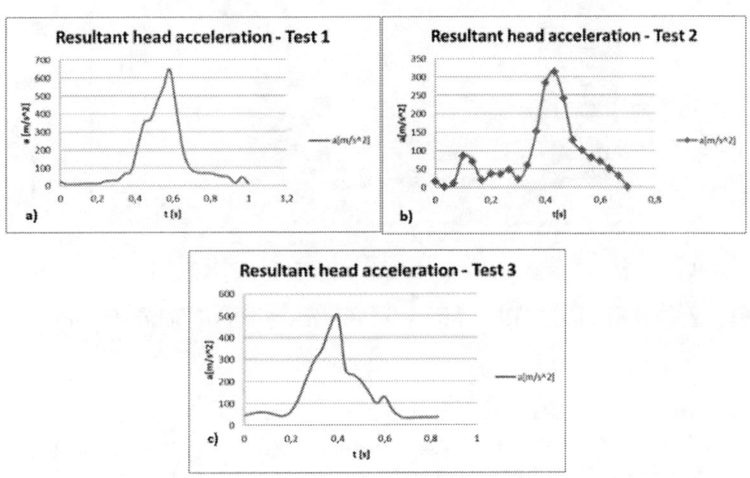

Figure 6. Resultant head acceleration for the first bench-test: a) the first test; b) second test; c) the third test

IOP Conf. Series: Materials Science and Engineering **568** (2019) 012095 doi:10.1088/1757-899X/568/1/012095

To quantify the injuries that may occur in the brain, various methods have been proposed to measure both kinematics and dynamic parameters. One of the most popular methods to determine head impact tolerance is HIC (Head Injury Criteria). The HIC value is determined by integrating the resultant acceleration, depending on the set time interval. Thus, the HIC calculation formula is [8], [9], [10], [11], [12], [13]:

$$HIC = \left\{ \left[\frac{1}{t_2 - t_1} \int_{t1}^{t2} a \cdot dt \right]^{2.5} \cdot (t_2 - t_1) \right\}_{MAX} \quad (1)$$

Where a represents the head acceleration, measured in m/s^2, while t_1 and t_2, is the time interval in which is calculated the HIC value and is measured in seconds. Usually, the area of interest of HIC calculation is where the peak accelerations are.

Therefore, using Mathcad software, we were able to calculate HIC values for each bench-test:

$$HIC_{test1} = \left\{ \left[\frac{1}{t_2 - t_1} \int_{t1}^{t2} a \cdot dt \right]^{2.5} \cdot (t_2 - t_1) \right\}_{MAX} = 500 \quad (2)$$

$$HIC_{test2} = \left\{ \left[\frac{1}{t_2 - t_1} \int_{t1}^{t2} a \cdot dt \right]^{2.5} \cdot (t_2 - t_1) \right\}_{MAX} = 127 \quad (3)$$

$$HIC_{test3} = \left\{ \left[\frac{1}{t_2 - t_1} \int_{t1}^{t2} a \cdot dt \right]^{2.5} \cdot (t_2 - t_1) \right\}_{MAX} = 281 \quad (4)$$

To summarize the obtained values, a diagram, together with the head impact locations was created and is presented in figure 7.

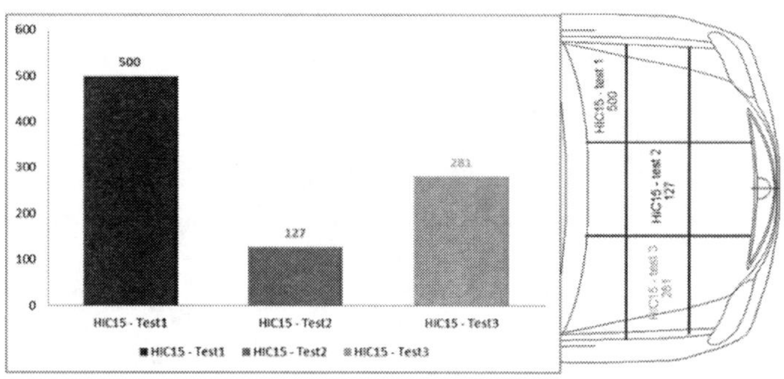

Figure 7. Obtained HIC values, in the function of head impact location

4. Conclusions

According to the obtained values, the highest chance for the pedestrian's to suffer a severe head injury is in the upper part of the bonnet (the part that connects with the lower part of the windshield), where we obtained a HIC value of 500. In the second scenario, where the impact was in the center of the

IOP Conf. Series: Materials Science and Engineering **568** (2019) 012095 doi:10.1088/1757-899X/568/1/012095

bonnet, the HIC value decreased with approximately 75%, while for the third bench-test scenario, the decreasing in comparison with the first crash test, was 56%.

Therefore, the vehicle's bonnet critical area is the upper part, where it connects with the windshield, while the safest part is in the center of the bonnet. To study the influence of the pedestrian passive safety systems (pop-up bonnet, pedestrian airbags), further tests will be performed on the designed test bench.

5. References

[1] ***CARE.2015.EUROSTAT for population data. http://ec.europa.eu/transport/road_safety/specialist/statistics_en, accessed in 12.03.2019

[2] Otte D 1998 *Verletzungsmechanik des Kopfanpralls von Fußgängernbei Windschutzscheibenkontaktan PKW*. Verkehrsunfall und Fahrzeugtechnik. 1998;36(11)

[3] Burg H and Moser A 2013 *Handbook of Accident Reconstruction: Accident Investigation-Vehicle Dynamics-Simulation. Basics*. Create Space Independent Publishing Platform; 2013

[4] Xu D, Zhu X, Miao Q, Ma Z, and Wu B 2010 *The research of reversible pop-up Hood for pedestrian protection*. In Proceedings of 2010 IEEE International Conference on Vehicular Electronics and Safety 2010 Jul 15 (pp. 42-47). IEEE

[5] Jakobsson L, Broberg T, Karlsson H, Fredriksson A, Gråberg N, Gullander C, and Lindman M 2013 *Pedestrian airbag technology–a production system*. In 23rd International Technical Conference on the Enhanced Safety of Vehicles (ESV) National Highway Traffic Safety Administration 2013 (No. 13-0447)

[6] ***European Enhanced Vehicle-safety Committee 1998 *EEVC Working Group 17 Report – Improved Test Methods to Evaluate Pedestrian Protection Afforded by Passenger Cars*. TN0 Crash-Safety Research Centre: Delft, The Netherlands. 1998 Dec: 6-8

[7] Kalliske I and Friesen F 2001 *Improvements to pedestrian protection as exemplified on a standard-sized car*. SAE Technical Paper; 2001 Jun 4

[8] Radu A I, Trusca D and Toganel G 2018 *Influence of tire pressure on the braking distance when driving on snow and asphalt*. In IOP Conference Series: Materials Science and Engineering (Vol. 444, No. 7, p. 072013), IOP Publishing, ISSN: 1757-8981

[9] Radu A I and Cofaru C. Study of the current state of crash testing. Bulletin of the Transilvania University of Brasov. Engineering Sciences. Series I. 2015 Jul 1;8(2):31

[10] Radu A I, Trusca D and Toganel G 2018 *Comparison of Passenger Vehicle Braking Distance When Traveling On Snow and Asphalt at Different Velocities*. AMMA - The International Congress of Automotive and Transport Engineering, AMMA 2018

[11] Oţăt O V, Dumitru N, Oţăt V, and Dumitru I 2016 *Determination of Kinematic and Dynamic Behavior in the Driver's Skull upon the Impact with the Steering Wheel*. In International Congress of Automotive and Transport Engineering 2016 Oct 26 (pp. 812-819), Springer, Cham, ISBN: 978-3-319-45447-4

[12] Oţăt O V, Marinescu G and Oţăt V 2017 *Analyses and statistics on the frequency and the incidence of traffic accidents within Dolj County*. The 11th edition of The International Congress of Automotive and Road Transport Engineering, University of Pitesti, Faculty of Mechanics and Technology, Scientific Bulletin Automotive series, year XXIV, no. 28, CAR 2017

[13] Oţăt O V, Nicolae D and Ilie D 2017 *Dynamic models to analyze the influence of the seat belt in a frontal collision*. In IOP Conference Series: Materials Science and Engineering 2017 Oct (Vol. 252, No. 1, p. 012017) IOP Publishing, ISSN: 1757-8981

IOP Conf. Series: Materials Science and Engineering **568** (2019) 012096 doi:10.1088/1757-899X/568/1/012096

Improving the engineering education system – a students' perception

C P Simion[1], G C Slusariuc[2], R D Stanciu[1] and O Bucovetchi[1]

[1]University POLITEHNICA of Bucharest, 313, Spl. Independentei, Bucharest, Romania
[2]University of Petrosani, address: Universitatii St, Petrosani, Romania

olga.bucovetchi@upb.ro

Abstract. Nowadays, we hear more and more complains about the quality of the Romanian educational system that decreased over the last decades. Therefore, one of the three pillars of the 2018-2030 strategy regarding the education system is "the adaptability of the educational system to external changes and trends of the future". As everybody may notice, trends change with every new generation. Those who are now students, the so-called members of the Z-generation are different from those before them. The authors consider that to identify the best teaching methods for increasing the efficiency of the educational process, it is necessary to address to the representatives of the Z generation. Starting from their expectations, a more efficient strategy can be developed. This will lead to an increase in the quality of the educational activities and a better integration of the graduates on the labor market. The authors chose two universities in Romania (University "Politehnica" of Bucharest and University of Petrosani) for the pilot sample and leaded several focus groups that aimed to identify students' dissatisfaction and expectations towards the educational process. After analyzing the results, there were identified several elements that can be implemented in the didactic process for improving it.

1. Introduction

Generation Z was born after 1995 and is part of the generation that does not conceive the world without internet. They are active, and Sparks & Honey [1] says that 60% of them want to change the world compared to "only" 39% of the Millennials. They hate hierarchies, and they feel able to take the boss's place anytime. Furthermore, they do not compromise on free time and personal life.

All these characteristics have produced a strong earthquake in the labor market. There are tens of thousands of available jobs, but not candidates. Studies show that one in four young people has no plans for the future and does not want to work. They do not want to work eight days a day in an office; they take a long time and work only if they have inspiration. To bring them to companies, there are employers who have decided to completely change the rules.

Given these, also the educational environment deals with a serious challenge in attracting these students into classrooms.

The anthropologist Vintilă Mihăilescu considers that "Digital natives are another type of human. When language becomes technically 90%, and social relationships include that much technology that can be called techno-relation, it already means that we deal with another type of human" [2].

Interactive teaching strategies promote active learning, involve sustained collaboration between students who, organized into micro-groups, work together to achieve pre-established goals. The teaching

IOP Conf. Series: Materials Science and Engineering **568** (2019) 012096 doi:10.1088/1757-899X/568/1/012096

environment places the emphasis, not on the role of speaker of informational messages but the roles of organizer, facilitator, and mediator of learning activities. The teaching approach is designed so that the center is the student, not the teacher. The role of the teacher remains a capital, but by renouncing the old rigid and uniform educational practices, he becomes the organizer of a learning environment tailored to the peculiarities and needs of the beneficiaries, facilitating the learning process and developing skills. The focus is on how assimilated information is processed, structured, interpreted, and used in various situations. Thus, students acquire solid skills, but also the confidence that they will prove operational and will serve them authentically in various life contexts.

The teacher influences the active and creative behavior of the student by asking for answers to a problem by organizing the information and training activities of the students by emphasizing the development of cognitive applicative processes,

The didactic strategy - in a postmodern vision - is thus the fruit of collaborative participation of the teacher with the pupils, completing their work plan with their interests, wishes for knowledge and intellectual activity. Thus, they can express their desire to learn through cooperation, teamwork, collectively or individually, they can opt for certain teaching materials to use, for specific methods, techniques, or working methods. Giving them the chance to make such choices, professor it contributes to increasing activism and developing the creativity of their pupils and teaching strategy, that comes from the harmonious combination of all stakeholders can successfully lead towards achieving desiderates primarily by providing learning.

2. Methods

The main objective of the research was to identify the students' perception toward the state of art of engineering education system starting the conclusions of the Presidency's project "Educated Romania". The authors consider that to identify the best teaching methods for increasing the efficiency of the educational process; it is necessary to address to the representatives of the Z generation. Thus, starting from their expectations, a more efficient strategy can be developed. This will lead to an increase in the quality of the educational activities and a better integration of the graduates of higher education in the labor market.

The definition of the variables of the research had in view their conception to reflect the researched problem and to be in line with the trends in the field of the literature [3,4]. This research has operated with several categories of variables: basic variables, descriptive variables, situational variables, independent variables, and dependent variables. Qualitative research was designed to answer questions such as "Why?" and "How?" to discover, examine, and understand the root causes of the phenomenon.

To achieve the objective of this paper, qualitative primary research was used, based on a focus group which brought together 12 students from the final years. The topic of the focus group was "students expectations towards the didactic process", the students responding to several 10 open questions. In this context, specific interactions were developed, on the one hand, between moderator and participants (questions and answers) and, on the whole, between the participants (debates, questions, answers). The focus group was practically a group interview, with people representing a specific and homogeneous audience. The main disadvantages of the focus group are the existence of a very close relationship of dependence between the quality of the moderators and the quality of the obtained results, respectively the fact that there is the risk of reaching hasty generalizations when they are not followed by more complex investigations. That is why the focus group sessions were recorded to be analyzed in detail, and the moderator created a relaxed atmosphere in which the members of the group felt comfortable and were able to express themselves freely. Two focus group events were organized in each university center to see if the results of the separate event analyzes were similar.

The second step within the present demarche was to identify several strategies to improve the detected quality level of engineering education system. For that, several semi-structured interviews were conducted. Those interviews were held with the most talkative subjects for the focus groups.

IOP Conf. Series: Materials Science and Engineering **568** (2019) 012096 doi:10.1088/1757-899X/568/1/012096

3. Results

When assessing the focus groups' activity, there where several elements that came up frequently in terms of complains towards the educational process or the educational infrastructure, as follows:

- The classes embed too it much theory and too little practice
- There is unaddressed/unspecific information or too general thus the examples are not helpful enough
- During laboratories are used obsolete computer programs
- There is a lack of correlation between theory and labor market demands
- There are few practical applications
- Lack of internet in some rooms, video projectors, interactive whiteboards

It is interesting to mention that no matter the university nor whether there was the master of bachelor students, the main complain emphasized was related to the fact that the teaching methods are conventional and they consider that their digital skills are not used.

Given these, one can draw the following proposals of improvement:

- the use of tablets and smartboards during seminary activities at first and afterward maybe even during lectures
- participating in debates or free discussions rather than solving exercises
- using simulation software for a better understanding of in vivo technical systems for applying more the theory into practice
- solving case studies
- use of modern software programs that are specific to engineering and management environment
- involving students in designing the training activity
- creating blended courses
- using teaching strategies based on action, application, research, experimentation

Upgrading exposure can be done by prior notice the plan and objectives pursued, the use of the genetic process, explanation, anchoring in the reality of time, the use of the teacher's personal research, rhetorical questions, positioning, making opinion polls, formulating some value judgments illustrate by public education presentation computer, exposing the opponent, exposing debate etc.

4. Discussion

When using an interactive methodology, the roles of the teacher are diversified, enriched, so that he becomes an animator, counselor, moderator, participant with his students to solve problems, even a member of the working teams. It increases the degree of activism and the student's involvement in the activity, from the simple receiver to the active participant. Interactivity also implies a positive attitude towards human relationships, the importance of teamwork, and openness to co-operation, an attitude of supporting the ideas that emerged through collaboration with others. It specifies interactive teaching methods and multi-relationship between teacher and pupils, between student and colleagues, on the one hand, pupils and content on the other.

Exposure methods do not require exchanges between educational agents, being unidirectional, the message that they only tear from the teacher to the student. The interactive methodology aims at achieving cognitive goals (stimulating higher cognitive processes, developing the capacity to link knowledge to each other, and creating conceptual networks, developing multiple intelligences, etc.) and achieving socio-emotional goals.If the decision of the teacher aims at presenting the information in an actively problematic way, requiring the direct participation of the pupils in the rediscovery of knowledge, the methodology used also involves active and interactive methods (problem, discovery, collaboration, case study, mutual learning, etc.).

The academic environment is currently focusing on promoting techniques and methods that focus on active student participation; the direct interest of its development training, when distance-based methods conditioning, memory, and repetition.

There are several ways of activation of students, which educational practice and pedagogical research have validated [5, 6,7]:

- the use of a system of activation training and activation methods, such as heuristic approach, problematization, discovery, modeling, experiment, assisted training / self-training computer, case study, role play, learning using teaching simulators
- using methods to develop students' creativity;
- the integration into the training and self-training activities of the educational means and the valorization of their stimulating function by ensuring the interdependence with the other components of the didactic strategies [8].

Games have been created with the purpose of entertainment for a long time. They have always attracted people because they provide fun and excitement. For this reason, game developers have tried very hard to invent and introduce new ways of enjoyment.

Thankfulness to the rapid advancement and vast expansion in computer hardware and visualization technologies, every time, the graphical detail and definition of games have become more realistic and accurate to the player. With the ability to mimic or simulate reality, some game designers started to adopt a purpose of playfulness to develop another kind of games, a more serious one. Those applications can be used for research, education, and training as if they are seen as fit or compatible with the study and finding objectives. When discussing the use of games or concept of gaming in education, there are at least four categories of an existing idea emerging in the literature.

- Serious Gaming: The term represents the use of games for specific purposes or other benefits, such as learning, researching, and training, instead of entertainment value only. It may look just like playing digital games, but its objective is to achieve something extraordinary.
- Simulations: Simulation technologies are quite similar to severe games unless they can replicate and visualize the real-world objects or environment for real-life training.
- Game-Based Learning: The concept refers to the use of games in the traditional classroom to strengthen the course objectives and to enhance the learning and teaching experience. Game-Based Learning is considered a branch of serious gaming that designed to deal with designated learning outcomes.
- Gamification: Gamification adverts the use of game design elements in non-game contexts to motivate or influence desired behaviors. Those attributes include elements, such as leaderboards, badges, levels, trophies, or any other rewards. A concept is a considerably new to scholarly society but a substantially developed in the business world.

During the past 20 years, there are numerous research and studies on the effectiveness and productiveness of using serious games as a learning tool or gaming concept as a teaching technique. For example, [9] quotes that one investigation from a psychological association dating back in the early 1980s has logically revealed that playing digital games, both video-based or computer-based, develops the reduction in responsive times, improvement in hand-eye coordination, and encouragement in players' self-esteem and self-respect. Considering this found evidence, Prof. Griffiths also concludes that video games or computer games have great benefits and positive potentials not only for their entertainment value but also non-entertainment purposes, and a future success of this initiative can be achieved when appropriate games selection and playing requirements are designed to address a particular problem clearly or to teach a specific skill specifically [9]

Many research and experimentations also suggest that some critical or unique skills may be developed or strengthened by playing severe games [9-12]. For instance, the abilities of spatial planning and visualization, such as creative and critical thinking, data allocation and management, and three-dimensional objects rotation and manipulation gradually can be evolved along with gaming experiences [13]. Given that, serious games may seem to be more productive and advantageous for young people, like children and teenagers, who started with relatively beginner skills. As a result, researchers, educators, and corporates are now using video games and computer games as an application or a tool for studying individuals, teaching students, and training personnel and staff.

The major objective of active and interactive pedagogy is to empower the student, to help them develop, shape their personality, and develop their self-confidence by resorting to pedagogical practices centered on an individual or collaborative learning.

5. Conclusions

Modernizing the means of learning and making them more efficient is an essential direction in pedagogical innovation. Information technologies are used for collecting, storing, organizing, processing, presenting, and communicating information. Modern technologies can also be used to teach disciplines to support and stimulate individual study. The diversity of teaching methods and tools used in student teaching and learning respond to a fundamental need for variation, differentiation, nuance and customization of educational activity, while at the same time broadening and accumulating considerable teaching experience and student learning experience, multiple action strategies appropriate to the many tasks and educational-educational.

References

[1] Gen Z, https://www.sparksandhoney.com/gen-z, accessed in May 2019
[2] Mihailescu V 2007 *Anthropology: Five introductions* (*Antropologie: cinci introduceri,* in Romanian*)* (Iasi: Polirom Publishing House)
[3] Danielyan N 2017 Internet Technologies and Trans-Science: Orientation For Future *European Proc of Social and Behavioural Sciences* **35** pp 246-253 DOI: 10.15405/epsbs.2018.02.28
[4] Stoian AC 2017 The Identification of the Needs of Teachers` Training for ICT *European Proc of Social and Behavioural Sciences* **23** pp 1067-1074 DOI: 10.15405/epsbs.2017.05.02.131
[5] Nicolescu BN, Macarie T and Petrescu TC 2017 Is Blended Learning a New Way to an Optimal Adult Education? *European Proc of Social and Behavioural Sciences* **23** pp 1380-1389 DOI: 10.15405/epsbs.2017.05.02.169
[6] Grecu I, Ghita I and Ionescu, S 2009 The Network Structure of a Virtual Design Organization Using a Social Network Approach *4th Int Conf of Management and Industrial Engineering (ICMIE 2009)* (Bucharest: Niculescu Publishing House) pp 283-287
[7] Arrosagaray M, Gonzalez-Peiteado M, Pino-Juste M and Rodriguez-Lopez B 2019 A comparative study of Spanish adult students' attitudes to ICT in the classroom, blended and distance language learning modes *Comput Educ* 134 pp 31-40 DOI: 10.1016/j.compedu.2019.01.016
[8] Badea D and Ispas L 2014 The integration into the collection we training of practical and theoretical approaches about game and simulation to design a laboratory for urban combat *10th International Scientific Conference on eLearning and Software for Education* (Bucharest: Carol I Natl Defence Univ Publishing House) p 319-323
[9] Griffiths, MD 2002 The educational benefits of videogames *Education and Health* 20 pp 47-51
[10] Michael DR and Chen SL 2006 *Serious games: games that educate, train, and inform* (Boston: Thomson Course Technology)
[11] van Eck R 2006 Digital game-based learning: It's not just the digital natives who are restless *EDUCAUSE Review* 41(2), pp 16-30, available online at https://er.educause.edu/articles/2006/1/digital-gamebased-learning-its-not-just-the-digital-natives-who-are-restless, accessed in May 2019
[12] Susi T, Johannesson M and Backlund P 2007 *Serious Games – An Overview Technical Report HS-IKI-TR-07-001*, University of Skövde, Sweden
[13] Subrahmanyam K and Greenfield PM 1994 Effect of video game practice on spatial skills in girls and boys *Journal of Applied Developmental Psychology* 15(1), pp 13–32

IOP Conf. Series: Materials Science and Engineering **568** (2019) 012097 doi:10.1088/1757-899X/568/1/012097

The implementation of an adaptive traffic light concept in regards to tram access in a complex intersection

S Bogdan[1], G Grebenişan[1], D A Ulica[1], M Ratiu[1]

University of Oradea

sandaromocea@yahoo.com

Abstract- Case study on the solution of the implementation of an adaptive traffic light concept in regards to tram access of in the intersection created between the Decebal Bvd-Aradului Path-Primariei Street of Oradea Municipality.Description of the proposed solution for the intersection Decebal Bvd-Aradului Path- Primariei Street

1. Introduction

The intersection of the Decebal Bvd-Aradului Path-Primariei Street is an intersection in a well-defined geometric form, being systematized from a topological point of view. The topological intersection of the Decebal B-Calea Aradului – Street Primariei is arranged as a roundabout. The elimination of conflicts between vehicle flows is solved classically by the location of the "Give Way" indicators located on the four entries of the roundabout. Once entered in the roundabout the vehicles have priority to rotate around it and engage on any of the exits.

2 Case study

This is on the solution of the implementation of an adaptive traffic light concept in regards to tram access in the intersection created between the Decebal Bvd-Aradului Path-Primariei Street in Oradea Municipality, adapted to the requirements of public transport.

The particularity of this roundabout is the path of the tramway and their circulation. It is in two different situations. The first refers to the entry of trams in the roundabout and the second resolution of the conflicts between tram movements and the circulation of vehicles inside the roundabout. In the first situation, the trams entering the roundabout are deprived of priority in relation to vehicles already engaged inside the roundabout. In the second situation, once penetrate in the roundabout, trams have priority over the rest of the vehicles. This regulation is flagged with " Attention Tram " indicators located in the intersection and presented in accordance with figure 1, 2, 4, 5, to be presented in the following.

It is clear that the need to prioritize tram access in the roundabout based on the general policy of prioritizing public transport with the other traffic participants. In this case, as outlined above, the prioritization of trams can only be achieved by the traffic lights of the conflicts between them and the rest of the participants, vehicles, and pedestrians.

As a result, we describe the circulation flows of vehicles that have access to this intersection as follows:

- Exit from the Primariei Street to Decebal Bvd has the right turn to the Train Station, heading forward to the Aradului Path with the possibility of turning left in the roundabout to the Decebal Bvd;

- The exit of the Aradului Path to Decebal Bvd to the right, after the vehicles cross the passage of pedestrians with the button on the Aradului Path, benefit from a strap, which removes from the regulation by traffic lights this relationship, the direction of walking forward is regulated by traffic lights;
- For the flow of vehicles entering the intersection on the meaning of walking the Decebal Bvd towards the Zoo the movement of vehicles is regulated at the roundabout entrance for both the forward direction and the right turn through the road signs, and at the level of the conflict point between the traffic flow of the tramways on the path of its running and the flow of movement of vehicles on the rolling track within the roundabout, the circulation is regulated by traffic lights both for the sense of movement Decebal Bvd towards the Zoo and for the meaning of walking Decebal Bvd towards the Train Station at the entrance of the roundabout.

The conflicts that occur at the level of pedestrian passage with the flow of vehicles and tramways generated by walking directions present the following characteristics: the passage of pedestrians from the Primariei Street, the passage of pedestrians from the Decebal Bvd and the passage of pedestrians from Aradului Path is flagged by road markings, with road signs being light by an adaptive system at the request of pedestrians, by actuating a pedestrian button equipped with a sensor for referral of applications-phase dedicated to crossing the passage of to pedestrians. [3], [4], [10].

3. Describing the proposed solution

To prioritize access to the roundabout and through it by trams, as we have already suggested, it is necessary to request the conflict points between trams and the rest of the traffic participants, as well as between pedestrians and the rest of the participants in traffic.
Not being general traffic lights, classical, which should have solved all the conflicts between the traffic participants, the function of the traffic light of this atypical intersection will be similar to that of the railway crossings. More clearly, only the conflicts between trams and the rest of the traffic participants and the conflict between pedestrians crossing the pedestrian crossings related to the intersection with the rest of the participants will be resolved. The rest of the circulation will be regulated, as outlined above, through the established prioritization of a roundabout.
The intersection traffic lights will work on request, that means a general state of calling for a request from the trams and the phase that will serve them access and crossing the roundabout, called the phase on demand, applications for the serving of pedestrian are related and conditioned by tram movements. In this way, the circulation of vehicles will be interrupted only at the appearance of trams and at the request of pedestrians, which will be conditioned by the circulation of trams.
Through this systematization and traffic lights, the ability to pass and retrieve traffic flows at the intersection level, increases, so the level of service increases by increasing the safety of road streams for vehicles, public transport and for pedestrians using modern traffic lights and by markings and indications.
The intersection traffic lights must resolve the following conflicts between the traffic flows of the tramways on their path and the traffic flows of vehicles on the driving path, as well as the conflicts occurring at the level of crossings pedestrians with the circulation flows intersecting them as follows:

- The exit of Primariei Street to go forward to Aradului Path road and turn right towards the train station;
- Exit from the Aradului Path direction forward and turn right towards the Decebal Bvd;
- Entrance to the intersection on the way to Decebal Bvd towards the Zoo on the forward direction and turning right towards the Aradului Path
- Entrance to the intersection on the way to Decebal Bvd towards the Train Station on the direction forward.
- Crossing the Bvd Stefan cel Mare by pedestrians;
- Crossing the Primariei Street by pedestrians of pedestrian crossings located at the exit of this street.

- Crossing by pedestrians of pedestrian crossings on the Aradului Path located in the proximity of its intersection with the roundabout on the Decebal Bvd
- Crossing by pedestrians of pedestrian crossings from the Decebal Bvd located in the proximity of the roundabout on the Decebal Bvd on the Zoo side.

These conflicts that occur in this intersection are the consequences of intersecting traffic flows that are generated by the road of motor vehicles on the one hand and the other side of the traffic flows that are generated by the tramway , will be solved by an atypical traffic light in the motion phases on demand generated by the demand of tram movements and pedestrian crossings:

3.1 Phase I

The calling phase of an application for crossing the intersection of tram or pedestrians, allows the movement of vehicles only on the front and right directions of the Primariei Street, the movement of the self-propelled both ways on the Decebal Bvd as well as turning right towards the Aradului Path, the movement of motor vehicles in the forward and right direction of the Aradului Path route on the Decebal Bvd;

In this phase or calling the state of a tram request or requested by pedestrians P5.1/P5.2. The traffic lights in the intersection will function as follows: Vehicle traffic lights V1.1, V2.2, V3.3, V5.1, V6.1, V6.2, V8.1 works on FLASHING YELLOW, Pedestrian traffic Lights P5.1, P5.2 on RED and all tram traffic lights T1.1, T2.1, T3.1, T4.1, T5.1 T6.1, T 7.1 and T8.1 indicate STOP (Figure1).

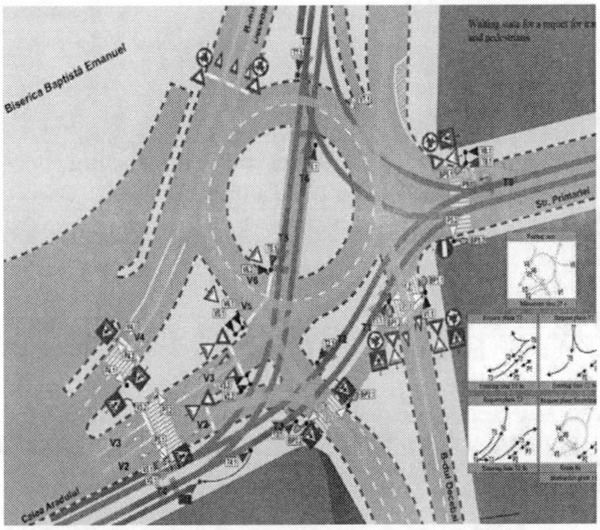

Figure 1. Calling phase of an application for crossing the intersection for tram or pedestrians

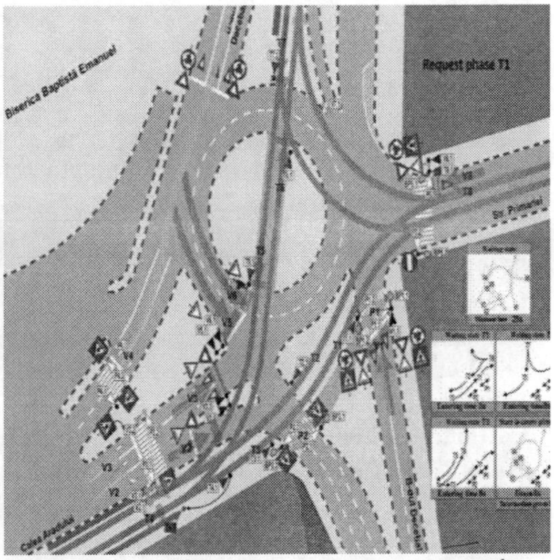

Figure 2. Phase on request crossing intersection for tram T1

3.2 Phase II,

On-demand phase for the T1 tram allows the movement only for the trams circulating on the Primariei Street direction towards the Aradului Path and those that drive to the station on Decebal Bvd, the traffic lights on all directions before and the right turns will see red.

In this phase on request of the T1 tram, which serves the demand of tram coming from Primariei Street, the traffic lights at the intersection will function as follows: Vehicle traffic lights: V1.1, V 2.2, V3.3, V5.1, V6.1, V6.2, V8.1 will be on the RED color, The traffic lights of pedestrians P5.1, P5.2 on RED, and tram traffic lights T1.1, T2.1, T3.1, and T8.1 indicate FORWARD while T4.1, T5.1 T6.1 and T7.1 indicate STOP. (Figure 2).

3.3 Phase III

IOP Conf. Series: Materials Science and Engineering **568** (2019) 012097 doi:10.1088/1757-899X/568/1/012097

On-demand phase for the tram T2, allows the movement of only trams that run on the Decebal Bvd to the left to Primariei Street and turn right towards the Aradului Path, the vehicle traffic lights on all directions forward and the turns to the right will have red.

In this phase on request of the tram T2 , serving tramways that circulate and have access to the intersection of the Primariei Street, the traffic lights in the intersection will function as follows: V1.1, V2.2, V3.3, V5.1, V6.1, V6.2, V 8.1 will be RED, pedestrians P5.1, P5.2 on RED, and tram traffic lights T 5.1 and T 7.1 indicate FORWARD while T1.1, T2.1 T3.1, T4.1, T6.1 and T8.1 indica STOP (Figure 3)

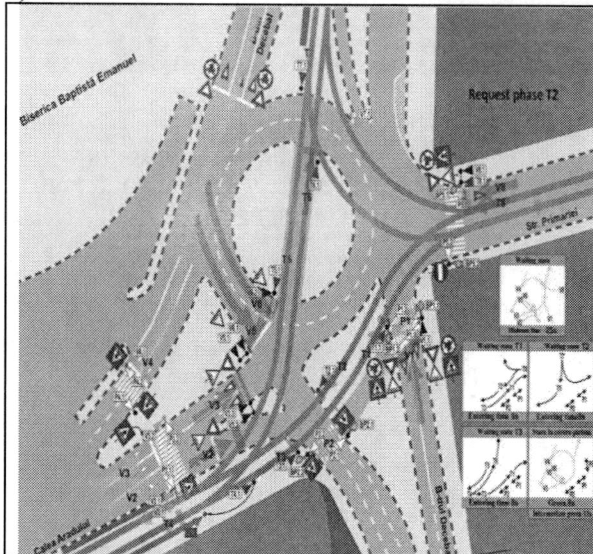

Figure 3. Phase on request crossing the intersection for the T2 tram

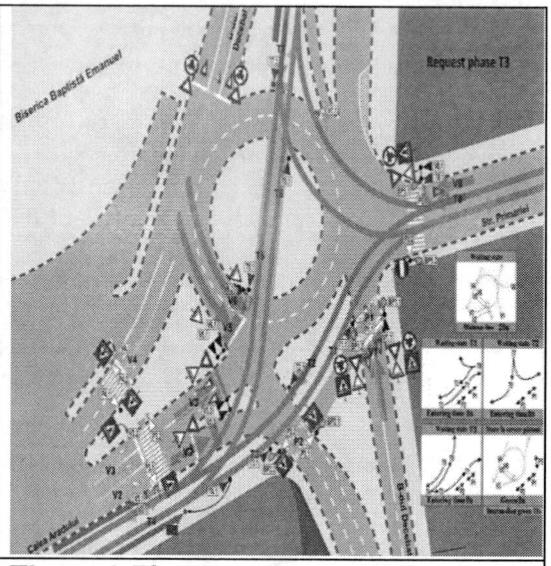

Figure 4. Phase on request crossing the intersection for the T3 tram

3.4 Phase IV

Phase on demand for the T3 tram, allows the movement of only trams that circulate on Aradului Path, turning left towards the Decebal Bvd and forward direction to the Primariei Street, the traffic lights on all directions forward and the cornering at the right will have red.

In this phase at the request of the T3 tram, serving the tramways coming from the Aradului Path, the traffic lights in the intersection will function as follows: pedestrians P5.1, P5.2 on RED, and tram traffic lights T1.1, T3.1, T4.1, T 5.1 and T6.1 indicate FORWARD while T2.1, T7.1, and T8.1 indicate STOP (Figure 4).

3.5 Phase V

Phase on demand for pedestrian traverses the Primariei Street, and, Decebal Bvd, allows the movement of pedestrians only requesting the demand to circulate on pedestrian crossings on the Street Primariei and Decebal Bvd , vehicle traffic lights and tramways on all directions forward and the corners to the right will have red.

In this phase on demand for the P5 pedestrians serving pedestrians crossing the Primariei Street, the traffic lights in the intersection will function as follows:V1.1, V2.2, V3.3, V5.1, V6.1, V6.2 will function on FLASHING YELLOW, V8.1 on RED color, pedestrians P5.1, P5.2 on GREEN, and tram traffic lights T1.1, T2.1, T3.1, T4.1, T5.1 T6.1, T7.1 and T 8.1 indica STOP (Figure 5).

4. Recommendation

I recommend the City Hall of Oradea to conduct a traffic study on which to design the regulation of the circulation in this intersection by reconfiguring the topological and extending the traffic lights system for solving all conflicts

IOP Conf. Series: Materials Science and Engineering **568** (2019) 012097 doi:10.1088/1757-899X/568/1/012097

Figure 5. Phase IV, on-demand phase for pedestrians crossing the Primariei Street

5. Conclusions

With an almost no interruption in the field of traffic lights from 1998, almost 21 years, I see that I have never encountered such a situation before. Traffic lights are made primarily to increase the safety of the participants in traffic, introducing clear restrictions materialized by the RED color of the traffic light or STOP in the case of trams. In this case, we have a mixed traffic light combined with regulation with indicators. For this reason, the signaling must be explicit, placed visibly in order not to create ambiguities.

The monitoring of the chosen solution implies selecting some measurable parameters; the performance index shall be analysed based on the result. These measurable parameters can be, for example, duration of time, " green" allocated to each phase. Based on this measurable parameter we can calculate the percentage, "green" used from the total maximum allotted time, as well as the optimum cycle duration and the distribution of the "green" times allocated to this cycle. Adjusting the parameters in the function can be done as a result of graphic representations and the repetition of the whole process, following the analysis of this aspect can be carried out a study that will lead to the optimal variant chosen. Measuring the efficiency of traffic crossings and analysis of implemented programs requires an integrated system for collecting traffic data, registering them for a reasonable period, transferring them through a network of Communication in computing and finally processing center, graphical representation, and analysis, making statistical reports.

To solve these deficiencies, the author proposed the solution to implement a concept of adaptive traffic light system presented in this work and which currently solves this problem.

References

[1] Romocea, S 2010 Contributions to the actual test samples and the existing structure of the signaling system, University of Oradea, Faculty of Managerial and Technological Engineering, DAI - Industrial Engineering
[2] *** SR 10144-4 / 1995 - Spatial intersections of streets, IRS 1995
[3] *** STAS 6926/11/1999 .: - Road vehicles. Test methods-time measurement minimum speed and acceleration
[4] *** VIP 19 "RACK, Software Handbook
[5] *** VICCOM / E Version V1.12 VIP 19 "RACK, User Handbook
[6] *** AS26-User Handbook
[7] *** www.traficon

IOP Conf. Series: Materials Science and Engineering **568** (2019) 012098 doi:10.1088/1757-899X/568/1/012098

Python-Based Data Analysis Tool for a 6-DoF Industrial Robot

A Alsabbagh[1], A N Nasser[1] and G Husi[1]

[1] Department of Mechatronics, University of Debrecen, 4028 Otemeto utca 2.-4., Debrecen, Hungary

Ammar-alsabbagh-eng@hotmail.com

Abstract. A newly developed tool for analysing and visualizing data based on python. The main aim of this tool is saving the data analysis time by automating the data analysis process and creating a Graphical User Interface (GUI) to facilitate the usage of this tool regardless how much they experience programming with Python. This paper describes the architecture of the developed software tool and provides a variety of examples to show the developed tool's advanced features and functionalities. Moreover, in this paper, an industrial problem related to data analysis is discussed, and then a solution of this problem is presented. This tool is developed using open source libraries (Pandas, Matplotlib, Tkinter, and SciPy) to manage multi data files at once using one tool, as well as visualizing and storing these data in one file in a clean and organized way.

1. Introduction

In R&D, process development department, one of the important steps while building the production line is simulating the assembly process machines in order to define and determine the configuration of the production line and the part managements. For this purpose, Battenberg Robot is used, which is a 6-DoF (Degree of Freedom) Robot that is integrated with 6-Axis Force and Torque sensors (ATI-Sensors) for measuring the required forces and torques of the assembly processes.

The described Robot exports the sensor measurements, the robot end effector position and orientation, and the angles of the robot motor's joints to text files, (one file for each single movement). To manage all those generated txt files, in an automated way, data analysis tool was developed using Python language.

2. The data analysis tool functions

2.1. The data analysis tool capabilities

The developed tool is capable of doing all the following functions using only the text files that are generated by the Robot (see the block diagram in Figure 1).

- Calculate the velocity and the acceleration (both linear and rotational) of the robot end-effector
- Visualize the data on two axis coordinates X, Y.
- Calculate and plot the average curve of all selected files, in addition to obtaining the maximum and minimum curves.
- Implementing the Savitzky Golay filter for smoothing the visualized data.
- Visualizing two or more different data sets on the same figure for an easy comparison.

- Save the graph for most of possible image extensions (png, jpg, pdf, etc).
- Obtaining the minimum and maximum in a specific region of the all graphs (regional min, max).
- Calculate the Process Capability Index Cpk, for a specific point of the process.
- Export the row data, filtered data, calculated average, maximum and minimum to sets to an Excel file xlsx and Comma Separated Value file csv

2.2. The tool steps of analyzing the data

The steps of data analysis tool can be seen in figure 3. It starts by opening the text files until the final GUI is open are:
'Open window' pops up for the user to choose the files that should be analyzed. then, a first user action is required for choosing the file/files that should be managed by the tool. After that, the chosen files are opened one by one, and the data sets are shown in the First GUI, allowing the user to select the required data sets of the experiment. A second user action then is required for selecting the data that is required to be analysed, visualized, saved to the data base. Then the data sets of all the selected files will be organized in one data frame (each data set is linked to the file's name) and the final GUI pops up. When the final GUI is showed, the user can do the data analysis will have all the data analysis option that is provided by the GUI.

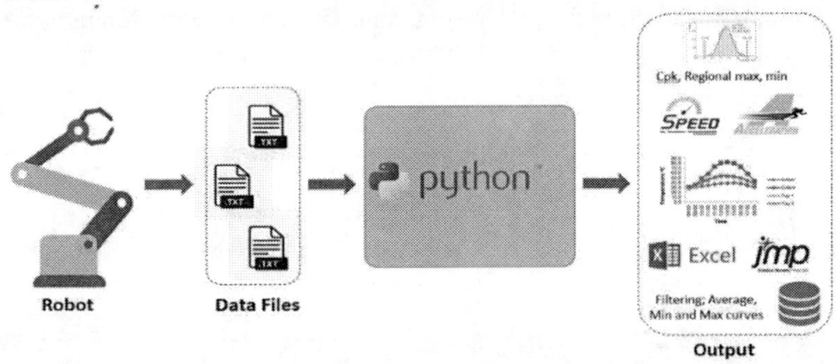

Figure 1. Block diagram of the data analysis tool.

When a user runs the program, the "open file window" pops up, allowing the user to select whatever the required files to be analyzed. With this window, the user can select one or more text files at the same time. Then the program opens those text files and checks automatically the available data sets in those files (the data sets are changeable according to the robot program and sensors configuration), then it shows the available data on the first graphical user interface

The first GUI allows the user to select among of the available data. When the user selects the required data to be analyzed, these data will be cleaned, organized and added to one data frame, taking in consideration the change of the text files according to the change of the time and type of assembly test (number of text lines, the size of the data, the type of the robot movement, the location of the data sets header, the number of the text lines at the end of the text file, and etc.). All the transitional velocity and transitional acceleration of the robot end-effector are calculated in this stage, or the rotational velocity and rotational acceleration of the robot end-effector are calculated in this stage according to the type of movement. Moreover, the set velocity and acceleration of the robot are read by the program to compare them with the actual velocity and acceleration set by the robot program.

2.3. Final GUI

Figure 2 shows the final graphical user interface for analyzing the data. This GUI contains all the tool Functions. Both sections "Choose Y axis" and "Choose X axis" are changeable according the selected data set in the Previous GUI (First GUI), while all the other section on this GUI (Statistical Options, Zooming, Labeling, and etc.) are graphically constants. The user can choose the output according to documentation requirements. The Visualization will be on a separate window, then the users can have as many figures as they demand.

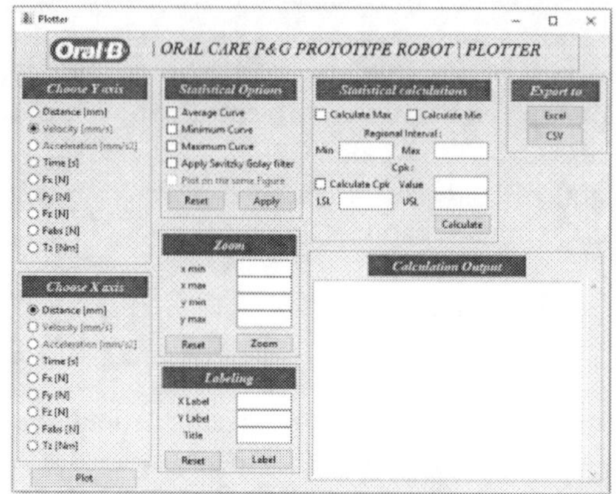

Figure 2. Final Graphical user interface for visualizing the data

The axes selection sections enable the user to choose the X axis and Y axis for visualizing the data. The data in black is the row data that come directly from the robot text files, while the data in green is the calculated data. The tool can calculate all the following data according to the robot movement type: transitional velocity, transitional acceleration, angular velocity, and angular acceleration.

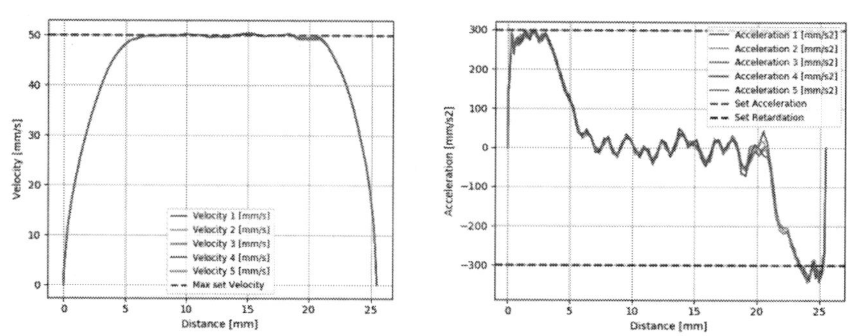

(a) (b)

Figure 3. The Robot end-effector velocity and acceleration

Figures 3 (a) and (b) show the comparison between the actual end-effector velocity with the set speed (a), and the same for the acceleration and the retardation (b). This feature is very significant for the assembly test, as setting the velocity of the robot should be exactly the same of the velocity of the assembly machine of the production line. With this feature the user can check the speed validity of the robot. And in case the robot is not reaching the desired velocity, an action should be taken, increasing the acceleration for instance.

3. Statistical Options

The "Statistical Options" section enables the user to have variety of statistical options for the visualization as well as for defining some of the most important curves. All can be done by selecting the required option among these section options. With this section the user can do all the following:

1. Calculate the average curve of all the selected text files, add it to the data frame and plotting it on the figure.
2. Define the minimum curve among all the curves that come from the selected files, add it to the data frame, and potting it on the figure.
3. Define the maximum curve among all the curves that come from the selected files, add it to the data frame, and potting it on the figure. Figure 6. Show an example of calculating the average of acceleration signals that are coming from different six test files, in addition to the finding the maximum/minimum of those six files, then visualizing them.

4. Smoothing the raw data using Savitzky Golay filter, add the filtered data to the data frame, and plotting it on the figure.

A Savitzky–Golay filter is a digital filter that can be applied to a set of digital data points to smooth the data, i.e. to increase data accuracy without distorting the tendency of the signal. This is achieved by fitting successive sub-sets of adjacent data points with a low-grade polynomial using the linear least squares method. When the data points are equally spaced, an analytical solution for the least square equations can be found in the form of a single set of "convolution coefficients" that can be applied to all data sub-sets to provide estimates of the smoothed signal (or smoothed signal derivatives) at the center point of each sub-set. The method was popularized by Abraham Savitzky and Marcel J. E. Golay, who published tables of convolution coefficients for various polynomials and sub-set sizes in 1964 [1][2], based on established mathematical procedures. Some tables errors have been fixed [3][4]. The method for handling 2-and 3-dimensional data has been extended [5].

figures 4 illustrate the acceleration of six different text files before (a) and after applying Savitzky Golay filter (b). The figure after applying SG filter, can show that the singles are smoothed and more understandable. That can be helpful for understanding the complete trend of the signals.

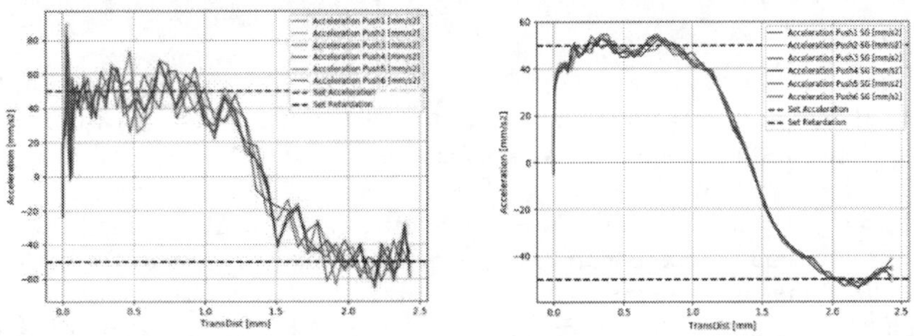

(a) (b)

Figure 4. the acceleration before and after applying Savitzky Golay filter

4. Statistical Calculation

A process capability index is a numerical summary that compares the behavior of a product or process characteristic to engineering specifications. These measures are also often called capability of performance indices or ratios; however, the generic term is capability index. A capability index relates the voice of the customer (specification limits) to the voice of the process. A large value of the index indicates that the current process is capable of producing similar parts, that will meet or exceed the customer's requirements. This term is widely used in the process because it reduces complex information about the process to a single number. Many customers ask their suppliers to record capability indices for all special product characteristics on a regular basis. These indices are used to communicate how well the process has performed. It is assumed that these indices also indicated expected future performance. Suppliers may also use Cpk for different characteristics to establish priorities for improvement activities. Similarly, the effect to process change can be assessed by comparing capability indices calculated before and after the change.

If the upper and lower specification limits of the process are USL and LSL, the target process mean is T, the estimated mean of the process is $\hat{\mu}$ and the estimated variability of the process (expressed as a standard deviation) is $\hat{\sigma}$, then commonly accepted process capability indices include:

$$\hat{C}_p = \frac{USL-LSL}{6\hat{\sigma}} \tag{1}$$

$$\hat{C}_{pk} = min\left[\frac{USL-\hat{\mu}}{3\hat{\sigma}}, \frac{\hat{\mu}-LSL}{3\hat{\sigma}}\right] \tag{2}$$

That is why calculating Cpk is a significant function in this GUI, especially for process development team. The data analysis tool can calculate the Cpk for any point of the selected signal, it is used for the most critical point during the executed tests. It should be mentioned, that in the assembly tests, few tests are executed for the same parts (less than a hundred). Therefore, it is mentioned on the output section of the data tool that the Cpk value may not give a conclusion of the basic distribution. Nevertheless, the Cpk feature is still giving a pre-description of the process before it goes for the production line. The Cpk value and all the other critical values will be showed on the "Calculation output" section.

Finally, one of the most useful features is exist in the "Export section". In this section there are two buttons for saving the data in two different file types. The First one is xlsx extension which is the extension of the Excel file, and the other extension is CSV Comma separated value that can be operated using different software like JMP, Excel etc.

With this feature the user would be able to save the selected data sets of all the files, the clean data, the calculated data (velocity, and acceleration), the filtered data, the maximum curve, minimum curve, and the average curve in an organized and accessible way, and this data will be linked to their corresponded files. The importance of this feature come from allowing the user to store the data in one SharePoint which is the ground of the database. Moreover, the user would be able to do more advanced data analysis which is not exist in the data analysis tool by using JMP program for example.

5. Conclusion

By using this data analysis tool, the time efficiency increased dramatically; more than 90% of data analysis consumed time is saved using the developed graphical user interface. Moreover, multiple tests on this tool were done, asking people who were never introduced to this tool before to do around fifteen tasks. All of them were able to finish all the fifteen tasks in less than 10 minutes, which shows how much this GUI is user-friendly, easy to be used, and sufficient for manipulating the data, moreover the same fifteen tasks had required more than three hours to accomplish them by the traditional way (copy data to Excel or JMP then analyze the data and visualize it). Finally, it should be mentioned that this tool was developed as a fast prototype for test and demo tool. Which means, more development is required, and all the cybersecurity issues should be taken into consideration for getting a safe and reliable version.

Acknowledgments

The publication is supported by the EFOP-3.6.1-16-2016-00022 project. The project is co-financed by the European Union and the European Social Fund.

References

[1] Whittaker E T and Robinson G 1940 Calculus of observations: *Treatise on Numerical Mathematics 3rd Edition* Blackie and Sons Limited, London.

[2] Guest P G, and Guest P G 2012 Numerical methods of curve fitting *Cambridge University Press.* pp. 147

[3] Savitzky A, and Golay M J 1964 Smoothing and differentiation of data by simplified least squares procedures. *Analytical chemistry,* **36(8)** 1627-1639.

[4] Pochinok N B. 2008 Contemporary Problems of Social Work. *Bulletin of Chelyabinsk State University,* **(4)** 88-97.

[5] Steinier J, Termonia Y, and Deltour J. 1972 Smoothing and differentiation of data by simplified least square procedure. *Analytical chemistry* **44(11)** 1906-1909.

[6] Larive C K, and Sweedler J V 2013 Celebrating the 75th anniversary of the acs division of *analytical chemistry*: A special collection of the most highly cited analytical chemistry papers published between 1938 and 2012

[7] Alsabbagh A, SZEMES P T, & Baki A 2018. Brushless DC Motor Modeling Using Bond Graph Method and Control using LabVIEW. *(RIim) Recent Innovation in Mechatronics,* **5(1)**

[8] Boyles R A 1991 The Taguchi capability index. *Journal of Quality Technology* **23(1)** 17-26

IOP Conf. Series: Materials Science and Engineering **568** (2019) 012099 doi:10.1088/1757-899X/568/1/012099

Comparative analysis of different heating systems based on the surface area of a household

C Cristea[1], M Cristea[2], R Tirnovan[3] and C Fagarasan[4]

[1,2,3] Electrical Engineering Faculty, Technical University of Cluj-Napoca, 26-28 Baritiu Street, Cluj-Napoca, Romania

[4] Machine Building Faculty, Technical University of Cluj-Napoca, 103-105 Muncii Avenue, Cluj-Napoca, Romania

[1]ciprian.cristea@emd.utcluj.ro

Abstract. The scope of this paper was to determine if the efficiency of a heating system depends on the surface area of a household and to identify the surface intervals for which a heating system is most appropriate than others. There were considered nine alternatives for the analysis: five different heating sources (wood boiler, electric boiler, gas boiler, pellet boiler and electric heating carpet) combined with two different heating terminals (radiant floor heating and the classic radiator), which were evaluated based on initial investment, fuel cost and the total annual expenditure for each heating system. The results indicated that the type of heating system for a household might vary depending on its surface, considering the three comparison factors mentioned before.

1. Introduction

When building or renovating a house, the occupant must decide between different heating systems to assure thermal comfort inside, as it can affect both his/her health and productivity [1]. Selecting the most low-priced system does not always mean that is the most efficient. Besides the raw material and installation costs, there are other factors that can differentiate between heating systems: fuel cost, lifecycle cost, impact on the environment, yield, system guaranty. There are other factors that depend on the area where the house is located, like fuel availability or system autonomy. These criteria can influence the decision maker's opinion in selecting the most appropriate heating system for his/her house.

In previous related research, there have been made some comparisons between different heating systems based on multiple factors like efficiency and cost. In [2] a comparative review of heating systems in EU countries, taking into consideration the seasonal efficiency and fuel cost was made. An analysis was conducted, and the lowest cost was established for the geothermal heat pumps; whereas, the highest cost was determined for the electric heating. A similar study was conducted in [3], where an analysis was developed for residential buildings in the Mediterranean climate, by using criteria like the equipment's efficiency, lifecycle cost and the buildings' location, keeping in mind also their environmental impact. The research indicated that the location of the buildings had a significant impact because, as the results showed, in the metropolitan area, natural gas is the most economical solution. For buildings situated in the urban area, the most economical and environmentally friendly solutions are condensing natural gas boilers. In [4], a new method was used to determine the best heating system in the residential sector of Turkey. This method is called the ex-ergo-environmental analysis, and it is a combination between an exergy analysis and lifecycle assessment. There were considered three heating systems: a conventional coal boiler, a condensing natural gas boiler, and a ground source heat pump.

IOP Conf. Series: Materials Science and Engineering **568** (2019) 012099 doi:10.1088/1757-899X/568/1/012099

The primary conclusion of this paper was that the condensing natural gas boilers were the best choice, considering all the circumstances taken into account.

In [5], different heating systems were analyzed and evaluated by using criteria like the basis of the investment, costs of operating as well as the emissions and energy consumption. The study was made on single-family dwellings during renovation works, and the main conclusion was that recovering the existing heating installation was the first step in finding the optimal solution. The second step was the installation of a gas condensing boiler. These two steps would also satisfy the environmental and economic criteria. Fuel cost and equipment efficiency were considered in [6] to compare the heating systems used within several residential areas in Greece; as results, it appears that the geothermal heat pumps have the lowest costs and the electrical heating has the highest cost. In [7], an analysis was made on existing Danish buildings with the following purpose: to provide a list of costs and benefits for low-temperature district heating. The results indicated that an investment in the improved heating system would reduce district heating temperatures. The payback for this investment is estimated to be between 0.3 years to 18.7 years for a customer and between 1.3 years to 4.2 years when considering the total energy system. Distributed and centralized heating systems were put under analysis in [1], in the context of a typical apartment situated in Hangzhou, China. These two heating systems were compared: air-source heat pump (distributed heating) and steam/hot water radiator heating (centralized heating). The result showed that the first one renders 34.5% less carbon dioxide emissions than the second one, and it is preferred if this would be the single decision criteria. The cost was not considered in this article.

In this paper nine alternatives were compared based on initial investment, accounting for criteria such as raw material price, boiler price and installation price, fuel cost and total annual expenditure, considering the system reliability. Also, a sensitivity analysis was made to determine whether the efficiency of a system depends on the surface of the household and if so to indicate which heating system is the most appropriate for each surface interval.

2. Heating systems and criteria descriptions

In this study were considered nine heating systems for a household located in Satu-Mare County, Romania, which were evaluated against five criteria, considering the information received from a company dealing with electrical installations work. The first alternative (A1) is represented by the wood boiler combined with radiators, and with radiant floor heating for the second alternative (A2). The wood boiler is most used in rural and mountain areas, due to lack of infrastructure, low fuel cost, and existing storage space. In the same areas, it is also used the pellet boiler, combined with radiators (A7) and radiant floor heating (A8). It is one of the most well-known biomass boilers, being used more and more in residential, industrial, and commercial buildings, which have storage space for the fuel.

The next two alternatives use as a heating source, the electric boiler, which is combined with radiators (A3) and radiant floor heating (A4). This type of boiler is often used in residential and office buildings located in the cities, where the infrastructure is well developed, and the possibility for fuel storage is limited. Another choice for the type of buildings mentioned above is the gas boiler, combined with radiators (A5) and radiant floor heating (A6). This type of boiler is fuelled by natural gas and is considered less expensive than the electric boiler and electric heating carpet (A9).

The alternatives described above are compared against five criteria, and their performances are presented in table 1. The first criterion is the raw material price ([RON/m^2]), representing the cost of the heating terminals, calculated per square meter. There are three different values (currently applied on Romanian internal market, more specific in Satu-Mare County): for radiators (60 RON/m^2), radiant floor heating (65 RON/m^2) and electric heating carpet (45 RON/m^2). Depending on the quality of the material, there are different prices on the market. For this paper were considered average values for each type of heating terminal. Another criterion is the boiler's price, which depends on the surface of the household (kW output) that needs to be heated and on the type of fuel. The smaller ones (smaller wattage) have a lower price, while the ones used to heat larger interior spaces are more expensive. Also, the boilers fuelled with natural gas are the cheapest, while the ones fuelled with pellets (an environment-friendly material) have the highest price. The A9 alternative does not need a boiler, electric heating carpets being connected to the house's electrical installation. The installation price ([RON/m^2]) is the amount needed

IOP Conf. Series: Materials Science and Engineering **568** (2019) 012099 doi:10.1088/1757-899X/568/1/012099

to install the entire heating system and depends on the surface of the house. Table 1 briefly presents the performances of the analyzed heating systems.

Table 1. Heating systems performances.

	Raw material price [RON/m²]	Boiler price [RON]	Installation price [RON/m²]	Fuel price [RON/m²/year]	System reliability [years] Raw material	System reliability [years] Boiler
A1	65	3000-5000	85	28.36	40	10
A2	60	3000-5000	80	28.36	20	10
A3	65	2700-3800	85	138.24	40	10
A4	60	2700-3800	80	138.24	20	10
A5	65	3000-8000	85	34.45	40	10
A6	60	3000-8000	80	34.45	20	10
A7	65	8000-18000	85	64.36	40	10
A8	60	8000-18000	80	64.36	20	10
A9	45	-	95	180.78	50	-

The fuel price ([RON/m²/year]) is another criterion used to evaluate the alternatives. For each type of heating system was calculated, the fuel consumption needed to heat a square meter per hour, which was multiplied by the number of hours of operation per year. The annual consumption was then multiplied by the fuel price for each consumed unit to obtain the fuel price per year for each heating system. The last criterion ([years]) represents the lifetime of each heating system, which differs based on the quality of the product. The system reliability is divided considering the two components: heating terminals (raw material) and a heating source (boiler type).

3. Comparison factors

To perform the sensitivity analysis, the performances of the nine alternatives (A1-A9) were evaluated against three factors: initial investment, fuel cost and total annual expenditure, which were calculated for various surface, ranging from 100 to 300 square meters. The first factor, initial investment (I), is calculated for each alternative considering the raw material price (p_{rw}), boiler price (p_b), installation price (p_i) and household surface (h_s), with the following formula:

$$I = p_{rw} \cdot h_s + p_b + p_i \cdot h_s \tag{1}$$

The fuel cost (C_f) depends only on the fuel price (p_f) criterion and the household surface (h_s):

$$C_f = p_f \cdot h_s \tag{2}$$

The last factor, total annual expenditure (C_T), was calculated based on raw material price (p_{rw}), boiler price (p_b), installation price (p_i), household surface (h_s), raw material reliability (r_{rw}), boiler reliability (r_b) and fuel cost (C_f) with the following equation:

$$C_T = \frac{p_{rw} \cdot h_s}{r_{rw}} + \frac{p_b \cdot h_s}{r_b} + p_i + C_f \tag{3}$$

4. Results

To select the optimal heating system for a household located in Satu-Mare County, Romania, a sensitivity analysis was performed to determine which alternative is more appropriate based on the surface of the household. Several heating systems were evaluated considering three significant factors: initial investment, fuel cost, and total annual expenditure.

The results obtained after the first simulation, where the performances of the analyzed alternatives were evaluated considering an initial investment, are presented in table 2. The evolution of each alternative, considering the household surface is described in figure 1. It can be noticed that for a household which has the surface area between [100-160] square meters, the most appropriate solution

IOP Conf. Series: Materials Science and Engineering **568** (2019) 012099 doi:10.1088/1757-899X/568/1/012099

is A4, followed closely by A2 and A6. The alternatives with the highest initial investment for the same surface are A7, A8, and A9. For the surface interval [160-290] square meters, the initial investment fluctuates, determining the change of places between the alternatives, until [290-300] square meters, where the ranking is: A4, A2, A3, A1, A6, A5, A9, A8, A7.

Table 2. Alternatives performances considering the initial investment.

	A1	A2	A3	A4	A5	A6	A7	A8	A9
100	18,000	17,000	17,700	16,700	18,000	17,000	23,000	22,000	19,500
110	19,500	18,400	19,200	18,100	19,500	18,400	24,500	23,400	21,450
120	21,000	19,800	20,700	19,500	21,000	19,800	26,000	24,800	23,400
130	22,500	21,200	22,200	20,900	22,500	21,200	27,500	26,200	25,350
140	24,000	22,600	23,700	22,300	24,000	22,600	30,000	28,600	27,300
150	25,500	24,000	25,200	23,700	25,500	24,000	31,500	30,000	29,250
160	27,000	25,400	26,700	25,100	27,000	25,400	33,000	31,400	31,200
170	28,500	26,800	28,200	26,500	28,500	26,800	34,500	32,800	33,150
180	30,000	28,200	29,700	27,900	30,000	28,200	36,000	34,200	35,100
190	32,500	30,600	31,500	29,600	32,500	30,600	38,500	36,600	37,050
200	34,000	32,000	33,000	31,000	34,000	32,000	40,000	38,000	39,000
210	35,500	33,400	34,500	32,400	35,500	33,400	42,500	40,400	40,950
220	37,000	34,800	36,000	33,800	37,000	34,800	44,000	41,800	42,900
230	38,500	36,200	37,500	35,200	39,500	37,200	46,500	44,200	44,850
240	40,000	37,600	39,000	36,600	41,000	38,600	48,000	45,600	46,800
250	41,500	39,000	40,500	38,000	42,500	40,000	50,500	48,000	48,750
260	43,000	40,400	42,000	39,400	45,000	42,400	52,000	49,400	50,700
270	45,500	42,800	43,500	40,800	47,500	44,800	55,500	52,800	52,650
280	47,000	44,200	45,000	42,200	49,000	46,200	57,000	54,200	54,600
290	48,500	45,600	47,300	44,400	51,500	48,600	59,500	56,600	56,550
300	50,000	47,000	48,800	45,800	53,000	50,000	63,000	60,000	58,500

The same procedure was used to compare the alternatives considering the fuel cost. The results are presented in figure 2 and indicate that the most appropriate heating source is the wood boiler (A1 and A2), followed by the gas boiler (A5 and A6). Also, it can be noticed that the alternatives A3, A4, and A9 have a more pronounced growth of the fuel cost than the other alternatives.

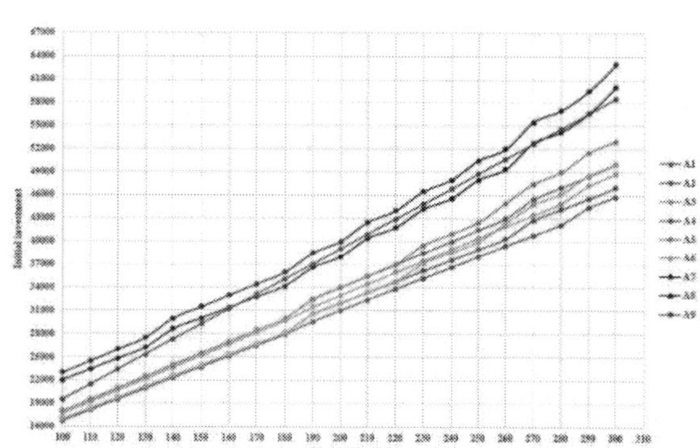

Figure 1. Sensitivity analysis based on the initial investment.

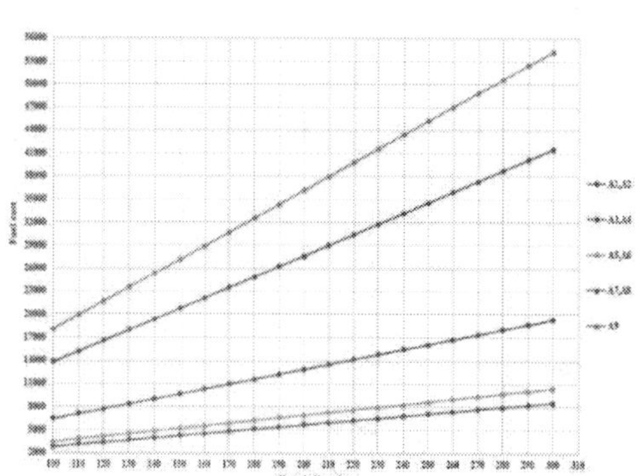

Figure 2. Sensitivity analysis based on fuel cost.

The results from the final simulation, regarding total annual expenditure, are described in figure 3. For the surface interval [100-150] square meters, there are four alternatives that can be considered appropriate for a household, and these are: A2, A7, A6, and A5. The total annual expenditure difference between these alternatives is relatively low for the interval surface mentioned above. For [150-200] square meters, it can be noticed a slightly increasing difference between the four alternatives, which becomes higher as the surface of the household grows ([250-300 square meters]).

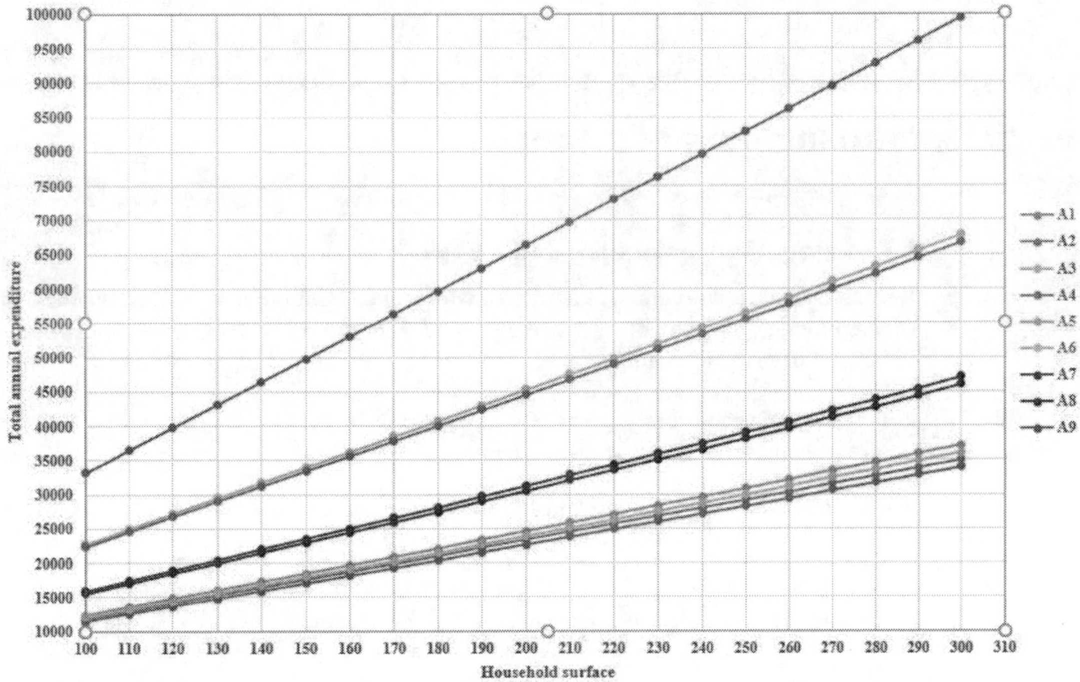

Figure 3. Sensitivity analysis regarding total annual expenditure.

5. Conclusions

In this study, a sensitivity analysis was performed to determine if the performances of the analyzed alternatives depend on the surface of the household and to identify the optimal solution for each surface interval for a household located in Satu-Mare County, Romania. The results obtained suggest that the type of heating system depends on the surface area of the household, considering the initial investment. The sensitivity analysis performed based on fuel cost and total annual expenditure are robust, but it can be observed an increasingly pronounced growth of costs between the best ranked alternatives and the worst ranked ones, in both cases.

6. References

[1] Zheng M, Fang R and Yu Z 2016 *Energy Procedia* **104** 287
[2] Martinopoulos G, Papakostas K and Papadopoulos A 2018 *Renew Sust Energ Rev* **90** 687
[3] Martinopoulos G, Papakostas K and Papadopoulos A 2016 *Energ Buildings* **124** 79
[4] Abusoglu A and Sedeeq M 2013 *Energ Buildings* **62** 268
[5] Obyn S and Moeseke G 2014 *J Energ Convers Manage* **88** 153
[6] Papakostas K, Martinopolous G and Papadopoulos A 2015 *17th Symposium on Thermal Science and Engineering of Serbia*
[7] Ostergaard D S and Svendsen S 2019 *Energy* **176** 718

IOP Conf. Series: Materials Science and Engineering **568** (2019) 012100 doi:10.1088/1757-899X/568/1/012100

Design of a device for testing and analyzing the friction coefficient during metal cutting

I Tamaşag[1], D Amarandei[1] and I Beşliu[1]

[1]Ştefan cel Mare University, Faculty of Mechanical Engineering, Mechatronics and Management, Department of Mechanics and Technologies, 13 University Street, 720229, Suceava, Romania

ioan_tamasag@yahoo.com

Abstract. This paper seeks to illustrate the development of a device for tribotesting and analysing of friction and normal forces in orthogonal cutting. In order to investigate the friction and normal forces, the objective of the device will be focused on the analyzing of the behavior of the tool-chip (rake surface) or tool-workpiece interface (flank surface) under dry metal cutting conditions. The device will simulate an orthogonal cutting process conditions that can be afterward modified for another cutting process basically with different geometries. The testing setup allows the friction coefficient to be estimated using the well-known Coulomb's friction law. Experimental procedures were conducted to validate the new device. The experiments were carried out using a tungsten carbide P20 as insert and C45 as the workpiece material, and the results showed that the coefficient of friction is dependent on the area of contact between the tool and workpiece.

1. Introduction

The friction due to the contact between the cutting tool and workpiece has been the subject to many research activities and studies in the past due to its high importance regarding the exit parameters of the process (tool life, chip formation, surface quality etc., overall sustainability).

With the increased development of new materials and tools, and the present work being insufficient to accurately describe the friction phenomena, it becomes mandatory that further study on this subject must be applied. The determination of friction coefficient using analytical relationships is complicated and often gives very approximate results due to the fact that the utilized formulas doesn't include all the influences or the interdependence of different factors. Such being the case, very often, for the determination of the exact values of the friction coefficients, scientists have resorted to empirical methods which utilises for those determinations the so-called "tribometers". This paper seeks to illustrate the development of a new device capable of simulating and modifying tool geometries in order to contribute to the studies on friction at the tool-chip-workpiece interface.

2. Background

The authors of paper [1] published a state of the art paper regarding the development of empirical models for friction in metal. According to his research, three different models can be used to investigate the friction during metal cutting: cutting force measurements, conventional tribometers and specially designed tribometers.

IOP Conf. Series: Materials Science and Engineering **568** (2019) 012100 doi:10.1088/1757-899X/568/1/012100

The first method consists in measuring the normal and tangential forces during orthogonal cutting with piezoelectric or strain-gauge dynamometers and estimating the macroscopic equivalent friction coefficients by comparing the values obtained for these output parameters. The second method consists in measuring the normal and tangential forces of a specially designed tribometer that adopt one of the classical arrangements pin-on-disc, ball-on-plate, pin-on-plate and pin on plane (horizontal) (fig.1). The results obtained with this kind of tribometers are usually unsatisfactory given the fact that the surface of the specimen is not refreshed and corresponds to the closed tribosystems. One example for this kind of method is presented in [2]. Another good example is presented in [3] where the authors measured the sliding coefficient of friction between flank wear and workpiece using the arrangement block-on-disk, the block being a CBN insert and the disk made of 60WCrV7 steel bar hardened to 55+2 HRc.

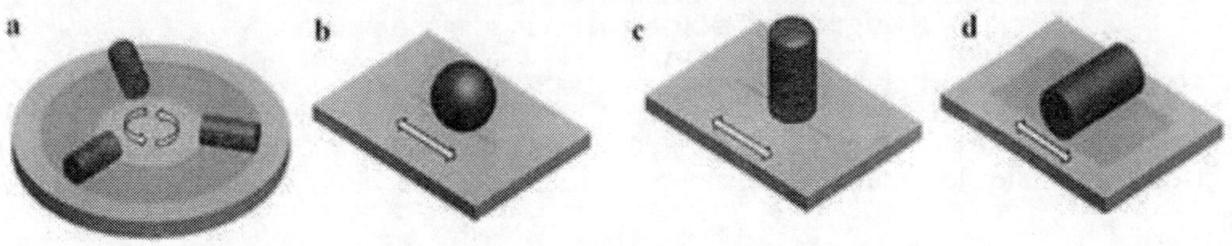

Figure 1. Most widely used tribo-testing geometric configurations: a. pin-on-disk, b. ball-on-plate, c. pin-on-plate, d. pin on plane (horizontal) [4]

The third method consists in specially designed tribometers that corresponds to the open tribosystems (newly refreshed surface) using a pin placed right after the cutting zone during machining. These kind of tribometers are used by many authors like [5] where he studied the friction between stainless steel and cemented carbide pin, [6] (fig. 2.b) where the authors developed a tribometer designed for frictional analysis at the tool/chip/workpiece interfaces based on a plane-sphere contact configuration between 27MnCr5 annealed steel and a carbide cutting tool under dry cutting conditions. There are many more authors like [7] (fig 2.c),[8] (fig. 2.e), that used specially designed open tribometers.

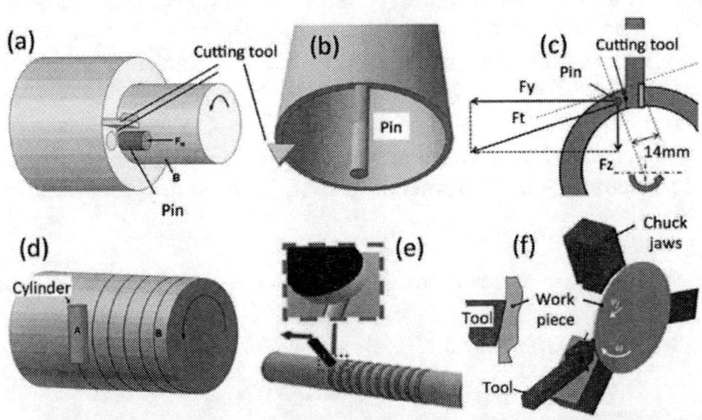

Figure 2. Tribometers for determining friction in cutting [1]

Other studies represent modified pin-on-ring systems, where the pin is in contact with the cylinder during a helical movement [9] (fig. 2.d). In the paper [10]. (fig. 2.f), the authors used a adapted type of tribosystem that simulates a large negative rake angle. The friction coefficient was determined for the couple of materials C45E+N (AISI1045) and uncoated cemented carbide insert.

3. Concept and theory of a new friction device

A new friction testing tribometer was designed to determine and analyze the friction coefficient simulating the normal cutting conditions. The tribometer and its components are shown in figure 3.a.

IOP Conf. Series: Materials Science and Engineering **568** (2019) 012100 doi:10.1088/1757-899X/568/1/012100

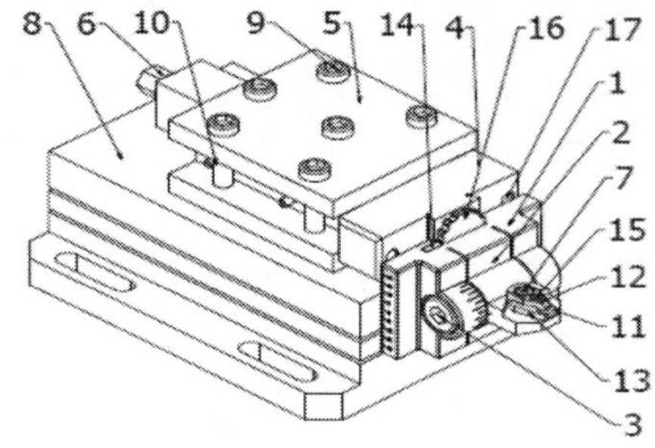

a) Device for friction coefficient determination

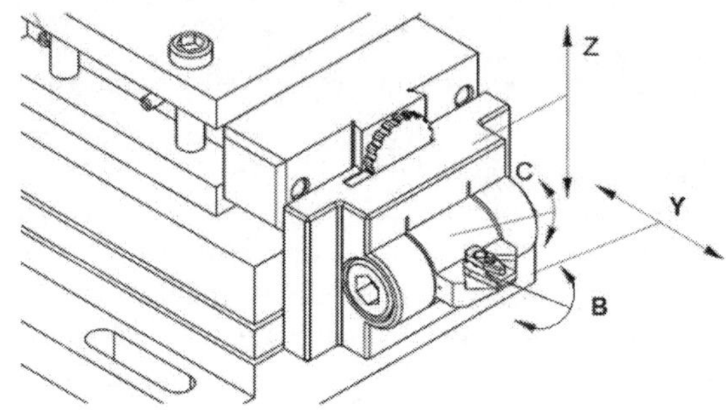

b) degrees of freedom of the device

Figure 3. The tribometer and its components

1. Rack body; 2. Insert head; 3. Lock bolt; 4. Tool body; 5. Dynamometer support; 6. Pinion cylinder; 7. Fixing screw; 8. Dynamometer; 9. Dynamometer support fixing screw; 10. Tool body locking screw; 11. Insert; 12. Bridle; 13. Intermediate support; 14. Rack; 15. Locking screw; 16. Pinion; 17. Locking screw.

The tribometer can simulate the rake and flank face and varying the cutting edge, rake and clearance angles. To measure the friction between the rake or clearance surface with the designed device, the device is fastened with the dynamometer support 5 on the Kistler dynamometer, who also is mounted on the SN 320x750 transverse slide

Figure 3.b. represents all the possible movements of the devices elements.

To adjust the device for a certain position (tool geometry), proceed as follows:

o Vertical position (OZ) – Moving the rack body 1 by rotating the pinion cylinder 6 which contacts the pinion 16 and the rack 14, the rotation being made through the end of the pinion cylinder by means of a standard wrench with the opening S = 12. Locking into the desired position is accomplished by tightening the locking screws 17 mounted in the standardized pockets of the tool body 4.

o Position in the OXY plane –Rotation over OX axis – axis B – The movement of the insert 11 on the axis of rotation B is done by means of the intermediate support 13. The insert is fixed to the support by means of the bridle 12, which in turn is blocked by the locking screw 15 on the intermediate support 13.

o Translation along the OY axis – The translation on OY axis didn't offer any possibility of movement because the length of the dynamometer. However, there is the possibility of moving on the OY axis of the whole assembly – dynamometer + device, by moving the transverse slide.

4. Experimental setup

In order to validate the device, a series of tests were carried out. Figure 4 represents the experimental chain as follows: 1. The device described at section 3; 2. Piezoelectric dynamometer Kistler, 9257B for turning operations ; 3. Kistler Amplifier 5070.; 4. Lathe SN320x750 .

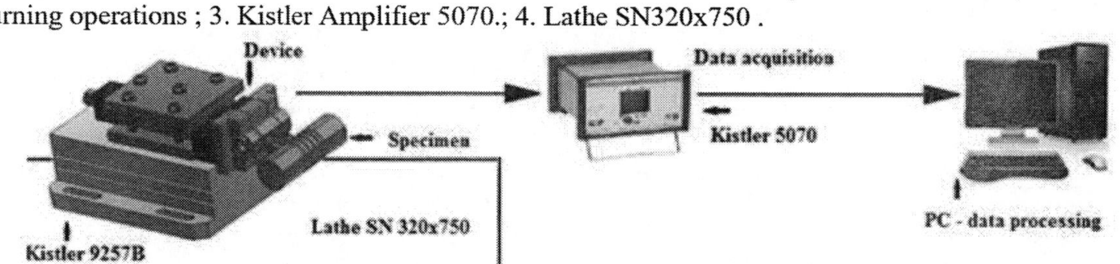

Figure 4. Measuring chain of the experimental setup

The device is mount on the Kistler dynamometer in order to determine the friction and normal forces during the process. The testing material is made of C45E and it has four rings with different widths as showed in figure 5.a.

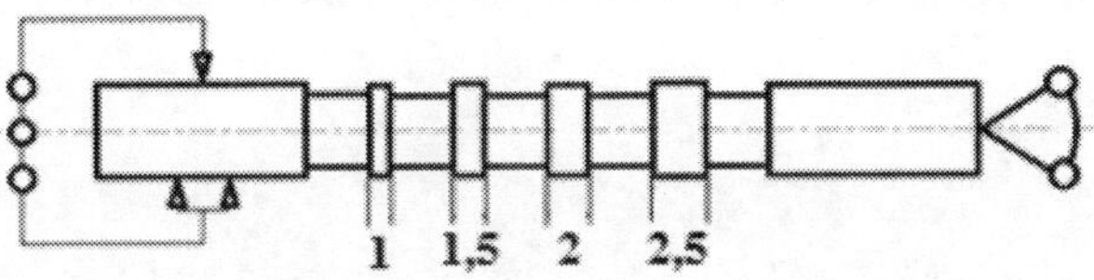

a) Specimen made for friction tests

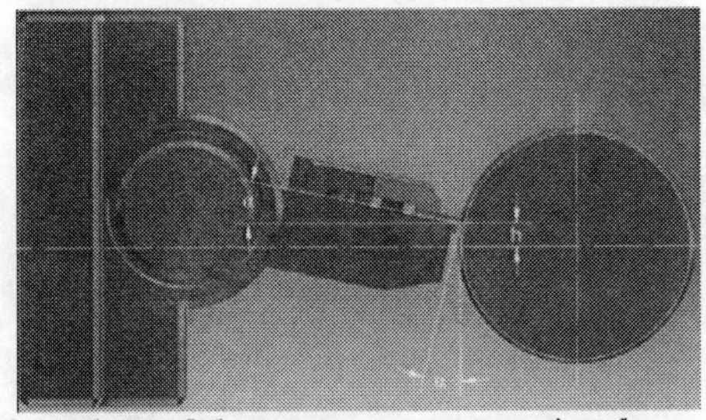

b) elevation of the contact zone to simulate clearance angle

Figure 5 The testing material

The tool insert used in the machining tests that were carried out was a P20 grade carbon steel. The device is set to have the clearance angle of 10° and the rake, inclination and attack angle 0°. In order to simulate orthogonal machining and the clearance angle, the contact of the insert with the workpiece is set at a given distance (fig. 5.b.) from the axes as presented in equation (1).

$$h = R \cdot tg(\alpha) \tag{1}$$

where R = radius of the workpiece and α = clearance angle

The depth of cut (pressure) was kept constant during all determinations. In this setup, the tool nose radius is absent or neglected. For all the determinations the sliding velocity is v_c = 20 m/min.

The friction coefficient is obtained using the well known Coulomb's friction law, equation (2).

$$\mu = \frac{F}{N} \tag{2}$$

where F is friction force and N is normal force.

5. Experimental results

The experiments were carried out on four different surface widths that simulates the depths of cut during normal machining. Figure 6 illustrates the experimental data obtain during the tests.

In the beginning, when the first contact is made, the plastic deformation of the material can be seen.

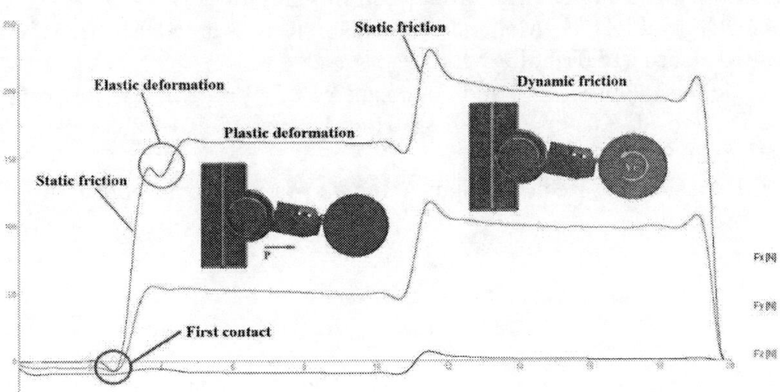

Figure 6. Graphic analysis of the simulation process

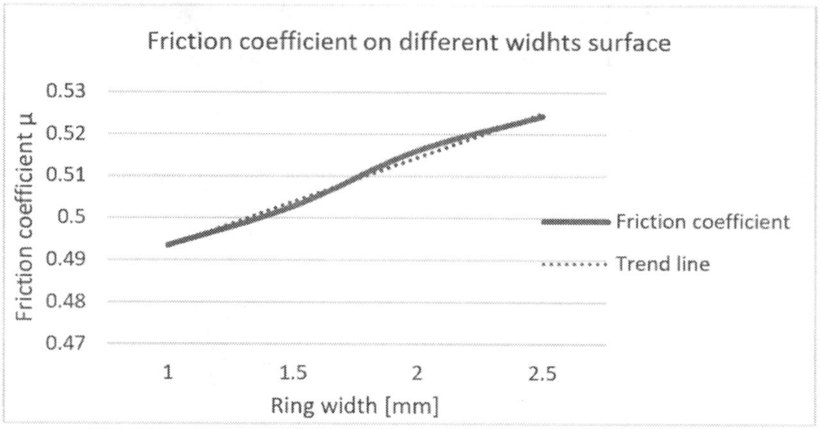

Figure 7. Variation of friction coefficient with contact length

After the contact, when the actual machining simulation occurs and the workpiece starts to rotate, we can measure the static and kinematic friction. Also, during the machining simulation, because the surface is not refreshed, the effect of peening appears where the workpiece surface smoothens. Even though the peening effect occurs, after several measurements in different parts of the graph, the friction coefficient seems to remain constant. Also, in figure 7, the obtained results show us the variation of friction coefficient with contact length. As seen in the graph, the friction coefficient increases with the increased width of the surfaces which states a true affirmation that the coefficient of friction increases with the increasing depth of cut in normal machining.

6. Conclusions and future work

A new friction device was developed that can measure the friction coefficient in the contact zone of the workpiece with the flank surface. The device is designed to be able to continue future works with different kinds of geometries, even with the implementation of cutting edge angle. The experimental results obtained after the tests validates that the device can be used for future studies.

References

[1] Claudin C, Mondelin A, Rech J and Fromentin G 2010 International Journal of Machine Tools and Manufacture **50** 681–688
[2] Melkote S, Grzesik W, Outeiro J, Rech J, Schulze V, Attia H, Arrazola J P, M'Saoubi R and Saldana C 2017 CIRP Annals - Manufacturing Technology **66** 731–754
[3] Hedenquist P and Olsson M 1991 Tribology International **23** 143–150
[4] Milutinović M and Tanović L 2016 Technical Gazette **23** 1373-1379
[5] Naisson P, Rech J and Paris H 2013 Engineering Transactions **61** 239–248
[6] Puls H, Klocke F and Lung D 2012 Production Engineering **6** 349-354
[7] Rehan Zahid, Masjuki H H, Mahendra Varman, Riaz Ahmad Mufti, Abdul Kalam Md and Mubashir Gulzar 2015 Tribology Letters **58** 1-28
[8] Schulze V, Michna J, Schneider J and Gumhsch P 2011 Procedia Engineering **19** 331–336
[9] Smolenicki D, Boos J, Kuster F, Roefolds H and Wyen C 2014 CIRP Annals - Manufacturing Technology **60** 97–100
[10] Zemzemi F, Bensalem W, Rech J, Dogui A and Kapsa P 2008 Tribotest **14** 11–25

IOP Conf. Series: Materials Science and Engineering **568** (2019) 012101 doi:10.1088/1757-899X/568/1/012101

Research on designing a manufacturing cell for the Disk Limiter

C I Indre[1,2], F S Blaga[1] and C Miculaş[2]

[1] University of Oradea, Faculty of Managerial and Technological Engineering, Universităţii Street No. 1, Oradea, Romania
[2] SC.FaistMekatronic.SRL, Nicolae Filipescu Street, No. 2, Oradea, Romania

indreclaudiuioan@yahoo.com

Abstract. The paper presents the results of the experimental researches carried out in order to increase the productivity of the processing process for part of the automotive industry. Thus, by increasing the volume of parts ordered by the recipient, research has been done to make a manufacturing cell having the ability to obtain a competitive unitary time. Manufacturing cells that are equipped with one or more industrial robots, machine tools with numerical control and auxiliary components (conveyors, equipment, etc.) have a rather old concept that has evolved and has reached a high level today. The machines integration in a flexible manufacturing cell is the way of making profitable the use of machines with numeric control. If they exist alone, the human users, that works on the cell, would be forced to make manually the download of machining programs, and to take decisions themselves, in real time. These decisions are related with the complex resources management, as machining time, machining tools and raw material.

1. Introduction

Manufacturing cells have been one of the main solutions to several problems caused by change in the world context in manufacturing systems [1]. Some of these changes are: reduction in lot sizes, increase in the product variety, improvement of quality, cost reduction, market competition, flexibility, relationship with customers (for example, consumer's satisfaction), etc. Manufacturing systems need a quick answer to the changing market, due to the variety that the customers bring to the market. They can look for other alternatives in product, quality, time and cost [2].
This paper presents the adoption of flexible manufacturing systems in industry. In recent years, advances in machining and new CNC tools has given rise to a stiff competition in the manufacturing industries. The economic and social justification of Cellular Manufacturing System (CMS) lies on the cost effectiveness over years.

In all situations, where traditional machines are used for operations on parts, concurrent formation of part-families and machine-cells is necessary [3]. This cell is a combination of microelectronics and mechanical engineering in manufacturing process that adapt automatically to random changes in every possible aspect. It combines the flexibility of job shop with the high-volume production of fixed automation. [4] In this paper a cost-volume-profit analysis is done to justify the initial cost of CMS. Flexibility in product design and manufacturing are necessary in today's ever-changing socio-economic society [5].This research paper proposes replacing a current manufacturing process served by an operator with a manufacturing cell serviced by the robot. Most of the researchers worked on major objectives as minimization of intercellular movements, machine/ load utilization and numerous cost considerations. There is need to consider research issues discussed in this paper to take complete benefit of CMS [6].

2. Current state of the manufacturing process

2.1 Presentation of the product

In various specialized works we define that the flexible manufacturing system is is a complete system that takes care of the process flexibility, product flexibility, production flexibility, machine flexibility, operation flexibility, volume flexibility, routing flexibility etc. In this paper I present an analysis and a method of designing a flexible manufacturing cell. The design analysis of the future manufacturing cell started from the existing machining process of one of the automotive parts.

Figure 1 shows the part after the pressure casting process. The particularity of this product lies in the fact that it is an ensemble of three elements. This product has as its base a pressure cast aluminium alloy piece, and two metal inserts, figure 2. Figure 3 shows the part after the CNC machining process.

a) b)

Figure 1. Disk Limiter–after the pressure casting process

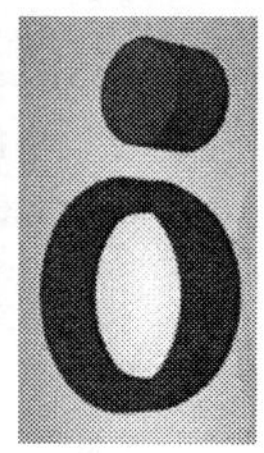

Figure 2. Metal inserts

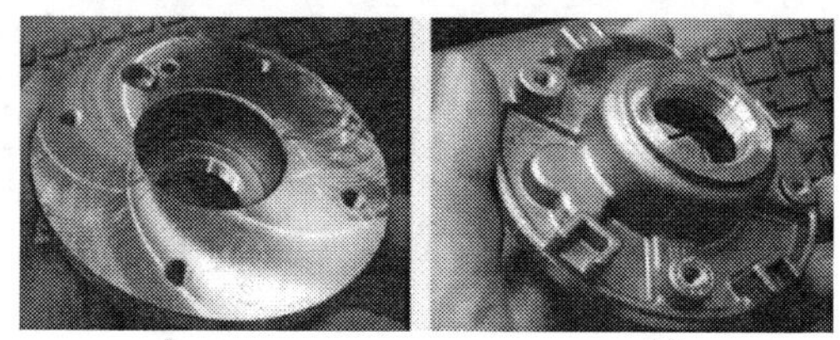

a) b)

Figure 3. Disk Limiter–after the CNC machining process

2.2 The current layout of the manufacturing flow in the machining area

Taking into account the requirements imposed in the execution drawings and the overall, for the manufacturing sequence for the CNC processing department, this product is made in three manufacturing stages, as follows: Operation 10-OP10 (CNC machining), Mechanical assembly two metal inserts) and Operation 20-OP20 (CNC machining).

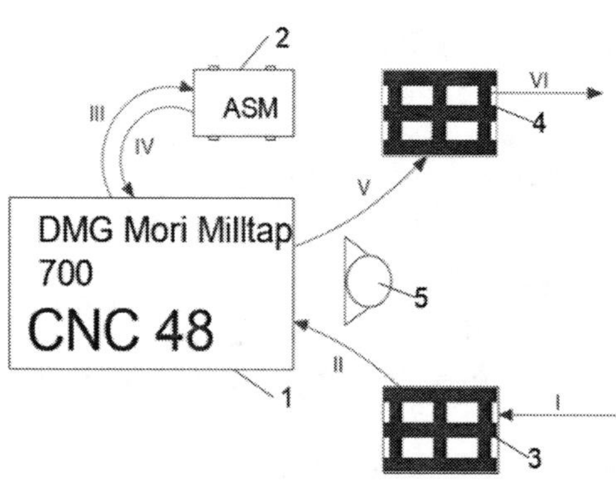

Figure 4. Layout Manufacturing Cell CNC Machining Area and CNC Assembly

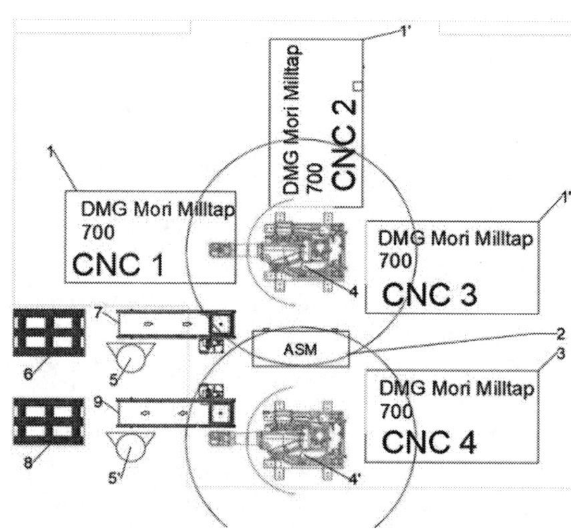

Figure 5. Layout manufacturing cell

In the present stage of machining, OP10 and OP20 are simultaneously processed on a vertical machining center, DMC MillTap700, from the company. The positioning, orientation and fixing of the workpiece is performed on a mechanically operated device with two positions (a position for OP10 and a position for OP20). This assembly of metal inserts is performed on a mechanically assembled mechanical device assembled manually by an operator.

The production sequences presented above are performed in the CNC processing department of SC FaistMekatronic SRL under the following structure, as shown in figure 6. For the execution of the product in the current manufacturing layout in the CNC department, we need technological resources that compete with the finished product execution.

2.3 Results of the current state of manufacture. Classic version.

The results of the current manufacturing stage are reflected in unit execution time of 135 sec /piece. This unit manufacturing time is obtained directly by the machining center on each operation on each cutting tool. The results of the mechanical processing are shown in table 1.

Table 1. Machining operations from the technological machining itinerary

OP	Operation	Tool		Machining parameter: Spindle (rot/min) /Feed (mm/min)	Machining time/tool (s)
10.1	Roughing the central zone	T101	Mill_D20	8000/3000	5
10.2	Rear piece milling	T7355	Mill_T	2500/500	9
10.3	Thread Milling	T2410	Thread-mill	7000/1000	11
10.4	Chamfer back- thread	T1645	Dovetail cutter	7000/1000	8
10.5	Roughing the hole d11.287	T634	Mill_D8	8000/1500	6
10.6	Reaming d11.287	T7329	Reamer_D11.29	1000/2000	4
10.7	Chamfer hole d11.287	T7056	Chamfer_D10	8000/1500	3
10.8	Roughing the hole d4.98	T480	Drill_4.8	8000/2500	3
10.9	Reaming D4.98	T7349	Reamer_D4.98	2000/1000	3
20.1	Drilling D5.5	T526	Drill_D5.5	10000/3500	6
20.2	Thread M6	T7760	Thread_M6	3000/3000	10
20.3	External milling D94.8	T101	Mill_D20	9000/2000	22
20.4	Milling of metal inserts	T7338	Mill_D8	4400/560	15
20.5	Face milling	T1140	Mill_D40	12000/4000	15
20.6	Marking part	T6063	Chamfer_D6	6000/1500	15
Machining time (s): 135= Time OP10 (s): 52+ Total time /OP20 (s): 82					

It is worth mentioning that the manual assembly process of the two metal inserts is 15 seconds, it is performed in parallel with the mechanical processing and does not influence the unitary manufacturing time according to the product operating standard in the fabrication classic

3. Researches on the design of the flexible manufacturing cell

Considering a significant increase in customer demand in terms of several parts, it is a question of identifying solutions for organizing the manufacturing process so that the production volume is about 1000000 pieces/year. To determine the actual time of manufacture of a piece under the possible constraints (restrictions) that may occur, a calculation of the processing time was made. The results of the calculation are: 134 parts / hour- 1 and 26.89 seconds/ part.

The processing variant, by equipping the classic multi-function device, solves to some extent the manufacturing time but does not allow the expected volume of products (1 million parts per year). To achieve this, research has been carried out to improve processing technology by modifying the cutting regimes and the use of new tools. Another step in optimizing the production of this piece was the new technologization of the manufacturing section for the CNC department.

In the scientific work [9] is presented a variant of improvement of the production sequence in the CNC department machining the piece with the help of special tools combined. The scientific paper presented the advantages of using combined tools. After the new technological itinerary, which also includes the new combined tools, the CNC machining time for the first OP10 operation is 67 seconds, and for the OP20 operation is 21 seconds.

To make these parts, it is necessary to provide hydraulically actuated guiding and fastening devices.

The design of the manufacturing cell was done starting from the machining technology specified in the above rows. From a relatively simple mathematical calculation, a manufacturing cell has been designed to meet customer requirements, figure 5.

After the improvements made to meet the volume of parts necessary for customer satisfaction, it was necessary to simulate and manufacture a new manufacturing cell that will include:

- Three CNC machines to carry out operation 10 (OP10) of three workstations;
- A machine with three automatic assembling posts for metal inserts;
- A CNC machine for operation 20 (OP20) with three workstations.

The technological itinerary for this manufacturing cell works on the same concept as the classic one presented above. In the case of the manufacturing cell, the machines are serviced by two industrial robots synchronized with the machining centers and the automatic assembly station.

4. Design and simulation of the robotic cell using the program - ROBOT STUDIO

For the design and simulation of the manufacturing cell, we used the ABB Robot Studio program. Robot Studio provides the tools to increase the profitability of your robot system by letting you perform tasks such as training, programming, and optimization without disturbing production. This provides numerous benefits including Risk reduction; Quicker start-up; Shorter change-over; Increased productivity.

Having the structure of the manufacturing cell and knowing the succession of its specific sequences, it is a question of studying the function of the cell as a whole. More specifically, the reduction of dead times, auxiliary times of time intervals due to defects and maintenance activities will be considered.

Starting from the proposed layout of the proposed manufacturing cell, a manufacturing cell assembly was made in modeling and simulation software. For the design and simulation of the robotic manufacturing cell, we used Robot Studio Robotic Cell Simulation Software.

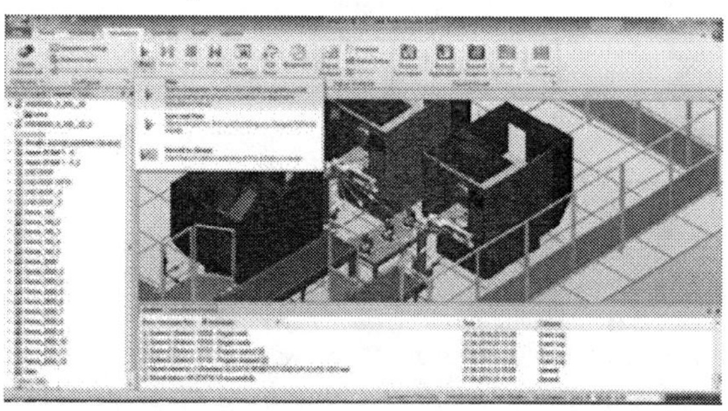

Figure 6. ABB Robot Studio - Synchronization of the manufacturing cell

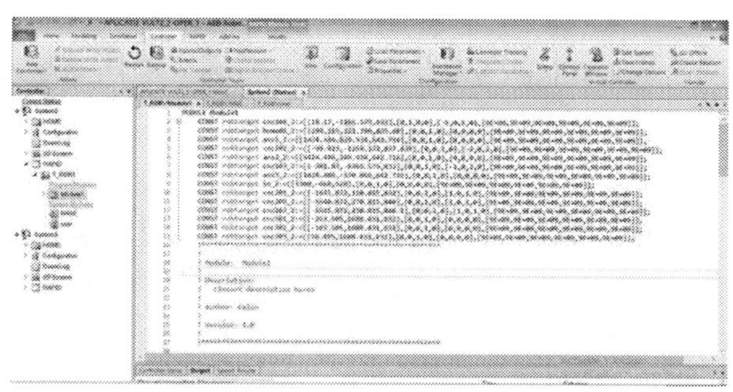

Figure 7. ABB Robot Studio - The manufacturing program of the manufacturing cell

An essential action in simulation is the synchronization of the created cell with the robot's virtual controller. An important step is to select the Simulate Run button from the ABB Robot Studio software menu, used to start cell simulation, where the robots will follow the paths, depending on the trajectories and the operating mode. This is shown in figure 6. The last sequence is the production of the manufacturing cell program. The result of the simulation is the generation of the manufacturing cell program shown in figure 7.

IOP Conf. Series: Materials Science and Engineering **568** (2019) 012101 doi:10.1088/1757-899X/568/1/012101

Offline programming is the best way to maximize return on investment for robot systems. ABB's simulation and offline programming software, Robot Studio, allows robot programming to be done on a PC in the office without shutting down production [10].

5. Conclusions

Manufacturing cells that are equipped with one or more industrial robots, machine tools with numerical control and auxiliary components (conveyors, equipment, etc.) have a rather old concept that has evolved and has reached a high-level today.

The manufacturing sector has become increasingly competitive as markets become more globalized. Consequently, there have been significant shifts in the design of manufacturing systems using innovative concepts. The adoption of cellular manufacturing (CM) has received considerable interest from both practitioners and academicians that offers several significant advantages, including a reduction in lead times and work-in-process inventories, and reduction of setup times due to the similarity of part types produced. Reorganizing the cell layout to meet the changing needs, however, may be time-consuming and costly. Further, if these changes occur very frequently, reconfiguration becomes impracticable or even infeasible. In such an environment, it appears that manufacturers tend to adopt a traditional job shop layout combined with the benefits of cellular manufacturing systems.

The machines integration in a flexible manufacturing cell is the way of making profitable the use of machines with numeric control. If they exist alone, the human users, that works on the cell, would be forced to make manually the download of machining programs, and to take decisions themselves, in real time. These decisions are related to sophisticated resources management, as machining time, machining tools, and raw material.

6. References

[1] Antonio B and Shahrukh A, 1996, Design of manufacturing cells: strategy, software and performance measurement, *Gest. Prod.* vol.3 no.3 São Carlos

[2] Arvindh B, 1994, Cell formation: the need for an integrated solution of the subproblems, *Int. of Prod. Re.* **32** **(5)** p. 1197

[3] Burgess A, Morgan, I and Vollmann T, 1993, Cellular manufacturing: its impact on the total factory, *Int. Journ. of App. Eng. Res.* **31** **(9)** p 2059

[4] Shivprakash B, Prakash M and Preeti P, 2011, Research Issues in Cellular Manufacturing Systems, *Int. Journ. of App. Eng. Res.* **6**, **(3)** p 291

[5] Chalapathi P, 2012, Complete Design of Cellular Manufacturing Systems, *Int. Journ. of Adv. Eng. Tech.* , K.L.University, Vijayawada, AP, India

[6] Jain KC, 2004, *Principles of Automation and advanced manufacturing systems, first edition*, Khanna Publishers

[7] Okpala A, 2018, Application of Break-Even Analysis In Industrial Development through Flexible Manufacturing Systems, *Proc. of the World Cong on Eng* Vol II, London, U.K.

[8] Costa A and Garetti M,1985, Design of a control system for a flexible manufacturing cell, J. of Man. Sys. **4** **(1)**, pp 65-84

[9] Indre CI, Blaga FS, Miculas C, and Negrau DC, 2019, Research to decrease the time manufacturing unit using the combined tools, *Innovative Manufacturing Engineering & Energy International Conference, The 23rd Edition*, Pitesti, Romania

[10] https://new.abb.com (accessed in 02.04.2019)

Acknowledgments

We, the authors of this scientific paper, wish to express our great appreciation to colleagues of the technical department at SC.FaistMekatronic.SRL Oradea for valuable and constructive suggestions in the planning and development of this research work.

IOP Conf. Series: Materials Science and Engineering **568** (2019) 012102 doi:10.1088/1757-899X/568/1/012102

Application of new data formats for electronic document management in government bodies

V Pysarenko[1], L Dorohan- Pysarenko[1], N Kantsedal[1]

[1]Poltava State Agrarian Academy (PSAA), Poltava, Ukraine

vyacheslav.pusarenko@pdaa.edu.ua

Abstract. Increasing the volume of information around the world and the rapid development of the latest information technologies led to the emergence of new opportunities for their use in the life of society, in the work of the authorities and local self-government. The introduction of computer technology for the processing, transferring, storage and use of information have led to the creation of documents on fundamentally new media, which caused the emergence of such concepts as an electronic document or document in electronic form. Analyzing the current state of the organization of processes of electronic document circulation in the activities of the public administration bodies of Ukraine as a component of e-government, it should be noted that there is a need for further study of the problem issues of the implementation of electronic document circulation, namely the order of application of electronic documents and the composition of their identity and legal registration, in order to provide legal force electronic documents and make it possible to use them on a par with paper documents. The development is devoted to the possibilities of using a new type of biometric digital signature in electronic documents of public administration bodies. The advantages and disadvantages of using this type of digital signature are analyzed. The directions of introduction of alternative means of identification of the person with the signing of electronic documents are indicated.

1. Introduction

Management acts as one of the most complex and most responsible areas of practical intellectual activity of people. It is always carried out to achieve specific goals and objectives, and their implementation is through the adoption and implementation of many managerial decisions that must be fully justified, timely, with the necessary completeness of the content and consistent with previously adopted decisions.

Ukraine's preparation for joining the EU requires changes to the current legislation and the development of new legal acts on the implementation of electronic document management.

In clause 9 of the International treaty of the Directive of 13 December 1999 No. 1999/93 / EC on the legal framework for the regulation of electronic signatures within the Community, it was stated that "... electronic signatures are planned to use in a wide variety of circumstances and applications that should expand the scope of the use of new services that are related to electronic signature, or use them. The definition of such goods and services should not be limited to the issuance and management of certificates. It should also include other goods or services that use or supplement the electronic signature, such as registration services, referral services, computer or consulting services related to electronic signature"[1].

The EU position on the legal regulation of electronic document circulation documents includes the achievement of some results - it is mainly the development of similar laws in the independent states that differ significantly in legal systems and traditions. The purpose of the European Community is not to

IOP Conf. Series: Materials Science and Engineering **568** (2019) 012102 doi:10.1088/1757-899X/568/1/012102

create the same legislation, but to promote the development of investment activity, trade, so, despite the presence of its harmonization, it is only one of the means of achieving the main goal.

The advantage of electronic document management systems for working with documents is among other things the ability to protect documents from unauthorized access by using an electronic digital signature.

Another issue for electronic document management and electronic document circulation is the need to secure the legal validity of electronic documents.

Due to the fact that the Law of Ukraine "On Electronic Digital Signature" [2], which regulates the use of electronic digital signature, was finally adopted, it became possible to provide electronic documents with legal status.

This Law defines the legal status of an electronic digital signature and regulates the relations arising from the use of an electronic digital signature, but the effect of this Law does not extend to relations arising from the use of other types of electronic signature, including the digitized self-handwritten signature image, or papillary fingerprint.

According to the Law an electronic digital signature is a props of an electronic document intended to protect it from forgery, a cryptographic tool that makes it possible to verify the absence of distortions in the text of an electronic document, and, where appropriate, to identify the person who created such a signature.

The digital signature infrastructure provides:

- the possibility of identification of the person who expressed his will through the preparation of an electronic document (the authorship of the document) or confirms the fact of familiarization with the document (visa on the document, notary's stamp, mark of inspection, etc.);
- the possibility of ensuring the confidentiality of the workflow.

An electronic digital signature in an electronic document is equivalent to a self-signed signature in a document on a paper carrier as confirmed by the current legislation.

An electronic digital signature can be obtained as a result of a cryptographic transformation of information using a private key. Of course, the complexity of the mathematical apparatus of two-key cryptography, through which the mechanism of electronic digital signature is implemented, is large.

The electronic document management system of the executive body must comply with the requirements of Clause 3.7 of the Technical Specifications (TU U 30.0-33240054-001: 2005) [Electronic document management executive authority. Specifications. TU 30.0-33240054-001: 2005], on the complex of means of protection of information from unauthorized access [3].

The key features of a digital signature are the ability to provide legal force by e-mail. The digital signature itself provides the status of an electronic document to files that contain informational content. In addition, digital signature allows you to transmit electronic documents in the same form and bring to court their authorship.

For the proper functioning of the digital signature, it is necessary to create a serious infrastructure that has received recognition in the world called "public key infrastructure" or PKI (public key infrastructure) - in the international terminology. All developed countries of the world, including Ukraine, form similar national infrastructures.

The creation of a fully-fledged digital signature infrastructure today requires the introduction of amendments to the digital signature law, which establish:

- the duty of state authorities, state organizations, their officials to accompany documents digitally and to admit them to circulation, to accept documents from citizens and their associations in electronic form, accompanied by a digital signature;
- requirements to certification centers that carry out the certification of public key digital signature, which are used to ensure the legal significance and / or confidentiality of document flow between the parties, if at least one of these parties is a state authority;
- the duty of the state authorities to certify the public key of a digital signature for use in the document circulation between this body and a citizen or organization at the request of that citizen or organization;

IOP Conf. Series: Materials Science and Engineering **568** (2019) 012102 doi:10.1088/1757-899X/568/1/012102

- the duty of public authorities to disclose and officially publish their public keys of digital signature;
- the responsibility of the certifying centers of the public keys of the digital signature for mistakes in the certificate of keys, as well as state authorities and their officials for compromising their own private keys of digital signature;
- standards that ensure the compatibility of the digital signature keys and standards that provide a specified level of stability of the keys of the digital signature;
- requirements for the development and implementation of alternative standards based on accepted algorithms and data formats in foreign countries and ensure compatibility with the existing international digital signature infrastructure;
- requirements for the development of open-source software that implements standards of digital signature in a standard manner.

In accordance with the Law of Ukraine "On Electronic Digital Signature" [2], in the absence of the possibility of overlaying and verifying an electronic digital signature with the help of an enhanced certificate (due to the lack of accredited certification centers that have the right to issue such certificates), to provide legal status to electronic document signed by electronic digital signature, the participant of the electronic circulation may act within the framework of an agreement on the recognition of an electronic digital signature, the validity of which is certified without the use of an enhanced certificate. Since the number of participants in electronic document management in public administration is large enough and the conclusion of agreements between the authority and each of them is a rather complex and inconvenient process, it may be appropriate to provide contractual basis for electronic document circulation on the basis of the accession agreement. Such a contract is made by one party - the initiator of the document circulation (coordinator), the rest of those who wish to join the system join it by providing the coordinator of the completed form-card (the mechanism of signing the contract may be determined separately). After the conclusion of the agreement the parties act in accordance with it and signed by electronic digital signature the electronic documents have a legal status. Thus a certain corporate network of electronic circulation is organized.

To provide subscribers of the corporate network of electronic document circulation with electronic digital signature services, each subscriber uses the services of the Center of Certification of keys. For this the subscriber concludes an agreement on the provision of electronic digital signature services in the form of an accession agreement with the Key Certification Center. At the same time, each subscriber (or his authorized person) must arrive at the Center of Certification of keys personally with documents confirming the information entered into the key certificate or transmitted by courier delivery (the order of certification and the list of documents are determined by the relevant documents of the Certification Center keys).

Thus for the implementation of secured electronic document circulation each subscriber must:

- Install on your computer a special software for generating electronic digital signature keys provided by the key certification center, and generate a pair of electronic digital signature keys and a couple of encryption keys. The personal digital signature key remains with the subscriber, and the public key in the form of the application for certification is provided to the Center of Certification of keys;
- to carry out the certification of its public key at the Center of Certification of keys, while providing all necessary documents confirming the information entered into the certificate of the key;
- install on your computer a special software for overlaying and verifying the electronic digital signature provided by the key certification center.

To sign and encrypt electronic documents, you need to use your private key, and to use for checking and decrypting documents received from other system participants, use the certificates of keys of these subscribers.

Each time, upon receipt of the signed and encrypted document, verify the validity of the certificate of the subscriber who signed it. To do this, you need to go to the particular resource of the Center of Certification of Keys - the table of valid certificates. If the signer's certificate is not valid at the time of the creation of the document - such a document is not valid.

IOP Conf. Series: Materials Science and Engineering **568** (2019) 012102 doi:10.1088/1757-899X/568/1/012102

In this methodology for the encryption and for decoding by the sender and the recipient the same key that they have agreed to use before the interaction is used. If the key has not been compromised, then the decryption automatically performs the identification of the sender, since only the sender has a key that can encrypt the information, and only the recipient has a key that can decrypt the information. Since the sender and the recipient are the only people who know this symmetric key, when the key is compromised, only the interaction of these two users will be compromised. The problem that will be relevant to other cryptosystems is the question of how to safely distribute symmetric (secret) keys.

2. Review of literature

Scientists interpret the concept of a digital signature - in different ways. A.V. Tkachev considers electronic digital signature as a legal category and believes that the use of such expressions in normative documents as an analog of a personal signature and equivalent to a personal signature of a person is undesirable, as it entangles the participants of the legal relationship and can create procedural difficulties in identifying persons, who used an electronic digital signature to certify computer documents [4].

The solution of the problem of the authorship of a paperless document can be achieved only with the use of an electronic digital signature, defined by L. A. Sysoevu as "... a tool that allows using cryptographic methods to reliably establish authorship and authenticity of a document" [5].

In the Law of Ukraine "On Electronic Digital Signature", which defines the legal status of an electronic digital signature and regulates the relations arising from the use of an electronic digital signature, the terms "electronic signature" are used as data in electronic form, which are added to other electronic data or logically from they are linked and intended to identify the signer of this data, and the "electronic digital signature" as the type of electronic signature obtained as a result of the cryptographic transformation of the electron collection s data included with this set, or logically combined with it and allows you to confirm its integrity and to identify signer. An electronic digital signature is superimposed using a private key and verified using the public key [2].

The reason for this is that an electronic digital signature allows you to set only the fact of its creation using a particular private key, as opposed to a personal signature that carries information about the distinctive signs of the author. The conclusion about the identity of the author of the document with the owner of the private key is based on the assumption that he is known only to its owner, but this provision can be refuted. Therefore, the idea of an electronic digital signature as an analog of a handwritten signature, according to many experts in the field of law, is based only on the similarity of the functions performed by these types of signature certificates.

The analysis of the scientific literature presented in this review revealed several unresolved issues regarding the effectiveness of the use of electronic digital signatures and their use when signing electronic documents in public administration.

3. The purpose and objectives of the study

The purpose of our research is to develop and use electronic digital signature documents, namely, the biometric digital signature, which continues to be of interest to scientists from different countries of the world and today. Despite the adoption of several legislative acts in this area, many problems remained unresolved and required additional regulation both at the legislative and regulatory-methodical levels.

4. Finding the best parameters for the process of using a biometric digital signature on electronic documents

The purposeful formation of the database of information resources necessary for the management of institutions and organizations has put the problem of document management, which is why modern document science pays great attention to studying the possibilities of advanced information technologies for their use in document management.

The basic concepts of document studies were borrowed from international standards and dictionaries, manuals issued by archivists from Australia, the United Kingdom, Canada, and the United States, as

IOP Conf. Series: Materials Science and Engineering **568** (2019) 012102 doi:10.1088/1757-899X/568/1/012102

well as from legislative acts regulating the organization of work with service documents in these countries and the theoretical works of international and national professional organizations.

A fundamentally new method of identity verification is offered at the signing of electronic documents based on biometric means of identification.

The advantages of biometric identifiers based on unique biological and physiological peculiarities of a person, which uniquely identify a person, have led to the intensive development of the appropriate means. Biometric identifiers use static methods based on the physiological characteristics of a person, that is, on the unique characteristics given to him from birth - drawing papillary lines of fingers. The general signs of patterns allow to differentiate the imprints by types and, in case of their differences, to conclude that there is no identity. To obtain the same definite conclusion about individual identity, in addition to the coincidence of standard features, it is necessary to coincide with a specific quantitative and qualitative set of distinct features, which include the beginning and end of the lines, the merging and splitting of lines, bridges, islets, hooks, points, specks, dashed lines, bends of lines etc.

The biometry allows uncontested identification of the person, and this information cannot be tampered or repaired.

Biometric identifiers provide very high rates: the probability of unauthorized access - 0.0001%, the identification time - a few seconds.

5. The strategy of solving the tasks of information security of electronic document circulation in the authorities

At the moment, reliable information security is one of the main criteria for selecting systems for storing and processing critical information. This is due to the probability of unauthorized access to such systems, as they have extensive information interaction with adjacent control systems through the Internet.

Therefore, providing information security should be the most crucial stage in their development.

Protection based on the biometric parameters of the human body, in particular, the fingerprint, has several indisputable advantages: ease of use, convenience, and reliability.

The analyzed characteristics of a person can not be lost, transmitted, forgotten, and extremely difficult to fake. They are practically not subject to wear and do not require replacement or restoration. The entire identification process takes a little time and does not require effort from those who use this access system. Research has also shown that the use of a fingerprint for personality identification is the most convenient of all biometric methods. The probability of error in identifying a user in this way is much smaller than in other biometric methods. Also, the fingerprint identification device does not require much space on the keyboard or in the mechanism. Nevertheless, it is necessary to carry out the corresponding work on the creation of a secure network for the transmission of biometric data and the infrastructure of informational interaction of state authorities among themselves, with local authorities, with consumers of management services.

6. Conclusion

As can be seen from the preceding, the method of identity verification when signing electronic documents based on biometric means of identification:

- more efficient - high reliability with use and low level of error;
- does not require - the creation of a severe infrastructure for the full functioning of the digital signature, the creation, and approval of standards that ensure the compatibility of digital signature keys and standards that provide a certain level of stability of the digital signature keys;
- easy to use - the lack of open and closed keys and the need for special software for generating EDS keys provided by the key certification center;
- convenience - there is no need for each of the subscribers to conclude from DSK an agreement on the provision of EDS services in the form of an agreement on the admission and directly to the CSCE personally documents confirming the information provided in the critical certificate and

access to the exclusive resource of the CSK each time upon receipt of the signed and encrypted document, in order to verify the validity of the certificate of the subscriber who signed it;
- cost-effectiveness - affordable software and fingerprint identification devices, no need to create and maintain a network of key certification centers.

In addition, it is necessary to clearly define at the legislative level what is a biometric digital signature, namely, a digital representation of the biometric characteristics of the person (fingerprint), a set of personal data collected on the basis of fixing its characteristics, which are of sufficient stability and substantially different from similar the parameters of other persons. To amend the Law of Ukraine "On the Uniform State Demographic Registry and Documents Affirming the Citizenship of Ukraine, Identifying a Person or Its Special Status" [6] regarding the possibility of its use.

This comparison shows a significant advantage of the method of identity verification when signing electronic documents using a biometric digital signature based on biometric identification means - a biometric digital signature.

References

[1] Directive 1999/93 / EC of the European Parliament and of the Council of 13 December 1999 on the system of electronic signatures within the Community.

[2] Law of Ukraine "On electronic digital signature" on May 22, 2003 № 852-IV. *Supreme Council of Ukraine*. 36 (2003), Art. 276.

[3] J. Chyrskyy Digital Signature: legal aspects of. *Handbook secretary and office manager*. 1 (2007), 26-31.

[4] L. Tkachev Pravovoy status of computer documents: Basic characteristics. Moscow, 2000, p. 8.

[5] L.L. Sisoeva Problems electron organization vyzyrovanyya documents in electronic document systems. *Deloproyzvodstvo*. 2 (1998), 47.

[6] The Law of Ukraine "On the Uniform State Demographic Registry and Documents Affirming the Citizenship of Ukraine, Identifying a Person or Its Special Status" of November 20, 2012 № 5492-VI. *Bulletin of the Supreme Council (BPD)*, 51 (2013),716

[7] M. Dutov, Legal problems of electronic document management. *Law of Ukraine*. – 6 (2002),122-124.

[8] Law of Ukraine on electronic documents and electronic document circulation. *Information from the Verkhovna Rada of Ukraine dated September 5*, 36 (2003), Article 275

[9] Shahverdov V.A. Digital signature in office automation and workflow systems / V.A. Shahverdov // *Records Management and Document Management at the Enterprise*. 8 (2003), 30-38.

[10] Electronic document management system of the executive authority Specifications TU U 30.0-33240054-001: 2005 *[Electron. resource]. Method: URL: <http://www.stc.gov.ua. - header. from the screen.*

IOP Conf. Series: Materials Science and Engineering **568** (2019) 012103 doi:10.1088/1757-899X/568/1/012103

The method of identity verification when signing electronic documents based on biometric means of identification

V Pysarenko[1], L Dorohan-Pysarenko[1], N Kantsedal[1]

[1]Poltava State Agrarian Academy (PSAA), Poltava, Ukraine

vyacheslav.pusarenko@pdaa.edu.ua,

Abstract. The development of global communications in business and everyday life has led to the emergence of a new plane of relationship, the subject of which is the electronic data exchange. In this exchange of data involved bodies of state power, local self-government, public administration, commercial and non-profit organizations, as well as citizens in their official and personal relationships. The problem of protecting electronic documents with the help of electronic-digital signature is extremely widespread in our time. As the era of the Internet and electronic technologies has come to an end, problems with the security of electronic document circulation, including those in public administration, are beginning to emerge. The issue of the preservation of electronic documents from copying and forging requires for its solution specific means and methods of protection. One of the means of such protection is a biometric digital signature, which undoubtedly confirms the authenticity of the information in the document, its details and the fact of signature by a particular person. The technology of electronic document circulation with the use of biometric digital signature in the public administration is not implemented today, which significantly reduces its quality and the development of electronic document circulation in general. The development is devoted to the possibilities of use and benefits of the introduction of electronic document circulation using the method of identity verification when signing electronic documents based on biometric identification tools.

1.Introduction

The essence of the problem and its solution. The purpose of this article is to solve the issues of organization of electronic document circulation in the authorities using the method of identity verification when signing electronic documents on the basis of biometric means of identification, namely, the biometric digital signature, the advantages of practical use of such a type of digital signature, and proposals for legislative and regulatory acts that determine the organizational principles for using a biometric digital signature in Ukraine.

Despite the existence in Ukraine of a certain legal and regulatory framework for electronic document management, there are many unresolved issues in improving the organization of electronic document management in government. Improvement and updating of the legal framework of Ukraine, creation of special legal norms will facilitate effective implementation and functioning of electronic document circulation.

Comprehensive use of electronic digital signature, namely biometric, will enable not only to simplify and speed up the circulation of documents between business entities, but also to create conditions for the development of export-import operations, e-commerce, electronic banking services, full-scale distance learning and provision of medical services from application of the latest information

IOP Conf. Series: Materials Science and Engineering **568** (2019) 012103 doi:10.1088/1757-899X/568/1/012103

technologies. In addition, the introduction of a biometric digital signature will establish an organizational and technical basis for the provision of electronic information services by public authorities, local governments, public administration, legal and physical entities with the use of the Internet. The reliable functioning of electronic document circulation will be a guarantee of Ukraine's information security.

This paper demonstrates the practical feasibility of organizing electronic document circulation in government using the method of identity verification when signing electronic documents on the basis of biometric of identification means - a biometric digital signature.

2. Review of literature

The management activity that is generally carried out and reproduced through and through the management of government records in government agencies is a dynamically functioning link in the process of ensuring interconnection and interoperability and the interconnection between the management and the managed systems.

The activation of the efforts of Western scholars at one time to develop scientific aspects of document management and coordination of scientists from different countries made it possible to recognize document management as a general management function, however, there was no complete agreement on understanding the management of documentation as a scientific category.

There was an understanding of the concept of managing documentation and information in their inextricable link, based on the combination of information technology capabilities and the benefits of traditional documentation.

In the scientific literature, the problematic aspects of the introduction of electronic document management systems are insufficient, which is due to the lack in Ukraine of the approved legislative framework for electronic documents functioning until 2004 in Ukraine. In the works of scientists, mainly only theoretical foundations of automated documentation are considered [1, 2, 3]. The adoption in 2003 of the laws of Ukraine "On electronic documents and electronic document circulation" [4] and "On electronic digital signature" [5] facilitated the introduction of integrated electronic document management systems. However, the theoretical and practical aspects of the implementation of such implementation need further development.

In order to regulate legal relations in the field of information technologies, the Verkhovna Rada of Ukraine adopted several Laws of Ukraine, which entered into force: "On mandatory copy of documents", which defines the legal principles of functioning of the system of mandatory copy of documents and regulates information relations related to replenishment national information fund of Ukraine [6]; "On the National Program of Informatization", which defines the general principles for the formation, implementation and adjustment of the National Informatization Program [7]; "On Telecommunications", which establishes the legal basis for activities in the field of telecommunications and defines the powers of the state regarding the management and regulation of the said activities, as well as the rights, responsibilities and principles of liability of individuals and legal entities involved in the activity or use telecommunication services [8]; "On the National System of Confidential Communication", which regulates social relations related to the creation, operation, development and use of the National Confidential Communication System [9]; "On Information Protection in Information and Telecommunication Systems", which regulates relations in the field of information security in information, telecommunication and information-telecommunication systems [10].

The Law of Ukraine "On the Uniform State Demographic Registry and Documents Affirming the Citizenship of Ukraine, Definition of a Person or Its Special Status," which defines the legal and organizational foundations for the establishment and functioning of the Uniform State Population Register and the issuance of identity documents, confirm the citizenship of Ukraine or a special the status of the person, as well as the rights and obligations of the persons to whom such documents have been issued [11].

Only in December 2017 the Cabinet of Ministers of Ukraine approved the resolution "On Approval of the Regulation on the National System of Biometric Verification and Identification of Citizens of

IOP Conf. Series: Materials Science and Engineering **568** (2019) 012103 doi:10.1088/1757-899X/568/1/012103

Ukraine, Foreigners and Stateless Persons" [12], which regulates the implementation and functioning of the national system of biometric verification and identification of citizens of Ukraine, foreigners and stateless persons, as well as defines its structure and purpose. The document states that this is an automated system created for the sake of national security, economic prosperity and human rights, which ensures the establishment of a person of a foreigner and stateless person who enter Ukraine, depart from Ukraine, exercise control over their observance of the rules of stay in the territory of our state.

The terms used in this Regulation are defined and have the following meanings:

- biometric data (parameters) - digitized fingerprints, digitized face image;
- data center - fault-tolerant integrated centralized system (software and hardware complex);
- workstation - a complex of hardware and software;
- fixing of biometric data (parameters) of a person - the process of collecting biometric data (parameters) of a citizen of Ukraine, a foreigner and stateless person (hereinafter - a person), making and storing in the departmental information systems of subjects of the national system;
- biometric identification - "one to many" search by recognizing and comparing one or two biometric data (parameters) of a person with biometric data (parameters) of persons in departmental information systems of subjects of the national system;
- biometric verification - search by principle.

3. The purpose and objectives of the study

The purpose of our study is to develop organizational measures for the implementation of electronic document circulation in public administration using a fundamentally new method of identity verification when signing electronic documents based on biometric identification tools. This issue will require further research by scientists from different countries, as existing legislation does not streamline the issue of electronic document circulation using biometric digital signatures.

4. Finding the best solutions for the implementation of electronic document and document management in public administration using a biometric digital signature

In the world, standards are becoming increasingly important. The main reason for this is the changes related to the globalization of the world financial market, the erosion of borders along the way of the movement of capital, goods, ideas and information. Scientific and technological progress, the rapid development of information technology and their active implementation - all this contributed to the process of development and implementation of international standards in all areas of human activity, in particular. in the management of documentation [17].

The use of electronic documents has been used in many areas of public administration, local government and public administration. The use of electronic document management systems can achieve enormous economic benefits, but the introduction of electronic document management systems can not be forgotten about security. One of the most important requirements for any electronic document management system is to ensure the security of electronic exchange of documents. Currently, systems of secure electronic document circulation are becoming increasingly popular. This is due to an increase in the number of confidential documents in public authorities, local self-government, other organizations of different forms of ownership and an active transition to electronic circulation of documents.

The legal protection of databases in the countries of the European Union is based on Directive 96/9 / EC dated March 11, 1996, which has received direct effect in the territory of the member states of the EU from 01.01.1998 [18]. In the European Union, databases are protected by a special legal institution - sui generis, that is, special protection if qualitative or quantitative investments are made in their selection, certification or submission.

The essence of special protection is that nobody can do:

- the transfer of a database or significant parts thereof to any other means of communication or the production of a copy (copy) in any form and at any time (paragraph A of paragraph 2 of Article 7 of the Directive);

IOP Conf. Series: Materials Science and Engineering **568** (2019) 012103 doi:10.1088/1757-899X/568/1/012103

- communication of the database to the public in any way or to enable the database to be used by various means (paragraph B, paragraph 2, Article 7 of the Directive);
- multiple systematic extraction or reuse of non-essential parts of the database and which compromise the intelligent interests of the compiler due to unforeseen use of the database, for example, with the help of the program (Clause 5, Clause 2 of Article 7 of the Directive).

In Ukraine, which is preparing for joining the EU, there is a need for regulatory documents regarding electronic document circulation and electronic digital signature, to bring the requirements of the European Union. However, the use and development of Ukrainian scientists is possible to ensure EU requirements. A new scheme of organization of electronic document circulation in the authorities is proposed using the method of identity verification when signing electronic documents based on biometric means of identification - a biometric digital signature.

5. Strategy for the development of organizational measures for the implementation of electronic document management in public administration with the use of a fundamentally new method of identity verification when signing electronic documents based on biometric means of identification

For the organization of electronic documentation in public authorities, local self-government using the method of identity verification when signing electronic documents based on biometric means of identification, it is necessary to prepare and issue joint orders of the chairman of the regional state administration and the regional council and, respectively, district state administrations and district councils for implementation in work institutions of electronic document circulation, using the method of identity verification when signing an e-mail Throne documents based on biometric identifiers, where:

- to indicate the specialists who will be obliged to include on the outgoing electronic documents, biometric data of the persons who signed the paper version of this electronic document and conduct identification of the biometric data of the persons who signed the incoming document (hereinafter the specialist);
- define one e-mail address in each institution from which the original electronic information will be sent and on which electronic documents will be received;
- oblige the relevant units to equip the automated workplace (APM) specialist with appropriate equipment and software.

5.1 Preparation for implementation

Provide an automated workplace for a specialist in the documentation department - a personal computer that needs to be password protected, a fingerprint scanner, such as ZK 6000-USB, connect the Internet and the e-mail of a Ukrainian provider.

Prepare a list of government officials with electronic addresses to which electronic documents will be sent, and their phone numbers for the contact.

Identify a list of positions and prepare lists of government officials who have the right to sign documents.

Install a fingerprint identification software on a personal computer, such as an automated dactyloscopic identification system, for example, "DACTO-2000".

Move the cursor over the scanner icon to the personal computer monitor and click on the left mouse button and prepare the scanner for work.

5.2 Creating a fingerprint database

Start preparing for the creation of a fingerprint database by filling out a fingerprint card specialist (hereinafter referred to as dactyloscopic card) on persons in the list, namely:

- registration number (identification number, card number);
- surnames;
- name;
- surname;
- positions;

IOP Conf. Series: Materials Science and Engineering **568** (2019) 012103 doi:10.1088/1757-899X/568/1/012103

- place of work;
- citizenship;
- date of birth;
- place of birth;
- information about registration at the place of residence.

The index finger of the left and then the right hand of the responsible person alternately placed on the scanner window. The reader scans your finger quickly and automatically. The fingerprint image is encoded before transferring to a computer via a usb port. For example, the ZK 6000 uses an optical fingerprint sensor for high quality and reliability. The fingerprint recognition algorithm has an unmatched ability to identify even the hardest fingers. The data stream received from the sensor is subjected to software processing to restore the fingerprint image and extract it from the necessary for further comparison with the information template. The restored image typically has a size of 25x14 mm, which is equivalent to the number of pixels of 500x280. The imprints of the index finger of the left and right hands are made in order to be able to use one of them in the case when the responsible person confuses his hands, or illness or other reasons will not allow the use of imprints from one hand.

The database will be completed when the dactyloscard is completed on all officials listed.

In order to ensure secrecy and due to the limited amount of memory, it is not desirable to store full fingerprints in the fingerprint recognition system. Of course, they can be stored in a safe place as a backup to access them in special cases, but for normal operation of this system, full-size image fingerprints are not required. In normal operation of the system from the image is extracted a unique set of data on the imprint. The extraction is performed using the pattern recognition process or using the principle of details (minutiae).

Authorities exchange their databases, creating in each institution a complete, common database, which is replenished with changes in staffing and the recruitment of new employees.

The database is stored on the server of the institution with the right of access only by the appropriate specialist.

5.3 Preparation and sending of outgoing electronic documents

An electronic document is prepared, the paper version of which is submitted and signed by the corresponding person in person.

In the absence of the head and signature on a paper document or the person who substitutes it, the paper document indicates his position and surname, and the electronic document supplements his electronic image of the papery patterns of fingers.

The signed paper document, together with the electronic document, is forwarded to the specialist for registration and dispatch.

The expert, having removed from the datacart of databases, the scanned fingerprint of the official who signed the paper document imposes it on the electronic document in the place where the signature is usually held and preserves it.

The electronic document is sent to the addressee by e-mail.

A paper document is stored in the institution in accordance with the requirements of normative documents regarding the terms of the storage of documents, and the electronic document is transferred to the electronic file of the control card of the document.

5.4 Procedure for processing received incoming electronic documents

Before sending an e-mail to the specialist e-mail address specified by the institution, an electronic document is copied to the file, and the specialist, using the automated fingerprint identification system, for example "DAKTO-2000", carries out the identification of the electronic image of the papillary fingerprints on the incoming electronic document with the database in order to identify the person who signed it.

If the print corresponds to the post and the name of the official who signed the document, it is printed on paper and handed over to the management for familiarization. Further work with a paper document and its electronic copy shall be carried out in accordance with the regulations and the document and control system in force in the institution.

In the event that the imprint does not correspond to the position or name of the official who signed the document, the electronic document is postponed to the "Unidentified" folder for detailed examination by the head of the documentation department and the identification of the reasons (failure of the system of identification or unreliability of the information).

Introduction of electronic document circulation using the method of identity verification when signing electronic documents based on biometric means of identification in comparison with electronic document circulation using electronic digital signature have significant advantages.

6. Conclusion

In the above studies, an analysis of the conditions and constraints that provide the necessary quality of electronic document circulation in public administration using the method of identity verification when signing electronic documents based on biometric means of identification, namely, a biometric digital signature, is carried out.

As can be seen from the foregoing, the method of identity verification when signing electronic documents based on biometric means of identification:

- more efficient due to high reliability at use and low level of error;
- does not require the creation of a serious infrastructure for the proper functioning of the digital signature, the creation and approval of standards that ensure the compatibility of the keys to digital signature and digital signature, and standards that provide a specified level of stability of the digital signature keys and digital signature;
- easy to use in the absence of open and closed keys and the need for special software for generating electronic digital signature keys provided by the key certification center;
- Convenience - there is no need for each of the subscribers to conclude an agreement on the provision of electronic digital signature services in the form of an agreement on subscription with the center of certification of keys, and to provide directly to the CSC personally the documents confirming the information included in the key certificate and access to the special resource of the key certification center each time receipt of a signed and encrypted document in order to verify the validity of the certificate of the subscriber who signed it;
- cost-effectiveness - affordable software and fingerprint identification devices, no need to create and maintain a network of key certification centers.

Consequently, the approval of the regulatory framework for electronic document circulation is a positive precondition for the use of primary media in electronic form. However, for the widespread introduction of these systems, further improvement of the normative and methodological base is required in accordance with the proposed recommendations.

Comprehensive use of electronic digital signature will enable not only to simplify and speed up the flow of business between business entities and to strengthen the competitiveness of domestic enterprises, but also to create conditions for the development of export-import operations, e-commerce, electronic banking services, full-scale distance learning and provision of medical services from application of the latest information technologies. In addition, the introduction of an electronic digital signature will establish an organizational and technical basis for the provision of electronic information services by public authorities and local self-government bodies to legal entities and individuals through the use of the Internet. The reliable functioning of electronic document circulation will be a guarantee of Ukraine's information security.

References

[1] A.N. Kurbatsky Automation of document processing. *Minsk,* (1999), 20.

[2] F.F. Butinets, S.V.Ivakhnenkov, T.V. Davidov, T.V.Shahraichuk Information Systems of Accounting: Textbook. *Ed. prof. FF Buttons; 2nd species, processing. and add – Zhytomyr.* (2002), 544.

[3] M. Bilukha Electronic Documents in Accounting. *Accounting and Audit.* 5 (2003).
[4] Law of Ukraine on electronic documents and electronic document circulation. *Information from the Verkhovna Rada of Ukraine dated September 5, 2003.* 36 (2003), Article 275.
[5] Law of Ukraine "On electronic digital signature" on May 22, 2003 № 852-IV. *Supreme Council of Ukraine.* 36 (2003). Art. 276.
[6] Law of Ukraine "On mandatory copy of documents" dated April 9, 1999 № 595 XIV. *Bulletin of the Verkhovna Rada of Ukraine dated June 11, 1999.* 22 (1999), Article 199.
[7] Law of Ukraine "On the National Program of Informatization" dated February 4, 1998 № 74/98-BP. *Bulletin of the Verkhovna Rada of Ukraine.* 27-28 (1998), 181.
[8] Law of Ukraine "On Telecommunications" dated November 18, 2003 № 1280-IV. *Bulletin of the Verkhovna Rada of Ukraine.* 12 (2004), 155.
[9] Law of Ukraine "On the National System of Confidential Communication" of January 10, 2002, № 2919-III. *Bulletin of the Verkhovna Rada of Ukraine.* 15 (2002), 103.
[10] Law of Ukraine "On Protection of Information in Information and Telecommunication Systems" dated July 5, 1994 № 80/94-VR. *Bulletin of the Verkhovna Rada of Ukraine.* 31 (1994), 286.
[11] The Law of Ukraine "On the Uniform State Demographic Registry and Documents Affirming the Citizenship of Ukraine, Identifying a Person or Its Special Status" of November 20, 2012 № 5492-VI. *Bulletin of the Supreme Council.* 51 (2013), 716.
[12] Resolution of the Cabinet of Ministers of Ukraine "On Approval of the Regulation on the National System of Biometric Verification and Identification of Citizens of Ukraine, Foreigners and Stateless Persons" dated December 27, 2017, № 1073. *Official Bulletin of Ukraine dated January 12, 2018.* 4 (2018), 152, article 178, code of the act 88680/2018.
[13] The Law of Ukraine "On Protection of Personal Data" of June 1, 2010 № 2297-VI. *Bulletin of the Verkhovna Rada of Ukraine.* 34 (2010), art. 481.
[14] Resolution of the Cabinet of Ministers of Ukraine "Some Issues of Documenting Management Activity" of January 17, 2018, № 55. *Official Bulletin of Ukraine dated 23.03.2018.* 23 (2018), 24, article 770, act code 89399/2018.
(Law of Ukraine "On electronic trust services" dated October 5, 2017, № 2155-VIII. *Bulletin of the Verkhovna Rada.* 45 (2017), 400.
[15] Resolution of the Cabinet of Ministers of Ukraine "On Approval of the Procedure for the Use of Electronic Trust Services in State Authorities, Local Self-Government Bodies, Enterprises, Institutions and Organizations of State Ownership" of September 19, 2018, № 749. *Official Bulletin of Ukraine dated October 5, 2018. 76 (2018), p. 23, article 2528, code of act 91594/2018.*
[16] Federal Law of 27.07.2006 № 149-ФЗ "On Information Based Information Technology on Information Protection". *Reference Legal System CONSULTAN.*
[17] A.A. Vaishnurs Present and prospects of legal protection of databases in Russia, the USA and the European Union. *Lawyer.* 12 (2003). 45.
[18] Electronic document management system of the executive authority Specifications TU U 30.0-33240054-001: 2005 *[Electron. resource]. - Method: URL: <http://www.stc.gov.ua>.- header. from the screen.*
[19] Directive 1999/93 / EC of the European Parliament and of the Council of 13 December 1999 on the system of electronic signatures within the Community.

IOP Conf. Series: Materials Science and Engineering **568** (2019) 012104 doi:10.1088/1757-899X/568/1/012104

Using "GUI" programming environments to calculate the level of service at intersections

I Dumitru[1], L D Racila[1], L Matei[1], D Tutunea[1], B S Florescu[1]

[1] University of Craiova, Faculty of Mechanics, Craiova, Romania

mateiclucian@yahoo.com

Abstract. Inside busy cities, road congestion is common due to the large number of vehicles. Modifying existing infrastructure only for certain periods of peak time or for specific events involves high costs, so it is desirable to use its full capacity. At the same time those in charge of traffic optimization level infrastructure and coordination are municipalities. Based on the above premises it can be seen that for a better optimization of traffic in a critical situation and require the use of unique software for calculating traffic light times, service level, etc. The paper aims at using software platforms in "GUI" applications to be used in the context of dynamic changes in road traffic.

1. Introduction

The level of service indicates the degree of use (load intersection, through the transit of vehicles and a degree of creating congestion) at an intersection, translated in time constant [1] [2].

These criteria of analysis are necessary for rapid examinations imposed a junction and offers some quick values that can identify in real time where the intersection and the problems studied.

The maximum service flow rate is defined as the maximum number of vehicles that can cross a uniform road section over a given time period (preferably 15 minutes) under certain road, traffic and control conditions to maintain a certain level of service projected [8].

They are highlighted as follows:

- Level A: characterizes free flow, with individual users, virtually unaffected by the presence of other traffic vehicles.
- Level B: characterizes stable flows with a high degree of freedom in choosing speed and operating conditions that have little influence on other road users.
- Level C: characterizes restricted flows that remain stable but interact with other participants in traffic flow. The overall comfort and safety level drops considerably.
- Level D: characterizes the high-density flow in which speed and maneuverability are severely restricted, and comfort and safety are low even if the flow remains stable.
- Level E: characterizes unstable flow at or near the capacity limitation, with the lowest level of comfort and safety.
- Level F: corresponds to conditional traffic in which the number of vehicles arriving at a point exceeds the service possibilities and therefore creates the conditions of tail formation (traffic jams), low comfort and increased accident risk.

IOP Conf. Series: Materials Science and Engineering **568** (2019) 012104 doi:10.1088/1757-899X/568/1/012104

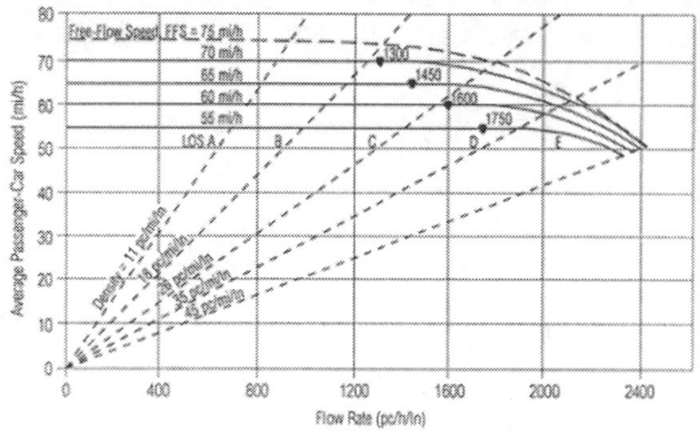

Figure 1. Level of service based on speed flow density curve [8]

2. Designing the graphical user interface (GUI)

Designing the graphical user interface was done using the MATLAB platform, GUI mode. Figure 2 shows how the buttons, windows, etc. were arranged in the design interface of the user interface, and also figure 2 represents the graphical interface in the running mode, i.e. as seen by the user [3] [4].

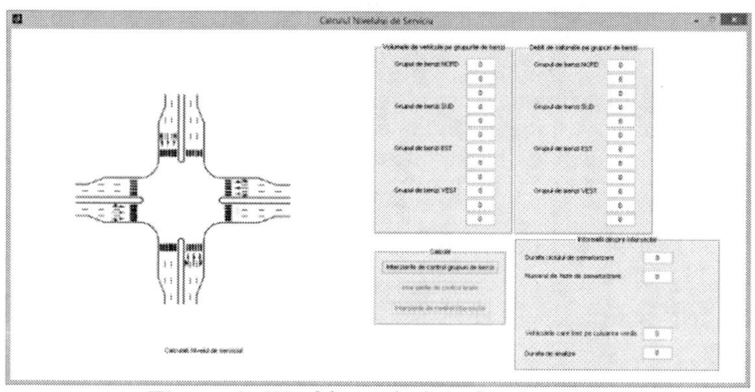

Figure 2. Working window for the GUI

The main window (Figure 2) is divided into four distinct zones [5] in order to enter the input data for the level of service calculation:

- Area 1 is the area where inputs will be entered for vehicle volumes for each lane of bands existing on that arm.
- Zone 2 represents the area where the saturation flow for each group of bands will be entered.
- Zone 3 is the area where all the values associated with the types of traffic lights are to be entered, depending on the signal plan configuration.
- Zone 4 is the group area of the control delay calculation buttons for the band groups, arms and intersection studied.

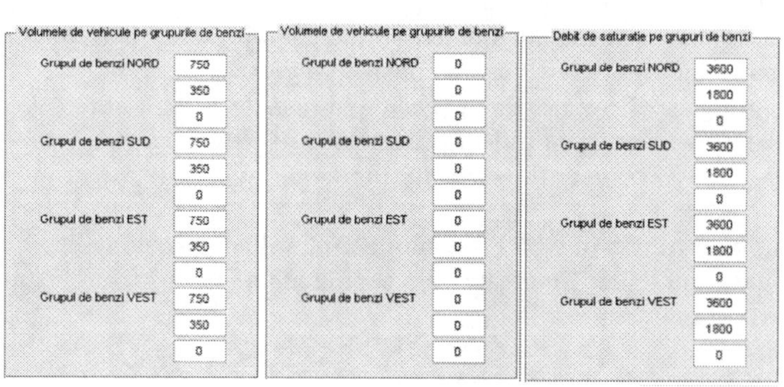

Figure 3. Area 1 and 2 of the GUI window

IOP Conf. Series: Materials Science and Engineering **568** (2019) 012104 doi:10.1088/1757-899X/568/1/012104

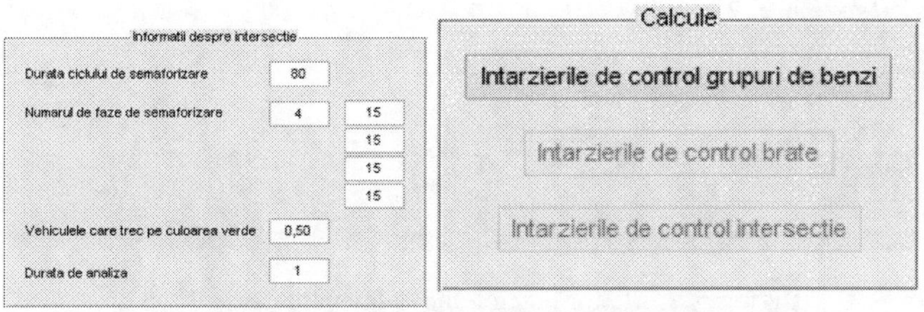

Figure 4. Area 3 and 4 of the GUI window

In order for the software not to be stuck in an infinite loop or loop due to errors that may occur when data is entered, codes are provided on all critical areas to provide error windows. For example, one of these errors is found at the introduction of traffic phases, so this window appears if less than two phases are entered [6].

2.1. Conceiving the software instruction code

This subchapter presents the main sequence of instructions, sequences that calculate, check, and display the service level values at the intersection studied.

Figure 5 shows the code for initializing the graphical interface, entering the text boxes with the default value of 0, and resetting the constants and the database to calculate the service level.

```
%Debit de saturatie pentru grupurile din Nord
set(handles.edit13,'string','3600');
set(handles.edit14,'string','1800');
%Debit de saturatie pentru grupurile din Sud
set(handles.edit16,'string','3600');
set(handles.edit17,'string','1800');
%Debit de saturatie pentru grupurile din Est
set(handles.edit19,'string','3600');
set(handles.edit20,'string','1800');
%Debit de saturatie pentru grupurile din Vest
set(handles.edit22,'string','3600');
set(handles.edit23,'string','1800');
```

```
for i = 2:3
   set(handles.(['pushbutton',num2str(i)]),'enable','off');
end;

for i = 42:45
   set(handles.(['edit',num2str(i)]),'visible','off');
   set(handles.(['edit',num2str(i)]),'string','0');
end;

for i = 13:25
   set(handles.(['text',num2str(i)]),'visible','off');
end;
```

Figure 5. Program code lines

Figure 6 is the procedure for displaying the error when the user enters the number of traffic cycles less than 2.

```
%% verificare pentru numar de faze de circulatie mai mare ca 2 sau egal cu 2
for i = 1:4
    if (str2num(get(handles.edit26,'string')) == 0 || str2num(get(handles.edit26,'string')) == 1)
         errordlg('numarul de faze nu este posibil','eroare');
         break;
    else
       for j = 42:(42+(str2num(get(handles.edit26,'string'))-1))
           set(handles.(['edit',num2str(j)]),'visible','on');
       end;
    end;
end;
%%
```

Figure 6. Program code lines for error syntax

The next step is to retrieves user input into the text boxes and converts them into input variables to be entered into the service level code. After the values are transformed from string to integer variables the code for calculating the traffic capacity that takes into account the intersection geometry and all the traffic volumes stored in the database when the software is opened, is called, as can be seen in the figure 7.

After the traffic capacity is calculated the code for the flow-capacity calculation is called. All variables used in the program code are retrieved from the database, created temporarily by the software. The next code called is for the calculation of the uniform delays and incremental delays at the intersection studied, and uses "for" and "if" instructions to be able to calculate the delays values as accurately as possible depending on the intersection geometry. The next step is the call of the code for the calculation of control delays for every group of movement, the aggregation of the control delay for

IOP Conf. Series: Materials Science and Engineering **568** (2019) 012104 doi:10.1088/1757-899X/568/1/012104

every junction arm and finally the control delay for the intersection and the placement of delays in a Level of Service structure [7].

```
if Dc < 10
    set(handles.text38,'string','Nivel de serviciu - A - Circulatie fluenta, fara cozi de asteptare, viteza libera de circulatie');
elseif 10 <= Dc && Dc < 20
    set(handles.text38,'string','Nivel de serviciu - B - Circulatie fluenta, fara cozi de asteptare, viteza mai redusa');
elseif 20 <= Dc && Dc < 35
    set(handles.text38,'string','Nivel de serviciu - C - Circulatie acceptabila, posibilitati pentru formarea cozilor de asteptare, viteza mai redusa');
elseif 35 <= Dc && Dc < 55
    set(handles.text38,'string','Nivel de serviciu - D - Circulatie acceptabila, cozi de asteptare reduse, viteza redusa');
elseif 55 <= Dc && Dc < 80
    set(handles.text38,'string','Nivel de serviciu - E - Circulatie dificila, cozi de asteptare permanente, viteza redusa');
elseif Dc >= 80
    set(handles.text38,'string','Nivel de serviciu - F - Circulatie foarte dificila, cozi de asteptare permanente, viteza redusa, opriri multiple');
end;
```

Figure 7. Code for placing the delay values in a LOS area

3. Case study for a junction in Craiova city

Verification of the LOS and at the same time the functionality of the computing application made in the MatLab platform will be accomplished with data taken from an intersection in Craiova. The intersection chosen is a transfer area between two neighborhoods and the city center, as well as a critical point thanks to the municipal hospital in Craiova, as can be seen from Figure 8.

Figure 8. Studied junction, aerial view

Based on the measurement the vehicle volumes for each movement arm at the intersection studied are formed by two groups each with calculated based on a cumulative volume of movements. At the same time, the saturation flows for the lane groups will be considered as 1800 vehicles.

Figure 9. Measured vehicles volumes

In order to be able to check the service level of the intersection studied, it is necessary to provide the software with the data on the intersection traffic lights. Thus, the greening times at the intersection reported for each phase (the northern phase has 32 seconds, the south phase has 29 seconds, the phase is 16 seconds, and the west phase has 15 seconds), were measured also. It can also be seen that the duration of the traffic light cycle is 112 seconds, and the intersection protection and release times at the end of each phase are 3 seconds of yellow and 2 seconds of red.

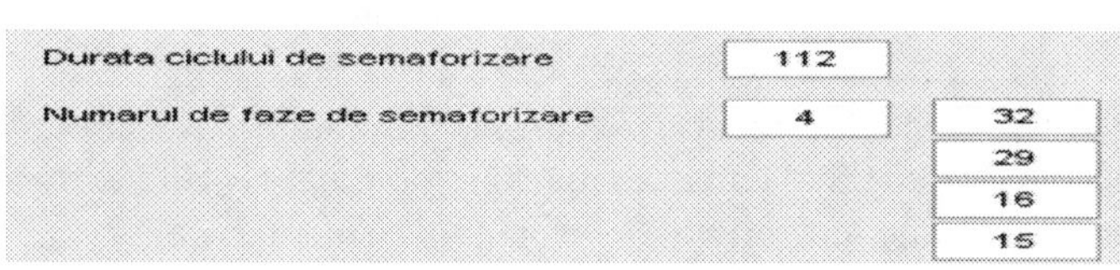

Figure 10. Traffic light signal plan for the studied intersection

After running the input data through the LOS calculation software, we obtained the specific intersection results studied divided into delays related to movement groups, arm and intersection.

IOP Conf. Series: Materials Science and Engineering **568** (2019) 012104 doi:10.1088/1757-899X/568/1/012104

We can draw the following conclusions by evaluating figure bellow:

- Delays for the North arm are 35.5 seconds
- Delays for the South arm are 55.3 seconds
- Delays for the East arm are 42.3 seconds
- Delays for the West arm are 43.3 seconds
- Delays for the study junction are 44.4 seconds, so the Level of Service is in Class D (approaching unstable operations where small increases in volume produce substantial increases in delay and decreases in speed [8]).

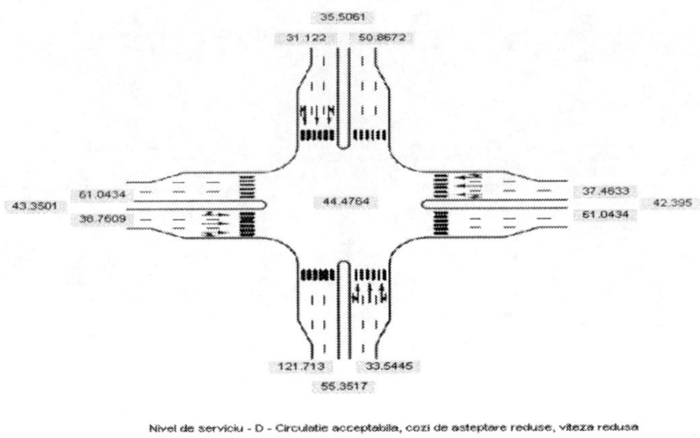

Figure 11. LOS representation based on the MATLAB application

4. Conclusions

The utility behind the LOS MATLAB application is being designed to help in the following aspects:

- A rapid verification tool for intersection delays;
- Graphical user interface as simple and intuitive as possible;
- Creating the programming procedures so as to use as few resources as possible;
- It is an application designed for rapid calculation so that it can be easily used by personnel without special training in the field;
- Can be used on the field or in the office by changing input data very quickly to see the differences in delay time between different traffic flows;
- The application can rapidly evaluate different types of signal plans based on the LOS results.

References

[1] Andrews C M, Mohammad S E, James E C 1998 *Traffic-Control System In Transportation Research Record*, p. 150-155,.

[2] Helbing T, Helbing M, Helbing D 2002 *Microsimulation of Freeway Traffic Including Control Measures*"

[3] Kenneth B D 1999 *Principles of Transportation Economics* (Addison Wesley Longman publishing house)

[4] Dowling R 2007 *Definition, interpretation, and calculation of traffic analysis tools measures of effectiveness* (Federal Highway Administration, Washington DC)

[5] Stewart J A, Aerde M V 1998 *An Assessment of Adaptive Co-ordination of Traffic Signal offsets within integration* Volume 39 (Traffic Engineering and Control)

[6] Otat O V, Dumitru I, Otat V, Matei L 2013 *The role of info-mobility systems in optimizing public transport. Case study Craiova* (Congress AMMA - Automotive Motor Mobility Ambient, 17-19 October, Cluj – Napoca)

[7] Otat O V, Dumitru N, Otat V, Dumitru I 2016 *Determination of Kinematic and Dynamic Behavior in the Driver's Skull upon the Impact with the Steering Wheel (*CONAT International Congress of Automotive and Transport Engineering) ed (Springer International Publishing Switzerland, 2017) pp 812-819,

IOP Conf. Series: Materials Science and Engineering **568** (2019) 012104 doi:10.1088/1757-899X/568/1/012104

[8] Transportation Research Board, American Association of State Highway and Transportation Officials (AASHTO) and Federal Highway Administration 2010 *Highway Capacity Manual*

IOP Conf. Series: Materials Science and Engineering **568** (2019) 012105 doi:10.1088/1757-899X/568/1/012105

Rapid filament prototyping of a bionic hand

L I Culda[1] E S Muncut[2] and A Komjaty[3]

[1,2,3]Department of Automation, Industrial Engineering, Textiles and Transport, Aurel Vlaicu University of Arad, RO, Arad, Romania

lavyy_99@yahoo.com

Abstract. No matter how we call it: additive manufacturing, fast manufacturing, fast prototyping, fabrication in layers or DMF (desktop manufacturing) manufacturing, 3D printing is rapidly developing due to the diversity of applicability areas. In this paper, we would like to introduce the way of realizing a bionic hand that imitates human hand movements with the help of the 3D printer. The actual function that can perform is the movement of the finger segments for closing and opening the fist. The imprint was made using the Creality 3D Direct Store 3D Ender 3 3D Printer Original. As components, we have the handprint, the print of servomotor holder, the Arduino uno board, the breadboard, 5 actuators, male jumpers and fishing thread for ligaments. For programming, the integrated development IDE environment based on the Processing project has been used, which includes support for the C and C ++ programming languages.

1. Introduction

The quick prototyping or additive manufacturing has been born to obtain rapid results. Prototyping has always been the expression of efficiency, as it is more advantageous to study and analyze a prototype and to launch at the production of the product rather than to simulate and the physical realization of the product to present problems.

In the present paper, it is desired to present a case study, namely obtaining a bionic hand on a fast filament printer.

There is an explosion of interest by the general public in 3D printing. 3D printing is also known under other names, more or less similar, such as additive manufacturing, fast manufacturing RM (rapid manufacturing), or rapid prototyping (RP). Other less used names are digital manufacturing, digital fabrication, layered manufacturing, or desktop manufacturing.

In recent years, engineering has taken a considerable dynamism, using increasingly powerful computers and mechatronic products becoming indispensable to the technique.

Increased competition has intensified, customer tastes can change very quickly, and the desire for a product to have an attractive look, to be innovative and with proper functionality is more significant than ever.

In obtaining prototype products, the basic is a CAD system that offers the possibility to design, simulate, and then print the desired product. A CAD system uses various data elements and structures to define a detail, but geometry is only part of the data being retained, followed by material, surface quality, production cycle, etc.

The rapid prototyping is a technological process in contrast to abrasive processes such as erosion, drilling, polishing and so on. This process is a generative or addition of material by which a product is

IOP Conf. Series: Materials Science and Engineering **568** (2019) 012105 doi:10.1088/1757-899X/568/1/012105

formed by the addition of bulking elements as opposed to abrasive products through which the product is made by removing the material, [1].

The term rapid prototyping is most used to describe this generative process, even if it is not the most correct in terms of semantics.

More precisely, rapid prototyping is processes that are added to 2D contours of constant thickness. The layer is contoured in an x-y plane on two dimensions, the third dimension will result from the set of layers deposited on each other, but there will be a constant coordinate z. Thus, the models are made up of 3D parts arranged exactly on the working plane in the direction x-y which more accurately copies the desired product as the advance in the z direction is made in several layers.

Rapid prototyping can produce both positive and negative patterns. Negatives are molds and dies to produce other positive ones, this process being called a generative process of tooling or rapid prototyping of tools.

2. Imprinting using two 3D prototyping techniques: powder printing and filament printing

The primary purpose of a prototype is to serve as a support for testing certain aspects or features without cost and time. It will not be developed into a final product, but it provides information needed for iterative design, performance evaluation, study of errors and improved product. To obtain the desired product several rapid prototyping procedures has been studied starting with a study of FDM print job with a Zprinter 650 printer.

2.1. Imprinting using the 3D Zprinter printer

Typically, 3D printers do not offer greater precision and durability of a completed prototype, but the mechanical properties of these prototypes are sufficient to cope with design and product development.

For the study of printing, it was taken into consideration the 3D printer from Z Corporation, one of the simplest "jet" printers - the so-called Fused Deposition Modeling (FDM), figure 2. The FDM idea is very simple - the dispensing head pushes the heated droplets of the thermoplastic onto a cooling platform (as virtually any industrial thermoplastic material can be used as a material). The droplets solidify and stick together, forming the layers of the future object (printing is also in layers). The FDM process technology allows the production of finished and ready-to-use (with a minimum thickness of 0.12 mm) precision parts with large gauge sizes (up to 600 x 600 x 500 mm). The basics of this technology were developed in 1988 by Scott Crump.

Z Corp's 3D printing technology supports all major 3D formatted files, including STL, WRL, ply, and SFX files which can be exported.

Minimum element thickness: 2.5 mm the objects printed 3D in composite powder has the consistency of the ceramic and its hardness is determined by the application in question. It is suitable for design objects, prototypes and for studying the shape. The thickness of the walls is recommended to start from 2.5 mm.

Figure 1 Zprinter 3D [2]

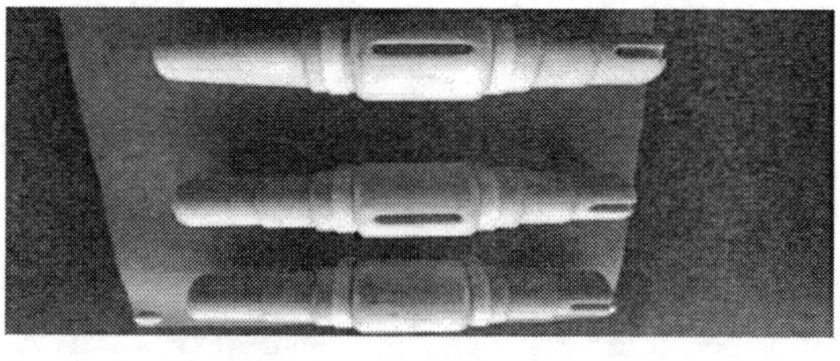

Figure 2 Case Study

IOP Conf. Series: Materials Science and Engineering **568** (2019) 012105 doi:10.1088/1757-899X/568/1/012105

We have done some projects on this printer to study the strength of the product and to see if we can print without the need for a support material on the geometry of the product. Due to the material from which the printed material is produced, we do not have the flexibility that we need to obtain a bionic hand. So, we gave up this system and switched to filament printing.

2.2. Printing using the Big Printing Bed High Precision print 3d Reprap Prusa i3 DIY 3d printer

Regardless of the print process, we always discuss composing the final object layer by layer. The process is slow, lasting at least several hours for an average complexity object. Under no circumstances, we are talking about fast printing in seconds, as is the case with a traditional printer. FDM models are plastics that are heated, melted, and laid in layers on the print platform using a three-dimensional digital model.

If we put the cost in the 3D printing in the balance of an ordinary object vs. purchase, we will most often find that the cheapest is to buy the product directly from the store. There are many steps to go before it is cheaper to print a plastic dish, so the advantages of 3D printing are quite different at the moment.

For our study, we used Reprap Prusa i3 DIY 3D Printer, figure 3.

When we chose it, it was taken into account:

- The volume and print area. The maximum print speed is expressed in width x depth x-height, with values in centimeters or millimeters. The volume is sufficient for the prototype.

- The print speeds. This depends not only on the machine's ability to heat and place the plastic filament on the surface of the object but also on its complexity (number of layers, corners, lines, and geometric details).

- The resolution - The quality of the final object is influenced by its number of layers and their thickness. A small resolution object will look like a 3D game from the beginnings of computers, meaning pretty pixelated. The higher the resolution is the resulting objects will have finer margins to touch, but the print speed decreases considerably.

- The extrudes - is the assembly of the melting nozzle, the motor that forces the filament through the nozzle, a temperature sensor and the cooling system. A 3D printer with more extruders will print faster and allow the use of different color materials in the same time.

- Connectivity

- The LCD display showing your machine's current status.

- The control software - is a very important component of the printer. Ease of use of the printer control program is very important because we usually talk about a complexity of the process higher than paper printing. Additionally, mistakes in the commands will be seen sometimes after hours of printing, when you will already have consumed a lot of material and precious time.

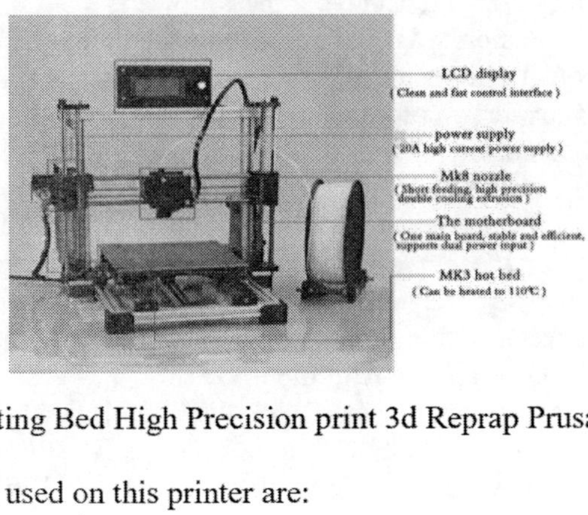

Figure 3 Big Printing Bed High Precision print 3d Reprap Prusa i3 DIY 3d printer

The filaments that can be used on this printer are:

- PLA (Polylactic Acid) is a thermoplastic material classified as polyester plastic. This is the most common 3D printing material. The PLA filament is easy to print 3D and is biodegradable. There are many different colors and varieties, and almost every filament manufacturer in the market deals with PLA production. Also, its properties allow the addition of other materials such as metal powders, hemp, coffee or wood. Acrylonitrile Butadiene Styrene (better known ABS) is the second most popular 3D printing filament. This thermoplastic is inexpensive, durable, lightweight, and easy to extrude - making it perfect for 3D printing. It's the same plastic used in LEGO parts and bicycle helmets. But there are drawbacks in using the ABS filament. This requires a higher temperature to reach the melting point, usually in the range of 210 ° C to 250 ° C. Moreover, a heated bed is required. This prevents the first layers of printing from cooling too quickly so that the plastic does not deform and collapse before the object print is finished. Another disadvantage of this 3D printer filament is the intense smoke that occurs during printing. The smoke can be dangerous for people (or pets) causing breathing difficulty.

3. Obtaining the bionic hand

3D printing is used in the medical industry to obtain 4D-printed artificial muscles, tissues with blood vessels, cheap prosthetic parts. The 3D printing opens a completely new way for people to advance their fields and provide patients with really useful alternatives for traditional medicine.

We did not want to get a prosthetic hand, but we just tried to get a bionic hand to study the conception, the print and the mobility of the product. In its realization we started from the examples on the internet, such as the baby's prosthetic hand in figure 4.

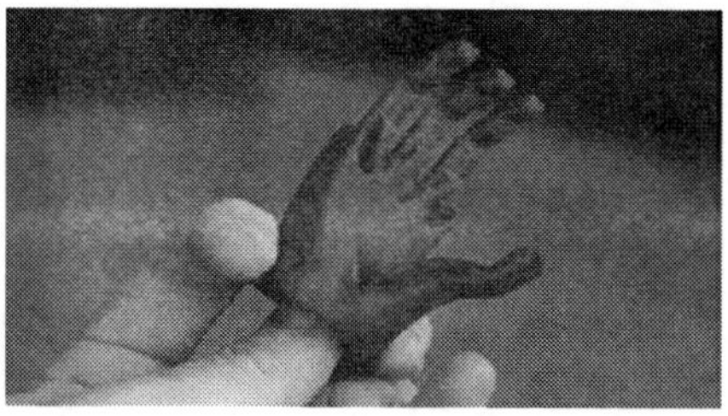

Figure 4 Prosthetic hand [3]

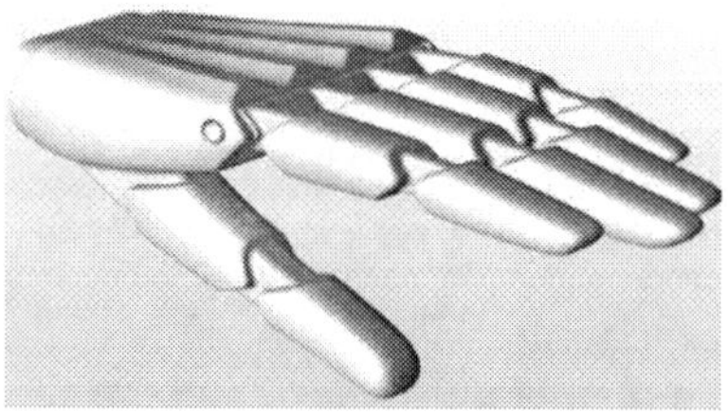

Figure 5 Hand drawing

It is known that the human hand has its bones, tendons, cartilage, nerves that work together for its mobility. In a robotic hand we have to replace all this with something that will allow us as natural as possible and as much mobility as possible.

For the 3D printing, parts of the hand: fingers, palm, wrist, and arm are all 3D printed.

The desired function is to move the finger segments to close and open the fist.

To achieve the goal, he following steps have to be followed:

- For this the drawing in Fusion was made and translated for the printer's software to make the imprint. The print lasted about 12 hours, because a better resolution was wanted, figure 5.
- Performing the support for servomotors using 3D printing too
- An Arduino plate has been used
- Breadboard
- 5 servomotors
- Male Jumper
- Fishing thread as tendons
- Making connections between Arduino uno, breadboard and servomotors
- Creating and loading the program for Arduino uno

The electric servomotor was chosen because it has the role of transforming an electric control signal into an electromagnetic torque, or in a rotating motion of its shaft, through which the mechanism that performs the desired operation is trained. Thus, they are included in automatic control systems as

IOP Conf. Series: Materials Science and Engineering **568** (2019) 012105 doi:10.1088/1757-899X/568/1/012105

execution elements, working in a wide variety of conditions both in terms of load character, speed, power, voltage and frequency. I got the hand out of the figure 6.

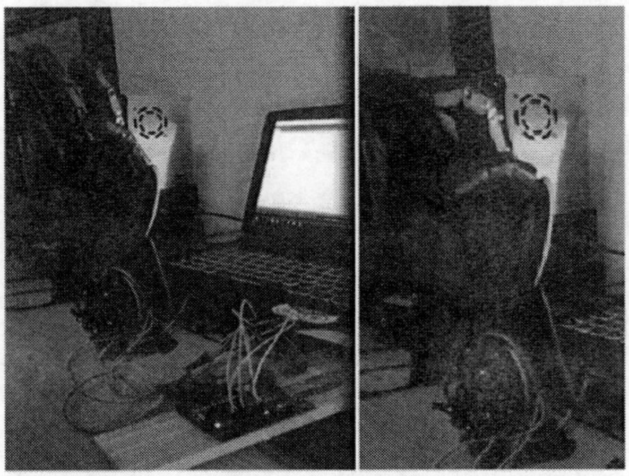

Figure 6 Bionic hand

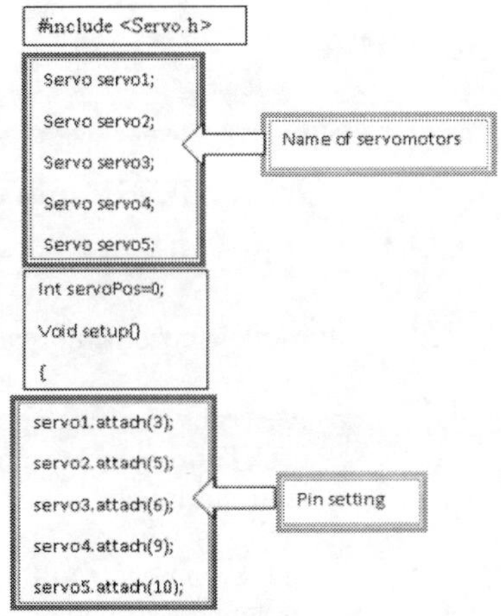

Figure 7 The current running program on Arduino Uno

The programming of the Arduino Uno board was done with the software offered by Arduino and can be seen in figures 7.The print quality was given by the number of layers in a print. The height of the layer is measured in microns. To achieve high resolution, thin layers should be used. With high-resolution printing, the subject's layers become difficult to observe, as they can be printed just as thin as a sheet of paper at a thickness of only 100 microns. The objects achieved have a lower resolution and thicker layers. They seem hard by touch and contain layers that are slightly more visible to the naked eye, similar to the rings of a tree or the texture of a vinyl disc. We have obtained a good resolution with a figure-like support that allowed the "tendons" to bind to the control plate.At the beginning satisfactory results were achieved, but the project is constantly developing.

4. Conclusion

The conclusion of our study to date is as follows: For the realization of a bionic hand, 3D filament printing is the most advantageous, as texture, resilience and mobility. Printing does not require support brackets and the combination of the printed model with the Arduino plates is a happy one, the connection is easy to make and the results are spectacular. The Arduino open-source programming allows appropriate programming for the intended purpose.

Even if we are still studying and wanting to get movements as close as possible to a human one, the phase we have reached gives us hope to use the prototype to obtain a functional didactic robot using the same procedure.

References

[1] Rapid Prototyping and Manufacturing: A Review of Current Technologies, Hugo I. Medellín-Castillo, Conference Paper January 2009, DOI: 10.1115/IMECE2009-11750

[2] Drd. Ing. Ciobota Nastase-Dan, Prof. Univ. Dr. Ing. Gh. Ion Gheorghe, CS I Alexandru Moldovanu, "Rapid prototyping" - Advanced technology implemented in the mechatronics industry (Tehnologie avansata implementata in industria de mecatronica), 021631, Bucuresti, Romani.

[3] Ciobota Năstase-Dan, Implementation of advanced adhesives of selective laser sinterization for applications in mechatronics and biomecatronics (Implementarea tehnologiilor avansate de sinterizare selectivă cu laser pentru aplicații în mecatronică și biomecatronică), aprilie - mai 2015, www.incdmtm.

IOP Conf. Series: Materials Science and Engineering **568** (2019) 012106 doi:10.1088/1757-899X/568/1/012106

Using virtual inductive loops to adapt the signal plans of junctions

L Matei[1], I Dumitru[1], L D Racila[1], D Tutunea[1], A C Oprica[1]

[1]University of Craiova, Faculty of Mechanics, Craiova, Romania

racila_laurentiu@yahoo.com

Abstract. Trafficking is a growing problem in cities around the world. Congestions occurring at peak hours have become common problems that all the major cities of the world are facing. Some metropolises, however, have found solutions to reduce traffic jams, especially in peak hours. Problem solutions are given by road traffic management by reducing waiting times and queues, in the same way as traffic safety increases. The first step in achieving coherent management is to achieve a method of collecting traffic data. Under the current infrastructure development conditions, one of the few solutions is given by virtual inductive loops. Inductive loop detectors have become the most utilized sensors in traffic management systems. The gathered traffic data is used to improve traffic efficiency (i.e., warning users about congested areas or planning new infrastructures). This paper aims to develop virtual intersection models by using primary data from virtual inductive loop video analyzers.

1. Introduction

The concept of an intelligent system to monitor and control traffic does not require integration to an existing system, but the combination of several equipments combined in a specific area studied and integrated with existing and prevailing in the area to form an intelligent system [1]. Thus, the current situation does not require identification of existing intelligent but identification of components that can be integrated into the requirements of Craiova, so use in combination lead to the creation of an intelligent system. At most intersections, to ensure effective crossing without significant safety delays, eliminating potential conflicts that may [2], [3] be caused by intersections of vehicle flows, traffic management is achieved through traffic lights.

The traffic light of an intersection must provide safety and efficiency for cars and pedestrians, and in designing a traffic light, several aspects need to be taken into account:

- Repeated traffic jamming on intersection access roads;
- Repeated vehicle routing with more than two minutes of waiting on secondary streets, the traffic density on the main street being too large to allow cross traffic or cross-filtering;
- Accidents due to non-observance of intersection priority rules.

You can see the following advantages of using a signal plan, assuming it was done according to the intersection characteristics [4]:

- Ensures the orderly movement of vehicles at the intersection;
- Increase the intersection capacity to serve the crossing requests;
- Reduce the frequency and severity of collisions between vehicles;
- Tries to maintain a decongested traffic, as fluent as possible;
- Are used to limit the waiting time to enter the intersection of vehicles;
- Improve safety on crossing the intersection.

IOP Conf. Series: Materials Science and Engineering **568** (2019) 012106 doi:10.1088/1757-899X/568/1/012106

2. Inductive loop data

For the collection of traffic data, "SmartLoop" video sensors were used to monitor traffic using presence sensors called virtual inductive loops as seen in figure 1.

Figure 1. "SmartLoop" equipment

This equipment can be used for:

- Presence: detecting the presence of the vehicle;
- Queue: queue detection in the studied area;
- Speed: Approximate circulation speed of vehicles.

2.1. Area of influence

Thus, the data gathering from the study area was carried out in three stages, namely:

- Metering of vehicles at the intersection of Calea Bucureşti - Carol I Boulevard - underground passage;
- Metering of vehicles at the intersection Alexandru Ioan Cuza Street - Carol I Boulevard - Arieş Street;

The measurements were made in the course of a day, thereby identifying four time zone intervals referred to in the literature as hourly zones, important for traffic by the intersection: Zone I (Early morning - in the first part of the day - for example: transit from residential to service - often in metropolitan areas); Zone II (mid-peak time - transit to central areas where there are administrative, banking, etc.); Zone III (Top time in the second part of the day - for example: transit from service to residential areas); Zone IV (Evening - Transit to the commercial and cultural areas of the city).

2.2. Metering of Vehicles in the intersection

A video sensor positioned in Figure 2 at the Calea Bucureşti intersection - Carol I Boulevard - the underground passage was used for vehicle counting. Following the assessment of the entry and exit zones and the intersection conflict points [6], [7], it was concluded that to best measure the vehicles crossing the intersection, there were 18 virtual inductive loops needed, as can be seen in Figure 3.

Figure 2. Catch with inductive loop positioning at the first intersection

Thus, to cover all the bands and the points of conflict, the virtual inductive loops were used in the following way:

- Calea Bucureşti direction west-east were used 5 virtual inductive loops (numbers 5, 6, 7, 8 and 12 in Figure 2)
- Calea Bucureşti direction east-west were used 4 virtual inductive loops (numbers 15, 16, 17 and 18 in figure 2)

IOP Conf. Series: Materials Science and Engineering **568** (2019) 012106 doi:10.1088/1757-899X/568/1/012106

- Carol I Boulevard in the north-south direction were used 4 virtual inductive loops (numbers 1, 2, 3 and 4 in figure 2)
- Carol I Boulevard in the south-north direction were used 5 virtual inductive loops (numbers 9, 10, 11, 13 and 14 in Figure 2)

After the introduction of the virtual inductive loops, the length of time for which the reports for each loop were to be made was introduced. As can be seen from Figure 4 and 5, for each virtual inductive loop, the reports are generated every hour.

View summaries on disk

Detector Id	From	To	Total
15	2016-08-30 15:54	2016-09-02 23:00	42
14	2016-08-30 15:54	2016-09-02 23:00	42
13	2016-08-30 15:54	2016-09-02 23:00	42
12	2016-08-30 15:54	2016-09-02 23:00	42
11	2016-08-30 15:54	2016-09-02 23:00	42
10	2016-08-30 15:54	2016-09-02 23:00	42
9	2016-08-30 15:54	2016-09-02 23:00	42
8	2016-08-30 14:55	2016-09-02 23:00	43
7	2016-08-30 14:55	2016-09-02 23:00	43
6	2016-08-30 14:55	2016-09-02 23:00	43
5	2016-08-30 14:55	2016-09-02 23:00	43

Figure 3. Report Verification Window

3. Signal plan adaptation

To be able to adapt the inductive signal plan based on the inductive loop and the mathematical algorithm as accurately as possible [8], the semaphore cycle was measured so that it was possible to identify and create the signal plans for the intersection.

For the intersection of Figure 3, the greening times for each of the four phases and all four arms of the intersection were identified. Thus, the measurements made in the field are as follows:

- North arm traffic: Two movement phases F1 and F2 were identified for this movement arm, namely the F1 movement before the Carol I Boulevard and F2 the left movement on the Carol I Boulevard. The right-hand movement on Carol I Boulevard is made on the bracelet separated from the islet and does not enter the traffic light cycle. Following the F1 measurements, it has 17 seconds of green and F2 has 20 seconds of green.

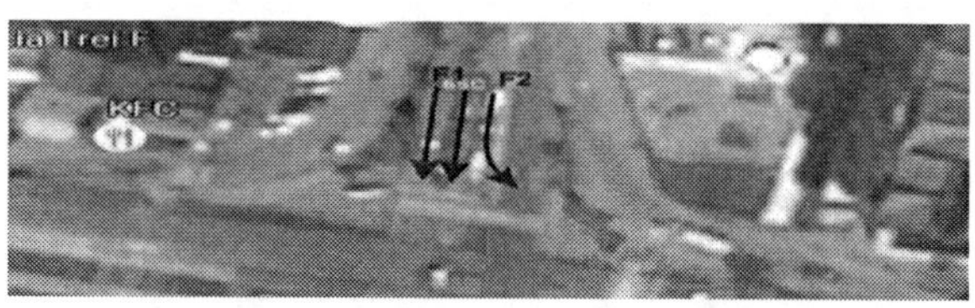

Figure 4. Selection of traffic phases on North arm of the intersection

- South arm traffic: Two movement phases F1 and F2 have been identified for this movement arm, which coincides with the phases of traffic lights on the north arm, namely F1 the forward movement both in the passage of the University and on Blvd. Carol I and F2 the left movement both from the passage of the University as well as from Carol I Boulevard. The right-hand movement in the passage is forbidden, and the right-hand movement on Carol I Boulevard is realized on a separate branch of the island. Following the F1 measurements, it has 17 seconds of green and F2 has 20 seconds of green.

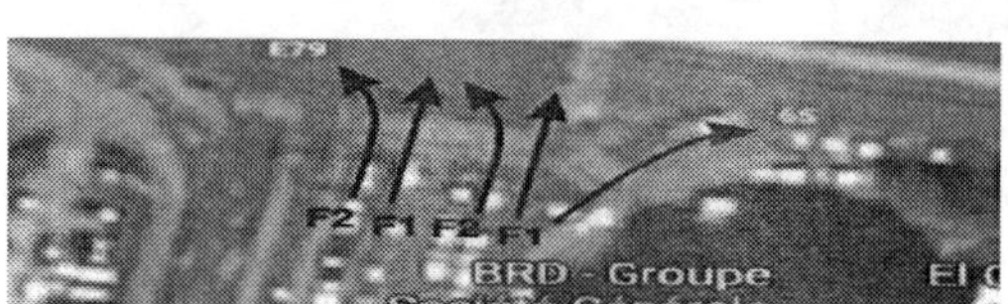

Figure 5. Selection of traffic phases on the South arm of the intersection

- West arm traffic: Two movement phases F3 and F4 were identified for this movement arm, namely F3 the left movement on Calea Bucureşti and F4 the forward and right movement in the passage of the University on Calea Bucureşti. The right-hand movement on Bucharest Avenue

IOP Conf. Series: Materials Science and Engineering **568** (2019) 012106 doi:10.1088/1757-899X/568/1/012106

to Carol I Boulevard is realized on a separate island from the island and does not enter the traffic light cycle. Following the measurements, F3 has 17 seconds of green, and F4 has 42 seconds of green.

Figure 6. Selection of traffic phases on West arm of the intersection

- East arm traffic: Two movement phases F3 and F4 were identified for this movement arm, circulation phases coinciding with the west arm, namely F3 the left movement on Calea Bucureşti and F4 the movement before Bucharest. The right-hand movement on Bucharest Avenue to Carol I Boulevard is realized on a separate island from the island and does not enter the traffic light cycle. Following the measurements, F3 has 17 seconds of green, and F4 has 42 seconds of green.

Figure 7. Selection of traffic phases on East arm of the intersection

3.1. Calculation of the new signal plan

According to the literature, the saturation flow for each group of bands is the flow of vehicles that can be taken over by the band group assuming that the phase with right of access for the particular group (green color) lasts 100% of the time. So, it will be calculated using the following formula:

$$s = s_0 \cdot N \cdot f_W \cdot f_{HV} \cdot f_g \cdot f_p \cdot f_{bb} \cdot f_a \cdot f_{LU} \cdot f_{LT} \cdot f_{RT} \cdot f_{LTp} \cdot f_{RTp} \quad (1)$$

Were:

- s - flow lane saturation; s_0 - the basic saturation flow; N - the number of bands in the group of movements; f - the adjustment factor for the following conditions (w - bandwidth; HV - heavy vehicles; g - declivity; p - parking; bb - buses blocking; a - the type of the studied area; LU - the use of bands; RT – right turn LT – left turn; Rpb - Interference of cyclists or pedestrians at the right turn, and Lpb - Interference of cyclists or pedestrians at the left turn).

The role of intermediate time (interchange and traffic phase protection) is to eliminate the potential conflict between vehicles that evacuate the intersection at the end of a traffic phase (after the end of the green phase of that phase) and the vehicles entering the intersection at the start of the greenback of the next phase. The mean time can be calculated with the formula:

$$L = C_g + C_r = t + \frac{V}{2 \cdot a + g \cdot G} + \frac{l + w}{V} \quad (2)$$

Were:

- L is the length of the exchange interval (or lost time) corresponding to a semaphore phase; Cg - is the time for yellow, (s); Cr is the time for full red, (s); T is the reaction time (can be considered 1 s); V - is the traffic speed at the intersection access points (m / s), 1 km / h = 0.27 m / s; A - is the deceleration of vehicles (can be considered 3m / s2); G - is the gravitational acceleration (9.81 m / s2); G - is the land gradient (%), positive on climb, negative at the descent; L is the length of the vehicle (m); W is the width of the intersection to be crossed (m).

IOP Conf. Series: Materials Science and Engineering **568** (2019) 012106 doi:10.1088/1757-899X/568/1/012106

The next step is to calculate the saturation flow based on the flow ratios and the total traffic light cycle.

$$Y = \sum_{i,critic} Y_i = \sum_{i,critic} \left(\frac{v}{s}\right)_i \quad (3)$$

$$C = \frac{1.50 \cdot L + 5}{1 - \sum_{i,critic} Y_i} \quad (4)$$

The final step is to calculate the length of the green light for each phase.

$$g_i = \left(\frac{v}{s}\right)_i \left(\frac{C_{ef}}{Y}\right) \quad (5)$$

Based on the formulas above and the calculation algorithm the new traffic light cycle is as follows: Phase 1 (F1) – 15s; Phase 2 (F2) – 18s; Phase 3 (F3) – 22s; Phase 4 (F4) – 20s.

4. Conclusion

Road traffic, in general, is divided into two areas: the peak area and the general traffic area. However, it can be seen from the data collected using virtual inductive loops that the traffic in the studied area has 4 intensive traffic areas: Zone I (Early morning - in the first part of the day), Zone II (mid-peak time), Zone III (Top time in the second part of the day), Zone IV (Evening).

Using the virtual inductive loops to adapt the signal plans of junctions is essential to high-level junction because it can account for every situation.

Another important aspect is that at the calculation of the traffic light cycle is based on how many measurements are in an hour. Based on this, the virtual inductive loop can generate reports based 5-minute increment so that the calculation of the traffic light cycle can be very accurate.

In the same tim, the virtual inductive loop can be used in creating a traffic signal plan for every zone identified as a peak hour.

References

[1] Andrews C M, Mohammad S E, James E C 1998 *Traffic-Control System In Transportation Research Record*, p. 150-155,.

[2] Helbing T, Helbing M, Helbing D 2002 *Microsimulation of Freeway Traffic Including Control Measures*"

[3] Kenneth B D 1999 *Principles of Transportation Economics* (Addison Wesley Longman publishing house)

[4] Dowling R 2007 *Definition, interpretation, and calculation of traffic analysis tools measures of effectiveness* (Federal Highway Administration, Washington DC)

[5] Stewart J A, Aerde M V 1998 *An Assessment of Adaptive Co-ordination of Traffic Signal offsets within integration* Volume 39 (Traffic Engineering and Control)

[6] Otat O V, Dumitru I, Otat V, Matei L 2013 *The role of info-mobility systems in optimizing public transport. Case study Craiova* (Congress AMMA - Automotive Motor Mobility Ambient, 17-19 October, Cluj – Napoca)

[7] Otat O V, Dumitru N, Otat V, Dumitru I 2016 *Determination of Kinematic and Dynamic Behavior in the Driver's Skull upon the Impact with the Steering Wheel* (CONAT International Congress of Automotive and Transport Engineering) ed (Springer International Publishing Switzerland, 2017) pp 812-819,

[8] Transportation Research Board, American Association of State Highway and Transportation Officials (AASHTO) and Federal Highway Administration 2010 *Highway Capacity Manual*

IOP Conf. Series: Materials Science and Engineering **568** (2019) 012107 doi:10.1088/1757-899X/568/1/012107

Designing and Implementing Control panel for Diesel generator

A Alsebai[1] **and S Afghan,**[1]

Department of Mechatronics, University of Debrecen, 4028 Otemeto utca 2.-4., Debrecen, Hungary

Ahmadalsebai@hotmail.com

Abstract. The Main goal of the study was to Build and implement a control panel to protect the diesel generator, and auto-starter with ability display the error using visual alarm and LCD, also, easily used by the operators. The purpose has been achieved as the test results show that it operated as intended. In this paper, we explain How the system works and How will respond to the error signals coming from the sensor (Thermostat, Coolant pump, Fuel, Lubricating, Stop emergency). We have Design and choose the components of the electrical circuit of the control panel to build a stable, reliable system and quickly responding system. It is also Protected from outside disturbance flaw such as over-voltage. The system can recognize the faults and initiate shut down on diesel generator also retain the fault indicator (LED and LCD notification) until the system is reset. Design and implementation of a microcontroller based automatic Diesel Generator controller. With the added ability to measure the amplitude and frequency of the generator's AC output, in addition to providing alarm state notification using LEDs and messages displayed on LCD.

1. Introduction

Diesel generators are increasingly being used in various applications, such as providing backup or standby power to various facilities and critical systems in case of power fail or emergencies, to avoid any power loss. Diesel generators are widely adopted in different industries due to their reliability, ability to generate adequate power for massive duty operations, portability, and efficiency. As a result, it is highly required to enhance both the efficiency and performance of the generator [1], [2], [3]. To address these concerns, current diesel generator control circuits are designed to monitor all parameters and sensors' data continuously and providing automatic on/off switch, if there is any power loss. All electrical machines have a set of nominal parameters under which they function at optimum levels, so any fluctuations in their parameters can cause these machines to run at lower efficiency. Power Generators are no exception to that; generators are typically designed to run most efficiently at or near sea level under standard conditions of temperature and press (STP) [4]. For proper ignition and functioning of generators, there must be appropriate ambient temperature and air press values. All generators, irrespective of the fuel that powers them, demand sufficient air for combustion. So, any deficiency in air density can lead to start-up fails. In diesel engines, air and fuel are infused together then the compressed air becomes hot, and when peak temperature and press values are achieved, diesel is injected, which then ignites under the given conditions [4].

For abnormal Circumstances, diesel generators must be equipped with a system to protect against costly damage. That protection system responds by shutting down the generator faster than a human operator when it detects unusual output or input values, which may cause damage. The quality and reliability of a protection system are paramount [3].

IOP Conf. Series: Materials Science and Engineering **568** (2019) 012107 doi:10.1088/1757-899X/568/1/012107

The protection system of a diesel generator set is aimed at preventing costly damage to the generator in the event of a mechanical or electrical emergency. The diesel engine is typically protected against low oil press, over-speed, and high oil temperature or water (coolant) temperature. The protection unit is designed to identify any abnormal condition within the engine through individual sensors and acts to shut down the engine [1].

The primary function of this smart Microcontroller based unit is to control and protect power generators in industrial applications; despite the sophisticated technology of the circuit, its interface is still user-friendly and straightforward to use by the operator. Thus, it does not require a professional nor a qualified worker to use it. The primary mode of operation is selected by a key switch, turning off the generator's control unit can be simply done by turning the key to "off" or by removing the key, and there is Push button for an emergency shutdown. Also, the unit allows you to control response time. The generator's control unit incorporates multi-function protection: an engine protection module, Auto starter, I/O monitoring Sensor signals, receiving analog or digital signals, monitoring the output voltage of the power generator, monitoring the output Frequency of the power generator, internal time clock, LCD provides helping messages which make using the unit intuitive.

2. Programming

2.1. Micro-controller programming

Microcontrollers are great when you need a few computer smarts in a physical system.[6] In this control unit, a higher-level language called MikroC with the integrated development environment (IDE) was used, it contains useful tools such as code editors, compilers, and debuggers. As the first step relates to programming, set data register manually to specialize Microcontroller for a specific task and activate some feats, this data register is unique each model or brand, so it must be checked from the datasheet. However, in our control unit, we selected two different microcontrollers ATMEGA 32/8. [2] In programming, We initialized the ports of microcontrollers as:

- Identify the input of the sensor and switches.
- Initialization of LCD ports.
- Activate the general Interrupt in the system
- Initializing of the timer setting.
- Identify the Output of the ports
- Initialization of external Clock Setting
- Activate the Emergency Interrupt
- Initializing of the counter with compare overflow set

The programming flowchart can be seen in Figure 1:

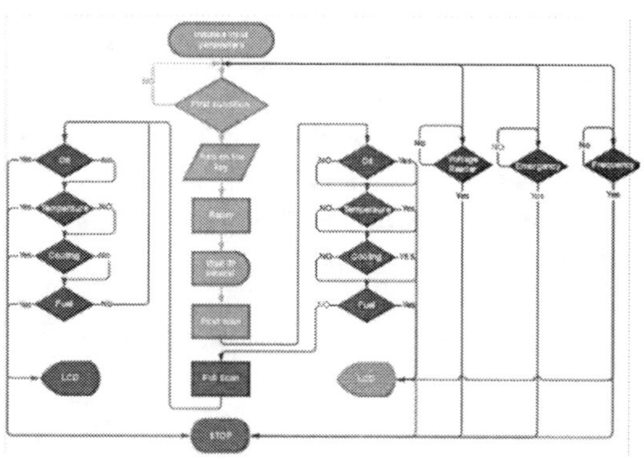

Figure 1 The flowchart of the program and How the system works

IOP Conf. Series: Materials Science and Engineering **568** (2019) 012107 doi:10.1088/1757-899X/568/1/012107

2.2. First scan

When the generator reaches a set point, the unit will scan for the first time; it will display all problems on the LCD if any. In the case of more than one problem at the startup of the generator, then, after displaying all error messages, the control unit will turn on the LED alarm, and shutdown the generator. As demonstrated in Figure 2:

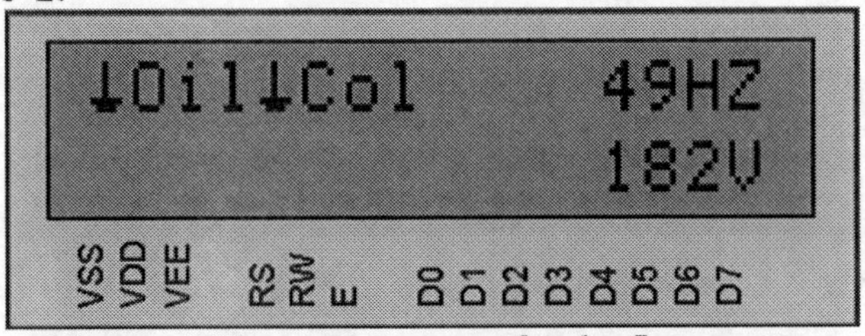

Figure 2 The error message for the first scan

3. Schematic project

3.1. Input

The optocoupler is an LED chip integrated into the same package with a light-sensitive device such as a photodiode or a phototransistor, the two components being electrically separate but optically coupled. For Microcontroller pins, as shown in Figure 3, we use optocouplers to provide isolation between pins and sensors to take precautions against input abuse or over-voltage coming from outside the control system. Also, it is a standard method used to achieve electrical isolation. Also, the LEDs are connected as a part of the sensors instead of being connected to the microcontroller's pins [5].

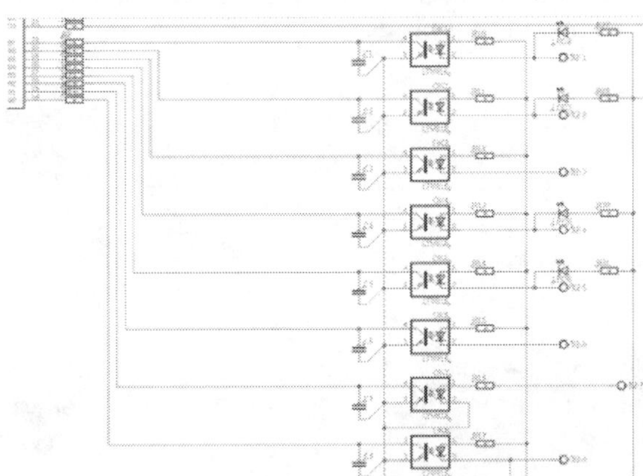

Figure 3 The schematic of input sensor connected to the Microcontroller pins

3.2. Auto starter and Coolant Pump

Until now microcontrollers can't drive electrical equipment directly. Consequently, we use relays. A Relay is just a switch that's opened and closed by an electromagnet, so they're the easiest to understand. Run a current through a coil, and it pulls a piece of metal into contact with another piece of metal and your switch closes. However, when the magnetic field collapses, it generates a reverse voltage, which can be quite large. To avoid the effects of that reverse voltage, we use Diode [6]. After Initializing the microcontroller and the system, it will scan the status of the sensors (First Condition), if all the sensors are on normal mode, you will be able to Switch the key to the "on" position. Meanwhile the microcontroller switches on the port which controls the NPN transistor to latch the starter relay, in addition, turning "on" the LED which is connected it.

After ignition engine start, it is required to wait for 30 seconds until the system reaches set point afterward it will scan the sensors. The microcontroller will provide the on/off signal to the base of this transistor. Which will be driven to saturation, and a current will start to flow. Obviously, in Figure 4, the microcontroller drives two relays:

One of them is used as a starter relay, the other one for turning on the coolant pump during high temperatures in the generator.

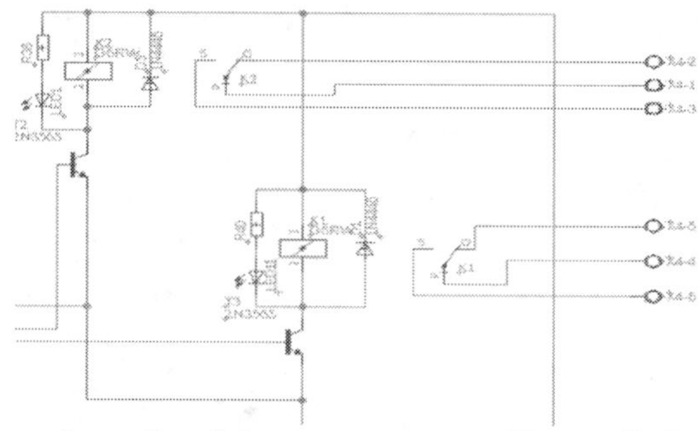

Figure 4 The schematic of the system output (Starter Relay, Coolant pump)

The Control panel typically protects the diesel Generator against High Temperature, Cooling system, Fuel System, Emergency, Overspeed and High Amplitude. Since about one-third of the heat is a result of fuel combustion inside the engine, it must be removed to the outside environment. Thus, a precise cooling system is needed to avoid high temperature, which might cause the generator to shut down [7].

The thermostat sensor is generally used in internal combustion engines, so it can be used to send the signal to the system when there is an over-temperature situation [1], when the system receives a signal from the thermostat, first of all, it will turn on a coolant pump, afterward, the response depends on settings put in the device and give the operator option to shut down immediately or wait (30/60/120) Seconds before shut down and during this period check the thermostat and other sensors. The flowchart in Figure 5 shows the overall workflow for such a case. Error messages will be displayed on LCD for temperature and coolant, as shown in Figure 6 and Figure 7.

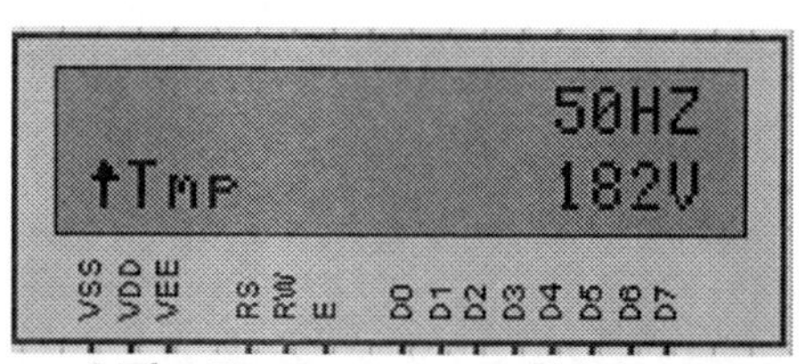

Figure 5 The Error message of high temperature

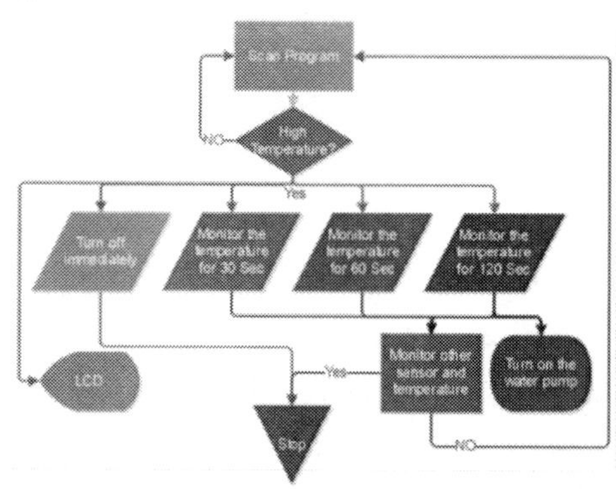

Figure 6 The flowchart of the program in case of high temperature and how it will respond

Generally, applied protection sensors are temperature switches and indicating gauges monitoring the temperature of the coolant. There is a useful protection feat by a low-press sensor positioned in the inlet to the engine's coolant system. It will detect any fail occurring in the coolant circulation with press sensitive device operating before a rise in temperature of the engine [1] when the system receives the signal from the sensor demonstrating that there is a problem in the cooling system, it will respond to this situation in the following steps :

- Shutdown the generator immediately.
- Turn on the alarm LED.
- Display the error on the LCD as shown in Figure 7

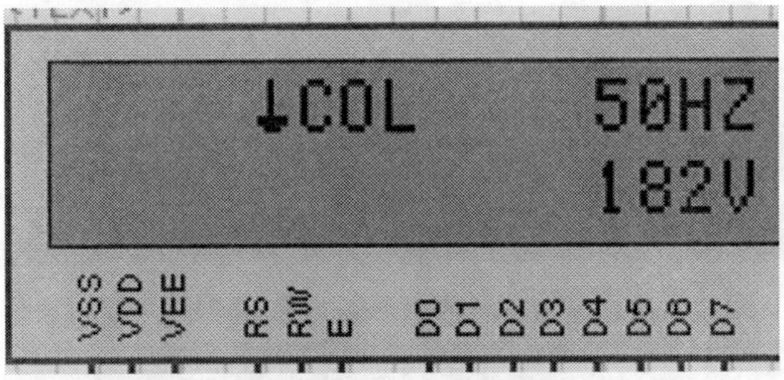

Figure 7 The error message of the cooling system

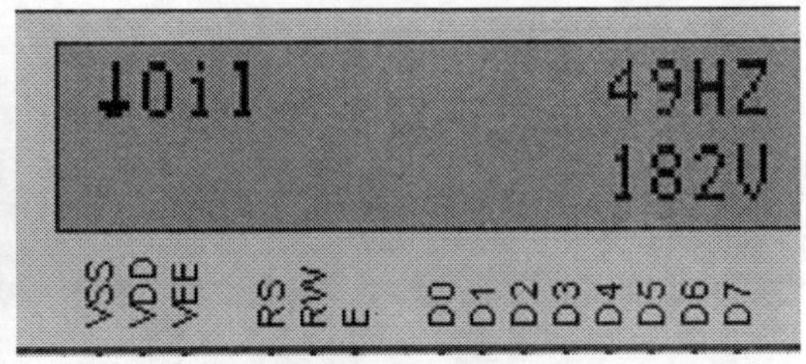

Figure 8 The error message of the oil system

4. Lubricating system

The lubrication system serves to lubricate and cool the pairings of the crankshaft drive, the valve train, and other moving components and to remove local contaminants, wear particles and combustion residues [7]. Sometimes oil press falls because of some reason such as too much tolerance between the piston ring and journals or overheating in the engine. So, any trouble in the system of lubrication will cause clattering sounds coming from the engine during operation due to high friction in the engine's components, causing damage or wearing the engine [1]. If the system receives a signal from the oil press sensor, it will respond to this situation in the following steps :

- Shutdown the generator immediately.
- Turn on the alarm LED.
- Display the error on the LCD as shown in Figure 8

5. Fuel System:

Sometimes due to a fault in Filters, Valves or Press controller, etc.. which affects the fuel flow rate and press to the injectors, this fault leads to a drop in the fuel system press, which is needed to cause ignition in the diesel engine. [7]

The system will respond to this situation in the following steps :

- Shutdown the generator immediately.
- Turn on the alarm LED.
- Display the error on the LCD as shown in Figure 9

6. Stop Emergency:

In emergencies, there is a button for the operator to press, which immediately shuts down the generator, Turn LED on, and display the error as it is shown in Figure 10.

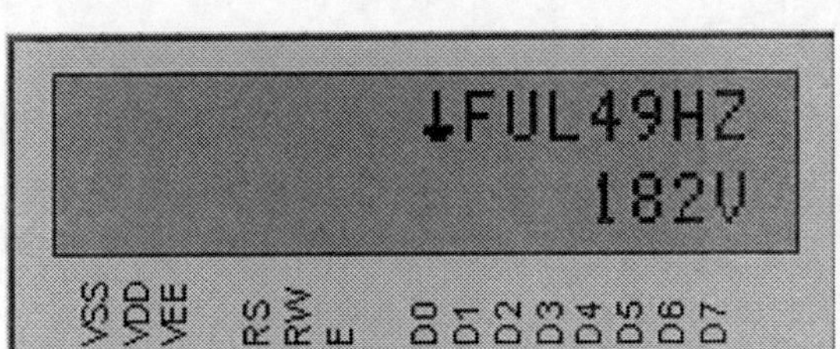

Figure 9 The error message of the fuel system

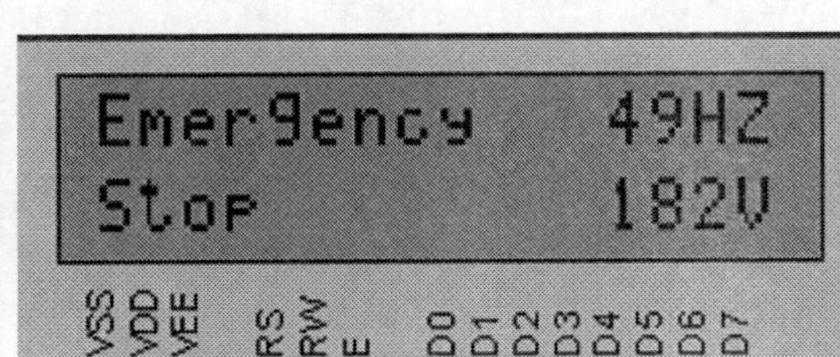

Figure 10 The error message of Emergency

IOP Conf. Series: Materials Science and Engineering **568** (2019) 012107 doi:10.1088/1757-899X/568/1/012107

7. Final Board:

In the end, we design with two options:

- ATmega 32 comes with LCD.
- ATmega 8 just necessary feats and without LCD, with only LEDs to show the errors.

Figure 11 The PCB for ATmega 32

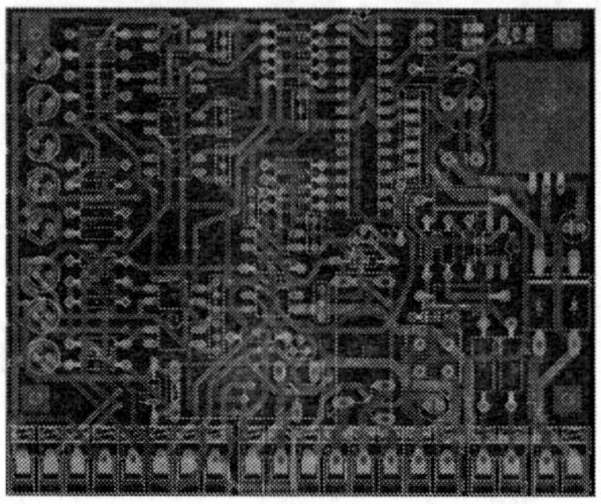

Figure 12 The PCB for ATmega 8

8. Conclusion:

The Main goal of the study was to Build and implement a control panel to protect the diesel generator, and auto-starter with ability display the error using visual alarm and LCD, also, easily used by the operators. The purpose has been achieved as the test results show that it operated as intended. In this paper, we explain How the system works and How will respond to the error signals coming from the sensor (Thermostat, Coolant pump, Fuel, Lubricating, Stop emergency). We have Design and choose the components of the electrical circuit of the control panel to build a stable, reliable system and quickly responding system. It is also Protected from outside disturbance flaw such as over-voltage. The system can recognize the faults and initiate shut down on diesel generator also retain the fault indicator (LED and LCD notification) until the system is reset.

Acknowledgments

This research was supported by EC4ST Company and Eng. Zohier Hamshow who provided insight and expertise to complete this work. I would like to express my very great appreciation to Syeda Adil Afghan for her valuable and constructive suggestions during the planning and development of this research work. Her willingness to give her time so generously has been very much appreciated also for her assistance in keeping my progress on schedule. I would like to thank Eng. Fares Atallah for his contribution to this research paper who provided comments that significantly improved the manuscript.

References

[1] Mahon L L J 1992 Diesel Generator Handbook
[2] A AVR Datasheet of ATmega 32
[3] Samuel I A, Awelewa A A, Emmanuel A, Agbetuyi F A and Katende J 2013 "An Electronic Protection Unit for Diesel Engine Standby Generators" **4(13) 2013.**
[4] "Diesel Service & Supply," Diesel Service & Supply, 1999-2019. [Online]. Available: www.dieselserviceandsupply.com.
[5] Wilson P 2018 The Circuit Designer's Companion, Kidlington: Elsevier Ltd.
[6] Williams E 2014 Make AVR Programming, Safari Books Online.
[7] Isermann R 2017 Combustion Engine Diagnosis, Darmstadt, Germany: Springer Nat.

IOP Conf. Series: Materials Science and Engineering **568** (2019) 012108 doi:10.1088/1757-899X/568/1/012108

Urban buses gearbox faults analysis

S C Dan, R D Mureşan and N Burnete

sebastiancdan@gmail.com

Abstract. Urban busses have to assure transport of thousands/tens of thousands/ hundreds of thousands passengers, on short distances and in extreme working conditions that stress the transmission. The multitude and diversity of buses that are used in public transportation in the cities of Romania, the structure of available fleets, the different road categories that are used for public transportation and the way that public transport is organised – are all factors that influence different faults of urban busses gearbox. In these conditions, reliability and durability of transmission composing elements are affected considerably. This paper is the result of a study that highlights the gearbox faults that appeared in the urban busses of one of Romania's important public transport company. The data analysed in this study has been collected from the order vouchers prepared in the workshop, containing information about the supposed fault that caused the bus to be called for repairs and the actual faults that needed intervention by the personnel, the last ones being considered as faults and noted as such.

1. Introduction

The multitude and diversity of buses that are used in public transportation in the cities of Romania, the structure of available fleets, the different road categories that are used for public transportation and the way that public transport is organised – are all factors that influence different faults of urban busses gearbox.

Compania de Transport Public Cluj-Napoca S.A. (CTP) – Public Transport Company of Cluj-Napoca is one of the most important public transport operators in the country, having to assure an annual transport of more than 200 million passengers [1]. CTP (delegated by Asociaţia Metropolitană de Transport Public Cluj – AMTPC – Cluj Metropolitan Association of Public Transport) ensures public transport services in the city of Cluj-Napoca and in the near-by settlements that are members of the AMTPC (Apahida, Baciu, Chinteni, Ciurila, Feleacu and Floreşti, region known as the Metropolitan Area of Cluj-Napoca) [2]. Established in 2014, CTP took over the actives and activities of until then Regiei Autonome de Transport Urban de Călători Cluj-Napoca (RATUC) – Administration of Urban Public Transport of Cluj-Napoca, an operator administrated by the municipality, but with a semiautonomous activity.

According to the Sustainable Urban Mobility Plan, the Cluj-Napoca Metropolitan Area was set up at the end of 2007 as an association, having as main objectives the sustainable development of its territory as a whole, the common economic and social development, the improvement of the quality of the environment and the quality of life. The metropolitan area is 1 630 km2, representing 24% of the total area of Cluj County, and includes 55% of the total population of the county. [3]

For providing the public transport services, in 2018 CTP used a fleet of 27 tramways, 257 buses, 81 de trolleybuses, 10 minibuses and 11 electric buses [4] and operates one minibus route, 3 tramway routes, 7 trolleybus routes, 43 urban bus routes and 24 metropolitan bus routes, summing a total length

IOP Conf. Series: Materials Science and Engineering **568** (2019) 012108 doi:10.1088/1757-899X/568/1/012108

of 650,45 km of routes serviced [5]. The present study took into consideration only the data that referred to the buses used for providing public transport services, the fleet used in 2018 sorted by maker brand and model is presented in Table 1 [6].

Table 1. The fleet of buses available for providing public transport services, segregated by producer make and model, and number of each model available.

	Renault Ares	Renault Agora Long	Renault Irisbus	Renault R312	Irisbus Agora	MJT	MAN	Scania Hess	Iveco Urbanway	Solaris Urbino 18	Mercedes-Benz Conecto G
No.	3	57	10	60	46	2	7	7	10	40	15

2. Study conditions

In case of buses that weight over 4 tons, the most common solution for the transmission is: rear drive, with the engine and the gearbox mounted in the back of the bus. This display has the advantage of a more simple construction, because of the smaller distance that appears between the engine and the driving axle, and the engine-gearbox assemble is positioned outside of the area destined for passengers, which ensures ease of access to the engine and the gearbox and offers the possibility to lower the floor for an increased accessibility. [7]

Causes of gearbox faults in urban busses can be accidental, or caused by the ageing of the components and auxiliary systems that are subjected to a very demanding exploitation regime. To implement an effective preventive and systematic maintenance in order to reduce and even anticipate fault developments, when considering such a large and vast fleet, considerable human, technological and financial resources are needed. Considering our country, where investments related to public transport, if even considered, are reduced to the renewal of the fleets due to the age of the vehicles, preventive maintenance is hard to implement because of the many shortcomings remembered that refer to technological resources and qualified personnel.

The faults identified in the present study can be divided by their technical nature in electrical faults, hydraulic faults, or mechanical faults. For the first category, the most representative are the faults that affected the wiring, electrical contacts and sensors. Representative for the hydraulic faults are the ones that occurred on hydraulic lines, while for the mechanical faults, the ones regarding shafts and gears are the most representative. The last ones, even if numerical there aren't many, they are the most serious faults, that couldn't be repaired and eventually led to the replacement of the gearbox, making them the most expensive faults registered.

The main way of collecting data for the analysis involved following a complex and somewhat rudimentary process, that relies on the deductions made by the drivers, who noticed an anomaly in the functioning of the bus, and reported that anomaly to the mechanics in the workshop. After the mechanics established the fault, they make a written request to the chief engineer to approve the repair and the removal from the storeroom of the needed components. Taking into consideration the fact that most of the buses that are in the studied fleet where acquired as second-hand vehicles so they didn't have the necessary equipment for maintenance, and being outdated from a technologically point of view, they don't have an electronic diagnosis, so diagnosing a fault relies only on the knowledge and experience of the personnel, but in order for the mechanics to evaluate the status of a gearbox, the bus needs to be recalled to the workshop and that leads to decreased productivity.

From the total of 257 buses studied, only 65 units have been bought first hand and have support from the producer and come with electronic diagnostic systems, making the maintenance easier and less time consuming. For three quarters of the studied fleet, the only way to establish faults depends on the drivers experience to notice changes in the functionality of the bus, and on the mechanics experience and knowledge to repair the faults in the workshop.

3. Urban buses gearbox fault analysis

From a total of 257 buses that have been used in 2018 for public transport services in Cluj-Napoca, 15 units were newly acquired and their delivery was spread out during the calendar year, so they weren't considered in this study. The remaining 242 buses that were used regularly in 2018 for providing public transport services in both urban and metropolitan area were analysed.

In 2018, in the bus maintenance and repair workshop, 14705 records were registered as faults in the order voucher. Of these, 359 interventions were planed according to the technical revision plan, 176 entries have been caused by road accidents, which leaves a total of 14170 records noted as faults. These faults are sorted in figure 1 based on the nature of the affected systems, as follows: 3331 electrical system faults, 2680 cooling system faults, 1742 braking system faults, 1546 faults were related to air system malfunctions, 1308 entries referred to engine fault, 977 body damages, 772 suspension system faults, 626 steering system faults, 589 transmission faults, 526 fuel system faults, 40 faults of the hydraulic system and 33 other interventions.

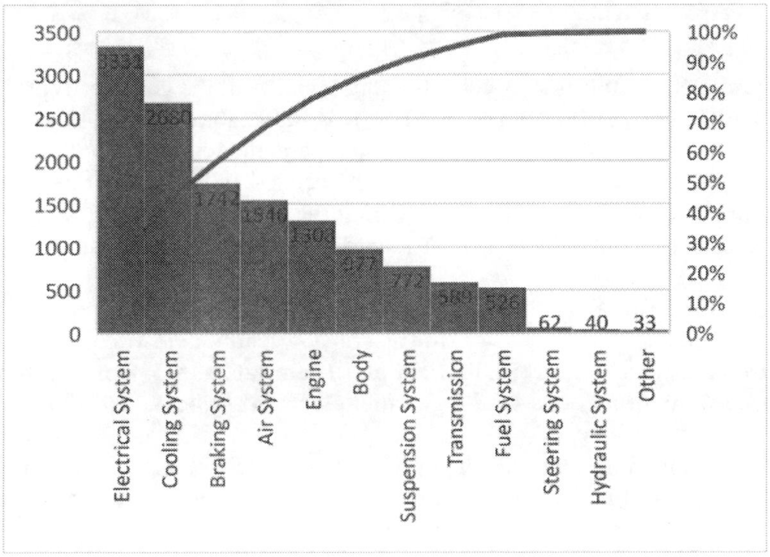

Figure 1. Pareto Diagram showing the number of faults recorded in the studied buses based on the nature of the affected systems

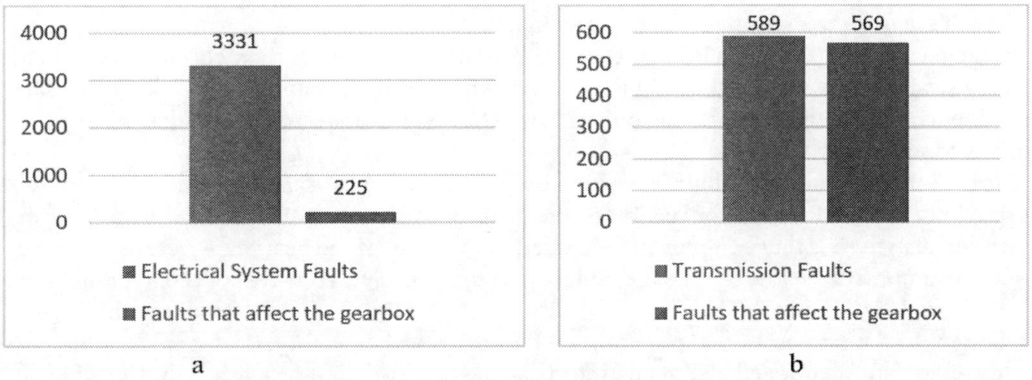

Figure 2. Number of faults that affect the electrical system (a) and the transmission (b) and how many of them are related to the gearbox the gearbox

Analysing the diagram in figure 1, one can observe that most of the fault, 3331, were registered as electrical system faults, which is explainable considering that all functional systems of a bus use electrical components. In order to analyse the faults that affect a particular component of the bus, all

IOP Conf. Series: Materials Science and Engineering **568** (2019) 012108 doi:10.1088/1757-899X/568/1/012108

of the systems faults that affect that component have to be considered. In this paper, the analysis that refers to the urban bus transmission, all faults that affected the functionality of the transmission were taken into consideration, including the ones registered in a general category named transmission faults and some faults of the electrical system.

Figure 2 shows the number of gearbox faults reported to the total faults registered in the electrical system and, respectively, the transmission.

Out of the 3331 faults registered in the electrical system, 225 directly influence the gearbox, representing 6,75% of total. But regarding the faults registered in the transmission, we have a completely different situation, where 569 out of 589 faults directly influence the gearbox, that is 96,6% of total.

Table 2. Gearbox faults segregated into categories corresponding to the affected component system.

	Lubrication System	Electrical System	Trans-mission	Preventive Verifications	Cooling System
No.	287	234	188	89	24
%	35	28	23	11	3

In total there have been recorded 823 faults that affected the functionality of the gearbox, representing 5,8% of the total number of faults registered on buses in 2018. It is noticeable that most of the faults were registered in the gearbox lubrication system – 287, meaning a percentage of 35% out of the total, but 130 entries refer to oil change adjacent operations that were needed in order to complete other repairs. Other than that, 69 entries refer to the necessity of filling the gearbox oil level, and other 67 times, an oil leakage needed to be sorted out. The rest of 21 recorded faults implied a change of a hose, a gasket, a grip ring or a pipe.

Gearbox electrical system faults in total of 235, registered only 2 operations, 121 entries note a necessary intervention on retarder electrical contacts, and the rest of 114 gearbox electrical system faults refer to the wiring of the gearbox. After the gearbox lubrication system faults, (35% of total) most of the faults recorded were caused by malfunctions of the gearbox electrical system, with 28% of total faults.

Out of the 89 preventive verifications, in 26 cases the purpose was an oil check, 52 times a general inspection of the gearbox was done, and in 11 cases that needed the bus to withdrawn from service, other components of the transmission were verified. It is worth mentioning that these checks were ordered by the bus drivers who noticed a different behaviour of the bus, so these operations are not part of the regular procedures scheduled according to the annual revision plan. These interventions represent 11% of the reasons to call a bus in the workshop.

In the case of the 24 interventions recorded that are related to gearbox cooling system faults, things are very straight forward, all of the 24 times there was a problem caused by the gearbox heat exchanger breakdowns. These faults represent only 3% of the total number of faults that have affected buses gearbox functionality.

Most of the mechanical failures sustained by the components of the transmission were related to the gearbox, not only to the selector section of the gearbox assembly, affecting the planetary gears and the shafts, but also to the other components included in the body of the gearbox – because of the complexity of such an automatic gearbox – like the clutch, part that suffered 13 faults, or the hydraulic retarder, part that recorded 23 failures. This category also contains the faults suffered by the gearbox shafts, diaphragm, or the rare cases that required a total repair of the gearbox or its replacement. Also, still in this category are mentioned the small interventions that implied a gearbox buffer change, or some gaskets, fastening collars and studs. Although most of the faults that affected the functionality of the gearbox were recorded in the hydraulic system, almost half of them were, in fact, oil change operations necessary in the process of other repairs. So, after analysing the urban buses gearbox faults, it is noticeable that the biggest issue was the electrical system of the gearbox, with the most faults recorded during a calendar year. The faults recorded as transmission faults, although not the most common, were

IOP Conf. Series: Materials Science and Engineering **568** (2019) 012108 doi:10.1088/1757-899X/568/1/012108

cause of the longest periods of inactivity, which translates into lack of productivity and implicitly loss of money, in addition to being the most expensive operations: shafts repairs, planetary gears repairs, and other components that needed special expert interventions.

The nature of these faults and their high number is explicable considering the age and the many years of service of most of the analysed buses, with only 50 of the 242 buses having less than 2 years of service in 2018, the rest of them, having more than 10 years of service in public transport. In addition, for the other 192 buses, being technologically obsolete, it is becoming more and more difficult to find the necessary spare parts and materials needed for their maintenance. Another big disadvantage is the fact that these buses, as they were second-hand acquisitions, no longer benefit from technological support and know-how from the producer, and repair manuals and workshop instructions, if any available, are at most incomplete and in a foreign language. In such conditions, repairing and maintain a bus in a functioning state relies only on the experience and knowledge of the personnel that operate and repairs the buses, fact that doesn't help reducing the service time and the period of inactivity in case of a breakdown. In these conditions, is in fact a remarkable achievement that there were 89 records of unscheduled checks, operations that prevented further damages and faults. And unfortunately, in these cases, also all the process depends only on the knowledge and experience of the drivers, who noticed and reported in a timely manner the possibility of an occurring breakdown of the gearbox.

On the other hand, it is worth mentioning that in 2018 there was no recording of a gearbox damage caused by road accident in which a bus was involved.

4. Conclusions

After analysing the gearbox faults of a fleet containing 242 urban buses destined for public transport over a period of a calendar year, 2018 respectively, a total number of 823 faults affected the functionality of the gearbox, representing only 5,8% of the total number of faults recorded in 2018. Taking into consideration the daily exploit conditions imposed by the urban area and the very busy traffic charts of these buses, which includes scheduled transports at regular intervals starting at 4:30 in the morning and ending at midnight, and considering that the average age of the fleet in 2018 was more than 15 years old, automatic planetary gearboxes prove once again, if there was any doubt, that they are the right choice for the urban buses destined for public transport, adding to the comfort and ease of the driving, the advantage of a reliability proven over time.

References

[1] ***http://www.insse.ro/cms/sites/default/files/field/publicatii/transportul_de_pasageri_si_marfuri_pe_moduri_de_transport_in_anul_2018.pdf, pg. 18;

[2] ***"Opportunity Study in order to delegate the management of the public local public passenger transport service through regular transports to the Metropolitan Public Transport Association Cluj for the member-administrative territorial units" („Studiu De Oportunitate în vederea delegării gestiunii serviciului public de transport public local de persoane prin curse regulate în Asociația Metropolitană De Transport Public Cluj pentru unitățile administrativ-teritoriale membre", in Romanian), pg. 13;

[3] ***"Sustainable Urban Mobility Plan Cluj-Napoca, Final Report, Version II, November 30, 2015" („Planul de Mobilitate Urbană Durabilă, Raport Final, Varianta II, 30 Noiembrie 2015", in Romanian);

[4] ***Compania de Transport Public Cluj-Napoca S.A., "Annual Activity Report" („Raport anual de activitate", in Romanian), 2018;

[5] ***Compania de Transport Public Cluj-Napoca S.A., "Exploitation Data" („Date de exploatare", in Romanian), 2018;

[6] ***Compania de Transport Public Cluj-Napoca S.A., "Bus Depot Annual Activity Report" („Raport anual de activitate Autobaza Autobuze", in Romanian), 2018;

[7] Naunheimer H., Bertsche B., Ryborz J., Novak W., *Automotive Transmissions. Fundamentals, Selection, Design and Application*, tr. A. Kuchle, Springer-Verlag Berlin Heidelberg, 2011;

IOP Conf. Series: Materials Science and Engineering **568** (2019) 012109 doi:10.1088/1757-899X/568/1/012109

Video analysis system of the gait cycle and posture of people wearing high heels

M. I. Baritz[1]

[1]University Transilvania Brasov, Product Design, Mechatronics and Environment Department, B-ul Eroilor, no.29, Brasov, Romania

mbaritz@unitbv.ro

Abstract. Gait cycle analysis is a comprehensive source of information, usable in the medical field, in locomotion, sporting recovery procedures or to increase performance in various sporting fields. Study methods of walking cycle parameters are very diverse, starting from biomechanical analysis, computer simulations and modelling, and completing with video analysis of kinematic and dynamic parameters. The use of optical and video methods is very widespread in all medical, sporting, ergonomic and even economic fields. Thus, the first part of the paper analysis the general aspects regarding the use of optical methods of acquisition and processing of video sequences of the normal gait cycle. Also, in the same context of the introductory part of the paper the aspects of the biomechanics of the normal and high heels cycle for the females are presented. In the second part of the paper is presented an experimental setup for investigating, with the sample of subjects chosen, the cycle of gait with shoes having different dimensions of the heel height. In the third part of the paper are presented the recordings made on *footscan* pressure plate and also the data acquisition on the especially sensory insole worn by the subjects. In the final part of the paper are presented the results of the processing of this information and the conclusions from the experiment.

1. Introduction

The posture of the human body in stability or in the gait cycle can indicate a number of issues related to the behavior of the subjects, their health or comfort behavior, and sometimes the level of adaptability to the environment. A number of scientific and research papers address various aspects of the whole body or component segments in order to identify and / or use functional parameters in order to achieve performance or to rehabilitate neuro-motor functions affected by various causes. As mentioned in the American Chiropractic Association documents [1], posture is defined as being "the position in which we hold our bodies while standing, sitting and lying down" or waking and running. "Good posture is the correct alignment of body parts supported by the right amount of muscle tension against gravity. Without posture and the muscles that control it, we would simply fall to the ground." From a functional point of view, posture is in turn related to the state of stability and balance of the human body, and the gait movement is related to the health of the skeleton, joints, muscles and the central nervous system. Analyzes and measurements to be performed on the human body cycle should identify as clearly as possible the causes that can alter the normal structure and not just the effects. From the point of view of rigid body dynamics there are two approaches, namely: the problem of direct dynamics in which the forces applied to a mechanical system are known and the main objective is the determination of the parameters of the movement, respectively the second one, which by the inverse dynamics the movement of the mechanical system is defined with initial details, and the main objective is to determine the forces

IOP Conf. Series: Materials Science and Engineering **568** (2019) 012109 doi:10.1088/1757-899X/568/1/012109

that generate this movement [2]. In general, all human body motion analyzes refer to a 3D coordinate system, from which, with the aid of center of mass (COM) movement trajectories or other markers positioned on different segments, one can obtain a full "image" of posture, stability or balance.

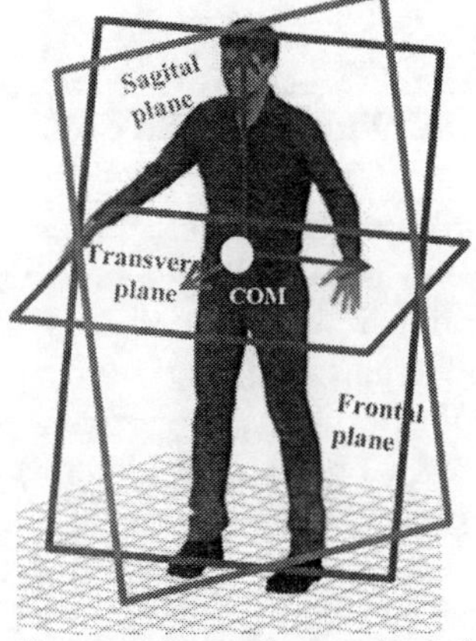

Figure 1. Coordinate system for analysis human body [3]

In biomechanical analyzes of posture and stability an extremely important issue is the shape and dimensions of the footwear of the subjects and the way (trajectory, duration) in which they move with them. As shown in a series of researches [4,5] in the sagittal plane, flexion of the knee is greater, and hip flexion is lower when walking with high heel shoes (HHS). These changes in the muscles activation system during gait with HHS are still not well documented and the assessment methods are varied (from electromyography, to video or sensory methods). In addition to these aspects, it is necessary to analyze and approach the HHS port by female subjects as a series of studies [6] show that approximately 86% of “American women wore shoes that were smaller than their feet, possibly for reasons of reducing the space for foot motion in the shoe. The inference from this evidence is that if arch height increases because of high heels, the foot will shorten, causing the shoe to become too long for the foot, which would alter fit, comfort, and stability”[6]. Interesting aspects are identified in HHS port analyzes not only in the normal walking cycle, but also in running, jumps or stepping, in more complex dynamic actions and leading to higher instability rates, with different studies showing that there is an interaction between speed travel and height of HHS [7]. Also other results from experiments indicate that subjects wearing HHS are moving more slowly, with a cadence that remains constant, with a base of support (BOS) (the distance between the planting surfaces of the feet on the ground) slightly enlarged but with the length of the steps reduced in comparison with no heels gait. All these observations indicate that the use of HHS with exaggerated dimensions to the anthropometric dimensions of the human subject is unbeneficial and causes inconvenient behavior, changes in the shape and structure of the foot, and an increase in instability that may generate over time her, lesions and loco-motor dysfunctions. A pertinent analysis of how the human subject moved with HHS, especially on the inclined plane (0^0-30^0), was able to highlight a number of aspects that change the comfort of the human subject during the walking cycle or even change the bipedal position by changing the position of center of pressure (COP) on the BOS [9]. As shown in [10] “Dynamic Systems Theory (DST) provides a non-linear framework to analyze human movement by representing inter-segmental interactions in angle-angle diagrams. DST offers an accurate solution to study coordination in human movement, but it also requires expensive hardware and very specialized biomechanical software.” However, some combined experimental systems (mechanical, video, optoelectronic) can overcome financial barriers, providing effective and feasible results for specific applications.

2. Theoretical aspects of gait cycle with or without HHS

The cycle of gait on horizontal surfaces as well as on inclined or stepped surfaces generates a series of changes in the osteo-muscular systems that are reflected in the walking trajectory, the trajectory of the

COM projection at the pressure center or the pressure distribution across the plantar surface.

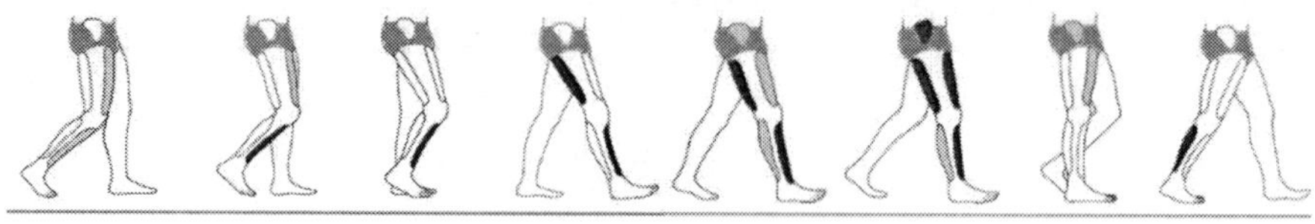

Figure 2. Gait cycle [2]

As mentioned in the fundamental biomechanical studies of gait [2] "traditionally the gait cycle has been divided into eight events or periods, five during stance phase and three during swing. The names of these events are self-descriptive and are based on the movement of the foot. The stance phase events are as follows: Heel strike initiates the gait cycle and represents the point at which the body's center of gravity is at its lowest position; Foot-flat is the time when the plantar surface of the foot touches the ground; Mid-stance occurs when the swinging (contralateral) foot passes the stance foot and the body's center of gravity is at its highest position; Heel-off occurs as the heel loses contact with the ground and push-off is initiated via the triceps muscles, which plantar flex the ankle; Toe-off terminates the stance phase as the foot leaves the ground. The swing phase events are as follows: Acceleration begins as soon as the foot leaves the ground and the subject activates the hip flexor muscles to accelerate the leg forward; Mid-swing occurs when the foot passes directly beneath the body, coincidental with mid-stance for the other foot; Deceleration describes the action of the muscles as they slow the leg and stabilize the foot in preparation for the next heel strike." Recently, the use of wearable sensor combinations (accelerometers, gyroscopes, force sensors) made it possible to measure the gait cycle parameters both in the experiment area and in the outdoor areas in which the human subject moves under real conditions, as well as continuously wireless recording and without large data loss [11]. The use of the heel shoes changes the gait cycle as well as the number of steps or the position of the whole body. The higher the heel, the more difficult the cycle to perform, and the more the movement of the ankle (of the planting surface) from the ground is changed. To simplify experimental instrumentation, it is considered that the foot movement is approximately flat and executed only in the sagittal plane [12]. But the most important data related to the HHS gait cycle is the ability to record planting pressure directly on the foot that is in the shoe with a possible option by using force sensors connected to an Arduino acquisition board [13,14].

3. Experimental setup

Experimental setup used in this research consists of a set of equipment that allows recording of planting pressures during a walking cycle and respectively recording the posture video and the trajectory of human body segments involved in locomotion (lower limb, joints) in the sagittal direction - along the pressure plate (Figure 4). Subjects participating in the experiment were selected according to initial criteria series that determined their choice in relation to their comparable anthropometric dimensions, without loco-motor dysfunctions, feminine gender and performing the same type of activity. In addition, the 4 subjects were chosen according to the following parameters: mean age (23 years), average height of 160.83 cm, mean body weight 51.5 kg and average shoe size 37.5. All the subjects provided informed consent prior to the test. The experimental protocol was conducted according to the Declaration of Helsinki [15]. They have been informed about the protocol of the experiment and the devices that were used. The recording of data on planting pressures and posture during the walking cycle was performed using the experimental structure of Figure 3 and four forms of footwear (small, medium 4 cm, large 8cm and very large 10 cm) (Figure 4.)

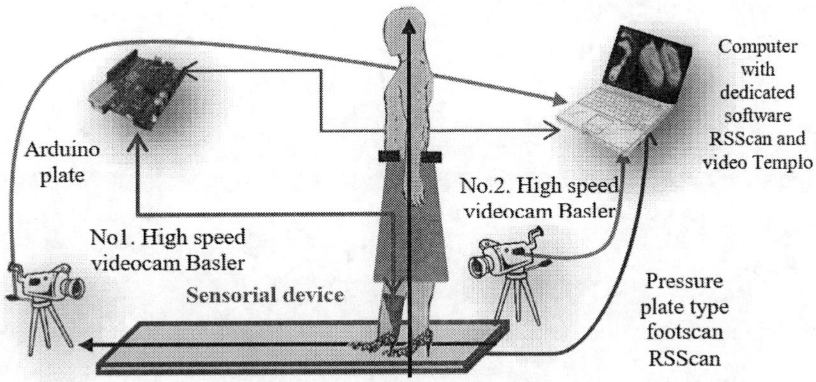

Figure 3. Experimental setup

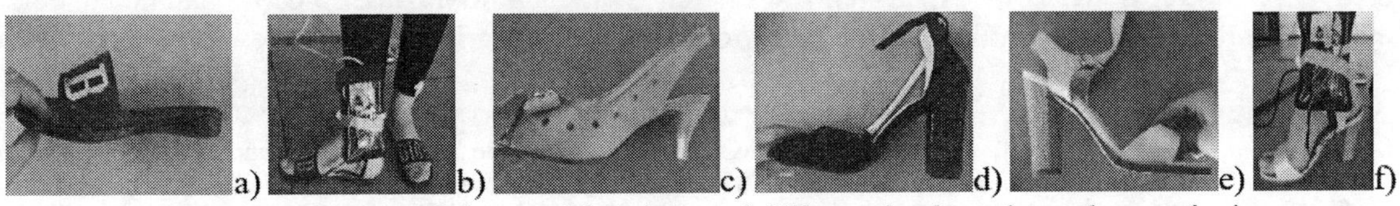

Figure 4. Types of shoes with heights of different heels and attachment device

4. Results and conclusions

The recorded data was processed by pressure plate software, *RSScan footscan 7.97 Gait 2nd generation*, and video sequences using Templo Contemplas and Kinovea dedicated software. As can be seen from Figure 6, the trajectories highlight the difficulty of moving with HHS (blue) and the comparable movement values in the version with the heel shoe of 4 cm (green), respectively with the values of the heelless shoe variant (red).

Figure 5. Video processing of knee trajectories in the cycle with and without HHS

When a sensorial device was attached to the foot and recorded simultaneously on the pressure plate, this similarity of variation was revealed in the case of the 4 subjects investigated and for which the four shoe variants were used (Figure 7 from - heelless, with a 4 cm, 8 cm and 10 cm high heel).

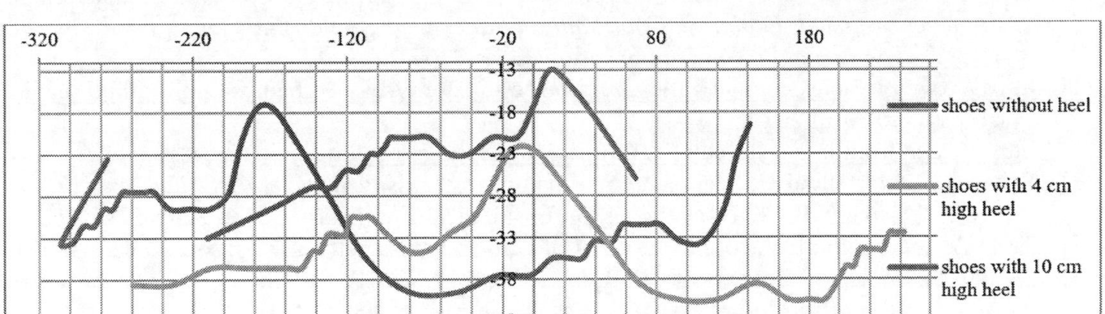

Figure 6. Example of knee trajectories during gait with and without HHS for the same subject

At the same time, video sequence software applications allowed the identification of subjects with difficulty in adapting to the HHS.

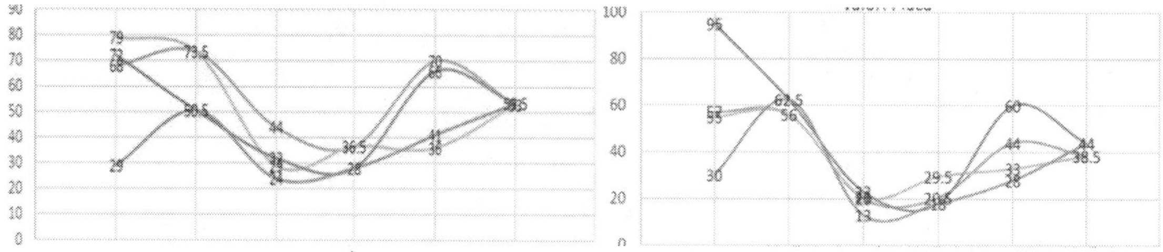

a) b)

Figure 7. Comparative values recorded on the sensor system (a) and on the pressure plate (b)

In this sense, the values of speed variations, acceleration, horizontal displacement (sagittal-axis Oy) or vertical recorded with the sensory and video system (simplified) can become basic parameters of the analysis and can be used in experimental research in the outdoor situation to evaluate, under any environmental or surface conditions, the behavior of subjects wearing HHS.

Acknowledgements
In these experiments we've developed the investigations with equipment from "Advanced Mechatronic Systems Research Center - C04" and Applied optometric Laboratory at University Transylvania of Brasov, with the students help from Medical Engineering study Program.

References

[1] Wyatt L.H, *Healthy Living, Patient Information from the American Chiropractic Association, Tips to Maintain Good Posture*, Texas Chiropractic College, 2006
[2] Vaughan Ch, *Dynamics of Human Gait*, Kiboho Publishers, South Africa, 1999
[3] www.makehumancommunity.org/ accessed in April 2019
[4] Pratihast M, Al-Ani A, Chai R, Su S, Naik G, *Changes in lower limb muscle synchronisation during walking on high-heeled shoes*, Healthcare Technology Letters, pp. 1–3, 2018
[5] Opila-Correia K.A., *Kinematics of high-heeled gait*, Arch Phys Med Rehabil, 71, (5), 1990
[6] Cowley E, Chevalier T, Chockalingam N, *The Effect of Heel Height on Gait and Posture, A Review of the Literature*, American Podiatric Medical Association Journal, vol 99, no6, 2009
[7] Rangra P, Santos D, Coda A, Jagadamma K, *The Influence of Walking Speed and Heel Height on Peak Plantar Pressure in the Forefoot of Healthy Adults*, Clin Res Foot Ankle 5, 239, 2017
[8] Emmanouil A, Rousanoglou N, *Effect of high-heeled shoes on postural control in the upright and the leaning body stance*, Physical Medicine and Rehabilitation Research, 2018
[9] Mishra E, Jenaa S, Bhoia C, Arunachalama T, Kumar Panda S, *Effect of high heel gait on hip and knee-ankle-foot rollover characteristics while walking over inclined surfaces*, The Foot, 40, 8-13, 2019
[10] Muñoz J.E, Villada J.F., Casanova S., Montoya M.F, Henao O.A., *Dynamic Systems Theory in Human Movement. Exploring Coordination Patterns by Angle-Angle Diagrams Using Kinect*, https://www.researchgate.net/publication/328138367 2018
[11] https://www.skysoles.com.au/galley-gossip/foot-anatomy-biomechanics-and-the-high-heel/, accessed in April 2019
[12] http:/ /naturalrunningcenter.com/2012/06/21/walking-vs-running-gaits/, accessed in April 2019;
[13] Sabatini A, Martelloni C, Scapellato S, Cavallo F, *Assessment of Walking Features From Foot Inertial Sensing*, IEEE Transactions On Biomedical Engineering, Vol. 52, No. 3, 2005
[14] Trica E., License project, coord. Baritz M.I., University Transilvania Brasov, Romania, 2018
[15] Donati, M, Cecchi, F, Bonaccorso, F, Branciforte, M, Dario, P, Vitiello, N, *A Modular Sensorized Mat for Monitoring Infant Posture*, Sensors 14, pp 510-531, 2014

IOP Conf. Series: Materials Science and Engineering **568** (2019) 012110 doi:10.1088/1757-899X/568/1/012110

Research on the relationship between chromatic vision and biomechanics of the posture and the gait cycle

M. I. Baritz[1]

[1]University Transilvania Brasov, Product Design, Mechatronics and Environment Department, B-ul Eroilor, no.29, Brasov, Romania

mbaritz@unitbv.ro

Abstract. Visual screening is a complex source of information, usable in the medical field, in vision recovery or sports, to increase performance in various sports fields that require the use of the visual system. For the study of the visual system a variety of methods are used, starting from visual parameters analysis, biomechanics, computer simulations and modelling and completing with optical and video analysis on kinematic and dynamic parameters. In many researches, a very close link between visual function and loco-motor function is identified, considering that the visual system is that which, when in dysfunction or overexploited, can cause changes in the characteristics of the gait cycle and posture of the human body. Thus, the first part of the paper analysis the general aspects of the behaviour of the visual system without dysfunctions and its connection with the loco-motor system. In the second part of the paper is presented an experimental setup for the analysis of the influence of chromatic stress on the way of movement and the posture of the human subjects, and in the third part of the paper the recordings made on footscan pressure plate and respectively with the perception assessment software application colour are presented. In the final part of the paper the results of the processing of this information and the conclusions from the experiment are presented.

1. Introduction

Chromatic vision is one of the most important features of the human visual system, providing the human subject with the ability to distinguish and visually perceive elements in the environment at a much higher level (theoretically about 16 million colors) than they would dyschromatopsia (view in gray levels - 256 levels). The ability of eyeball to have good chromatic visual acuity is especially important for assessing the behavior of subjects in various activities in which continuous, intermittent, light colored beams are being used. Changing the biomechanical behavior of subjects in these situations is important to be evaluated through various procedures that meet current standards to prevent musculoskeletal dysfunction, avoid motor dysfunction, and even improve ergonomics of working place. The posture of the human body in stability or in the gait cycle can indicate the status of the subject's behavior and can be related to their health or comfort level and sometimes to the level of adaptability to the environment.

A number of scientific and research papers address various aspects of the whole body or component segments in order to identify and / or use functional parameters in order to achieve performance or to rehabilitate neuro-motor functions affected by various causes. In these cases a visual function at very low levels of its parameters may additionally affect the adaptability level, change the duration and level of motor recovery (where appropriate), or even change the overall perception of the environment [1]. In addition to the chromatic characteristics of the visual function, 3D binocular vision is another parameter that can influence the spatial perception of the depth of the environment, the hierarchy of the dimensions

IOP Conf. Series: Materials Science and Engineering **568** (2019) 012110 doi:10.1088/1757-899X/568/1/012110

of the objects in the visual field and, last but not least, the discrimination of the movements and their positions in relation to posture or movement of the human body.

As shown in a series of research, [2] the quality of life in all its components is severely impaired in these visual dysfunction conditions, and its modifications due to dysfunctions are difficult to remove or recover. Also, as stated in World Health Organization documents (WHO) "visual impairment is an increasingly prevalent public health problem, particularly in aging populations. An estimated 1.64 million individuals in Japan and 3.22 million in the US have impaired vision. This number is expected to increase in developed countries because of the increased incidence of age-related eye diseases, such as glaucoma, macular degeneration, and diabetic retinopathy. Vision is a key element of health and quality of life (QoL). Blindness and visual impairment are among the 10 most common causes of disability in the US and are associated with both diminished QoL and a shorter life expectancy. Individuals with impaired vision have an increased risk of road traffic accidents and are more likely to cease or curtail driving. Visual impairment has also been associated with emotional distress and infrequent socialization. Thus, visually impaired individuals may have functional and psychosocial burdens that affect various aspects of their daily lives. Further, visually impaired older adults have difficulties with physical functioning, which increases the risk of falls and fractures. A fear of falls and perceived risks of mobility have been reported to limit physical activity in visually impaired patient." Color vision is achieved by the ability to transform various wavelengths of light radiation from the photosensitive cells of the retina, although the photochemical process taking place at the level of each cell is not yet well known. The most interesting research on visual color perception is the investigation of the chromatic sensations of human subjects in low illumination levels or the psychological and physiological effects of colors on the quality of life [3,4].

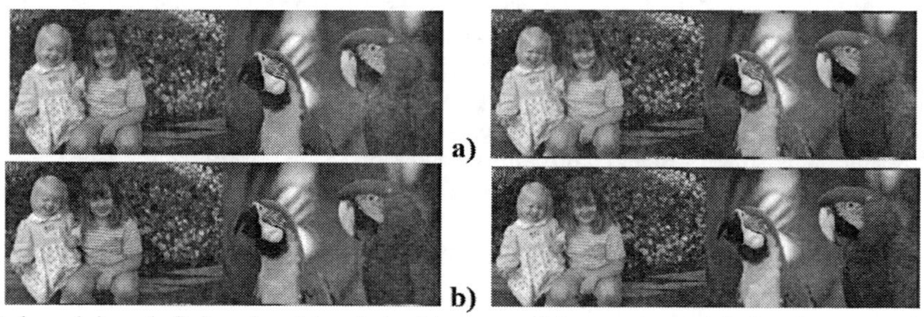

Figure 1. Color vision deficiencies (a) original images, (b) protanop, (c) deuteranope, (d) tritanope [5]

As shown in the paper [6], the authors identified, following research carried out on a sample of 14 totally blind subjects, an intrinsic connection between the postures, respectively the gait cycle and the visual system in dysfunction, concluding that the visual aids can also help improve movement, motor activity, or precision. Later on, these observations have been developed in the field of constructing robotic systems [7,8] with artificial vision for the creation of autonomous locomotion systems in the form of bipedal humanoid [9,10,11]

2. Experimental setup

Experimental setup used in this research consists of a set of equipment that allows the recording of *footscan* planting pressures during a walking cycle and simulation of chromatic stress. The area where the experiment was conducted is sound and optical isolated to avoid the influence of some audio-video parasitic stimuli in the environment. The sample of subjects selected for the experiment were chosen following similar general criteria of the same type: same age, same type of basic activity, average height 1.67 m and body weight 52.4 kg, footwear and leg clothing and the like. All of these conditions were followed throughout the experiment in order not to hurt the measurements made. In addition, the subjects were in a proper metabolic state so that chromatic stress did not affect the well-being they must manifest during the experiment. All the subjects provided informed consent prior to the test. The experimental protocol was conducted according to the Declaration of Helsinki [12]. They have been informed about the protocol of the experiment and the devices that were used. The Chromatic Stress Simulation Device (CSSD) is built to allow exposure of the subject's visual system to illuminations of varying intensity and illumination with different wavelengths for different periods of time before recording the cycle of movement on the pressure plate (see Figure 2).

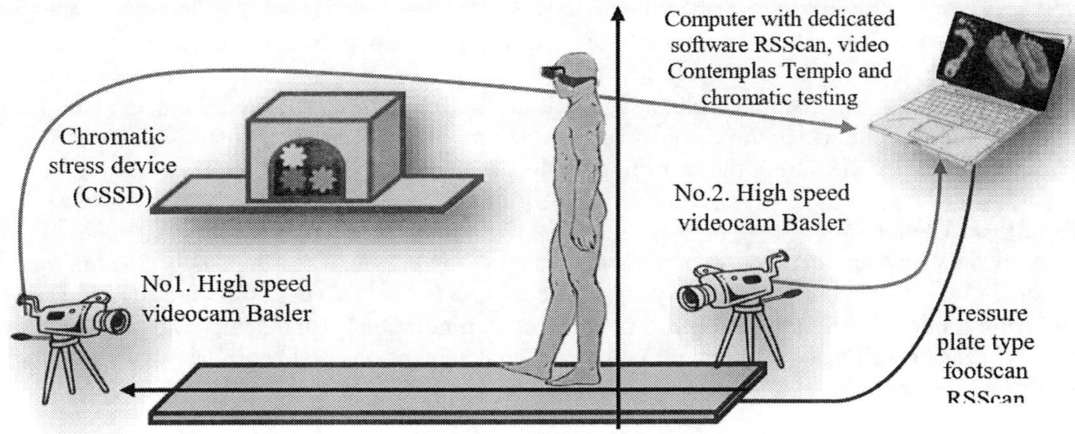

Figure 2. Experimental setup

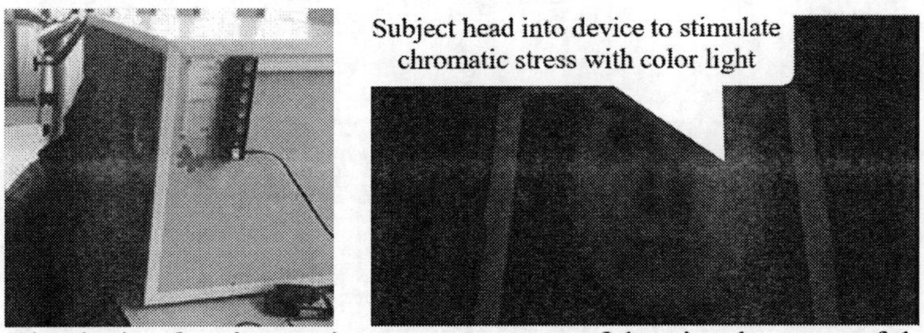

Figure 3. The device for chromatic stress exposure of the visual system of the subjects

The registration and evaluation procedures were carried out with all the subjects in the sample and they required at their onset to perform initial tests in order to identify the level of normality in all the subjects involved. The initial tests were based on measurements on two systems: biomechanical by measurements of the normal (no stress visual) cycle on the *footscan* pressure plate and the second, the history of the visual system and the evaluation of the chromatic performance.

All determined values are the initial values that will be compared with those obtained following the application of chromatic stresses by CSSD. During the 20 minutes of the chromatic stress procedure, the subjects were followed up with a video camera (Figure 3) mounted inside this device. The investigator's attention during the exposure was focused on the position and the way the subjects keep their eyes open by watching their visual behavior.

The system is equipped with 5 types of colored LEDs: red, green, blue, yellow and white, and the maintenance time on the stage was set at 20 min, since the test subjects are very young (23 years) and accommodated it occurs very quickly. In the procedure, it was chosen that the subjects were tested in turn in exposure to 3 types of consecutively colored light. The first time was red (5 min), then blue (5 min), and the last type of stimulation was done with all types of colored LEDs on, so the stimulation was strong (10 min).

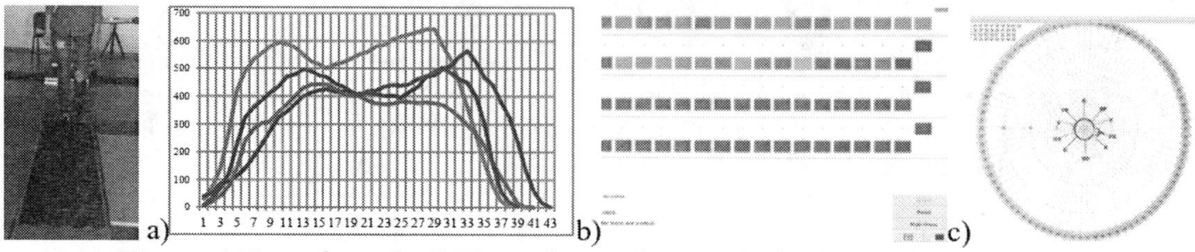

Figure 4. Records on the RSScan plate and respectively chromatic performance

The second stage was identical, changing only the wavelength of the light sources used (green, yellow and white). To enhance the chromatic effect, the subjects also wore glasses with the same filter at the exit from the CSSD during the recording on the pressure plate [13].

3. Results and conclusions

The recorded data was processed by software dedicated to the pressure board, namely *RSScan footscan 7.97 Gait 2nd generation*, and chromatic performance through the *Visiontest* software. At each stage of the procedure, immediately after exposure to chromatic stimulation, visual acuity was evaluated in all subjects to track the influence of bright radiation on visual perception and to correlate with the gait cycle typology (see Table no.1).

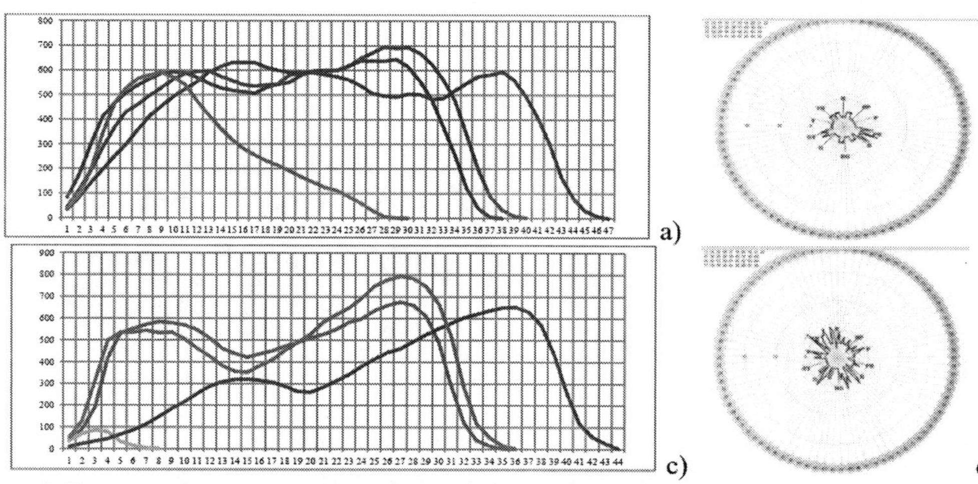

Figure 5. Pressure plate pressure (*a* and *c)* and chromatic performance (*b* and *d*) at all wavelengths

Table no.1. Visual acuity (VA) for right (RE) and left eye (LE) in different color light

Human subject	Visual acuity VA Red color light			Visual acuity VA Blue color light			Visual acuity VA White color light			Visual acuity VA Ambiental light (300lx)		
	RE	LE	B	RE	LE	B	RE	LE	B	RE	LE	B
1.	0.9	0.9	1	1	0.9	1	0.9	0.8	0.9	1	0.9	1
2.	0.9	1	1	0.9	1	1	0.8	0.9	0.9	1	1	1
3.	0.9	0.8	0.9	0.8	0.9	0.9	0.8	0.8	0.9	1	0.9	1
4.	0.9	0.8	0.9	0.8	0.8	0.9	0.8	0.8	0.9	1	1	1

The graphical representation of this feature indicates a decrease in visual acuity from normal to white light exposure of about 20% for right eye vision, a decrease of nearly 10% for the left eye, and in binocular decrease to 5%. After all the records, the following aspects were highlighted: the biomechanical behavior of the subjects after exposure to chromatic radiation (all categories) changed substantially, subjects doing on average a larger number of steps on the same plate length (by 10%

more); their movement was much slower (duration increased by an average of 26%); the posture of the subjects has changed many times many times down to the ground at a small distance; they used the arms to compensate for the slight imbalances observed both in movement and in plantar pressure values. In addition to these aspects of posture and movement, subjects were temporarily diminished monocular and binocular visual acuity, causing uncertainty in movement and discomfort in the perception of the environment.

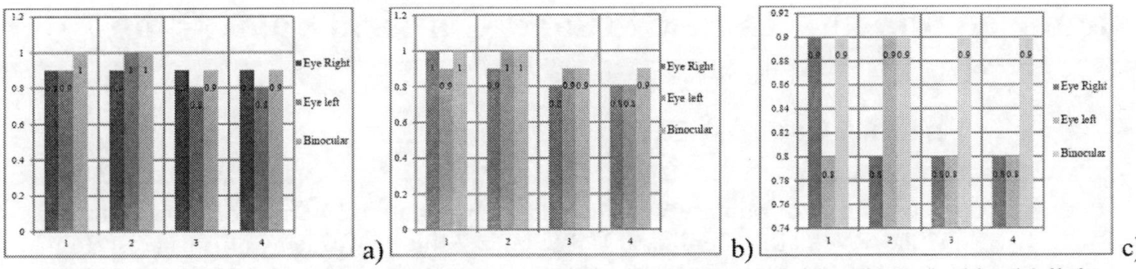

Figure 6. Visual acuity for right, left eye and binocular, in red (a), blue (b) and white (c) light

The issues emerging from this experiment, chromatic visual perception of subjects correlates with displacement parameters in the walking cycle, confirms the hypothesis that visual function can be considered an important component of the locomotion process.

Acknowledgements
In these experiments we've developed the investigations with equipment from "Advanced Mechatronic Systems Research Center - C04" and Applied optometric Laboratory at University Transylvania of Brasov, with the students help from Optometric study Program.

References
[1] Vaughan Ch, *Dynamics of Human Gait*, Kiboho Publishers, South Africa, 1999
[2] Inoue S et al., *Assessment of physical inactivity and locomotor dysfunction in adults with visual impairment,* Scientific Reports, www.nature.com/scientificreports/ 8:12032, 2018;
[3] https://www.getbodysmart.com/nervous-system/retina accessed April 2019;
[4] Lennie P, *The Physiology of Color Vision*, New York, The Science of Color, Elsevier Ltd, 2003
[5] Fairchild M. D., *Human Color Vision, Color Appearance Models*, Second Edition John Wiley & Sons, Ltd, 2005
[6] Miller J, *Vision, A Component Of Locomotion,* Journal of the Chartered Society of Physiotherapy, London, 1967
[7] Logan D, Kiemel T, Jeka J, *Using a System Identification Approach to Investigate Subtask Control during Human Locomotion*, Front. Comput. Neurosci. 10:146, 2017
[8] Logan D, Ivanenko Y, Kiemel T, Cappellini G, Sylos-Labini F, Lacquaniti F, Jeka J, *Function dictates the phase dependence of vision during human locomotion*, Journal of Neurophysiology, April 2014
[9] Reimann H, Fettrow T, Thompson E, Jeka J, *Neural Control of Balance During Walking,* Front. Physiol. 9:1271, 2018
[10] Muñoz J.E, Villada J.F, Casanova S, Montoya M.F, O.A. Henao, *Dynamic Systems Theory in Human Movement. Exploring Coordination Patterns by Angle-Angle Diagrams Using Kinect*, https://www.researchgate.net/publication/328138367, 2018
[11] Sabatini A, Martelloni C, Scapellato S, Cavallo F, *Assessment of Walking Features From Foot Inertial Sensing*, IEEE Transactions On Biomedical Engineering, Vol. 52, No. 3, 2005
[12] Donati, M, Cecchi, F, Bonaccorso, F, Branciforte, M, Dario, P, Vitiello, N, *A Modular Sensorized Mat for Monitoring Infant Posture*, Sensors 14, pp 510-531, 2014
[13] Fogoros R., *License project (Proiect licenta),* coord. Baritz M.I., Universitatea Transilvania Brasov, 2017

IOP Conf. Series: Materials Science and Engineering **568** (2019) 012111 doi:10.1088/1757-899X/568/1/012111

Research of CFR SI Engine and Dacia single cylinder SI engine equipped with LASER and Classical Spark Plug

B G Done[1], I Copae[1]

[1] Department of Military Vehicles and Transportations, Faculty of Aircraft and Military Vehicles, "Ferdinand I" Technical Military Academy of Bucharest, Romania.

bogdan.done@yahoo.fr,

Abstract. Latest investigation has shown that laser ignition offers many potential benefits compared to the classical spark plug ignition due to the improvement of the performance of internal combustion engines. This paper aims to show the progress made in recent research on laser ignited internal combustion engines and also discusses the advantages and control opportunities, while considering the challenges faced and the prospects for its future implementation. It has been demonstrated that laser ignition can be used to improve the control of the engine in areas such as improved combustion stability, reductions in emissions, lower idle speed and leaner operation. Moreover, laser ignition not only that is non-invasive, but it also has a great flexibility in terms of the ignition position. Therefore, it allows the possibility of multi-point ignition.On the other hand, spark plugs offer limited possibilities for optimising engine efficiency, as their fixed position within a cylinder and the protrusion of electrodes disturbs the cylinder geometry and can also quench the flame kernel. As a result, this paper explores several results of the recent research focusing on: indicated, mediated pressures and their wave dispersion .

1. Introduction

A potentially better technological solution for gasoline injection compression Ignition engines, however, continues to depend on fuel Cetane Number, which should not be in excess of CN15[1] causing longer Ignition delays. In other words, faster Ignition and combustion over very lean mixtures, have considerable benefits concerning fuel efficiency and emissions, due to the extremely high-power Ignition sources such as LASER Ignition and offer several advantages in comparison with classical Spark Plug Ignition [2,3]. This type of Ignition was tested initially in 1978, when a single-cylinder research engine was ignited by a CO_2 LASER [5].

In 2008, advanced Q-switched Nd:YAG LASERs were used to ignite a four-cylinder engine [4] but T. Taira et al. [5] made public the first gasoline engine ignited only by LASER. Following the research, Nd:YAG/Cr^{4+}:YAG LASERs which were either side pumped [7] or longitudinally pumped and that delivered one beam [7] or multi-beam output [7] proved to be feasible for engines.

The following main advantages of LASER ignition are:

- a choice of arbitrary positioning of the ignition plasma in the combustion cylinder
- absence of quenching effects by the spark plug electrodes
- ignition of leaner mixtures than with the spark plug => lower combustion temperatures => less NO_x emissions

IOP Conf. Series: Materials Science and Engineering **568** (2019) 012111 doi:10.1088/1757-899X/568/1/012111

- no erosion effects as in the case of the spark plugs => lifetime of a LASER ignition system expected to be significantly longer than that of a spark plug
- high load/ignition pressures possible => increase in efficiency
- precise ignition timing possible
- exact regulation of the ignition energy deposited in the ignition plasma
- easier possibility of multipoint ignition [6,7]
- shorter ignition delay time and shorter combustion time [4,7]
- fuel-lean ignition possible

The disadvantages of LASER ignition are:

- high system costs
- concept proven, but no commercial system available yet.

Current research is focusing on the improvement of reliability and miniaturization of LASER Ignition (LI) systems [7], also on the influence of delivered energy on a single point [7].

2. Experimental investigation

Experimental investigations were carried out in a research laboratory of the University POLITEHNICA of Bucharest.The experimental research was developed on an experimental single cylinder SI engine, equipped with LASER Ignition on two engines. The Dacia Single Cylinder SI Engine, operating regime was 2800 rev/min, 90 % load and the CFR Octane Rating Unit Engine, operating regime was 900/min. Both experimental engines were single-cylinder and were mounted on test beds adequately instrumented. The LASER spark plug used in the experiments was provided by INFLPR, Laboratory of Solid-State Quantum Electronics, Magurele, Romania. A photo of the LASER spark is shown in figure 1.

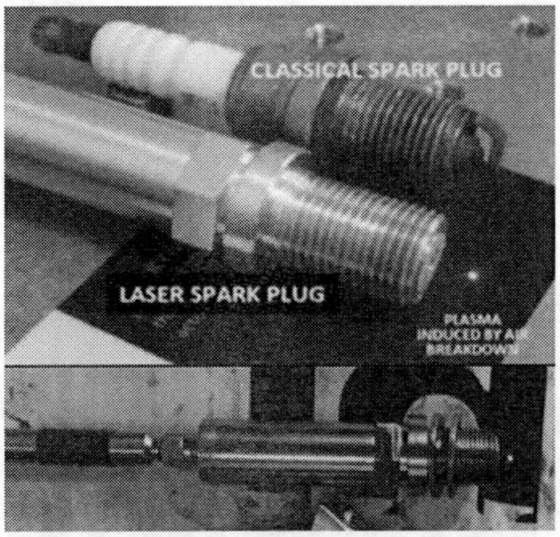

Figure. 1. A photo of a LASER spark plug is shown in comparison with a classical spark plug. The plasma induced in air by optical breakdown is visible.
(Courtesy of INFLPR, Laboratory of Solid-State Quantum Electronics, Magurele, Romania)

3. Results and discussion

As the graphs show, the dispersal of curves is higher in the case of the classical spark plug at narrow and wide β angles and increases in the case of the LASER spark plug up to a value of β=364°CAD. Detail A in figure 2, shows that the maximum values reached by the curves are 363°CAD for the classical spark plug and 364 °CAD for the LASER spark plug. Detail B highlights the fact that at an β=357.7 °CAD, the curve of the LASER spark plug is above that of the classical spark plug.

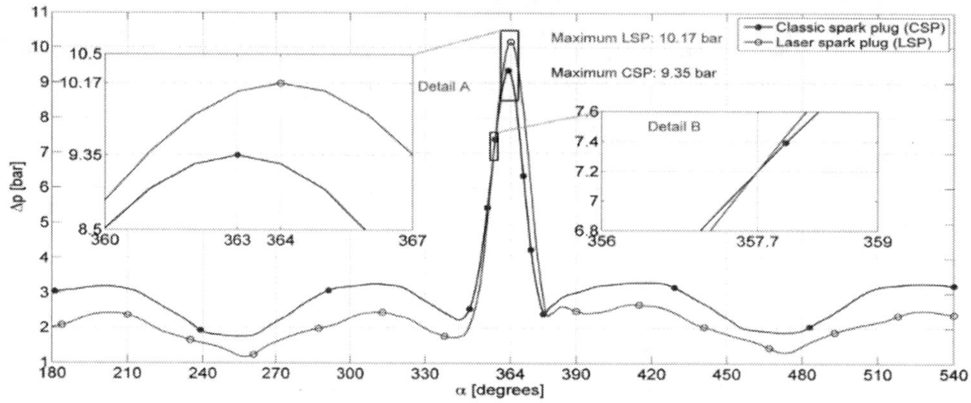

Figure 2. Dispersion of the cycle pressure curves for the classic spark plug and laser spark plug, β = 28.5°CAD, λ = 1.1

Thus, figure 3, presents the pressure curves indicated for the λ =0.9 classical spark plug case while figure 3b for the LASER spark plug, for all the 50 measured cycles; figure 3c presents the distances between the external outlines of the curves, at each crankshaft rotation angle. As figure 8c, shows, cyclic dispersion is higher for narrow angles in the case of classical spark plugs and higher for wider angles in the case of LASER spark plugs. Moreover, the classical spark plug registers the maximum value (33.1 bar). In exchange, figure 4, shows that in the case of λ =1.0, cyclic LASER spark plugs have a higher cyclic dispersion in the case of narrow angles while classical spark plugs have a higher cyclic dispersion in the case of wider angles. Furthermore, LASER spark plug register the maximum value (32.2 bar).

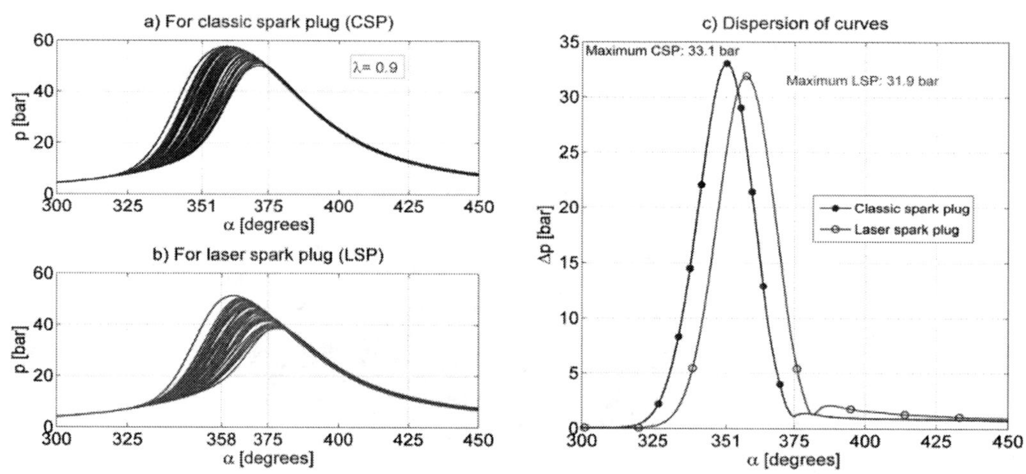

Figure. 3. The indicated pressures and the dispersion of their curves (details), Dacia monocillium, for: 50 cycles; n = 2800 rpm; κ = 90%; λ = 0.9

Moreover, the classical spark plug registers the maximum value (33.1 bar). In exchang,e, figure 4 shows that in the case of λ =1.0, cyclic LASER spark plugs have a higher cyclic dispersion in the case of narrow angles while classical spark plugs have a higher cyclic dispersion in the case of wider angles. Furthermore, LASER spark plugs register the maximum value (32.2 bar).
The analysis of both engines has resulted therefore, in families of curves for the indicated pressure, related to a given number of functional cycles. Based on them, obviously, it is possible to establish the median pressure curve, through mathematical calculation of the average value for each crankshaft angle.

IOP Conf. Series: Materials Science and Engineering **568** (2019) 012111 doi:10.1088/1757-899X/568/1/012111

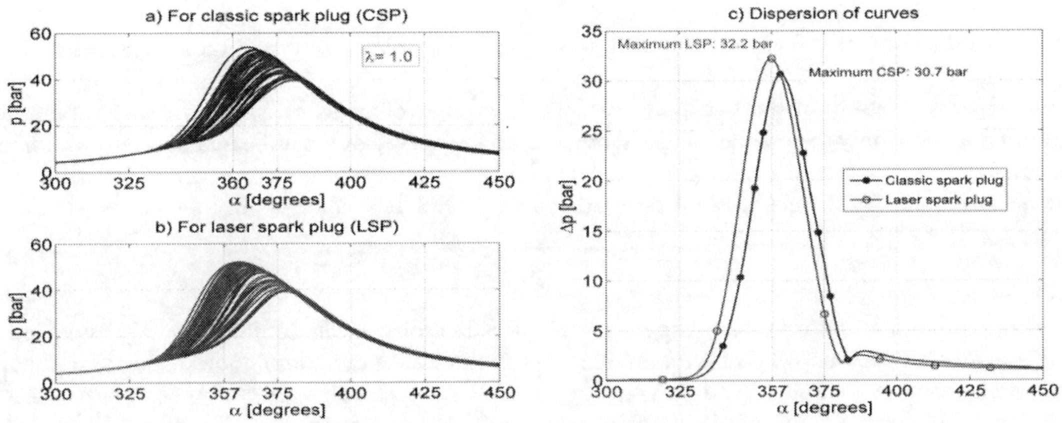

Figure 4. The indicated pressures and the dispersion of their curves (details), Dacia monocillium, for: 50 cycles; n = 2800 rpm; κ = 90%; λ = 1.0

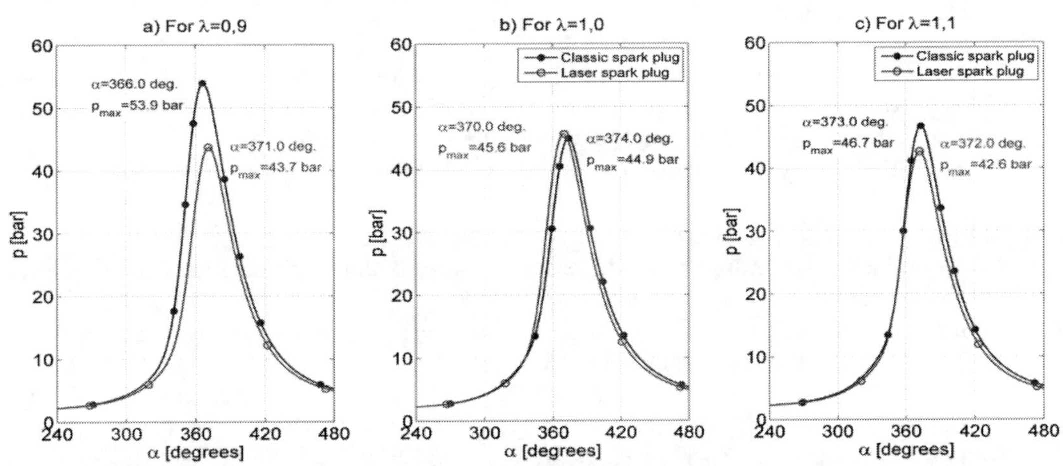

Figure. 5. The indicated pressure for the Dacia single-cylinder curve for: 50 cycles; n = 2800 rpm; κ = 90%; λ = [0.9; 1.0; 1.1]

For example, figure 5, indicates the median pressure curves for the Dacia mono-cylinder engine, for three values of the excess air coefficient λ. The graphs also show the maximum pressure values p_{max}, as well as the rotation angle β related to them. Also noticeable in figure 5, in the case of the stoichiometric mix (λ =1.0), the maximum pressure values indicated are the closest to the two types of spark plugs (44.9 bar and 45.6 bar respectively). Also, for this excess air coefficient value the maximum pressure value is obtained in the case of the LASER spark plug. In regard to the above however, it should be specified that operation based only on the median pressure of cycles represents a simplification which actually conceals the real phenomenon. For this reason, a live study should operate with all the indicated pressure curves, with the concept of uncertainty, the algorithms related to the uncertainty theory and highlighting the non-linear phenomena which accompany cyclic dispersion.

4. Conclusions

The experimental results obtained from the LASER spark plug used in the spark Ignition engine compared to the classic Ignition system have led to several important conclusions presented as follows:

IOP Conf. Series: Materials Science and Engineering **568** (2019) 012111 doi:10.1088/1757-899X/568/1/012111

- research carried out utilizing all the experimental data (the study presents only a part of it) confirms the well-known fact that cyclic dispersion, performance levels and pollutant emissions are influenced by the Ignition advance and the quality of the air-fuel mix (through the excess air coefficient) resulting in the need to quantify them by resorting to dispersion analysis, information analysis and sensitivity analysis.
- a comparative study of the two types of spark plugs, as close as possible to the real phenomenon, should target not only the median diagram but also all the curves of the indicated diagram which involves resorting to the uncertainty concept as well as highlighting the non-linear phenomena that accompany cyclic dispersion by means of bi-spectral analysis in frequency, analysis in time-frequency, the chaos theory, etc.

Acknowledgements
In order to prepare the necessary experimental investigations, I was offered the help and guidance of some respected persons, who deserve my deepest gratitude.As the completion of this assignment gave me much pleasure, I would like to show my gratitude to the research team comprises of Prof. Constantin Pana, Prof. Niculae Negurescu, Lecturer Alexandru Cernat, Lecturer Dinu Fuiorescu, Lecturer Cristian Nutu, Faculty of Mechanical Engineering and Mechatronics, Department of Thermotechnics, Engines, Thermal Equipments and Refrigeration Installations, University POLITEHNICA of Bucharest, for giving me good guidelines throughout numerous experimental investigations and for the use of the laboratory and its instruments.

The experiments were performed with a LASER spark device developed at National Institute for LASER, Plasma and Radiation Physics, Laboratory of Solid-State Quantum Electronics, Magurele, Ilfov, 077125, Romania. The authors would like to thank to Mr. Pavel Nicolaie, Mr. Dinca Mihai and Mrs. Croitoru Gabriela for their help and assistance during the experiments.
Also, the authors address special thanks to the AVL GmbH for providing the necessary eqiupments.

References

[1] Concawe Fuels and Emissions Management Group. Exploring a Gasoline Compression Ignition (GCI) engine concept. Report no. **13/14**(2014).
[2] Dearden G. and Shenton T LASER ignited engines: progress, challenges and prospects. Opt. Express **21**(S6) (2013): A1113-A1125.
[3] Dale J D Dale, Smy P R and Clements R M LASER ignited internal combustion engine: An experimental study. SAE International, paper 780329 (1978); DOI: 10.4271/780329.
[4] Mullett J, Dickinson P, Shenton A, Dearden G, and Watkins K G Multi-Cylinder LASER and Spark Ignition in an IC Gasoline Automotive Engine: A Comparative Study. SAE International, paper 2008-01-0470 (2008); DOI: 10.4271/2008-01-0470.
[5] Taira T, Morishima S, Kanehara K, Taguchi K, Sugiura A, and Tsunekane M World first LASER ignited gasoline engine vehicle. 1st LASER Ignition Conference (LIC'13), Yokohama, Japan, April 23-25, 2013, paper LIC3-1.
[6] Kroupa G, Franz G and Winkelhofer E Novel miniaturized high-energy Nd:YAG LASER for spark Ignition in internal combustion engines. Opt. Eng. **48**(10) (2009): 014202.
[7] Pavel N, Tsunekane M and Taira T Composite all-ceramics, high-peak power Nd:YAG/Cr4+:YAG monolithic micro-LASER with multiple beam output for engine Ignition. Opt. Express **19**(10) (2011): 9378-9384.
[8] Saito. T Influence of LASER incident energy and focal length on multi-point LASER Ignition in an internal combustion engine at N_2 dilution in LASER Ignition Conference 2017, OSA Technical Digest (online) (Optical Society of America, 2017), paper LWA4.2.

IOP Conf. Series: Materials Science and Engineering **568** (2019) 012112 doi:10.1088/1757-899X/568/1/012112

Comparative research results of CFR Octane Rating Unit Engine and Dacia Single Cylinder SI Engine equipped with classical Spark Plug and LASER Ignition

B G Done[1], I Copae[1]

[1] Department of Military Vehicles and Transportations, Faculty of Aircraft and Military Vehicles, "Ferdinand I" Technical Military Academy of Bucharest, Romania.

bogdan.done@yahoo.fr,

Abstract. Nowadays, research has developed new technologies for a better control of the combustion process. Among the new technologies used for Ignition and combustion control, the LASER spark plug system is defined as an innovative technology which could overcome several limitations of classical spark plug. The purpose of the paper is the experimental research of the LASER spark plug used in the spark ignition(SI) engine. Improved performance, ensuring rapid and robust combustion, depends on how the Ignition stage is achieved. An Ignition system that could provide such an enhanced combustion process is the one based on plasma generation using a Q switched solid state LASER that delivers pulses with high MW peak power. The LI device used in the current research was a LASER medium Nd:YAG/Cr4+:YAG ceramic structure (Baikowski Co., Japan) that consisted of a 8.0-mm long, 1.0-at.% Nd:YAG ceramic, optically-bonded to a Cr4+:YAG ceramic with saturable absorption. It was designed, integrated and built to resemble a classical Spark Plug and could be mounted directly on the cylinder of a CFR Octane Rating Unit Engine as well as on a Dacia Single Cylinder SI Engine which led to several results among which: indicated, mediated pressures and their wave dispersion.

1. Introduction

Recently, a large number of engine manufacturers have been directing their efforts towards enhancing electric mobility, in particular, the internal combustion engine. It is currently widely accepted that Ignition in most engines is produced by electrical system of the engine at a certain moment of the operating cycle and at a specific position in the combustion chamber where the spark plug is located.

Spark plug engine operation, unfortunately, still has several areas requiring improvement such as poor efficiency in the case of low loads, knock tendency in the case of high loads, nanoparticles emissions and elevated level of the global pollutant emissions. Future combustion technologies applied to heat engine will need to solve the problems of ignitability of lean mixtures and the stability of combustion in lean and diluted mixtures i.e. with high EGR, in spark ignited engines [1]. In other words, faster Ignition and combustion over very lean mixtures have considerable benefits concerning fuel efficiency and emissions, due to the extremely high-power Ignition sources such as LASER Ignition and offer several advantages in comparison with classical Spark Plug Ignition [2,3]. This type of Ignition was tested initially in 1978, when a single-cylinder research engine was ignited by a CO2 LASER [4]. In 2008, advanced Q-switched Nd:YAG LASERs were used to ignite a four-cylinder engine [5] but T. Taira et al. [6] made public the first gasoline engine ignited only by LASER.

IOP Conf. Series: Materials Science and Engineering **568** (2019) 012112 doi:10.1088/1757-899X/568/1/012112

Following the research, Nd:YAG/Cr4+:YAG LASERs which were either side pumped [7] or longitudinally pumped and that delivered one beam [7] or multi-beam output [8] proved to be feasible for engines.

2. Experimental investigation

Experimental investigations were carried out in a research laboratory of the Faculty of Mechanical Engineering and Mechatronics, Department of Thermotechnics, Engines, Thermal Equipments and Refrigeration Installations, University POLITEHNICA of Bucharest.

The experimental research was developed on an experimental single cylinder SI engine, equipped with LASER Ignition on two engines. The Dacia Single Cylinder SI Engine (figure 1) operating regime was 2800 rev/min, 90 % load and the CFR Octane Rating Unit Engine (figure 2) operating regime was 900/min. Both experimental engines were single-cylinder and were mounted on test beds adequately instrumented.

The LASER spark plug used in the experiments was provided by INFLPR, Laboratory of Solid-State Quantum Electronics, Magurele, Romania. A photo of the LASER spark plug is shown in figure 3. The LASER medium was a Nd:YAG/Cr4+:YAG ceramic structure (Baikowski Co., Japan) [9,10,11].

Figure 1. Dacia Single Cylinder SI Engine equipped with LASER Ignition

Figure 2. CFR Octane Rating Unit Engine equipped with LASER Ignition

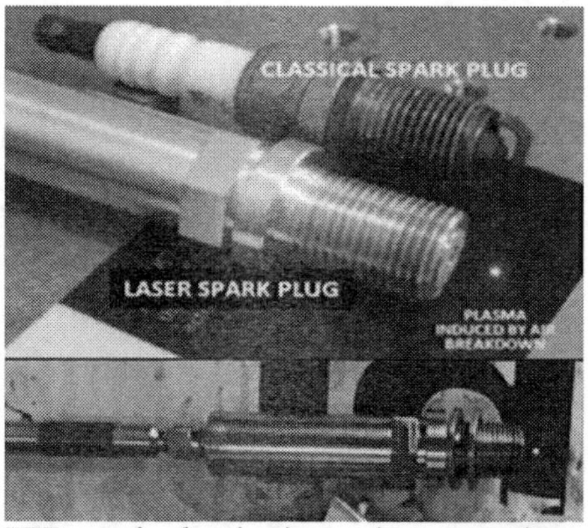

Figure. 3. A photo of a LASER spark plug is shown in comparison with a classical spark plug. The plasma induced in air by optical breakdown is visible.
(Courtesy of INFLPR, Laboratory of Solid-State Quantum Electronics, Magurele, Romania)

3. Results and discussion

Figure 4 shows the pressure values of the CFR engine for 491 cycles, n=400 rev/min, Ignition advance β=28.5 °RAC and λ=1.1 excess air coefficient.

The graphs also indicate the external outlines of the 491 cycles as well as the related maximum values for the classical spark plug (figure 4a) and the LASER spark plug (figure 4b). As noticeable in the 2 graphs, the maximum pressure values range was 22-26.3 bar in the case of the classical spark plug while the range for the LASER spark plug was 22.4-26.9 bar. As seen and expected, figure 4, shows the existence of the cyclic dispersion phenomenon. According to scientific literature, cyclic dispersion is defined, as a rule, by the coefficient of variation (COV), which for a given value *x*, represents the ratio between the standard deviation σ_x and the average value m_x:

$$COV_x = \frac{\sigma_x}{m_x} \qquad (1)$$

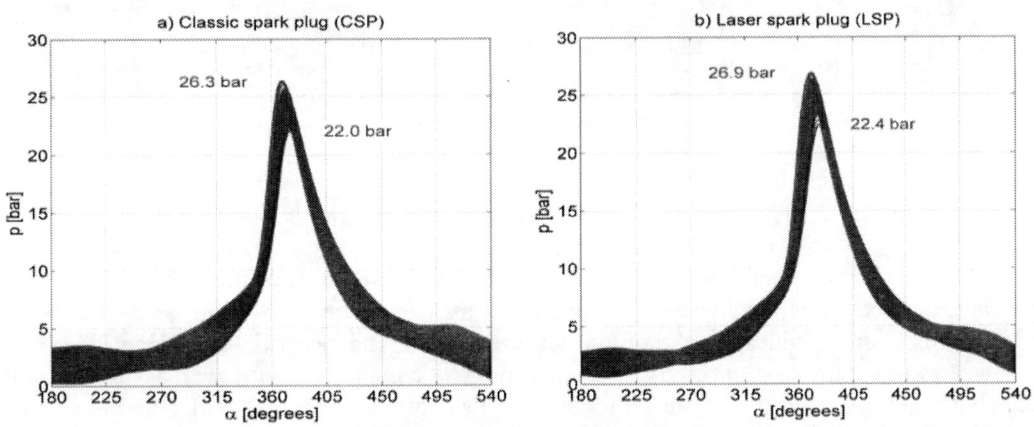

Figure. 4. The indicated pressure, instantaneous values and cycle tires the CFR engine for 491 cycles, n=400 rev/min, β=28.5 °CAD and λ=1.1.

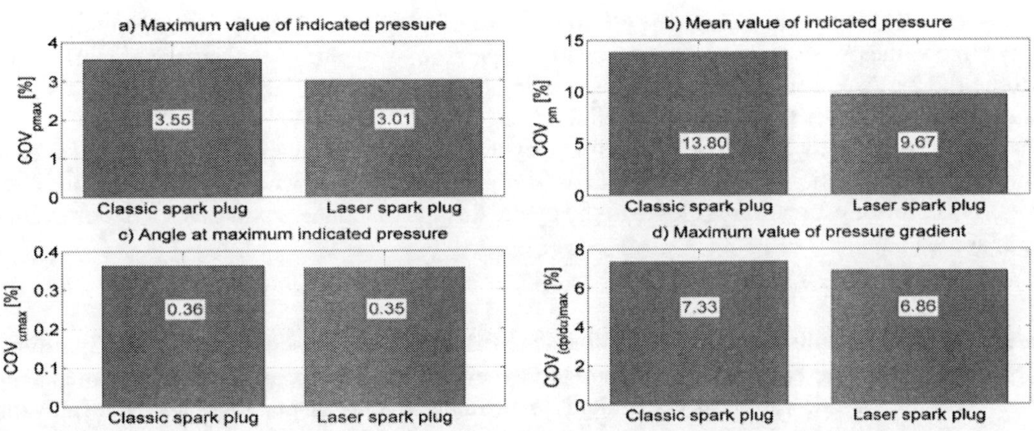

Figure 5. Coefficients of variation for maximum pressure, mean pressure, maximum pressure angle, maximum pressure gradient, CFR engine for: 491 cycles; n = 900 rev/min; β = 28.5°CAD; λ = 1.1

Figure 5 presents the variation coefficient values for the pressure curves in figure 4 and for the four values mentioned in the graph: the maximum and average pressure values indicated, the rotation angle of the crankshaft(CAD) related to the maximum pressure and the maximum value of the pressure gradient. As the graphs show, the LASER spak plug variation coefficient has lower values by

IOP Conf. Series: Materials Science and Engineering **568** (2019) 012112 doi:10.1088/1757-899X/568/1/012112

comparison with the classical spark plug, for all the four above-mentioned values. Consequently, in this case, based on the variation coefficient, the scale of cyclic dispersion is smaller than in the case of LASER spark plugs.

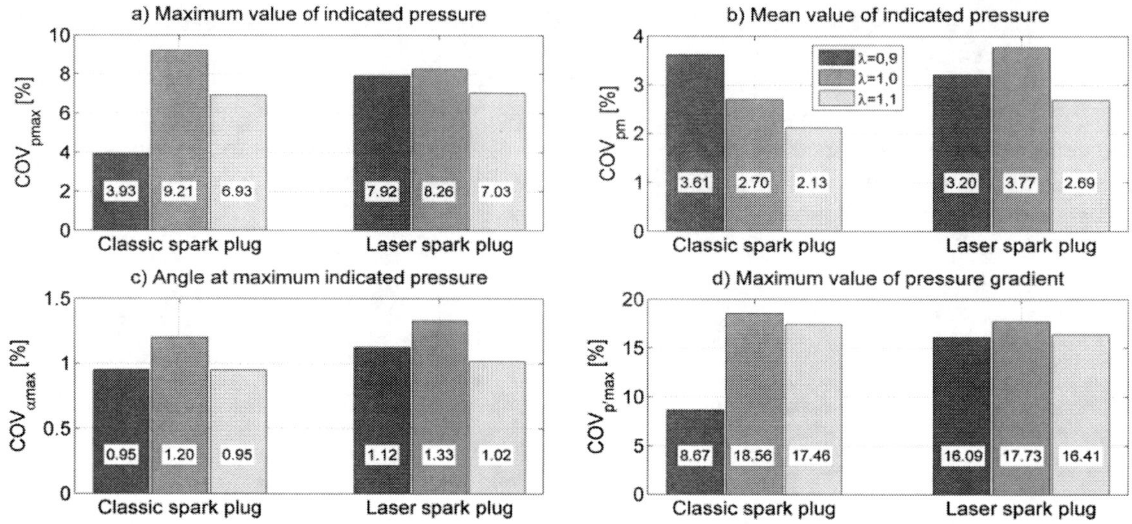

Figure. 6. Coefficients of variation for maximum pressure, mean pressure, maximum pressure angle, pressure gradient, Dacia monociller for: 50 cycles; n = 2800 rev/min; κ = 90%; λ = [0.9; 1.0; 1.1]

Similarly, Fig.6 presents the variation coefficient values in the case of the indicated pressure curves of the Dacia monocylinder engine, for three values of the excess air coeffcient λ as well as for the same four before-mentioned values. As the graphs show, this time the variation coefficient does not always have lower values in the case of the LASER spark plug by comparison with the classical spark plug, for all the four values mentioned and for all the excess air coefficient values. With reference to the above statements, one particular aspect related to the variation coefficient should be mentioned. As resulting from the relationship (1), this coefficient represents a ratio of two values: standard deviation and average value.

Consequently, COV can be higher both in the case of a high standard deviation, and in the case of a low average value. As a result, estimating cyclic dispersion by means of the variation coefficient proves not to be the best solution. In this sense, estimating cyclic dispersion based only on dispersion (square of standard deviation) is ultimately a better solution.

In conclusion, it is recommended to estimate cyclic dispersion by dispersing the curves at each angle of the crankshaft rotation, in other words based on the distance between the external outlines of the curves. This type of estimation is closer to the cyclic dispersion concept and gives a better evaluation of the phenomenon.

4. Conclusions

The experimental results obtained from the LASER spark plug used in the spark Ignition engine compared to the classic Ignition system have led to several important conclusions presented as follows:

a. the use of the variation coefficient to estimate cyclic dispersion has the disadvantage of representing the ratio of two values and as a result cannot lead to veridical conclusions;
b. the use of the distance between the external outlines of the curves of the diagram to estimate cyclic dispersion allows the highlighting of the real phenomenon, the study thus showing that cyclic dispersion varies depending on the crankshaft rotation angle and is not constant, as resulting from the variation coefficient;

IOP Conf. Series: Materials Science and Engineering **568** (2019) 012112 doi:10.1088/1757-899X/568/1/012112

Acknowledgements

In order to prepare the necessary experimental investigations, I was offered the help and guidance of some respected persons, who deserve my deepest gratitude.As the completion of this assignment gave me much pleasure, I would like to show my gratitude to the research team comprises of Prof. Constantin Pana, Prof. Niculae Negurescu, Lecturer Alexandru Cernat, Lecturer Dinu Fuiorescu, Lecturer Cristian Nutu, Faculty of Mechanical Engineering and Mechatronics, Department of Thermotechnics, Engines, Thermal Equipments and Refrigeration Installations, University POLITEHNICA of Bucharest, for giving me good guidelines throughout numerous experimental investigations and for the use of the laboratory and its instruments.

The experiments were performed with a LASER spark device developed at National Institute for LASER, Plasma and Radiation Physics, Laboratory of Solid-State Quantum Electronics, Magurele, Ilfov, 077125, Romania. The authors would like to thank to Mr. Pavel Nicolaie, Mr. Dinca Mihai and Mrs. Croitoru Gabriela for their help and assistance during the experiments.

Also, the authors address special thanks to the AVL GmbH for providing the necessary equipments.

References

[1] Kalghatgi G T (2015) Developments in internal combustion engines and implications for combustion science and future transport fuels. Proceedings of the Combustion Institute 35(1): 101-115.

[2] Rooney P D (2013) LASER versus conventional Ignition of flames. Opt. Eng. **33**(2) (1994): 510-521.

[3] Dearden G. and Shenton T LASER ignited engines: progress, challenges and prospects. Opt. Express **21**(S6): A1113-A1125.

[4] Dale J D, Smy P R and Clements R M (1978) LASER ignited internal combustion engine: An experimental study. SAE International, paper 780329; DOI: 10.4271/780329.

[5] Mullett J, Dickinson P, Shenton A, Dearden G, and Watkins K G (2008) Multi-Cylinder LASER and Spark Ignition in an IC Gasoline Automotive Engine: A Comparative Study. SAE International, paper 2008-01-0470; DOI: 10.4271/2008-01-0470.

[6] Taira T, Morishima S, Kanehara K, Taguchi K, Sugiura A, and Tsunekane M 2013 World first LASER ignited gasoline engine vehicle. 1st LASER Ignition Conference (LIC'13), Yokohama, Japan, April 23-25, paper LIC3-1.

[7] Kroupa G, Franz G and Winkelhofer E (2009) Novel miniaturized high-energy Nd:YAG LASER for spark Ignition in internal combustion engines. Opt. Eng. **48**(10): 014202.

[8] Pavel N, Tsunekane M and Taira T (2011) Composite all-ceramics, high-peak power Nd:YAG/Cr4+:YAG monolithic micro-LASER with multiple beam output for engine Ignition. Opt. Express **19**(10): 9378-9384.

[9] Dascalu D, Croitoru G, Grigore O, Pavel N (2016) High-peak power passively Q-switched Nd:YAG/Cr4+:YAG composite LASER with multiplebeam output. Photonics Research **4(**6): 267-271.

[10] Done B., Alexandru C., Fuiorescu D., Cristian N. 2017 Aspects of the cycle variability study of a SI engine with LASER plug Ignition system." U.P.B Sci. Bull. Series D, **Vol. 79**, Iss. **2**, ISSN 1454-2358.

[11] Done B. (2018) Experimental Investigations of LASER Ignition Use at Spark Ignition Engine 11th International Conference Interdisciplinarity in Engineering, INTER-ENG, Procedia Manufacturing **22** 659–666. 6

IOP Conf. Series: Materials Science and Engineering **568** (2019) 012113 doi:10.1088/1757-899X/568/1/012113

Analyzing the sustainability of an automotive component using SolidWorks CAD software

C Torcătoru[1], D Săvescu[2]

[1,2] Product Design, Mechatronics and Environment Department, Transilvania University of Braşov, B-dul Eroilor 29, 500036, Romania

[1] constantin.torcatoru@unitbv.ro;

Abstract. Design stages have been widely recognized as a milestone in the life cycle of a product in sustainability implementation, and this is of major importance for leading organizations that produce cost-effective and environmentally friendly products. This paper focuses on analyzing some aspects of environmental impact and ecological sustainability of design models using SolidWorks CAD design software. The material from which the piece was made was evaluated, with emphasis on its strength, durability and environmental pollution in terms of carbon footprint, energy consumption, air acidification, and eutrophication. Consideration was given to the whole life cycle of the product under consideration, from raw material extraction, processing, product production, assembling and use to the end of life, taking into account the mode of transport and the distance between these stages. In the case study is presented the virtual model of the analyzed product and the sustainability report that the SolidWorks software provides us and which can be analyzed later by the whole organization for the adoption of the right decision.

1. Introduction

The ideas of durability and sustainable design are an increasingly important part of today's design discussions. But what is a sustainable design and how can a more eco-friendly product be created? In this paper, there are some answers to these questions.

The World Commission on Environment and Development defines sustainable development as "*development that meets the needs of the present without compromising the ability of future generations to meet their own needs*."

Elkington [1] indicated that the concept of sustainability should cover economic, social and environmental aspects. This concept is also known as the "triple bottom line" where the profit, the planet, and the people are considered simultaneously. Therefore, the scope of sustainable design covers all the spectrum of product life.

Sustainable design is generally the process of developing a product that performs successfully, generates profits for the company, is socially acceptable, and uses a minimum amount of energy, that is, a minimal amount of material that produces hazardous waste.

One of the major challenges of sustainable development is encouraging the search for ecological ways and discouraging ways that harm the environment (air, water and soil or subsoil).

IOP Conf. Series: Materials Science and Engineering **568** (2019) 012113 doi:10.1088/1757-899X/568/1/012113

Contemporary design engineers should integrate the principles of sustainability into their work for at least one of the following reasons: *personal interest, professional growth, the intention of the company, industry rules.*

2. Literature Review

To analyze the sustainability of a product, there are dedicated software programs for these types of analysis, which can do much more in detail than those in the CAD range, for example. Such a dedicated software, which we can mention, is the GaBi Life Cycle Assessment.

Lately, global awareness of environmental issues has never been more acute. Consumers are increasingly looking at the environmental impact of the products they buy. It does not just look at recyclability. They want to know about the materials used and how they were produced. Producers, in turn, are trying to respond to rising demand. [6]

Why SolidWorks Sustainability?

Having implemented in SolidWorks and LCA mode as an engineer or product designer, there is the possibility to dramatically change how we interact with our environment. The question is how. Many designers do not know about the life cycle assessment process (LCA) or how they can guide them to more sustainable models. [5] Others may think that the process is too complex and consumes a lot of time or is the responsibility of someone else.

SolidWorks Sustainability provides this intelligent analysis module to make informed decisions about what materials to use. It can also be analyzed how the supply, manufacture, use, and disposal of region-specific materials will affect the life cycle of the product before starting the manufacturing process.

2.1. The concept of sustainability

Sustainable design analyzes how the product will evolve [7] from concept to recycling, focusing on four key factors to report on environmental impact::

- *Acidification of air* - combustion fuels creates sulfur dioxide, nitrogen oxides, and other acidic air emissions. This causes an increase in the acidity of the rainwater, which in turn acidifies lakes and soil, subsequently making the soil and water toxic to plants and aquatic life. Also, acid rain can slowly dissolve hand-made construction materials such as concrete.

This impact is usually measured in kilograms of sulfur dioxide equivalent (SO2).

- *Carbon footprint* - Carbon dioxide and other gases from combustion of fossil fuels accumulate in the atmosphere, with the consequence of increasing the average temperature of the earth. Also known as "global warming potential" (GWP), the carbon footprint is measured in carbon dioxide equivalent units (CO2e).
- *Total Energy Consumption* - A measure of non-renewable energy sources associated with the piece's life cycle in megajoules (MJ) units. This impact includes both the electricity or the fuels used during the life cycle of the product as well as the upstream energy required to obtain and process these fuels as well as the built-in energy of the materials that would be released in the event of combustion. Total energy consumed is the net calorific value of primary energy demand from non-renewable resources (for example, oil, natural gas, etc.). Efficiencies in energy conversion (for example, energy, heat, steam, etc.) are also factors.
- *Water Eutrophication* - Appears when a supernatant of nutrients is added to a water ecosystem. Nitrogen and phosphorus in wastewater and agricultural fertilizers cause overlapping algae that flourishes, which then diminishes oxygen water and results in the death of both plants and animals. This impact is usually measured either in kilograms equivalent to phosphate in kg (PO4) or nitrogen equivalent (kg).

IOP Conf. Series: Materials Science and Engineering **568** (2019) 012113 doi:10.1088/1757-899X/568/1/012113

3. Case study

In this chapter, for the case study, a component of a rotating nut jack-screw was selected for analysis. In Figure 2 shows the structural diagram and the main components of the assembly, these being: 1 - the cup, 2 - the screw, 3 - the rotating nut, 4 - the body, 5 - the actuator attached to the rotating nut. From this diagram it is possible to notice the operating mode, namely by rotating the nut 3 with the mechanism of action 5, the screw 2 will perform only a translational motion in the direction of load displacement Q, the rotation is prevented by the translation coupling made between the screw movement and body. [3]

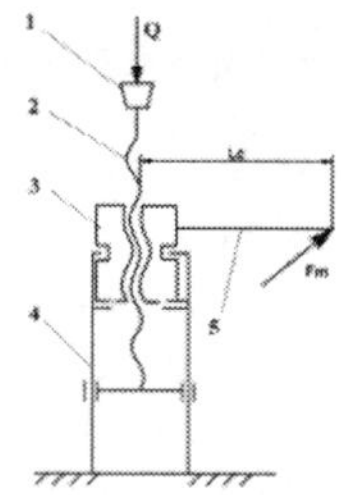

Figure 1. Structural scheme for jack-screw with rotating nut

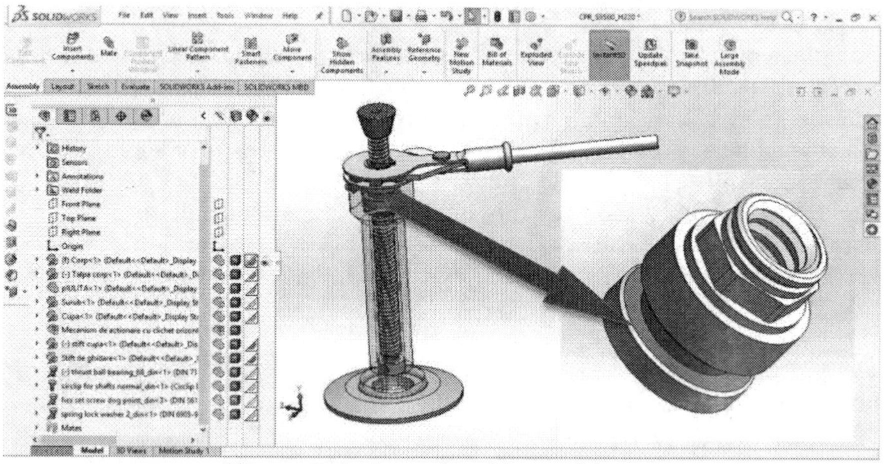

Figure 2. Jack-screw with rotating nut

In Figure 2, is presented SolidWorks CAD software interface, in which the 3D model of the jack-screw with the rotating nut has been made, having the input data a maximum stroke, Hmax = 220 mm and a maximum load, Qmax = 9500 N. For a result with a major impact, the sustainability analysis must be done for each benchmark with the necessary modifications, but in this paper only the rotating nut.

After finalizing the modeling, the Sustainability Study of the "rotating nut" was started using the "Sustainability" menu (Figure 3). By accessing the "Material Set" tab, the database of materials provided by the CAD software was launched. Figure 4 being sorted material quality carbon steel range. In this database, some features are shown for each material, for example, "Modulus of elasticity", "density", etc.

As mentioned in Chapter 2, besides the type of material used to manufacture the product, the impact on the environment is influenced by other factors such as the territory where it is manufactured and then used, the transport between the two territories, the manufacturing process, painting piece, etc. In the present case, a location in Europe was chosen as the manufacturing area (Figure 5), and as the location of use of the product, South America (Figure 7). The greater the distance between the two areas, the less environmentally friendly the product. [8]

In the first step, the manufacturing process used (Figure 6) is a machining process, after which a hot forging process has been chosen.

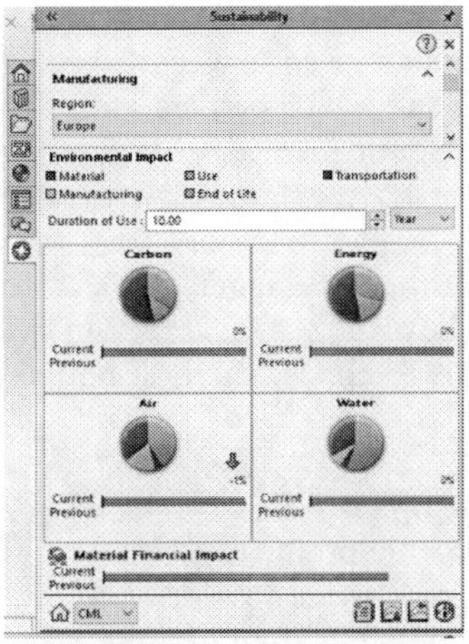

Figure 3. The product "Sustainability" menu

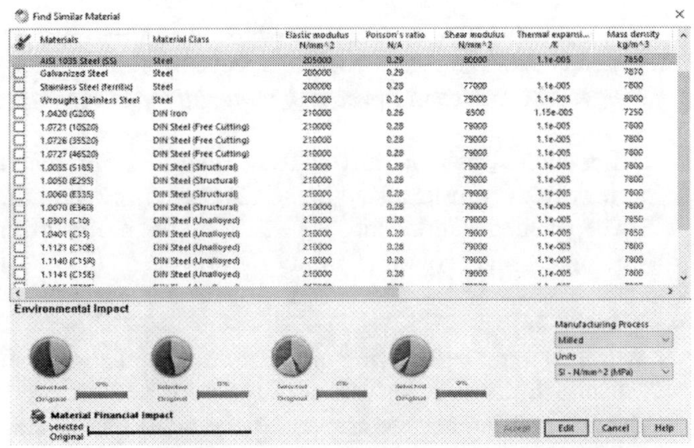

Figure 4. The steel materials datebase

In the first step of the production process used (Figure 7) is a cutting machining process, was subsequently machining process especially a hot forging.

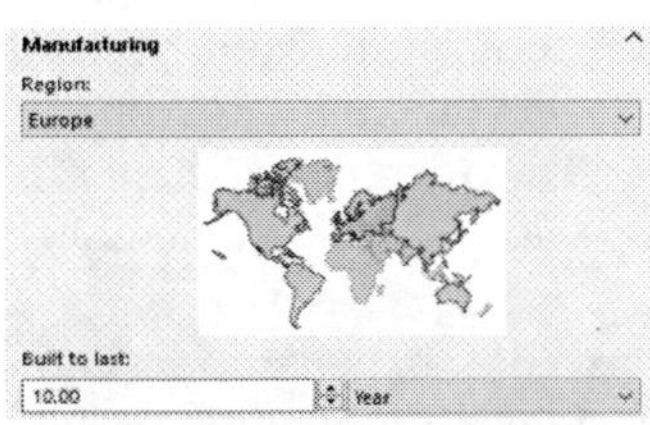

Figure 5. Manufacturing tab

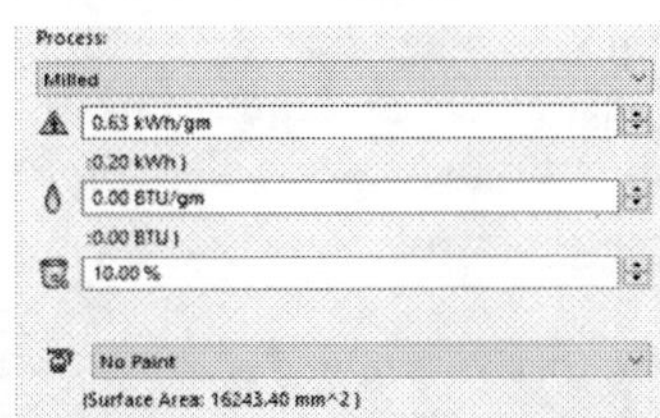

Figure 6. Process menu

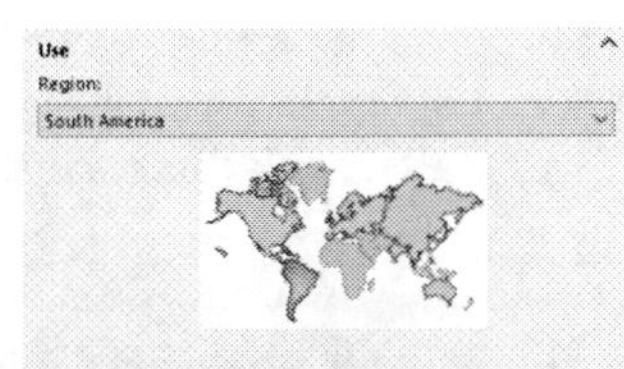

Figure 7. Use tab

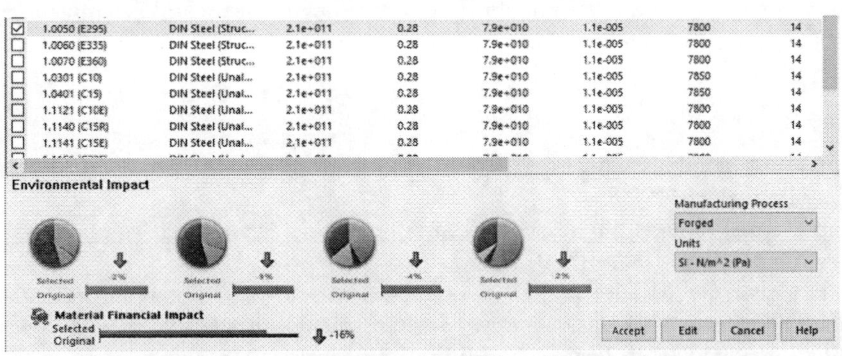

Figure 8. After changing the material and processing process

As can be seen in Figure 8, after changing cutting manufacturing process in hot forging and changing the material with similar characteristics, were obtained:

- A decrease in *carbon footprint* by 2%. If the final product is used in a region other than the one mentioned above, for example in Europe, this factor will shrink by another 6 percent, and if North America chooses, the factor will decrease by another 12 percent from South America.
- A decrease in *total energy consumption* by 3%. If the final product is to be used in a region other than the one mentioned above, for example across Europe, this factor will shrink by 8 percent, and if North America chooses, the factor will decrease by another 9 percent from South America.

IOP Conf. Series: Materials Science and Engineering **568** (2019) 012113 doi:10.1088/1757-899X/568/1/012113

- A decrease in *air acidification* by 4%. If the final product is to be used in a region other than the one mentioned above, for example in Europe, this factor will shrink by 16 percent, and if North America chooses, the factor will decrease by another 14 percent from South America.
- A decrease in *water eutrophication* by 2%. If the final product is used in a region other than the one mentioned above, for example in Europe, this factor will shrink by 16 percent, and if North America chooses, the factor will decrease by another 19 percent from South America.
- A decrease in the *cost of material* by 16%.

After such a sustainability analysis conducted by the Design Department, designers can get various reports with the results tested in the CAD environment, then they can be analyzed in a meeting with the other departments and using, for example, the brainstorming technique to choose the best option from an ecological and economic point of view.

4. Conclusion

New trends in manufacturing organizations focus on developing an organic product.

Sustainability analysis was performed for a single product, and the result was obviously diminishing environmental impact factors, and if the analysis is made across the whole and taking into account a small series production of approximately 1000 pcs / month, shrinking percentages for factors that impact on the environment will increase.

After such an analysis, the enterprise can decide which regions can export and sell the products, which would be the cheapest way of transport, etc.

Many organizations that do not have product execution capabilities but are highly focused on design and because the region where the company's headquarters is located has some sustainability legislation they can delegate for manufacturing elsewhere and later, to sell the products in another region.

In parallel with this analysis, design engineers can also approach the DFMA redesign method, from which you can get a reduction in material requirements, machining operations, lower assembly costs, and a lower production cost.

I believe that in the near future, because a company/enterprise that has a design department and wants to keep up the market and face the competition, you need to use this sustainability module provided by CAD software or to have dedicated software for these analyzes.

In the following papers, several sustainability case studies from several organizations will be dealt with for various parts made from both ferrous and non-ferrous materials in order to reduce the impact on the environment.

References

[1] Elkington, J., "Cannibals with Forks: the Triple Bottom Line of 21st Century Business," New Society Publishers, 1997.

[2] Sustainability of Products, Processes and Supply Chains Theory and Applications, 36, Elsevier, 2015 Accessed on 10 April 2019

[3] A. Jula, E. Chisu, Gh. Moldovean, D. Velicu, D. Savescu, Gh. Mogan, R. Velicu, E. Eftimie, Cl. Pozna, D. Oprean, M. Lates, " Screw Mechanisms, Design Guidelines" ("*Mecanisme Surub-Piulita, Indrumar de proiectare")*, Editura Lux Libris, 1976)

[4] A. D. Patel, et al, ESI or Energy & Environmental Science, (2012) Accessed on 15 March 2019

[5] Chowdary, B.V. and Phillips, C.M. (2014), "*Integration of computer-aided design and DFE principles for the development of a tennis ball retriever concept: A case study*", Asian Journal of Management Science and Applications, Vol. 1, No.4, pp.333-354

[6] Mutingi, M., Dube, P. and Mbohwa, C. (2017), "*A modular product design approach for sustainable manufacturing in a fuzzy environment*", Procedia Manufacturing, Vol.8, pp.471 478.

[7] "Introducing Sustainability in Value Models to Support Design Decision Making", Marco Bertoni, Vol. 9, Issue 6, 2017 Accessed on 25 March 2019

[8] Sustainable Design Guide in SolidWorks Accessed on 10 April 2019

IOP Conf. Series: Materials Science and Engineering **568** (2019) 012114 doi:10.1088/1757-899X/568/1/012114

Design and development of a virtual model of an electric vehicle of category L7

C M Pappalardo[1*], N Lombardi[2], P V Dašić[3] and D Guida[1]

[1] University of Salerno, Via Giovanni Paolo II, 132 - 84084 - Fisciano (Salerno) - Italy
[2] Spin-Off MEID4 S.r.l., Via Giovanni Paolo II, 132 - 84084 - Fisciano (Salerno) - Italy
[3] High Technical Mechanical School of Professional Studies, Trstenik, 37240, Serbia

cpappalardo@unisa.it

Abstract. This study is aimed at creating an accurate mathematical model, developed in Matlab-Simulink through the Simscape library, concerning an electric vehicle of class L7. To this end, a mechanical model with three degrees of freedom is used. The degrees of freedom are advancement, vertical displacement, and pitch. Two degrees of freedom are also added to this dynamical model in order to model the stiffness and the damping of the tires, whose behavior is described by the "Magic Formula". The propulsion unit is modeled in a standard design configuration. The vehicle can be controlled by supplying the driver's inputs, by controlling the cruising speed, or by ensuring that the vehicle has a driving cycle. The proposed model allows for identifying the critical aspects of the selected design solution, thereby reducing the time required in the design phases.

1. Introduction

This paper deals with the virtual design and the computer development of a dynamical model of an electric vehicle of category L7. This section provides background information on the problem addressed in this paper and the organization of the manuscript. Currently, one of the technologies with the greatest potential in the field of sustainability is electric propulsion [1,2,3]. Electric Vehicles (EV) do not emit polluting substances. However, the energy that feeds them generates different emissions globally based on the way in which the same amount of energy is produced. If this energy is obtained through the use of fossil fuels, the result is simply to transfer the emissions from the city to the production plant. On the other hand, if the energy required to power electric vehicles is produced using any renewable form of energy generation, a significant impact on the sustainability of the means of transportation can be achieved. It is, therefore, necessary to globally enhance the synergy between EV and renewable energy production. If these systems were used globally, there would be ideally zero emissions. The remaining part of this manuscript is structured as follows. Section 2 provides a short literature review on the topic addressed in this paper. In Section 3, the background materials and the computational methods employed in this research project are described. Section 4 and 5 proposes the developments of the virtual prototype of the electric vehicle of category L7 that represents the main object of the present paper. Section 6 offers a summary of the work and the conclusions reached in this paper.

2. Literature Review

Several recent research works are devoted to the virtual prototyping of electric vehicles. To obtain a general overview of electric vehicles, one can consult the journal paper [4], which illustrates the most used technologies, comparing the models on the market with the impact of electric vehicles on the environment, the economy, and the electricity grid. The theme of the impact of electric vehicles is dealt with in the detailed paper, in which the in-depth analysis proposed concerns vehicles of category L [5]. For the study of vehicle performance, an excellent reference is the vehicle dynamics textbook by Guiggiani [6]. Also, useful references are the journal publications [7,8], in which the half-body model of an electric vehicle is analyzed in particular. The longitudinal tire model was developed using Pacejka's well-known "Magic Formula" described in the

IOP Conf. Series: Materials Science and Engineering **568** (2019) 012114 doi:10.1088/1757-899X/568/1/012114

book [9]. The model developed in this work uses the Simulink model as a basis [1]. Furthermore, as far as the vehicle models developed in Matlab/Simulink is concerned, the reference papers [10-12] are recommended. For a review on the electrical motor drivelines in electric vehicles, the literature survey [13] is recommended. The suspension system suitable also for the vehicle of category L7 considered in this work are studied in the journal paper [14].

3. Materials and Methods

In the development of the present work, a systematic approach is used. Firstly, the object of study is analyzed, therefore analyzing the various existing components for EVs, such as batteries, charging stations, engines, and the different possible configurations. Subsequently, for the creation of the mathematical model, a systematic approach is used, dividing the whole system into simpler problems. Once solved the individual problems, the complete system is assembled creating a general vehicle model. However, the main focus of the paper is on the development of a structured dynamical model which couples the mechanical aspects and the electric aspects of the vehicle. For the development of the vehicle model, the Matlab/Simulink software was used, an application included in the Matlab calculation program and which constitutes, in large part, a powerful and intuitive graphic interface aimed at simplifying its use by the user. Using Simulink it is possible to program the execution of calculations in the Matlab environment in a much faster and error-free way compared to the writing of the long and complex m-files that are necessary, for example, to program the numerical integration of a system of high order differential equations. Through the visual tools available in the Simulink environment it is possible to simulate even very complex systems with an effort by the user that is limited to the tracking, on an electronic worksheet, of a block diagram representative of the system under examination. Simulink consists of a series of libraries that contain elementary blocks, which, properly interconnected, form the block diagram representing the desired functionality. The library most used to conduct this study is Simscape, in which the blocks represent physical elements having different levels of fidelity relative to the different levels of the system being analyzed, and the lines, which connect these components, correspond to physical connections in the system real that transmits power. This method thus allows us to describe the physical structure of a system rather than the underlying mathematics. The electrical and mechanical connections are represented by different colors that indicate the physical domain. This software was used in order to carry out a preliminary simulation of the behavior of the electric vehicle in a more economical and efficient way compared to hardware prototypes.

4. Electrical Model of the Vehicle

The electrical system is powered by a lithium-ion battery consisting of 10 parallels composed of 50 series of cells each. In the representation of the battery, the equivalent circuit of the cell is created, and by multiplying it appropriately, the entire generator is obtained. There is a DC-DC converter to bring the current from the battery voltage to the motor voltage. The electric motor is synchronous with permanent magnets with closed-loop torque control, which means that the input voltage and power adjust according to the torque required by the control system. Therefore, having at the motor input the electric current, regulated by the controller, the rotation of the rotor and the heat generated are obtained. Furthermore, in the case of negative torque and positive speed, the electric machine operates like a current generator recharging the batteries. The electric system is shown in Figure 1.

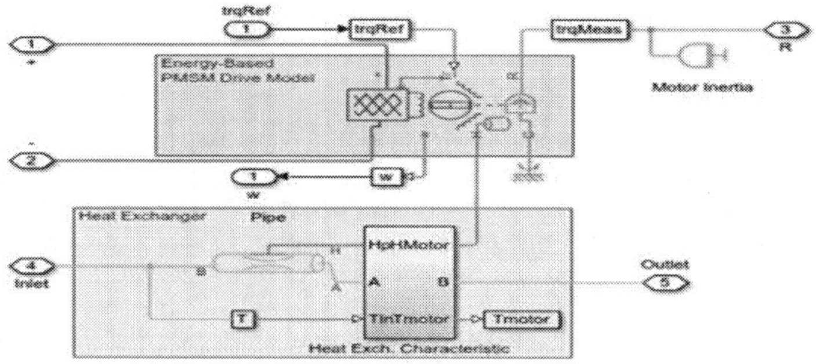

Figure 1. Electric motor.

IOP Conf. Series: Materials Science and Engineering **568** (2019) 012114 doi:10.1088/1757-899X/568/1/012114

5. Mechanical Model of Vehicle

The mechanical part of the electric vehicle of interest is modeled considering a simplified multibody approach [15,16]. To this end, the system uses a half-car vehicle model with three degrees of freedom: advancement, vertical displacement (bouncing), and pitching, in order to consider the different forces acting on the car such as viscous friction with the air, the elastic ones and the damping ones due to the suspensions [17]. The behavior of the tires was also examined using the well-known Pacejka "Magic Formula" and considering their stiffness and damping [18]. The propulsion unit has been modeled according to a standard configuration consisting of a central engine, with a fixed transmission ratio and a mechanical differential that distributes the driving torque to the wheels [19,20]. In the design configuration chosen, the rotation obtained by the motor is first reduced by a fixed gear, then distributed by a mechanical differential to the two driving wheels, from whose model, finally, the longitudinal force acting on the axis is calculated, and which affects the dynamic model as shown in Figure 2. In electric vehicles, it is possible not to use the speed change, since the electric motors, compared to traditional internal combustion engines, have a high starting torque, much higher maximum RPMs and good efficiency for a wide speed range.

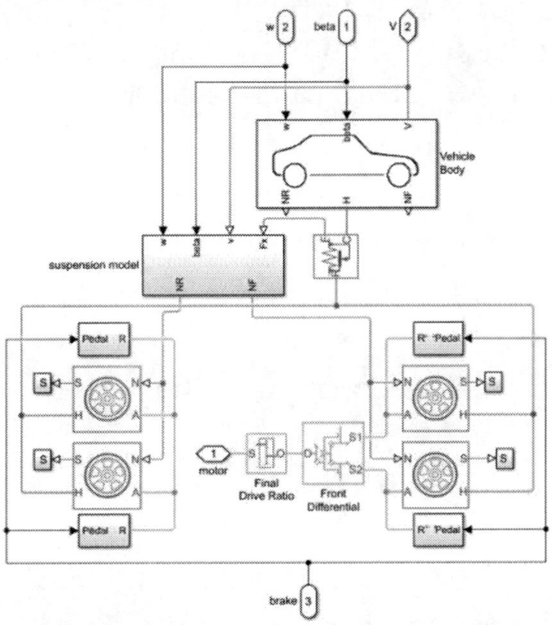

Figure 2. Matlab/Simulink model of the design configuration.

A dynamic half-car model with three degrees of freedom was used: advancement, vertical displacement (bouncing), and pitching, to which two degrees of freedom are added, giving stiffness and damping to the tires as shown in Figure 3.

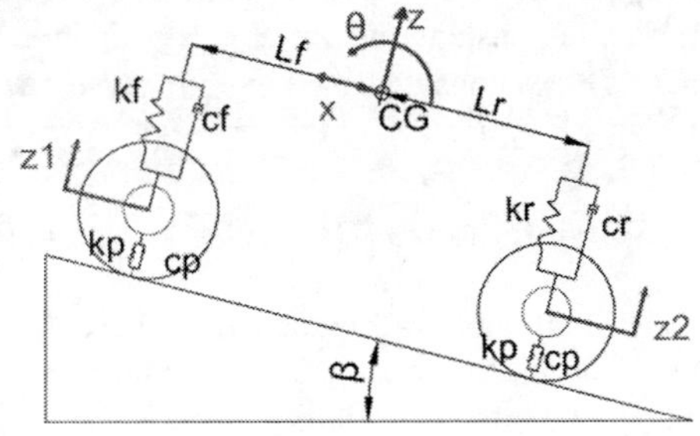

Figure 3. Scheme of the dynamical model.

$$\begin{cases} m\ddot{x} = F_x - F_{d,x} - F_{g,x} \\ m\ddot{z} = F_{d,z} + F_{sz,F} + F_{sz,R} - F_{g,z} \\ I_{yy}\ddot{\theta} = -L_f F_{sz,F} + L_r F_{sz,R} + hF_x - M_{d,y} \\ m_w \ddot{z}_1 = -F_{sz,F} + F_{s1} \\ m_w \ddot{z}_2 = -F_{sz,R} + F_{s2} \end{cases} \quad (1)$$

In these equations:

- F_x is the longitudinal force applied on the wheel axis;
- $M_{d,y}; F_{d,z}; F_{d,x}$ are respectively the drag moment, the longitudinal drag force, and the vertical drag force;
- $F_{sz,F}; F_{sz,R};$ are the elastic and damping force of the front and rear axle suspensions;
- $F_{gx}; F_{gz}$ are the components of the gravity force;
- $L_f; L_r$ are respectively the longitudinal distances from the center of mass to the front and rear axle;
- h is the normal distance between the CG and the wheel axis.

The viscous friction force between the air and the vehicle is therefore directly proportional to the square of the difference between vehicle speed and wind speed:

$$\begin{cases} F_{d,x} = \frac{1}{2} C_d \rho A(\dot{x} - w)^2 \\ F_{d,z} = \frac{1}{2} C_l \rho A(\dot{x} - w)^2 \\ M_{d,y} = \frac{1}{2} C_{pm} \rho A(\dot{x} - w)^2 (L_r + L_f) \end{cases} \quad (2)$$

The elastic and damping force of the suspensions and of the tires is given by the equations:

$$\begin{cases} F_{sz,F} = -2k_f(z - z_1 - L_f\vartheta) - 2C_f(\dot{z} - \ddot{z}_1 - L_f\dot{\vartheta}) \\ F_{sz,R} = -2k_r(z - z_2 + L_r\vartheta) - 2C_r(\dot{z} - \ddot{z}_1 + L_r\dot{\vartheta}) \end{cases}, \quad \begin{cases} F_{s1} = -k_p x_1 - c_p \dot{x}_1 \\ F_{s2} = -k_p x_2 - c_p \dot{x}_2 \end{cases} \quad (3)$$

The advancement equation is independent of the other ones, therefore it is resolved independently with the use of the "Vehicle body" block. The remaining equations are instead coupled and are solved with the use of Matlab Simulink blocks. The values used for the simulation refer to a generic vehicle of category L7, in particular the damping and stiffness coefficients of the suspension, and of the tires, are those present in the paper [15] in which a vehicle with a mass comparable to that chosen by us is analyzed. To model the tire using the Simulink pakage "Tire block (magic formula)", which describes the longitudinal behavior, using an empirical formula based on four adaptation coefficients. The block models the dynamics of the tire considering it a rigid set with the wheel, both in constant and variable road conditions and evaluating its inertia, rolling resistance and slip. Moreover, this block considers the longitudinal stiffness and damping of the tire. The model used is based on the Pacejka Magic Formula [9], obtained through experimental tests. The general formula is:

$$F_x = F_z D \sin(C \arctan\{B\kappa - E[B\kappa - \arctan(B\kappa)]\}) \quad (4)$$

The coefficients B, C, D, E are obtained graphically where the function passes through the origin of the Cartesian system, reaches a maximum and then tends to a horizontal asymptote. The slip is given by the ratio between the speed of slip and the speed of progress in the case in which there is no slip, therefore:

IOP Conf. Series: Materials Science and Engineering **568** (2019) 012114 doi:10.1088/1757-899X/568/1/012114

$$\kappa = \frac{V_{sx}}{|V_x|} = \frac{r_w\Omega - V_x}{|r_w\Omega|}, \quad \kappa = \frac{2V_{sx}}{V_{th} + \frac{V_x^2}{V_{th}}} \tag{5}$$

where the second equation is valid at low speeds.

6. Summary and Conclusions

The authors focus their research mainly on the dynamics of multibody systems [21-25], on the optimal control techniques [26-30], and on the algorithms of system identification [31-35]. Based on this research background, this paper deals with the design and the developments of a virtual prototype of an electric vehicle. This work is focused on the design of a mathematical model representing the engineering relevant aspects of an electric vehicle of class L7. In order to be able to carry out an analysis that takes into account the different forces acting on the system, a standard dynamic model with three degrees of freedom referred to as the half-car model is implemented. This mechanical model is further extended including also the longitudinal displacement of the vehicle in order to simulate appropriate driving cycles. The computer simulations are performed in Matlab-Simulink environment, using the Simscape library. A proportional-integrative control is used for the system motion, which allows the vehicle to follow the standard WLTP class 1 driving cycle. The electric system of the vehicle is also modeled in Matlab-Simulink environment in order to determine the energy consumption. The proposed dynamic model allows for effectively performing the virtual prototyping of the electric vehicle of interest.

References

[1] Electric vehicle configured for HIL [Online]. Available at: https://it.mathworks.com/help/physmod/sps/examples/electric-vehicle-configured-for-hil.html

[2] Ratti C and Claudel M 2016 *The city of tomorrow: Sensors, networks, hackers, and the future of urban life* (Yale University Press)

[3] Offenhuber D and Ratti C 2014. *Decoding the City: How Big Data Can Change Urbanism* (Birkhäuser Verlag GmbH)

[4] Yong J Y, Ramachandaramurthy V K, Tan K M and Mithulananthan N 2015 A review on the state-of-the-art technologies of electric vehicle, its impacts and prospects. *Renew. Sust. Energ. Rev.* **49** 365-85

[5] Santucci M, Pieve M, and Pierini M 2016 Electric L-category vehicles for smart urban mobility *Transp. Res. Proc.* **14** 3651-60

[6] Guiggiani M 2014 *The science of vehicle dynamics* (Pisa, Italy: Springer Netherlands)

[7] Cao J T, Liu H H, Li P, Brown D J and Dimirovski, G 2007 A study of electric vehicle suspension control system based on an improved half-vehicle model. *Int. J. Automat. Comp.* **4**(3), 236-42

[8] Brown D J, and Dimirovski G 2007 A study of electric vehicle suspension control system based on an improved half-vehicle model *Int. J. Automat. Comp.* **7** 56-61

[9] Pacejka H B 2005 *Tyre and vehicle dynamics* (Elsevier)

[10] Vo-Duy T, and Ta M C 2016 A signal hardware-in-the-loop model for electric vehicles. *Robomech. J.* **3**(1) 29

[11] Androne C F and Rao S B *Vehicle Simulation for Powertrain System Testing* (Chalmers University of technology)

[12] Thanh V D, Thinh D Q, Huy N B and Minh T C 2016 Design of Hardware-in-the-loop Model for Electric Vehicles *Proc. Pub. H. Sci. Tech.***1**(1)

[13] De Santiago J, Bernhoff H, Ekergård B, Eriksson S, Ferhatovic S, Waters R and Leijon M 2012 Electrical motor drivelines in commercial all-electric vehicles: A review. *IEEE Trans. Veh. Techn.* **61**(2) 475-84

[14] Gawai N, Yadav D, Chavan S, Lele A and Dalvi S 2016 Design, modelling & analysis of double wishbone suspension system *Int. J. Mech. Eng. Rob.* **4**(1) 58-62

[15] Pappalardo C M 2015 A natural absolute coordinate formulation for the kinematic and dynamic analysis of rigid multibody systems *Nonlinear Dyn.* **81**(4) 1841-69

[16] Pappalardo C M, and Guida D 2018 A time-domain system identification numerical procedure for obtaining linear dynamical models of multibody mechanical systems *Arch. Appl. Mech.* **88**(8) 1325-47
[17] Gonc M A, Lenz W B, Ribeiro M A, Tusset A M and Balthazar J M 2019 Magneto rheological damper in semi-active vehicle suspension system using SDRE control for a half car model *J. MESA* **10**(1) 75-91
[18] Cabrera J, Castillo J, Pérez J, Velasco J, Guerra A, and Hernández P 2018 A procedure for determining tire-road friction characteristics using a modification of the magic formula based on experimental results *Sensors* **18**(3) 896
[19] Hibbeler R C 2001 *Engineering mechanics* (Pearson education)
[20] Azman M, King P D and Rahnejat H 2007 Combined bounce, pitch, and roll dynamics of vehicles negotiating single speed bump events. *P. I. Mech. Eng. K-J Mul.* **221**(1) 33-40
[21] De Simone M C, Rivera Z B and Guida D 2018 Obstacle Avoidance System for Unmanned Ground Vehicles by Using Ultrasonic Sensors *Machines* **6** 18
[22] De Simone M C, Russo S, Rivera Z B and Guida D 2017 May, Multibody model of a UAV in presence of wind fields *Int. Conf. on Control, Artificial Intelligence, Robotics & Optimization (ICCAIRO)* (Prague, Czech Republic: IEEE) p 83-88
[23] De Simone M C and Guida D 2018 Identification and Control of a Unmanned Ground Vehicle By using Arduino *UPB Sci. Bull. Ser. D* **80** 141-54
[24] De Simone M C and Guida D 2017 On the development of a low-cost device for retrofitting tracked vehicles for autonomous navigation *Proc. of the 23rd Conf. of the Italian Association of Theoretical and Applied Mechanics* vol 4 (Salerno, Italy: Springer) p 71-82
[25] De Simone M C and Guida D 2018 Control design for an under-actuated UAV model *FME Trans.* **46** 443-52
[26] De Simone M C and Guida D 2018 Modal coupling in presence of dry friction *Machines* **6**(8)
[27] De Simone M C, Rivera Z B and Guida D 2018 Finite element analysis on squeal-noise in railway applications. *FME Trans.* **46** 93-100
[28] De Simone M C and Guida D 2019 Object Recognition by Using Neural Networks For Robotics Precision Agriculture Application *Eng. Lett.* In press
[29] Concilio A, De Simone M C, Rivera Z B and Guida D 2017 A new semi-active suspension system for racing vehicles *FME Trans.* **45** 578-84
[30] Quatrano A, De Simone M C, Rivera Z B and Guida D 2017 Development and implementation of a control system for a retrofitted CNC machine by using Arduino *FME Trans.* **45** 565-71
[31] Ruggiero A, Affatato S, Merola M and De Simone M C 2017 Sep, FEM analysis of metal on UHMWPE total hip prosthesis during normal walking cycle *Proc. of the 23rd Conf. of the Italian Association of Theoretical and Applied Mechanics (AIMETA 2017)* vol 2 (Salerno, Italy: Springer) p 4-7
[32] Ruggiero A, De Simone M C, Russo D and Guida D 2016 Sound pressure measurement of orchestral instruments in the concert hall of a public school *Int. J. Cir. Syst. Sign. Proc.* **10** 75-81
[33] De Simone M C and Guida D 2015 Dry friction influence on structure dynamics *Proc. of the COMPDYN 2015 - 5th ECCOMAS Thematic Conf. on Computational Methods in Structural Dynamics and Earthquake Engineering* (Crete Island, Greece) p 4483-91
[34] Iannone V and De Simone M C 2019 Modelling of a DC Gear Motor for Feed-Forward Control Law Design for Unmanned Ground Vehicles. *Actuators* Submitted
[35] Rivera Z B, De Simone M C and Guida D 2019 Modelling of Mobile Robots in ROS-based Environments. *Robotics* Submitted

IOP Conf. Series: Materials Science and Engineering **568** (2019) 012115 doi:10.1088/1757-899X/568/1/012115

Modeling techniques for kinematic analysis of a six-axis robotic arm

R Guida[1], M C De Simone[2*], P Dašić[3] and D Guida[2]

[1] MEID4 S.r.l., Fisciano 84084, Italy
[2] The University of Salerno. Via Giovanni Paolo II, 84084 Fisciano, Italy
[3] High Technical Mechanical School of Professional Studies, Trstenik, 37240, Serbia

mdesimone@unisa.it

Abstract. The purpose of this analysis is to evaluate the effectiveness of different modeling techniques for a COMAU six-axis robot arm. The robot manipulator represents one of the most used robots in the mechanical industry. For robots, with a high number of degrees of freedom, to obtain a kinematic model, Denavit-Hartemberg parameters allow representing geometric transformations in the Euclidean space through the minimum number of parameters. Furthermore, the use of detailed multibody models and the use of sophisticated tools like the Robotics Toolbox System, allow performing, in a very efficient way, inverse kinematic analysis and trajectory planning to control the robot move from one configuration to another.

1. Introduction

According to the International Federation of Robotics (IFR), the increased speed of development of robotics is mainly beneficed by the progress of off-the-shelf, low-cost equipment, software packages or codes almost ready to use, developments expected by the industry and active communities in the field [1]. To design a robot, it is essential to establish the optimal configuration by analyzing its kinematic behavior. There are many optimization techniques in literature for the design of optimal configurations. As reported in Cesconeto and Perondi (2019), optimum strategies must be evaluated to evaluate the optimum length of single links and joint angular limits to design an arm with the best possible kinematic performance within a desired design-defined workspace [2]. It is important for developers and implementers, to be aware of the available platforms, methods, algorithms, hardware components, that are most used, underlying physical and numerical paradigms [3-4]. All these aspects have an impact on safety and goal reaching motion; that is why we are interested in a good understanding of modeling techniques for kinematic and dynamic analysis and expertise on software architectures for robot application, as demonstrated by the growing number of developers of open-source systems for robotics as in the case of the ROS-Gazebo robot platform [5-6]. This paper represents a preliminary study to evaluate the effectiveness of different modeling systems for a subsequent experimental activity on the retrofitting of an industrial robot COMAU Smart Six6-1.4 of the Applied Mechanics laboratory [7-8].

2. The COMAU Smart Six 6-1.4

The COMAU Smart SiX 6-1.4 is a manipulator robot produced by the company Comau (COnsorzio MAcchine Utensili), an Italian multinational company that is part of FCA Group. It develops and realizes the automation process and solutions to meet specific manufacturing needs for industries ranging from railway, heavy industrial, renewable energy, and automotive [9]. The manipulator consists of 6 links, that form an open kinematic chain. Every joint represents a cylindrical hinge (shaped by a revolute joint). The motion is led by brushless DC electric motors (BLDC), also known as electronically commutated motors. The BLDC motors present various advantages: they have a high power to weight ratio, high speed, and electronic control. The motion transmission takes place by means of gears (for the axes 1, 2, 3, 4) and using a belt (for the axes 5, 6). The belt transmission also has a Harmonic Drive, a strain wave gear. The joints of the robot have end stops, that can be software or mechanical [10]. In Fig. 3, the technical specifics of the robot are reported. Before using the robot is necessary to calibrate it. The calibration consists in bringing the joints in a known position to guarantee the correct repetition of the programmed cycles [11]. The two possible different kinds of calibration are:

- Precise calibration, used before cycles that need to be very accurate;
- Reference notches calibration, used in other cases, this kind of calibration is faster but more inaccurate.

The robot is governed by an algorithm that performs a position control. The control system uses PID controllers (Proportional-Integral-Derivative control) that uses the encoders of the 6 axes to compare the actual position of the joints to the reference position [12].

The robot has its programming language, the PDL2, language is very similar to Pascal, an object-oriented programming paradigm. The code can be uploaded by using C4G, a controller given by the company, or using a PC, linked thanks to a LAN connection. As default every motion law is built using a trapezoidal trajectory, but this setting could be changed.

3. Robot Arm Kinematics

To proceed with the analysis of the anthropomorphic manipulator, the reference frame 0 is chosen with the origin at the intersection of z_0 and z_1 ($d_1 = 0$), and z_1 and z_2 are parallel as reported in Fig. 1a). Choosing x_1 and x_2 along the direction of the respective link in order to make null every d_i.

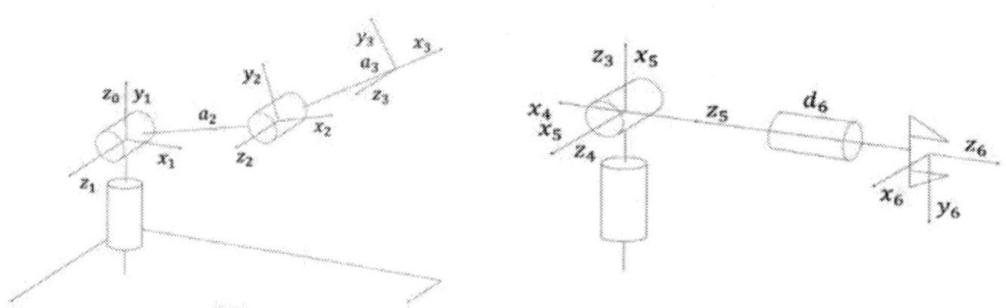

Figure 1. Reference frames used for the kinematic analysis: (a) frame references used for links 1,2 and 3, (b) frames references used for links 4, 5, and 6.

In (1), (2) and (3), are reported the transformation matrices for link 1, 2, and 3, respectively.

$$A_1^0(\theta_1) = \begin{bmatrix} c_1 & 0 & s_1 & 0 \\ s_1 & 0 & -c_1 & 0 \\ 0 & 1 & 0 & 0 \\ 0 & 0 & 0 & 1 \end{bmatrix} \quad (1)$$

$$A_2^1(\theta_2) = \begin{bmatrix} c_2 & 0 & s_2 & 0 \\ s_2 & 0 & -c_2 & 0 \\ 0 & 1 & 0 & 0 \\ 0 & 0 & 0 & 1 \end{bmatrix} \quad (2)$$

$$A_3^2(\theta_3) = \begin{bmatrix} c_3 & 0 & s_3 & 0 \\ s_3 & 0 & -c_3 & 0 \\ 0 & 1 & 0 & 0 \\ 0 & 0 & 0 & 1 \end{bmatrix} \quad (3)$$

The direct kinematic function is reported below:

$$T_3^0(q) = A_1^0\,A_2^1\,A_3^2 = \begin{bmatrix} c_1c_{23} & -c_1s_{23} & s_1 & c_1(a_2c_2 + a_3c_{23}) \\ s_1c_{23} & -s_1s_{23} & -c_1 & s_1(a_2c_2 + a_3c_{23}) \\ s_{23} & c_{23} & 0 & a_2s_2 + a_3s_{23} \\ 0 & 0 & 0 & 1 \end{bmatrix} \tag{4}$$

with $q = [\theta_1 \quad \theta_2 \quad \theta_3]^T$ the first, second and third joint variables. The second part of the kinematic study consist in considering exclusively the spherical wrist. In this case the first link is indicated with the number 4. Since the three axes of rotation intersect all in the same point, once identified z_3, z_4 and z_5 and defined x_3, the direction of x_4 and x_5 are indeterminate. Such directions are chosen as reported in Fig. 1b).

$$A_4^3(\theta_4) = \begin{bmatrix} c_4 & 0 & -s_4 & 0 \\ s_4 & 0 & c_4 & 0 \\ 0 & -1 & 0 & 0 \\ 0 & 0 & 0 & 1 \end{bmatrix} \tag{5}$$

$$A_5^4(\theta_5) = \begin{bmatrix} c_5 & 0 & s_5 & 0 \\ s_5 & 0 & -c_5 & 0 \\ 0 & 1 & 0 & 0 \\ 0 & 0 & 0 & 1 \end{bmatrix} \tag{6}$$

$$A_6^5(\theta_6) = \begin{bmatrix} c_6 & -s_6 & 0 & 0 \\ s_6 & c_6 & 0 & 0 \\ 0 & 0 & 1 & d_6 \\ 0 & 0 & 0 & 1 \end{bmatrix} \tag{7}$$

Considering the transformation matrices reported in (5), (6) and (7), the direct kinematic function assumes the following form:

$$T_3^0(q) = A_1^0\,A_2^1\,A_3^2 = \begin{bmatrix} c_1c_{23} & -c_1s_{23} & s_1 & c_1(a_2c_2 + a_3c_{23}) \\ s_1c_{23} & -s_1s_{23} & -c_1 & s_1(a_2c_2 + a_3c_{23}) \\ s_{23} & c_{23} & 0 & a_2s_2 + a_3s_{23} \\ 0 & 0 & 0 & 1 \end{bmatrix} \tag{8}$$

R_6^3 matrix, which can be extracted from T_6^3 , coincides with the rotation matrix of the Euler angles, ie θ_4, θ_5 and θ_6 compose the set of Euler ZYZ angles with respect to the reference frame O_3-x_3,y_3,z_3. Finally, for describing the COMAU Smart Six configuration the kinematic function is defined from link 0 to link 6.

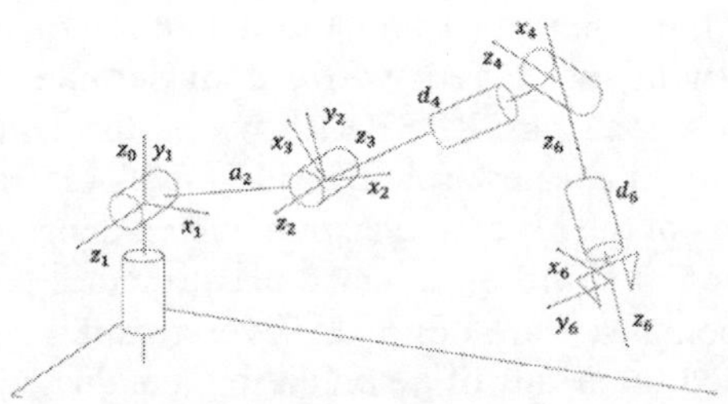

Figure 2. Reference frames used for the kinematic analysis of the anthropomorphic robot.

It cannot be obtained by product of the two matrices T_3^0 and T_6^3 since the 3rd frame of the manipulator cannot coincide with the 3rd frame of the wrist. The resolution of this problem follows two paths, the first is to impose a constant transformation between the two reference frames in order to overlap them, the second is to follow the procedure of D-H for all reference frames (see Fig. 2).
Doing so, we obtain the following parameters: rows 3 and 4 are different from the previously highlighted parameters, so the matrices of homogeneous transformation change, as reported in eq. (9) and (10).

$$A_3^2(\theta_3) = \begin{bmatrix} c_3 & 0 & s_3 & 0 \\ s_3 & 0 & -c_3 & 0 \\ 0 & 1 & 0 & 0 \\ 0 & 0 & 0 & 1 \end{bmatrix} \quad (9)$$

$$A_4^3(\theta_4) = \begin{bmatrix} c_4 & 0 & -s_4 & 0 \\ s_4 & 0 & c_4 & 0 \\ 0 & -1 & 0 & d_4 \\ 0 & 0 & 0 & 1 \end{bmatrix} \quad (10)$$

The other matrices instead, remain the same. After calculating the direct kinematics function, the position and the orientation of the tool axis can be expressed by defining the values of p^0, n^0, s^0 and a^0.

$$p^0 = \begin{bmatrix} a_2c_1c_2 + d_4c_1s_{23} + d_6(c_1(c_{23}c_4s_5 + s_{23}c_5) + s_1s_4s_5) \\ a_2s_1c_2 + d_4s_1s_{23} + d_6(s_1(c_{23}c_4s_5 + s_{23}c_5) + c_1s_4s_5) \\ a_2s_2 - d_4c_{23} + d_6(s_{23}c_4s_5 - c_{23}c_5) \end{bmatrix} \quad (11)$$

$$n^0 = \begin{bmatrix} c_1(c_{23}(c_4c_5c_6 - s_4s_6) - s_{23}s_5c_6) + s_1(s_4c_5c_6 + c_4s_6) \\ s_1(c_{23}(c_4c_5c_6 - s_4s_6) - s_{23}s_5c_6) - c_1(s_4c_5c_6 + c_4s_6) \\ s_{23}(c_4c_5c_6 - s_4s_6) - s_{23}s_5c_6 \end{bmatrix} \quad (12)$$

$$s^0 = \begin{bmatrix} c_1(-c_{23}(c_4c_5c_6 + s_4s_6) + s_{23}s_5c_6) + s_1(-s_4c_5c_6 + c_4s_6) \\ s_1(-c_{23}(c_4c_5c_6 + s_4s_6) + s_{23}s_5c_6) - c_1(-s_4c_5c_6 + c_4s_6) \\ -s_{23}(c_4c_5c_6 + s_4s_6) - c_{23}s_5c_6 \end{bmatrix} \quad (13)$$

$$a^0 = \begin{bmatrix} c_1(c_{23}c_4c_5c_6 - s_{23}c_5) + s_1s_4s_5 \\ s_1(c_{23}c_4c_5c_6 + s_{23}c_5) + s_1s_4s_5 \\ s_{23}(c_4c_5c_6 - s_4s_6) - s_{23}s_5c_6 \end{bmatrix} \quad (14)$$

4. Numerical Analysis

The aim of this paper, is to test the effectiveness of the methods presented for performing inverse kinematic analysis of the COMAU Smart Six 6-1.4. As a reference, we chose to create a multibody model of the robot by importing a three-dimensional CAD model of the robot under analysis into SimScape, the multibody environment of Simulink [14]. The use of detailed geometries allows to have a good model in which mass distribution is considered making the dynamic study much more accurate. In this paper the multibody model has been used in the kinematic analysis as a reference target. The Robotics Toolbox (RTB), developed by Peter Corke, is a MATLAB toolbox which allows to simulate the operation of terrestrial robots and of robotic arms. The original release dates back to the early 90s and supported exclusively the simulation of robotic arms, represented as one serial structure of links and to perform direct and inverse kinematic and dynamic analysis [13]. The Toolbox is based on a general method of representation of the kinematics and the dynamics of robots. The robot, as previously explained, is a manipulation with six degrees of freedom definite through six revolute joints. Such configuration is described by the string "RRRRRR" where the letters define the type of joint: (R) revolute or (P) prismatic. To define a new robot inside the Toolbox you must first of all define the Denavit-Hartenberg parameters of the robot in question, which depend not only on the type of robot but on the structural parameters reported in Fig. 2. Once the structural parameters have been selected, the various links of which it is composed are defined. Several values were used for testing the robotics toolbox. In Fig. 4.a) is reported the result of a simulation, conducted by using the Robotics Toolbox. Such toolbox, offers the possibility to take into account the external forces acting on the joints. To do so, it is possible to use the itorque, coriolis and gravload instructions which allow, respectively, to evaluate the inertial, the Coriolis and the gravitational terms.

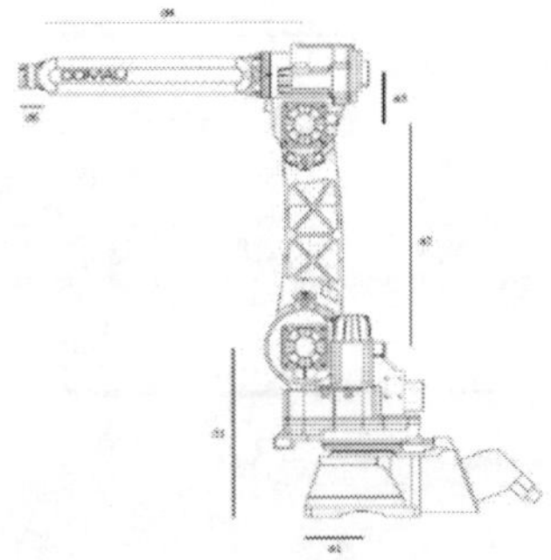

Table 1. D-H parameters

j	theta	d	a	alpha	offset
1	0.15	0.45	0.15	1.5708	0
2	0.59	0	0.59	0	0
3	0.13	0	0.13	1.5708	0
4	0.007	0.674	0	-1.5708	0
5	0.45	0	0	1.5708	0
6	0.64707	0.095	0	0	0

Figure 3. Geometric parameters and values of the D-H parameters used for the robotics toolbox.

The use of such function allows to create a mathematical model more coherent with reality, to create a more precise control of motion and position, but these details do not affect the study of the kinematics. Robotics Toolbox also provides tools for calculating differential kinematics, such as the tr2jac function, which outputs the Jacobian matrix to obtain the velocity between two links connected by a homogeneous transformation matrix. In Fig. 5, is reported the behaviour of the robot for the configuration reported previously. Such configurations, will be compared with multibody model described in the following section. The SimScape multibody simulation environment is a multi-domain environment in which it is possible to make different systems interact in a simple way. Starting from a three-dimensional geometry it is possible to obtain a multibody model in which the mass distribution is much more accurate and therefore the answer will be much more precise in simulation. For our application, the geometry of the anthropomorphic robot was replicated by using Solidworks 3D CAD modeller and thanks to the SimMechanics Link Tool, imported into the SimScape multibody simulation environment [14]. In this environment it is possible to model the actuators and the sensors by an electrical and electronic point of view in order to achieve a much more accurate model of the system [15]. In Fig. 4.b), is reported the simulation conducted on the multibody model for the same configurations.

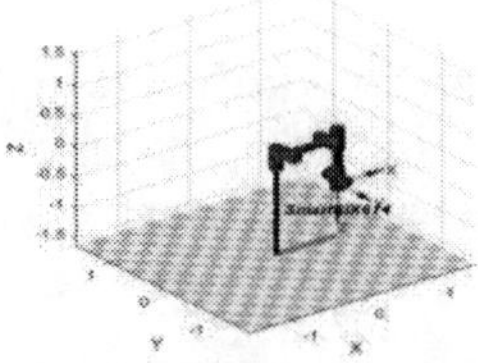

(a)

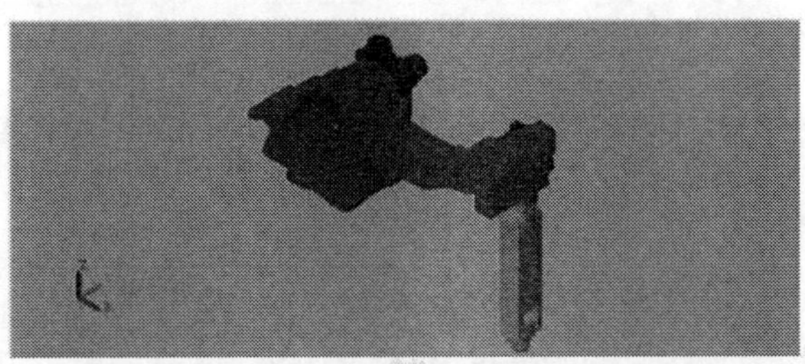

(b)

Figure 4. Kinematic simulation of the Comau Smart SiX 6-1.4: (a) robotics toolbox (b) Simscape.

In table 1, are reported the position of the end effector for the models with the percentage error.

5. Conclusions

The goal of the paper is to test different modelling techniques in order to evaluate the effectiveness of the several models for describing a specific anthropomorphic manipulator, the COMAU Smart SiX-6.14 for a subsequent retrofitting activity [16]. The analytical model and the multibody model were evaluated by using Matlab software, which allowed us to perform various tests in order to evaluate the consistency of the results very quickly [17]. For this study we chose to test the modelling techniques by using the Denavit-Hartenberg parameters, the robot toolbox and a rigid multibody model obtained using Solidworks and the Simscape environment. The errors made by the models used for this analysis ranged from 5% to 8%, however, remaining below 10% and thus showing a validity of all the proposed models.

Table 2. Comparison of the proposed methods by using the multibody model as a reference

Configuration	Multibody model	Denavit-Hartenberg	error	Robotics Toolbox	error
$q_z = [0 \ \ 0 \ \ 0 \ \ 0 \ \ 0 \ \ 0]$	[0.87 0 -0.27]	[0.87 0 -0.27]	5%	[0.87 0 -0.27]	5%
$q_r = [\frac{\pi}{2} \ \ 0 \ \ 0 \ \ 0 \ \ 0 \ \ 0]$	[0.87 0 1.17]	[0.892 0 1.17]	5%	[0.892 0 1.17]	5%
$q_s = [0 \ \ \frac{\pi}{2} \ \ 0 \ \ 0 \ \ 0 \ \ 0]$	[1.465 0 0.58]	[1.482 0 0.58]	5%	[1.482 0 0.58]	5%
$q_n = [\frac{\pi}{4} \ \ 0 \ \ 0 \ \ \frac{\pi}{4} \ \ 0 \ \ 0]$	[1.19 0 0.501]	[1.12 0 0.506]	5%	[1.211 0 0.501]	6%

References

[1] Cammarata A 2015 Optimized design of a large-workspace 2-DOF parallel robot for solar tracking systems *Mechanism and Machine Theory*. **83** 175-186.

[2] Cesconeto E M and Perondi E A 2019 Development of a hydraulic robotic arm – Determination of the kinematic parameters *Lecture Notes in Electrical Engineering*. **505** 537-544.

[3] Sahu P K *et al.* 2019 A Heuristic Comparison of Optimization Algorithms for the Trajectory Planning of a 4-axis SCARA Robot Manipulator *Advances in Intelligent Systems and Computing*. **711** 569-582.

[4] Fenucci A *et al.* 2016 An off-line robot motion planning approach for the reduction of the energy consumption. In *IEEE International Conference on Emerging Technologies and Factory Automation, ETFA, 2016-November*.

[5] Colucci F, De Simone M C and Guida D 2020 TLD Design and Development for Vibration Mitigation in Structures *Lecture Notes in Networks and Systems*. **76** 59-72.

[6] Živković M, Dašić P and Petrović Z 2019 Influence of New Technologies on Higher Energy Efficiency of Hydrostatic Devices and Systems *Lecture Notes in Networks and Systems*. **42** 386-396.

[7] Pappalardo C M and Guida D 2018 On the Computational Methods for the Dynamic Analysis of Rigid Multibody Mechanical Systems *Machines*. **6(2)** 20.

[8] De Simone M C, Rivera Z B and Guida D 2018 Obstacle avoidance system for unmanned ground vehicles by using ultrasonic sensors *Machines*. **6 (2)** 18.

[9] Knox J 2017 Comau leads the digital transformation in the industrial world together with Microsoft and ICONICS *Automotive Industries AI*. **197 (4)**.

[10] Nagpal N *et al.* 2019 Tracking control of 4-DOF robotic arm using krill herd-optimized fuzzy logic controller *Advances in Intelligent Systems and Computing*. **697** 629-643.

[11] Lin Y *et al.* 2018 A computationally efficient and robust kinematic calibration model for industrial robots with kinematic parallelogram. In: *2017 IEEE International Conference on Robotics and Biomimetics, ROBIO 2017, 2018-January*

[12] Trapani S and Indri M 2018 Task modeling for task-oriented robot programming. In: *IEEE International Conference on Emerging Technologies and Factory Automation, ETFA*. 1-8.

[13] Corke P I 1996 A robotics toolbox for MATLAB *IEEE Robotics & Automation Magazine*. **3(1)** 24-32.

[14] Concilio A *et al.* 2017 A new semi-active suspension system for racing vehicles *FME Transactions*. **45(4)** 578-584.

[15] Gao G *et al.* 2018 Structural parameter identification for 6 DOF industrial robots *Mechanical Systems and Signal Processing*. **113** 145-155.

[16] Pappalardo C M, De Simone M C and Guida D 2019 Multibody modeling and nonlinear control of the pantograph/catenary system *Archive of Applied Mechanics*.

[17] Francesco P and Paolo G G 2017 AURA: An Example of Collaborative Robot for Automotive and General Industry Applications *Procedia Manufacturing*. **11** 338-345.

IOP Conf. Series: Materials Science and Engineering **568** (2019) 012116 doi:10.1088/1757-899X/568/1/012116

Ensuring the adequacy of neural network models based on the optimization of test volumes

S Kovalevskyy[1], O Kovalevska[1], I Tasić[2], R Cvejić[3] and A Koshevoy[1]

[1]Donbass State Engineering Academy (DSEA), Kramatorsk, Ukraine,
[2]University of Novi Sad, Technical Faculty "Mihajlo Pupin", Zrenjanin, Serbia
[3]University UNION - Nikola Tesla, Faculty of Information Technologies and Engineering, Belgrade, Serbia,

e-mail: kovalevskii@i.ua

Abstract: Ensuring the accuracy and adequacy of mathematical models based on neural networks is an insufficiently studied area of modeling complex multifactor models based on a limited amount of input data. The article presents the results of studies of randomly organized test samples that complement the training set to the total amount of set of precedents provided for the construction of a mathematical neural network model of the object. It is shown that the adequacy of the neural network model is achieved by agreeing on the size of the set and the expected accuracy of the model on a given set of procedures. The formulation of the optimization problem for adequate neural network models is formulated. The results obtained significantly expand the approaches to increasing the reliability of neural network models.

1. Introduction

Mathematical modeling of objects is a necessary and, often, the only possible means for studying and applying for practical purposes knowledge about the features of simulated objects. The central concept of the theory of mathematical modeling is the concept of adequacy [**Error! Reference source not found.,Error! Reference source not found.,Error! Reference source not found.**]. An argumentated adequacy check ensures that good and practically significant results are obtained. Adequacy of the mathematical model is the correspondence of the results of the computational experiment to the behavior of the real object. This correspondence should be evaluated in terms of the research objectives. Therefore, various approaches to assessing the adequacy of different models are possible. To find out this correspondence for processes characterized by measurable quantities - parameters - it is necessary to compare the parameters of the model and the original under the same conditions. The question of the necessary and sufficient degree of conformity of the object - the original or the adequacy of the model is among the most important in the field of modeling methodology, and the answer to this question characterizes the efficiency of modeling, which reflects the practical utility of the model. Therefore, only the corresponding parameters must be compared to each other and only in the area of the functioning of the object in which it is supposed to be investigated. To verify the adequacy of the mathematical model to the real process, it is necessary to compare the values observed in the course of the experiment with the predictions of the model with certain process parameters [**Error! Reference source not found.**,5,**Error! Reference source not found.**].

In modeling, the researcher should always strive for the most complete and accurate reproduction in the model of the properties and characteristics of the object. The consequence of this is the increase in the complexity of the model, which manifests itself in the number of variables, the number of considered links and influences, and increasing the requirement for the accuracy of the source data.

In mathematical terms, the best combination of completeness-accuracy of the model being created on the one hand and simplicity on the other, almost never succeeds because of the non-formalizability and ambiguity of most of the factors to be taken into account. The deciding factor of efficiency is the mathematical apparatus. At the same time, the best quality or efficiency of any model is achieved as a reasonable compromise between the proximity of the model to the original (adequacy) and the simplicity that provides the possibility and convenience of using the model for its direct purpose, and excessive accuracy of the model in practice is no less harmful than its incompleteness [**Error! Reference source not found.**].

Consistency implies the identical nature of the change of the corresponding parameters, i.e. identical form of the basic properties of functional dependences. With a deeper consideration of this concept, the manifold of possible criteria for checking consistency becomes apparent. Since the comparable parameters in the object's area of operation can take many different values, so that any conclusions about the conformity of their behavior can be made only on the basis of statistical processing of such sets. Therefore, the adequacy is verified using statistical criteria, which can with a certain probability indicate the conformity of the results of the computational experiment to the behavior of the real object in the appropriate conditions.

Mathematical models of mechanical systems and processes are constructed basically as similar deterministic models possessing a common mathematical description with the original. Therefore, for the adequacy of the mathematical model of the behavior of the original - the mechanical system - it is enough to be convinced of the performance of two properties: accuracy and consistency. Accuracy in the tasks of mechanics means that the generalized characteristic of the mismatch of the corresponding parameter of the model and the original should not be greater than the predetermined value of the acceptable error. As such, the meanings of the mismatch, the mean value of the mismatch, or the statistical estimate may be the highest modulus.

When identifying difficult formalizable complex systems, when artificial neural networks are used to construct mathematical models, the use of statistical criteria is difficult and sometimes impossible [**Error! Reference source not found.,Error! Reference source not found.**].

2. Basic part

An approach based on a neural network basis is a mathematical structure with a variable structure dependent on input data. The structure of the elements of such structures is rebuilt using custom threshold elements whose level of excitation depends on the input data. The verification of the adequacy of such models is based on satisfactory testing on a test set that is not included in the training set, and the volume of the source data arrays is directly related to the quality of the models formed on the basis of these data. The purpose of the research: to establish the relationship between the teaching and test sets in the creation of adequate models of objects of different nature.

In general terms, there is a vector set *X*, which may be represented n by parameters of table 1.

In accordance with the goal set to explore the features of the choice of the trainee *Y* and test *Z* sets, when $X \ni Y, Z$ a table of precedents was formed (table 2).

The 32 tuples of the set *X* (table 2), test samples were created, representing 9 random combinations with 9 levels of volumes of test sample arrays C_Y^Z :

$$C_{32}^{3};\ C_{32}^{6};\ C_{32}^{9};\ C_{32}^{12};\ C_{32}^{15};\ C_{32}^{18};\ C_{32}^{21};\ C_{32}^{24};\ C_{32}^{27}. \quad (1)$$

These are tables of combinations of subsets *Y* and *Z*.

Table 1 - Vector set of source data for modeling

№	X_1	X_2	X_3	X_4	X_5	X_6	...	X_n

IOP Conf. Series: Materials Science and Engineering **568** (2019) 012116 doi:10.1088/1757-899X/568/1/012116

1	x_{11}	x_{21}	x_{31}	x_{41}	x_{51}	x_{61}	...	x_{n1}
2	x_{12}	x_{22}	x_{32}	x_{42}	x_{52}	x_{62}	...	x_{n2}
3	x_{13}	x_{23}	x_{33}	x_{43}	x_{53}	x_{63}	...	x_{n3}
...	...	...	...	...	...	...	...	...
m	x_{1m}	x_{2m}	x_{3m}	x_{4m}	x_{5m}	x_{6m}	...	x_{nm}

Table 2 - A set of precedents used for neural network simulation

№	x_1	x_2	x_3	x_4	x_5	x_6	x_7	x_8	x_9	x_{10}	x_{11}	x_{12}	Y
1	1	0	1	1	0	0	1	0	1	0	0	0	2
2	0	0	0	0	1	0	0	1	1	0	0	0	1
3	1	1	1	0	0	0	1	1	1	0	1	0	1
4	1	1	0	0	1	0	1	0	0	0	1	0	1
5	1	1	0	1	0	1	0	0	0	1	0	0	2
6	1	0	0	1	0	0	1	1	0	0	0	0	1
7	1	0	0	1	0	0	1	0	1	0	1	0	2
8	0	0	0	0	1	0	0	0	0	0	0	0	1
9	0	0	1	0	1	0	0	1	1	0	0	1	2
10	0	0	0	1	0	1	0	1	1	0	1	0	2
11	0	1	0	0	1	0	1	0	0	0	0	1	1
12	1	1	0	0	1	0	0	0	0	0	1	0	1
13	1	1	0	0	0	0	0	1	0	0	0	1	1
14	1	1	1	1	1	0	1	0	0	0	0	0	2
15	0	0	0	0	1	0	0	1	0	0	0	0	1
16	1	0	0	1	0	0	0	1	1	0	0	0	2
17	0	1	1	0	1	0	1	1	0	1	0	0	1
18	1	1	0	0	0	0	1	0	0	0	0	0	1
19	1	1	0	0	1	1	0	0	1	0	0	1	2
20	0	1	0	0	1	1	1	1	0	0	1	0	1
21	1	1	0	0	1	1	1	1	0	0	1	0	1
22	1	1	0	0	1	0	1	1	0	0	1	0	1
23	1	1	1	0	1	0	0	1	0	0	0	0	1
24	1	0	0	1	0	0	1	0	0	1	0	1	2
25	0	1	0	0	1	0	0	0	0	0	1	0	1
26	1	1	0	0	0	1	0	0	0	0	0	1	2
27	0	0	0	0	1	0	1	0	0	0	0	0	1
28	1	1	1	1	0	0	1	1	1	0	0	0	2
29	0	0	0	0	1	0	1	1	1	0	0	0	1
30	1	1	0	1	1	0	0	0	0	1	0	0	2
31	0	0	1	1	1	1	0	0	0	1	0	1	2
32	0	1	1	1	1	1	0	1	1	1	0	0	2

3. Methodology of experimental studies

The research methodology involves conducting neuro-net model accuracy-adequacy estimates based on the variation of the random selection of training and test subsets formed from the same data set available for the creation of a non-grid model. The method of comparative analysis included the following steps:

1. Build neural network models based on neural network modeling and data analysis Neuro Pro 0.25 [**Error! Reference source not found.**] based on data from training subsets *Y*.

2.Creation of the number of inputs of the neural network model x1..x12, leaving only significant values for the training subsets to be obtained for the given accuracy.

3. Testing of neural network models obtained as a result of the implementation of item 2 on test sets *Z* at the same time assessing the accuracy of testing.

4. Building the dependencies of the given training accuracy and the obtained accuracy of testing on the sets of samples of subsets of *Y* and *Z* arrays (1).

5. Analysis of the obtained results and development of recommendations on the method of assessing the adequacy of neural network models.

Thus, 81 neural network models with a given simulation error of ±1.0% were created and analyzed. The results of the studies are presented in table 3.

Figure 1 shows a graphical interpretation of the error *S* of the reproduction of the test set *Z*.

Table 3 - Error of testing 81 samples of neural network models

Accuracy testing	Error of testing with share p-test set from the size of the data set,%									Total error
	10	20	30	40	50	60	70	80	90	S
10,00%	0,335	0,031	0,669	0,002	0,337	0,008	0,004	0,335	0,008	1,727
20,00%	0,282	0,006	0,337	0,195	0,330	0,170	0,171	0,174	0,339	2,003
30,00%	0,118	0,198	0,008	0,340	0,401	0,117	0,113	0,170	0,005	1,472
40,00%	0,202	0,124	0,173	0,170	0,179	0,173	0,090	0,171	0,257	1,539
50,00%	0,401	0,536	0,138	0,378	0,138	0,213	0,177	0,140	0,205	2,325
60,00%	0,173	0,447	0,113	0,501	0,214	0,225	0,348	0,560	0,325	2,906
70,00%	0,146	0,656	0,287	0,435	0,170	0,226	0,351	0,101	0,312	2,684
80,00%	0,500	0,737	0,253	0,169	0,337	0,336	0,379	0,424	0,382	3,517
90,00%	0,408	0,371	0,151	0,334	0,446	0,335	0,721	0,150	0,482	3,398

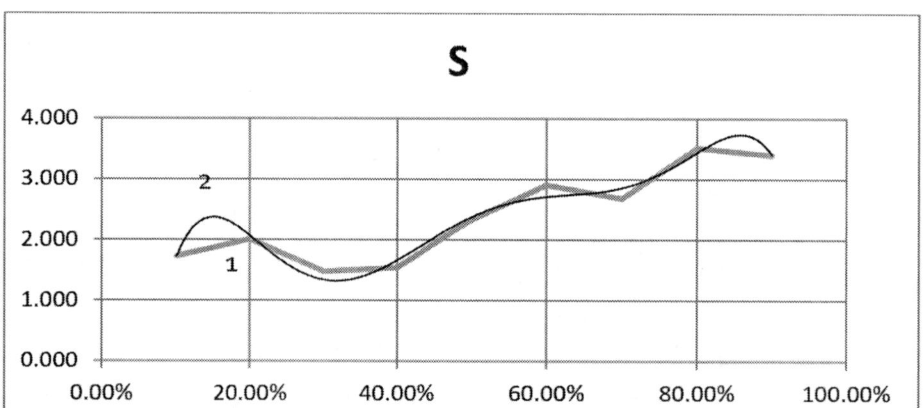

Figure 1 - Total Error 1 and its approximation by the polynomial of 2 test variants with a fraction of the test set from 10% to 90%.

Analysis of simulation and testing errors of models presented by the trend curve approximated by the polynomial dependence of the form:

$$S = -2219{,}9p^6 + 6889{,}8p^5 - 8388{,}4p^4 + 5041{,}9p^3 - 1535{,}2p^2 + 216{,}97p - 8{,}898 \quad (2)$$

allowed to determine the minimum of error.

IOP Conf. Series: Materials Science and Engineering **568** (2019) 012116 doi:10.1088/1757-899X/568/1/012116

The verbal description [**Error! Reference source not found.**] of the optimized model is represented by the following description (3):

Database fields (initial symptoms):

X4
X5
X8
X9
X12

Database fields (end syndromes):

y

Preprocessing the input fields of the database for supplying the network:

X4=(X4-0,5)/0,5
X5=(X5-0,5)/0,5
X8=(X8-0,5)/0,5
X9=(X9-0,5)/0,5
X12=(X12-0,5)/0,5

Functional Converters:

Sigmoid 1(A)=A/(0,1+|A|)

Level 1 syndromes:

Syndrome1_1=Sigmoid 1(0,5175867*X4-0,2253383*X5-0,7707164*X8+0,167638*X9-0,1165028*X12)
Syndrome1_2=Sigmoid 1(0,1908794*X8-0,09852738*X9+0,1385741*X12-0,1664897)
Syndrome1_3=Sigmoid 1(0,3572189*X4+0,3209354*X12)
Syndrome1_4=Sigmoid 1(0,6674076*X4-0,115299*X5-0,8144038*X8+X9+0,1843456*X12+0,7063498)

End-syndromes:

y=0,6864882*Syndrome1_1+Syndrome1_2-0,4366252*Syndrome1_3+Syndrome1_4+0,3727925

Post-treatment of end-syndromes:

Y=((y*1)+3)/2)

The minimum of the total error, which was estimated with respect to randomly formed subsets *Y* and *Z* came upon setting the test set as 1/3 of the randomly selected portion of the common data array. This recommendation, of course, differs from the recommendations of 80% / 20% for the fraction of subsets *Y* and *Z* [**Error! Reference source not found.**] and should be taken into account when constructing neural network models.

4. Conclusions

The studies, the results of which are presented in the article, showed that for neural network models, adequacy and accuracy are achieved under the condition of formation of the training and test sample, proceeding from the achievement of satisfactory accuracy at the maximum volume of the test sample. Therefore, the minimum of the total error at 10% of the fraction of the test set indicates a high probability of recognizing the test set. The minimum of the total error of the test set must correspond to condition (3):

$$\mathbf{S} \rightarrow \min; \ \mathbf{p} \rightarrow \max \qquad (3)$$

Applying this rule can ensure the accuracy and adequacy of the neural network model, based on the optimal amount of training and test sets. Neglecting this rule will result in the loss of the adequacy of the model with the ultimate achievable accuracy. It should be noted that the accuracy of the neural network model will be limited by the volume of the set of precedents.

References

[1] U.B. Filik, M. Kurban. A new approach for the short-term load forecasting with autoregressive and artificial neural network models. *International Journal of Computational Intelligence Research*. 3 (2007) 66-71.

[2] O.S. Kovalevska, S.V. Kovalevskyy. Application of acoustic analysis in control systems of robotic machine tools. *Naukovij zhurnal «Radioelektronika, informatika, upravlinnya» - «Radio Electronics, Computer Science, Control»*. 2 (45) (2018) 51-59.

[3] S. Kovalevskyy, O. Kovalevska, R. Turmanidze. Acoustic diagnostics of lever mechanisms with subsequent processing of data on neural networks. *Lecture Notes in Networks and Systems (LNNS)*. 42 (2019) 202-210.

[4] I.O. Konovalova, Yu.A. Berkovich, A.N. Erohin. Optimization of the LED lighting system vitamins space greenhouse. *Aviakosmicheskaja i jekologicheskaja medicina*. 50 (3) (2016) 17-22.

[5] A.E. Kononyuk. *Obobshhennaya teoriya modelirovaniya*. Nachala. K.1.Ch.1 :"Osvita Ukraïni", 2012. - 602 pp.

[6] S. Kovalevskyy, O. Kovalevska. Resource optimization with systemic design of robotized technological equipment. In: *Proceedings of the World Convention on Robots, Autonomous Vehicles and Deep Learning, Singapore*, 10-11 September 2018. 2018, pp. 50. doi: 10.4172/2168-9695-C3-0216th.

[7] V.A. Kaladze. *Mathematical models of casual processes with stationary increments and the non-uniform information dynamic processing*. Monograph. Lorman (MS - USA): Science Book Publishing House, 2012. - 136 pp.

[8] E.A. Ganceva, V.A. Kaladze, A.M. Polyakov. Intellektual'ny'j kriterij kachestva matematicheskix modelej slozhny'x sistem: ideologiya, perspektivy' razrabotki. *Vestnik Voronezhskogo gosudarstvennogo texnicheskogo universiteta*. 9 (5.1) (2013) 52-56.

[9] V.G. Caregorodcev. Izvlechenie yavny'x znanij iz tablic danny'x pri pomoshhi obuchaemy'x i uproshhaemy'x iskusstvenny'x nejronny'x setej. In: *Novy'e informacionny'e texnologii : materialy' vtorogo nauch.-prakt. seminara*. Moscow: Mosk. gos. in-t e'lektroniki i matematiki (MGIE'M), 1999. pp.. 40-50.

[10] A.N. Gorban', D.A. Rossiev. *Nejronny'e seti na personal'nom komp'yutere*. Kiev (Ukraine): Nauka, 1996. - 276 pp.

Using wave signatures for identifying mechanical objects

S Kovalevskyy[1], O Kovalevska[1], A Koshevoy[1], I Tasić[2]

[1]Donbass State Engineering Academy (DSEA), Kramatorsk, Ukraine
[2]University of Novi Sad, Technical Faculty "Mihajlo Pupin", Zrenjanin

Email: olenakovalevskaya@gmail.com

Abstract. The article provides a rationale for identifying objects based on wave signatures. Wave signatures in the article are limited to the acoustic range of vibration. The results of experimental studies based on modeling the dynamic processes of propagation of complex excitations in samples of various configurations and with the presence of various defects using the signature of acoustic oscillations excited in the objects of study are presented. As such objects are presented parts of engineering production. These studies confirm the possibility of complex diagnostics of objects with their subsequent use in product quality control systems, in particular, to reduce the level of peak values of residual stresses in body parts during multi-frequency vibro-processing.

1, Introduction

The creation of identification objects implies, most often, their use in control systems. Requirements for such models ensure compliance with the criteria of accuracy and accuracy of the objectives of the application [1,2]. Control by reference models allows control systems to provide a corrective effect on the control object, thanks to which the tasks of increasing the accuracy of the performance of the control function by the object or working out the optimal control in accordance with the objective function are solved [3,4].

Creation of identification models is preceded by the formation of an array of precedents. The reliability of the object depends on their volume and quality.

One of the features of the identification of objects is the statement about the dependence of the response function on the complexity of the disturbing influence on the object. The essence of this statement is that the complexity of the nature of the disturbing influence is proportional to the volume of the individual properties of the object, which can be fixed as a function of the response to this disturbing influence. From the theory of automatic control, it is known that single functions make it possible to identify the transfer functions of typical links of dynamic models of objects. This becomes possible due to the formation of the natural oscillation functions of the object, including its elements.

2. Basic part

The aim of the presented work is to substantiate and confirm the expediency of identifying objects based on wave signatures (in particular, acoustic), excited by disturbing influences in the object, whose amplitude-frequency characteristic has the same amplitudes of wideband oscillations in the range available for fixation ("white noise"). The signature of such signals should be understood as the amplitude-frequency characteristic of a certain state of the object. Thus, each identified state of the

IOP Conf. Series: Materials Science and Engineering **568** (2019) 012117 doi:10.1088/1757-899X/568/1/012117

object corresponds to a certain signature. Many such signatures are capable of describing a variety of object states. (1).

1-st state of the object – F(S)1;
2-nd state of the object – F(S)2; (1)
… … … …
n-th state of the object – F(S)n.

Since technical and technological objects can be described by a set of material M, energy E and informational I components, then it is the signature of objects having a wave character that contain the most complete amount of information about the object. It can be assumed that the range of the spectrum of signatures is directly proportional to the completeness of the identification models of objects. Moreover, the discreteness of signatures can solve the problem of boundedness of the array of source data.

3. Methodology of experimental studies

To confirm the applicability of wave signatures in the diagnostic systems of objects in order to identify any defects, the following conditions have been adopted. The database of sample signatures consists of valid (reference) and defective products, the type of which is shown in Figure 1 in the following sizes (Table 1).

Table 1. - Dimensions of samples, where L - sample length, mm; d1, d2 and d3 - diameter at the initial, middle and final cross sections of the samples, mm.

Sample number	Sample sizes			
	L, mm	d1, mm	d2,mm	d3, mm
1	200,10	Ø30,01	Ø30,23	Ø30,19
2	200,05	Ø30,23	Ø30,20	Ø30,17
3	199,95	Ø30,22	Ø30,24	Ø30,28
4	201,00	Ø30,25	Ø30,32	Ø30,29

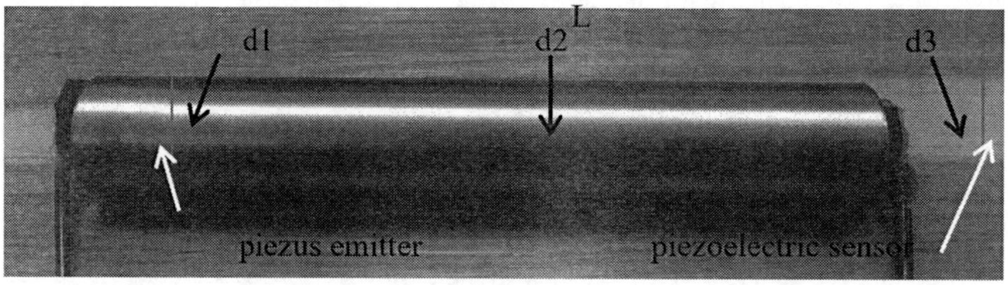

Figure 1. - Example of experimental samples with dimensions L, d1, d2 and d3 mm.

One sample is defective, has a specially made groove with a depth of 0.95 mm and a width of 2.75 mm at a distance of 27.4 mm.

Wave (acoustic) signatures created with the aid of a piezo emitter of a wave packet in the form of a broadband acoustic signal in the range of 20 Hz-20 kHz of constant amplitude ("white noise"), which is received and converted into an analog-digital device into an array; part which in the form of spectrographs is presented in Figure 2.

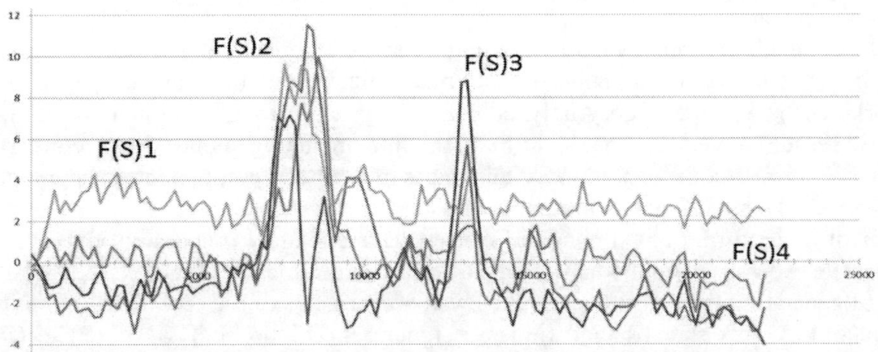

Figure 2. - Wave signature of experimental samples F(S)1, F(S)2, F(S)3, F(S)4.

Based on the data set of digitized signatures, neural network identification models were constructed for diagnosing the validity of samples by analogy with the authors' works [5-8]. The peculiarities of neural network modeling were that the number of implementations of signatures was taken by 10 for each sample. The difference in signatures was that for each measurement of signatures, the installation of a piezo emitter and piezo sensor was reinstalled. Thus, the invariance condition was ensured - the recognition of suitable and defective samples at any position of the emitter and sensor. To ensure the constancy of the clamping force of the emitter and sensor to the end surfaces of the samples, magnetic substrates of piezoelements were used (Table 2).

Table 2. The result of the analysis of spectral characteristics (wave signatures) of samples

Quality sign (1-quality, 0 defective)	Network forecast	Quality sign (1-quality, 4 defective)	Network forecast	Quality sign (1-quality, 4 defective)	Network forecast
1	1,028	1	1,036	1	1,036
1	1,007	1	1,039	4	3,859
1	1,034	4	3,859	1	1,039
0	0,013	1	0,961	1	0,961

Testing of samples with a random sequence of test sample presentation showed 100% recognition of the defective sample. Another example illustrates the capabilities of signatures for diagnosing technological regimes for improving the quality of engineering products. Production experience has shown that in case designs made of simple low-carbon steels and having a fairly high precision after welding, after further machining or stacking within two to three weeks, the tolerance limit was changed and they required additional processing. The main reason for such changes was the presence of residual stresses [9,10]. Changes in geometric sizes are increased under the influence of mounting, transport and operating loads, as well as with increasing temperature. In order to increase the stability of the geometric dimensions of foundry and welded structures, they are often subjected to general heat treatment, which requires additional energy costs. The urgent task of solving the problem of improving accuracy is the process of reducing the peak residual stresses by means of vibration stabilization treatment at resonant frequencies as this process is less energy intensive, but in most cases it is carried out at the resonant frequencies of the whole case. Vibration processing is carried out by means of excitation in the construction of low-frequency mechanical oscillations. In terms of energy consumption, vibration stabilization is tens of times lower than heat treatment, and in terms of performance significantly exceeds it. Approximately an order of magnitude decreases capital costs for the introduction of such processing technology. Vibration treatment is subjected to the construction not only of carbon steels, but also made of aluminum and titanium alloys. Widely used vibration processing of cast iron and steel castings [9,11,12].

Despite the presence of many discrete publications on the problem of vibration processing, summary information is inadequate [10,13,14]. Therefore, there is a need to propose a method for more efficient

IOP Conf. Series: Materials Science and Engineering **568** (2019) 012117 doi:10.1088/1757-899X/568/1/012117

vibration stabilization processing by simultaneously processing individual parts of the body part at their own resonant frequencies.

Example of method implementation: using piezo elements and software "Signal Generation" reproducing "white noise", the signal turns into mechanical oscillations using a piezar affects the body of the workpiece [15]. Spectrum Analyzer software is used to fix resonant oscillations [16]. The piezoelectric sensor converts mechanical vibrations into an analog signal, which converts into a digital signal of the resonant frequencies of individual parts of the body component using an analog-to-digital converter (ADC). This ADC is a computer sound card.

Excitement in the form of "white noise" causes wave signatures separately for each element of the complex profile. Converted by the hardware part of the experimental complex, the "white noise" signal is capable of generating the amplitude-frequency characteristics of the spectra of individual elements of the body detail in a tense state (diagnostic wave signatures) (Figure 3, F(S)1, F(S)2, F(S)3).

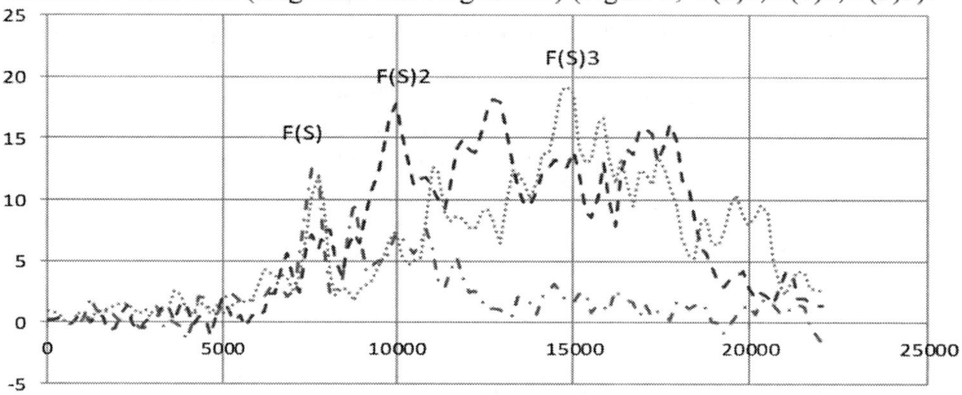

Figure 3 - Amplitude-frequency characteristics of the spectra of elements of the body detail in a stressed state

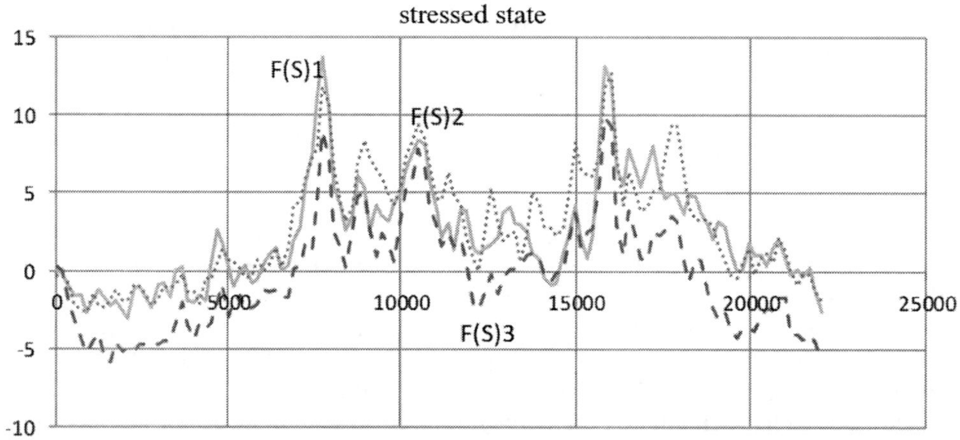

Figure 4 - Amplitude-frequency characteristics of the spectra of elements of the body detail after 30-minute processing at resonant frequencies

The graph of the amplitude-frequency characteristics of the spectra of plates in the stress state confirms that the signatures of each element are individual, despite the coincidence of some auxiliary ones.

The experimental booth is designed in such a way that it can excite each piezoelectric element on the surface of the body part with a spectrum of frequencies in accordance with its own diagnostic signature. Figure 4 shows the amplitude-frequency characteristics F(S)1, F(S)2, F(S)3 of the casing after 30-minute multifrequency machining.

IOP Conf. Series: Materials Science and Engineering **568** (2019) 012117 doi:10.1088/1757-899X/568/1/012117

Polyfrequency, according to signatures, simultaneous processing of all elements of body parts allows to significantly reduce the required vibration time and reduce the amplitudes of residual stresses. The technical result is an increase in precision due to vibrational stabilization of residual stresses in body parts, which increases productivity and reduces energy costs.

4. Conclusions

The simulation of dynamic processes for the propagation of complex excitations in samples of different configurations and the presence of various defects confirmed the possibility of complex diagnostics of objects. This means that, based on the data of one implementation of the diagnostic procedure, one can draw conclusions about the size, physical and mechanical properties, microheometry of the surface layer, the presence and nature of defects in products. Such products can have separate parts, and such knots and mechanisms of machines.The practical implementation of the use of wave signatures is the diagnosis of product quality and diagnostics with subsequent multifrequency effects on the product in order to change its complexity of stress.

References

[1] A.M. Vendrov. Proektirovanie programmnogo obespecheniya e'konomicheskix informacionny'x sistem. [2-e izd., pererab. i dop.]. Moscow: Finansy' i statistika, 2006. – 544 pp.
[2] V.M. Tomashevs'kij. Modelyuvannya sistem. Kiev: Vidavnicha grupa VNV, 2007. – 352 pp.
[3] A.A. Zhilenkov, S.G. Chyorny'j. Modelirovanie adaptivnogo upravleniya v slozhny'x raspredelenny'x sistemax s identifikaciej parametrov. *Visnik Xmel'nic'kogo nacional'nogo universitetu*. 6 (2013) 253-260.
[4] S. Mirjalili, S.M. Mirjalili, A. Lewis. Grey wolf optimizer. *Advances in Engineering Software*. 69 (2014) 46-61. doi: 10.1016/j.advengsoft.2013.12.007.
[5] O.S. Kovalevska, S.V. Kovalevskyy. Application of acoustic analysis in control systems of robotic machine tools. *Naukovij zhurnal «Radioelektronika, informatika, upravlinnya» / «Radio Electronics, Computer Science, Control»*. 2 (45) (2018) 51-59.
[6] S.V. Kovalevskij, E.S. Kovalevskaya, V.I. Tulupov. *Razvitie metodov akusticheskoj diagnostiki v mashinostroenii*: Monografiya. Kramatorsk : DGMA, 2014. – 91 pp.
[7] S.V. Kovalevskii, O.S. Kovalevska. *Diagnostics of technological systems and engineering products* (*using neural network approach*). Scientific Monography. Vrnjačka Banja: SaTCIP, 2016. – 169 pp.
[8] S.V. Kovalevs'kij, O.S. Kovalevs'ka, Є.O. Korzhov, A.O. Koshevoj. *Diagnostika texnologichnix sistem i virobiv mashinobuduvannya* (*z vikoristannyam nejromerezhevogo pidxodu*). Monografiya. Kramatorsk: DDMA, 2016. – 186 pp.
[9] R.G. Rizvanov, A.M. Fajrushin, D.V. Karetnikov. *Vliyanie parametrov vibracionnoj obrabotki v processe svarki na svojstva svarny'x soedinenij*. Kramatorsk: DDMA, 2016.
[10] V.M. Semenov, A.Yu. Den'shhikov, S.V. Podlesny'j. Issledovanie vliyaniya termicheskoj i vibracionnoj obrabotki na deformacii svarny'x konstrukcij. *VISNIK Donbas'koï derzhavnoï mashinobudivnoï akademiï*. 2 (19) (2010) 260-263.
[11] A.F. Bulat, G.A. Shevchenko, V.A. Lenda. *Povy'shenie e'ffektivnosti texnologii vibracionnoj stabilizacii ostatochny'x napryazhenij i geometricheskix razmerov v svarny'x konstrukciyax*. Kiev: Institut geotexnicheskoj mexaniki im. S. Polyakova NAN Ukrainy', 2012, pp. 84-87.
[12] G.I. Lashhenko. Texnologicheskie vozmozhnosti vibracionnoj obrabotki svarny'x konstrukcij. *Avtomaticheskaya svarka*, 7 (754) (2016) 28-34.
[13] P.I. Yashhericy'n, E'.V. Ry'zhov, V.I. Averchenko. *Texnologicheskaya nasledstvennost' v mashinostroenii*. Minsk: Nauka i texnika, 1977. – 256 pp.
[14] V.A. Kolot, S.V. Kovalevskij. «*Sposob stabilizacii ostatochny'x napryazhenij*» *Progressivnaya texnologiya v mashinostroenii*. Tezisy' dokladov. – Simferopol', 1994. – 121 pp.
[15] Web page: https://www.tek.com/signal-generator-software. Available on the 23.04.2019.
[16] Web page: https://www.consultixwireless.com/. Available on the 23.04.2019.

Increase of operational properties of tools and machine parts nitriding the powder mixture

V O Kostyk[1], K O Kostyk[2], V D Kovalov[1], R Turmanidze[3], P Dašić[4,5]

[1] Donbas State Engineering Academy, Akademichna 72, 84313, Kramatorsk, Ukraine
[2] National Technical University «Kharkiv Polytechnic Institute», Kyrpychova, 2, 61002, Kharkiv, Ukraine
[3] Georgian Technical University (GTU), 0175 Tbilisi, Georgia
[4] SaTCIP Publisher Ltd., 36210 Vrnjačka Banja, Serbia
[5] High Technical Mechanical School of Professional Studies, 37240 Trstenik, Serbia

Email: vikakostik777@gmail.com

Abstract: The effect of diffusion layers and alloying elements on surface hardness, micro-coarseness, and wear resistance of hardened tool steels in a nitrogen-containing medium is Studied. The experiments showed that nitriding in the powder mixture strengthens the surface layers in 2–3.5 times without reducing the volume strength of the material in question, which contributes to the preservation of high structural strength of tools and machine parts. It was found that nitriding in the dispersed powder mixture provides increased wear resistance of 1.3–2.5 times of steel, which can significantly improve the reliability and durability of products. Mathematical models of wear resistance of the investigated steels after the proposed hardening technology of nitriding are obtained.

1. Introduction

The creation of high-quality products with high structural strength with economical use of materials contributes to the reliability and durability of products, which is an urgent task in engineering today.

Structural strength depends on the design features of the part (shape and size); mechanisms of various types of destruction of the part; the state of the material in the surface layer of the part; processes occurring in the surface layer of the part, which leads to failures when working [1].

The hardening of the surface layers provides long-term and reliable operation of the material in the operating conditions of the products, which is associated with a complex of strength properties that are in the greatest correlation with the service properties of this product [2-5]. Reliability and durability of tools and machine parts are determined by the structural strength of the materials from which they are made. The performance of many products-wear resistance, corrosion resistance and others – are determined by the properties of the surface. To obtain high structural strength characteristics of the surface layers, various methods of surface hardening are often used, which allow to increase the service life and reduce the cost of repair of worn products [6]. Given the need to improve the performance of tools and machine parts, is an important task of research aimed at further improvement and development of surface hardening technology products. This suggests that the subject of research on the directional regulation of the processes of phase and structure formation of hardened layers on steel products through the use of effective methods of impact on the surface layers is relevant.

IOP Conf. Series: Materials Science and Engineering **568** (2019) 012118 doi:10.1088/1757-899X/568/1/012118

The Strengthening of the surface layers on the principle of interaction with the surface is carried out by changing the chemical composition of the surface (chemical and thermal treatment), applying a new material to the surface (spraying, deposition, surfacing, application of enamels and paints), etc. [7]. Types and methods of coating of metals are diverse and are classified as diffusion [8], thermomechanical [9], chemical[10] , vacuum plasma [11], etc. The choice of a method of saturation is carried out in accordance with the type of production, the required thickness of the surface layers, the necessary performance properties of the product. At present, there is a wide variety of processes of chemical and thermal treatment associated with the saturation of the surface of metals with various elements [12-19]. Of all the chemical-thermal hardening processes, nitriding is the most versatile and suitable for tool steels. However, the existing methods of nitriding (gas, liquid, ion-plasma, etc.) are a fairly long process (up to 50-100 hours) or require expensive equipment [17]. In this regard, the development of new nitriding technologies that provide accelerated saturation of the metal surface with nitrogen while maintaining the strength of the base material is of great practical importance. At the same time, it is advisable to use new developments of chemical and thermal hardening without additional costs for expensive equipment. Therefore, there is reason to believe that the issue of using a new technology of nitriding in the powder mixture to speed up the process of saturation of the surface layers of steel, increasing the performance properties of tools and machine parts, necessitates research in this direction. The conducted research was aimed at obtaining high performance properties of tools and machine parts by nitriding in a powder mixture.

To achieve this goal, you need to solve the following tasks:

- to determine the impact of the developed nitriding technology on the degree of hardening of the surface layers and the volume strength of tools and machine parts;
- to study the effect of chemical-thermal hardening in a nitrogen-containing medium on the wear resistance of steels;
- to obtain mathematical models of wear resistance of steels after the proposed hardening nitriding technology.

2. Materials and Methods

Materials for the study selected tool steels for different purposes, such as 3Cr3Mo3V, 5CrNiMo, 9CrSi and W6Mo5Cr4V2, this allowed to study in detail the structure of diffusion layers and the effect of alloying elements on wear resistance of hardened alloys during nitriding in the powder mixture.

The sample container was placed in a thermal furnace. The content of atomic nitrogen in the powder mixture, namely melamine (66.67 %) is 1.7 times higher than ammonia (up to 40 %), which allowed to speed up the process of saturation of the metal surface compared to traditional gas nitriding. Nitriding of steels was carried out at a temperature of 550 °C for 5 hours.

3. Results of research of operational properties of steel by nitriding in powder mixture

The microstructures of the diffusion layers of tool steels obtained after hardening in a nitrogen-containing medium are shown in Figure 1 and Figure 2.

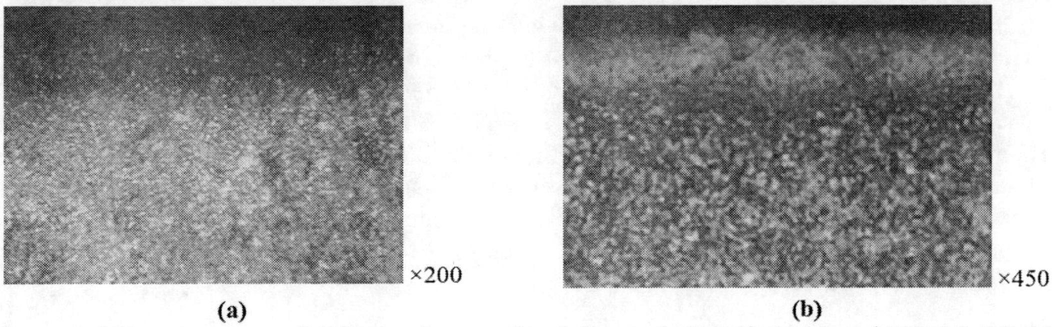

(a) (b)

Figure 1. Microstructures of diffusion layers of tool die steels 5CrNiMo **(a)** and 3Cr3Mo3V **(b)** after nitriding

IOP Conf. Series: Materials Science and Engineering **568** (2019) 012118 doi:10.1088/1757-899X/568/1/012118

The investigated steels were subjected to qualitative phase analysis after nitriding, which revealed, as expected, the presence of nitrides, carbides and carbonitrides of iron and alloying elements in the investigated steels.

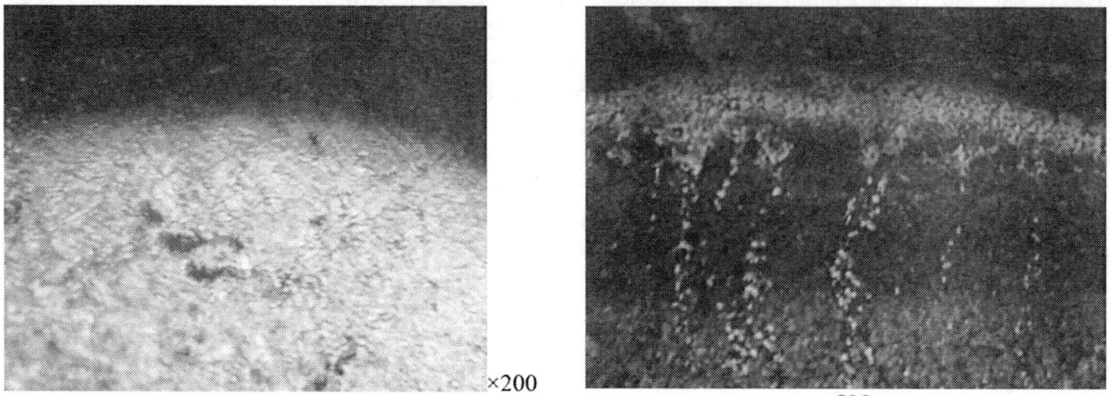

×200 ×500

(a) (b)

Figure 2. Microstructures of diffusion layers of tool steels 9CrSi **(a)** and W6Mo5Cr4V2 **(b)** after nitriding of drills

It should be especially noted that the analysis of the obtained diffractograms showed the formation of nitride phases of different stoichiometric composition in the surface layer (Fe_2N, Fe_4N). Thus, in high-speed steel after nitriding according to the proposed technology, iron nitrides were detected ε-Fe_2N, γ'-Fe_4N, nitrides of alloying elements Cr_2N, W_2N, Mo_2N, VN, carbides Fe_3C, WC, W_2C, MoC, Mo_2C, VC, V_2C, Cr_3C_2, Cr_7C_3, $Cr_{23}C_6$, oxides FeO, Fe_2O_3, Fe_3O_4, WO_3, WO_2, MoO_2, MoO_3, Mo_2O_3, CrO_3, Cr_2O_3, V_2O_3, V_2O_5 and a solid solution of nitrogen in α-iron.

For testing, smooth (surface roughness $R_a \leq 1{,}25$ μm) cylindrical specimens without corrosion and scale were used. Before the test, the samples were subjected to burnishing to achieve a complete fit of the samples to the skin under conditions identical to the test conditions. Wear of the tested samples was obtained under the same conditions, i.e. the samples passed the same friction path within the entire test series.

The essence of abrasive wear is the destruction of the metal with hard abrasive grains during plastic deformation and micro-cutting of rubbing surfaces. For tools and machine parts, abrasive wear is most often prevalent compared to other types of wear. Therefore, the work carried out tests on samples of this type of wear to assess and predict the resistance to abrasive wear of finished products.

Studies to determine the degree of resistance to abrasive wear were carried out on the machine for abrasion testing AP 40.613.20 R 43/82 (Germany). The degree of wear was studied by controlling the weight loss of the tested samples. with the set pressing force was carried out on the abrasive material at a constant speed and along a certain path.

It has been shown that the hardness of the surface layers after nitriding increased by 2–3,5 times depending on the steel grade. Thus, the surface hardness of tool steel 9CrSi increased in 2,5 times (up to 10.1 GPa), high speed steel – 2 times (12.5 GPa) and die steels 3Cr3Mo3V and 5CrNiMo – 2.5–3.5-fold (from 11.2 to 11.7 GPa) respectively. This is due to the influence of nitrides, carbides and carbonitrides of iron and alloying elements of the investigated steels formed in the surface layers.

Nitrided surface with high hardness, high wear resistance. Its greatest values are the continuous layers of ε-phase ($Fe_{2\text{-}3}N$) with a minimum nitrogen content and carbonitride phase type $Fe_{2\text{-}3}(N,C)$, characterized by high hardness and plasticity. Almost the same wear resistance is characterized by (ε+γ')-phase and γ'-phase.

After nitriding, the wear resistance of high–speed steel W6Mo5Cr4V2 and die steel 3Cr3Mo3V in the first test cycle increases by 2.0-2.5 times, and then with the continuation of wear tests at 2–5 cycles, the wear resistance is slightly reduced and remains at a high level. Thus, the zone of internal nitriding in steels increases wear resistance by 1.7–2.0 times.

In steels 5CrNiMo and 9CrSi after chemical and thermal treatment wear resistance in the nitride zone increases by 1.5–1.7 times on the friction path of 40 m (the first cycle), and in the internal nitriding zone the increase in wear resistance occurs in 1.3–1.5 times on the friction path of 80–200 m (2–5 cycles), which is provided by a nitrided solid solution with dispersed Fe_4N nitrides (γ'-phase).

On the basis of the obtained equations, graphs of the dependence of wear resistance before and after nitriding are constructed (Figure 3).

The graphs show that after nitriding, the wear resistance of steels increases during the first cycle (40 m) of tests, and then remains almost at the same high level regardless of further cycles (2–5) of tests, i.e. does not depend on the friction path (from 80 to 200 m). This is due to the presence of nitrides and carbonitrides of iron and alloying elements in the surface diffusion layers of tool steels, which is confirmed by x-ray phase analysis.

4. Discussion of the research results

The obtained results of the study of phase and structure formation, micro-coarseness, surface hardness, wear resistance of the hardened layers, thickness of diffusion layers on tool steels 3Cr3Mo3V, 5CrNiMo, 9CrSi and W6Mo5Cr4V2after nitriding in the powder mixture indicate that the proposed method of surface hardening of alloy steels can significantly increase the hardness of the surface layers and in turn contributes to a significant increase in the service life of tools and machine parts.

Indeed, after saturation with nitrogen at a temperature of 550 °C for 5 hours on the surface of alloyed tool steels formed a diffusion layer consisting of a nitride zone and a sublayer of α-solid solution. This, in particular, can be seen in Figure 1 and Figure 2 – microstructures of the strengthened layers confirm the presence of classical diffusion layers with clearly expressed nitride and transition zones. Such a change in microstructures leads to a change in the properties of steels, namely, the hardness of the surface layers after nitriding increased by 2–3.5 times (up to 10.1–12.5 GPA), and the depth of the hardened layer was 0.18–0.32 mm depending on the steel grade.

In this sense, of particular interest is the interpretation of the results of qualitative phase analysis after nitriding in the powder mixture which confirms the establishment of the presence of nitrides, carbides and carbonitrides of iron and alloying elements in the studied steels.

It should be especially noted that in contrast to traditional nitriding methods, the proposed method of diffusion saturation of the surface of steels using a nitrogen-containing powder mixture is favorably different and allows us to assert the following:

- processing speed compared to conventional nitriding increases 5–10 times due to the acceleration of diffusion processes and reduce the number of preparatory operations,
- the possibility of combining technological operations (leave with nitriding),
- no warping of tools and machine parts due to low nitriding temperatures (550 °C),
- reduction of materials and energy costs,
- the processes are non-toxic and environmentally friendly.

That is why the proposed method of improving the performance of tools and machine parts should be recognized as promising and possible for use in enterprises and in the laboratory.

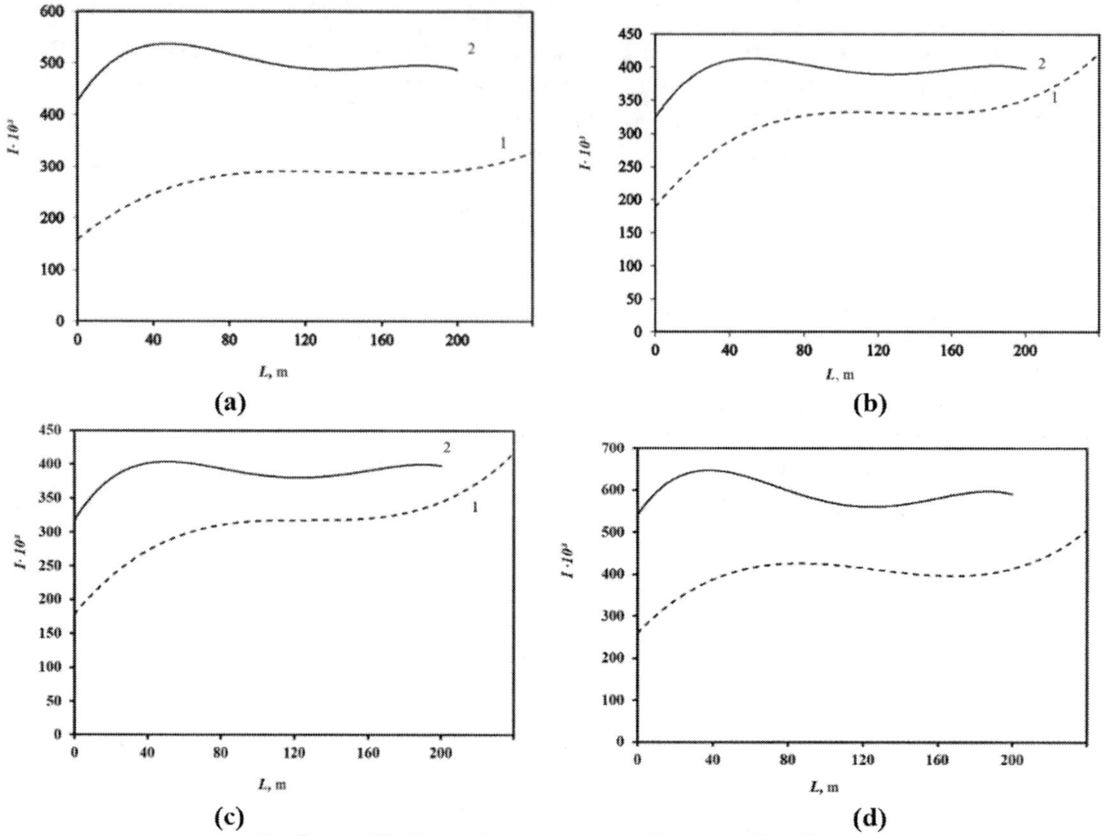

1-after preliminary heat treatment, **2** - after nitriding

Figure 3. Graphs of wear resistance before and after nitriding of tool alloy steels 3Cr3Mo3V **(a)**, 5CrNiMo **(b)**, 9CrSi **(c)** and W6Mo5Cr4V2 **(d)**

5. Conclusions

Studies have shown that the developed technology of nitriding of tool steels allows to strengthen the surface layers in 2–3.5 times without reducing the volume strength of the material in question, which contributes to the preservation of high structural strength of tools and machine parts.

It is experimentally proved that the chemical-thermal hardening in a nitrogen-containing medium allows to increase the wear resistance of steels by 1.3–2.5 times, which in turn contributes to a significant increase in the durability of products.

Mathematical models of wear resistance of the investigated steels after the proposed hardening technology of nitriding are obtained.

References

[1] V. Kovalov, Y. Vasilchenko, P. Dašić. Development of the integral complex of optimal control of heavy machine tools adaptive technological system for wind-power engineering parts. *Procedia Technology*, 19 (2015), 145-152. doi: 10.1016/j.protcy.2015.02.022.

[2] V. Kovalov, Y. Antonenko, P. Dašić. Method of structural design of heavy machine tools. *Procedia Technology*, 22 (2016), 146-152. doi: 10.1016/j.protcy.2016.01.023.

[3] V. Khoroshailo, V. Kovalov, P. Dašić. Improving of vibration resistance of boring tools by big diameter holes tooling on lathe. *Procedia Technology*, 22 (2016), 153-160. doi: 10.1016/j.protcy.2016.01.038.

[4] V. Kovalov, Y. Vasilchenko, M. Shapovalov, R. Turmanidze, P. Dašić. Impact of a pulsed magnetic field on a hard alloy during machining on heavy machine tools. *International Journal of Industrial Engineering and Management (IJIEM)*, 10 (1) (2019), 125-130. doi: 10.24867/IJIEM-2019-1-125.

[5] V. Kovalev, Y. Vasilchenko, D. Riznić. Evaluation of a level of quality of manufacturing process on heavy engineering enterprises. *Annals of The University of Oradea Fascicle of Management and Technological Engineering*, 14 (3) (2014), 1-7.

[6] J. H. Ai, H. M. Ha, R. P. Gangloff, J. R. Scully. Hydrogen diffusion and trapping in a precipitation-hardened nickel-copper-aluminum alloy Monel K-500 (UNS N05500). *Acta Materialia*, 61 (9) 2013, 3186-3199. doi: 10.1016/j.actamat.2013.02.007.

[7] C. Sun, Q. Xue, J. Zhang, S. Wan, A. K. Tieu, B. H. Tran. Growth behavior and mechanical properties of Cr-V composite surface layer on AISI D3 steel by thermal reactive deposition. 148 (2018), 158-167. doi: 10.1016/j.vacuum.2017.11.015.

[8] X. Xu, G. Mi, L. Chen, L. Xiong, P. Jiang, X. Shao, C. Wang. Research on microstructures and properties of Inconel 625 coatings obtained by laser cladding with wire. *Journal of Alloys and Compounds*, 715 (2017), 362-373. doi: 10.1016/j.jallcom.2017.04.252.

[9] A. M. Lazar, W. P. Yespica, S. Marcelin, N. Pébère, D. Samélor, C. Tendero, C. Vahlas. Corrosion protection of 304L stainless steel by chemical vapor deposited alumina coatings. *Corrosion Science*, 81 (2014), 125-131. doi: 10.1016/j.corsci.2013.12.012.

[10] J. Yang, H. Zhao, X. Zhong, F. Shao, C. Liu, Y. Zhuang, S. Tao. Thermal cycling behavior of quasi-columnar YSZ coatings deposited by PS-PVD. *Journal of Thermal Spray Technology*, 26 (1-2) (2017), 132-139. doi: 10.1007/s11666-016-0491-8.

[11] D. T. M. Thanh, P. T. Nam, N. T. Phuong, L. X. Que, N. Van Anh, T. Hoang, T. Dai Lam. Controlling the electrodeposition, morphology and structure of hydroxyapatite coating on 316L stainless steel. *Materials Science and Engineering*, 33 (4) (2013), 2037-2045. doi: 10.1016/j.msec.2013.01.018.

[12] W. A. R. Dhafer, V. O. Kostyk, K. O. Kostyk, A. Glotka, M. Chechel. The choice of the optimal temperature and time parameters of gas nitriding of steel. *East European Journal of Advanced Technologies*, 3 (5) (2016), 44-50. doi: 10.15587/1729-4061.2016.69809.

[13] P. N. Belkin, S. A. Kusmanov. Plasma electrolytic hardening of steels. *Surface Engineering and Applied Electrochemistry*, 52 (6) (2016), 531-546. doi: 10.3103/s106837551606003x.

[14] V. V. Budilov, R. D. Agzamov, K. N. Ramazanov. Ion nitriding in glow discharge with hol-low cathode effect. *Metallography and Heat Treatment of Metals*, 7 (2007), 33-36. doi: 10.1007/s11041-007-0065-y.

[15] M. K. Mohanad, V. Kostyk, D. Demin, K. Kostyk. Modeling of the case depth and surface hardness of steel during ion nitriding. *East European Journal of Advanced Technologies*, 2 (5) (2016), 45-49. doi: 10.15587/1729-4061.2016.65454.

[16] V. O. Kostyk, K. O. Kostyk, A. S. Dolzhenko, S. V. Nikiforova. High-speed method of nitro-cementation of alloy steel. *Bulletin of the National Technical University Kharkiv Polytechnic Institute. Series: New Solutions in Modern Technologies*, 14 (2015), 35-41.

[17] P. Dašić. Approximation of cutting tool wear function using polynomial regression equation. *Journal of Research and Development in Mechanical Industry*, 3 (3) (2011), 171-180.

IOP Conf. Series: Materials Science and Engineering **568** (2019) 012119 doi:10.1088/1757-899X/568/1/012119

The technique of designing high-power CNC lathes for enterprises of the heavy engineering industry

V Kovalov[1], Y Vasilchenko[1], R Turmanidze[2], P Dašić[3,4,5], T Sukova[1], M Shapovalov[1]

[1] Donbas State Engineering Academy (DSEA), Akademichna 72, 84313, Kramatorsk, Ukraine
[2] Georgian Technical University (GTU), 0175 Tbilisi, Georgia
[3] High Technical Mechanical School of Professional Studies, 37240 Trstenik, Serbia
[4] SaTCIP Publisher Ltd., 36210 Vrnjačka Banja, Serbia

[5]E-mail: dasicp58@gmail.com

Abstract. Proposed is the method of heavy machine tools constructions' structural-parametric synthesis based on their components processing parameters statistical study. The process of designing high-power metal cutting equipment that most completely meets the customer's essential requirements consists in developing its technological structure in the form of a functional-structural model with the main technological parameters (accuracy class, size of working space, necessary power parameters, cutting modes) substantiation.

1. Introduction

Enterprises of the heavy machine-building industry are specific with a wide range of products, those unique products sizes, and weight, a long cycle of parts manufacturing, the unique and small-scale nature of manufacturing, technological process arranged to base on the principle of operations concentration. The necessity to process various pieces predetermines this wide equipment versatility. Manufacturing parts weighing tens and hundreds of tons requires the use of different machines, which represent a characteristic feature of this machine building branch. That equipment nomenclature and quantitative composition should ensure the ability to carry out tasks in conditions of continuously changing production facilities. In this connection, acute is a need to increase the production of the following character at a given output volume.

In this case, the specificity of parts machining with heavy machine tools is not sufficiently considered, due to the scale factor (Figure 1) (machined parts up to 24000 mm in diameter and up to 5000 mm in diameter, weighing up to 250 tons), tool design, large dispersion of processing parameters, increased probability of tool destruction (up to 45%), and the need to consider tool reliability in the processing of extensive and costly parts.

IOP Conf. Series: Materials Science and Engineering **568** (2019) 012119 doi:10.1088/1757-899X/568/1/012119

Figure 1. Large rolling shaft

Developed is the knowledge base on massive metal cutting equipment operation. Analysis of high-load CNC lathes operation showed that often technological saturation with these machines does not correspond to the structural and technological characteristics of the parts. The machine tools working space size sometimes significantly exceed the overall dimensions of parts, the number of shaping movements and tools in the store are also superfluous. This leads to an increased material and energy intensity of machine tools, and produced pieces' overpriced production cost.

It was evident that the development of justified requirements for high-load machine tools with adaptive control is possible only with generalization and investigation of a particular set of characteristics specific to parts manufactured by heavy engineering enterprises. The application of computer numerical control (CNC) in the processing at the lathe is shown in the following papers [1-6] and in other processing is shown in the paper [7].

2. The technique of structural-parametric synthesis of heavy lathes

Established is that the technological basis for designing CNC machines can refer to parts grouping. Developed and in practice tested are several features and criteria for selecting parts in groups. Based on the cluster analysis of the overall dimensions and the number of necessary form-forming motions, the processed parts are analyzed for the conditions of heavy machine-building enterprises [8-17].

Each piece is characterized by a set of features:

$$D=\{P_1,\ldots,P_i,\ldots,P_n\} \tag{1}$$

where P_i – i-th feature of the piece;
n – features a total number.
The CNC machines' parameters are characterized by the set:

$$Y=(y_1,\ldots,\ y_i) \tag{2}$$

The set of classification features $X=(x_1,\ldots,\ x_i)$.
Multivalue representation of CNC lathe's parameters set Y within classification features set X:

$$\begin{gathered} v\colon B\rightarrow A \\ v\,(bi\,) = \{\,Ai\}\ \in A \\ Un \in Nv(bi)=\{ai/3n(n\ \in N)\ \wedge\ ai \in v(bi)\} \end{gathered} \tag{3}$$

where N – set of elements' ($b1$, ... bi,) images is B when elements of the set N.
Total set of classification features X:

$$A=\{(Un \in Nv\{bi)\ \in A)\}\ \in A\kappa i \tag{4}$$

IOP Conf. Series: Materials Science and Engineering **568** (2019) 012119 doi:10.1088/1757-899X/568/1/012119

Classification features values region:

$$Ako = \{ ako \}$$
$$Akl = \{akl1, akl2, ..., aklj\} \tag{5}$$
$$Aki = \{ aki1 , aki2, ..., akij\}$$

where $\{ako, \ldots , akij \}$ is Ak – set of classification features values representing areas of structural features' names values $Ako, \ldots , Aki$.

Developed is the algorithm of classification kinds reduction in compliance to the pieces' grouping process. That process consists of transforming the hierarchic structure of graph Lg using the graph grabbing operation. The practical realization of convolutional operation for graph Lg consists of combining the classification features $Akij$, for which processing required is the same composition and number of shape-forming coordinates Fk. A functional model of the primary process as to the metal-cutting equipment functional-structural model construction has been developed, that made it possible to establish all processes entering the primary process, their interrelation, incoming and outgoing information. Mathematical models have been developed that allow us to establish the functional dependencies between the processed parts complexity, equipment functions and the functional units they realize. Machine functions:

$$r: S \times T \rightarrow F \tag{6}$$

where r is the set mapping; $S = \{Si\}$ is the set of the complex detail representative of the complexity Si; $T = \{Te\}$ is the basic set of technological operations realized by the machine system; $F = \{Fi\}$ is the set of machine system functions.

$$r\ (Si \times Te) = \{Fi\!: (U{\cdot}Si\ x\ Te){\cdot}(Si\ x\ Te;\ Fi) \in r\} \tag{7}$$

Functional units determining:

$$B \in x\ \{B_i : i \in I\} \tag{8}$$

where B is the machine system in the form of a set of functional units; × the symbol of the Cartesian product; $Bi : i$ is the i-th functional unit; I set of functional units.

$$In \in: B_1\ x\ B_2\ x \ldots x\ B_n \tag{9}$$

Displaying a set of machine functions in a set of functional units

$$d: F \rightarrow B \tag{10}$$

Functional units B_i, providing the given function:

$$d\ (Fi)\ \text{-}\ (\{\ Bi\} : (\ni Fi)\ (Fi;\ Bi) \in d) \tag{11}$$

The finite subset of functional units $\{Bi\}$, required to implement the whole machine tool functional model:

$$U\ d\ (Fi)\ \text{-}\ \{\ Bi\ /\ \ni (m \ni M)\ \wedge\ Bi \in d\ (Fi)\} \tag{12}$$

where M – functions' images set $(Fb \ldots,\ Fi) \in F$ when representation d; m – an element of set M. Unified composition of functional units for the needed machine tools creation:

$$Bi = [U{\cdot}(Ud(Fi)) \in B\ \forall (Fi) \in F] \tag{13}$$

IOP Conf. Series: Materials Science and Engineering **568** (2019) 012119 doi:10.1088/1757-899X/568/1/012119

Correlation between different levels functions sets:

$$F=\{F0\}x\{F1i\}x\{fij\} \qquad (14)$$

Units possible composition:

$A \mid \mid Bi\ x\ fij \mid \mid$	*(15)*

where aij – matrix elements; i - l, n, j - 1, n – number of auxiliary functions and units.

Proposed are dependencies for determining the unified composition of functional blocks for the entirely new equipment spectrum construction on a single element base. A general view of the CNC machine functional-structural model is shown in Figure 2.

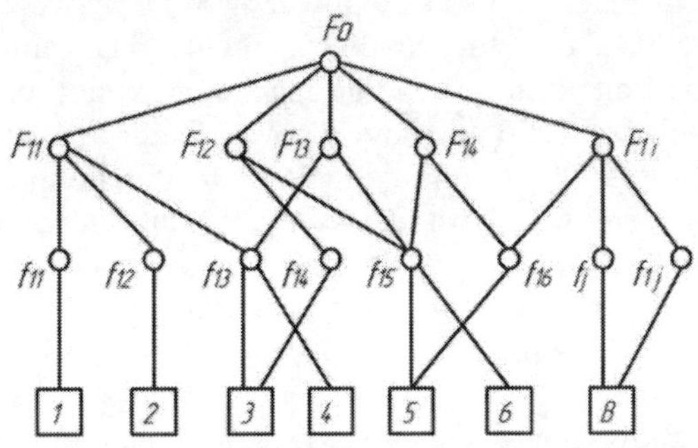

Figure 2. CNC lathe functional-structural model general scheme

Fo - system-wide function; F_{11} .-. *Fij* main functions;
Fi .. -fj auxiliary functions; *B* - the functional units of the machine.

The principal, essential, and auxiliary functions of the machine are formulated, and its generalized functional model is developed. As the basic machine functions, its shape-forming coordinates are taken, which composition determines the equipment efficiency. Developed is the machine functional-structural model is developed, which coordinates functions with functional blocks. Formulas for the evaluation of the new machine tool functions significance are proposed, and the functionally justified expenses for their implementation are determined. Building the functional-structural model allows us to establish a set of necessary and sufficient (optimal) functions implemented by the machine tool, and the executive mechanisms that provide them (assemblies, units, etc.).

3. High-load CNC machines operation in substantial engineering enterprises: results of the analysis

The data bank on the operation of heavy machines has been analyzed, that served to establish that the most probable lengths of parts lie in the range from 3 to 9 m, they slightly increase with increasing D_0 over 1,250 mm and remain constant. With the growth of D_0 from 1250 to 5000 mm, the maximum lengths of the details decrease from 24-27 m to 12-15 m. The most probable weight of parts with a change in D_0 from 1250 to 3200 mm rises from 0-10 tons to 25-50 tons and remains with a further growth D_0 approximately constant. The maximum weight of parts increases from 50-75 to 200-250 tons. Since on machines, D_0=5000 and 6300 mm mainly welded parts are processed, it is recommended to provide their load capacity of not more than 200 tons. For machines with D_0=1250, 1600 and 2000 mm, having L_0 25 m, and D_0=2500 mm with L_0=20 m, it is recommended to complete lengths of the front and two tailstocks for fuller use of the bed length.

It is established that a screw for tapping is necessary on the machine D_0=1250 mm, and on machines D_0=1600-2000 mm, the possibility of mounting the screw should be provided. Also, it is required to provide the possibility of increasing the thread pitch to be cut. Obtained are data on the applicability of processed and tool materials. Over 80% of all machining is carried out with a carbide tool. The average cutting depth for roughing operations with carbide cutters on machines D_0=1250-3200 mm was 20 mm and on machines D_0=4000-5000 mm - 25 mm. The maximum cutting depths are up to 45 mm. The feed at roughing with a carbide tool did not go beyond the limits 0.6-2.0 mm/rev. The average feed rate on the machines D_0=1250-1600 mm was 1.1mm/rev, on the machines D_0=2000-3200 mm - 1.2 mm/rev and on the machines D_0=4000-5000 mm - 1.5 mm/rev. At finishing operations, the feed rates were in the range of 0.05 mm/rev to 40 mm/rev.

It is established that in the overwhelming majority of cases, the cutting force vertical component does not exceed 80-90 *kN*. The most likely forces are about 30*kN*. Taking into account the increase in cut cross section due to the growth of feeds in the future, the calculated force values for machine tools of different models are determined. It is established that with increasing D_0 up to 2000 mm, the most significant actual value of *Mcr* increases sharply, but does not exceed 700 *Nm*. The most probable value of *Mcr* does not exceed 100 *Nm*.

The average coefficient of machine used by power with an increase in D_0 from 1250 to 5000 mm decreases from 0.8-0.45 to 0.05-0.1; the smaller values refer mainly to finishing machines. The maximum values of this coefficient on most large machines do not exceed 0,8-0,5; at the same time, many roughing machines with D = 1250 mm have a power factor of more than -1.0. The average cutting power with an increase in D_0 to 2000-2500 mm increases for finishing machines to 12*kW* and roughing machines to 19*kW*, and then decreases. By the same law, the cutting power maximum value will change. The average power consumed by the machine does not exceed 90*kW*, and the maximum 150*kW*. Based on reference sources' data mathematical processing concerning machines of different producers, power-law dependences of several machine parameters (*Qo, nmax, nmin, Mcr, N, etc.*) on *D* were obtained. A method for analyzing the machines use in time was developed and tested based on their state data at each moment. This method allows many times to reduce the time required to determine the average data on the equipment used. A statistical survey of the multi-purpose equipment use and analysis of machined parts requiring its application show that no unified solutions can be found here. It is established that due to the wide variety of factors influencing the expediency of using multi-purpose/ universal equipment on heavy lathes, the vast majority of equipment should be supplied only on a particular order. These types of equipment include open, closed, and semi-closed lunettes of all sizes, milling, drilling, boring, and grinding devices.

4. Conclusion

As a result of the conducted research, the actual scientific and technical problem of key importance industrial significance has been stated that consists in the need to develop scientific bases for designing of machine tools aimed to process pieces at heavy machine building industry. The technological bases of designing new metal-cutting equipment with NC control, specializing in the manufacture of commonly used components for the production of economically feasible complexity with an operative selection of rational equipment from the existing type machine tools range for certain production conditions have been elaborated and developed. The stated problem solution allowed to increase productivity by 30-45%, 3-4 times augmenting accuracy and rendering the substantial machinery parts manufacturing reliability more certain by 20-25%.

References

[1] P. Dašić. Comparative analysis of different regression models of the surface roughness in finishing turning of hardened steel with mixed ceramic cutting tools. *Journal of Research and Development in Mechanical Industry (JRaDMI)*, 5 (2) (2013), 101-180.

[2] Z.C. Du, J.G. Yang, Z.Q. Yao, B.Y. Xue. Modeling approach of regression orthogonal experiment design for the thermal error compensation of a CNC turning center. *Journal of Materials Processing Technology*, 129 (1-3) (2002), 619-623. doi: 10.1016/S0924-0136(02)00668-4.

[3] C. Natarajan, S. Muthu, P. Karuppuswamy. Investigation of cutting parameters of surface roughness for brass using artificial neural networks in computer numerical control turning. *Australian Journal of Mechanical Engineering*, 9 (1) (2012), 35-46.

[4] M. Taka, S.P. Raygor, R. Purohit, V. Parashar. Selection of tool and work piece combination using multiple attribute decision making methods for computer numerical control turning operation. *Materials Today: Proceedings*, 4 (2) (2017), 1199-1208. doi: 10.1016/j.matpr.2017.01.138.

[5] W.-H. Sun, S.-S. Yeh. Using the machine vision method to develop an on-machine insert condition monitoring system for computer numerical control turning machine tools. *Materials*, 11 (10) (2018), Article no. 1977. doi: 10.3390/ma11101977.

[6] Y. Shrivastava, B. Singh. Stable cutting zone prediction in computer numerical control turning based on empirical mode decomposition and artificial neural network approach. *Transactions of the Institute of Measurement and Control*. 41 (1) (2019), 193-209. doi: 10.1177/0142331218757285.

[7] B. Naithani, S. Chauhan. Mathematical modelling approach for determining optimal machining parameters in turning with computer numerical control (CNC) machines. *International Journal of Computer Aided Engineering and Technology*, 4 (5) (2012), 403-419. doi: 10.1504/IJCAET.2012.048838.

[8] K.-C. Wang, H.-C. Shen, C.-H. Yang, H.-Y. Chen. Temperature sensing and two-stage integrated modeling of the thermal error for a computer-numerical control swiss-type turning center. *Sensors and Materials*, 31 (3) (2019), 1007-1020. doi: 10.18494/SAM.2019.2225.

[9] Y. Rodichev, H. Soroka, V. Kovalov, Y. Vasilchenko, I. Savelova. Predicting the fatigue strength of hard alloys, treated with Pulsed magnetic field. In: *Proceedings of the 13th International Conference «Research and Development in Mechanical Industry» (RaDMI-2013)*, 12-15 September 2013, Kopaonik, Serbia, pp. 254-259.

[10] M. K. Mohanad, V. Kostyk, D. Demin, K. Kostyk. Modeling of the case depth and surface hardness of steel during ion nitriding. *East European Journal of Advanced Technologies*, 2 (5) (2016), 45-49. doi: 10.15587/1729-4061.2016.65454.

[11] V. Kovalev, Y. Vasilchenko, P. Dašić. Adaptive optimal control of a heavy lathe operation. *Journal of Mechanics Engineering and Automation (JMEA)*, 4 (4) (2014), 269-275.

[12] V. Kovalev, Y. Vasilchenko, D. Riznić. Evaluation of a level of quality of manufacturing process on heavy engineering enterprises. *Annals of The University of Oradea Fascicle of Management and Technological Engineering*, 14 (3) (2014), 1-7.

[13] V. Kovalov, Y. Vasilchenko, P. Dašić. Development of the integral complex of optimal control of heavy machine tools adaptive technological system for wind-power engineering parts. *Procedia Technology*, 19 (2015), 145-152. doi: 10.1016/j.protcy.2015.02.022.

[14] V. Kovalov, Y. Antonenko, P. Dašić. Method of structural design of heavy machine tools. *Procedia Technology*, 22 (2016), 146-152. doi: 10.1016/j.protcy.2016.01.023.

[15] W. A. R. Dhafer, V. O. Kostyk, K. O. Kostyk, A. Glotka, M. Chechel. The choice of the optimal temperature and time parameters of gas nitriding of steel. *East European Journal of Advanced Technologies*, 3 (5) (2016), 44-50. doi: 10.15587/1729-4061.2016.69809.

[16] V. Khoroshailo, V. Kovalov, P. Dašić. Improving of vibration resistance of boring tools by big diameter holes tooling on lathe. *Procedia Technology*, 22 (2016), 153-160. doi: 10.1016/j.protcy.2016.01.038.

[17] V. Kovalov, Y. Vasilchenko, M. Shapovalov, R. Turmanidze, P. Dašić. Impact of a pulsed magnetic field on a hard alloy during machining on heavy machine tools. *International Journal of Industrial Engineering and Management (IJIEM)*, 10 (1) (2019), 125-130. doi: 10.24867/IJIEM-2019-1-125.

Institute of Physics
Dirac House, Temple Back
Bristol BS1 6BE UK

CURRAN ASSOCIATES INC.
proceedings.com

ISSN: 1757-8981
ISBN 978-1-5108-9811-0

9 781510 898110